DATE DUE

DEMCO 38-296

Desert
Geomorphology

Desert Geomorphology

Ron Cooke
University College London

Andrew Warren
University College London

Andrew Goudie
University of Oxford

UCL
PRESS

First published in 1993 by UCL Press

UCL Press Limited
University College London
Gower Street
London WC1E 6BT

The name of University College London (UCL) is a registered
trade mark used by UCL Press with the consent of the owner.

ISBN
1-85728-016-4 HB
1-85728-017-2 PB

A CIP catalogue record for this book
is available from the British Library.

Cover photograph by Peter L. Kresan
Crest of a star dune, Gran Desierto, Northwest Sonora, Mexico.

Printed in England by Clays Limited, St. Ives plc.

CONTENTS

PART 1 Global perspectives

PART 2 Desert surface conditions

PART 3 The fluvial domain in deserts

PART 4 Wind in desert geomophology

PART 5 The evolution of desert landforms: a global perspective

PREFACE

When we presumptuously wrote *Geomorphology in deserts* in 1970, our aim was to provide a convenient and original synthesis of desert geomorphology that focused on the dynamics of contemporary processes. Two decades later, our aim is modified, and our presumption is greater, for in the intervening period there has been an explosion of desert research. There are several reasons for this explosion, as follows.

The advent of satellite imagery has provided a revolutionary new perspective on deserts. It has made many unstudied, remote areas easily and relatively cheaply "accessible" for the first time, at least for reconnaissance observations. It has also allowed landforms to be viewed at a scale that was impossible before, an opportunity that has revealed new landforms and new landform patterns.

In addition, the rapid emergence of extraterrestrial research and the study of planetary landscapes has led to the search for terrestrial analogues in deserts. Thus the advance of planetary geomorphology has stimulated desert research on Earth and laboratory simulation of desert processes.

At the same time, the environmental issues arising from the contrasting problems of poverty and plenty in deserts have created new applied contexts in which geomorphological research has flourished. The conscience of the world has become alerted to the perils of famine, and to the processes of desertification and environmental change in drylands. Public concern has generated a library of literature on desertification, and studies of geomorphological processes, not surprisingly, figure prominently within it. The poverty accompanying desertification in some regions contrasts starkly with the prosperity in those deserts blessed with oil resources and a rising income from oil revenues since the early 1970s. Massive and rapid urban and industrial development often encountered unexpected environmental hazards, many of which have their physical basis in geomorphology. Thus there has grown a need to apply geomorphological research to the solution of planning and engineering problems, especially in the Middle East, and there has been substantial progress.

A fundamental consequence of these developments has been an exponential growth over the past two decades in the volume of literature we need to consider. For example, in 1970 the aeolian literature was quite small and still heavily dependent on Bagnold's (1941) seminal work, *The physics of blown sand and desert dunes*. Today, there are ten or more new books on aeolian processes and forms, and hundreds of new articles, most of which contribute something to our understanding.

In confronting the daunting task of synthesis, we have retained the three major parts of the original book – on surface conditions, the fluvial domain and the aeolian domain. This division is not perfect, but it is robust and convenient. In the earlier book, our emphasis was more on process than on landform. Here we have moved towards a more balanced view of forms and processes in which there is a better evaluation of temporal change and inheritance. In addition, we have sought to demonstrate briefly some of the ways in which geomorphological knowledge can be applied to planning, engineering and agriculture. Beyond the three systematic parts we have introduced a new final part, written by Andrew Goudie, which attempts a regional integration and review of the geomorphological history of the world's deserts. The result, we believe, completely supersedes the first book but, like it, provides a synthesis of generalizations, a compendium of unanswered questions, and a new baseline for future research.

The literature of desert geomorphology is spatially very biased, reflecting cultural prejudices and scientific predilections, sponsorship and problems of field access. These variations deserve a book to themselves, and we touch upon some of them in Part 1. But some features are clear. Research in the southwestern USA is still of the greatest importance and continues unabated. The recent flowering of research in Israel is strongly linked philosophically to American work. Franco-German efforts, following different objectives, still contribute substantially to progress especially, but not only, in the Sahara and the Middle East. In Eurasian deserts, Soviet and Chinese research is very extensive, but unfortunately much of it is not accessible to us. Elsewhere, in the huge drylands of South America, southern Africa and Australia for example, the landscapes are vast and the labourers few. These spatial variations are also often delimited by language barriers that still reduce communications between geomorphologists from different cultures. We have tried to overcome some of these artificial and unhelpful barriers here, especially by using more German and French work than in our first volume. Where possible we have referred to publications by Europeans in English, to make them more accessible to our readership.

But our prejudices remain. We are Eurocentric geomorphologists steeped in the traditions of G. K. Gilbert, W. M. Davis, R. A. Bagnold and their successors. We only visit deserts, and our efforts have become increasingly directed towards using geomorphology to help understand and solve desert problems related to human activity. These prejudices no doubt explain some of the weaknesses of the volume; they are also its justification.

Since 1970, we have greatly extended our experience. To the research Cooke and Warren had done at that time in North America, Chile and Peru, northern Africa and Pakistan, we have now added studies in many countries of the Middle East, and further work in the American Southwest. Andrew Goudie brings to Part 5 of the volume experience of many deserts, especially in Africa, Australia and the Middle East.

Various parts of the volume have been reviewed by colleagues. We are particularly grateful to A. D. Abrahams (Buffalo, USA), A. J. Parsons (Keele, UK), and C. M. Clapperton (Aberdeen, UK).

The diagrams were prepared by Tim Aspden, Louise Saunders and Robert Bradbrook in the Cartographic Unit, Department of Geography, University College London. Harfiyah Haleem, Olivera Markovic and especially Dr Suse Keay tirelessly helped to prepare the text. Finally, we are deeply indebted to friends and colleagues, too many to mention individually, who have helped in so many ways to support our fieldwork. We extend our sincere thanks to them all.

Ron Cooke   Andrew Warren   Andrew Goudie
JANUARY 1992

ACKNOWLEDGEMENTS

We are most grateful to all those authors and publishers who have kindly given us permission to reproduce their diagrams in this volume. Unfortunately, we have been unable to contact a small number of authors for permission, but we hope they will approve the way we have used their work. The sources of all work from other authors are fully referenced in the Bibliography. Specific acknowledgements are due to the following: the American Meteorological Society for Figure 20.12; Blackwell Scientific Publications Limited for Figures 18.8 and 26.33; Elsevier Science Publishers BV for Figure 11.18; the Geological Society Publishing House for Figure 14.12; the Royal Society of Edinburgh for Figure 20.16; the University of Chicago Press for Figures 5.4, 10.30, 12.15 and 19.6; John Wiley & Sons Ltd for Figures 11.4, 11.5, 11.7, 11.16, 11.23, 12.12, 12.14 and 16.2; and Williams & Wilkins for Figure 6.4.

PART 1
GLOBAL PERSPECTIVES

CHAPTER 1

The desert realm

1.1 Definition and distribution of deserts

To some, deserts are simply barren areas, barely capable of supporting life forms. Many places meet this criterion: Mangin's *The desert world* (1869) embraced environments as diverse as the waste heaps of the china clay quarries in Cornwall, the steppes of Tartary, the Dead Sea and the Arctic wildernesses. But most deserts are areas of aridity and they are usually defined scientifically in terms of some measure of water shortage. Such measures, **indices of aridity**, are commonly based on relationships between water gained from precipitation and water lost by evaporation or transpiration. There are plenty of indices to choose from, the differences between them reflecting different objectives of classification (e.g. Meigs 1953, Wallén 1967, Howe et al. 1968, Budyko 1974, Thompson 1975).

An excellent study of aridity was published by UNESCO in 1979. Its criteria take account of a growing network of meteorological stations, climatological advances, and new information on soils and vegetation. The UNESCO map (Fig. 1.1) is based partly on aridity indices, and partly on consideration of soil, relief and vegetation data. The aridity index is *P/ETP* (where *P* is the annual precipitation, and *ETP* is the mean annual potential evapotranspiration, based on the Penman formula). Aridity increases as the value of the ratio declines, and four zones of aridity are delimited:

(a) subhumid zone $(0.50 < P/ETP < 0.75)$
(b) semi-arid zone $(0.20 < P/ETP < 0.50)$
(c) arid zone $(0.03 < P/ETP < 0.20)$
(d) hyper-arid zone $(P/ETP < 0.03)$

In addition, the map shows the seasons of dryness, the number of dry months, and the mean temperature of the coldest month in the year.

The areas shown in Figure 1.1 constitute the warm desert realm. Within it, there are five major regions of aridity: the deserts of North and South America, North-Africa–Eurasia, southern Africa and Australia. They cover a third of the Earth's land surface and are the context for this study of desert geomorphology. There are also arid areas in polar latitudes but, geomorphologically, they are very different from the subtropical deserts (e.g. Péwé 1974) and they are excluded from this review.

1.2 Causes of aridity

There are five major climatic causes of aridity. First is distance from marine or other moisture sources (or **continentality**), which means that the largest deserts are near to the centres of the largest landmasses. Continental aridity commonly extends to the western margins of the subtropical landmasses, chiefly because the dominant trade winds lose their moisture travelling over the continents and arrive at west coasts as dry, offshore airstreams (Thompson 1975; Fig. 1.2).

Secondly, and certainly as importantly, the maintenance of global circulation requires atmospheric subsidence in the subtropics that is associated with high-pressure cells. Here, **dynamic anticyclonic subsidence** normally prevails, and is accompanied by adiabatic warming and low relative humidity (e.g. Thompson 1975, Barry 1978). The patterns of the subtropical high-pressure systems vary seasonally, and occasional rain-bearing low-pressure troughs may penetrate the zone. Such rain-bearing systems are associated, for example, with easterly waves in the trade winds and depressions originating along the polar front. The penetration of the monsoons behind

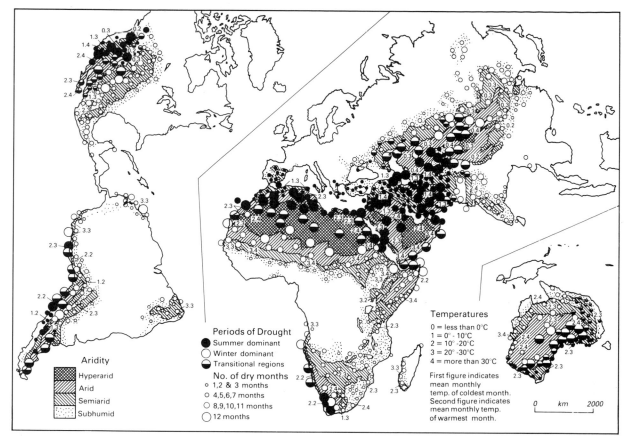

Figure 1.1 The world's arid lands (after UNESCO 1979).

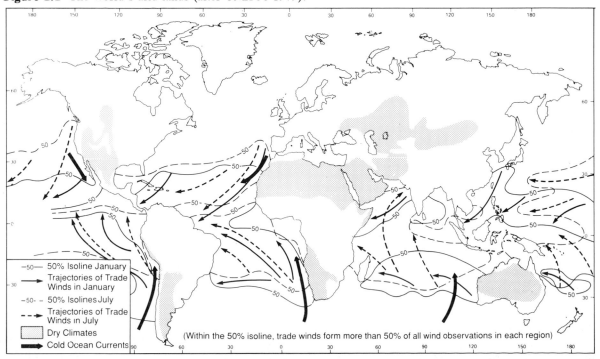

Figure 1.2 Distribution of dry climates and trade-wind predominance (after Crowe, from Thompson 1975).

3

the intertropical convergence zone on the equatorward sides of the high-pressure cells is more persistent and regular (Thompson 1975).

Orographic influences are a third major control of, or at least an accentuating influence on, aridity. Mountain ranges occlude marine influences and rain-bearing winds, and on descending the lee slopes are warmed and dried adiabatically. The Great Divide of Australia cuts off the interior from trade winds, the Andes shelter the Argentinean drylands, the Western Ghats deprive the Deccan Plateau of rainfall, and the Rockies shield the plains of Alberta.

Fourthly, aridity is also reinforced on the dry western littorals of continents by coastal **upwelling** of cold water associated with cold, equatorward-flowing offshore currents (Fig. 1.2), creating advection fogs and accentuating atmospheric stability (Thompson 1975).

Finally, the aridity of subtropical deserts may also be reinforced by the **high reflectivity (albedo)** of desert surfaces themselves. This may cause net loss of radiative heat, create a horizontal atmospheric temperature gradient along the desert margin, and induce circulation systems that induce subsidence or reinforce prevailing subsidence (Charney 1975, Barry 1978).

The varied causes of desert climates result in an immense diversity of climatic conditions. This diversity profoundly influences desert geomorphology. Thus, the diversity of temperature and humidity conditions underlies the spatial complexity of weathering processes (Chs 4–8); the varied patterns of rainfall are reflected in the distinctiveness of fluvial activity (Chs 9–15); and wind regimes generated by these weather systems determine the variety of aeolian movements (Chs 16–29).

1.3 Changing distributions

A fundamentally important fact in the study of desert geomorphology is that desert climates have changed. The changes, which are considered in Part 5 and elsewhere, are evident at several different scales. First, *drought* is a seasonally and annually varying short-term hazard. Droughts impose a temporary regime of greater aridity on areas that are otherwise more humid, and may be accompanied by contemporaneous changes of geomorphological processes. Secondly, the desert realm (and especially its margins) has been substantially changed in recent decades as a result of many environmental pressures arising from

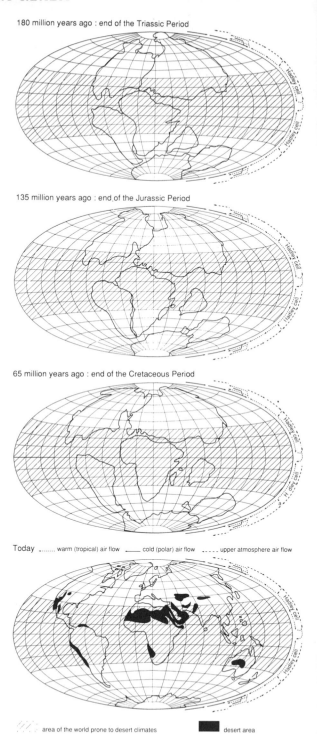

180 million years ago : end of the Triassic Period

135 million years ago : end of the Jurassic Period

65 million years ago : end of the Cretaceous Period

Today .... warm (tropical) air flow ____ cold (polar) air flow ___-___ upper atmosphere air flow

area of the world prone to desert climates   desert area

Figure 1.3 Continental movements relative to the global zone of aridity (from various sources, after Warren 1984).

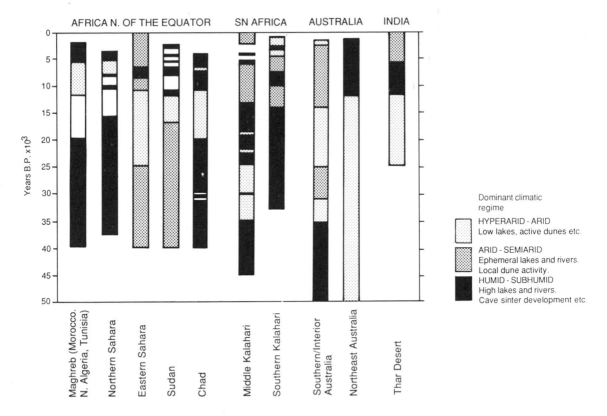

Figure 1.4 Simplified chronologies of late Quaternary expansions and contractions of tropical and subtropical arid zones, based mainly on geomorphological and sedimentological data (from various sources, after Thomas 1989b).

both drought-creating climatic trends and from increased degradation. The result of these complex changes has in places led to *desertification*, a problem that is of global significance (e.g. Warren & Agnew 1988).

Finally, desert climates have evolved over much longer time periods. The dry belts associated with the subtropics have expanded and contracted and moved over the landmasses, either as a result of global climatic shifts (accompanying glaciations, for example), or as a result of the movement of land-masses, or a combination of both. The ways in which the landmasses have moved with respect to the tropical desert zone are shown in Figure 1.3: for instance, dune fields covered northern England in the Permian, whereas the arid core of Australia was once attached to the more humid Antarctic continent, as is clear from its heritage of ancient soils.

Since the Cretaceous, when most desert landscapes have been formed, and especially during the Quaternary, the margins of aridity have contracted and expanded on several occasions. Some areas that are

today arid contain abundant evidence of pluvial conditions, such as former lakes and rivers. Lake Mega-Chad, in west-central Africa, covered an area as large as the Caspian Sea in the late Quaternary, but is now occupied in part by active dune fields.

Other, rather larger areas were once more arid, as can be seen by the distribution of now stabilized sand seas, which cover more of the semi-arid realm than active sand seas cover the present deserts. The dune patterns of these ancient sand seas survive in a vegetated landscape underlain by deep-leached soils. The northern shore of modern Lake Chad, within the limits of Lake Mega-Chad, is a delicate pattern of these ancient dunes, now even cultivated. Nevertheless, some areas have remained arid throughout much of the post-Cretaceous period, perhaps only varying between hyper-arid and semi-arid. Such *Kernwüsten*, or core deserts, include parts of the Chilean and eastern Saharan deserts.

Many climatic shifts in the deserts and on their margins were associated with the same forcing factors that generated the glaciations in high latitudes. In general, the glacial periods were dry in the tropics and the interglacials wet, although the relationships were not everywhere simple. Whereas late Pleistocene arid phases coincided in East and West Africa, Rajasthan, Australia and the Arabian Gulf (e.g. Williams 1975, Doornkamp et al. 1980), some of the Australian arid periods were out of phase with this pattern (e.g. Bowler 1975). There are also anomalies in the southwestern USA, where high lake levels appear to have corresponded with times of greater precipitation in early or late glacial phases (e.g. Morrison 1968). Some of the possible complexities are shown, for Australia and the Sahara, in Table 1.1 (Rognon & Williams 1977), and a simplified chronology of late Quaternary climatic fluctuations of desert climates is provided in Figure 1.4 (Thomas 1989b).

Thus the margins of the desert realm, and the patterns of climate within it, are constantly in motion. Change over space and time is the norm, and such change is fundamental in fashioning the desert landscape.

Table 1.1 Summary of late Quaternary climates in Australia and North Africa.

Interval (1,000yr BP)	Tropical margins		Temperate margins	
	Southern Sahara	North Australia	Northern Sahara	South Australia
40–20	wet	dry	wet, cool	wet, cool
20–13	dry, cold	very dry, cold	dry, cold	dry, cold
12–7	wet	wet, warm	dry generally, warm	wet, warm
7–0	drier	drier, cooler	dry, cooler	drier, cooler

Source: Rognon & Williams (1977).

CHAPTER 2

Diversity and distinctiveness of deserts

2.1 Ingredients of diversity

The immense diversity of environmental conditions in deserts conspires to create a complex pattern of landforms. It is quite false to believe that lack of water engenders a simple distinctive geomorphology. Climatic variability over space and time has lead not only to a large range of landform-producing processes, but also to a complementary diversity of soils and vegetation which, in turn, influences geomorphological activity.

Desert soils, as discussed in Part 2, are distinctive in several respects. Although the underlying rocks are as varied in deserts as in any other global environment, the loose surface sediments, from which most arid soils have formed, are either coarser alluvium than is found in humid climates or, more importantly, great blankets of aeolian sand and silt, laid down in various periods in the past. Because soil development in arid areas is slower than erosion, on all but the gentlest gradients, soils with distinctive horizons can survive only on slopes much gentler than in humid climates. Most desert soils are thin and poorly horizonated, and have a low clay content; many are reddened and include subsurface horizons enriched in salt (in the drier regions), gypsum (in moderately dry regions) or calcium carbonate (in semi-arid areas). In the limited areas where the water table is shallow (as on alluvial plains), water may be drawn up to the surface by capillarity and so create salinized surface horizons. Where gradients and erosion rates are low (especially in the very arid deserts), many soils have features inherited from previous wetter climates.

Vegetation is also distinctive and it exhibits a great diversity in deserts, reflecting individual climates, soils and evolutionary histories (Noy-Meir 1973,

Evenari 1985). In some deserts there is great species diversity, reflecting an enormous range of adaptations to drought, and in some cases isolation over millennia. In most, the species complement is very small. At the regional scale, biomass is virtually zero over huge swathes of the most arid deserts such as the Sahara, and it increases slowly and fitfully towards the desert margins, responding to latitudinal and altitudinal gradients. At the local scale, biomass is characterized by much greater contrasts in arid than in humid climates, responding principally to water availability (being extraordinarily dense in some hollows to which water drains), but also to fire and exploitation. Where biomass is low, the vegetation is characteristically much more clumped than vegetation in mesic environments (Shmida et al. 1986)

From the geomorphological perspective, these vegetation patterns have several important features. Crucially, sparsity of vegetation means that a greater proportion of the ground surface is directly exposed to the action of wind and water. Thus, there should be a simplified and more direct relationship between climatic phenomena and the surface. Raindrop impact and splash erosion may be more effective, thermal changes tend to be large, and winds may act with little hindrance on surface material. The clumped pattern of vegetation, where it occurs, encourages an intricate pattern of wind and water erosion and deposition.

These generalizations require some qualification. For example, some xerophytic desert plants have extensive, near-surface rooting systems that serve to bind the surface and limit the full impact of surface erosion – despite a sparse above-ground biomass. Moreover, vegetative protection varies significantly in a seasonal or more erratic fashion. Litter fall may be considerable in some seasons and may provide temporary surface protection, for breakdown may be slow, even in wet conditions. Finally, in the few areas where moisture is abundant, vegetation can play a crucial rôle in geomorphological processes. It is a surprise to discover woodlands of *Prosopis* in parts of the Wahiba Sands of Oman: the tap roots of these trees can allegedly penetrate to water tables 10m below the surface. The trees act to trap sand and build up large dunes. Dense stands of riparian vegetation are not uncommon in gullies in the southwestern USA for the same reason, and in these locations exert a strong influence on contemporary processes, for example by hindering channel change.

Wind and water, as exogenetic processes, work on landscapes driven also by endogenetic forces, such as tectonic and volcanic activity. Deserts have experienced great temporal and spatial variability of these endogenetic processes. On the one hand, there are apparently tectonically stable areas, such as central Australia, the Kalahari and the western and eastern Sahara; on the other, there are patently unstable areas such as central Asia, Iran, Chile and California; in between there are regions that are temporarily quiescent, but which show signs of activity in the recent past, such as parts of North Africa and the central Sahara.

The interplay of endogenetic and exogenetic processes determines the nature of topography. In arid areas experiencing intense tectonic activity, the topography is dominated by the direct consequences of earth movements, such as fault scarps, disrupted, tilted and folded surfaces, and the blocked and diverted drainage of the southern part of the Great Basin Desert in California (Fig. 2.1). Contrasted with this are the plains and complex suites of aeolian features in the arid, but tectonically stable, northeastern Sahara. The increase of precipitation and erosional activity southwards through the tectonically unstable Atacama Desert may help to explain why landforms produced by endogenetic processes are so much more clearly displayed in the extremely arid northern part of the area. The chief reason for the diversity of desert landforms is the variation in the relationship between the groups of endogenetic and exogenetic processes, making the task of realistic generalization even more difficult.

2.2 Desert domains and landform inheritance

2.2.1 Desert domains
In so far as simple generalizations about desert geomorphology are sustainable, it is convenient to

think in terms of two principal process domains. The **aeolian domain** is dominated by the erosional and depositional activities of the wind. The first of these – wind erosion – is pervasive, although not always easy to recognize, and only dominant in extremely dry areas. Deposition by the wind is also widespread, but thick accumulations of dust and sand are widespread throughout the moisture spectrum and are easy to identify: aeolian sand seas shroud some 25% of the arid realm, and (though now stabilized) up to half of some semi-arid areas, such as the West African Sahel. Desert dust, in the form of loess, also blankets vast areas of the desert margins, as in other parts of the Sahel and in parts of North Africa and the southern Levant.

The **fluvial domain**, where the work of water is dominant, is to be found especially in the less arid areas, such as the Great Basin of the USA and parts of North Africa and the Middle East. Here, streams erode valleys and spread the debris in alluvial fans and valley trains. However, only a few locations are exclusively the product of either fluvial or aeolian processes, and most areas reflect a subtle interplay between them. For instance, aeolian dust deposits commonly occur in places where the landforms are largely of fluvial origin, and aeolian and fluvial processes are constantly trading sediment. There are many more areas where the interplay of domains is dominant than areas where one process holds exclusive sway.

Because of climatic changes, the areas dominated by one or other of the domains have varied in size and location over time. Thus, the geomorphological record varies greatly within the desert realm, from those few locations where either aeolian or fluvial activity has exclusively produced the landscape, to much more numerous locations where the present-day forms reflect the frequently changing relationships within and between the two groups of processes. This not only means that many of today's deserts reveal evidence of wetter conditions of the past, but it also means that former desert conditions are preserved in landforms beyond the present areas of aridity.

2.2.2 The past in the present
Of paramount importance in interpreting desert landforms is the possibility of inheritance from previously different climatic conditions. Many deserts are today more arid than they were during the pluvial conditions of the Pleistocene, so that they are commonly experiencing reduced fluvial activity and, perhaps, enhanced aeolian activity. In contrast, many landforms

Figure 2.1 A recent fault scarp on an alluvial fan, Panamint Valley, California.

of the desert margin preserve evidence that there was much greater aridity during the time when glaciers were more extensive in the higher latitudes than they are today. The single most important and pervasive consequence of this fact is that landforms produced under the earlier, different conditions have a good chance of surviving in the new environment. Thus the problem of "fossil", "stabilized" or "relict" landforms has been central to studies of desert geomorphology.

There are at least three main types of landform preservation: burial, abandonment and preservation by protective veneer.

(a) Burial

An alternating sequence of climatic changes in deserts may be accompanied by an alternating sequence of erosional and depositional episodes, so that landforms created under one set of circumstances may be buried under sediment laid down under the other. The changes are very evident along drainage lines, where fluvial deposits survive from former fluvial episodes, and where aeolian sediment may bury former fluvial sediments. It is even more evident where sand seas have blanketed whole fluvially derived landscapes. Burial may be succeeded by exhumation, as is apparently the case with the very ancient glacial landscapes of the central Sahara (Rapp 1975) and parts of the Arabian Peninsula (Fig. 2.2).

(b) Abandonment

A more common form of preservation occurs where

Figure 2.2 A striated pavement, overlain by glacial diamictite and created by a Permo-Carboniferous glacier, now in very arid conditions in Central Oman.

a landform or a landscape produced under a given climatic regime is merely abandoned by the processes which formed it with the advent of aridity. The degree of abandonment may vary greatly. In a river system, for instance, drainage may change from perennial to ephemeral or intermittent, or ephemeral drainage may become even more sporadic: the channel system may appear to retain its form, although effective drainage area and drainage density may be reduced, and there may be appropriate changes of channel geometry.

There is extensive evidence of fluvial systems and landforms on a scale too great to have been created or maintained under the present climatic conditions. A reconstruction of the drainage system of the south-western Great Basin Desert in the USA during the Wisconsin period is shown in Figure 2.3. At this time, or for a short period within it, rivers from the Sierra Nevada and ranges to the south flowed through a series of lakes into Death Valley, and there were also many small, isolated lakes. Today the system is

Figure 2.4 An overflow channel between playa basins in the Mojave Desert, California. "Flow" was from Silurian Basin northwards to Searles Basin, which is in the distance. This is a fossil feature inherited from earlier, pluvial conditions (see Fig. 2.3).

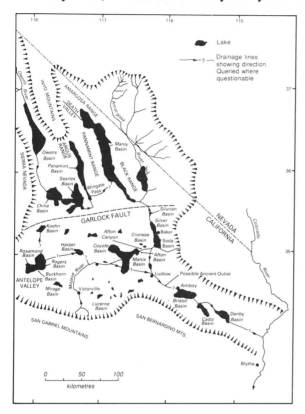

Figure 2.3 The drainage network in the southwestern Great Basin during the Wisconsin period (after Black-welder 1954 and others).

clearly preserved with many examples of shoreline features, overflow channels (Fig. 2.4) and terraces, but drainage rarely extends into the desert beyond the Owens and Silver basins.

(c) Protection

The preservation of landforms by protective veneers is both common and distinctive. On a large scale, the formation of duricrusts – which may be composed of silica, gypsum or calcium carbonate (Ch. 6) – may serve to protect associated surfaces from further erosion. In these circumstances, no climatic change is necessary to produce the protective effect, but such a

change is often involved. Protection may be on a massive scale. For example, the widespread occurrence of siliceous duricrusts in deserts contributes to the preservation of extensive surfaces (e.g. Summerfield 1983a), as in parts of Australia. However, such preservation is also possible locally and ephemerally. For instance, some of the dunes in the Great Eastern Sand Sea of the Tunisian Sahara are preserved and stabilized by a thin, delicate gypsum crust.

In addition to preservation, there are two types of response by landforms in deserts to a climatic change towards aridity: continued development without apparent change of form, and changes leading to superimposition of landforms.

(d) Continued development

Landforms or erosion systems may respond to climatic change merely in the *rate* of their development. For example, an erosion system composed of mountains, pediments and alluvial plains may develop more slowly under conditions of increased aridity and may experience no distinguishable change of form. It is true that such systems do often show obvious evidence of adjustments to climatic change, but adjustments are by no means universal. In the case of the pediplain landscape of the high Andes in the southern Atacama Desert, for example, it is possible that mountain-front retreat has continued in the undissected areas throughout much of Pliocene and Quaternary time. In such circumstances the landforms may be descended from features formed under different conditions, which may display no characteristics diagnostic of those conditions or, for that matter, of contemporary conditions.

Examples of continued, although accelerated or decelerated, development occur also in soils and aeolian landforms. The rate of development of calcretes (calcium-rich subsurface soil horizons), depending as it does on the supply of calcium dust and of rainwater, is in constant adjustment as these inputs respond to climate change, but the processes continue to produce much the same end-product. The rate of adjustment of most mega-dunes must have decelerated since the windy conditions of the Pleistocene but, in many, development is undoubtedly continuing, albeit at a slower rate.

(e) Superimposition

Where one erosion system is replaced by another, landforms characteristic of the new system may be superimposed upon them. The most blatant examples of this process are the replacement of fluvial by aeolian systems or vice versa. The fluvial landscape upon which the Saharan sand seas have been so obviously superimposed may have been inherited from Miocene times, as is shown in the northwestern Sudan (Breed et al. 1982) and in eastern Egypt (e.g. McCauley et al. 1986, McHugh et al. 1988). In the opposite sense, the stabilized dunes of the Sahel are now overlain by a developing fluvial drainage network, with gullies and alluvial fans (Talbot & Williams 1978).

To conclude, an important qualification is necessary. Implicit in much of this discussion of landform responses to changing climate has been the idea that landforms may acquire a state of equilibrium with one set of environmental circumstances, and that a change in those circumstances causes a period of readjustment followed by the establishment of a new equilibrium. This view is especially apposite to channel changes.

There are, however, two further possibilities. First, it is quite possible that some landforms may develop an equilibrium condition which is not only relevant to the initial circumstances, but is sufficiently secure to be maintained when the situation changes. For example, stone pavements might develop to a stable condition under a semi-arid climate, but retain their form virtually unchanged when the climate becomes arid. Some ancient soils have become so leached of mobile constituents, or have developed such durable horizons, that it would require extreme climatic changes to provoke their destruction. Mega-dunes may create their own local wind patterns which help them to survive a changed climatic pattern.

Secondly, it is equally possible that adjustment to new conditions may intervene before full equilibrium has been reached. This could easily be the case with mega-dunes, which demonstrably need millennia for growth to full size.

CHAPTER 3

Geomorphologists in deserts

3.1 Exploration and exploitation

Most deserts are frontiers, and from a very early period they invited exploration and presented potentials for exploitation. Partly as a result, they have generated conflict, and much of the scientific work in deserts has been accomplished within this context.

Exploration is an amalgam of romance, curiosity, escape, obsession, courage and cupidity, and the scientific reports from early desert explorations are memorials to some or all of these drives. Witness the reports of Blake (1858), Powell (1875), Gilbert (1875), McGee (1896) and more recently Bryan (1925) in the southwestern USA. Huntington (1907) invested experiences in the Gobi Desert with his consistent philosophy of environmental determinism; and Hedin's (e.g. 1904–5) curiosity is recorded in the accounts of his Asian forays. In the Sahara, successive French, British and Egyptian expeditions brought a wealth of knowledge and speculation to metropolitan scientific societies' thirsty for knowledge. Such were the expeditions of Gautier & Chudeau (e.g. 1909), Flammand (e.g. 1899), Tilho (e.g. 1911), Newbold (e.g. 1924), Hassanein Bey (1925) and Bagnold (e.g. 1931). An almost apostolic succession of gifted explorers infiltrated Australian deserts from the middle of the nineteenth century into the early twentieth century, several of them contributing important scientific information (e.g. Spencer 1896, Wells 1902, Madigan 1936). Usually, exploration was accomplished by group expeditions; but lone travellers also played a significant rôle, as with Thesiger (e.g. 1949) and Monod (e.g. 1958).

A surprising number of the early desert travellers were motivated by high romance. The desert west of the Nile was repeatedly scoured for lost oases such as Zerzura (e.g. Bagnold 1931, Lancey-Forth 1930, Hume 1925). Passarge (1930) investigated the geomorphology of southern Tunisia as an aside in his attempt to ascertain the location of Atlantis. And in India, the search for the Vedic Sarasvati, and for reasons behind the decay of the ancient Indus civilization, gave the initial impetus to scientific enquiry into deserts of the subcontinent (Raverty 1878).

Great hardship in desert exploration is a thing of the past. Although there are those who choose deliberately to confront the harsh conditions (for example, by wind-sailing alone across the Sahara), they rarely contribute seriously to scientific research. The balance between science and survival – a crucial problem for early geomorphologists and other scientists – has now tipped decisively in favour of science. A desert research team can now use reliable off-road vehicles, fix its position accurately by satellite, travel with the aid of good maps or satellite images, and be in constant communication with the outside world. This may reduce the romance but it improves the science, which is all to the good, for major scientific challenges remain.

This is not to say that carefully organized expeditions are no longer needed, and there are many examples of excellent scientific research from this kind of exploit. The expeditions to the central Sahara from the Freie Universität Berlin have established a research station at Bardai in Chad (e.g. Jäkel 1974); the Royal Geographical Society has recently sponsored scientific projects in Oman (Dutton 1988) and Australia (Goudie & Sands 1989); and multinational expeditions are increasingly common (e.g. El-Baz et al. 1980, El-Baz & Maxwell 1982).

Until the mid-twentieth century, the influence of national groups has largely been confined to particular deserts: German-, French-, Italian- and English-speakers (from at least two very different national traditions) each had, as it were, their own territories. Three aspects of this segregation are important. First, scientific research reflected the prejudices and intellectual attitudes of the groups concerned. Secondly, communication was restricted, by barriers of language and competition. Thirdly, and perhaps most importantly and perniciously, much of the work was accomplished by people working in unfamiliar environments.

Through the efforts of UNESCO, UNEP, the International Association of Geomorphologists, INQUA and new journals such as the *Journal of Arid Environments*, *The Journal of Desert Research* and *Problems of Desert Development*, these antiquated barriers are thankfully crumbling, and fresh traditions are being created in desert science.

The exploitation of deserts has been generated by a variety of pressures: imperialism, colonialism and military adventurism have certainly played prominent if dubious rôles. They have been replaced in many areas by the search for agricultural land and mineral resources. Much scientific intelligence has come from research intended to plan exploitation and development. As the pressures on land have increased, so this form of research has assumed a more important rôle.

Frontier movements into arid areas within established nation–states have also been important in extending scientific knowledge. The new settlers have had to accommodate themselves to alien environments, and scientific knowledge has been acquired in the process. The settlement of the western USA is a good example of this, and there have also been settlement movements in Israel, Australia, South Africa and China. In the USA, the latter half of the nineteenth century saw an enormous number of resource evaluations in desert areas. Some of the earliest were the surveys for railroad routes (e.g. Blake 1858, Parke 1857). Powell's (1878) more general (but classic) *Report on the lands of the arid region of the United States* was the precursor of a library of resource evaluations. Amongst more recent evaluations are the regional surveys of the CSIRO in Australia (e.g. Perry et al. 1962), and environmental appraisals for military purposes (e.g. Neal 1965b, Mitchell & Perrin 1967). Many resource evaluations incorporate valuable geomorphological information, and some soil surveys have also laid important foundations for later geomorphological research.

Today, the growing scientific interest in deserts has led to the establishment of many indigenous centres of desert research, such as the University of Arizona, the Negev Institute of Desert Research in Israel, several universities in the Arabian Gulf (such as the Kuwait Institute of Science and Technology, the Research

Institute of the King Fahd University of Petroleum and Minerals at Dhahran and the Sultan Qaboos University in Oman), the Research Station at Gobabeb in Namibia, the Universidad del Norte (in Antofagasta, Chile), the Desert Institute of the Turkmenistan Academy of Sciences, the Desert Department of the Academia Sinia at Lanzhou in China and the Central Arid Zone Research Institute, Jodhpur.

3.2 Desert geomorphologists and geomorphology

3.2.1 Declining description
For many decades, most desert geomorphology emanated from short expeditions, in which science was rarely the main objective, and which therefore generated rather ephemeral, largely descriptive (as opposed to analytical) accounts. Processes were rarely observed in operation, and most conclusions were drawn from observations of forms and materials.

It is easy to understand why mere description occupies such a high proportion of the desert geomorphological literature. Desert landforms, denuded as they are of vegetation and artefacts, present a bizarre landscape to the explorer from more temperate areas. Some of the responses to this strangeness were original and imaginative, and have had implications for geomorphology well beyond the confines of the desert. The two outstanding examples are Gilbert's *Report on the geology of the Henry Mountains* (1877), the product of a fertile mind and two months field work, and Bagnold's *The physics of blown sand and desert dunes* (1941), which began in creative observations on resourceful expeditions to the Western Desert in Egypt. These two books have become canons of the geomorphological literature. A common response was to focus on individual, and often spectacular or exotic, features. This kind of desert geomorphology came to be characterized by concern for specific landform types (such as **zeugen**, **yardangs**, **dreikanter** or **barchans**) which may be neither common nor of great significance.

3.2.2 The ascendency of indigenous research
The rôle of description has mercifully declined, to be replaced by observation and analysis of desert processes by scientists in more sustained contact with the desert. Much of the new research requires permanent and expensive installations, regular visits and,

in the highly variable environment of the desert, observations over many years. Semi-permanent monitoring stations have been established in several deserts in recent years, and they have begun to provide important data. They include those at Sde Boqer and Nahal Yael and Avdat in Israel (e.g. Evenari et al. 1971), Tombstone in Arizona (e.g. Abrahams et al. 1988), Arroyo de los Frijoles in New Mexico (e.g. Leopold et al. 1966), the Repetek site in Turkestan, some in southeast Spain (e.g. Thornes 1976), and several sites in Australia, as at Fowler's Gap in New South Wales.

Another contemporary trend in desert geomorphology is practical application, in the context of terrain evaluation, and land exploitation for agriculture and urban development. The problems of sand drift, salt weathering and alluvial-fan flooding, for example, have all attracted contributions from geomorphologists (e.g. Cooke et al. 1982) and they have encouraged support for further and less applied research.

Coincidentally, the moves towards the study of processes and application have been accompanied by a belated realization that traditional, indigenous groups often had a more secure understanding of desert environments than many of those with alien perspectives. Much has been learned, for example, about water harvesting, from the activities of the Nabateans and Israelites in the southern Levant (e.g. Evenari et al. 1971) and the Papago and Hohokami in Arizona. And it is salutary, although it should not be surprising, to discover that the nomadic Bedouin have a clear and detailed understanding of landform, such as the dunes in the Wahiba Sands of Oman (Webster 1988).

3.2.3 A broadening perspective

Aerial photographs have for decades been the basis for much geomorphological mapping in deserts. They still are. Their perspective, detail and stereoscopic potential provide a valuable basis for field mapping, and opportunities for monitoring environmental change. Many geomorphological studies in deserts are based on maps that could not have been created without their aid (e.g. Ehlen 1976). Similarly, land resource appraisals and terrain classifications in deserts, such as those in Australia by CSIRO (e.g. Finlayson 1984) and by military agencies (e.g. Perrin & Mitchell 1969, 1971), and various hazard studies in desert cities (e.g. Cooke et al. 1982) depend heavily on aerial photographs.

Satellite imagery has now extended the opportunities provided by aerial photographs. Imagery is now available from many satellites, including the Apollo-Soyuz test project and similar missions (e.g. El-Baz & Warner 1979, El-Baz & Maxwell 1982); general-purpose satellites such as LANDSAT and SPOT; (e.g. Millington et al. 1987); shuttle imaging radar (e.g. McCauley et al. 1986); Meteosat (e.g. Mainguet et al. 1980a); NIMBUS-3, with its high-resolution infrared scanner (e.g. El-Baz 1979); and the large-format camera products carried by some manual flights (Walker 1985).

Satellites provide, in ways that complement aerial photography, images from different and selected parts of the electromagnetic spectrum, repetitive images, and images that are either in hard copy form or are on digital tapes that can be analyzed and enhanced in many different ways on computers. They are particularly useful in deserts (ERIM 1982, Jones 1987) because there is usually a relatively low proportion of cloud cover, the lack of vegetation reveals landforms in all their detail, and areas of wetness are usually clearly portrayed. Such imagery allows large areas to be mapped, and it can reveal previously unrecognized large-scale features such as former drainage systems in the eastern Sahara (McCauley et al. 1986), sand drifts (e.g. Jones et al. 1986) and structural lineaments (e.g. El-Baz & Warner 1979).

While their perspectives are being broadened, geomorphologists have also been acquiring new and much more powerful tools for investigating deserts. There are many such tools: the electron microscope for investigating particle surfaces and fabrics for diagnostic signs of their history; much more effective sensors to monitor flows of air, water and sediment; controlled hardware models of surfaces, soils and flows in the laboratory; and more accurate and longer-period dating. However, one of the new tools, the computer simulation model, is having an especially large impact. Desert geomorphological and soil systems are turning out to be much more complex than the pioneers had ever supposed. The mathematical model can piece together what is known of the processes and point to the gaps in knowledge. Slopes, soils, streams and dunes have now all been modelled in this way, to the extent that this kind of model is becoming the essential core of geomorphological research.

3.2.4 The search for synthesis

Just as some early desert exploration was inspired by the hunt for lost oases and mineral resources, many of them little more than mirages, so desert geomorphologists have sought, often unsuccessfully, to discover simple, comprehensive explanatory models of desert landforms. The better known of these models still provide stimulating reading, although they have been largely superseded by more complex ideas arising from more detailed data.

One group of ideas focused on the view that desert landscapes were produced largely by mechanical weathering and wind action (e.g. A. Penck 1905a, Passarge 1904, Walther 1924, Keyes 1912). The "aeolianists" believed that field observations in arid and semi-arid areas demonstrated "beyond all shadow of doubt that as a denuding, transportive, and depositional power, the wind is not only fully competent to perform such work, but that it is comparable in every way to water action in a moist climate" (Keyes 1912: 541). A cornerstone of the aeolianist model was the concept of desert-levelling or plain formation, by wind action, a land-lowering process that was restricted only by groundwater tables.

The aeolianists' views carried conviction for some years, but their acceptance rapidly declined in the face of evidence, mainly from North America, that fluvial activity was of vital importance in deserts. The weakness of their case lay in the failure to demonstrate unequivocally the formation of extensive wind-eroded plains. However, the aeolianist ideas are not dead. Recently, satellite imagery has revealed wind-eroded surfaces, including giant grooves (Ch. 21.3.2c) to be much more widespread than had been previously thought (Mainguet et al. 1974); modern studies of dust deflation by wind unequivocally show it to be very extensive (Ch. 22.1); analysis of ocean-core sediment reveals that dust storms were more common in Pleistocene dry phases; and an apparent example of a base-level plain of deflation controlled by the groundwater table has been discovered in Oman (Gardner 1988). So the pendulum of evidence is again swinging, slightly, towards the aeolianist view.

However, it has a long way to go before it re-establishes a new balance between the perceived rôles of wind and water in deserts. The evidence for fluvial activity is compelling and pervasive, at least in the semi-arid deserts. An unequivocal fluvialist, W. M. Davis (1905, 1980) was among the first to codify a model of landform evolution in arid climates (Fig. 3.1). Davis's views were persuasively extended and embellished by disciples such as Cotton (1947). Although Davis and Cotton did not ignore the action of wind, they believed that fluvial processes domin-

ated landform evolution in what they considered to be the "climatic accident" of aridity: "the agent chiefly responsible for the development of the major features of most desert landscapes by erosional sculpture and by transportation and deposition of waste is flowing water, which acts in association with the disintegration by weathering of the material of slopes and of all salient features, and with the downhill gravitational transfer of the resulting debris. . . . only some minor features of the desert surface can be explained as resulting directly from the effects of erosion by wind" (Cotton 1947: 11).

The aeolianists and the fluvialists might expect to provide convincing models within the exclusively acolian and fluvial domains respectively but, as discussed above, the areas of such domains are relatively small compared with the enormous areas within deserts both where they co-exist today and where they have alternated through time. The dilemma of the extremists is neatly illustrated by the case of the large enclosed basins in many deserts, for which the only acceptable model must be one of interaction, over long periods of time of aeolian, fluvial and, probably, tectonic activity (Ch. 21.3.4).

Another model that attempts comprehensively to describe the nature and evolution of desert landscapes is the continental European view of "climatic geomorphology" in which associations are sought between climate, landforms and landform evolution, and which recognizes distinctive evolutionary sequences or "cycles" in arid lands (e.g. Tricart & Cailleux 1960, 1989, Birot 1968). Furthermore, distinctive evolutionary models have been developed for semi-arid areas by Cotton (e.g. 1947) and Birot (1968); King's (1962) "standard epigene cycle of erosion" drew its initial inspiration from semi-arid lands, and owes much to the earlier ideas of Walther Penck (e.g. 1953). Beyond the semi-arid environment, "savanna cycles" have been formulated by Cotton (1947) and Büdel (1957).

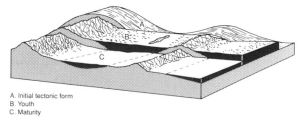

A. Initial tectonic form
B. Youth
C. Maturity

Figure 3.1 W. M. Davis's arid cycle in a mountainous desert (after Cotton 1947).

The grand schemes of the early part of this century are either virtually untestable or fail when tested against the field evidence from more than a few areas. Of necessity, many were based on little more than deductive guesswork, and reflections stimulated by the views from the tops of hills. The main reaction to such schemes has been a new search for empirical evidence on processes, a search that has focused attention on smaller areas of enquiry, thus making broad generalizations rather difficult.

However, the exploration of general models is not a thing of the past; indeed, it is as fresh today as ever it was, as the move to simulation modelling shows. The ultimate and legitimate aim of much desert geomorphology is still the formulation of acceptable generalizations. The huge amount of effort in recent years to understand desert processes and geomorphological systems, which is the basis of this book, has yet to yield comprehensive models, but it is clear from what follows that rapid progress is being made towards understanding desert landforms.

PART 2

DESERT SURFACE CONDITIONS

CHAPTER 4

Introduction

4.1 Topographic settings of desert surfaces

The architecture of desert topography may be conveniently classified into two main types: *shield and platform deserts* and *basin–range deserts*. The former dominate areas of relative tectonic stability. They are principally associated with the vast, stable plainlands of Africa (including much of the Sahara and southern Africa), with much of the Saudi Arabian Peninsula, and with parts of Central Asia, central India and Australia. These areas are dominated by erosional surfaces that are often cut across basement igneous rocks (as in Western Australia and southern Africa), by huge shallow basins (such as the drainage area of Lake Eyre Basin in Australia, and the Lake Chad Basin in Africa), or by extensive surfaces on horizontally bedded and gently warped sediments (as in the Colorado Plateau, Central Asia, or the Nubian sandstone areas of North Africa). The plains, whose origin may owe little to contemporary aridity, are occasionally surmounted by isolated hills or mountains that are often associated with more resistant rocks or the eroded rumps of former mountains (Fig. 4.1a).

In places, such broad plains may be entrenched with canyons, providing a distinctive mixture of plateaux and valley-side escarpments, as in the canyon country of the Colorado Plateau, and in northern Namibia (Fig. 4.1b).

Basin–range deserts are dominated by alternating mountains and basins, and often by enclosed drainage systems (Fig. 4.1c). They are most common in areas

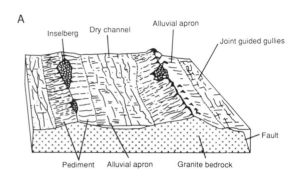

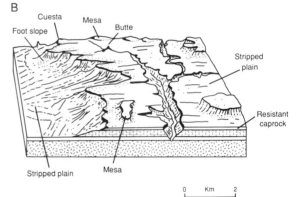

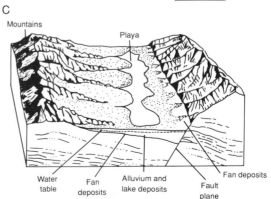

Figure 4.1 Characteristic fluvial dryland landscapes. (A) Inselberg and pediment landscape; (B) canyon and escarpment country; (C) basin–range topography with alluvial fans and playas (after Goudie & Wilkinson 1978).

18

of tectonic activity, such as the southwestern USA, the dry littorals of Chile and Peru, the majority of Iran, Afghanistan and Pakistan, and parts of the deserts of central Asia. In a sense, it is possible to imagine that in basin–range deserts characterized by enclosed drainage "une charactéristique de la morphogenèse désertique est l'ensevelissement du relief sous ses propres débris" (Dresch 1982: 43).

Both shield and basin–range deserts include mountains and plains, although their relative proportions differ greatly. In basin–range country, the mountain:plain ratio may approach one, whereas in shield deserts it may approach infinity. Typically, the mountains are characterized by steep-sided valleys and they are locally terminated by relatively steep slopes at the mountain/piedmont junction (the **piedmont angle**; *angle du piémont*). Isolated mountains are called **inselbergs**. Typically, too, the piedmont plain may be subdivided into different and distinctive landforms, the most important of which are plains cut in bedrock (**pediments**) and plains formed of alluvium

(**alluvial plains**) (Fig. 4.2). The latter may include **bajadas** (**tapis alluvial**: alluvial zones, often of considerable stratigraphic and topographic complexity), **alluvial fans** and **base-level plains** (such as river-valley floors or **playas**). In addition, aeolian deposits in the form of **sand-seas** (**ergs**) occur on the plains, usually in topographic lows.

Much of the diversity of desert landforms lies in the infinitely varied ways in which these different features are combined. Figure 4.2 provides some examples. The distribution and relative spatial importance of these fundamental landform types has not yet been consistently or quantitatively described throughout the desert realm, but the very approximate data in Table 4.1 suggest some of the regional differences.

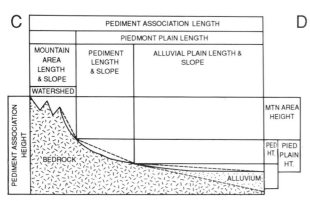

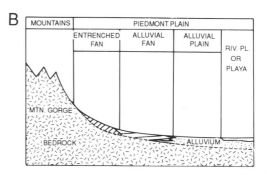

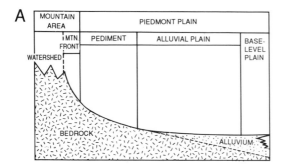

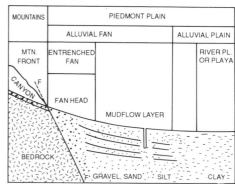

Figure 4.2 The main components of desert topography. (A) The basic mountains and piedmont plain model; (B) The main morphometric characteristics; (C) Variation 1 – alluvial fans and buried piedmont; (D) Variation 2 – alluvial plains, no pediment and a faulted mountain front.

Table 4.1

(a) Proportions of landform types in deserts.

Landform type	Southwestern USA	Sahara	Libyan Desert	Arabia
Desert mountains	38.1	43	39	47
Playas (base-level plains)	1.1	1	1	1
Desert flats	20.5	10	18	16
Bedrock fields (including hamadas)	0.7	10	6	1
Regions bordering throughflowing rivers	1.2	1	3	1
Dry washes	3.6	1	1	1
Fans and bajadas	31.4	1	1	4
Dunes	0.6	28	22	26
Badlands and subdued badlands	2.6	2	8	1
Volcanic cones and fields	0.2	3	1	2
	100.0	100.0	100.0	100.0

Source: Clements et al. (1957). Figures are percentages of area.

(b) Extent of desert landform types in Australia.

	% of arid zone
Mountain and piedmont desert	17.5
Riverine desert	4.0
Shield desert	22.5
Desert clay plains	13.0
Stony desert	12.0
Sand desert	31.0
	100.0

Source: Mabbutt (1971).

4.2 The nature and pattern of desert surface conditions

The nature of desert surfaces is determined by several factors and processes. Pre-eminent among the factors is *position* within the topographic model described above. Beyond that, the character of the surface material (e.g. bedrock or alluvium) and soils is fundamental. Petrov (1976) recognized this in his classification of Asian deserts, in which he differentiated desert surfaces within climatically and structurally defined regions in terms of ten **litho-edaphic types**:

(a) sand, on loose deposits of ancient alluvial plains;
(b) sand–pebble and pebble, on gypsiferous Tertiary and Cretaceous structural plateaux and piedmont plains;
(c) gravel–gypsiferous, on Tertiary plateaux;
(d) gravel, on piedmont plains;
(e) stony, on low mountains and hillocks;
(f) loamy, on weakly carbonate black loam;
(g) loess, on piedmont plains;
(h) clayey takyrs, on piedmont plains and in ancient river deltas;
(i) clayey badlands, on low mountains made up of saliferous marl and clay of different age;
(j) solonchak, in saline depressions and along coasts.

From these categories he created a classification based on the combination of litho-edaphic types with topographic locations (Table 4.2).

In fact, the relations among surface material, surface form and topographic position are immensely variable, and comprehensive generalizations are rather difficult because so many of the categories overlapping. Thus, in Table 4.2, for example, sand–pebble deserts may

Table 4.2 Classification of Asian deserts based on "litho-edaphic types".

1. Typical sand deserts on desert-sand **serozems** and grey-brown soils of ancient alluvial and coastal plains

2. Sand-pebble deserts (**gobi**) on gypsiferous grey-brown soil of Tertiary and Cretaceous plateaux

3. Gypsiferous gravel deserts on grey-brown soils of Tertiary plateaux

4. Gravel deserts on piedmont plains

5. Stony and gravel deserts on low mountains and hillocks

6. Loamy desert on weakly carbonate soils of stratified plateaux and plains of northern deserts

7. Clay-loess desert on **serozems** on piedmont plains

8. Takyr (clay-pan) desert in piedmont plains and ancient river deltas

9. Badland deserts

10. **Solonchak** (salt soil) deserts of salt-lake depressions and sea coasts

Source: Petrov (1976).

be associated with gypsum in Asia, but elsewhere gypsum may be replaced by, or co-exist with, calcium carbonate.

The processes and features of desert surfaces are ultimately dictated by climate, geology and the ages of the surfaces, as Petrov (1976, Table 4.2) recognized. Beyond that, it may be useful to attempt to summarize the diversity of desert processes and features with respect to their locations in the mountain–piedmont transect in the form of a speculative, predictive model (Fig. 4.3). The local differentiation predicted in this model is determined by many considerations such as the occurrence and weathering susceptibility of rocks and alluvium, and the fact that the size of debris normally decreases down slope, as does slope inclination. Similarly, surface-water discharge, infiltration capacity and surface wetness are all spatially variable, and related in part to the nature of surface debris, slope and soil development. The nature of such local differentiation is explored fully in the following sections.

The processes at work on desert surfaces may be classified into three main groups: soil-forming and weathering processes, fluvial processes and aeolian processes. These groups are considered in Parts 2,3, and 4 respectively. In the context of soil-forming and weathering processes, there are four main groups of interrelated processes, each of which is associated with distinctive surface features (see Ch. 6): rock disintegration and weathering, soil development and duricrust formation, surface particle sorting and concentration, and volume change and related patterned-ground phenomena. Although the features created by these processes occur in many different deserts, it cannot be assumed that the occurrence of the same land-surface feature in different areas means that the same processes were responsible for it. Meckelein's (1965, 1974) studies of geomorphological similarities between polar and hot deserts, for example, makes it clear that such an assumption is unjustified. Different processes or combinations of processes may create similar features in different areas from different initial conditions.

Surface *stability* is fundamental to the interplay between processes and factors in determining the character of desert surfaces, and surface stability is spatially varied. Areas of natural relative stability tend to be those where slopes are gentlest, where rock exposures are resistant to weathering, and where surfaces are characterized by crusts or stone pavements. Soil development may be both a consequence and a cause of stability. Soil development is possible only on relatively stable surfaces, but some soil-forming processes may tend to reinforce that stability. For example, weathering crusts (such as carbonate or sulphate crusts produced in semi-arid or arid soils respectively, or silcretes, ferricretes and other crusts that are often inherited from earlier different climates), are commonly surface-stabilizing features. In many deserts, soils are often relatively primitive in terms of their profile differentiation and depth, compared with more humid areas. In part this may reflect the fact that the surfaces on which they are forming are unstable or young; but it may also be an expression of the relatively slow pace of pedogenesis in deserts where organic matter and water are relatively rare. Nevertheless, as Peterson (1981) has explained, the mountain–plain model (Fig. 4.2) provides an excellent basis for identifying units for soil mapping, and is capable of considerable elaboration to accommodate specific pedological circumstances.

In contemporary deserts the stability of many desert surfaces is increasingly disrupted by human processes of ground-surface disturbance, sometimes on a huge scale. Perhaps of greatest concern is the widespread damage caused by wheeled vehicles, agricultural practices, overgrazing, civil-engineering construction, and the concentration of human activity on the desert fringes of settlements. The patterns of these processes are markedly different from those suggested in Figure 4.3, although the disturbance is undoubtedly concentrated on piedmont plains.

The fact that desert surface conditions are, nevertheless, to a considerable extent predictable, in terms of the model shown in Figure 4.3, has substantial practical benefits. For example, Fookes & Knill (1969) showed that both engineering soil types and other geotechnical features are systematically related to the main mountain–piedmont zones (Fig. 4.4).

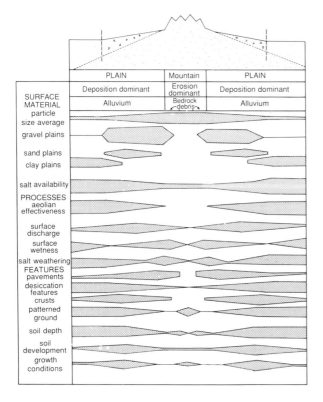

Figure 4.3 A characteristic mountain–piedmont profile, and the common relationships of various surface conditions to it. This is a speculative rather than an authoritative summary diagram. The maximum development of all features is shown by the same bandwidth.

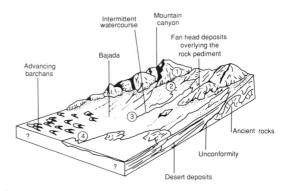

ZONE	1	2	3	4
PRINCIPAL ENGINEERING SOIL TYPES		Rock fans	Silty stony desert and sandy stony desert. Some evaporites	Sand dune loess and evaporites
SLOPE ANGLE OF DESERT SURFACE		2-12°	0.5-2°	0-0.5°
PRINCIPAL TRANSPORTING AGENT OF THE ENVIRONMENT		Gravity and as 3	Intermittent stream flow and sheet floods	Wind and evaporation
GEOTECHNICAL FEATURES		Good for foundation and fill	Generally very good foundation and fill material	Erratic behaviour to load bearing. Migrating dunes Metastable loess. Saline Absence of coarse material

Figure 4.4 Some relations between geomorphological zones of the mountain–piedmont model and features of engineering significance (after Fookes & Knill 1969).

CHAPTER 5

Weathering forms and processes

5.1 Weathering forms

Weathering forms in deserts – many, distinctive and varied – are associated with both bedrock outcrops and surface debris. The forms themselves are often not diagnostic of the processes that formed them, despite the fact that the assumption of process from form is very common. It simply is not yet possible, for instance, to see a split rock and to deduce accurately that it was produced by a specific process: there are almost always several possible explanations.

Desert weathering has two pre-eminent characteristics: *superficiality* and *selectivity*. Its superficiality is shown by the shallowness of soil profiles and weathering mantles, and the preponderance of surface crusts and patinas. It results largely from the fact that the zone of water penetration and temperature change tends to be shallow in deserts. The selectivity of desert weathering is demonstrated by the localization of such features as weathering pits and caverns. Selectivity results mainly from two causes. First, as bedrock is commonly exposed, structural and lithological properties of rocks strongly control the effectiveness of superficial weathering processes. In particular, joints, faults, crystal boundaries and other lines of weakness may be exposed, and they tend to be the loci of greatest weathering. Secondly, extreme local variations of temperature and humidity across uneven surfaces are an important cause of selective weathering. For instance, shaded crevices may be weathered more rapidly than exposed, dry surfaces.

Rock outcrops and superficial debris are commonly affected by **flaking**, **spalling**, **splitting** and **granular disintegration**. Flaking, spalling and splitting (the *Kernsprung* of German authors) all involve the loosening of fragments from parent particles, often without any apparent chemical alteration. Examination of split, flake and spall phenomena affecting superficial boulders in Chile and California (Cooke 1970b) suggested that they occur on most rock types, but less

commonly on coarse-grained igneous rocks; the particles affected vary in their size and shape, and in the extent of their immersion in finer material; and the orientation of split planes to the surface, the width of spaces between splits and the number of splits also vary greatly. A distinctive form of splitting, with the cleavage of a fragment along two or more subparallel fractures roughly normal to the ground surface, sometimes known as "sliced-bread weathering", is illustrated in Figure 5.1. **Split rocks** have been described in most deserts. Sometimes salt crystals or calcrete are present along split planes (Yaalon 1970). Debris resulting from spalling, splitting and flaking is characteristically angular and coarse, and it often provides a large proportion of the detritus on the desert surface.

A number of mechanical weathering processes have been invoked by various authors to explain these phenomena, of which insolation weathering (contraction cracking on cooling, Evans 1971) and salt weathering are the most common (e.g. Peterson 1980a,b). Chemical weathering processes have also been impli-

Figure 5.1 Parallel splitting of a pebble in the salty zone of an alluvial fan in Death Valley, California.

cated. One point should be emphasized. The splitting of rocks, especially those embedded in superficial detritus, may involve *two* phases: the establishment of the split, and its enlargement; the same process is not necessarily responsible for both phases.

The **small-scale exfoliation** of individual silicate-rock boulders, sometimes called "onion-skin weathering", is a form of rock splitting. It was qualitatively analyzed in a classic paper by Blackwelder (1925). He argued convincingly that the peeling of skins involved expansion and that theoretically this could be effected by temperature changes (caused by fire, lightning or insolation), expansion of foreign substances (e.g. ice, salt and rootlets), or chemical alterations (e.g. absorption of water by colloids; hydration or oxidation of silicates). He favoured an hypothesis that involved the establishment of cracks (by undefined processes) that may act as channels for percolating water. Chemical alteration is then effective along the cracks, even when the exposed surface of the boulder is dry and, as a result, the undersurface of the plate expands more than the exposed surface, so that the plate is warped. The cracks might be established along boundaries between concentric zones of rock surrounding dense cores which solidified at an early stage in the consolidation of the magma (Schattner 1961).

In non-siliceous rocks an alternative explanation of small-scale exfoliation to that of Blackwelder is required. LaMarche (1967) offered one hypothesis. He observed a contrast between the angular weathering of unmetamorphosed dolomite and the spheroidal weathering of metamorphosed dolomite marble in the White Mountains, California. The unmetamorphosed dolomite weathers to polycrystalline fragments and fine debris, whereas the marble weathers by granular disintegration into the constituent mineral grains of the rock. The reason suggested for this contrast is that the marble has developed intergranular stresses as the result of the large drop in temperature since crystallization. In addition, stress relief, for example by fracturing along grain boundaries, may promote corrosion.

Exfoliation of spheroidally weathered rocks may depend on the fact that porosity increases near the margins of the marble blocks because of intergranular corrosion. An expansion mechanism related to the porosity differences would explain concentric fractures. In the White Mountains, repeated freezing and thawing of pore water provides such a mechanism. Some small-scale exfoliation, for example on sandstones, may arise from microbiological processes (Ch. 6.2.5).

Exfoliation can also occur at a *large scale*, perhaps involving rock sheets several metres thick and curved failure planes. This phenomenon is not the result of weathering processes, and it is not confined to deserts, but it is particularly common in granite. It has been attributed to expansion by pressure release along joint planes in granite, following erosion of superincumbent rocks, and to the creation of curved joint planes under conditions of radial compressive stress.

Exfoliation may play a rôle in the development of natural sandstone arches, a distinctive feature in southeast Utah (Blair 1987). Here, a sandstone rock **fin** is a prerequisite for arch formation, and is formed where joints are closely spaced, or the retreat of canyon walls removes sufficient interchannel volume of rock. The fin may develop an arch where sandstones are horizontally bedded and vertically jointed, there are some less competent horizons, weathering continuously weakens surface materials and there is systematic release of stress through exfoliation.

Granular disintegration – the breakdown of rocks into individual crystals – is chiefly found in coarse-textured igneous and metamorphic rocks. Figure 5.2 shows an example where a rounded, granitic boulder has been disintegrated until it is nearly flush with the desert surface. Such weathering has been attributed to both mechanical and chemical processes, and different explanations may apply in different circumstances. Granular disintegration usually seems to affect rocks in which at least some minerals have been chemically altered.

Two widespread weathering phenomena in deserts that have attracted considerable attention are **tafoni**

Figure 5.2 A granite boulder, partly weathered by granular disintegration on its exposed surface in the Atacama Desert, Chile.

(and the associated, but smaller, honeycomb or alveolar weathering) and **gnammas** (weathering pits). Neither is exclusively a desert phenomenon.

5.1.1 Tafoni and alveoles

Tafoni are cavernous weathering features that typically have volumes of several cubic metres, arch-shaped entrances, concave inner walls, overhanging margins (**visors**), and fairly smooth, gently sloping, debris-covered floors. They usually occur in groups, prompting an analogy between the rock outcrops and Swiss cheese. They are common in many different climates and coastal environments (e.g. Martini 1978), and they have been described in several deserts including the Mojave (Blackwelder 1929), the Sonoran Desert (Bryan 1925), the Colorado Plateau (Mustoe 1983), Antarctica (Calkin & Cailleux 1962), the Atacama (Grenier 1968), the Sahara (Cote 1957, Smith 1978), Saudi Arabia (Chapman 1980) and South Australia (Dragovich 1969).

Several generalizations about tafoni appear to be justified. First, they occur in many different rock types. (Examples in the literature include argillite, porphyry, rhyolite, conglomerate, lava, limestone, sandstone, gneiss and granite.) Secondly, their formation includes at least two phases: disintegration of the walls, and removal of debris from the floors. Thirdly, once they are established, the air and surfaces of the shaded caverns are often cooler and moister, and the ranges of temperature and relative humidity tend to be smaller than those outside, so that the caverns may be loci of relatively rapid weathering. Fourthly, cavern walls are generally crumbly, puffy and friable. They are often broken up by flaking or granular disintegration, and surface impregnations of salts are common. Finally, tafoni may be either fossil or active. The former may be preserved by lichen covers, desert varnish, crusts such as dust and gypsum, or hardening. The disintegration of carved names and cave drawings, or loss of a varnished surface, may reveal active tafoni. Most of the published descriptions appear to be of active caverns, although Grenier (1968) recorded "dead" tafoni in Chile, and mentioned similar features in the heart of the Sahara.

The origin of tafoni is a matter of controversy. Often the features appear to have been initiated along pre-existing joints, fractures, bedding planes and other rock structures, in randomly distributed cavities resulting from general surface weathering, or at points of mineralogical weakness, especially where water naturally accumulates. Schattner (1961) suggested that

Figure 5.3 Tafoni in the fanglomerates of Papago Buttes, Tempe, Arizona.

in Sinai some tafoni on granite may have originated by the detachment of "core" boulders. Some authors recognize a distinction between tafoni developed on the underside of boulders and rock sheets ("basal tafoni") and those developed on the exposed slopes of outcrops ("sidewall tafoni") (e.g. Bradley et al. 1978, Smith 1978). In South Australia, Dragovich (1969) established a relationship between basal tafoni and ground level. She suggested that cavern development accompanies "ground-level weathering" in which there is a disparity between the rates of disintegration above and below the surface. Above-surface weathering is more rapid because changes of temperature and moisture conditions are more pronounced. This is especially true on the downslope side of boulders and rock outcrops. Tafoni on breccia outcrops in Tempe, Arizona (Fig. 5.3) are clearly initiated along bedding planes that dip steeply. However, the tafoni appear to develop floors that are approximately parallel to the surface and cut across bedding planes. This feature might also be explained by ground-level weathering, although some of the tafoni are high above the present surface today. In places, the development of caverns at ground level may arise from spring sapping, or from weathering in the capillary fringe above the water table (e.g. Doornkamp et al. 1980).

Many authors consider that the disintegration of tafoni walls proceeds by flaking and granular disintegration. Some consider that the main work is by hydration weathering of rock minerals such as feldspars, promoted perhaps by the relative dampness of the cavity environment. Other processes that have been invoked include wind scouring, insolation weathering, wetting and drying, frost action, burrowing animals, solution of calcareous cements in sandstones, case hardening and salt weathering. All of these are subaerial processes. It is also possible that some tafoni or similar weathering pits are produced by preferential moisture attack beneath the ground surface, and subsequently exhumed (e.g. Twidale & Bourne 1976). Any one or any combination of these processes might be responsible for tafoni in particular circumstances. *Removal* of debris from the floor of tafoni may be by wind, soil creep, rainwash, or channelled flow of water.

Of all the tafoni-forming processes, salt weathering is the most commonly invoked. For example, along the foggy littoral of the Atacama Desert, where salts compose the condensation nuclei of fogs, salt weathering appears to be particularly significant. Rögner (1987) attributed a major rôle to salt crystallization in tafoni formation in Israel. Similarly, Bradley et al. (1978) showed that small-scale flaking in the surfaces of active granitic tafoni in South Australia is associated with high concentrations of halite and gypsum that are probably derived from the rocks themselves (Fig. 5.4). This interpretation was challenged by Winkler (1979) who argued that the outward migration of moisture containing dissolved minerals, leading to case-hardening of the surface and weakened bonding of the material beneath, is a primary cause of tafoni, with salts playing only a secondary or negligible rôle. Mustoe (1983) suggested that in the Capitol Reef, Utah, tafoni occurred mainly at sites where soluble salts had accumulated.

Most observers attribute the salt effect to crystal growth or hydration pressures (see Ch. 6.2). But Young (1987) offered an alternative interpretation, based on observations of tafoni in quartzose sandstones in contrasted arid environments in Australia. She found no evidence for weathering due to crystallization pressures, but strong evidence of solutional etching of quartz grains and clays in soils led her to conclude that caverns evolve by the loosening of quartz grains as voids are enlarged by solution of silica, a process that is accelerated in the presence of sodium chloride.

In places, the outer surfaces of tafoni, especially the visors, may be **case-hardened**. The hardening, which tends to preserve the outer surface while the walls of the cavern are progressively developed, takes various forms. It may, for example, be a reddening akin to desert varnish, or a cementation with silica, calcite, gypsum, hydrated calcium borate (colemanite) or other salts (e.g. Conca & Rossman 1982). But, as

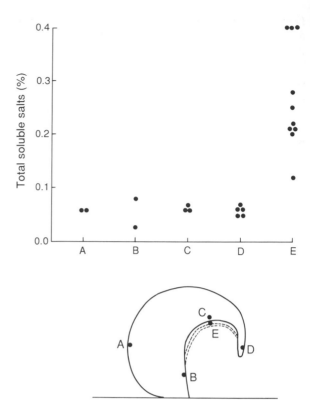

Figure 5.4 Total soluble salts versus sample location in a tafoni, South Australia. (A) The outside of the boulder; (B) the solid (non-flaking) tafoni wall; (C) the massive rock behind the flakes; (D) the visor; (E) tafoni flakes (sample with 0.12% salts graded into massive rock with 0.07% salts) (after Bradley et al. 1978).

Dragovich (1969) pointed out, case-hardening is not a prerequisite for tafoni formation. Occasionally, too, the *inner* surfaces of tafoni are encrusted, for example with gypsum (e.g. Mustoe 1983).

Two variants on the general tafoni form deserve mention. Flared lower slopes of some granite inselbergs in South Australia and elsewhere are somewhat similar features, and indeed, they merge with tafoni in places (e.g. Twidale 1982a). Such steepened, sometimes overhanging, slopes which sweep smoothly upwards and outwards, are occasionally found beneath the alluvial mantles fringing the inselbergs. The flaring is best explained by subsurface weathering around the margins of inselbergs and at the piedmont junction, where runoff is concentrated, followed by erosion of debris to expose the slopes.

Alveoles are a second, extremely widespread variant

(Fig. 5.5). They are small hollows (e.g. *c.* 5–50cm in diameter) that occur in clusters in apparently homogeneous rocks to create a "honeycomb" appearance. The hollows are morphologically similar to small tafoni, and occasionally the two forms occur together, although this does not imply a common origin (Mustoe 1982). Alveoles do have many of the general characteristics of tafoni. For example, they occur widely in different climates; they are common in coastal locations; they occur in many different rock types; and they may be recently formed or fossil. The origin of alveoles is similarly controversial, but there is also a convergence in the literature towards a recognition of the importance of salt weathering, particularly because the process causes the disaggregation of mineral grains without chemical decomposition (Mustoe 1982). Individual alveoles are

Figure 5.6 Weathering pits with flared rims in limestone near Wupatki, Arizona.

commonly separated by thin partitions which may be relatively resistant to weathering if they are strengthened by "case-hardening" features (such as exuded precipitates) or protected by algal growths. Other processes invoked to explain alveoles include wind erosion, frost shattering, exfoliation, and solution of cements.

Weathering pits, which are sometimes known by the Australian aboriginal word *gnammas*, occur widely in relatively bare, horizontal or gently sloping surfaces of granite, sandstone, quartz porphyry and other coherent and massive rock types (Fig. 5.6). They are by no means confined to deserts, but they are most common on unvegetated surfaces. Gnammas may have diameters of up to about 15m, and depths of up to some 4m. Shapes in plan are varied, sometimes reflecting detailed structural patterns. The pits are often elliptical or circular on horizontal surfaces, and asymmetrical on sloping surfaces. In cross section, gnammas are usually flat-floored or semicircular, and they occasionally have overhanging sidewalls. The pits appear to be caused by differential weathering of the rock, possibly along a weathering front which is beneath the surface, or possibly under

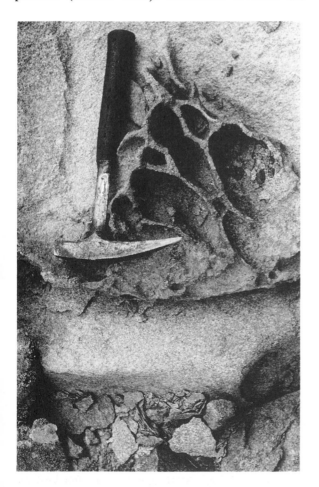

Figure 5.5 Alveolar weathering in granite in the moist, salty coastal zone of the Namib Desert, near Luderitz.

a soil cover or a weathering mantle, and subsequent removal of the resulting debris (e.g. Twidale 1980).

In a study of granite weathering pits in South Australia, Twidale & Corbin (1963) concluded that the depressions were formed by differential disintegration concentrated at points of physical weakness or along joints, by flaking of granite surfaces, or by lichen disintegration. Once established, the depressions become sites of water accumulation, and chemical alteration – especially hydration – becomes important. Material is lost from the pits by periodic flushing or deflation. Overhanging sidewalls might also be due to the fact that water remains longer on the lower part of a pit, so that weathering may continue to extend into the rock at that level (Twidale & Corbin 1963). The possibility of granular disintegration occurring beneath a case-hardened rock surface, especially where the pits flare out beyond a small entrance, has been suggested by Ollier (1984). In a study of weathering pits (*Opferkessel*) in the Navajo Sandstone in the central Colorado Plateau, Schipull (1978) attributed their formation to solution by periodic wetting, and algal and lichen weathering (see also Ch. 6.2.5).

5.2 Weathering processes

Recognition of weathering debris which has apparently been produced without chemical alteration – such as cracked and split rocks, and sharp, angular material – has led to the conclusion that mechanical weathering processes are important in deserts. Indeed, mechanical processes have been said to be more important than chemical and biological processes in arid and semi-arid lands, because the former are enhanced by the direct relationship between rock surfaces and the atmosphere, whereas the latter are reduced by limited plant and soil cover, and low humidities. In this section the main mechanical weathering processes in deserts are reviewed; chemical processes are considered in the context of soil development (Ch. 7).

Before discussing each process, six introductory observations are justified. First, desert weathering processes are likely to be distinctive because of the distinctive features of the daily and seasonal march of temperature and relative humidity. Secondly, and contrary to popular belief, moisture for weathering is widely available – from rainfall, dew and fogs (especially in the deserts of dry western littorals); and in certain locations, such as valley floors and caverns (e.g. Goudie 1989). Thirdly, relative humidity is often high at night, a fact relevant to several weathering processes. Dew is common. Fourthly, although it is difficult, as elsewhere, to determine unequivocally the mixture of processes responsible for particular forms, and physical processes are probably significantly more important than elsewhere, the rôle of chemical processes should not be ignored. Some mechanical processes, such as salt weathering, are chemically based. Fifthly, and despite the emphasis commonly placed on distinctive weathering landforms in deserts, the most important rôle of weathering is certainly to provide debris for the fluvial and aeolian systems. Sixthly, some of the debris and weathering features seen today may well be inherited from different climatic conditions in the past.

5.2.1 Insolation weathering

The concept of "insolation weathering" involves the idea that rocks and minerals are ruptured as the result of large diurnal and/or seasonal temperature changes, and steep temperature gradients from the surface into the rock. Such changes are said to cause differential expansion and contraction of rocks and minerals, depending on coefficients of thermal expansion and other properties of the affected materials, that can lead to fracture.

The idea is an old one, often defended by reports of fusillades of heard but usually unseen rocks cracking in the desert night (e.g. Whitaker 1974). It was particularly cherished because it evidently provided a mechanism for initiating a split, whereas many other processes were thought to be able only to exploit established lines of weakness. Theoretically, it provided a mechanism of instant rupture (although stresses may be cumulative). It had the appeal of simplicity and an apparent sympathy with the desert environment, where diurnal temperature ranges are commonly high.

For many years, evidence has apparently accumulated that discredits the process, although few have paid great attention to the key groups of variables that probably determine its efficacy: the magnitude, frequency and rate of diurnal and seasonal temperature change; the coefficients of rock and mineral expansion; the size and nature of rocks; and rock "spalling resistance". The crucial evidence has for decades been the laboratory studies of Blackwelder (1925, 1926, 1933), Griggs (1936) and Birot (1962), all of whom concluded that temperature change by itself is unlikely to cause cracking. But this prevailing view, already opposed by some (e.g. Ollier 1984), seems set for reappraisal, and the process for rehabilitation.

In the first place, the experimental evidence is open to question. Aires-Barros et al. (1975), for example, showed in a set of laboratory experiments that, while "dry" weathering under conditions of controlled "insolation" produced much less weathering than when water and saline solutions were present, weathering did in fact occur under "dry" conditions. Microfissures were created, enlarged and polished mineral surfaces became pitted and clouded, and weight was lost. They believed that the humidity of the air under "dry conditions" was sufficient to cause the damage.

Again, Griggs's work was based only on *small* rock samples that were *unconfined*. Rice (1976) showed that the larger the sample, the larger the thermal stress and the greater the likelihood of insolation weathering. The tendency for *spalling* he described as follows:

$$kx \left[\frac{\alpha}{\sqrt{k} \, \epsilon_S} \right] \propto 1/S \qquad (5.1)$$

where x is some characteristic of dimension, ϵ_S = the breaking strain, k contains the temperature difference and the time elapsed since that difference was applied and S is the "spalling resistance index", equal to $k\sigma_S/\alpha\epsilon$, (where k is thermal diffusivity, σ_S is the tensile strength, α coefficient of thermal expansion, and ϵ is the modulus of elasticity).

According to Rice, it is quite possible for insolation to aid weathering processes in, say, granite, provided that x is large and the time elapsed is long and/or the frequency of "quenching" (i.e. rapid temperature reduction) is high. This view was reinforced by Garner (1976), who pointed out that the linear expansion of gneiss outcrops c. 100m long in Brazil under ambient temperature conditions was calculated many years ago to be considerable (Table 5.1) and sufficient to cause exfoliation. For convenience, this process can be called **dry insolation weathering**.

In addition, Rice (1976) and Bauer & Johnson (1979) argued that "dry" insolation can induce **microcracking**, even in small rocks, thus increasing their permeability. This paves the way for exploitation by other processes, as suggested by Birot (1962) and Whalley et al. (1982a). Winkler & Rice (1977), following the lead of Birot (e.g. 1962), maintained that, when moisture is associated with high diurnal

Table 5.1 Estimated linear expansion of a gneiss surface in Brazil.

Depth (m)	Temperature change (°C)	Expansion (mm)
Surface	57.0	47.09
1	8.3	6.80
2	5.6	4.57
3	3.0	3.20
4	2.8	2.28
5	2.2	1.82

Source: Garner (1976).

temperature changes, water entrapped in rock capillaries may (because of its high coefficient of expansion) generate sufficiently high internal pressures eventually to fracture rock under a temperature range of 10–50°C. This process of **moisture insolation weathering** depends on rock pore structure, temperature range and moisture availability. Beyond this, there is general agreement that the presence of moisture in association with high temperature ranges can accelerate weathering (e.g. Griggs 1936, Aires-Barros et al. 1975).

There have been several recent studies of desert air/surface/rock temperatures. In general, daily *air* temperature ranges are often high in deserts, up to as much as 54°C (in the Atacama Desert, Keller 1946). But in the context of rock weathering, surface and subsurface rock temperatures are fundamental. Figure 5.7 shows some examples of daily ranges in different rock types, and their changes with depth. The daily surface temperature range is often greater than that in the air above or in the rock beneath. In Figure 5.7, the ranges are up to 40°C, and the maximum recorded temperature is over 80°C.

Several points deserve emphasis from these and other similar data. First, temperature *gradients* may also be steep (in the order of $1°C \, cm^{-1}$ here, but as much as $7°C^{-1}$ has been recorded elsewhere (e.g. Ravina & Zaslavsky 1974)). Time-lags in temperature change with depth. The effects of rain showers may steepen these gradients and significantly influence stress patterns (Peel 1974). Secondly, seasonal and aspect differences may be important. For example, Smith (1977) argued that mechanical weathering on the limestone hamadas of the Sahara is likely to be greatest where moisture is available and tends to be retained, but temperature ranges and gradients are high: that is to say, on east- and west-facing slopes

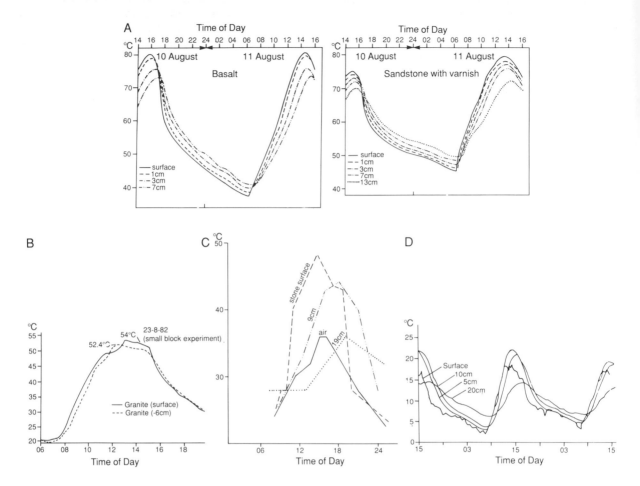

Figure 5.7 Temperature ranges and daily cycles in deserts. (A) Examples from Wour, Tibesti: basalt has a higher thermal conductivity and shows the largest range (after Peel 1974). (B) Curves for granite in southeast Morocco (after Kerr et al. 1984). (C) On and in a quartz monzonite boulder in the Mojave Desert, California (after Roth, from Winkler & Rice 1977). (D) An example from a limestone boulder in Tunisia (after Smith 1977).

below cliffs. Thirdly, the fabric of a rock may be weakened by many small volume changes over protracted periods, rather than by spectacular daily changes, and "rock fatigue" may be important. Fourthly, some rocks such as granite may have considerable initial *internal* stresses caused by crystallization, which are an important supplement to external forces of rock disintegration. Fifthly, rock thermal properties are fundamentally important. Kerr et al. (1984) and McGreevy (1985) emphasized that the variations reflect, *inter alia*, differences in such properties as colour/albedo, specific heat capacity and thermal conductivity. Basalt, for example, shows high

surface-temperature ranges and typically has low albedo, specific heat capacity and thermal conductivity. In terms of actual cracking, coefficients of thermal expansion may be critical, however, and these are lower for basalts than for, say, quartzites, which are therefore likely to be more *affected* by the changes (e.g. Evans 1971).

Finally, not all rises in rock temperature are necessarily the direct result of atmospheric heating. For example, Rögner's (1987) monitoring of Israeli tafoni revealed a temperature rise in the sheltered rear wall of the feature shortly after sunrise that could not be attributed to shadow, insolation or air temperature change. He saw it as a result of salt crystallization from highly concentrated solutions at a time when air temperature was rising and relative humidity was falling.

Fire provides an exceptional opportunity to create unusually high temperature changes in rock and to cause cracking and spalling. Forest fires are in fact quite frequent in many semi-arid environments (e.g. Cooke 1984). Blackwelder (1926) first attributed rock spalling in forested areas of the southwestern USA to this process. Ollier & Ash (1983) demonstrated in Australia, where bush fires are common, that fires adjacent to granodiorite outcrops and boulders caused a variety of features, including spalling, and cracking parallel, radial and tangential to rock surfaces. These features were possibly formed during cooling, and affected granodiorite because it has a relatively high coefficient of thermal expansion. Fire effects also were noted on limestones in semi-arid Western Australia (Spath 1977).

Thus, the evidence in favour of insolation weathering appears to be growing, even if the case for it is not conclusive. The fact is that this evidence confirms the intuition of many desert geomorphologists that at least some cracking and spalling features on certain rocks can reasonably be attributed to insolation effects.

5.2.2 Frost weathering

Frost weathering – the result of pressures exerted by crystal growth or by the volumetric expansion of water when it is frozen in confined spaces – has received little attention in deserts. It was mentioned by Tricart & Cailleux (1989), emphasized by Coque (1962), experimentally investigated by Schumm & Chorley (1966) on sandstones from the Colorado Plateaux, and invoked by Melton (1965a) as a mechanism which produced detritus during colder phases at higher elevations in Arizona.

Laboratory studies of frost weathering (e.g. McGreevy 1981) indicate a need for detailed climatological data on moisture conditions, and the frequency and extent of ground and soil temperature fluctuations about freezing point. Although such data

Figure 5.8 In high-altitude areas, salt weathering may combine with frost action to produce debris rapidly, as here in the Karakoram Mountains, Pakistan.

are rarely available in deserts, three generalizations seem reasonable. First, frost action is present today in those high-altitude deserts where temperatures regularly fall below freezing point, such as much of the Great Basin, the Atlas Mountains in Africa, and the many mountain ranges in Iran, Afghanistan, Mongolia, and throughout Eurasian deserts (Fig. 5.8). Secondly, and especially in such areas, there is a high probability of effective frost action during cooler, moister periods of the Quaternary. This view has, for example, been built into hypotheses to explain sedimentation phases on alluvial fans (Ch. 12). Thirdly, there is some experimental evidence to support the idea that the presence of salts in solution, a common feature of desert weathering environments, may either inhibit (if salt supply is limited) or enhance (if salt supply is plentiful) frost damage (McGreevy 1982), depending on whether or not the minimum rock temperature falls below the appropriate eutectic point (Jerwood et al. 1987).

5.2.3 Wetting and drying weathering

Goudie et al. (1970) showed that wetting and drying of fresh black shale in distilled water caused the rock to disintegrate. Fahey (1986) similarly demonstrated a minor rôle for wetting and drying in the weathering of shales. Ollier (1984) reported unpublished experiments by Condon in which fine-grained rocks such as shale disintegrated by flaking and splitting when repeatedly wetted and dried. A possible explanation of this process is in terms of "ordered-water" molecule pressure. Ollier noted that, as water is a "polar" liquid, with two positively charged hydrogen atoms at one end of a molecule and a negatively charged atom at the other, the positively charged end is attracted to the negatively charged surface of clay or other materials, and other molecules line up to form an ordered-water layer. Wetting and drying may allow the molecules to become increasingly ordered and thence perhaps to exert an expansive force that thrusts apart the confining surfaces. In argillaceous rocks, this adsorption process has been described as a form of hydration weathering (e.g. Fahey 1983).

Desert surfaces may be wetted by rainfall, fog or dew. Wetting by dew may be very important, often allowing a daily wetting and drying cycle. Dew is common in some deserts. For example, the number of days with dewfall in Jordan often exceeds 50 per annum, and falls of over 20mm yr^{-1} have been recorded in several places. In Israel, Yaalon has noted that wetting by dew can extend up to several milli-

metres below a bare surface and up to several centimetres under trees and bushes. Since the dew in Israel is saline, weathering of surface material could be promoted there by both wetting and drying and salt.

A related process was postulated by Ravina & Zaslavsky (1974). They argued that cracking could be produced by pressures arising from the very high electrical gradients in the water of the electrical double layers at rock/water interfaces. Such pressures are said to be most effective where the double layers of different rock surfaces interact, and the resultant cracking may occur in both large rock outcrops and small particles. The pressures are said to vary with relative humidity and diurnal temperature ranges, both of which are often high in deserts.

5.2.4 Salt weathering

The nature and consequences of salt weathering in deserts have been examined extensively in recent years in the field and the laboratory (e.g. Evans 1970, Cooke 1981, Goudie 1985b). Salt weathering undoubtedly comprises a major set of geomorphological processes that are more effective in both hot and cold deserts than in temperate areas (Davidson 1986).

(a) Introduction to salt weathering processes

Rock disintegration by salt may result from either chemical or physical processes. Several *chemical* processes of salt weathering, as Fookes & Collis (1977) have outlined, may result in volume changes that can cause disintegration. For example, chemical exchange between sulphate radicals in solution and calcium aluminium hydrate in cement may produce reactive products of larger volume. Reactions between cement and gypsum may produce expansive portlandite (calcium hydroxide). Accelerated corrosion of ungalvanized steel embedded in concrete and in the presence of chlorides may disrupt the concrete, and chloride–aluminate reactions may also create expansive forces. In addition, as French & Poole (1976) have described, reactions between silica-reactive aggregate and cement paste may produce a gel that can absorb water and swell to many times its volume to initiate cracking. This process may be exacerbated by salts contaminating the aggregate.

Cooke & Smalley (1968) recognized three main *physical* processes by which salts can cause rock disintegration through expansion or growth in confined spaces, especially pores: (a) the growth of salt crystals in confined spaces as a result of evaporation

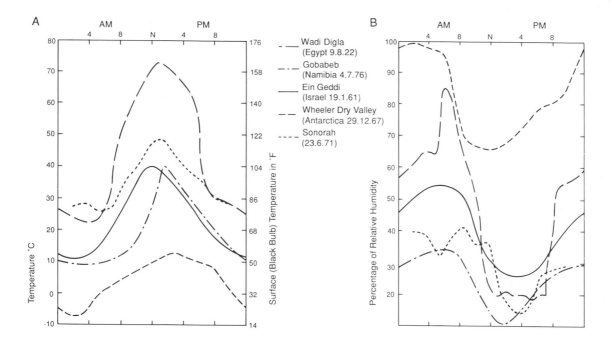

Figure 5.9 Ground-surface temperature and relative humidity conditions at selected desert locations (derived from Williams 1923, Warburg 1964, Cameron 1969, Friedmann 1980).

and/or cooling of saline solutions; (b) change in volume of salt crystals through hydration; and (c) the thermal expansion of salt crystals as a consequence of pronounced daytime heating, especially where the salts have high coefficients of thermal expansion that exceed those of rock minerals.

(b) The salt weathering system

An understanding of these processes requires the components of the weathering system to be analyzed. In its simplest form the system comprises three main groups of variables: the *environmental conditions* (especially climate and ground-surface characteristics); the *materials subject to salt attack*; and *the salts, the solutions and their chemical properties*.

ENVIRONMENTAL CONDITIONS The climatic conditions conducive to salt weathering are pre-eminently those characteristic of hot arid environments, where during the day relative humidities are usually low and temperatures and evaporation rates are high; and where diurnal temperature and relative humidity ranges are also normally high. In such conditions, saline solutions are regularly and frequently subjected to evaporation and temperature changes that may

promote crystal growth; extreme vapour pressure changes are favourable to hydration processes; and salts with high coefficients of thermal expansion may undergo substantial volume changes as a result of high temperature ranges.

However, these characteristic climatic conditions do not prevail uniformly throughout the world's hot deserts. The fundamental climatic properties of temperature and relative humidity vary greatly, especially daily and at or near ground level where most salt weathering occurs (Fig. 5.9). Although it is still difficult to document this climatic variability precisely, there can be little doubt that it is sufficient to influence the spatial variability of salt weathering processes in deserts.

Climate also plays an important rôle in determining the nature of ground-surface conditions that are so fundamental to much salt weathering, for it promotes surface and near-surface evaporation and saline concentration of groundwaters, and capillary migration of water above the water table. The capillary fringe, the zone of saturated and partially saturated material above the water table, varies in thickness in any given environment chiefly with the particle size of the host sediments. In very fine sediments and extreme desert

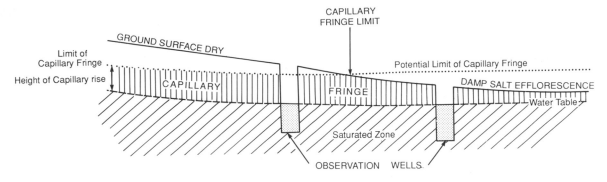

Figure 5.10 The capillary fringe, and related ground-surface and groundwater phenomena.

conditions, it may be over 3m thick (Fig. 5.10). The fringe may be confined below the ground surface, where its upper limit may be marked by crystal growths, often of gypsum to produce desert sand roses. Alternatively, it may intercept the surface to create a zone of salt efflorescence and crusting, and cause salt weathering in features that rise out of the surface (the so-called "wick effect", e.g. Goudie 1986).

MATERIALS In any given set of environmental circumstances, the response of rocks to salt weathering will vary with rock properties. Various experimental studies (e.g. Goudie et al. 1970, Cooke 1979, Whalley et al. 1982a, McGreevy & Smith 1984, Smith et al. 1987) suggest that of these properties, porosity and particularly microporosity (i.e. the proportion of micropores), together with water absorption capacity, saturation coefficient, potential for crack

Figure 5.11 A saline playa in the salt-dome topography of the Cerro de la Sal in the Atacama Desert, Chile.

propagation, and tensile strength are the most important. In addition, such factors as the surface texture, the presence of clay minerals, bedding-plane orientation, the area:volume ratio and shape may all be of significance. Experimental work has shown that igneous and metamorphic rocks are more resistant to salt weathering than are sandstones or limestones (e.g. Goudie 1974).

SALTS At or near desert surfaces, salts may be derived from one or more of several major sources: from the deposition of saline aerosols that have served as condensation nuclei in clouds and are chiefly derived from the oceans (such salts usually show a coastal concentration that declines inland); from dried-out water bodies such as lakes and lagoons; from other atmospheric sources such as dust or volcanic gases; from the exposure of fossil salt deposits by erosion, or by diapiric movements that produce salt plugs (Fig. 5.11); from groundwater, which in turn derives its salt either from rainwater solutions or stratified salt deposits; from subsurface penetration of saline sea water into coastal flats and adjacent areas, sometimes as a result of over-extraction of adjacent freshwater bodies; and, perhaps most importantly, from the chemical weathering of rocks and from the transfer of solutions in flowing water to locations where they are concentrated by evaporation.

As Goudie & Cooke (1984) have shown, the types of salt found in deserts vary greatly from place to place, partly in response to the spatial variability of these processes. For example, in Australia, NaCl is preponderant; in the Chilean salars (Stoertz & Ericksen 1974), NaCl is commonly associated with calcium and sodium sulphates; and in East African salt lakes, nahcolite ($NaHCO_3$) is widespread. The combinations of salts found at any one locality fundamentally influence the local potential for salt weathering.

Normally, salts are disseminated and are only found in small quantities. As Mueller (1968) stated, for example, the mean concentration of nitrate in the desert soils of Chile, the Gobi Desert, the Sahara and the Mojave Desert, is in the order of 0.1–0.2%. What distinguishes the Chilean nitrates from those in other areas is that they are concentrated in certain locations and are somewhat separated from other salts. *Such mechanisms of salt concentration and segregation are fundamental to the creation of loci for salt weathering.*

GLOBAL ZONES OF SALT CONCENTRATION On a global scale, three main environments of salt accum-

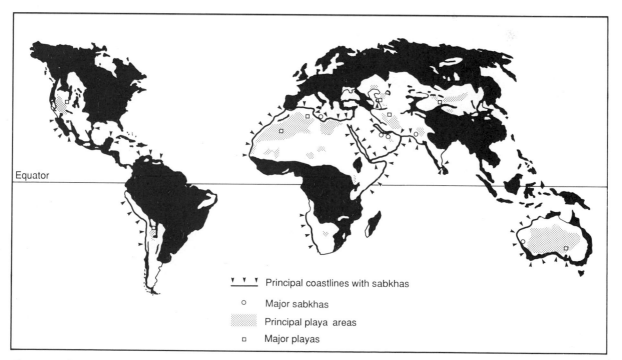

Equator

Principal coastlines with sabkhas

Major sabkhas

Principal playa areas

Major playas

Figure 5.12 Sabkhas and playas: major environments of salt concentration in dry environments (after Cooke 1981).

35

ulation are evident: humid deserts in dry western littorals, playas and sabkhas.

In *the humid deserts of dry western littorals*, including the deserts of Chile, Peru, Namibia and western Australia, ocean salts may accumulate at rates of up to 150kg ha^{-1} yr^{-1}. A coastal origin for inland desert salt accumulations may apply elsewhere, as in the Thar desert, where it was estimated that the annual supply of sea salt exceeded 200,000t (Holland 1911).

Salts also accumulate within and around the margins of the *floors of enclosed drainage basins*, the features known variously as **playas**, **salt lakes**, **chotts**, **pans** and **kavirs** (Fig. 5.12, Ch. 14). The salts in playas may be fossil, they may be washed or blown in, or they may be derived from groundwater discharge. Locally, conditions similar to playas may occur in and adjacent to drainage channels where flow is concentrated and wetting and drying is most common.

In *areas of coastal aggradation*, especially adjacent to relatively land-locked seas of high salinity, such as the Arabian Gulf and the Red Sea, **sabkhas** may form (Ch. 14). These broad intertidal flats are loci of extensive salt accumulation, chiefly as a result of sea water being drawn to the surface by capillary action (e.g. Fookes et al. 1985).

These environments are not only highly saline; they are amongst the most humid and wettest locations in deserts in terms of runoff, sea water and groundwater. Thus, they are all significant regions of salt weathering.

LOCAL SALT CONCENTRATION MECHANISMS   Within these zones, local salt concentration and segregation mechanisms are important in determining spatial variability of salt weathering. The most important of these relates to the *evaporation of solutions*, solutions provided by surface runoff, the sea, the capillary fringe above the water table, or percolating rainwater. In the highly evaporative conditions, *vertical and spatial zonation* of salts may arise because of differences in the solubilities of salts and the mobility of anions in mixed solutions. Thus, one may find annular rings around a basin floor dominated by a central zone occupied by the most soluble chlorides, an intermediate zone of sulphates, and an outer zone of the least soluble carbonates, as, for example, in Death Valley, California (Figs 5.13 & 14) and on Bonneville Salt Flats in Utah (e.g. Hunt 1975, Lines 1979). Similarly, a contrast between crusts of halite surrounded by a rim of sulphates is common in Chilean salt pans (Stoertz & Ericksen 1974).

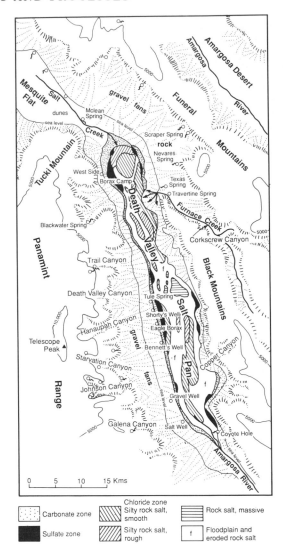

Figure 5.13   The Death Valley salt pan (after Hunt 1975).

Again, in a soil profile, salt concentrations may decline with depth, and sulphates may be deposited higher than the more mobile and more soluble chlorides. Finally, Mueller (1968) convincingly demonstrated that the occurrence of nitrates in commercial concentrations in alluvial slopes away from the floor of the central depression of the Atacama Desert was due to the capillary migration of residual solutions up slope from the saturated zone, and that the vertical and horizontal segregation of salts reflects the relative mobility of anions in solution. Sulphate ions migrate slowest and concentrate near the surface adjacent to

Figure 5.14 Highly weathered alluvial-fan debris in the salty sulphate zone adjacent to the floor of Death Valley, California.

the saturated zone; nitrates occupy an intermediate position; and chlorides migrate farthest and fastest.

The "**salt cycle**" is another local mechanism of salt movement and concentration (Fig. 5.15). Salts are slowly dissolved from rocks and soil, and are then transported haltingly in solutions across the intervening zones to depositional areas, where they are concentrated. The resulting salt crystals are then blown back up slope by the wind. Young & Evans (1986) described an example of this cycle in Nevada.

A. T. Wilson (1979) has suggested an entirely different mechanism of salt segregation in the extremely arid McMurdo region of Antarctica. Here, precipitation falls as snow that does not melt but is lost directly by sublimation, leaving behind salts. The more deliquescent of these, such as $CaCl_2.6H_2O$, percolate deeper into the soil and farther down slope. This fractionation is determined by the relative humidity and deliquescence of salts in the region.

TYPES OF SALTS   From the point of view of salt weathering, a final important consideration is the types and mixtures of salts present in any location. Particularly common are calcium carbonate (especially in semi-arid areas), various forms of calcium sulphate (mainly in arid regions) and sodium chloride (chiefly in coastal locations). Otherwise the occurrence of salts is extremely variable. Nitrates, carbonates, chlorides and sulphates, especially of sodium, magnesium and potassium, are by far the most important. The weathering effectiveness of the salts varies with their chemical properties, such as solubility (Fig. 5.16). Experimental studies (e.g. Goudie et al. 1970, Goudie 1985b) suggest that by far the most effective (in isolation) are sodium sulphate and magnesium sulphate (Table 5.2). Equally, the mixture of salts appears to be important in determining weathering effectiveness. The principal reason for the varying effectiveness of salts lies in their crystallization and hydration properties.

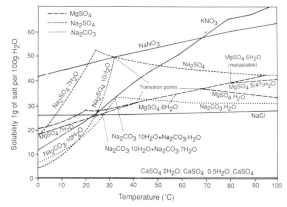

Figure 5.16 Relations between solubility and temperature for some relatively common desert salts (from various sources, after Cooke 1981).

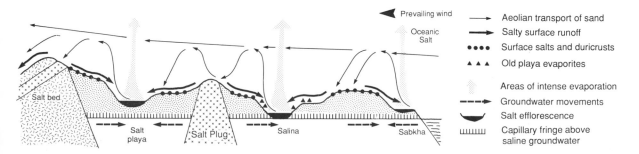

Figure 5.15 The sources, movements and destinations of salt in deserts – a generalized model.

Table 5.2 Experimental ranking of the power of different salts (most effective salt at top of column).

Pedro (1957)	Kwaad (1970)	Goudie, Cooke, Evans (1970)	Goudie (1974)	Goudie (1985)
$NaNO_3$	Na_2SO_4	Na_2SO_4	Na_2SO_4	Na_2CO_3
Na_2SO_4	Na_2CO_3	$MgSO_4$	Na_2CO_3	$MgSO_4$
$Mg(NO_3)_2$	$MgSO_4$	$CaCl_2$	$NaNO_3$	Na_2SO_4
K_2SO_4	$NaCl$	Na_2CO_3	$CaCl_2$	$NaCl$
KNO_3	$CaSO_4$	$NaCl$	$MgSO_4$	$NaNO_3$
Na_2CO_3			$NaCl$	$CaSO_4$
K_2CO_3				
$MgSO_4$				
$CaSO_4$				
$Ca(NO_3)_2$				

Source: Goudie (1985b).

Sodium sulphate, for example, is believed to be a particularly effective agent, because hydration of thenardite (Na_2SO_4) to mirabilite ($Na_2SO_4.10H_2O$) involves volume changes; its solubility is relatively sensitive to temperature change; it is highly soluble, so that substantial quantities are available for crystal growth when saturated solutions are evaporated; and its crystal habits are prismatic or acicular, so that forces of crystal growth are concentrated along one axis and are thus likely to be more effective in promoting disintegration in pores.

(c) Crystal growth

Experimental studies have demonstrated that, of the three main mechanical processes of salt weathering, *crystal growth from solution* is the most important (Goudie 1974, Cooke 1979, Sayward 1984). Crystal growth is promoted by a decrease in solubility as temperature falls, by evaporation of solutions, or by mixing of different salts in solution. Atmospheric humidity is fundamental to the crystal-growth process, because crystallization will occur only when ambient relative humidity is lower than the equilibrium relative humidity of the saturated salt solution (Goudie 1985b), and such equilibrium humidities vary with both salt type and temperature.

Correns (1949) concluded that salt crystals can continue to grow against a confining pressure when a film of solution is maintained at the salt/rock surface. The maintenance of this solution depends on the

interfacial tensions at the salt/rock, salt/solution and solution/rock interfaces. When the sum of the last two is smaller than the first, the solution can penetrate between the salt and the surrounding rock. The pressure exerted by a growing crystal depends mainly on the degree of supersaturation of the solution and may be described by the following equation (e.g. Evans 1970):

$$P = \frac{RT}{V} . \ln \frac{C}{C_S} \qquad (5.2)$$

where P is the pressure on the crystal, R is the gas constant, T is the absolute temperature, V is the molar volume of crystalline salt, ln is the natural logarithm, C_S is the concentration at saturation point without pressure effect, and C is the concentration of a saturated solution under external pressure P.

(d) Hydration

Disruptive stresses may also be exerted by **anhydrous salts**, dehydrated in high desert temperatures, which become *hydrated from time to time*. The presence of both anhydrite and gypsum in deserts suggests that some conversion of one to the other may occur and, in fact, the presence of other salts tends to promote the dehydration reaction. There is certainly a volume change associated with the hydration of $CaSO_4$. When a test-tube is partly filled with plaster of Paris and water and allowed to set, the tube is often broken by the apparent expansion, yet there is an overall diminution in volume of about 7% when the reaction, represented by the equation,

$$2(CaSO_4.1/2H_2O) + 3H_2O = 2(CaSO_4.2H_2O) \qquad (5.3)$$

proceeds to completion. It seems that, although there is a theoretical decrease in volume of 7% during the hardening of a $2(CaSO_4.1/2H_2O)$–H_2O paste, there is in fact a 0.5% expansion (Chatterji & Jeffery 1963).

Thus, emplaced salts which have become even partially dehydrated can be expected to exert a force on the walls of the containing fissure or pore when they are wetted. The various forms of hydrated and anhydrous $CaSO_4$ have widespread occurrence in deserts. The weathering mechanism proposed involves, first, deposition of a hydrated form in a rock fissure or pore. This is followed by slow dehydration in high desert temperatures and in the presence of dehydration-assisting salts. Then a relatively sudden wetting causes rapid hydration, with consequent tensile stresses on the fissure walls.

It is only recently that the efficacy of the hydration process in disintegrating rocks has been demonstrated experimentally. Sperling & Cooke (1985) showed that the hydration of sodium sulphate under laboratory-simulated desert conditions was capable of breaking up rocks independently of, but not as successfully as, the process of crystal growth. Fahey (1986) also came to the conclusion that hydration plays only a minor rôle in the break-up of shales by sodium sulphate under experimental conditions. He showed that salt crystallization and hydration together were no more effective in generating fines (>2mm) from the shale than crystallization alone, but that in combination they showed a greater internal transfer amongst existing size grades (Fig. 5.17) and brought about a 40% reduction in mean size (compared with only a 24% reduction for crystallization acting alone). He attributed this apparent contradiction to two facts: first, that the production of flakes larger than 2mm was due mainly to crystal growth on the surface of the samples; and, secondly, that the penetration of salt solutions into exploitable discontinuities within the shale, where both processes could operate under conditions of confined stress, led to the production of particles of different size and to the size transfers shown in Figure 5.17.

Field evidence of the effect of hydration is very limited indeed, but Smith & McAlister (1986) believed that some hydration salt weathering in the Kenyan Rift Valley might be associated with the presence of thermonatrite ($Na_2CO_3.H_2O$) and kieserite ($MgSO_4.H_2O$).

Hydration forces are certainly potentially important (Kwaad 1970, Winkler & Wilhelm 1970). Pressure created by the hydration of entrapped salts can be repeated many times during a season or perhaps even during a single day. The precise mechanism of anhydrous salt formation and its translation into higher hydrates may be approximately as follows. When soil temperatures during the day are high, capillary-rising salt solutions may yield crystals without, or low in, water of crystallization. Then, during the night, temperatures fall, and the anhydrous salts or lower hydrates absorb water vapour from the atmosphere when the pressure of the atmospheric aqueous vapour exceeds the dissociation pressure of the hydrate involved. Higher hydrates are then formed. The transformation is therefore related to relative humidity and temperature conditions in the atmosphere and to dissociation vapour pressures of salts. In addition, some salts have transition temp-

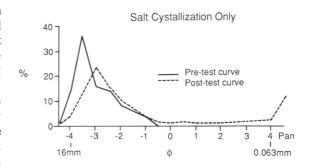

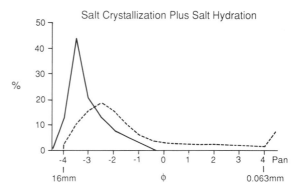

Figure 5.17 Pre- and post-test grain-size frequency distributions for two shale aggregate samples, one subjected to salt crystallization alone, and the other to salt crystallization and hydration (after Fahey 1986).

eratures above which only the anhydrous form or a lower hydrate is stable (Kwaad 1970).

Hydration pressure can be calculated using an equation that takes these and other related variables into account. One equation is that used by Winkler & Wilhelm (1970):

$$P = \frac{(nRT)}{(V_h - V_o)} 2.3 \log \left[\frac{P_w}{P'_w} \right] \qquad (5.4)$$

where P is the hydration pressure in atmospheres; n is the number of moles of water gained during hydration to the next higher hydrate; R is the gas constant; T is the absolute temperature (degrees Kelvin); V_h is the volume of hydrate (molecular wt. hydrated salt) density (gm/cc); V_o is the volume of original (salt molecular wt. orig. salt) density (gm/cc); and P_w is the vapour pressure of water in the atmosphere (mm mercury at given temperature, P'_w is the vapour pressure of hydrated salt (mm mercury at given temperature).

The pressures actually produced depend on the environmental conditions and the salts involved. Thus, for example, the transitional temperature in a pure solution for sodium sulphate (Na_2SO_4, mirabilite) to thenardite ($Na_2SO_4.10H_2O$) is 32.4°C. This transition falls as the proportion of NaCl rises, to 17.9°C in an environment saturated with NaCl (e.g. Obika et al. 1989). In contrast, NaCl does not normally have a hydrated form.

(e) Thermal expansion

The extent to which *salts may expand when they are heated* depends on their thermal characteristics and the temperature ranges to which they may be subjected. Many of the salts that commonly occur in deserts have coefficients of volumetric expansion that are high, and are higher than those of many common rocks, such as granite (Fig. 5.18; Cooke & Smalley 1968). Figure 5.18 shows the thermal expansion characteristics of

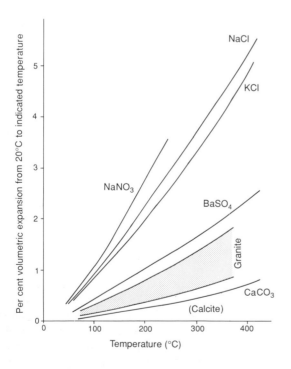

Figure 5.18 The relationship between temperature and the volumetric expansion of various salts and granite (Cooke & Smalley 1968). Johannessen et al. (1982) postulated slightly lower volume changes than those shown here.

five of these salts over a range of about 300°C. The temperature–volume relationships are approximately linear in each case, and it is assumed that this linearity extends to atmospheric temperatures, and that the percentage expansion which the salts undergo in natural conditions can be estimated from this diagram. The potential of salt expansion was also noted by Hunt & Washburn (1966), who stated that the thermal coefficient of linear expansion of NaCl per 1°C at 40°C is 40.4×10^{-6}, which approaches that of ice (47.4×10^{-6}).

The nature of diurnal temperature ranges in deserts has been outlined above. The effect of such temperature changes may be considerable on salts precipitated from solution in crevices and pores in rocks near the surface. The stress-causing system in rock associated with salt in a small crack is efficient in that the stresses caused by thermal expansion of the salt produce tensile stresses in the rock which are concentrated at the tip of the crack, the point at which further failure must occur. If the rock is granite, with its considerable internal stresses caused by crystallization, then the combination of these internal stresses with the stresses caused by salt expansion may produce fracturing. Of course, the cracking produced by salt expansion provides avenues for the further penetration of saline solutions, and the process may continue until the rock disintegrates.

Until recently, the salt-expansion process seemed only to be a theoretical probability (Cooke & Smalley 1968). But empirical observations of cliff weathering under different conditions of insolation (Johannessen et al. 1982) pointed strongly towards its effectiveness. They showed that, in the mid-latitude coastal environment of Oregon, the rate of cliff weathering was at least ten times faster on sunny south-facing cliffs than on shaded north-facing cliffs. This contrast is attributed to the expansion of NaCl on heating in the spray zone on the south-facing cliffs. The most important factor is probably absolute range of temperature differences. Johannessen et al. illustrated the effect for NaCl operating on quartz crystals (SiO_2) in a sandstone with a temperature change of 50°C. The basic equation of potential pressure, they suggested, is:

$$K_{NaCl} \quad P + K_{SiO2} = T(E.NaCl + E.SiO_2) \quad (5.5)$$

where K is the linear compression a crystal undergoes when pressured, P is the pressure in kilobars, T is the change in temperature, and E is the linear thermal expansion of a crystal during a heating episode (when:

$K_{NaCl} = 0.0014$/kbar, $K_{SiO2} = 0.0009$/kbar, ENaCl = 0.000040/°C, E.SiO$_2$ = 0.000005/°C, and T = 50°C). In these circumstances:

$$0.0014P/°C + 0.0009P/°C$$
$$= 50°C\ (0.00004 + 0.000005)$$
$$P = 0.97\ \text{kbar} = 970\ \text{bars}.$$

The pressure of 970 bars, they argued, is sufficient to break quartz crystals from their surrounding cement in the sandstone.

(f) Salt weathering effects

The notion that salts in rocks can cause weathering in deserts is long established, although not always proven (e.g. Jutson 1918, Walther 1912, Mortensen 1933). In recent years the evidence has mounted for the weathering efficacy of salts (e.g. Goudie 1985b). It has been held responsible for rock flaking, granular disintegration, and splitting (Ch. 6.1). It can produce silt-size quartz, feldspar and other particles that may subsequently be moved as loess (e.g. Goudie et al. 1979, Pye & Sperling 1983, Smith et al. 1987); and it is probably crucial in forming cavernous weathering features, including tafoni and alveoles (Ch. 6.1) in places as far apart as Utah (Mustoe 1982), various parts of Australia (Bradley et al. 1978, Young 1987) and near Lake Magadi, Kenya (Smith & McAlister 1986). Salt weathering at ground level, in the capillary zone above the water table, may also be responsible for fashioning pedestal rocks or **zeugen**, features that are traditionally attributed to wind abrasion (Fig. 5.19). Mass segregation of salts may lead to the creation of mounds and ground heaving.

Salt weathering may also contribute to the creation of pitted cobbles, as in Death Valley (Butler & Mount 1986) and the Atacama (Wright & Urzúa 1963). It may help the disintegration of alluvial fan gravels in Death Valley (Goudie & Day 1980), and the generation of debris in sandy limestones of eastern Saudi Arabia (Chapman 1980), in the dry Karakoram Mountains of Pakistan (Goudie 1984), and around Tunisian sabkhas (Goudie & Watson 1984). It is certainly a major process of debris comminution (Peterson 1980b). It may also help the development of pans in southern Africa (Goudie & Thomas 1985).

Salt weathering also has a fundamentally important *applied context*, because it can cause damage to buildings, roads and engineering structures in deserts, especially as a result of the ingress of groundwater (Fig. 5.20B, 5.21). This problem has now been exten-

Figure 5.19 A pedestal rock in Bahrain that probably owes more to weathering in the zone of capillary rise than to wind erosion.

sively reviewed (e.g. Fookes & Collis 1977, Netterberg & Loudon 1980, Cooke et al. 1982). Suffice it to say here that its control and management depend crucially on knowing the composition of the groundwater, the depth of groundwater, the thickness of the capillary fringe, and the relations between these features and the depth and composition of foundations. For example, several studies in the Middle East have suggested that the nature of the salt weathering hazard is directly proportional to the shallowness of groundwater and its salinity, and that it can therefore be predicted through the spatial analysis of these variables (Fig. 5.20A; Doornkamp et al. 1980, Cooke et al. 1982).

5.2.5 Weathering by lichens and algae

Lichens and algae occur widely in deserts, no doubt partly because they are distributed by the wind (e.g. Friedmann & Galun 1974). The number of species is often large. For example, over 150 lichen species have been recorded in the Atacama, over 130 in the Sahara, and over 50 in the Negev (Friedmann & Galun 1974). They occur on the soil, in the soil, on the lower surfaces of stones partly buried in the soil, and on and in rocks. They have been found to create weathering features in both hot and cold deserts (e.g. Krumbein 1969, Friedmann 1980).

From the weathering perspective, there is a significant difference between **endolithic** and **epilithic**

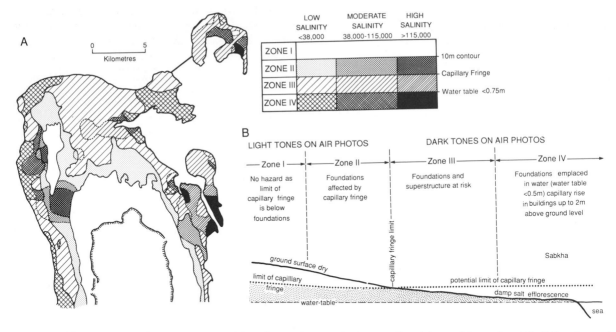

Figure 5.20 (A) Aggressive ground conditions in northern Bahrain: predicted hazard intensity. (B) Saline hazard zones associated with saline groundwater – a general model (Doornkamp et al. 1980).

organisms. The former grow *beneath* the rock surface, in one of two forms. **Cryptoendolithic** organisms, such as the blue–green algae (e.g. *Glaeocapsa*) and the lichen *Caloplaca alociza*, grow *beneath* the surface in zones parallel to it of distinctive colours (Fig. 5.22A). They grow in protected, relatively isolated and favourable pore-space habitats up to about 40mm deep. Such growth is limited to initially porous rocks, and rocks which are light-coloured and to some extent translucent, such as marble, granite and granodiorite. In contrast, **Chasmoendolithic** organisms colonize pre-existing minute cracks in the rock that are open to the surface. Epilithic organisms, such as the widespread crustose lichens, are firmly attached to rock surfaces by their thallial tissues.

There is now substantial evidence of the weathering effectiveness of lichens and algae (e.g. Danin et al. 1982, Friedmann 1982). In general, their effects are to embroider or modify the micromorphology of rock surfaces, through several processes. *Chemical* processes are said to include the aqueous dissolution of respiratory carbon dioxide, the formation of water-soluble lichen compounds which are able to form soluble metal complexes, and the excretion of oxalic and other acids by growing lichens (Jones et al. 1980). Of these processes, the last appears to be effective in chemically etching feldspar, ferro-magnesian minerals and quartz, and in degrading hydrothermally altered minerals (e.g. Hallbauer & Jahns 1977, Jones et al. 1980). Lichen activity has also been held responsible for concentrating minerals at or near to stone surfaces, especially through iron mobilization, and therefore for contributing towards desert varnish formation (Ch. 6).

Physical processes are associated with the volume changes to gelatinous and mucilaginous substances, for example in thallial tissues, that contract on *withdrawal* of water, under certain circumstances exerting a pressure on the rock to which they are attached, and causing a fragment to be broken off (in much the same way as drying gelatin may chip flakes off the surface of a glass plate). Disintegration may also be caused by the *wetting* of thalli lodged in confined spaces, which swell and may exert sufficient pressure on the rock for it to fracture. Fry (1924, 1927) demonstrated the effectiveness of these processes on various rock types. Their efficacy in deserts is not substantiated, but the presence of epilithic lichens, and the wetting and drying of lichens (especially in such "high-humidity" deserts as the Atacama and the Negev) suggests a potential for physical weathering.

Lichens and algae are probably responsible for

Figure 5.21 Salt efflorescence and destruction of a paved surface within the capillary fringe zone, Bahrain.

several weathering features. Friedmann (1982) showed that cryptoendolithic organisms, of both filamentous fungi and blue–green algae which form a symbiotic lichen association up to 8mm thick (Fig. 5.22A), probably created **exfoliation** of sandstone in Antarctica. Here, the horizontal stratification of organic growth (Fig. 5.22A) presumably leads to chemical, volume and other changes that cause plates of the relatively impermeable crust to flake off. This, in turn, causes the coloured zones to extend deeper into

the rock, and a new surface crust to form (Fig. 5.22B). In some circumstances, the cryptoendolithic growth may be replaced by the epilithic growth of crustose lichens (Fig. 5.22A). This form of exfoliation is very similar to that observed in hot deserts, where it may have a similar origin.

Endolithic organisms, such as *Caloplaca alociza*, are capable of dissolving limestone in deserts. For example, Danin et al. (1982) showed that on limestone surfaces in Israel dissolution tends to be greatest where two lichen colonies meet. Here, grooves 0.1–0.2mm deep are formed at the line of contact. Such grooves make a complex fretwork pattern. This is revealed when the lichens die or are removed. Danin et al. (1982) also observed weathering **pits**, 10–20mm across and 5–15mm deep, that appeared to be formed by colonies of blue–green algae under arid conditions. They also showed that lichen weathering may be an indicator of *changing* weathering conditions. At one site, for example, they argued that broad pits reflected an arid climate; that subsequent grooving by endolithic lichens was produced under "Mediterranean" conditions; and present-day aridity is associated with micro-pitting by blue–green algae.

In contrast, epilithic crustose lichens are probably capable of modifying surfaces through the penetration and modification of thallial tissues, as described above. Such change may cause granular disintegration or flaking, or help to create smooth surfaces. One problem is to determine how far the presence of a crumbly weathered surface is a prerequisite for lichen

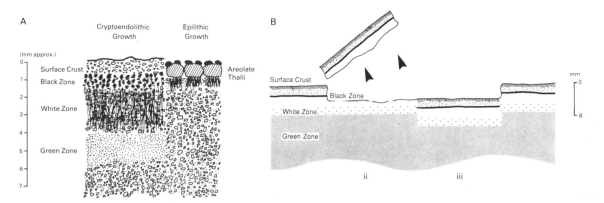

Figure 5.22 Diagrams of lichen weathering in sandstone. (A) The structure of cryptoendolithic growth and its relationship to the epilithic areolate growth form. The areolate thalli appear in protected areas after breaking of the surface crust by biogenous weathering. (B) Exfoliative rock weathering, showing, from left to right: the initial level of lichen growth; exfoliation of the surface crust due to biological activity; the site of an earlier exfoliation, with the lichen growing deeper into the rock substrate; and the formation of new surface crust and a portion of the old surface crust at the initial level of lichen growth (after Friedmann 1982).

colonization, and how far it is the result of lichen activity.

5.2.6 Solution processes and limestone: karst

From the preceding discussion, it is clear that some limestone weathering in deserts is accomplished by lichen and algae (Fig. 5.23). However, it is not clear how far evidence of limestone solution in deserts arises from such microbial activity or from solution by acidified rainfall. For example, in Israel, studies of limestone tombstones and other dated monuments by Klein (1984) in Haifa and Danin (1983) in Jerusalem led to a common agreement on an *average* limestone weathering rate of $5\mu m\ yr^{-1}$, but to a difference of opinion on the processes responsible. In Haifa, solution in an urbanizing atmosphere is at least a partial probable explanation; in Jerusalem, it seems that limestones were influenced only by microbial activity.

There is undoubtedly evidence of solutional karstic phenomena in deserts, despite the fact that lack of water and vegetation might lead to a general expectation that such features would be relatively scarce (Jennings 1983, 1985). In fact, although they are not common, most solutional karst features have now been described in deserts, including small solution hollows and pits, limestone pavements with clints and grykes, cave systems, dolines and dayas, and certain tafoni (e.g. Jennings 1985, Smith 1987). Solutionally sculptured flutes and grooves, such as **rillenkarren** and similar features, have been reported from Australia (Dunkerley 1979), the Namib (Sweeting &

Figure 5.23 Limestone pitting as a result of lichen growth, near Avdat, Israel.

Lancaster 1982) and the Sahara (Smith 1987). However, these features should not be confused with somewhat similar wind-abrasion fluting (Ch. 21).

Some karst features in deserts are almost certainly inherited from more humid conditions; but others, such as the vertical cylindrical tubes found near limestone cliffs in southern Morocco, may be formed under contemporary aridity. Castellani & Dragoni (1987) ascribed these tubes to dew condensation associated with air movement through fractures in the limestone. Their model suggests that the solution features may form over 10^5yr under present conditions.

CHAPTER 6

Desert surfaces and soils

6.1 Introduction

Soils in deserts are much more closely implicated in geomorphology than they are in humid climates. This is partly a consequence of greater exposure, but it is also a result of the toughness of many subsurface soil horizons. This section examines these near-surface features and soils in arid and semi-arid regions, and briefly considers their relations with geomorphology.

Surfaces and soils in deserts differ from those in humid zones in seven significant respects, all to do with scant moisture:

(1) Less weathering and leaching, which has four consequences: coarse textures, shallow soils, retention of soluble substances, slow soil formation.
(2) Large areas of bare rock and coarse weathered mantle.
(3) Great expanses of patinated surface.
(4) Major inputs of aeolian material.
(5) Low rates of erosion, especially in ancient shield deserts such as the Sahara and in Australia, permitting the survival of extensive areas of very old soils.
(6) Abrupt soil boundaries as a consequence of abrupt thresholds in soil behaviour: between slopes with angles that can and those that cannot maintain soils long enough to develop horizons; between well drained and poorly drained soils; between sandy and non-sandy parent materials (as explained in Chs 16.2.3 & 4, and 27.3.1); between large barren tracts and small patches of biologically active soil beneath bushes, which accumulate nutrients and aeolian sediments (some of them "nabkha" – Ch. 25.2.3a); between the extensive barren areas where moisture is scarce, and much smaller areas with dense vegetation where it collects, as at the bases of slopes.

(7) As a consequence of the critical nature of these thresholds, degradation is more easily provoked in arid than in humid soils.

Although there is unquestionably less biological activity in desert than in humid soils, research is showing, against expectation, that organisms are important to some desert soil processes, as will be shown.

There are four major distinctive groups of loose mantles, soils and surfaces in contemporary deserts (and some minor, peripheral types):

– bare patinated rock;
– loose, coarse, weathered mantles;
– free-draining, horizonated soils;
– soils of depressions.

The distribution of these soil groups is a function of climate and topography (Fig. 6.1). In very arid areas, bare rock and coarse weathered mantles prevail, even

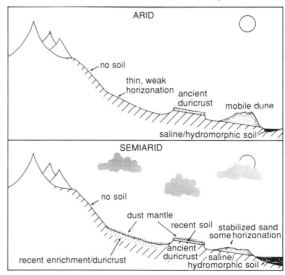

Figure 6.1 Generalized toposequences of soils in arid and semi-arid areas.

45

on gentle slopes; many ancient soils are preserved; and hydromorphic soils are limited to a few oases and to the valleys of exogenous rivers. In semi-arid environments, coarse mantles are found only on the steepest slopes; gentler slopes support free-draining, horizonated soils, many of them containing an abundance of aeolian material; most ancient soils are buried by aeolium or alluvium; and hydromorphic soils in valley bottoms are much more extensive than in very arid environments.

The formation of coarse weathering mantles is discussed in Chapter 5, and the character of stone pavements in Chapter 7. The red colour of many desert surface soils is examined in Chapter 22.2.5a, for redness is more typical of aeolian than alluvial soils. The soils of stabilized dunes are examined in Chapter 29.2.4d, these too being distinctive. This section provides an examination of rock patinas and contemporary, hydromorphic and ancient soils.

6.2 Rock patinas

Rock patinas are not confined to deserts, occurring also in high mountains (Whalley 1984), near the Poles (Glasby et al. 1981, Douglas 1987), near coasts and close to river channels in many climates (Boussingault 1882, Whalley 1983). Nevertheless, patinas are more extensive and conspicuous on the immense exposures of bare rock in deserts than anywhere else on Earth.

6.2.1 Rock varnish

(a) Characteristics
Rock varnish (sometimes called "desert varnish") is the most widespread and prominent of the patinas. In the deserts of the southwestern USA, it is estimated to cover 75% of all bare rock (Engel & Sharp 1958). It is a hard rind, 5μm to about 100μm thick (rarely 500 microns), above and sometimes just below the soil surface. Thickness varies greatly within short distances, being greater in small indentations and cracks, on dry, sun-facing exposures which are not wind-abraded or where there are no higher plants (Dorn 1983). The greatest thickness is at the surface of the soil abutting a bare rock outcrop or pebble: the "ground-line band".

Most above-ground varnish is black or deep red (Munsell 10R 3/3 or 3/4). Beneath soil, the colour is paler red (10R 4/8, 2.5YR 4/6 to 5/6, 5YR 7/6; Fig.

Figure 6.2 Rock varnish on limestone in northern Sinai. The discontinuous cover suggests colonization by fungi or bacteria. Note the pitting of the limestone (cf. Fig. 5.23).

Figure 6.3 A varnished pebble from the central Libyan desert. The rock in the centre of the pebble has weathered to incoherence and been removed by wind when exposed by the breakage of the pebble. The tough varnished carapace has survived.

6.2). Higher manganese contents give darker hues. Most varnishes contain clay minerals (about 10%), particularly mixed-layer illite–montmorillonite (Potter & Rossman 1977), and they "scavenge" many minor elements (Whalley 1983). Layering, picked out by varying amounts of manganese is common (Hooke et al. 1969, Perry & Adams 1978), the layers dying out laterally and merging into varnishes with other structures (such as botryoidal). Rock varnish is tougher than many of the rocks beneath, which may weather to incoherence while varnish survives (Fig. 6.3; Tricart & Cailleux 1964, Hooke et al. 1969).

(b) Processes

Most early workers believed varnishing to be a physicochemical process, associated with the high Eh (dry, oxidizing) and pH (unleached, alkaline) conditions in deserts. The occasional incoherence of the underlying rock suggested to some that iron and manganese had been drawn in solution from the rock beneath and precipitated by evaporation (Tricart & Cailleux 1960, 1964). The differentiation of iron from manganese in varnishes, and the greater concentration of manganese in varnish than in parent rock, was explained, in this model, by the fractionation of the more soluble manganese from the less soluble iron as solutions were concentrated by evaporation. Ground-line bands were seen as the result of the preferential evaporation of soil solutions in these positions.

Engel & Sharp (1958), while still assuming physicochemical fixation, discovered that many of the elements in varnishes were derived from dust. The importance of dust is not now in dispute, for it has repeatedly been demonstrated that most varnish constituents could not have come from underlying rocks, many of which have little iron or manganese. Moreover, most varnishes overlie the rock with a sharp boundary (Potter & Rossman 1977). The poor development of varnish on the surfaces of some minerals, such as quartz, was explained in the physicochemical model by the absence of catalysts, seen as necessary to precipitate the manganese (Whalley 1983).

The physicochemical model was challenged at an early stage by biological models of manganese fixation (Whalley 1983), and this explanation is now in the ascendancy, albeit with new evidence and a more precise notion of the organisms involved (Krumbein & Jens 1981, Dorn & Oberlander 1981a, Taylor-George et al. 1983, Dorn 1989a). In the new biological model, micro-organisms, probably micro-colonial lichens and bacteria, play the major rôle in fixing manganese. They are adsorbed to the surface of clay dust which settles on rock surfaces and they obtain energy by oxidizing incoming manganese, when occasionally activated during brief wet periods. The manganese is then strongly adsorbed to the clay minerals, especially to the illite–montmorillonite,

where it is held between layers in the crystal lattice. The varnish then protects the bacteria from the harsh conditions at the desert surface. Once begun, manganese-rich varnish formation is said to experience positive feedback, trapping more clay and manganese. In this model, ground-line bands are the result either of high concentrations of manganese in soil solutions, or of the greater number of wetting cycles at the ground-line than on the exposed rock.

Evidence for bacterial involvement comes from the spatial patterns and speed of growth of manganese-rich crusts (Krumbein & Jens 1981). The strongest evidence is contained in laboratory experiments in which manganese can be fixed with a suitable mix of clays and manganese, but not without micro-organisms (Dorn & Oberlander 1981a), and in which respiration can be recorded from moistened varnishes (Taylor-George et al. 1983). The greater concentration of manganese than of iron in above-ground varnishes is seen in this model to be the result of the preference for manganese as an energy source.

For all the excellent evidence for the biological model, recent studies show that manganese varnish formation may be achieved in a number of ways, including purely physicochemical processes in some environments (Glasby et al. 1981, Douglas 1987). The physicochemical and biological models are not in dispute about iron-rich varnishes: they agree that iron is fixed physicochemically. Red, iron-dominated varnish occurs where manganese-fixing bacteria cannot survive, as on the undersides and buried portions of stones and boulders. In the biological model, intermediate shades of varnish are the consequence of intermittent survival by manganese-fixing microbes (Dorn & Oberlander 1981a).

(c) Environments

According to Dorn's (1983) biological model, manganese-rich varnish will not form where: Eh fluctuates wildly (from oxidation to reduction); pH is too high; manganese-fixing organisms are outcompeted by mosses, other lichens or higher plants; rock surfaces are too smooth; organic matter deters oxidation; there is continuous wetness or dryness (inhibiting adhesion); or there is little manganese. There are neither red nor black varnishes on surfaces destabilized by weathering (as on shales) or abrasion (as in exposed, windy conditions). Established varnishes are destroyed if there is a change in Eh, pH, wetness, or the erosion rate (by wind or water) (Dorn & Oberlander 1981b).

Engel & Sharp (1958) believed that varnishes were confined to areas with less than about 150mm of rainfall and, although varnish has been reported from other environments, confusion between these and desert varnishes is unlikely. Moreover, Whalley (1983) reported experiments which showed that varnish soon disappeared when exposed in a wet climate, so that varnish is never likely to survive long enough to be an indicator of desert conditions in the past.

(d) Rock varnishes and dating

Dark, manganese-rich varnish has been seen to appear in 13 years in southern Israel (Krumbein & Jens 1981), but there is no doubt that more heavily varnished surfaces are associated with greater age, for example on alluvial fans (Dorn 1983), artefacts (Dorn et al. 1986) and rock engravings (Dragovich 1986). However, thicknesses and colour allow only relative dating, because these properties are too dependent on factors other than time to act as measures of absolute age. Layering in varnishes has also been claimed as a dating tool (Perry & Adams 1978), though other authorities doubt whether it is a regularly periodic phenomenon (Dorn & Oberlander 1981a, Dragovich 1984).

This does not mean that varnishes are not datable. The ratio of the contents of titanium to those of potassium and calcium in manganese-rich varnishes (Mn concentration > 0.1) has been claimed as a method that, if calibrated with other methods such as ^{14}C, can give absolute ages. This is because titanium is chemically stable, whereas potassium and calcium are mobile, and are apparently removed at a logarithmically steady rate (Dorn 1983, Dorn et al. 1987). The accuracy of this "cation-ratio" technique is constrained by the fact that varnish appears in recognizable form on desert surfaces only after about a hundred years, and it does not appear to work in varnishes that may be young (Dragovich 1988). Dragovich has also speculated that some varnishes may grow patchily, and so destroy neat pseudo-stratigraphic patterns. If the cation-ratio method does work, it is unaffected by the probable cessation of varnish formation in some deserts in the most recent climatic phase (Knauss & Ku 1980).

The technique has shown the minimum age of some Australian varnishes to be over 5,000yr (Dragovich 1986), and of some in the southwestern USA to be over 10,000yr (Dorn 1982). Other dating methods also suggest great age: in southeastern Utah, Knauss

& Ku (1980) believed a thick varnish to be over 300,000yr old; and in Nevada, Harrington & Whitney (1987) found some varnishes to be a million years old. The most remarkable aspect of these discoveries is the immense age and stability even of this fragile outer shell of some parts of the desert landscape.

(e) Rock varnishes as indicators of climatic conditions

Though layering may not be periodic, it may be characteristic of sites or periods with high ambient levels of dust, as in deserts, whereas botryoidal structures indicated less dusty, presumably more humid, environments (Dorn 1986). Dorn & DeNiro (1984) also suggested that the $\delta^{13}C$ value of varnishes could distinguish arid from humid environments.

6.2.2 Desert glaze and case-hardening

"Desert glaze" was a term used by Fisk (1971) to describe lustrous, silica-rich, colourless and almost ceramic coatings on pebbles in the central Sahara. The glaze was about 0.01mm thick and it occurred only on silica-rich rocks such as flint, agate, fossilized wood and quartzite. Fisk's glaze closely resembles coatings found in the semi-arid parts of Hawaii by Farr & Adams (1984) and Curtiss et al. (1985). These appear to thicken with age and to darken as they trap reddish or greyish dust. Curtiss believed them to be derived from siliceous, locally derived dust, which was dissolved by dew and fog once deposited on a rock surface, and later precipitated by evaporation. Desert glaze seems to be very common and deserves much more attention than it has received.

Case-hardening is not confined to deserts but, like rock varnish, it is more conspicuous on the extensive rock exposures of desert than of other climates (Goudie 1983c). Case-hardening occurs on fine sandstones in the southwestern USA, where it varies in thickness from 0.5 to 5mm, with an average of about 1.5mm (Conca & Rossman 1982). It is paler in appearance and much harder than the underlying rock, from which it is separated by an abrupt boundary. It has a calcareous surface horizon, over a zone of kaolinite enrichment, which is poorer in quartz and haematite than the underlying rock. Conca & Rossman believed that the kaolinite had been brought to the surface by evaporating water, and later cemented by dew and occasional rain.

6.3 Freely drained soils

Most free-draining soils of arid and semi-arid regions are classified in the US Department of Agriculture (USDA) system as either *entisols* (with minimal horizonation) or *aridisols* (dry soils with subsurface horizons) (Dregne 1976).

6.3.1 Surface horizons

Because they are so intimately involved in geomorphic processes, the properties of surface horizons are dealt with under several headings in this book. The surface horizons of soils underlying stone pavements are discussed in Chapter 7. It is shown there that many of these soils are silty, stone-free and have a vesicular structure. This section examines other properties of immediately subsurface horizons.

(a) Texture

Most desert surface horizons (except those of alluvial soils in major river valleys) are coarse-textured (Jenny 1941, Weinert 1965). This is because weathering is generally slow.

ADDITIONS OF AEOLIAN MATERIAL Another property is not so visible, but probably more significant to soil genesis: huge additions of aeolian material.

It is shown in Chapter 29.2.1 that as much as 50% of the tropical semi-arid lands are covered with ancient dune sands. Soil development in these sands is described in Chapter 29.2.4d. Sand added in smaller quantities to finer-textured soils may be incorporated by inwashing down cracks and by biological processes (Blume 1987). An example of a soil with significant additions of aeolian sand is the clayey *qardud* soil of semi-arid central Sudan (Warren 1970).

In the Sahel, southern Tunisia and southern Israel, dust additions have been enough to form great depths of loess (Ch. 22.1.5c), providing deep, silty soils. These aside, few upper horizons of semi-arid soils have escaped significant enrichment with dust of one kind or another (Yaalon & Ganor 1973, Chartres 1983a, Stahr et al. 1989; Ch. 29.2.4b). The addition of dust may be insidious and hard to detect in many soils, but research is showing it to be very widespread, as the discussion below will show.

The added dust ranges in size from fine clay to silt, being generally coarser near to sources (Smith et al. 1970). Some clays are added as sand-size pellets (Ch. 25.3.2; Chartres 1982). Rates of addition are discussed in Chapter 22.1.5, where it is shown that the highest rates of deposition are in semi-arid desert margins where large quantities are filtered out by rain and vegetation. Local patterns of dust deposition are determined by plumes down wind of sources, such as playas (Kolm 1982, Chadwick & Davis 1990), and by patterns of airflow in the vicinity of hills (Queiroz et al. 1982; Ch. 20.4.7). Dusty soils in arid deserts are rarer than those in semi-arid areas; the North African "fech-fech" is one example. It consists of a surface horizon of loose dusty material, only lightly protected by a surface crust, which is easily disturbed (Rognon 1990).

Where dust is relatively coarse and insoluble, it remains in the upper soil and is only slowly mixed with lower horizons by animals, or by penetration down cracks. Surface horizons are progressively clogged by dust, and increasing impermeability permits less and less of the dust to be incorporated (Amit & Gerson 1986, Wells et al. 1987). Where the dust is soluble it is redistributed through the profile in solution in infiltrating waters (Sec. 6.3b). Additions of salts such as sodium chloride are significant only within a few hundred kilometres of coasts (Dan & Yaalon 1982). Others, such as gypsum, are significant in areas close to and down wind of coastal lagoons and inland lakes (Coque 1962).

(b) Cations

Different cations are leached from surface horizons according to their solubility. Even with mean annual rainfalls of as low as 50mm, most sodium salts are washed out of surface horizons (Blume et al. 1985). In increasingly wetter climates, the sequence of decreasing solubility (and therefore of soil constituents retained in progressively wetter soils) is K^+; SO_4^{2-}, Mg^{2+}, Ca^{2+} (Drever & Smith 1978). Thus sodium salts are retained only in very arid topsoils, gypsum only in fairly arid soils, and carbonates in slightly wetter ones. This sequence is discussed in several sections below.

(c) Clay minerals

Weathering is generally slow in desert soils, and the clay mineralogy of even quite old soils bears a close resemblance to that of their parent material. The major factor differentiating alluvial soils, for example, is their sedimentary character rather than any pedogenic feature (features due to soil-forming processes, defined in Sec. 6.3.2c) (Western 1972).

When a soil has developed over thousands of years, a characteristic suite of clay minerals appears. In the

Mojave, Holocene soils do not have this suite: late Pleistocene ones do (Amundson et al. 1988). In mature arid soils one might expect rather different clay minerals from those in humid soils, for there is less leaching in arid soils, but it is not always easy to distinguish between the effects of climate and those of parent material (Atkinson & Waugh 1979), amount and character of added dust ((a), above), topographic position and environmental history (below, *passim*).

When these other factors can be controlled, it appears that there is more smectite (montmorillonite), illite, chlorite and palygorskite in arid than in humid soils (Birkeland 1974: 226–30, Singer & Amiel 1974, Singer 1984, Amundson et al. 1988, Reheis 1990). Some palygorskite may be inherited from dust, itself derived from playas and sabkhas where it forms in large quantities (Coudé-Gaussen et al. 1984a), but palygorskite in arid soils may also be an authigenic (*in situ*) transformation of montmorillonite in the presence of an enhanced supply of magnesium, and in moderately saline (high pH) and perhaps temporarily waterlogged or brackish conditions (Singer & Norrish 1974, Yaalon & Wieder 1976, Singer 1979). Palygorskite also forms in calcretes (Sec. 6.3.5), probably where calcium carbonate precipitation is favoured over magnesium carbonate, and where therefore magnesium is concentrated in the remaining soil solution (Hassouba & Shaw 1980, Watts 1980, Smith & Whalley 1988). Illites are found in large quantities in the topsoils of some arid soils, though are rare in others. They also appear to have formed authigenically, for they occur in soils over igneous rock. It may be that they synthesize more quickly where they are wetted and dried frequently; their potassium may come as dust (Singer 1989).

(d) Biological activity

Organic matter contents of very arid soils are very low, building up slowly as rainfall increases, perhaps at a rate of about 0.3–1.1mg of dry matter for each gramme of water in the 38 to 170mm rainfall zone (Whittaker et al. 1968, Gile 1977, Crawford & Gosz 1982). The main constraint on the production of organic matter is the low productivity of desert ecosystems and the generally low root:shoot ratios of desert plants (Orians & Solberg 1977), but a countervailing factor is slow (though variable) decomposition, allowing some litter to lie on the surface for many years before it disappears. Experiments show that intense brief rain showers do not accelerate decomposition, probably because of the absence of decom-

posing organisms. Thus breakdown rates are related to long-term average rainfall conditions, for it is these that control the composition of the bacterial fauna (Santos et al. 1984).

The effects of invertebrates are minor in very arid soils, and increases in importance towards the semiarid and then humid areas (Yair & Rutin 1981). Termites are present in the soils even of some very arid areas, but play a small rôle in soil formation (Chhotani 1988). In semi-arid deserts such as Arizona, termites may be locally effective, bringing some 744kg ha^{-1} yr^{-1} of clay- and nutrient- (particularly nitrogen) rich subsoil to the surface, and thus making a significant, though patchy contribution to soil formation and nutrient cycling (Noy-Meir & Harpaz 1976). Termites have a major effect only in soils in quite humid environments (Pullen 1979).

Organic matter in arid and semi-arid soils may have a higher concentration of nitrogen than organic matter in humid soils, partly because there are proportionately more leguminous plants in arid than mesic environments (Crawford & Gosz 1982), and partly because some algae in soil-surface crusts are nitrogen fixers (Fuller et al. 1960, MacGregor & Johnson 1971). On the other hand, organic matter contents are very low, and biological denitrification can be fast, so that nitrogen levels are generally low (Noy-Meir & Harpaz 1976).

Nitrogen and phosphorus accumulate close to the top of the profile in arid soils, both because of fast turnover when plants and soil micro-organisms are briefly active (Charley & Cowling 1968), and of the low rate of leaching. Potassium, however, is as efficiently circulated as in humid soils and may be deficient in some arid topsoils (Hanawalt & Whittaker 1977). The concentration of nutrients is spatially very inhomogeneous for two reasons: the patchy nature of desert vegetation, itself determined by variable water supply related to micro- and macro-topography (Crawford & Gosz 1982, Hunter et al. 1982); and the concentration of nutrients at wetting fronts as water moves through the soil; these can be stranded when the water evaporates (Hunter et al. 1982).

6.3.2 Subsurface enrichment in general

(a) Introduction

The formation of subsurface horizons is so slow in arid and semi-arid climates that they survive only where erosion is also slow. This confines them almost wholly to soils on low-angle slopes. They are formed by two processes whose products are not always readily distinguishable in the field: sedimentation and pedogenesis.

Three types of sedimentary deposit occur in arid soils: colluvium, brought down by slope processes (mainly surface wash); alluvium, near to stream channels; and aeolium. Episodicity in these processes (for many reasons, but often because of climatic change) varies the rate of deposition and produces strata which can be confused with horizons caused by pedogenesis. Not only do some sedimentary strata closely resemble pedogenic horizons, but pedogenesis and sedimentation are commonly superimposed (Chartres 1983b). The sedimentary processes are examined in other parts of this book, leaving pedogenesis to be examined here.

(b) Characteristics

Enrichment ranges from zero to nearly 100%. In early pedogenesis, enriched horizons inherit the characteristics of the host material, the introduced material being dispersed and fine-grained where the host was fine-grained (= fine-pored, as in clays), and massive where pores in the host were coarse (as in gravels) (S. H. Watts 1978a, Wieder & Yaalon 1982). In later stages, when they have been replaced and displaced, host materials lose their influence; fragments of pre-existing material "float" in a massive matrix of the enriching material, or disappear completely (N. L. Watts 1978, S. H. Watts 1978a, Summerfield 1983a).

Very tough enriched horizons are known as duricrusts. Most duricrusts are more than 10^4yr old, and only survive in very stable sites. Though well displayed and extensive in arid and semi-arid areas, not all duricrusts were formed in these environments and many are relics of wetter climates. Individual duricrusts – some pedogenic, some non-pedogenic – commonly occur in complex profile sequences, stacked one on another, in alluvium or in weathered profiles (themselves composed of layers with different textures). The sequences may contain horizons derived from the erosion and re-cementation of remnants of previously enriched duricrusts. These profiles may be 200m deep, and include up to 45 individual horizons, many truncated (Goudie 1983c). Although many sequences of horizons may indicate only that the variable textures in an alluvial profile have encouraged variable contemporaneous precipitation, they may also be evidence of repeated deposition and erosion, precipitation and pedogenesis through complex climatic changes over thousands, if not millions, of years.

(c) Modes of formation

The chief distinction between enrichment processes is between those that are pedogenic and those that are not. Pedogenic enrichment occurs in four main ways: **illuviation** (the commonest), in which material is removed from upper and deposited lower horizons; **authigenic processes**, whereby incoming or *in situ* constituents are disaggregated and recombined within the horizon to form new substances; **selective removal** of constituents from a horizon; or **greater weathering** in a particular horizon (or any combination of these four processes). A significant number of enriched horizons have been formed by non-pedogenic processes, many associated with fluctuating water tables; these are defined and discussed in various sections below.

(d) Distribution

On a regional scale, the main control on movement to the subsoil is the amount of water available for leaching. This is a function of rainfall, evaporation and the texture of the soil.

A climatic succession of subsurface enrichment follows the solubility progression discussed in Section 6.3. Even in very arid deserts, showers of 50mm can wet a loamy soil down to 30–50cm and a sandy soil to 60–80cm, moving chlorides, carbonates and sulphates to the subsoil (Blume et al. 1985). With mean annual rainfalls above about 50mm, sodium salts are moved well down or out of the soil, and gypsic subsoil horizons are more common (Watson 1983). Carbonates are not generally removed from upper horizons under these conditions, but above 175–300mm mean annual rainfall (depending also on evaporation), gypsum is mobilized, and carbonate is the commonest form of subsurface enrichment. Carbonate is mobilized as much by the higher soil CO_2 levels associated with greater organic activity as by more infiltrating water. In turn, carbonates retreat beneath the surface at higher rainfalls, and are totally removed at rainfalls above 400–600mm, whereafter clay enrichment comes into its own.

The depth at which redeposition occurs is commonly believed to be the depth to which rainwater normally penetrates (Fig. 6.4; Arkley 1963, Sehgal & Stoops 1972, Gile 1975, Nettleton et al. 1975, Dan & Yaalon 1982, Rabenhorst & Wilding 1986b). Measures of infiltration after extreme storms, or in very wet years, may be more relevant parameters than the mean annual rainfall in this respect (Marion et al. 1985). Other factors controlling the depth of enrich-

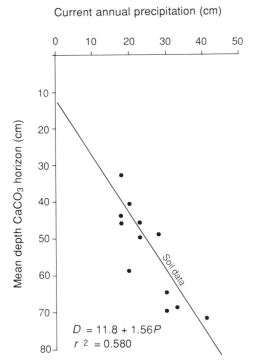

Current annual precipitation (cm)

$D = 11.8 + 1.56P$
$r^2 = 0.580$

Figure 6.4 The relationship between the mean annual rainfall and the depth to maximum calcium carbonate concentration in Arizonan soils (after Marion et al. 1985).

ment include: the texture of the host material, enrichment tending to occur where a porous layer overlies an impermeable one; the depth of disturbance in shrink–swell soils such as vertisols (Blockhuis et al. 1968); the varying partial pressure of the soil CO_2, itself dependent on temperature and soil organic activity (Marion et al. 1985); decreasing pore space with compaction; plugging of the horizon with the early stages of enrichment (McFadden & Tinsley 1985); and the relative mobility of various enriching substances in particular environments. Thus horizons enriched in different substances can be superimposed in the same profile, even without climatic change.

At the local scale, topography and soil texture are the main controls on the type and depth of enrichment. There are two opposed topographic tendencies: the water-based and the wind-based. As a general rule, the more soluble, more mobile substances are taken farthest down slope by leaching processes (Glazovskaya 1968). Thus halite travels farther down slope than gypsum, which itself travels farther than carbonate.

However, once precipitated as evaporites on the surface of playas, some salts, particularly halite and gypsum, are removed as dust, redistributed and deposited in patterns determined by wind directions and local topography (Ch. 22.1). In some wind regimes, dust may be spread all over the basin whence its constituents had been taken by leaching – Coque's (1962) "*cycle de gypse*". Where winds are more directional, the salts are distributed down wind in the zone of dust plumes. When washed down into the soil, the mobile constituents of the dust create various forms of enrichment.

Soil texture controls the rate infiltration. In general coarser-textured soils experience deeper leaching, and enriched horizons are deeper in these than in fine soils.

(e) Forms of enrichment and climatic change
Though many enriched horizons within the same profile may form in an unchanged climate, as explained in (d) above, superimposition is more likely to be the result of climatic change. Climatic changes can trigger a number of changes in soil-forming factors. They may control the filling and emptying of lakes, and so the covering and uncovering of sources of dusts of various kinds. They may also, rather more obviously, trigger changes in the amount of water infiltrating soil profiles, and this in turn brings about either a change in the type of material being washed down the profile or a change in the depth they reach in the profile.

With all these controls, most profile sequences are rather ambiguous evidence for climatic change. For example, a clay-enriched horizon underlying a carbonate horizon might indicate desiccation, or it might indicate an earlier period in which large amounts of clay dust were released to the atmosphere. The older of two superimposed clay-rich horizons could be interpreted as evidence either that the earlier period of clay movement had lasted long enough to move the clay well down the profile; or that it had been especially wet, or that it was simply older (Dan et al. 1973). Some clay-enriched horizons may have been destroyed after emplacement by a change to wetter conditions (Gile 1966) and there are always problems of distinguishing pedogenic from sedimentary processes.

There are, however, some greater certainties. Few forms of enrichment survive changes to wetter conditions, though massive silica, carbonate or clay enrichments are only slowly weathered and removed

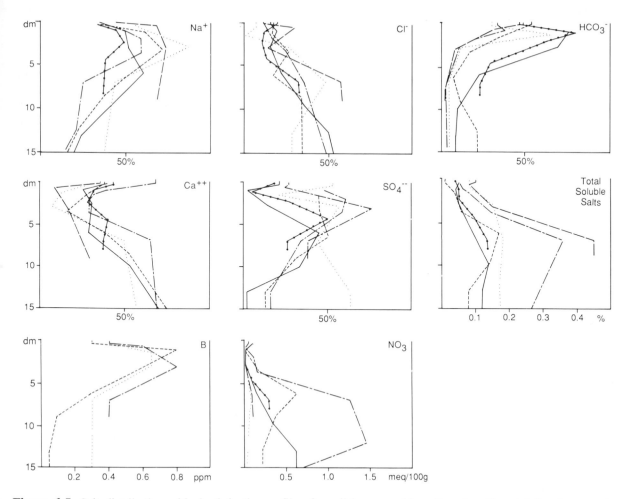

Figure 6.5 Salt distribution with depth in the profile of a soil in very arid southwestern Egypt (after Smettan & Blume 1987).

from the profile. Gypsum and halite enrichment quickly disappear. On the other hand, almost all forms of enrichment survive changes to a drier environment; the discussion below contains many examples of this. With silcretes, it has been suggested that a change in climate was necessary even to bring about any precipitation.

6.3.3 Subsurface salt enrichment

Chlorides, sulphates and nitrates are washed from the subsurface horizons of all except the soils of very arid areas. Only in extremely arid deserts, such as northern Chile, do they survive at the surface. The extensive sodium nitrate deposits of northern Chile and nearby Peru occur in horizons of between a few centimetres to a few metres thick. The great depth of these

nitrate deposits points to long continued extreme aridity, perhaps for as long as 10 to 15 million years (Ericksen 1981).

In what is apparently a slightly wetter area in southwest Egypt (though it is still very dry), borates and sodium carbonate occur in the topsoil; sodium sulphate in middle horizons; succeeding downwards to sodium chloride, calcium chloride, calcium nitrate and sodium nitrate, as might be expected from the solubility series (Fig. 6.5; Smettan & Blume 1987). In both of these areas, saline horizons are deeper in the lower parts of the landscape, where there is presumably more infiltrating water. Smettan & Blume believed that this pattern of salt enrichment was also controlled by temperature, which could become very high.

The rôle of past climates in creating the pattern in southwest Egypt is hard to evaluate: subsurface salt enrichment could have occurred either in the slightly wetter climate of the Neolithic, or after occasional heavy showers (the interpretation favoured by Smettan & Blume). Horizons enriched in soluble salts such as sodium chloride develop quickly in loose material once a surface has been stabilized (Dan & Yaalon 1982), though subsurface horizons can continue to accumulate salts over 70,000 years in very arid areas (Dan et al. 1982). As with all enrichment, the depth to maximum mobile salt enrichment is believed to reflect the depth of penetration of the mean annual rainfall. In one estimate, nitrates are leached down between 1mm and 5mm for each millimetre of rain in loamy soils (Dan & Yaalon 1982, Eisenberg et al. 1982; Fig. 6.6).

As with all other enriching constituents, the main source of salts in surface and subsurface saline horizons is thought by most authorities to be dust, for few underlying deposits contain the concentrations of salts found in the soils that overlie them (Hallsworth & Waring 1964, Smettan & Blume 1987, Wells et al. 1987). Salt inputs decline down wind away from sources such as coastal zones or playa lakes (Eriksson 1958). The source of the nitrates in Chile is mysterious: it may have come from seaweed, guano, bacterial decay of organic matter, or from volcanic rocks (Ericksen 1981).

6.3.4 Gypsum enrichment

Horizons enriched in gypsum are up to 5m thick (Watson 1983). Gypsum concentration is never as great as in some carbonate or silica-enriched horizons (the gypsum content is usually 75–97%), and gypsum cements are never as tough as calcretes or silcretes. Horizons enriched in gypsum are known as "gypsic", and the soils in which they occur as "gypsiorthids" in the USDA system. Hardened gypsic horizons are known as "gypcrete". Watson (1979), synthesizing earlier systems of classification, proposed that gypsic horizons formed in four distinct conditions: in well drained soils, as buried evaporites, in hydromorphic soils, and by the exposure of subsurface horizons by erosion.

(a) In well drained soils

These are gypsic horizons formed by pedogenesis. Though pedogenetic gypsum may derive from gypsum-rich parent material, by far the greater amount is derived from dust, itself traceable ultimately to playa evaporites (Jessup 1961, Coque 1962, Reheis 1987b), deposited on the surface and then illuviated to a subsurface horizon. Two pieces of evidence suggest this: first, this form of gypsum-enriched horizon occurs in soils over gypsum-poor parent materials in positions far above regional water tables; and second, there is sufficient gypsum in dust. In Wyoming, gypsum dust is deposited at a rate of

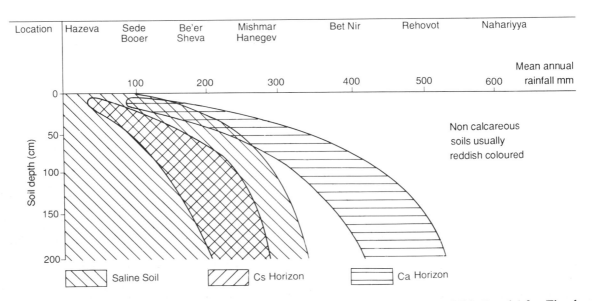

Figure 6.6 The depth of horizons enriched in different salts related to mean annual rainfall in Israel (after Eisenberg et al. 1982).

Figure 6.7 "*Rose de sable*", a twinned gypsum crystal formation from a "*croûte de nappe*" in southern Tunisia.

26×10^{-6} to 60×10^{-6}g cm^2 yr^{-1}, whereas the rate required to produce the existing gypsum enrichment in soils is 1.1×10^{-6} to 120×10^{-6}g cm2 yr^{-1} (Reheis 1987b).

The gypsum in these horizons is usually mesocrystalline. The crystals have grown by displacing the soil above, rather than other soil constituents in the same horizons. The only signs of replacement are of calcite.

Some pedogenic gypsum may be formed by the reaction of rainwater (or groundwater) rich in hydrogen sulphide with calcium carbonate in the soil. This is thought by some authorities to occur in the Namib Desert, the sulphate here supposedly being derived from offshore diatom oozes. Another source of sulphate for authigenic gypsum could be iron pyrite (Watson 1979).

(b) In hydromorphic soils
In soils near the water table, gypsum crystallizes from groundwaters rich in calcium and sulphate to form what French workers call "*croûte de nappe*" (Coque 1962). Some of these horizons are distinguished by large crystals (large enough, in southern Tunisia, to be sold to tourists as "*roses de sable*" or "sand roses") (Fig. 6.7; Watson 1985). This form of gypsic horizon occurs in sediments close to playas or sabkhas, such as the great Chotts in southern Tunisia and nearby Algeria and on the shores of the Persian

Gulf. It commonly reaches 1.5m, but may reach 5m thick: 1m thick crusts of this type have formed in less than 20 years in the Souf area of Algeria after water tables have been lowered by intensive groundwater extraction for irrigation (Bellair 1958).

Even in such high osmotic pressures, plant roots may play a rôle in precipitating the gypsum, initially as anhydrite (Gunatilaka et al. 1980). This mechanism is accelerated by high halite salinity (Ali & West 1983) and perhaps also by sulphate-reducing bacteria (Watson 1983). In some cases, the growth of this kind of horizon may displace the upper horizons, and produce nearly pure gypsum. In others, particles of the parent soil are included in the gypsum crystals.

(c) Burial of evaporites
Gypsum deposited on the shores of shallow playas (Ch. 14.4), may be buried by other sediments and appear as a soil horizon. Gypsum deposited in this way is usually bedded and graded by size of crystal, commonly contaminated by many other salts, and may be massive and nearly transparent (Watson 1988).

(d) Erosional exposure of originally subsurface gypsic horizons
Any of the first three types of gypsic horizon may be exposed if the loose surface horizons are stripped off by erosion. On exposure, the gypsum becomes columnar, powdery or cobble-like, depending on the character of the original horizon and the state of disintegration. Many of exposed gypsic horizons underlie a stone pavement, which is good evidence of exhumation. The exposed surface of a gypsum crust is frequently cracked polygonally with units 25cm to 1m in diameter (Tucker 1978).

(e) Distribution
The rate of development, thickness, concentration and geographical distribution of any of these kinds of gypsic horizons, as with calcic horizons (Sec. 6.3.5e), is a complex function of rainfall, dustfall, topographic position, characteristics of the parent material, age and climatic history. Gypsic horizons are better developed near to sources of gypsum, from which they may be moved locally by groundwater flow, subsurface throughflow or by the wind. Gypsum is more mobile and remains in the soil for a shorter period of time in wetter climates. Topography can have two effects: in controlling the pattern of dustfall; and in concentrating leaching solutions down slope. Thus gypsic horizons should have a distinctive pattern

around hills that affect dust deposition (Ch. 22.1.5d); and should be better developed in soils of lower rather than upper slopes (Connacher 1975, Watson 1983).

(f) Age

Gypcretes in relatively humid southern Tunisia appear to date from the first injections of gypsum dust from newly desiccated chotts, 8,000 to 9,500yr ago (Watson 1988). Gypsic horizons in the Dead Sea rift valley in Israel, also near to sources of gypsum dust but in a much more arid environment, are thought to be more than 70,000yr old (Dan et al. 1982). Because gypsic horizons are both too soluble to survive for long after a change to a wetter climate, and too weak to withstand wind erosion when exposed at the surface, they are not as persistent features of the landscape as calcic or silica-enriched horizons (Watson 1979).

6.3.5 Carbonate enrichment

(a) Introduction

Carbonate enrichment is the commonest and most prominent form of subsurface enrichment in semi-arid soils (Fig. 6.8). Carbonate-enriched horizons may underlie as much as 13% of the total land surface of the Earth (Goudie 1983b, quoting Yaalon 1981), and soil carbonate accounts for more carbon storage than living terrestrial vegetation, and about half the amount in soil organic matter (Schlesinger 1985). Carbonate-enriched horizons also occur in many ancient rock sequences. They have been given several general and local names, the best known being *caliche* and *kunkar*.

"Calcic" horizon is a broad, general term that will be used here for all soil horizons more than 15cm thick that are visibly enriched with carbonate. Arid-zone soils with such horizons are known as *calciorthids* in the USDA system, or calcisols.

Some calcic horizons in semi-arid areas are striking features. Tough duricrusts of pure carbonate, up to 3m thick, form prominent caprocks and scarps. These are **"calcretes"** or "petrocalcic horizons". Earlier this century, most geologists saw calcretes as marine or lacustrine limestones (Passarge 1930). Research since then has shown conclusively that most were formed in terrestrial conditions, and that many are pedogenic. This is not to say that there are no shallow calcretes that are of lacustrine or other sedimentary origin (Hutton & Dixon 1981).

When rapid evaporation produces supersaturated solutions of calcite (Suarez 1981), and where there is also a high magnesium:calcium ratio in the waters, high-magnesium carbonates are formed (Watts 1980). Where large quantities of magnesium are available, and where fresh or brackish surface waters mix with saline groundwaters, giving alternately very different conditions (so-called "schizohaline" conditions), carbonate-rich horizons, both pedogenic and non-pedogenic, may be dolomitic (Stalder 1975, Watts 1980, Maizels 1988); when hardened, these are termed **"dolocrete"** (Fig. 6.9; Netterberg 1979).

(b) Pedogenic enrichment

Most unindurated calcic horizons and many calcretes are pedogenic. In pedogenesis, carbonate is mobilized in surface horizons with high CO_2 contents, through

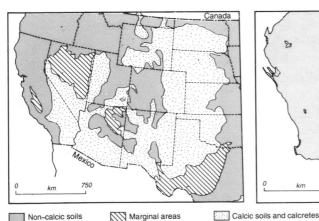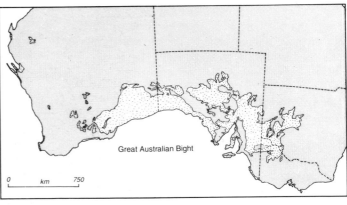

| Non-calcic soils | Marginal areas | Calcic soils and calcretes |

Figure 6.8 The distribution of calcic soils in the western USA and southern Australia (after Machette 1985, and Milnes & Ludbrook 1986, quoting Northcote).

Figure 6.9 Dolocrete developed beneath raised river channel gravels in north-central Oman.

which rainwater infiltrates, and deposited lower down the profile where CO_2 levels are lower, or from which water is lost by evaporation. It is claimed that deposition due to changing CO_2 levels creates carbonates with higher $\delta^{13}C$ values than those deposited because of evaporation. In most pedogenic carbonates this criterion indicates precipitation due to changing CO_2 levels, rather than because of evaporation (Salomons et al. 1978, Dijkmans et al. 1986, El Hinawi et al. 1990). However, the $\delta^{13}C$ value is controlled by several factors, including the species of local vegetation, and evaporative loss and CO_2 pressure are not entirely separate processes, so that the diagnostic value of this criterion is not yet certain (Bowler & Polach 1971, Schlesinger 1985, Amundson et al. 1990). The presence of sulphate ions (usually in gypsum) also encourages carbonate deposition (Lattman & Lauffenburger 1974) and, furthermore, it is likely

that there are important biological mechanisms of carbonate fixation in soils, the active organisms being lichens, algae, bacteria and mycorrhizas (Klappa 1978, Phillips et al. 1987, Phillips & Self 1987).

Pedogenic calcic horizons in semi-arid soils are not necessarily or even most commonly derived from weathering of parent soil, for many contain far more carbonate than could have come from this source. Hence, it is postulated, most of the calcium in many of these horizons comes from dust or rain. This hypothesis is corroborated by measurements of carbonate inputs (Machette 1985). At one site in the southwestern USA, only 1% of the soil carbonate can have come from bedrock; modern carbonate dust flux at the site was estimated at 0.5×10^{-3}g cm^{-2} (Mayer et al. 1988). In general, present rates of deposition of calcium in dust and rain are consistent with the age and amounts of calcium in calcic horizons throughout the southwestern USA (Schlesinger 1985). Local patterns of calcic horizon distribution probably reflect the availability of carbonate dust, as near to the carbonate dune sands on the South Australian coast (Milnes & Ludbrook 1986). The world distribution of carbonate horizons, with concentrations on the poleward rather than the tropical margins of the main deserts, may reflect the distribution of calcium inputs in rain and dust, or some temperature-related effect on calcium solubility.

In early pedogenesis, calcic horizons are little more than dispersions of carbonate nodules or powdery filaments. These horizons can develop quickly in recent alluvium if there are adequate supplies of infiltrating water. Recognizable enrichment has been reported in soils no more than 75 years old in northern Algeria (Boulaine 1958b), but in dryer conditions and over non-calcareous parent materials, even early Holocene alluvium shows only slight enrichment (Shlemon 1978, Nettleton et al. 1990).

Over thousands of years, pebble coatings merge and carbonate begins to take up more and more of the inter-particle space in the horizon of strongest accumulation. A pebble coating may take 3,000yr to develop fully (Schlesinger 1985, quoting Ku & Liang 1984). Eventually, a much more complex profile evolves. Such sequences are generally less than 2.5m thick (Watts 1980). There is little doubt that such profiles are pedogenetic, for they occur well away from streams, high above the water table, over calcium-poor groundwaters, or in coarse soils through which there is no capillary rise (Machette 1985).

Calcrete commonly occurs in complex profiles

which show signs of multiple phases of sedimentation, carbonate accumulation, replacement, solution, break-up and recementation. They may be more than 200m thick, and must have developed over many thousands if not millions of years (Gile & Hawley 1966, Leeder 1975, Watts 1980, Arakel 1982, Goudie 1983c, Machette 1985).

Three features of calcic horizons attract more than their share of controversy. First is the pure, commonly banded laminar layer. Early discussion of this layer was summarized by Coque (1962), who himself believed it to have originated as a blanket of aeolian sediment, a view shared by Brown (1956). Later authorities saw the laminar layer variously as an algal limestone, a freshwater lake deposit or a travertine. In one contemporary model, the laminar layer is formed when an impermeable hardpan calcrete is exhumed by erosion. Surface wash, aided by algae and lichens, produces the banded, pure carbonate, and subsequent reburial restores the laminar layer to its place in an apparently normal soil profile (Lattman 1973, Klappa 1979, Blümel 1982). The same model has been proposed for laminar horizons in dolocretes (Stalder 1975). Although stripping of the topsoil is undoubtedly common, other authorities have pointed to structures that suggest that banding is associated with a mat formed by roots unable to penetrate the hard calcrete (Wright et al. 1988). The most widely accepted model has lamination formed beneath the surface as throughflow is channelled over a plugged hardpan (Gile et al. 1965, 1966, Machette 1985).

A second controversy is over replacement and displacement, for though these are the only ways in which fragments of the host material could have been separated by pure carbonate, it is not always easy to decide which process took place in particular calcretes. Replacement has undoubtedly occurred in the most developed calcretes and dolocretes, where few of the original constituents or structures are recognizable, and where the edges of original grains are corroded (Amiel 1975, Hay & Reeder 1978, Watts 1980). Some authorities believe that replacement is possible only in markedly seasonal climates (Halitim et al. 1983). Replacement is less certain where more fragments of the host survive and where host grains have fresh surfaces, and here displacement is more likely (N. L. Watts 1978, Klappa 1979). Displacement implies that the soil expanded (Machette 1985), and contorted structures such as pseudo-anticlines may be evidence of this (Blank & Tynes 1965, Watts 1977), though this too has been disputed (Klappa 1979).

The third enigma in calcretes is fragmentation (at scales from a few millimetres to several centimetres, even metres), the fragments having either rounded edges (boulder calcrete) or angular edges (breccia calcrete). Hypotheses here include: solution and re-cementation (for boulder calcrete) (Goudie 1983c); root activity (though perhaps only in wetter climates where there are many deep-rooted plants) (Klappa 1980); expansion of the matrix and displacement of the original grain fabric (especially for the smaller forms of brecciation) (N. L. Watts 1978); and erosion and redeposition of fragments of an earlier calcrete (giving large angular fragments) (Lattman 1977). Individual boulder or breccia calcretes could have been formed in more than one of these ways.

ENVIRONMENTS OF PEDOGENIC CALCIC HORIZONS
Calcic horizons can form in high-rainfall areas in carbonate-rich parent material, or where there is concentrated carbonate-rich lateral wash (Braithwaite 1983). Examples of these include: "nari", developed on limestones and chalks in central and northern Israel (Yaalon & Singer 1974), and other horizons in North Africa (Riche et al. 1982), Florida (Kahle 1977) and Western Australia (Semeniuk 1986). Similar calcic horizons occur over limestones in arid and semi-arid areas (Blank & Tynes 1965, Rabenhorst & Wilding 1986a), but in semi-arid and arid zones calcic horizons also occur in soils in carbonate-poor parent and host materials.

In semi-arid and arid areas six factors influence the state of development of calcic horizons (Machette 1985, McFadden & Tinsley 1985):

(1) The texture of the host material, determining in turn its porosity, bulk density and water-holding capacity; dense accumulations cannot develop in soils with little void space.

(2) Partial CO_2 pressures in the soil air. CO_2 pressures presumably depend on plant and animal activity in the soil. As is pointed out below, this is one of the least understood features of calcic horizon development.

(3) The rate of supply of calcium. The calcium content of rainfall may be relatively invariant geographically, but the calcium carbonate content of dust appears to have a marked spatial variation (Goudie 1973, Machette 1985). Variations in the carbonate content of parent material and of lateral throughflow are further factors.

(4) The amount of infiltrating water available for leaching, itself dependent on rainfall and

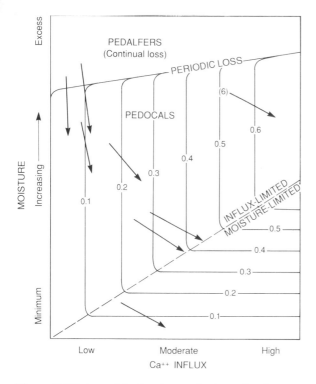

Figure 6.10 A model of carbonate horizon development in relation to Ca^{++} flux (largely from dust) and moisture flux (largely from rainfall). Arrows show paths of individual soils through different sequences of climatic change (after Machette 1985).

evaporation (and therefore also on temperature). If CO_2 partial pressures and the porosity of the host material are assumed to be constant, or related to climatic factors (admittedly large assumptions), carbonate and moisture supply can be built into a model of carbonate accumulation (Fig. 6.10). In this model, moisture is seen to be the limiting factor in some cases, calcium flux in others. When moisture is limiting, but calcium flux sufficient, surface horizons remain relatively carbonate-rich, and subsurface horizons are not well developed. When calcium flux is limiting, but moisture sufficient, surface horizons are leached of carbonate, and again subsurface accumulation is limited. The maximum development of calcic horizons occurs where dust and moisture flux are balanced.

Two further factors, however, are needed to explain the state of development of calcic horizons, namely:
(5) The age of the profile. Even in dry areas, there may be enough infiltration to develop calcic horizons, given sufficient time. In the now-classic model of Gile et al. (1966), calcic horizons proceed through a sequence from light cementation, to the development of the "hardpan" calcrete, which prevents further infiltration and so allows the development of the laminar layer above. Following this model, one can view the total amount of carbonate in a single-stage profile as a measure of its age (Machette 1985). Using this criterion, it is seen that full development of the calcic horizon sequence is achieved in New Mexico soils only in profiles dating from the mid-Pleistocene (Gile & Hawley 1966).

Various dating techniques confirm that many calcretes are very ancient: 30,000–15,000 BP in Namibia and southeast Spain (Blümel 1982); from "at least 1 million years" (Shlemon 1978); 83,000yr ($\pm$10,000) (Ku et al. 1979) to as far back as the late Miocene or Pliocene in the southwestern USA, though most of these calcretes probably date from the last glacial period (Brown 1956, Gile & Hawley 1968, Gardner 1972, Schlesinger 1985, Machette 1985); and greater than 1 million yr in South Africa (Netterberg 1979). Some of these dates suggest rates for calcrete accumulation: Schlesinger's (1985) review of rates in the southwestern USA showed them to vary between 1.0 and 12.00g CaCO$_3$ m^{-2} yr^{-1}.

Nevertheless, there is yet one more factor that influences calcrete development, namely:
(6) The climatic history of the site.

CLIMATIC CHANGE AND PEDOGENIC CARBONATE ENRICHMENT It is probable that most well developed calcic horizons are relics of climates rather different from the present. In the southwestern USA, calcic horizons are apparently too deep to have formed in present rainfalls (Marion et al. 1985).

Climatic fluctuation brings about two changes in relation to the development of calcic horizons: varying amounts of infiltration, itself depending on rainfall and temperature; and varying amounts of calcium input (Ch. 22.1). Inputs of calcium might rise with rainfall if it has a constant concentration in rainwater (Marion et al. 1985); or may fall, if more rainfall discourages wind erosion (Machette 1985). Subtle climatic shifts, as between winter and summer rainfall, could also cause changes, for winter rain is more effective, coming as it does at a time when evaporation is less (Marion et al. 1985).

Depending on their positions in relation to these two factors in Figure 6.10, calcisols could have been affected in various ways by climatic change. It is evident that the same profile characteristics should have developed as a result of more than one climatic sequence (McFadden & Tinsley 1985).

In some areas (for example, New Mexico), calcium supply and rainfall may have varied in such a way as to maintain a steady rate of calcic-horizon development over long periods of time: drier conditions produced more carbonate dust, but less rain to move it down profile; wetter conditions provided the leaching medium, but less carbonate to be leached. In more arid areas, where moisture was more limiting during the driest phases (for example, western Arizona), climatic change to more humid conditions might have dramatically increased the rate of calcic horizon formation (Machette 1985). There are clearly other distinctive paths of carbonate-horizon development through the diagram on Figure 6.10. Moreover, calcretes could experience a change to wetter conditions (and be presumably slowly redissolved), or be stranded and inactivated in climates drier than those in which they formed. A relic calcrete in a dry climate occurs in eastern Saudi Arabia (Chapman 1974). Netterberg (1969) believed that most southern African calcretes were also inactive.

Climatic scenarios can be tested by modelling water and calcium flux (McFadden & Tinsley 1985, Marion et al. 1985, Mayer et al. 1988). McFadden & Tinsley's model produced soils not unlike their natural counterparts, but Mayer found that one modelled soil had far less carbonate than its real counterpart, suggesting that, even in the past 6,000yr, there had been a period with higher carbonate flux (more dust), and more effective leaching (more rain). These models, like all good models, are valuable in highlighting inadequacies in the understanding of the natural system. In calcisols, the models of Marion and Mayer showed that the behaviour of soil CO_2 was a very weak link in the understanding of pedogenic calcretes, and Mayer's model showed in addition that dust influx was another important unknown.

(c) Non-pedogenic calcretes
Field relations and characteristics show that many calcretes are non-pedogenic, though the distinctions between these and pedogenic calcretes are not always easily drawn, and considerable debate surrounds the interpretation of individual profiles. The debate is made no simpler by the occurrence in the same profile of calcretes of manifestly different origins and by the pedogenic modification of some non-pedogenic cal-

cretes (Machette 1985). The thickest carbonate-enriched profiles develop in accreting land surfaces, the upper surface of the calcrete building up as the surface accretes, and in some cases being pursued upwards by a water table from beneath (Gardner 1972, Machette 1985).

WATER-TABLE CALCRETES Many calcretes are said to have been deposited near a water table. It is widely believed that the main mechanism is evaporation from the capillary fringe, but it has been found that calcium carbonate deposition can be induced below a water table by the changing CO_2 contents of the water; the calcite formed in this way is generally more coarsely crystalline than the carbonate formed above the water table (Land 1970, Thortensen et al. 1972). Most authorities use the absence of meniscus cements round and between particles to distinguish these "phreatic" (water-table-related) from "vadose" or pedogenic calcretes (Warren 1983). Like horizons enriched pedogenically, vadose calcic horizons may be massive, laminar or concretionary, but many are in thinner sheets than are found in pedogenic calcrete, and rhizoconcretions are rarer (Semeniuk & Searle 1985). They are less restricted to semi-arid areas than pedogenic calcic horizons, forming in wet climates where there are strong concentrations of carbonate, as in limestones and chalks. High magnesium calcite may be more common in this form of calcrete (Watts 1980). Water-table calcretes develop especially well in valley alluvium, where water tables are near the surface and where the matrix is coarse.

FLOODWATER CALCRETES Water-table calcretes may be difficult to distinguish from calcretes thought to have formed in yet another way – by the infiltration of carbonate-rich floodwaters. Individual horizons formed in this way are usually much thicker (often > 10m) than other forms of calcrete, occur only in alluvium close to stream channels, and show evidence that the alluvium was fresh when cemented (Bogoch & Cook 1974, Lattman 1977, Arakel & McConchie 1982). In northern Oman, where large quantities of magnesium are available from ultrabasic ophiolites, floodwater dolocretes and calcretes cement the alluvium in nearly every valley, ranging in character from cemented cobbles (Fig. 6.9), to profiles in which dolomite has replaced materials of other lithologies and destroyed the original sedimentary structure (Stalder 1975, Maizels 1988). The cement in these calcretes is usually as strong as, or stronger than, the parent cobbles. Floodwater calcretes and dolocretes are commonly interstratified with (and in turn hard to distinguish from) travertines, tufas or lacustrine marls (Vogt 1982, Ashour & Abd-el-Mogheith 1983, Goudie 1983c).

The age of the alluvium into which some of this last type of calcrete has been emplaced suggests that it can develop quickly (Machette 1985, Jones et al. 1988). Though this may be so, many of these calcretes are of great age. In Oman and Western Australia some valley calcretes date back to the early Tertiary (Stalder 1975, Arakel & McConchie 1982). As with pedogenic calcretes, many valley calcretes may now be relics of periods in the past in which conditions were different. For example, valley calcretes incorporating palaeolithic tools have been reported from the northwest Sudan, whereas today there is neither flow in wadis, nor a water table anywhere near (Szabo et al. 1989)

6.3.6 Clay enrichment

Clay-rich subsurface horizons, known as "argillic" in the USDA system, are usually denser and redder than the horizons above or beneath (Gile 1967). When they are so weakly developed that clay enrichment is hard to detect, but where there is a difference in colour and structure from horizons above and below, as they often are in arid climates, these horizons are called "*cambic*" (Gile 1966).

Clay enrichment is achieved by three, not always easily distinguishable, processes:

(1) Downwashing of clay from the topsoil following decalcification and acidification. This process is more typical of desert-margin, high-rainfall areas than of arid areas (Gile 1977, Felix-Henningson 1984, Honeycutt et al. 1990). Carbonate-rich soils do not develop argillic horizons, because carbonate inhibits the dispersal of clay.

Thin sections of clay-enriched horizons may exhibit clay coatings on the surfaces of coarser grains and aggregates when seen under the microscope, and these clay coatings are presumed by some workers to indicate clay illuviation (McFadden et al. 1986), though this is difficult to prove. Another indication of clay illuviation is a comparable degree of weathering in surface and argillic horizons (Nettleton et al. 1975). Clay-rich lamellae may be better indicators of illuviation (Peterson 1980). In the USDA classification, arid-zone soils with argillic horizons are known as *argids* (or *haplargids* if they are only slightly

developed – as is commonly the case).

(2) Authigenic (*in situ*) clay may form in a subsurface horizon if more moisture is retained there. In arid conditions such a process is more likely to produce a cambic than a fully argillic horizon. Cambic horizons seem to develop by this process even in very arid parts of the Sahara (Blume et al. 1985).

(3) Clay may be dispersed and washed from the upper soil in the presence of high concentrations of sodium. Both carbonate and silicate clay, and even silt, can migrate down the profile under these conditions. This type of process is most common in low positions in the landscape, as near playas, to which sodium migrates (Peterson 1980). Sodium-rich soils can also occur elsewhere in the landscape if sodium is being blown around as dust (Hallsworth & Waring 1964), or at saline seeps. Soils with subsurface clay and sodium enrichment are known as *natrargids* in the USDA system or, more commonly, as *solonetz*.

Rainfall is the chief regional control of the distribution of argids. The local distribution of haplargids and natrargids is a function of the amount of local run-on. Permeability, related mainly to soil texture, is another important factor at the local scale: clay illuviation is more marked in pebbly and sandy than in finer soils (Brewer 1956). The clay may come from weathering of the parent material, or may be washed in from floodwater (Walker et al. 1978). The occurrence of clay enrichment in soils where weathering might be expected to be slow suggests that dust may be an important source (Dan et al. 1982, Peterson 1980).

Where there is little sodium in the topsoil, clay enrichment is not yet detectable in Holocene or even late Pleistocene alluvium or aeolium in arid areas, being found only in older deposits (Yaalon & Dan 1974, Dan et al. 1982, Felix-Henningson 1984, Nettleton et al. 1990). In the arid parts of the southwestern USA, cambic horizons may appear in 5,000yr (Gile 1966), but detectable clay enrichment does not appear in soils younger than between 12,000–18,000yr (Nettleton et al. 1975, McFadden et al. 1986). Of course, enrichment is much faster where there is more rainfall on the margins of deserts (Gile 1975).

The great length of time that is apparently needed for clay enrichment in relatively acid, arid conditions increases the likelihood that many, if not most, clay-rich horizons are relics of wetter climates in the past (Gile 1966, Gile & Grossman 1968, Nettleton et al.

1975, Reheis 1987a). Further evidence of inheritance comes from clay enrichment superimposed on carbonate enrichment (Nettleton et al. 1990).

Clay enrichment is much faster in some soils that are sodium-rich. In arid southern New Mexico these natrargids appear in less than 2,000yr (Peterson 1980), and in Nevada, with rainfalls between 100mm and 150mm annually, in less than 6,600yr (Alexander & Nettleton 1977).

6.3.7 Silica enrichment

(a) Types of silica enrichment
The first necessary distinction in a discussion of silica enrichment is between horizons enriched in silicate clay minerals (where silica enrichment occurs, but in association with alumina and bases), and silica enrichment *per se*, in which enrichment is in silica alone, and which is the main (though not only) concern here.

The second distinction is not as clear-cut: it is between enrichment that occurs in semi-arid areas, much of it quite geologically recent, and very ancient silcretes, thought by most authorities to have formed in wetter environments (Hutton et al. 1978, Ollier 1978). An attempt is made below to separate these two categories.

After oxygen, silicon is the most abundant element in continental rocks, but in rock minerals it is either held in massive, poorly soluble forms such as quartz, or bound to bases or aluminium. Because the oxide, silica, comes rapidly into solution at high pH, it is tempting to suggest that it is most easily mobilized in unleached arid soils. Aridity or semi-aridity would also encourage evaporative precipitation of silica (Smale 1978). The association of many silcretes with semi-aridity encourages this school of thought, but there is plenty of evidence that silica is mobilized and precipitated in soils under many different climatic conditions. For example, it is being removed in large quantities from soils in cool and very humid Wales (Crompton 1960), apparently because of the high rate of leaching, and it has also been precipitated in the lower horizons of some of these soils (Bridges & Bull 1981). Moreover, silica solubility is not only controlled by pH and the leaching rate, but also by the presence of other ions in solution (Summerfield 1983a).

The main geomorphological interest in silica enrichment is in **silcrete** duricrusts, rather than in weakly enriched horizons. The term "silcrete" was coined to describe small outcrops of hard silica-cemented rocks in southern England, specially sarsen stones (Lamplugh 1902), but silcrete has subsequently been found to be much more extensive in arid, semi-arid and tropical subhumid environments. Even further research has shown that there are smaller outcrops under many climatic regimes (Dury & Habermann 1978).

Silcretes are very hard, brittle, usually grey, intensely indurated and very silica-rich (commonly 90%; Hutton et al. 1978, Summerfield 1983a). Most of the cement is microcrystalline quartz, but some is opaline, including commercial opal. Many authors describe the loud, ringing sound and characteristic smell when silcrete is hit with a geological hammer. Horizons are between 1m and 3m, and rarely 5m thick, and many silcretes, like other duricrusts, are prominent caprocks in desert landscapes. They occur in weathered profiles, in sediments and even in unweathered rock. Their textures, as in other duricrusts, reflect that of the host material. Like other duricrusts, there is commonly more than one silcrete in a profile; they may occur in the same profile with, and superimposed on, calcrete, ferricrete or gypcrete (Summerfield 1983a).

(b) Silica enrichment in aridisols
There are three conditions in which silica accumulates in aridisols: in shallow, recent profiles; in association with calcretes; and in playa-margin sediments.

SILICA ENRICHMENT IN SHALLOW, GEOLOGICALLY RECENT SOIL PROFILES Silica enrichment and cementation have been reported in the western USA in Quaternary soils that have inputs of volcanic glasses (Flach et al. 1969, Chadwick et al. 1989). Silicification in these soils is in the form of "durinodules" rather than completely cemented horizons, and occurs at the depth of minimum pH. Rather more extensive silicification in Pleistocene aeolian sands and alluvium is reported from Botswana by Summerfield (1982), who speculated that the source of the dissolved and reprecipitated silica in aeolian soils might be the silt- and clay-size fragments of quartz broken from sands in aeolian transport (Ch. 22.1.3). He believed that the silica had been precipitated from slowly moving solutions by evaporation, or by changes in pH. Hence, he believed that silicification occurred only in low-relief landscapes. Summerfield's (1982, 1983b) use of titania (TiO_2) content as a criterion for separating arid-zone from humid zone silcretes has been challenged (S. H. Watts 1978b). His observation that arid-zone silcretes never occur in deeply weathered profiles is also of dubious diagnostic value, for neither do many silcretes said to have developed in humid climates ((c), below).

SILICA ENRICHMENT ASSOCIATED WITH CALCRETES

This form of silica enrichment produces some hard and thick silcretes, but appears to be a much more recent process than the ones that produced the ancient silcretes described below. Silcrete is closely associated with calcretes in the Kalahari (Watts 1980, Summerfield 1982, 1983a), and in South Australia (Ambrose & Flint 1981). When silcretes and calcretes are found in the same profiles in the same area, silcretes occur in coarser sediments and in pre-existing calcretes, while calcretes occur where there are gentler hydraulic gradients or fine sediments (Shaw & De Vries 1988). Watts suggested that silica had been released when silicates had been replaced by calcite in the calcrete-forming process, and that the silica had been precipitated in zones in the soil profile where pH was lower, or where CO_2 partial pressures were higher.

SILICA ENRICHMENT IN PLAYA-MARGIN SEDIMENTS

The last of Summerfield's (1982) types of arid-zone silcretes occur in playa-margin sediments, where presumably they have been released by high pH conditions, and precipitated when silica-rich solutions come into contact with saline solutions. In these positions diatoms may have been the source of some of the silica. Ancient silcretes have also been found in playa-margin situations (c, below)

(c) Ancient silcretes

All ancient silica enrichment is in the form of silcrete duricrusts rather than mere silica-enriched horizons. The accepted wisdom is that most of these silcretes were the products of warm, humid climates, though just how wet is still under discussion (Dury & Habermann 1978, Langford-Smith 1978, Ollier 1978, Wopfner 1978, Summerfield 1983a, Young 1985, Twidale & Hutton 1986). Only a climate considerably wetter than the semi-arid would permit the thorough weathering necessary to release silica from silicate minerals in the quantities required. Although pH values may have been low in these conditions, two other restraints on silica solution were probably effectively removed, these being alumina and iron. Organic compounds, which are likely to be in larger quantities in wet conditions, may also have played a

part. Thus silica could have been dissolved, albeit slowly, and large quantities could have been released from siliceous parent rocks, such as sandstone.

In some silcretes, high contents of low-mobility titania (up to 3%, well above that of the parent rock; Butt 1985) and signs of solution on quartz grains of the parent rock, indicate derivation from deeply weathered profiles, from which iron and mobile cations have been lost. A pedogenic model for silcrete formation in these profiles has silica being released near the surface and moving down to lower horizons where it is precipitated near a fluctuating water table (Summerfield 1983a). In other silcretes, titania contents are very low and grains of the host material are unweathered, signifying infiltration of silica taken far from source. However, the significance of titania is still a point of some debate (Hutton et al. 1978, S. H. Watts 1978b, Summerfield 1983a, Milnes & Twidale 1983). Both types of silcrete may therefore be genetically associated with deep alumina-rich, silica-poor weathered profiles (so-called "pallid zones" in "laterite profiles"), so that they may be parts of the same process in which ferricrete developed (Dury 1966a). Both ferricrete and silcrete probably formed in stable landscapes with low rates of erosion.

In many ancient silcretes, the silica must have migrated long distances to its place of deposition – another indication of derivation in a wet climate – for migration would also need more water than could have been supplied in a semi-arid climate (Stephens 1971, Ollier 1978). Among the ways in which this far-travelled silica could have precipitated are a change in pH, some organic process and evaporation. Evaporation, of course, would be more likely in a dry climate or a strongly seasonal one (Stephens 1971, Summerfield 1983a), but the environments of silica deposition are in as much dispute as those of its release (Wopfner 1978, Young 1985). Yet another reason for precipitation could be a change in climate from the humid one in which the silica was mobilized to a more arid one in which it was immobilized (Ambrose & Flint 1981, Callen 1983). Precipitation could also occur only in the absence of alumina and iron, which must therefore have been themselves leached out of the host profiles (Summerfield 1983a), or in pockets of low pH (Ambrose & Flint 1981). Precipitation could have occurred in basins of interior drainage, such as the Lake Eyre Basin in central Australia (Ollier 1978). The discovery of ancient silicified shorelines of a huge Miocene lake southwest of Lake Eyre suggests quite a complex mode of precipitation, probably from gels. Silicification may have occurred as silica-rich groundwaters met saline lake waters, as in the playa-margin silcretes described above (Ambrose & Flint 1981).

Another form of silica-enriched duricrust has recently been described from Australia. In these duricrusts, the silica is accompanied by alumina. These duricrusts are apparently very widespread, though not as tough as silcretes proper, disaggregating much more easily on exposure (Butt 1985).

(d) Age

It is therefore fairly certain that, where massive, extensive silcretes now occur in deserts, as they do in parts of Australia, southern Africa and the Sahara, they are relics of another climate. Silcretes began to form in Australia in the late Jurassic, on low-relief, stable landscapes, and in a hot, humid environment (despite Australia's more southerly position at that time). Dates of older than 23 million yr have been obtained for the most prominent silcretes in eastern Australia, some of which have been folded. There were at least two major subsequent phases of development in the Australian silcretes, the last in the early Pleistocene (Langford-Smith 1978, Ollier 1978, Ambrose & Flint 1981, Callen 1983, Twidale 1983a). In the Sahara, silcrete is thought to date from wet periods in the Oligocene (Busche 1983), and in southern Africa it may date back to the Cretaceous (Summerfield 1983a).

6.4 Soils of depressions

The soils of desert depressions fall into three categories: saline soils (solonchaks), hydromorphic soils and takyrs.

6.4.1 Solonchaks

Solonchaks are a very common soil type, covering as much as 7% of Earth's land surface, nearly all of this being in the arid zone (Dudal & Purnell 1986). They are soils with an electrical conductivity of greater than 15dS m^{-1} in the 0.75–1.25m horizon. They are known, broadly, as halorthids in the USDA system. Most of the salts are marine in origin; others come from marine sedimentary rocks, or from other rocks, particularly volcanics. Most solonchaks occur in fine alluvium at the distal ends of desert drainage systems. Some natural solonchaks occur in "slick-spots" where there are saline seeps.

Artificial solonchaks are much more extensive than natural ones. They are created in two types of site. First, they appear when artificial irrigation, usually of alluvial soils, raises the water table to the point at which its capillary fringe intersects the surface, so that salts are taken to and concentrated at the surface. These soils develop first in low-lying sites where the soil surface is near the water table. They are specially extensive in South Asia (nearly 50% of Earth's solonchak surface, Dudal & Purnell 1986), but they are common in irrigation systems everywhere. Some are very ancient, dating in Mesopotamia from as early as 6,000–7,000yr ago (Hardan 1971). Second, solonchaks appear where water tables are raised by forest clearance (so-called "dryland salting"). This form of salinization has accelerated alarmingly, from a low level at the start of the past decade to a very serious problem in parts of Australia and the northwestern Great Plains in North America (Connacher 1975). Because these soils pose major problems for agriculture, they are covered by a large literature (Bresler et al. 1982). This section discusses only natural solonchaks.

Most solonchaks have white surface crusts, a few centimetres thick, composed of crystallized salt. The commonest salt is sodium chloride, but salts composed of other cations, particularly calcium, and other anions such as sulphate, carbonate, nitrate and borate, occur in local concentrations. If a number of cations and anions are available in a soil, they are distributed in the profile and on the surface according to their solubility. Fractionation occurs in the capillary fringe above the water table. NaCl usually reaches the surface, being the most soluble; it is succeeded down profile by $CaSO_4$, $CaCO_3$, and FeOOH in turn. Also, because of its greater solubility, NaCl is taken farther down slope, and it characterizes the soils nearest to a playa, while the other salts are deposited at higher altitudes.

In some solonchaks, organic matter, remaining undecomposed in these conditions of high osmotic pressure, stains the surface dark, giving a "black alkaline soil". Where salts are deliquescent (such as CaCl and MgCl), the surface remains damp, and dust incorporated into it stains it yellow or pale brown. The surface crust may overlie a subsurface horizon darkened by water held by the salts.

The balance between capillary rise (bringing salts to the surface) and leaching (removing them down slope or down profile), creates intricate spatial patterns in solonchaks. Capillary movement occurs fully only when the capillary fringe reaches the surface. Because the depth of this fringe is a critical function of soil texture and structure, there is commonly a sharp line between non-saline soils with a water table just too deep, and very saline soils where it is just shallow enough, to intersect the capillary fringe. Leaching, on the other hand, occurs preferentially in places where surface water is concentrated and in sandy rather than clay soils. Thus there is usually an intricate pattern of sharply defined patches of saline and non-saline soils. High salinity occurs on slightly higher (but not too high) positions, and in clay soils (Wiegland et al. 1966, Williams 1968).

Most plants cannot withstand the high osmotic pressures experienced by roots in saline soils, and many solonchaks are unvegetated. A few halophytes are adapted to very saline conditions, and may themselves even exacerbate salinity. Some, notably *Tamarix* spp., bring salt up through their roots from the water table and add it to the surface as litter (Hutton 1968). Others have quite a different effect: the roots of *Prosopis glandulosa* can draw water from a shallow saline groundwater, but not salts, and this action concentrates salts in the groundwater (Jarrel & Virginia 1990).

6.4.2 Hydromorphic soils

The lower profile of solonchaks, being invaded at least seasonally by the water table, is commonly gleyed. Hydromorphism can also produce other subsoil features. In the Dallol Bosso in Niger, nodules of goethite occur between 30cm and 50cm below the surface, and maghemite–haematite occur near the water table. The formation of these clay minerals is thought to occur where the water table is only about 1m below the surface and there is intense summer evaporation. High contents of $NaCO_3$ are also found in these soils, giving very high pH values, and the high mobility of silica in these conditions permits the formation of beidellite (Bui & Wilding 1988, Bui et al. 1990).

6.4.3 Takyrs

"Takyr" is a Russian term for silty soils found in some wide depressions in central Asia. Other takyrs have been described from the Middle East and North Africa (Boulaine 1958a). Takyrs are carbonate-poor, and often saline at depth. They have a hard, very alkaline, crust in the dry season, which may break into polygonal patterns. The crust swells and absorbs water in wet periods, and becomes sticky. Takyr-filled

depressions often have a marked nabkha microtopography (Ch. 25.2.3a).

6.5 Vertisols

Vertisols are soils of semi-arid rather than arid areas. They have high contents of montmorillonite or, more generally, smectite, these being high-silica clay minerals which can form in low positions in the landscape to which silica is leached, or near rocks, such as basalts, where weathering rapidly releases large quantities of silica. Vertisols are characteristic of the semi-arid Deccan Plateau in southern India, and the Gezira and neighbouring clay plains in eastern Sudan (Williams et al. 1981).

Smectites have great capacity to swell on wetting and shrink on drying, and this characteristic produces soils that are extremely active in two ways. The first is the production of marked micro-topography (Ch. 8.1). The second is a churning action whereby soil falls down widening cracks in the dry season and is incorporated into the subsoil as the soil swells again in the wet season. Thus, surface soils are continuously renewed, subsurface enrichment is prevented, at least in the churning horizons, and the subsoil is characterized by slickenslides on cracks and joints where soil structural units have rubbed together in the shrink/swell process.

6.6 Ferricretes

In deserts, ferricretes are undoubtedly relics of another climate. Lamplugh's (1902) term "ferricrete" is now preferred by many authorities, because the older term "laterite" has become ambiguous (Ollier & Galloway 1990). A ferricrete is a duricrust cemented with iron, the structures of which may be massive, vesicular, pisolithic, nodular, slabby or vermiform. Iron concentrations may be up to 80%. Individual ferricretes are between 1m and 10m thick.

Ferricretes and silcretes have much in common. Both are found in weathered profiles and in sediments. Both have formed in repeated phases; in both the oldest exposures are very ancient; and in both younger ones are reworkings of older material (Coque 1976, Bourman et al. 1987, Spath 1987). Like silcretes and other duricrusts, there has been debate about whether particular ferricretes are pedogenic, colluvial, alluvial, lacustrine or phreatic (Beaudet et al. 1977a). Ferricretes and silcretes occur together in some exposures, and while this does not in itself mean a strong genetic connection, many silcretes and ferricretes may ultimately be derived from the same deeply weathered profiles, though usually transported over different distances and deposited or precipitated in different conditions (Ollier 1978, 1988). Another explanation could be that silcretes derive from acidic rocks, and ferricretes from basic rocks in the same climatic environment (Butt 1985). Both are better developed in the lower parts of the landscape.

Older authorities followed Walther (1915) in believing that most (or at least "typical") ferricretes had formed as part of a deeply weathered profile, but Ollier & Galloway (1990) claim never to have seen such a profile. In their view, all ferricretes are disturbed upper profiles in which there has been soil movement or redeposition, the evidence being an unconformity between the weathered rock and the ferricrete (often a stone line). In most ferricretes, the iron was released by deep weathering and then probably travelled towards the surface where it was oxidized and immobilized (Ollier 1988). Distances of travel to the point of first precipitation varied greatly.

Unlike silcretes, however, there has never been much doubt that ferricretes are products of warm, wet, though probably highly seasonal, environments. Where ferricretes are today found in desert environments, they are therefore reliable evidence of climatic change. Ferricretes are best developed in the southern Sahel and northern savanna zone of West Africa, and become thinner, and more discontinuous, northwards into the Sahara (Coque 1976, Beaudet et al. 1977a, Nahon 1986). They are best developed on the margins of the Australian deserts (Fig. 6.11; Twidale & Hutton 1986).

6.7 Soils and geomorphology

6.7.1 Upper horizons
Surface horizons are very transient features of the desert landscape. The removal of bonding agents, such as clay and carbonate, weakens their resistance to both wind and water erosion. Water erosion is further accelerated by increased runoff, itself aggravated by three processes: the plugging of surfaces

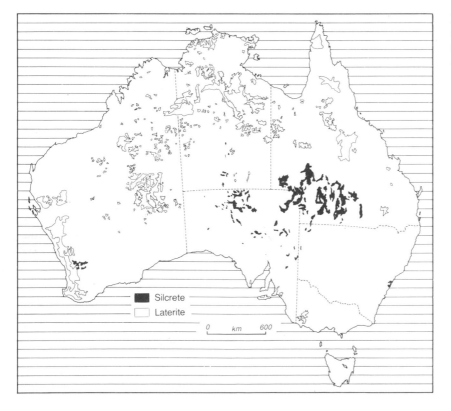

Figure 6.11 The distribution of ferricretes ("laterites") and silcretes in Australia (after Twidale & Hutton 1986).

Silcrete

Laterite

0   km   600

with dust, the dispersal of clays in surface horizons with sodium, and the development of impermeable shallow subsurface horizons (Amit & Gerson 1986, Wells et al. 1987). Wind and water erosion often truncate soil profiles, but they are as frequently replaced with aeolian or colluvial sediment (Chartres 1983a, Akpokodje 1984, McFadden et al. 1986, Wells et al. 1987)

6.7.2 Enriched horizons and duricrusts

Incision into a landscape underlain by argillic or gypsic horizons produces steep slopes on gully or yardang sides, but the new landforms in these weak materials are ephemeral. Lightly cemented dunes (as with gypsum) may be eroded to form yardangs, but these too survive only for short periods (Ch. 21.3.2g).

Duricrusts have a much more profound effect on the land surface and they provide the main relief features of low-relief deserts in central Australia, southern Africa and the central Sahara. In general, duricrust caprocks preserve ancient landforms, be they ancient geomorphic surfaces, valleys or shorelines. An example is the extensive Ogallala caprock in the High Plains of the USA, which is a calcrete (Brown 1956). In high-relief deserts such as those of the southwestern USA, it is alluvial fan surfaces that are preserved, again almost exclusively by calcrete (Gile & Hawley 1966, Lattman 1973, Ruellan et al. 1979). In these fans there is a relationship between the rate of accretion and the rate of calcic horizon development. Full calcic horizon sequences can develop only on slowly accreting fans (Leeder 1975).

Erosive incision into a major duricrusted landscape, as in Australia, can create a complex of second-order landforms. Because most duricrusts preserve the deposits of valleys, rather than of deeply weathered (and weakened) interfluves, incision may create **inversion of relief**, whereby the ancient valleys are preserved as elongated plateaux, while the ancient divides become valleys (Jessup 1961, Ollier 1988). In Oman, Saudi Arabia, the central Sahara and parts of central China, narrow, often markedly meandering channels were duricrusted and have survived as "raised channels" (Maizels 1988). Unlike the larger-scale inversion of relief, which is thought to have been caused by fluvial erosion, most of the incision that created these raised channels is thought to have been aeolian (Ch. 21.3.5). Other commonly inverted features are shallow depressions in which evaporitic duricrusts (of many kinds) have formed.

Third-order landforms are produced at the actual sites of incision into duricrusts. These include steep vertical scarps or "breakaways" in the duricrust itself, caverns in the softer material beneath, and blocky talus slopes onto which fragments of the duricrust collapse (Blume & Barth 1979). In some places groundwater seepage may undermine duricrusted scarps, as in Botswana (Shaw & De Vries 1988). Where duricrusts cover an undulating landscape, incision produces complex "flat-iron" topography (Everard 1963). The most remarkable landforms in duricrusted landscapes are karstic: solution hollows, collapse features and cave networks. These occur in calcretes, where they might be expected, but also in ferricretes and silcretes (Goudie 1983c, McFarlane & Twidale 1987). It has been claimed that some of the large depressions in the central Sahara, in which oases such as Sabhah in Libya are found, are karst-like dolines and poljes dissolved from an ancient silcrete (Busche & Erbe 1987).

6.7.3 Soil–landform interaction

Over the long periods of time during which some desert landscapes have survived (e.g. in parts of the Sahara and Australia), soil (and duricrust) formation has at times outstripped denudation, while at other times the balance has been reversed. It has been tipped one way and the other both regionally and locally, by tectonism or changes in climate. When the local balance is tipped in favour of erosion after a long period of stability, the soils developed in stability are eroded and their materials spread down slope. As the balance tips back to stability, new soils form on the new surfaces, developing in the debris of older ones (e.g. Arakel 1982, Gile & Hawley 1966, Brewer et al. 1972).

CHAPTER 7

Surface particle concentrations: stone pavements

7.1 Pavement form and structure

Stone pavements are armoured surfaces composed of angular or rounded fragments, usually only one or two stones thick, set on or in matrices of finer material comprising varying mixtures of sand, silt or clay (Fig. 7.1). They occur on weathered debris mantles, alluvial deposits and soils in many different environments, ranging from periglacial and mountain areas to recently dug gardens. However, they are extremely common in hot deserts, where they adopt a variety of forms and have acquired many local names such as **gibber plains** or **stony mantles** in Australia, **saï** in central Asia, **desert pavement** in North America, and **hammada** and **reg** in north Africa and the Middle East. Hammada is a term used both for boulder-strewn outcrops or pavements as discussed here, and for rock surfaces with a residual scatter of stones. Reg is generally applied to pavements on alluvial surfaces that consist of gravel or smaller particles.

Stone pavements are not only ubiquitous desert features; they are also of fundamental importance, for at least six reasons. First, they exert an exceptionally important control on *surface stability*, and they provide *protection* for desert soils. They act as a barrier to the processes attacking the surface, and in this sense they can be seen as a substitute in aridity for vegetation. Secondly, as recipients of the attack by desert processes, they are a most valuable record of the *activities of processes at desert surfaces*. Thirdly, once disrupted they often do not heal quickly, so that accelerated erosion of formerly protected material becomes possible. Disruption comes in many forms, but one of the most important is the activity of "off-road vehicles" – for example, pleasure vehicles in the southwestern USA, and four-wheel-drive pick-ups used as camel substitutes in Saudi Arabia (e.g. Webb & Wilshire 1983).

Fourthly, pavements have provided an excellent source of material for stone tools and a basis for graphic art in several areas. Artefacts are common on pavements, many of them produced from local work sites (e.g. Brakenridge & Schuster 1986); and, no doubt partly because once a pavement is cleared it rarely recovers fully or quickly, pavement pictograms created by the removal of stones are enduring "art" forms. For example, in California, stone alignments on pavements in Panamint Valley may be as much as

Figure 7.1 A stone pavement on an alluvial fan in the Mojave Desert, California.

1,000 years old. Gravel clearings and intaglios on terraces of the Colorado River appear remarkably fresh, with only a partially reconstituted pavement, although they may be several hundred years old (e.g. Davis & Winslow 1965). Similarly, the famous geometrical "scrapings" or pictograms, up to 7km in length, on pavement surfaces near Nasca, southern Peru, remain clear today, although they may have been created up to 2,000 years ago. Fifthly, pavements may provide a fundamental but little appreciated *store* of loose sediment in transit by the wind. Each pavement particle has both a small upwind and a larger downwind storage area that can contain loose debris temporarily until winds of appropriate velocities remove it. Finally, stone pavements probably fundamentally influence surface infiltration and runoff characteristics (see below).

Pavements are best developed where vegetation is absent. There may be an excellent reason for this: pavement areas may have characteristics that adversely influence plant growth. In a study in Yuma County, Arizona, for example, Musick (1975) showed that, compared with adjacent non-pavement areas, pavement soils had *relatively* high salinity, relatively high exchangeable sodium percentage (i.e. high alkalinity) and, no doubt partly because of their high ESP which aids deflocculation of soil colloids, relatively *low* infiltration capacity. Of these factors, Musick thought that the last was most important (Table 7.1) because low infiltration capacity would tend to keep water near the surface, where it does little good to plants. Pavements do not even have salt-tolerant or alkali-tolerant plants. They are xeric habitats that shed water to promote growth in adjacent rills and washes. Low infiltration capacity may be related in many areas to surface sealing intensity, a feature arising from surface crust formation under raindrop impact (see below; and Poesen 1986). Abrahams et al. (1991) found a negative correlation between infiltration rate and stone cover on shrub-covered hillslopes in Arizona.

Lack of vegetation also promotes particle concentration processes. Pavements are particularly common where coarse and fine particles co-exist in a deposit and where *generally* unsorted alluvial deposits occur, such as in the areas of smooth, gently sloping surfaces between channels on the upper part of alluvial fans, and on the debris aprons on and adjacent to mountain fronts. Pavements occur in areas ranging in size from a few square metres to several hundred hectares.

The stone mosaic is composed of coarse particles which are either on, or lie embedded in, the finer matrix. Even in the most developed pavements there are gaps between particles. The density and spacing of particles is extremely variable, reflecting the original composition of the deposit, its degree of superficial disintegration, and the extent to which particle concentration has progressed. In California, Cooke recorded densities of up to 1.36 particles per cm^2 and noted that up the 17% of the ground surface at sample pavement sites was not covered by coarse debris in areas of well developed pavement; particle density increased and the spacing of particles decreased with an increase in the slope (Cooke 1970b). In Australia, silcrete gibber cover ranges from "complete" to 50% (Dury 1968). Gaps between coarse particles are occupied by finer material. If this contains a proportion of clay or salt, as it often does, the fine material is usually indurated to form a thin, strong carapace. If there are no binding agents, the material is loose.

The coarse particles are either *primary* or *secondary*. Primary particles are similar in almost all respects to the coarse particles found in the underlying materials. These are particles which have always been at the surface or have subsequently appeared on the surface. Secondary particles are derived from primary particles by processes of disintegration. Split boulders and granular disintegrates are common on pavements, and the weathering products are usually angular, often lying freely on the surface. They are produced by the weathering processes described in Chapter 5.2. Artefacts are a form of secondary particle common in pavement areas that were formerly suitable human habitats or sources of stone for implement manufacture (Ascher & Ascher 1965, Cooke 1970b). Superficially weathered fragments are often moved down slope from their parent particles.

A coarse particle may be veneered with a desert varnish (Ch. 6). The patina is a dull black rind on the exposed portion of fragments; on the buried portion it is usually an orange–brown stain; and at ground level

Table 7.1 Infiltration characteristics, Yuma County, Arizona.

Cumulative infiltration for 50 minutes	Amount	Rate	Depth
1. Non-pavement bajada	8.9cm	9.6cm h^{-1}	33cm
2. Runnel	7.7cm	6.0cm h^{-1}	31cm
3. Stone pavement	1.0cm	0.84cm h^{-1}	5cm

Source: Musick (1975).

it is a pronounced black band.

An important distinction may be drawn between pavements that are *underlain by soils* (such soils are regs, see Ch. 6) and those that are not. The former are the more common (Mabbutt 1965, Cooke 1970b), but in such circumstances is not necessarily correct to conclude that the pavement results from development over a long period of time on a relatively stable ground surface by processes acting within the soil profile.

A fairly common feature of many soils beneath pavements is the absence of coarse material in the upper part of the profile (Fig. 7.2). Often, there is a vesicular A horizon of mainly silt–clay-size particles. The fines may be of aeolian or pedological origin, and the vesicles are probably due to trapped air that expands as soils dry out in rapidly rising temperatures following rainfall (e.g. Evenari et al. 1974). However, it must be emphasized that, although this stone-free zone is common, coarse-particle distribution can vary from being uniform throughout the profile, to the extreme situation shown in Figure 7.2. In addition, where a significant soil profile has had time to develop beneath a pavement, the common A_v horizon is also often underlain by a reddened B horizon and calcium carbonate rich B_{Ca} horizon (see Ch. 6).

Stone pavements are not always associated with soil profiles. Pavements may be produced without soil development on relatively unstable ground surfaces by superficial processes, such as deflation and runoff, that import pavement particles. In some cases, further-

more, there may have been an inadequate period of surface stability to enable soils to form beneath the stone layer.

Pavements are immensely varied in their forms, for these and other reasons. However, such variety, which can be of substantial practical importance, has only rarely been recorded. For example, Doornkamp et al. (1980) recorded the distribution of pavement on the limestone dipslopes in Bahrain which varied according to the amount of sand available, the extent of wind abrasion, particle density and packing, the salinity of the surface and subsurface, the presence of gypsum and the proximity of bedrock to the surface. Again, in a survey of the wind erodibility of surfaces in Saudi Arabia, Jones et al. (1986) described in one region five different pavement types which were classified according to their particle size, density, stability characteristics, potential for storing mobile sediment, degree of abrasion and varnishing.

7.2 Processes of particle concentration

Pavements may arise from several different inputs of material, from several different surface processes, from changes in primary particles, and from changes related to soil profile differentiation. Figure 7.3 provides a comprehensive model of pavement formation. There are three main groups of particle-concentration processes: (a) deflation of fine material by wind; (b) removal of fines by surface runoff and/or

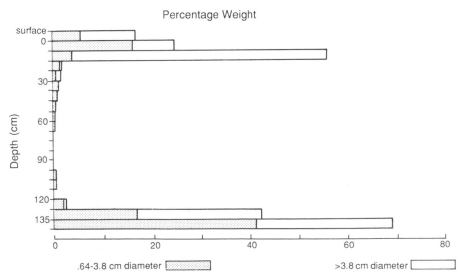

Figure 7.2 The stone distribution in pavement soils from central Australia (after Mabbutt 1965).

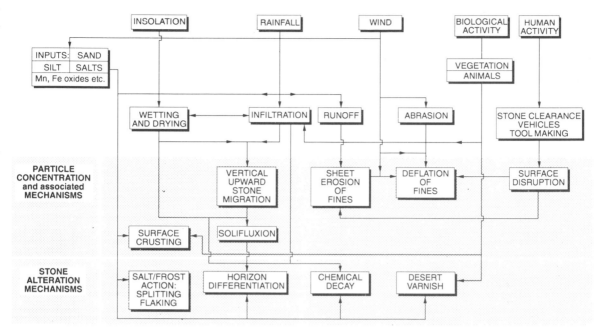

Figure 7.3 A model of stone pavement development.

creep; and (c) processes causing upward migration of coarse particles to the surface. In addition, pavements may evolve in close association with aeolian deposition and soil-profile differentiation.

7.2.1 Deflation

Stone pavements in hot deserts have usually been almost universally explained by the deflation of fine material from the surface, which leaves a residue of coarse particles. The concentration of coarse particles is seen as a function of their distribution in the original sediment and the extent of deflation. It is assumed that the accumulation of coarse particles at the surface represents a layer of soil similar in depth to a layer of uneroded soil with a similar quantity of coarse particles.

Simple and persuasive though the deflation hypothesis is, conclusive evidence of deflation is not always to be found. The proportions of sand, silt and clay in superficial sediments compared with the proportions for the sedimentary material as a whole may give an indication of the extent of deflation. For instance, Lustig (1965) showed that, in certain areas of Deep Springs Valley, California, silt:clay ratios are unusually high, and he suggested that these areas are related to the paths followed by winds which have removed much of the clay material. Evidence of wind abrasion on pebbles in pavement areas may also

suggest the probability of local deflation.

There is no doubt, of course, that loose, fine material *can* be removed by the wind, as experimental work has demonstrated (e.g. Symmons & Hemming 1968). Indeed, it is possible that, in circumstances in which deflation is the only process responsible for the superficial concentration of coarse particles, the distribution of coarse particles in a stable pavement surface may be expressed in terms of Chepil's (1950) "critical surface-roughness constant". This is defined as the ratio of the height of non-erodible surface projections above the surface to the distance between projections which will barely prevent movement of erodible fractions by the wind.

Arguments can be advanced *against* the ubiquitous application of the deflation hypothesis. Although desert winds are probably strong enough to move most loose fine material, in semi-arid areas scrubby vegetation may reduce winds to below the requisite threshold velocities, and surface vegetable litter may protect loose fines. Even more important is the fact that *loose* fine material is *unusual* in most deserts. Fine material is abundant, but it is usually formed at the surface into a thin **carapace** or **crust** (the **Staub-haut** of Mortensen 1927), *which is strong, dense, relatively impermeable and protects the underlying fines from erosion, especially by the wind.* The crust is up to about 3mm thick and is composed of a "washing-in layer", and a thin upper coating of clay particles orientated parallel to the surface. There is also evidence that surface crusting is associated in many deserts with the growth of micro-organisms between surface particles (Campbell et al. 1989). Mineral grains are commonly bound together by extracellular metabolic products (sheaths) or by the motile behaviour of filamentous cyanobacteria (especially *Microcoleus vaginatus*). Once established, the microbial crust increases the water-holding capacity and organic content of the soil and thus promotes a successional sequence; it also helps to increase infiltration capacity and reduce runoff and erosion, and thereby to stabilize the surface (see also Ch. 5.2). Beneath the crust, a vesicular layer is often developed.

Crusts formed by rainfall develop as follows. Wet soil aggregates are first broken down by raindrop impact, and fines are washed into surface pores, reducing their volume. After aggregate destruction, raindrop impact causes compaction of the surface, and produces the thin coating which is extremely impermeable. The compaction is a function of the size and terminal velocity of raindrops, and hence it might be expected to be extensive in desert areas experiencing intense rainfall. Laboratory studies by Poesen (1986) suggested that the crusts play a significant rôle in *surface sealing* and hence in reducing the infiltration capacity. Poesen showed that the surface sealing intensity in loose sediments is inversely proportional to the slope angle. On steeper slopes, the sealing is obstructed by intense rainwash and rill erosion, so that infiltration and percolation rates are reduced. On gentler slopes, where pavements are often best formed, the reverse is true. In addition, Poesen discovered that the *position* of surface stones affects the sealing intensity, and hence the infiltration capacity. For some soils the sealing intensity is lowest where stones lie *on* the surface, and it is higher where the stones are embedded *in* the surface.

The crust has been observed in many deserts, such as Australia, Chile and Israel (e.g. Mabbutt 1965, Mortensen 1927, Sharon 1962). In pavement areas it commonly shows little sign of having been deflated. Deflation will occur only if the carapace is broken by, for example, prior wind abrasion or vehicular traffic.

A further possibility is that pavements are sites of aeolian *deposition*, and that the silts of the surface crust and the A$_v$ horizon beneath the pavements are actually loess *deposits*. The **parna** deposits in Australia are of this origin. Such deposition may continue while particles remain at the surface (Sec. 7.2.4).

7.2.2 Water sorting

Experimental observations show beyond doubt that some stone pavements are often composed, at least in part, of coarse particles that remain after finer materials have been dislodged by raindrop erosion and removed by running water. Sharon (1962) demonstrated that surface runoff is the dominant process of pavement formation at sites in Israel. At cleared pavement sites in California, Cooke showed that surface fines and some buried coarse particles were removed by surface runoff and collected in sediment traps down slope; after only four years, stone pavements were becoming re-established at these sites. Experimental work on stony soils by Lowdermilk & Sundling (1950) revealed that the amount of fine-sediment removal under conditions of uniform rainfall intensity increased directly with gradient. Furthermore, they showed that, as the experiments proceeded, the amount of material removed diminished and a stone pavement emerged, first on steeper slopes and later on gentler slopes, until the test sites were ultimately covered with a stable pavement surface. In the absence of vegetation, stone pavements are an important influence on hydraulic roughness and may actually tend to disperse runoff, or at least retard flow concentration.

The frequency of runoff, and the extent of sediment dislodgement and removal, is largely a matter of speculation in most desert areas. However, two points deserve emphasis. First, the effect of raindrops on particle detachment may be considerable in arid and semi-arid areas where rainfall intensity and raindrop momentum are occasionally high, and there is little vegetation to inhibit direct access of raindrops to the ground. Secondly, runoff necessary for the removal of dislodged particles may be promoted in pavement areas by the relative impermeability of the surface crusts. These two features are not unrelated, however, for although raindrops cause detachment, they also promote crust development which, as rainfall continues, reduces soil splash (e.g. Farres 1978). Reduced permeability related to crusts may be offset partly by increased permeability due to cracking in clay materials. As indicated above, crust impermeability relates to surface sealing intensity (Poesen 1986).

The occasional saturation of pavements may cause fine material at the surface – and especially in stone-free upper soil horizons – to *flow* slightly. For example, if coarse particles are removed from a saturated pavement, the depressions which are left tend to fill with mud, adjacent particles jiggle slightly, and the surface settles down to a new equilibrium (e.g. Denny 1967). This settling phenomenon probably explains the *overall smoothness* of many pavement surfaces.

7.2.3 Upward migration of coarse particles

The concentration of coarse particles at the surface and at depth, and the relative scarcity of coarse particles in the upper soil profile (Cooke 1970b, Mabbutt 1965, Springer 1958; Fig. 7.2) prompts the suggestion that stones may have moved upwards through the soil to the surface. Two main mechanisms might be involved in such movements: cycles of freezing and thawing, and cycles of wetting and drying.

Laboratory experiments have shown that alternate freezing and thawing of saturated mixed sediments causes coarse particles to migrate upwards. However, there is some disagreement as to the precise mechanism. Corte (1963) demonstrated that the movement of particles may depend on the amount of water between the ice/water interface and the particle, the rate of freezing, the distribution of particles by size, and the orientation of the freeze/thaw plane to the surface. Inglis (1965) described a simple mechanism as follows (Fig. 7.4). In a slurry frozen from above

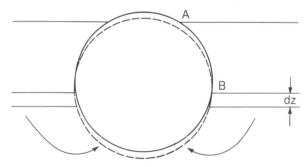

Figure 7.4 The vertical movement of a spherical stone in a partially frozen material (after Inglis 1965). (For explanation, see text; copyright 1965 by the American Association for the Advancement of Science.)

down to line A, the adhesion of the ice to the top of the large sphere will be strong enough to support its weight; subsequent freezing of a layer of thickness z causes its expansion by the amount dz (where d is the coefficient of expansion of water and sediment) which

lifts the frozen slurry and the sphere upwards. Because of the rigidity of the sphere, a cavity would be left beneath it except for the fact that unfrozen slurry will move in to fill the cavity. With thawing from above down to level B, the sphere will be supported by frozen slurry beneath it. As the thawing surface advances a distance z, contraction permits the whole mass of material above it to fall by the amount of dz. But the sphere, still unmoved, protrudes into this descending material. The net effect of both movements of the freezing plane is for the large sphere to move upwards through the slurry.

Although sorting by freeze/thaw action may seem improbable in hot deserts today, it cannot be entirely ruled out, especially in high-altitude deserts such as those in Bolivia and central Asia. It may have been more effective in some places during cooler, moister periods of the Quaternary. In addition, it is conceivable that, in very saline localities, the rôle of water, ice and a freezing plane in bringing coarse particles to the surface is replaced by water, salt and a desiccation plane (Fig. 7.5).

Figure 7.5 The large particles on this pavement surface in Bahrain may have been moved upwards to the surface by processes at work within the highly saline, damp puffy ground that lies within the capillary fringe.

A much more effective and widespread migration

mechanism in deserts is thought to be related to wetting and drying of the surface soil. Laboratory experiments by Springer (1958) on pavement soils from Nevada showed that the distance between the top of the soil and the tops of some particles placed in it decreased by up to 2.2cm after 22 wetting and drying cycles. Jessup (1960), using stony table-land soils from Australia, demonstrated maximum upward movement of coarse particles, after 22 wetting/drying cycles, of one centimetre; and Cooke induced four pebbles to emerge at the surface from a depth of 1.2cm in a pavement soil sample from the Mojave Desert, California after only four wetting/drying cycles. The precise nature of this process is uncertain, but Springer (1958) logically suggested the following explanation. When the soil is wetted, it expands and a coarse particle is lifted slightly; as the soil shrinks on drying, cracks are produced around the particle and within the soil; and because of its large size the coarse particle cannot move down into the cracks, whereas finer particles can. The net effect is an upward displacement of the coarse particle. An important prerequisite for the operation of this process is the presence of expanding clay minerals in the fine material, and the process is probably most effective where such minerals are most abundant. McFadden et al. (1987) also pointed to the efficacy of wetting and shrinking of columnar peds in well developed A_v horizons in displacing surface clasts. If upward migration alone concentrated coarse particles at the surface, the density of surface particles would be a function of the density of particles within the wetting/drying zone, and the length of time the process had been at work.

7.2.4 Interplay of processes through time

The relative importance of, and interplay between, these processes in creating the pavements at a site is far from clear, although it is probable that all can operate independently in suitable circumstances. However, it would be wrong to assume that the processes described are the only ones responsible for pavements. Recent studies by McFadden et al. (1987), for example, have thrown the whole matter into a new light. They argued that on the aeolian-mantled basalt lavas of the Cima volcanic field in the Mojave Desert, *the pavements are born at the surface*, and the individual pavement stones were *never* buried. The pavements in this area were said to evolve as follows.

(1) Basalt is broken up by weathering into clasts, a process accelerated by the ingress of aeolian silt into cracks and its subsequent volume changes with wetting and drying.

(2) The clasts move from higher areas to lower areas that are already filled with aeolian silt and clay, by colluvial and alluvial processes (such as creep and wash).

(3) Deposition of aeolian debris continues.

(4) The aeolian mantle is altered pedogenically, to create, for example, the vesicular A horizon.

(5) Finally, the clasts are maintained at the surface, while (3) and (4) continue, by processes that inhibit burial.

In this explanation, therefore, *no aeolian deflation or upward migration is necessary*, although it is possible that both, and especially the former, may occur from time to time. *Both aeolian deposition and the development of the A_v horizon are crucial.* McFadden et al. (1987) and Wells et al. (1985) also believed that pavement development was associated over long periods with phases of aeolian deposition; for example, in the late-Pleistocene and early Holocene, when nearby pluvial lakes dried out (Fig. 7.6).

This model may well have wider applications. Certainly many (but not all) pavements in different parts of the desert realm are associated with soil-profile development, and some (but not all) are associated with silty A_v layers that could be of aeolian origin. It may well apply, with local modification, where these features exist together with an initially irregular surface (such as a bar-and-swale topography on an alluvial fan).

Another perspective on the nature of pavement evolution over soil profiles was well illustrated by Amit & Gerson (1986) on a series of 15 Holocene alluvial terraces in the Dead Sea region, where the degree of pavement development (measured by per cent surface cover of stones and sorting, for example) is most pronounced on the oldest surface and weakest on the youngest surface (Fig. 7.7). Such changes are associated over 14,000yr or so with surface weathering, and contemporaneous soil-profile differentiation. The whole group of changes may provide a relative and approximate means of dating otherwise undatable surfaces, although the relative periods between the formation of the different surfaces may be unclear. In this area, as in the Mojave, the aeolian input of silt and salt is fundamental, and the creation of a full stone cover helps to reduce infiltration capacity and promote runoff.

Elsewhere, it is not uncommon to find pavements on surfaces developed before the last glaciation (e.g.

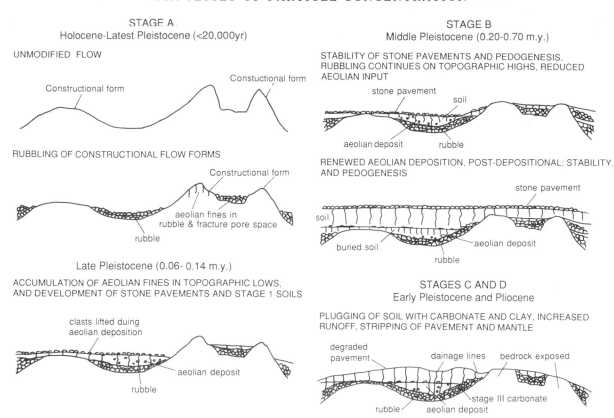

Figure 7.6 Schematic topographic and stratigraphic sections of late Cenozoic basaltic lava flows in the Mojave Desert, California, illustrating stages of flow-surface modification. The evolution of basaltic flow-surfaces is as follows. Stage (A): accretionary mantle deformation on flow-surfaces by weathering and colluviation of bedrock topographic highs and by aeolian deposition in topographic lows. Stage (B): flow-surface stability; decreased weathering, colluviation, and aeolian deposition; increased soil development; renewed aeolian deposition and soil burial; long-term surface stability; and the development of thick argillic B and carbonate horizons in soils. Stage (C): decreased permeability and increased runoff; increased surface erosion, soil stripping, and bedrock exposure. Stage (D): petrocalcic fragments (caliche rubble) exposed on deeply dissected flow-capped mesas. (After Wells et al. 1985.)

Wells et al. 1985) or in the past few years (e.g. Sharon 1962). Whatever the processes and their relative importance may be, it remains clear that once established, stone pavements protect the underlying surface from erosion by wind or water, and permit the progress of soil profile development. Indeed, many mature pavements may be both "wind-stable" and "water-stable".

7.2.5 Disruption and recovery: applied aspects
Stone pavements protect desert surfaces. This is a fact of great potential value to those seeking to limit surface erosion and to stabilize surfaces. The message from the previous discussion is clear: *protect surfaces by not disturbing pavements; stabilize surfaces by creating pavements.*

As noted, pavements have relatively low infiltration capacities, so that runoff from them may be relatively high. This is a phenomenon discussed further in Chapter 10.2. The disturbance of pavements by off-road vehicles (ORVs), through the shearing and compressive action of wheels, leads to a reinforcement of this characteristic. The vehicles disrupt pavements, compact surface soil and subsoil, strip protective vegetation, and cause erosion directly. Penetrometer studies of this disruption in the Mojave Desert (e.g. Wilshire & Nakata 1976) showed that *compaction* was a main consequence, along with a consistent increase in the *variability* of compaction. This led to reduced infiltration capacity and thus to increases in runoff (e.g. Eckert et al. 1979), together with either a reduction or an increase of wind erosion, an increase in the time the surface is vulnerable to erosion, reduced seed germination and a reduction in the small mammal population. The ability of vehicles to cross pavements can change dramatically. When dry, pavements often provide excellent driving surfaces; when wet, they can become impassable quagmires.

Once the damage has been done, recovery may be slow and difficult. Pavement intaglios, some of them over 500 years old, still have not fully recovered, and are today marked by an immature pavement, often of smaller particles. ORV tracks in the Mojave Desert, in contrast, seem to have recovered in places within a few decades (e.g. Elvidge & Iverson 1983). Elsewhere, recovery may never occur. Much depends on the availability of stones, and the effectiveness of compaction, which may restrict pavement-producing processes.

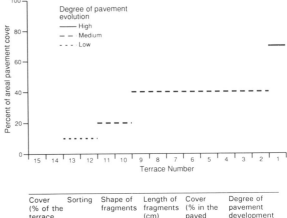

Cover (% of the terrace surface)	Sorting	Shape of fragments	Length of fragments (cm)	Cover (% in the paved segment)	Degree of pavement development
70+	good	platy	2–5	70	high
40–20	medium	platy & rounded	2–15	40–50	medium
20	bad	cobbles & pebbles angular & rounded	10–15	30	low

Figure 7.7 The degree of desert pavement evolution on terraces in the Dead Sea region, Israel (after Amit & Gerson 1986).

CHAPTER 8

Volume changes: patterned ground phenomena

Patterned ground phenomena – circles, nets, polygons, steps and stripes – have been widely reported in high latitudes and at high altitudes, where they have generally been explained in terms of frost action and solifluction. However, similar features are found in other climatic environments, and they are quite common in some deserts. Although frost action cannot be completely disregarded at present or in the past as a cause of patterned ground in some arid and semi-arid lands, ground patterns in deserts appear to be generally attributable to two different, and often related, groups of processes, both of which involve changes at or near the ground surface: wetting and drying, and solution and recrystallization of salts. The effects of the first group of processes are most apparent in fine sediments or "swelling" soils, especially where they contain a high proportion of swelling clays such as montmorillonite. Such soils occur commonly on the lower plains of deserts (Fig. 4.2). The effects of the processes associated with salts are manifest in zones of salt concentration, such as around the margins of ephemeral water bodies, along drainage channels, and where the capillary zone of water movement in the soil intercepts the surface.

8.1 Wetting and drying phenomena

8.1.1 Gilgai

The term **gilgai**, an Australian aboriginal word meaning a small waterhole, is applied to certain small-scale surface undulations in clay-rich soils. Most gilgai appear to be related to the differential expansion and contraction of the soil with periodic wetting and drying. Although gilgai are usually associated with fine-grained alluvial soils, they are also found in stone

pavement areas where particle sorting may occur (e.g. Ollier 1966). Gilgai have been reported from arid and several semi-arid areas, including Coober Pedy (Ollier 1966) and New South Wales (Hallsworth et al. 1955) in Australia, the central Sahara (Meckelein 1959), and the Middle East (Harris 1959, White & Law 1969), and on tropical black earths in east and central Africa (Stephen et al. 1956), South Dakota (White & Bonestall 1960) and elsewhere (Verger 1964).

Elements of gilgai topography often include puffs or mounds, depressions ("crab-holes" or "melon-holes") or channels, and the shelf areas between them. Not all elements are always present in any one area. Verger (1964) constructed a simple classification of puffs and depressions according to their shape, and their grouping in plan. Round gilgai and some network gilgai occur on flat ground; lattice, wavy and some network gilgai occur on gently sloping ground, and their puffs or depressions are generally orientated parallel to slope contours.

The dimensions of the different elements are extremely variable. The vertical interval between puff, crest or depression floor and shelf may vary from a few centimetres to about 3m; the diameter of non-linear positive or negative features may be up to about 50m; linear features may be many metres long and as much as 12m wide.

The many explanations of gilgai and related features extend to the bizarre and the ridiculous. It is improbable that all gilgai result from the same process, but many areas of gilgai display diagnostic characteristics. A feature common to most gilgai areas is the presence of soils that expand and contract differentially on wetting and drying. Beyond the recognition of this simple fact, the analyses of Hallsworth & Beckmann (1969) and Hallsworth et al. (1955) provide several generalizations which are probably of wide relevance.

(i) The vertical interval in well developed gilgai soils is often significantly, directly and principally related to the swelling capacity of the clay (which provides the force for soil movement), and to the sodium saturation of the exchange complex (the sodium ions adsorbed to the clay complex produce larger, more resistant clods, and the larger the clods moved by swelling, the greater the amplitude of the undulations).

(ii) Gilgai usually occur in soils where the clay content, the montmorillonite content and the proportion of sodium on the exchange complex all increase with depth, and where, in consequence, subsoil swelling is significantly greater than surface swelling (which is itself often over 10%). (However, not all gilgai soils include montmorillonite as a major clay mineral.) The type of gilgai that forms is related to the depth of the layer of maximum expansion and to the variations in expansion between different layers. The surface layer should be sufficiently coherent to allow subsoil cracks to extend to the surface, and thin enough to allow the drying front to extend into the swelling horizon.

(iii) In addition to soil properties, gilgai formation depends on an external set of variables, of which the climatic conditions (controlling the nature of wetting and drying) and the soil moisture conditions are probably the most important. For instance, if the swelling layers never dry out, movement of the wetting front will be minimal and gilgai would be unlikely to form. Gilgai formation is most likely in areas with a marked alternation of wet and dry seasons.

(iv) A key consideration in explaining the patterns of gilgai development is the *differential* nature of the process. Hallsworth et al. (1955: 25-6) offered the following explanation:

> As the soil dries out, the topsoil dries first, shrinking and cracking as drying proceeds. This cracking aids the drying out of deeper layers, which, as drying continues, also shrink, ultimately to a considerably greater extent than the topsoil . . . When re-wetting takes place . . . the whole soil swells again. If the drying out and subsequent shrinkage had been uniform, and had decreased uniformly from surface to subsoil, re-wetting and consequent swelling would merely have restored the soil level to what it was before.

In a deeply cracking soil, however, uniformity of drying and shrinkage is unlikely, since the moisture content will show a horizontal gradient from the face of each crack into the interior of the block, as well as the general vertical gradient from surface to subsoil. Re-wetting will also be uneven. Light showers will merely wet the surface, whilst heavy rains . . . will wet and swell the subsoil at the base of the cracks long before the rest of the subsoil at the same level has had any change in moisture content. Uneven re-wetting is consequently likely to be the usual occurrence on such soils.

The cracks themselves also contribute indirectly to the uneven distribution of pressures on re-wetting. During the dry period pieces of surface soil fall down the cracks, and the fact that the subsoil shrinks more than the topsoil, giving cracks that are wider below than they are at the surface, makes it easy for large pieces of the surface soil to break off and to slip down into the crack below. Subsequent re-wetting and rehydration would lead to excess pressure causing the fracturing of blocks of soil from that lying below. Repetition of this process would cause the blocks to move upwards towards the surface, since subsequent cracking and fracturing would start to occur again about the position of the earlier cracks, which would have become lines of weakness.

On the basis of this argument and the similar one of Verger (1964; Fig. 8.1) it is probable that the production of isolated pressure zones and the subsequent development of gilgai may be closely related to the prior formation of cracks and crack patterns.

Patterns of interlocking gullies in enclosed drainage depressions on the Mesopotamian plain of Iraq and on the Nile flood basins in the Sudan appear similar in certain respects to network gilgai described in Australia, and a similar origin has been attributed to some of them (Harris 1959). However, White & Law (1969) identified significant differences between the two areas of gully patterns (**tabra** channels), and pointed out that in places they lacked some of the features noted by Hallsworth et al. (1955) in Australia, such as a high expanding-clay content and an

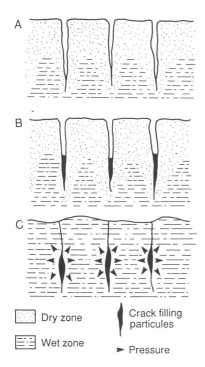

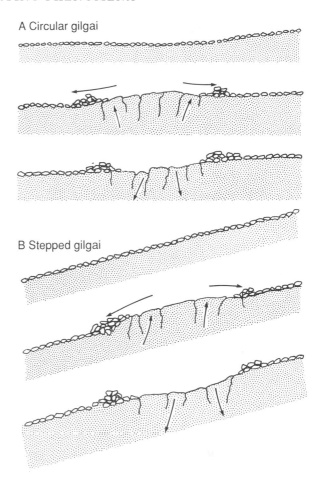

Figure 8.1 The formation of gilgai by cycles of wetting and drying, crack filling and crack expansion (after Verger 1964).

increase of exchangeable sodium with depth. They suggested that in Iraq and the Sudan the patterns have formed entirely by the local intensification of compaction caused by water standing for long periods in pre-existing surface depressions, rather than by heaving of soil in the areas between channels. The pre-existing depressions were rills which have become inactive, abandoned irrigation ditches, or features associated with desiccation cracks. As with other explanations, periodic wetting and drying out is essential to the effectiveness of the process. This alternative explanation yet again illustrates the fact that similar phenomena in different areas are not necessarily produced by the same processes.

Ollier (1966) recognized two types of gilgai in gibber plains of the central Australian Stony Desert, both of which involve a degree of particle sorting: **circular gilgai** and **stepped gilgai** (Fig. 8.2). Circular gilgai consist of bare clay topsoil shelves surrounded by slightly raised rims of pebbles. They are up to 9m wide and they occur on flat surfaces. Stepped gilgai occur on slopes, run parallel to slope contours, are up to 3m wide, and are composed of bare clay and

Figure 8.2 The development of (A) circular and (B) stepped gilgai in the central Australian Stony Desert (after Ollier 1966).

topsoil shelves with pebble accumulations on their upslope and downslope sides. Both features are associated with soils containing considerable quantities of swelling clays. Figure 8.2 summarizes the evolution of these features. If a patch of soil is wetted, it expands, and a small dome is formed; pebbles on the surface of the dome move towards its edges and settle there; when the soil dries, the clay contracts, and the clay shelf is slightly lower than the pebble rims; when dry, wind may remove some of the clay particles from the shelf. On flat surfaces (Fig. 8.2A) pebbles accumulate around the dome; on sloping surfaces (Fig. 8.2B), stones move down slope, and even the step itself may move *en masse* (with a gypsite layer at the base of the soil profile perhaps acting as a lubricating layer).

It is not clear, on present evidence, whether these or other gilgai are being continuously formed and destroyed, whether they are being continuously modified over a long period of time, whether they soon reach a condition of equilibrium, or whether they are fossil forms. It is known, however, that in some areas where gilgai have been eliminated by cultivation they will re-form in several years, so that the processes are apparently active at present (Hallsworth et al. 1955).

Features similar to stepped gilgai are the small steps which interrupt generally smooth but sloping stone pavement surfaces. They are characterized by risers of pebbles, and treads of finer material which is often capped by a thin veneer of stone pavement. In some areas, such as in Panamint Valley, California, the steps appear to be produced by the accumulation of fine material up slope of large boulders that extend through the soil profile. Elsewhere, in Death Valley, California, for example (Denny 1965, 1967), downslope creep of fine, relatively stone-free material *beneath* the pavement when it is saturated may lead to the development of steps. Hunt & Washburn (1960) indicated that the treads of some steps had up to ten times as much salt as stable ground around the steps, which might indicate that the salts may play a rôle in developing the features.

8.1.2 Desiccation cracks

As a saturated fine-grained, cohesive sediment dries out, it may pass through liquid, plastic and brittle-solid phases, and its volume is reduced until its **shrinkage limit** is reached. Volume reduction up to the shrinkage limit caused by evaporation of water from the sediment is often accompanied by sufficient tensional stress for rupture to occur and cracks to be formed. In general, the morphology of rupture patterns depends mainly on the intrinsic conditions of the material (such as moisture content, structure and degree of packing) and on extrinsic conditions of the environment (temperature, humidity, rate of desiccation, etc.) (Corte & Higashi 1964). There has been surprisingly little experimental study of desiccation cracks, and the comments that follow here rest largely on the work of Corte & Higashi (1964), Lachenbruch (1962), the early work of Kindle (1917), and on Maizels' (1987) study.

The microtopography of a desiccated sediment is composed of patterns of *cracks* and of polygonal *blocks* (flakes or cells). Cracks are usually fairly straight or smoothly curved in plan; but their lengths,

Figure 8.3 Concave mud flakes west of Chuquicamata in the Atacama Desert, Chile.

depths, widths, numbers and patterns vary greatly (e.g. Chico 1963). The plan shape of blocks is determined, of course, by the crack pattern. The profile of blocks may be flat, convex, concave, or irregular (Fig. 8.3). Many of the morphometric properties of cracks and blocks may be significantly related. Corte & Higashi (1964), for instance, in a series of experiments involving stone-free tills of different initial degrees of compaction, demonstrated relationships between the mean size of blocks and the total length of cracks and the thickness of the material. The mean size of the blocks is proportional to a power function of the thickness of the material, depending on the material at the base of the experimental trough and the dry density of the material; and the total length of cracks is also proportional to a power function of this thickness, depending on the same factors.

Maizels (1987) argued convincingly that the processes of crack formation in sediments are similar in periglacial and arid environments because, in both, contraction is the result of *effective volumetric loss of free moisture* (and it may be relevant that the coefficients of contraction and expansion of salt (arid) and ice (periglacial) are similar). Thus, her laboratory studies of crack development in *different sediments* under similar freeze/thaw conditions provide, together with the results of studies by others, some useful general ideas that are relevant to the study of desiccation cracks in deserts.

(i) The *number* of cracks is inversely proportional to sediment size, being 40–50 times greater in silt and clay than in sand and gravel (Fig. 8.4).

(ii) The cracking *pattern* is often established quickly, with 70% of the final cracks being initiated in the first three freeze/thaw cycles.

(iii) The *mean* crack *length* ($\bar{L}_c$) is directly proportional to sediment size (*d*) in the form:

$$\bar{L}_c = 10(1.41 - 0.29 \log \bar{d})^{-1} \qquad (8.1)$$

This is because texture influences sediment moisture content and retention, and therefore internal moisture stresses. *Total* crack length is inversely proportional to sediment moisture content.

(iv) Crack length shows a tendency to *decrease* with time, because new, short cracks continue to be formed during each new cycle.

(v) The proportion of "linked" cracks (*L*) *increases* over time for each sediment type, a feature that is dependent on the relative change of the crack length and frequency (*density, P*), such that

$$L = f(\alpha P) \qquad (8.2)$$

where α is constant for sediment of a given texture.

(vi) Research by Kindle (1917) suggested that the *spacing* of cracks may increase with the *rate* of desiccation and the *proportion* of clay present in the material. The *type* of clay present (Chico 1963) and the *cohesive properties* of the material may also influence the extent of contraction. For instance, montmorillonite-rich sediment is likely to contract more than a sediment with a comparable proportion of kaolinite.

(vii) The occurrence of *stones* in and on fine-grained sediments affects the nature of cracking in several ways (Corte & Higashi 1964). First, when they are appropriate distances apart, surface stones act as starting points for cracks. Secondly, the shape and porosity of stones significantly affect cracking patterns. For instance, Corte & Higashi (1964) showed that porous cubes developed cracks radiating from their corners, whereas impervious and semi-porous cubes often became surrounded by semi-circular cracks. A third interesting phenomenon noted by Corte & Higashi is "habituation" – the repetition of crack patterns when wetting and drying are repeated. Habituation occurred when there were layers of shale on the surface, but not when there was a surface cover of gravel. A possible explanation of this difference is that under the influence of wind and rain the shale pieces

A Coarse/Medium Sand

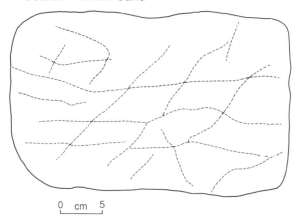

0 cm 5

B Clay

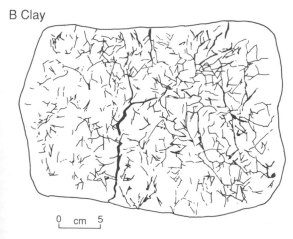

0 cm 5

Figure 8.4 Surface cracking patterns produced after 12 freeze/thaw cycles under experimental conditions in (A) coarse-medium sand and (B) clay (after Maizels 1981).

moved only a little into the cracks and were left tilted on the depressed surface of the old cracks, whereas the gravel moved farther down into the cracks. The mixture of clay and gravel is probably more cohesive than the clay, and the next generation of cracks therefore developed in new, weaker positions. The clay beneath the shale responded to the next desiccation in the same way as before. Finally, Corte & Higashi demonstrated that wind and water spray on unvegetated, cracked surfaces can lead to particle concentration in the cracks to produce stone nets etc.: some stone patterns in deserts may result from this process.

(viii) The surface of *blocks* between cracks may in profile be concave, convex, flat or irregular. The concave surface is attributed to the more rapid drying and shrinking of the surface layer than the material immediately beneath. In its extreme form, the concave upper layer may break away from the underlying sediment and produce a **mud curl** (Fig. 8.3). Occasionally, the cleavage line may represent a boundary between unlike sediments. Optimum conditions for mud curl development appear to be the rapid drying of a very thin mud layer.

Kindle (1917) demonstrated that convex block profiles could be formed when a considerable proportion of salt is present in the fine-grained matrix. When such a mixture dries out, a contraction-crack network develops, but there is also slight expansion which produces up-arching and which is sometimes accompanied by salt efflorescence on a "puffy" surface. Field observations suggest that these blocks are often moister than flat or concave blocks nearby. One possible explanation may lie in the hygroscopic nature of some salts.

A possible explanation of flat blocks is suggested by the previous comments on other profiles: they may be formed when relatively thick salt-free material is slowly dried.

(ix) Lachenbruch (1962) recognized two common patterns of cracks: an **orthogonal** system, in which cracks meet at right angles, and a **non-orthogonal** system which is characterized by tri-radial intersections commonly forming obtuse angles of about 120° (Fig. 8.5). Both systems may occur in a single area. Crack systems of different dimensions may also be found together. On the surface of Panamint Playa, California, for example, networks of small mudcracks coexist with "giant" desiccation polygons.

Desiccation patterns are developed according to the principle of least work in which polygons are compelled to have a maximum area (Hewes 1948). The area of polygons probably varies directly with the total work done and inversely with the square of the tension (Hewes 1948):

$$\frac{W}{A} = \frac{T^2}{E} \qquad (8.3)$$

Where W is the total internal work, E is the modulus of elasticity, T is the tension, and A is the area.

According to Lachenbruch (1962), orthogonal systems are probably characteristic of inhomogeneous or plastic media in which stress builds up gradually, with cracks forming first at loci of low strength or high stress concentration. As cracks do not form simultaneously, a new crack tends to join a pre-existing one orthogonally.

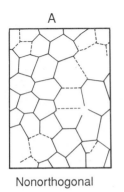

Nonorthogonal

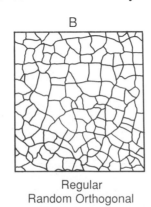

Regular
Random Orthogonal

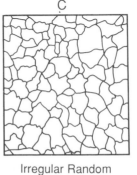

Irregular Random
Orthogonal

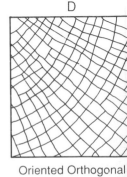

Oriented Orthogonal

Figure 8.5 Common systems of desiccation cracks (after Lachenbruch 1962, Neal 1965b).

Orthogonal systems are of two types (Fig. 8.5): those showing preferred orientation ("oriented orthogonal"), and those without preferred orientation ("random orthogonal").

Non-orthogonal systems probably form in very homogeneous, relatively non-plastic media which are dried uniformly. Lachenbruch suggested that under these circumstances "cracks propagate laterally until they reach a limiting velocity . . . and then branch at obtuse angles. The branches then accelerate to the limiting velocity and bifurcate again" (Lachenbruch 1962). All elements in this network are thus generated virtually simultaneously. Corte & Higashi (1964) independently reached a similar conclusion: cracks in orthogonal systems develop sequentially, whereas those in non-orthogonal systems form virtually instantaneously.

Geometrically regular crack patterns, as in a hexagonal system, are unusual in reality because uniform conditions of material and uniform desiccation are

rare. It is probably more realistic to assume a random distribution of high-stress points or weak points from which complex, irregular polygonal patterns develop.

Giant desiccation fissures are found in some alluvial sediments, and especially on playas. Their suggested causes have included seismic events; piping erosion (Ch. 8.2); tensional cracking, due to subsidence resulting from differential compaction caused by hydrocompaction or groundwater withdrawal for irrigation (Ch. 8.2); salt mobilization within playa sediments; shrinkage due to desiccation; or some combination of these.

On playas, which afford attractive, flat surfaces for vehicular movement, these giant fissures are dangerous, so that considerable effort has been expended in studying them (e.g. Neal 1965a, Neal 1968).

Giant desiccation features have been recorded in detail in playas of former Quaternary lakes in the western USA that are usually characterized by little groundwater discharge from the capillary fringe, rigorous desiccation, and hard, dry compact crusts composed largely of clays experiencing considerable volume changes on drying (Ch. 14.2). The fissures may be up to a metre wide, are sometimes over a metre deep, and may be up to several hundred metres long. They may occur in isolation, or they may form irregular random orthogonal, striped or other patterns (Fig. 8.6). Fissures may be continuous, discontinuous or merely a series of aligned holes. Cavities beneath the surface between holes suggest that the fissures may be initiated beneath the surface.

There can be little doubt that many of the giant fissures on playas result from the shrinkage by desiccation of playa material, but their great size suggests that high-magnitude forces are responsible for them. It seems probable that the large stress required to produce the fissures builds up gradually, perhaps over several years, and is concentrated beneath the surface in a slowly changing capillary fringe. Intense evaporation extending deep into the playa material, arising from desiccation following the last pluvial period, may have played an important part in lowering the capillary fringe and promoting fissure formation (Fig. 8.7). In some areas of the USA, the water table may have fallen in recent years as a result of human exploitation of underground water resources. As most fissured playas in the western USA are located in tectonically unstable areas, it is also possible that earthquakes may trigger the release of tensile stress.

Another feature described by Neal (1965b; Fig. 8.7B) is the "ring fissure", which is produced around

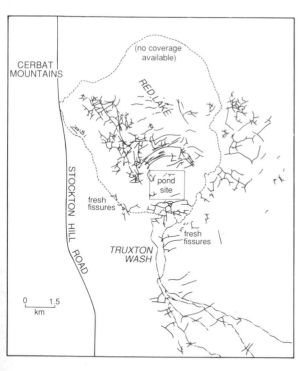

Figure 8.6 A generalized map of giant desiccation polygons, earth fissures and areas of recent fissuring, Red Lake Arizona, based on the interpretation of aerial photographs. The fissures are attributed to long-term desiccation and local subsidence due to water extraction and/or subsurface salt flow (after Lister & Secrest 1985).

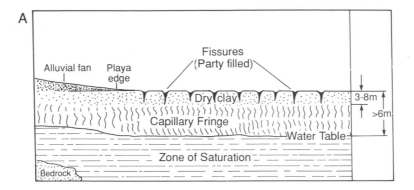

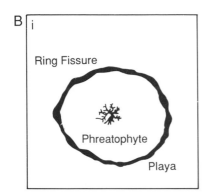
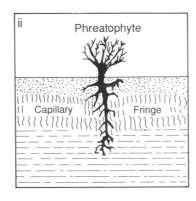
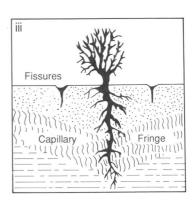

Figure 8.7 (A) Hypothetical cross section of a giant fissure environment, showing the relative positions of the water table, capillary fringe and fissures (after Neal 1965b). (B) Ring fissures: (i) plan view; (ii) before and (iii) after a sequence leading to fissure formation, showing the lowering of water levels with increased growth of the phreatophyte and associated drying of the superficial sediment (after Neal 1965b).

a phreatrophytic plant that lowers the water table and capillary fringe as it grows, dries out surface material, and causes shrinkage and cracking.

Once they are formed, cracks are usually modified fairly rapidly. A crack may suffer wind abrasion, and gradual infilling by windblown material and by debris washed into the crack when the surface is flooded or runoff occurs. Cracks may be zones of slightly greater moisture, and plants may grow in them. Eventually, cracks may be completely eliminated and their only trace may be lines of bushes or slight surface discolorations.

8.1.3 Applied contexts

Wetting and drying phenomena, and especially ground-heave associated with swelling soils, and ground-cracking accompanying desiccation at both small and large scales, pose potentially serious problems for construction. For example, Camuti et al. (1986) recognized substantial areas in Addis Ababa, Ethiopia, with high ground-swelling potential, where special foundation designs and other responses are essential for safe construction. The sudden creation of giant fissures in playas can seriously damage runways and roads, as in the western USA (e.g. Neal 1965a).

8.2 Piping and subsidence phenomena

8.2.1 Piping

Piping (or *suffossion* in French) is a term used to describe subterranean channels formed by water moving through and eroding incoherent and insoluble sediments. It is associated with subsurface ephemeral discharge in many drylands, including those in the USA (e.g. Fig. 8.8; Parker 1963, Parker & Jenne 1967), northeast Victoria, Australia (Downes 1946), the Hoggar area of the Sahara, Israel, southeast Spain and Saskatchewan (Bryan & Yair 1982) and the drier loess areas of Asian deserts. Piping poses a subsidence/collapse hazard in drylands, threatening roads and railroads, and other engineering structures such as dams, bridges and retaining walls. The most recent major survey is by Bryan & Yair (1982), who showed that piping is often associated with intensely dissected, vegetation-free areas known as "badlands".

Pipes are commonly formed adjacent to steep "freefaces", such as alluvial channel banks, and in alluvial slopes as a result of rapid water movement down steep hydraulic gradients through a permeable, easily dispersed clay, silt or other sediment. Piping often occurs where high infiltration capacities are found in material of low intrinsic permeability, which allows concentrated infiltration (Bryan & Yair 1982). Initially, water seeping through the alluvium carries with it dispersed and disaggregated clay and silt particles, and a small hole develops. The hole is enlarged by this process, but as it becomes larger it acquires more water. The increased flow leads to more rapid development of the pipe by corrasion and wall caving.

Figure 8.8 Piping in salty, sodic soils, from central Tunisia.

Eventually, the roof of the pipe collapses, creating a sinkhole down which surface-water may be funnelled, augmenting the flow and the erosion of the pipe still further. Ultimately, the sinkholes merge and a ragged gully pattern is created. Parker (1963) identified four basic criteria for this type of piping: sufficient water to saturate part of the material above base level; a hydraulic head to permit subterranean water movement; permeable and erodible soil above local base level; and an outlet for flow.

Pipes also occur in materials which may not appear, at first sight, to be very permeable, such as clays, silts, loess and volcanic ash. All of these materials may contain swelling clays, usually montmorillonite, but often illite or bentonite (Parker 1963). Swelling clays have two important qualities in this connection: when they are dried they shrink and create cracks, thus rendering impermeable materials more permeable; and when wet, they become highly dispersed, non-cohesive and, therefore, easily removed in suspension by water, even by slow-moving water. For both of these effects to be realized, wetting and drying are essential, and this is common in many arid and semi-arid areas.

Local processes that assist or are associated with piping include the removal of clay coatings and iron, manganese oxide, calcite and silica cements that bind stable aggregates and the consequent reduction in cohesive strength; clay-mineral swelling; and the physical dispersion of clay minerals and clay aggregates. In addition, secondary breakdown and subsidence of low-density earth materials may occur, following wetting of alluvium, colluvium and loess, in which there may be volume decreases of up to 13% and surface lowering by up to 5m. This is called **hydrocompaction** (e.g. Cooke et al. 1978). Piping intensity tends to increase with the Na:Ca–Mg ratio, soluble salt content, montmorillonite and other swelling clay-mineral content, degree of desiccation, rates of surface runoff and denudation, available relief, and hydraulic gradients between land surface and gully floors. In addition, piping may increase runoff dramatically by concentrating subterranean flow.

Piping is often a precursor of gully development. For example, Harvey (1982) found in southeast Spain that some gully development was promoted by piping in three circumstances: where soils have a light surface crusting over a less consolidated layer in thick, poorly consolidated alluvium; where similar soils occur over relatively weak bedrock with differential strength or porosity; and where deep trenching has

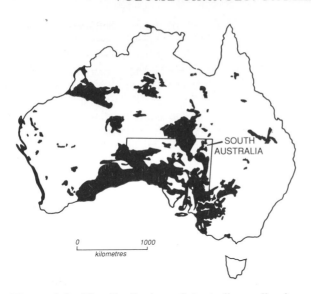

Figure 8.9 The distribution of Australian collapsing soils (Selby 1982d).

produced near-vertical walls that lead to tension cracking.

Piping, and consequent surface collapse and gully development, can cause damage to human activities, such as disruption of banks and fields (e.g. Baillie et al. 1986). In such circumstances, specific conservation responses may be required, including decreasing the volume and concentration of throughflow, the disruption of pipes, and the reduction of soil dispersability (Baillie et al. 1986, Colclough 1973).

8.2.2 Subsidence phenomena

Hydrocompaction is, in places, a regional phenomenon associated with the initial irrigation of alluvial sediments. In the Central Valley of California, for example, subsidence of over 3m and covering hundreds of square kilometres has resulted from the compaction of deposits as the clay bonds supporting voids within them are weakened by substantial quantities of irrigation water passing through them for the first time since they were laid down (e.g. Bull 1964a,

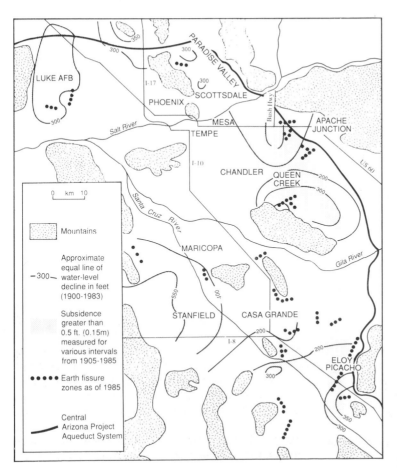

Figure 8.10 Water-table falls, land subsidence and earth fissures in south-central Arizona (Péwé 1989).

Lofgren & Klausing 1969). Similar subsidence is associated with so-called "collapsing soils" – soils that decrease in volume when wetted – in several deserts. In Australia, for example, such low-density, often sandy or silty soils that obtain their dry strength from calcareous cements, or clay or silt matrix, are widespread (Fig. 8.9). If structures are built on such soils and the foundations become wet, significant damage can result (Selby 1982d).

Ground-surface subsidence can also arise from groundwater withdrawal by pumping (e.g. Cooke & Doornkamp 1990). In general, compaction of sediments and thus surface subsidence is proportional to the thickness and compressibility of de-watered sediments, and the extent of water-table fall, once a threshold of extraction is crossed that is related to the natural preconsolidation stress in the sediments before pumping began. Thus, in central Arizona (Fig. 8.10), where over 8,000km^2 are now affected by subsidence and the maximum vertical change in elevation is

approaching 4m in places, the water table fell by *c.* 30–45m in alluvium before subsidence began (e.g. Holzer 1981). In this area, the ratio of subsidence to groundwater fall is in the order of 1:100, and the rate of surface decline is up to 0.170m yr^{-1} (Larsen & Péwé 1986). Similarly, Lofgren & Klausing (1969) described over 2,000km^2 of subsidence in the Central Valley of California, where maximum subsidence in 1964 was 3.6m, and the maximum groundwater decline was 61m; the ratio of subsidence to head decline was from 0.5×10^{-2}m to 5.0×10^{-2}m.

In certain circumstances, such ground-surface subsidence is accompanied by ground cracking that can lead to the formation of giant fissures – some of them over 10km long, 3m deep and 4m wide – which are similar to those created by desiccation (Ch. 8.1; Fig. 8.11). As Bell (1981) showed, the fissures are initiated in alluvium by subsurface tension cracking. They are enlarged by piping until they are revealed at the surface. They are then developed subaerially, especially by bank collapse, and eventually they fill with debris (Fig. 8.12). The fissures appear to be formed in areas where groundwater has fallen substantially (e.g. by over 60m in Central Arizona), and where the alluvium is over 60m thick. In these areas, fissure initiation occurs at three major loci (Fig. 8.13): over a buried bedrock hill, along a hinge-line of surface subsidence, and above a steeply sloping buried bedrock surface (Larsen & Péwé 1986). In other words, they occur where the thickness of alluvium changes abruptly (Jachens & Hölzer 1982).

Land subsidence, and fissuring due to irrigation or groundwater withdrawal, are major hazards to agricultural and urban areas in drylands, such as central California, Arizona and Australia. Irregularities in the ground surface can seriously disrupt drainage, sewerage and irrigation systems. Wells may fail, roads and linear services can be broken, and damage can be caused to buildings, bridges and other structures.

Figure 8.11 A ground fissure due to groundwater extraction, south of Chandler, Arizona.

8.3 Salt and patterned ground

Although discussion of the rôle of salts in forming patterned ground can be traced back at least as far as Huntington's (1907) observations in the Lop desert, the literature on this theme is rather thin. Perhaps one reason is that patterned ground is commonly associated with wetting and drying, as well as with salts, and the former offers a more obvious and better understood explanatory mechanism. Nevertheless,

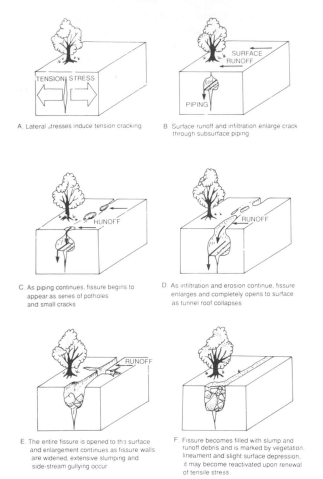

Figure 8.12 Generalized stages of fissure development (modified after Bell 1981).

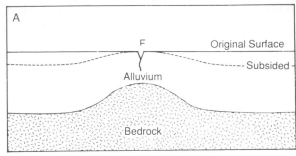

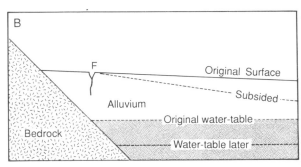

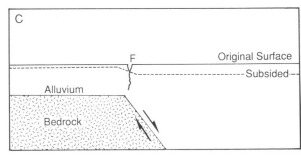

Figure 8.13 The geological settings of areas of potential fissuring. (A) a buried hill; (B) a hinge-line; (C) a buried fault-scarp (F = fissure; after Larsen & Péwé 1986).

salt-dominated patterned ground is common in playas and sabkhas, and in other salty locations, and it is associated with some very distinctive features, including sorted and non-sorted nets, polygons, steps and stripes. As discussed in Chapter 6.2, salt occurs at desert surfaces mainly through surface evaporation either of rainfall and runoff, or of solutions raised to the surface by capillarity. The segregation of salts in stratification and annular rings that reflect crystallization sequences (Ch. 6.2) means that patterned-ground phenomena associated with different salts may show similar segregation.

Hunt & Washburn (1966) and Hunt (1975) provided detailed accounts of salt-related patterned ground in the salt pans of Death Valley and the gravel fans marginal to them, offering a valuable general model for understanding such phenomena.

On the floodplain of the salt pan in Death Valley, the patterned ground is ephemeral and it reflects the hydrological regime. When a salty brine evaporates, a salt crust develops, and polygonal cracking is ultimately created by contraction. The cracks narrow downwards, and a wedge of salt grows within each crack, usually fed by evaporation from groundwater. The salt wedges may widen the crack, and grow to produce **ramparts** around the polygons. The ramparted polygons can then become miniature salt pans when flooding next occurs. In addition, groundwater evaporation *between* the surface crust and the underlying muds can cause salt-crystal growth that imposes

blisters on the polygonal surface. Open pools are formed where salt layers in mud have been dissolved and the surface has *collapsed*.

In the chloride zone beyond the limit of flooding, which is characterized by varieties of rock salt, a salt crust is developed on damp mud. Smooth surfaces of silt saturated with salt have polygonal patterns of desiccation cracks, and solution pits are common at crack junctions. Nets and polygons have developed in the gypsum deposits of the sulphate zone. In the *wet* sulphate zone, rampart polygons are formed. Polygons in *dry*, massive gypsum consist of primary cracks that extend through a thin surface layer and the underlying anhydrite to the crumbly, porous gypsum beneath, and of secondary cracks confined to the surface layer. Sorted and non-sorted nets occur in the carbonate zone towards the edge of the salt pan. Debris has been washed into cracks to produce the sorted nets. Sorted polygons occur where there is a layer of rock salt beneath the surface (Fig. 8.14 & 15). The salt is polygonally cracked, and the cracks are reflected in shallow surface troughs in the overlying silt, in which coarse debris accumulates. On older gravel fans there are sorted steps, sorted polygons and stone stripes.

Hunt & Washburn's conclusions are of general significance, and they may be summarized briefly as follows. The patterned-ground features occur in an area characterized by salt crystallization, clearly suggesting a genetic relationship. Although the precise processes are not well understood, their effects are very similar to those of ice in high latitudes. Salts may affect the form of patterned ground in several ways. The geometry of cracks is influenced by the nature and amount of the salts found in the muds. Salt accumulating in cracks tends to widen and accentuate them. By retarding evaporation, salt crusts favour upward salt migration in the capillary fringe, and this

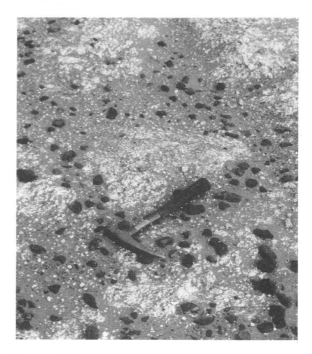

Figure 8.15 Stone polygons in gypsum, near Tabuk, Saudi Arabia.

produces heaving. Volume reduction of salts, and hence cracking, result mainly from either drying or cooling. Volume expansion, and hence some heaving features, occur as the result of crystallization, thermal expansion or hydration. The areal patterns of patterned-ground phenomena associated with salts depend first on the nature, and vertical and horizontal disposition of salts and, secondly, on important climatic and hydrological variables such as the frequency of wetting and drying, the depth of the capillary fringe, the relative importance of surface and capillary water, and

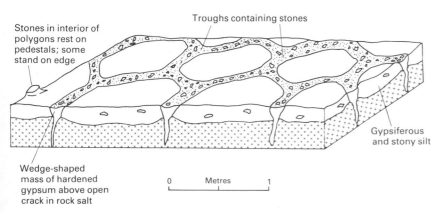

Stones in interior of polygons rest on pedestals; some stand on edge

Troughs containing stones

Gypsiferous and stony silt

Wedge-shaped mass of hardened gypsum above open crack in rock salt

0   Metres   1

Figure 8.14 Sorted stone polygons related to ridges and cracks in an underlying layer of rock salt in Death Valley, California (after Hunt & Washburn 1966).

A

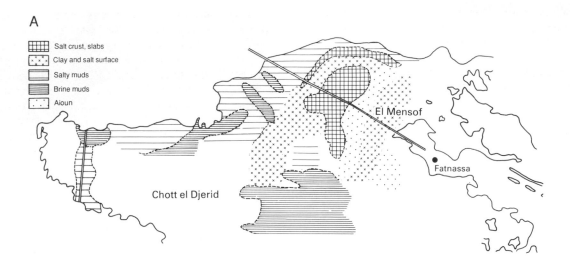

B

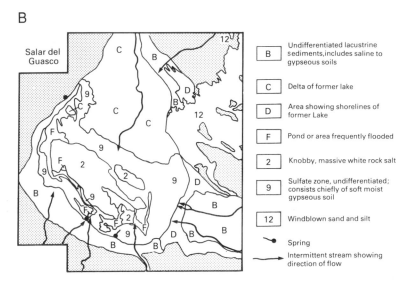

Figure 8.16 Zonation of salty ground on playas. (A) Chott el Djerid, Tunisia (after Coque 1962); (B) Salar del Guasco, Chile (after Stoertz & Ericksen 1974).

the nature of temperature changes.

The model of patterned-ground phenomena in Death Valley can be applied elsewhere. For example, Krinsley (1970) recorded similar patterned-ground phenomena in the salt crusts of the Great Kavir, Iran. Here, groundwater is the main source of salt and it reaches the surface through desiccation cracks in the crusts. In areas near the centre of the basin, where polygonal cracks persist between inundations, salt accumulates more rapidly than it is removed, and fresh brine introduced along the desiccation cracks is less saline than the indigenous brine and thus it

evaporates more easily. Fresh brine may bubble out along the cracks during the day, and its rapid evaporation may result in salt blisters or "blossoms" along the cracks. The salt cells or plates between cracks grow by salt crystallization on their borders, and the growth may be most rapid on their windward sides. Windward edges may eventually overlap the leeward edges of adjacent plates, giving the illusion of overthrusting. In addition, black saline muds may be extruded through the cracks, perhaps as the result of thermal expansion or of pressure from the overlying crust. The muds may quickly solidify to form dykes along the cracks. The uneven escape of muds may be responsible for tilting salt plates and for the creation of an extremely irregular surface.

Similarly, Coque (1962) recognized two main contexts of patterned-ground development in the Chott el Djerid, Tunisia: a 20 × 10km zone of salt-crust slabs and, surrounding it, an extensive area of salty muds (Fig. 8.16A). Features within the former area, which is dominated by halite and some gypsum, include raised-rim polygons surrounding often curved plates, puffy efflorescences, "boils" of deliquescent salts, wind-generated streaks of salt, and **aiouns** (Ch. 14.2). In the same way, Stoertz & Erickson (1974) analyzed the microtopography of three major playa surface types in northern Chile (e.g. Fig. 8.16B). **Hard evaporite crusts**, dominated by halite and fed by open-water evaporation and capillary migration, are associated with six major surface types: slabby, nodular, layered rock salt; rugged gypsum or anhydrite; massive, coarsely crystalline rock salt; smooth rock salt; and silty, nitrate-bearing crust. In contrast, there are **soft playa surfaces** that are often moist, puffy, smooth or hummocky, and sulphate-bearing (mainly gypsum); and there are also **hard clay playa surfaces**. Stoertz & Erickson (1974) also described several distinctive features arising from the **corrosion** of rock salt due to surface flooding: salt saucers, salt nodules, salt pinnacles, salt cusps, salt channels, salt pseudo-barchans and salt tubes. Constructional salt features, often associated with springs and capillary rise, include gypsum buttresses, salt veins, salt stalactites and salt cones.

Patterned ground associated with salt is not confined to playas. For example, crack polygons in gypsum crusts, often overlain by stone pavements, have not only been described in Death Valley, but also on pediments in northern Iraq (Tucker 1978), northern Saudi Arabia (Jones et al. 1986) and central Asia (Hörner 1933).

PART 3

THE FLUVIAL DOMAIN IN DESERTS

CHAPTER 9

Desert drainage systems

9.1 Introduction

The accelerating interest in fluvial geomorphology, growing from the seeds planted by the pioneer studies of G. K. Gilbert, R. E. Horton and others, has yielded a rich harvest of successful textbooks. Some of these attempt to cover the whole field (e.g. Leopold et al. 1964, Gregory & Walling 1973, Schumm 1977); others deal more specifically with slopes (e.g. Carson & Kirkby 1972, Selby 1982a) or channels (e.g. Richards 1982, Knighton 1984). There are also several substantial edited volumes on significant fluvial themes (e.g. Morisawa 1973, Gregory 1977, Rhodes & Williams 1979, Abrahams 1986, Gardiner 1987). Only one major, recent book addresses drylands exclusively: Graf's (1988a) study of *Fluvial processes in dryland rivers.*

The study of fluvial geomorphology owes much to dry environments, ever since G. K. Gilbert's early studies in the Henry Mountains of Utah (Gilbert 1877), and his later work in California (Gilbert 1914). The subsequent work of the US Geological Survey, such as that by Kirk Bryan (e.g. 1925) and especially more recently by L. B. Leopold and his associates, has had very strong links with the deserts of the American Southwest. Significant fluvial research still emanates from American drylands: as shown for example, in some of the outstanding studies by S. A. Schumm and W. L. Graf and their associates, by W. B. Bull, T. Maddock and V. R. Baker, and by those working at Walnut Gulch in Arizona (e.g. Abrahams et al. 1988). A singular feature of some of this work has been its emphasis on empirical field studies of contemporary runoff processes and their associated

Table 9.1 Status of drainage-basin variables during timespans of decreasing duration.

Drainage-basin variables	Status of variable during designated timespans		
	Cyclic	Graded	Steady
1 Time	independent	irrelevant	irrelevant
2 Initial relief	"	"	"
3 Geology (lithology, structure)	"	independent	independent
4 Climate	"	"	"
5 Vegetation	dependent	"	"
6 Relief (or volume of system above base level)	"	"	"
7 Hydrology (runoff and sediment yield per unit area within system)	"	"	"
8 Drainage network morphology	"	dependent	"
9 Hillslope morphology	"	"	"
10 Hydrology (discharge of water and sediment from system)	"	"	dependent

Source: Schumm & Lichty (1965).

landforms. The monitoring of process–form relations in the drylands of the USA has stimulated similar studies elsewhere, notably by A. P. Schick, A. Yair, the late R. Gerson and their colleagues in Israel, by J. B. Thornes and his collaborators in Spain, and by J. A. Mabbutt, D. H. Pilgrim and others at Fowler's Gap in Australia. The collective result of such relatively recent work has been to transform understanding of fluvial activity in deserts. The purpose of this Part is to review the main conclusions of these studies and to explore the ways in which they reveal the diversity and distinctiveness of fluvial systems in deserts.

9.2 The fluvial system: introduction

A basic tenet of most studies in fluvial geomorphology is that the drainage basin is an open system characterized by input, throughput and output of energy and materials; and by self-regulation among the component variables. Some of the variables are independent, others are dependent, and the relations among them vary according to the time span that is considered. Table 9.1 suggests the status of drainage-basin variables at three different time spans: "cyclic" time (the time encompassing an erosion cycle), "graded" time (the time when grade and a condition of dynamic equilibrium exists) and "steady" time (a fraction of graded time). The variables in the table are listed in approximately increasing degrees of dependence.

Time, initial relief and geology are not, of course, intrinsically distinctive in deserts. Climate constitutes the pre-eminent variable in deserts, influencing the nature of fluvial activity. In large measure, it is precipitation that determines the distinctiveness. Similarly, precipitation greatly influences vegetation, and both influence runoff and sediment yield. Precipitation, vegetation, runoff and sediment yield in deserts are examined in the following sections.

9.3 Precipitation

Deserts occur mainly in areas associated with the presence of high-pressure cells around the 30th parallels, where air is subsiding, the atmosphere is relatively stable, and divergent air flows characterize low altitudes (see Part 1). The western and poleward margins of these fundamentally dry areas are occasionally penetrated by rain-bearing low-pressure sys-

Table 9.2 Rainfall data, Avdat, Negev, Israel, 1 March 1960.

	Rain gauge	Rainfall (mm)	Rain gauge	Rainfall (mm)
Automatic	1	7.3	8	4.4
Automatic	2	3.4	9	2.4
Standard		7.8		
Small	1	7.0	10	2.9
	2	6.8	11	4.2
	3	7.3	12	3.0
	4	7.0	13	4.8
	5	5.6	14	5.8
	6	2.2	15	5.0
	7	4.4	16	5.5
			17	4.6

Source: Evenari et al. (1968).

tems that may seasonally dominate precipitation patterns. Eastern and equatorward margins are occasionally penetrated by tropical easterly waves and tropical cyclonic depressions. Where continentality reinforces the general circulation patterns, precipitation is also commonly associated with small, convective cells that can produce intense, short-duration storms. Thus, for example, at the Walnut Gulch Research Watershed in southern Arizona, USA, winter precipitation is caused by depressions passing eastwards from the Pacific Ocean, whereas small convective thunderstorms, associated with moist air moving north from the Gulf of Mexico, dominate summer rainfall (e.g. Renard 1970). Similarly, winter rainfall in Sinai tends to be associated with convective cells formed in conditions of atmospheric instability at the front of a belt of low-pressure (Schick & Sharon 1974, Yair & Lavee 1985).

Precipitation in deserts, as elsewhere, is variable in frequency, intensity, duration and "spottiness', so that generalizations are difficult. Nevertheless, certain features of desert precipitation are important in the context of fluvial processes.

(i) Precipitation can be very varied *spatially* (Fig. 9.1). Table 9.2 shows the rainfall collected by 20 rain gauges during a single storm over 10h in Avdat, Israel. The mean value was 3.07mm, the standard variation 1.75, and the range 7.8 to 2.4. Figure 9.1A reveals the important fact that spatial variability also has a temporal context. Such spatial and temporal variability is not unique to

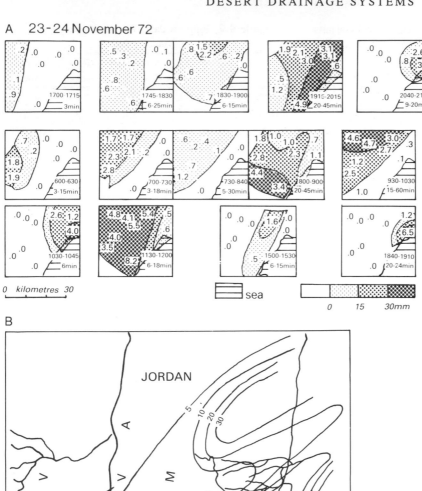

Figure 9.1 (A) Spatial and temporal variability of rainfall for a storm period in the Nahal Yael area of southern Israel based on studies by D. Sharon (after Schick 1974b); (B) the pattern of rainfall in the area of southern Israel and Jordan affected by the storm of March 1966 (after Schick 1971).

deserts, but it is exceptionally important because it means that commonly only *parts* of drainage systems are affected by rainfall and activated by runoff at any one time. Equally important, the *location* of runoff often varies from event to event. This is especially true when, as is often the case, rainfall is not topographically controlled. A corollary of this is that in large drainage systems it may be rare for the whole system to be activated at any one time, and it may be exceptionally difficult to predict where runoff will occur. On the other hand, the location of precipitation and runoff may be predictable where the cause is fixed. For example, snowfall in the Andes of northern Chile, although temporally variable, tends to occur at similar altitude-determined locations and to generate runoff seasonally which nourishes the streams that cross the Atacama Desert. Similarly, the escarpment at Taif (Saudi Arabia) provides a firm local topographic control of rainfall.

(ii) Individual *storm precipitation* totals can be very high compared with mean annual figures. As Table 9.3 shows, single-storm precipitation can exceed mean annual precipitation, resulting in a typically positively skewed distribution of annual totals (Fig. 9.2), and a standard deviation of mean rainfalls exceeding the overall mean (Slatyer & Mabbutt 1964). In the Saharan extremely arid areas, the 24-hour rainfall can reasonably be expected to exceed the mean annual rainfall by 3–4 times once in every 20–30 year period, and

rainfall extremes often exceed the annual mean (Schick 1987, Dubief 1953; Fig. 9.3). Thus, rainfall, and its consequent runoff, may often be highly concentrated and, in comparison with other environments, possibly more effective in the geomorphological work they do. A corollary of the extreme concentration of precipitation is the occurrence of long dry spells that serve, *inter alia*, to reduce vegetation cover and thus perhaps to promote the erosional effectiveness of rainfall when it occurs.

(iii) *Rainfall intensities* within storms can be very high (Tables 9.3, 9.4; see also Dubief 1953). How-

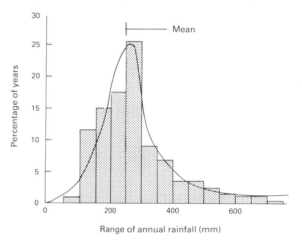

Figure 9.2 Annual rainfall distribution, 1874–1959, at Alice Springs, Northern Territory, Australia (after Slatyer & Mabbutt 1964).

Table 9.3 Extremes of precipitation.

Location	Date	Mean annual precipitation (mm)	Storm precipitation (mm)
Chicama (Peru)	1925	4	394
Aozou (Central Sahara)	May 1934	30	370 /3 days
Swakopmund (SW Africa; Namibia)	1934	15	50
Lima (Peru)	1925	46	1524
Sharjah (Trucial Coast)	1957	107	74 / 50min
Tamanrasset (Central Sahara)	Sept. 1950	27	44 /3hr
Bisra (Algeria)	Sept. 1969	148	210 /2 days
El Djem (Tunisia)	Sept. 1969	275	319 /3 days
Themed (Sinai)		30	142 /1 day
Djanet (Sahara)		19.1	42.4 /1 day

Table 9.4 Rainfall per rainy day in arid areas.

Area	Number of stations in arid zones	Range of mean annual rainfall (mm) (30yr average)	Range of number of rainy days >0.1mm per annum	Average rainfall per rainy day, over 30 years (mm)
USSR (former)	12	92–273	42–125	2.56
China	6	84–396	33–78	4.51
Argentina	11	51–542	6–155	5.41
North Africa	18	1–286	1–57	3.82
West Africa	20	17–689	2–67	9.75
Kalahari	10	147–592	19–68	9.55
COMBINED	77	1–689	1–155	6.19

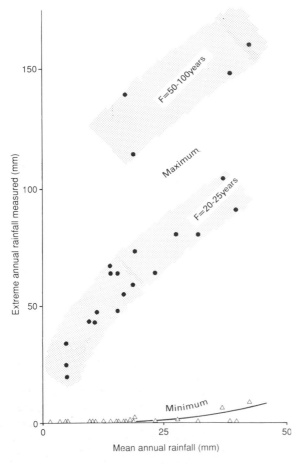

Figure 9.3 Extreme annual rainfall in the extremely arid parts of the Sahara (mean annual rainfall <50mm) for stations with a record of 20yr or more. Broad dotted bands indicate possible rainfall frequencies. (Data from Dubief 1953, after Schick 1977.)

ever, rainfall intensity is also spatially variable within storms and, when measured as rainfall per rainy day, the average figures may not be especially high (Table 9.4). From the perspective of geomorphological work, it is the occasional locally intense storm that may be the most effective agent of change, a situation that contrasts with temperate environments where it is the medium-sized event that is often most significant.

(iv) Rainfall in deserts characteristically varies through time. If temporal variability, V, is expressed as

$$V\% = \frac{mean\ deviation\ from\ the\ average}{the\ average} \times 100 \quad (9.1)$$

these values can be mapped (Fig. 9.4). Often variability exceeds 30% and in places, such as the central Sahara, it may be over 100%.

Longer-term fluctuations have been discerned from rainfall records in drylands. Sometimes these show rises and falls, but no consistent temporal trend; elsewhere, geomorphologically significant secular trends may be evident. In southern Arizona, for example, there do appear to be statistically significant secular trends (Fig. 9.5), notably in the early part of the record, when heavy rains were more frequent, and light rains were reduced in frequency. Such trends may be extremely significant in influencing the nature of runoff and erosion. For example, it could be that in the 1860–90 period, erosion was increased because a higher frequency of heavy, runoff-producing storms coincided with a lower frequency of grass-sustaining (and thus surface-protecting) small rainfalls: the heavy storms may have attacked a particularly vulnerable surface (Cooke & Reeves 1976).

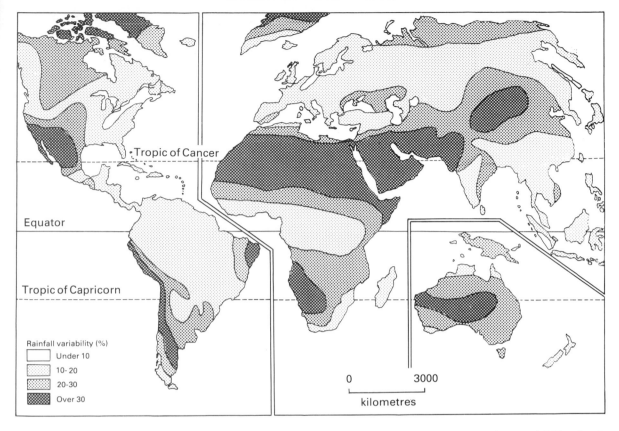

Figure 9.4 World map of rainfall variability. Note the way in which the major desert belts show variability that is often in excess of 30% (after Goudie & Wilkinson 1978).

Predicting precipitation, its quantity, frequency and duration, is a major problem. Several techniques have been developed (e.g. Kappus et al. 1978) and an example, based on the Probably Maximum Precipitation model, is provided in Figure 9.6 for the Iranian Gulf coast. Possibly of greater geomorphological value, however, are the techniques for predicting runoff (see Sec. 9.5).

9.4 Rôles of vegetation

Vegetation is, of course, normally sparser in deserts than in areas of higher rainfall, and as a result it is relatively less influential in acting as an intermediary between rainfall and runoff. However, it is a mistake to assume that its rôle is insignificant. Although, as Part 2 showed, there are large areas in some deserts that are relatively free of vegetation, most areas have some vegetation, and it can play a major rôle in

influencing fluvial processes. In deserts, as elsewhere, that rôle is especially important in protecting the surface, influencing infiltration, retarding wind and water flow, and consuming water that might otherwise be available as runoff, and thus in influencing the nature and rates of erosion on slopes and in channels (e.g. Lee 1981).

Desert plants are characteristically either perennial xerophytes (woody shrubs or small trees) or perennial grasses and rhizomes (e.g. Goodall & Perry 1979). Low densities of desert shrubs are common on most surfaces, excluding those that are excessively saline, free of a debris mantle, or covered with mobile sand. In many areas, quick-growing perennials respond opportunistically to rainfall. In semi-arid regions, perennial grasses provide a major additional component of the natural vegetation; and where surface water is present in channels, or can be reached at depth by tap roots, riparian vegetation may be strongly developed. Overall, the occurrence of plants on a desert

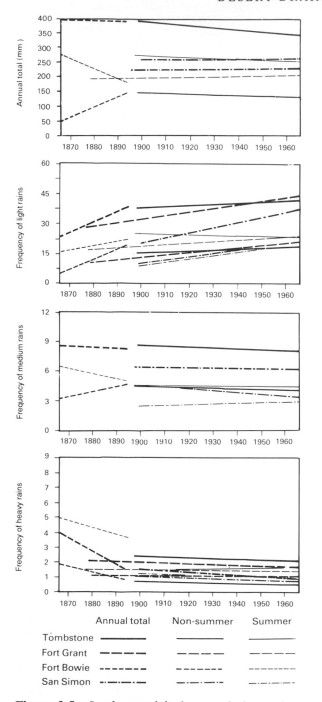

surface is probably the single most sensitive indicator of water availability.

Both types of desert vegetation are often distinctively adapted to aridity in many various ways. Some of these adaptations may significantly influence surface-runoff processes. For example, desert plants often have low leaf areas in order to reduce transpiration, so that the surface protection they provide may be low compared with that afforded by leafy shrubs in wetter climates. Similarly, desert plants require relatively extensive root networks in order to survive, with the result that plant density is low and a greater proportion of the surface is left unprotected and vulnerable to erosion than in wetter environments. At the same time a relative sparsity of *shallow* roots may leave near-surface and surface sediment more vulnerable to surface erosion. Conversely, the roots, even of dead or dormant plants, may help to bind near-surface materials and to promote infiltration capacity. Vegetation litter, which is often present, may also enhance surface protection and promote infiltration (Thornes 1985). Vigorous riparian vegetation in channels often draws upon water at depth through phreatophyte tap roots, and may significantly influence the nature and effectiveness of channel flow (e.g. Graf 1978, 1981).

The influence of vegetation on the crucial factor of infiltration capacity is illustrated by Figure 9.7, based on field studies in southern Spain by Thornes (1985) and others, where the relatively sparse vegetation significantly increases infiltration rates. A somewhat different influence has been clearly demonstrated in areas where semi-arid Mediterranean vegetation is destroyed by fire. In the mountains of southern California, for example, organic matter accumulates on the surface, and the upper soil layers become water-repellant due to the intermixing of partially decomposed organic matter and mineral soil. During a chaparral fire, the surface litter is burned, organic particles coat, and become chemically bonded to, the cooler soil particles below, and vaporized substances condense on mineral soil particles at depth (Fig. 9.8; DeBano 1981). The new, deeper, thicker and usually discontinuous hydrophobic layer reduces infiltration capacities. The direct consequence is that when the burned surface is attacked by rainfall, the runoff coefficient is increased and soil erosion is accelerated (Cooke 1984).

The implication of the preceding discussion is that vegetation influences runoff, but less so and in different ways compared to more humid areas. This may

Figure 9.5 Secular precipitation trends in southern Arizona, represented by regression lines showing precipitation totals and annual and seasonal frequencies of rainfalls in different categories. Light rains (0.1–12.7mm), medium rains (12.7–25.4mm) and heavy rains (25.4+mm). (After Cooke & Reeves 1976.)

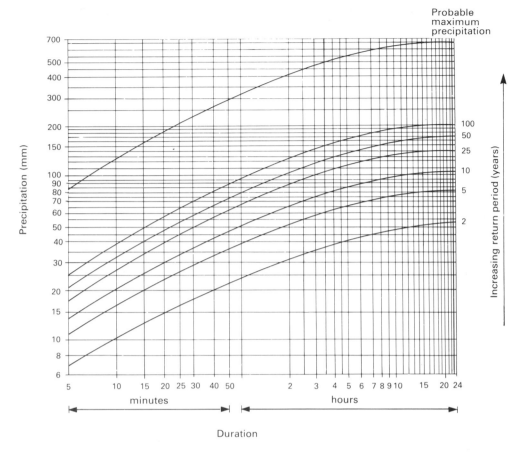

Figure 9.6 Precipitation depth–duration–frequency curves for part of the Iranian Gulf coast (after Kappus et al. 1978).

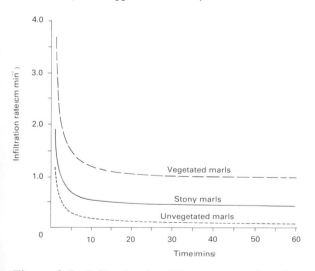

Figure 9.7 Infiltration in different types of surface materials in semi-arid soils (after Thornes 1985).

not be an entirely accurate generalization. For example, *experimental* studies by Moss (1979) of erosion rates on surfaces with different plant and litter covers led him to suggest that in passing from dense plant covers to none, a crucial threshold is crossed below which the little remaining vegetation has no significant effect on limiting erosion. Between these two extremes, Moss suggested, are areas where vegetation cover varies so that the threshold would be crossed. Such areas, which probably include semi-arid regions and desert margins, would be the most sensitive to disturbance and change, and to management efforts; they are the major areas of "desertification".

The geomorphological importance of vegetation in deserts is considered below, at two scales: at the scale of the drainage basin, in which its influence is chiefly seen in sediment yield (Sec. 9.6); and at the scale of the specific slope or channel (see Chs 10 and 11).

101

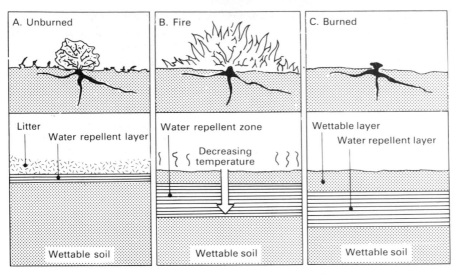

Figure 9.8 Chaparral-induced hydrophobic soil. (A) Before fire, hydrophobic substances accumulate in the litter layer and the mineral soil immediately beneath it; (B) fire burns the vegetation and litter layer, causing hydrophobic substances to move downwards along temperature gradients; (C) after fire, a water-repellent layer is present below and parallel to the soil surface on the burned area (after DeBano 1981).

9.5 Runoff

9.5.1 The organization of runoff: introduction

A geomorphologically useful distinction can be drawn in deserts between **endogenic** and **allogenic** drainage, and between **endoreic**, **areic** and **exoreic** drainage (de Martonne & Aufrère 1928). Endogenic drainage is that derived from precipitation within the desert area. Allogenic drainage originates outside the desert and flows into it. An example of an allogenic stream is the Mojave River, in California, which rises in the San Bernardino Mountains and flows north into the

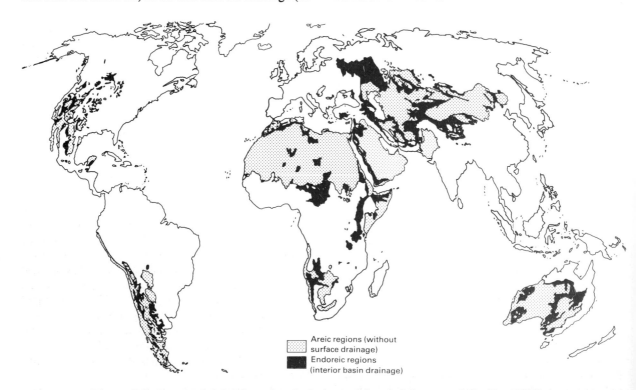

Figure 9.9 Areas of endoreic and areic drainage (after de Martonne & Aufrère 1928).

102

Mojave Desert; another is the Oued Saoura, which flows south to the Sahara from the Atlas Mountains in Algeria. Other examples include the Senegal, Gambia, Niger, Lagone Chavi and Nile rivers which flow north from humid to arid areas in central Africa. Geomorphologically, allogenic streams may be associated with landforms and deposits in deserts which are alien to the desert environment and reflect landforms more common in humid areas, such as with well defined terrace sequences.

Much desert drainage, whether it be endogenic or allogenic in origin, is endoreic; that is to say, it does not reach the sea, mainly because of high water losses due to evaporation and surface infiltration, and because much drainage is within enclosed basins. In areic areas, including sand seas, surface drainage is practically nil, or is at best ephemeral, and drainage channels are poorly connected. De Martonne argued that endoreic drainage characterized semi-arid areas, and areism, arid areas; but such a distinction is too bold because the nature of local drainage reflects more than climatic conditions.

Some deserts do have areas of exoreic drainage, where flowing water reaches the sea. In the southern Atacama Desert, for example, three perennial allogenic streams flow from the Andes across the desert to the Pacific. The approximate world pattern of endoreic, exoreic and areic areas is shown in Figure 9.9. Geomorphologically, these distinctions are important, for characteristic landforms may be associated with the different types. Changes of sea-level may be reflected in the landforms developed by exoreic drainage, and much sediment may be removed to the sea from such basins; enclosed basins, on the other hand, may experience progressive accumulation of sediment, and may remain unaffected by changing sea-level, although they may have experienced changing lacustrine base levels in the Quaternary Era.

Within this broad hydrological context, several different types of runoff can be recognized. On *slopes*, surface flow is commonly in the form of anastomosing threads of linear flow; occasionally, deeper flow may create a thin, semi-continuous film of flowing water. In addition, *throughflow* in soils, although probably less important than in humid regions, may be significant. These processes are examined more fully in Chapter 10.2.

Runoff *within channels* is usually **perennial**, **intermittent** or **ephemeral**. Perennial flow is rare in deserts, except when the drainage is allogenic. Intermittent flow – in which there is an alternation of dry and flowing reaches along a drainage channel – is more common. Most channels in deserts, however, only carry water during storms, and flow in them is therefore ephemeral. As a result, flow in most river systems is only rarely, if ever, integrated. Furthermore, the location and size of the sector affected by runoff within a basin may vary from storm to storm. In addition, streams may be "underfit"; that is, they may be too small for the valleys in which they flow. Underfit streams are common in many environments; in deserts, streams may be underfit perhaps because of the contemporary variability of runoff events, or in part because the valleys and channels are inherited from wetter conditions in the past.

For these reasons, there is a fundamental distinction to be drawn within desert drainage systems between the *actual* drainage area and the *effective* drainage area within it. For example, in a desert basin which has suffered a degree of desiccation since the last pluvial period, the effective drainage area may only occupy the mountains and higher piedmont plains near the watershed of the basin. A particularly striking example of such a situation is the Dalol Mauri drainage system, which extends from the Air Mountains in Niger south to the River Riger: flow in this system rarely extends far beyond the mountains. Because the actual drainage area is larger than the effective drainage area, and the ratio varies both spatially and temporally, drainage area *may* be a less useful predictor of drainage characteristics such as runoff than it is elsewhere.

9.5.2 Some global generalizations

McMahon (1979), in a study of available streamflow records from 72 stations distributed across the arid regions of six continents, derived several new generalizations concerning the flow volumes and peak discharges of desert streams. He demonstrated that, just as desert precipitation varies in its seasonal distribution and frequency, so does runoff. Thus, he showed that seasonality of runoff is particularly high, as might be expected, in the Mediterranean region and southern Africa. Again, his data revealed that the coefficient of variation of runoff (C_v = standard deviation of runoff divided by the mean) in arid continental areas, such as Australia, is about double that for continental areas as a whole (Table 9.5); and that C_v is generally higher in areas of low precipitation and runoff.

Furthermore, the frequency distribution of annual

Table 9.5 Seasonal flows in arid zones (% of mean flow per season).

Zone	Autumn	Winter	Spring	Summer	Coefficient of variation (C_v), annual flows
Australia	29	21	14	36	0.67
East Mediterranean	2	81	16	1	
North America	17	28	35	20	0.3
Southern Africa	25	18	9	48	
Europe					0.2
Asia					0.2
ARID ZONE					0.99

Source: McMahon (1979).

flows in the arid zones is considerably more skewed than for more humid regions; and persistence (the effect of one runoff event on subsequent events,

measured by the serial correlation coefficient), is only 0.03, considerably lower than the world average for non-arid zones of 0.15. This difference no doubt reflects the relatively low persistence effects of ephemeral runoff. For peak discharges, McMahon revealed, *inter alia* that specific mean peak discharge $(\bar{q})$ is predictably related to mean annual runoff (*MAR*) in the form:

$$\bar{q} = 3.3 \times 10^{-3} (MAR)^{0.83} \qquad (9.2)$$

but that there are substantial regional variations. For example, the eastern Mediterranean produces higher mean annual floods and the Australian zone produces smaller floods than the mean floods for other arid zones; and specific mean peak discharge correlates well with drainage area ($N=63$, $r^2=0.44$). However, the slope of the regression line (Fig. 9.10) is unexpectedly steep, being about 1.5 times that for non-arid zones, suggesting that flood magnitude is affected by drainage area more in arid zones than elsewhere.

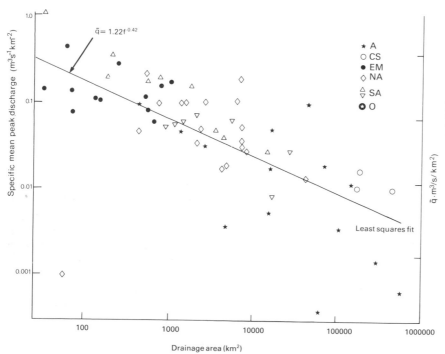

Figure 9.10 Effect of drainage area on specific mean peak annual discharge (after McMahon 1979). A, Australian zone; CS, Caspian Sea zone; EM, East Mediterranean zone; NA, North American zone; SA, South American zone.

$\bar{q} = 1.22 f^{-0.42}$

★ A
○ CS
● EM
◇ NA
△▽ SA
◉ O

Least squares fit

Specific mean peak discharge ($m^3 s^{-1} km^{-2}$)

$\bar{q} \cdot m^3/s/km^2$

Drainage area (km^2)

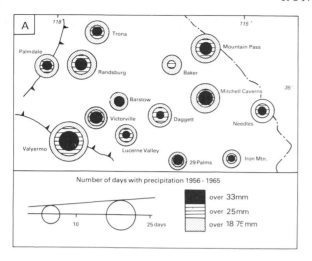

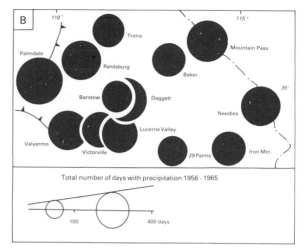

Figure 9.11 Precipitation in the Mojave Desert, California. (A) Number of days in a ten-year period with precipitation totals above 19mm, 25.4mm and 34mm; (B) total number of days with precipitation.

9.5.3 Predicting runoff from precipitation

The distinctive characteristics of desert runoff led McMahon (1979: 122) to several conclusions concerning the problems of predicting runoff.

(1) *Arid zone data* Because of inadequate arid zone data, it is essential that available data be subjected to detailed analysis.

(2) *Rainfall–runoff models* In view of (1) above, the development of rainfall–runoff models for synthesizing arid zone streamflow data is essential.

(3) *Stochastic data generation* It was concluded that, for arid zone streams, difficulties with stochastic data generation procedures will be severe.

(4) *Peak discharge analysis* Although more variable, peak discharges of arid streams present no unique features for analysis.

(5) *Humid to arid zone extrapolation* Hydrological characteristics based on humid region data should not be extrapolated to arid zones.

Such problems certainly make runoff prediction in deserts difficult. One way forward is to base runoff predictions on empirically tested models that use climatic variables available from published records. Such models are often crude because they inevitably ignore the fact that runoff is a precipitation residual affected by infiltration capacities, vegetation, water deficiencies in surface material and topography, most of which are commonly unknown.

A very crude, general impression of runoff frequencies may be gained by examining the frequency of daily precipitation totals above certain limits. Figure 9.11 shows, for the Mojave Desert, California, the number of days in a 10-year period with precipitation totals above 19mm, 25.4mm and 34mm (the last threshold was selected because it was associated with runoff in 1965). If runoff occurs when daily precipitation exceeds 19mm, its frequency will be once a year in many places and almost everywhere once in two years; if the 25.4mm level is taken, the frequency would be approximately once in every three years; and if there is runoff only when daily precipitation exceeds 34mm, then it will be experienced only once or twice in a decade. Analysis of the precipitation record in the Mojave Desert yields other points of geomorphological interest. Runoff is more likely in winter because most heavy rainfall occurs at this time, when temperatures and evaporation are lower. In this area precipitation, and presumably runoff associated with summer thunderstorms, is relatively unusual, for an average of only 15% of rain days in which precipitation exceeds 19mm occur from July to September. Another feature is that winter precipitation usually affects a large area, a fact which may be attributed to its origin in Pacific air masses. In contrast, heavy summer precipitation not only occurs infrequently, but is also more localized. Total precipitation declines, and the proportion of summer precipitation increases towards Arizona in the east. Thus, it would seem likely that across the Mojave Desert there is an eastward decline in the amount and frequency of runoff, and an increase in its localization.

Nemec & Rodier (1979) outlined a more sophisticated approach in which "climatic runoff", f, can be computed from:

$$f = r - E\,(R,L,r) \qquad (9.3)$$

where r is the precipitation, R is the radiation balance of the land surface, L is the latent heat necessary for evaporation of r (r, R, and L are annual averages per unit of surface) and E is the evapotranspiration. In addition, the runoff coefficient (f/r) will be:

$$f/r = 1 - E(R,L,r)/r \qquad (9.4)$$

Figure 9.12 summarizes the relations between precip-

itation and runoff, subdivided in terms of R, and the absolute values of runoff with respect to R and R/Lr (Budyko's radiation index of aridity, in which 2–3 = "arid" and 3+ = "desert") are summarized in Figure 9.12.

Such predictions ignore surface conditions, but there are some models which attempt to elaborate rainfall–runoff models in terms of surface conditions. Davy et al. (1976), for instance, created a family of rainfall–runoff curves for different surface conditions in the Sudano-Sahelian zone of West Africa (Fig. 9.13). Local, small-scale rainfall–runoff models have been developed empirically in a few areas. For example, in the "summer convective storm" desert of

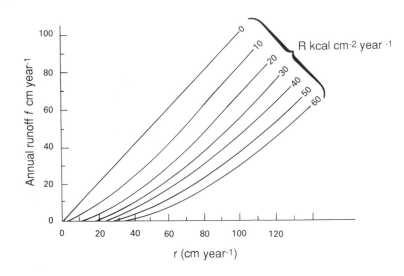

Figure 9.12 (A) Annual runoff versus annual precipitation, as a function of R (kcal cm^{-2} yr^{-1}); (B) absolute values of annual runoff in different climatic zones (after Budyko, from Nemec & Rodier 1979).

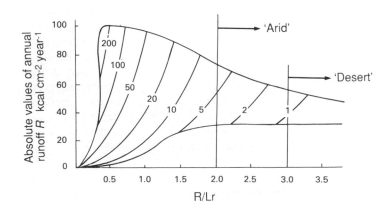

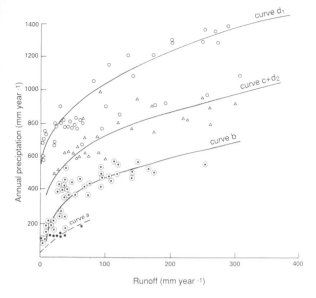

Between these two extreme approaches – prediction based on climatic data alone and empirical models developed for specific sites – there are other practical approaches, such as the Cook Method, and the Rational Method (Chow 1964, Graf 1979b), that do not require runoff data. The Cook method determines discharge at a given location for various return intervals, by drainage area, taking into account a dimensionless runoff factor W (which varies between 0 and 100, and is the sum of factors for relief or slope, vegetation cover, infiltration and surface storage; values of these variables can be derived from tables; see Chow 1964). The Rational Method is similar, in which peak rate of flow, qp, is determined by:

$$qp = Cia \qquad (9.5)$$

where C is a constant ranging from 0 to 1 according to watershed conditions, including infiltration rates, i is rainfall intensity for the given frequency, and a is drainage area (Chow 1964).

Figure 9.13 Relationship between annual precipitation and runoff as a function of vegetation in the Sudano-Sahelian zone of West Africa. Curve a (●) bare rock (at least 30% of the basin) and steppe for the remainder of the basin; curve b (⊙) steppe and thorny steppe with less than 50% crop fields; curves c and d_2 (△) thorny steppe, aboraceous steppe and aboraceous savannah in Chad (north of 10°N); curve d_1 (○) arboraceous savannah with less than 50% crops (except Chad) (from Davy et al. 1976, after Nemec & Rodier 1979).

southern Arizona, Schreiber & Kincaid (1967; Fig. 9.14) showed from a multiple regression study of small runoff-plot data that average runoff was proportional to precipitation quantity; that the taking into account of crown spread of vegetation improved runoff prediction, but only slightly; and that antecedent soil moisture was relatively unimportant. The three independent variables accounted for 72%, 3% and 0.5%, respectively, of the explained variance. Their general equation indicates that 6.6mm of rain must fall before runoff begins. In a later study, Osborn & Lane (1969) extended the investigation to small watersheds (0.2–4.4ha) in the same area. Their multiple regression analysis showed that runoff volume was again most strongly correlated with total precipitation; that peak rate of runoff related most closely to the maximum 15-minute depth of precipitation; that flow duration was best correlated with watershed length; and that lag time was most strongly correlated with watershed area. The independent variables explained approximately 70%, 70%, 50%

Figure 9.14 Runoff plot at the Walnut Gulch Experimental Watershed, Tombstone, Arizona.

Runoff prediction in ungauged catchments can be based on several other empirically derived, approximate methods which permit sensible calculations. For example, Begin & Inbar (1984) determined discharge-frequency curves for part of the Negev using estimates of shear stress, and its relationship to bed particle size and channel hydraulic geometry (both of which can readily be measured in ungauged catchments). Other methods involve extrapolation from established runoff–altitude relations and drainage area (Moore 1968), and prediction based on channel geometry (e.g. Griffiths 1978, Riggs 1979).

9.5.4 The runoff coefficient

The previous discussion reveals that a key relationship in the fluvial system is that between rainfall and runoff, which is usually expressed as the **runoff coefficient** (%) (e.g. Table 9.6a). From first principles, and indeed from some empirical data, it might seem to be likely that runoff coefficients will be relatively high in deserts, compared with more humid areas, mainly because of the relatively low infiltration capacities on vegetation-free surfaces, the occasionally high rainfall intensities, and the widespread occurrence of relatively impermeable rock surfaces. All that is generally true.

Table 9.6a Relationship between runoff coefficients and slope angle at experimental plots of Avdat farm, Israel.

Slope angle (%)	Runoff coefficient (%)
10.0	27.1
13.5	22.9
17.5	18.5
20.0	14.1

Source: Evenari et al. (1968).

However, such factors may be offset by the fact that the ground surface is often dry prior to precipitation so that a high proportion of initial rainfall is absorbed, and by high evaporation rates. In addition, infiltration will vary locally and greatly with rock type and alluvial and/or soil covers. Thus, local variability in runoff coefficients can be high. For example, experimental runoff studies at Avdat (Israel) by Evenari et al. (1968) showed that the runoff coefficient can vary between 14% and 27%, and that – in contrast to more humid areas – this variation is inversely related to slope inclination, because the factors that promote

infiltration (e.g. roughness and grain size) are strengthened on steeper slopes (Table 9.6a; Evenari et al. 1968, Yair & Klein 1973). Commonly, it seems that the proportion of runoff is better correlated with precipitation intensity, with intensities of at least 1mm/min and totals of some 10mm at these intensities normally being required to initiate channelled runoff (Table 9.6b). These matters are considered further in Chapters 10.2 & 4.

Table 9.6b Published data on the intensity of rainfall generating runoff in arid areas.

Area	Intensity generating runoff	Total
Mojave	34mm/24hr	
Negev	1.7mm/3min 3.4mm/10min	18mm
Ahaggar	25.4mm/10min	
Central Australia (bankfull Q)		12.7mm 50.8mm
Piedmont areas		25–50mm
Nahel Yael	1.5mm/3min	8–30mm
Walnut Gulch		>6mm
Alice Springs	44.6mm/50hr	

Source: Slatyer & Mabbutt (1964), Renard (1970), Schick (1971), Cooke & Warren (1973).

9.6 Sediment yields and budgets

9.6.1 Sediment yield

One useful, integrated measure of fluvial activity in deserts is sediment yield. It is usually expressed in tonnes or $m^3 km^{-2} yr^{-1}$. This index broadly measures the amount of material removed by fluvial processes from a given drainage area per unit time. In reality, it is rather difficult to measure precisely, especially in areas of ephemeral flow, because bedload and suspended load present particular monitoring problems, and some sediment may be lost by wind erosion. Thus, estimates of sediment yield tend to be based either on suspended sediment monitoring or on measuring the extent of sediment accumulation in reservoirs. Sediment yield data do not reveal anything about the nature and relative importance of different fluvial processes at work in desert drainage basins, and to ignore wind erosion and solution loads may significantly affect the value of the data (see Ch. 11.2 and Schick 1977). Nevertheless, they do provide a useful comparative yardstick.

Traditionally, there have been two major, apparently partly contradictory, generalizations about desert sediment yield. The first asserts that sediment yield is normally low in deserts. For example, Corbel (1964) estimated that for tropical arid areas with a precipitation of less than 200mm per annum, a rate of $1.0m^3$ km^{-2} yr^{-1} characterizes mountain areas, whereas in the plains a rate of $0.5m^3$ km^{-2} yr^{-1} is a reasonable figure. These predicted rates are lower than those for any other region he considered, and compare, at the other extreme, with $2,000m^3$ km^{-2} yr^{-1} in humid, glaciated polar regions. Corbel also estimated that the total erosion from warm arid lands amounted to only about 7% of erosion from all arid lands, and that erosion from all arid lands (warm, temperate and cold) accounted for only some 9% of erosion in unglaciated lands and about 4% of total world (fluvial) erosion. Although all of these figures are approximate, as are the figures of Fournier (1960) and others, the overall generalization does seem to be reasonable.

The second, more important and more controversial generalization derives from Langbein & Schumm's (1958) study relating annual sediment yield (based on sediment-station data for approximately 3,900 km² drainage basins, and on reservoir sedimentation data from approximately 78km² drainage basins in the USA) to mean annual effective precipitation (the amount of precipitation required to produce a known amount of

runoff under specified temperature conditions). Figure 9.15 shows a family of curves based on extrapolation from the original curve for an average annual temperature of 10°C. It should be emphasized that the original curve is itself based on only six group averages of sediment data. The curves have an extremely distinctive and much quoted form. Sediment yield reaches a maximum between 254mm and 625mm (depending on annual temperature) and yield falls off sharply towards areas of more or less effective precipitation:

> "This variation in sediment yield with climate can be explained by the operation of two factors, each related to precipitation. The erosive influence of precipitation increases with its amount, through its direct impact in eroding soil and in generating runoff with further capacity for erosion and for transportation. Opposing this influence is the effect of vegetation, which increases in bulk with effective annual precipitation."

(Langbein & Schumm 1958: 1079–80).
These opposing factors can be represented by the equation:

$$S = aPm\left[\frac{1}{1+bP^n}\right] \tag{9.6}$$

where S is the annual sediment yield, P is the effective annual precipitation, m and n are exponents and a and b are coefficients. aP^n describes the erosive factor and $(1/1+bP^n)$ represents the vegetarian-protection factor.

This equation can be solved graphically and, converting to metric units, the approximate result is:

$$S = \frac{1.631(0.03937P)^{2.3}}{1+0.0007(0.03937P)^{3.3}} \tag{9.7}$$

where S is the sediment yield in m^3 km^{-2} and P is the effective precipitation in mm (Douglas 1967). Most deserts occur on the rising limbs of the curves. In these areas, as precipitation rises from zero, sediment yield increases at a rapid rate, because more runoff becomes available to move sediment and there is still bare ground susceptible to rainwash erosion. As the peak is approached, the desert scrub progressively gives way to grasses which reduce the vulnerability of the surface to erosion. These semi-arid areas also tend to have high channel densities (e.g. Abrahams 1972) which might promote more efficient sediment removal. If these curves are typical of sediment yield-climate relationships on a world scale, it would seem reasonable to draw a conclusion which accords well

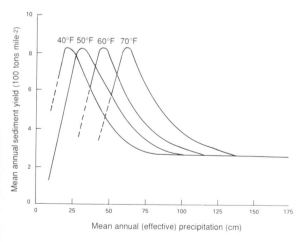

Figure 9.15 The effect of temperature on the relationship between mean annual sediment yield and mean annual precipitation (after Schumm 1965).

with the previous observations on the diversity of deserts: while sediment yield in deserts is generally low, deserts include a wide range of sediment yields from virtually nil to over 300m^3 km^{-2} yr^{-1} and they include both the lowest and the highest yields.

Langbein & Schumm's work emphasizes the relationship between sediment yield and climate. Their data need to be qualified by reference to other factors affecting the yields. First, the data are derived from relatively small drainage basins of about 3,900 km^2 or less; yield from larger basins may be lower per unit area. Secondly, erosion rates may often increase with the ratio of drainage basin relief to drainage basin length. Thirdly, the correlation between sediment yield and precipitation is improved if the seasonality of precipitation is considered. Erosion rates tend to be higher in those areas of seasonal rainfall such as north-east Queensland (Douglas 1967), where the erosive impact of intense storms is increased by the fact that the vegetation is relatively sparse because of seasonal drought. In some semi-arid areas of seasonal precipitation, therefore, very high sediment yield may be expected.

Fourthly, interference by man in natural ecosystems often leads to increased sediment yields, so that the data used in contemporary estimates of erosion may be unnaturally high. In particular, as Douglas stated (1967), the peak of Langbein & Schumm's curve at an effective precipitation of about 304mm (at 10°C) – in the areas of transition between desert scrub and grassland – may be exaggerated, because in these areas removal of vegetation may cause great erosion by the few storms which produce significant runoff. Finally, sediment yield data will clearly be affected by the erodibility of materials and the infiltration characteristics of the areas under consideration. For example, Yair & Enzel (1987) showed that within the Negev Desert there is evidence to suggest that sediment yield varies *inversely* with precipitation because it is locally dominated by high runoff and yield from steep, *rock*-covered slopes with low infiltration rates that prevail in the *drier* areas, whereas soil-covered slopes, which are more extensive in slightly wetter areas, have higher infiltration rates and relatively lower rates of runoff erosion. They argued that the important control is, for a given rainfall, the efficiency with which rainfall is converted into runoff.

A review by Walling & Kleo (1979), based on the analysis of sediment yield for 1246 stations throughout the world, served significantly further to qualify such generalizations of the Langbein–Schumm type . They

initially concluded that the various "global" studies of sediment yield do not produce a clear or consistent picture (Fig. 9.16B); and they doubted the value of the Langbein–Schumm curve (and its derivatives) in evaluating the relative magnitude of sediment yields from all arid and semi-arid areas, partly because it is based on group-average data, and on data drawn only from a restricted range of climatic and topographic conditions in the USA.

Their analysis of mean annual precipitation and sediment yield (Fig. 9.17) shows no clear relationship, strongly pointing to the relative importance of *other* factors, such as rainfall seasonality, relief, human impact, and soil erodibility. They concluded that sediment yield does *not* show a tendency towards a maximum in areas of low precipitation, although such areas often show a peak (Fig. 9.16); and that sediment yield in areas of low precipitation is *not* distinctively highly variable annually compared with the other areas. On the other hand, rivers draining areas of low precipitation *are* distinctive in having mean suspended sediment concentrations that are an order of magnitude or more higher than in other areas, mainly because of low runoff volumes and low vegetation cover. Sediment yield in arid and semi-arid areas *is* probably also relatively distinctive in so far as a large proportion of it is produced in very short periods. Furthermore, sediment-delivery ratios (the ratio between gross erosion and sediment yield) are probably lower in dryland catchments because of the localization of runoff in ephemeral flows, and many possibilities for local deposition within the system arising from, for example, high channel-bed infiltration, alluvial fans and braided channel networks. If sediment yield is plotted against drainage area, an inverse relationship is observed, as expected (Fig. 9.18; Renard 1972). However, in areas of low precipitation (such as the southwest USA) the inverse relation shown by the slope of the regression line, is possibly greater, reflecting a more rapid decline in sediment-delivery ratios as drainage area increases, the absorption of ephemeral flows in channel beds, and local sediment storage within drainage systems.

A final generalization may be appropriate. It is a widely held belief that arid, or more especially semi-arid, ecosystems are especially fragile and susceptible to human activity (e.g. Walling & Kleo 1979). If this is so, and it is certainly open to debate, then human interference could lead to relatively rapid acceleration of erosion in deserts. Equally, *recovery times* may be longer in such environments, perhaps as much as four times longer than in humid areas (Walling & Kleo 1979). In such circumstances, human activities could not only increase sediment yield, but they could also have a longer-lasting impact than elsewhere.

9.6.2 Sediment budgets

Sediment yield is a useful guide to rates of debris removal by fluvial processes from desert drainage basins, but more data are required for the preparation

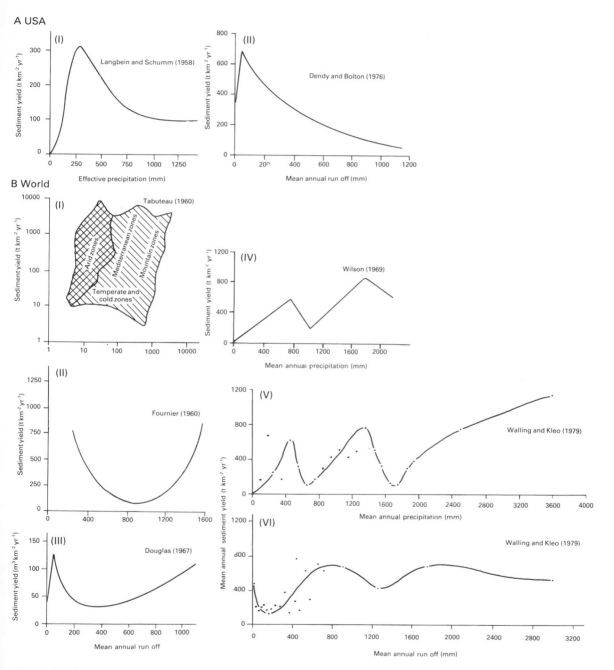

Figure 9.16 Examples of generalized relationships between sediment yield and mean annual precipitation or runoff: (A) USA; (B) the world. (V) and (VI) show subjectively fitted curves drawn through group average values of sediment yield in drainage basins of less than 10,000km² area (after various sources, from Walling & Kleo 1979).

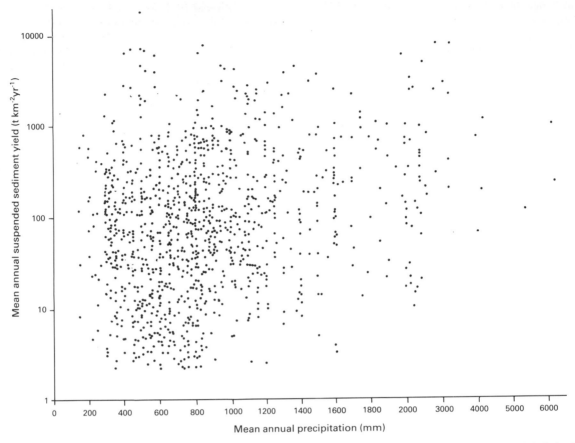

Figure 9.17 Plot of mean annual suspended sediment yield versus mean annual precipitation for a global database (after Walling & Kleo 1979).

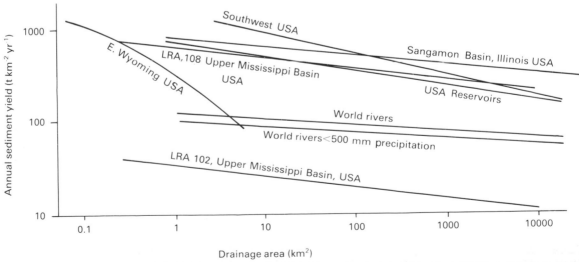

Figure 9.18 The relationships between sediment yield and drainage basin area, based on Walling & Kleo's global database and other regional studies (based on various sources, after Walling & Kleo 1979).

of sediment budgets and to determine sediment-delivery ratios. Very few sediment budgets have been calculated successfully for desert catchments, mainly because the important events are rare, many years of observation are required, and field monitoring is often difficult (e.g. Schick 1977). Two studies stand out. The first detailed attempt to create a dryland sediment budget was made by Leopold et al. (1966). They monitored erosion, transportation and deposition rates near Santa Fé, New Mexico, over seven years. They measured changes in channel dimensions; changes in scour and fill in channel beds (using scour chains); the extent of slope erosion (using nails inserted into the ground through washers); soil creep (by means of carefully located pipes); the rate of headcut erosion; and the rate of sediment accumulation in reservoirs. Valuable results were obtained from each set of measurements. For example, the scour-chain evidence showed convincingly that the period 1958–64 was in general a time of net channel aggradation. More importantly, it is possible to use data from the individual studies to build up a picture of the sediment budget in the area. Extrapolation of experimental-plot data to the whole contributing area was difficult, and in any case the observations were not comprehensive. However, by making realistic estimates, Leopold, Emmett and Myrick compiled the sediment budget shown in Table 9.7. Clearly, sediment production in this area is dominated by surface erosion (slopewash), and only a small proportion of the material is being trapped in channels and reservoirs.

A second long-term process monitoring study that used methods similar to those in the New Mexico project was undertaken by Schick (1977) for over 10 years in the 0.5 km² drainage basin of the Nahal Yael (southern Negev, Israel). The catchment is rugged, with many slopes over 25° on schists and granites. The main channel, with an overall slope of 0.05 merges onto a small alluvial fan. Schick estimated that 99% of the erosional work during the 10-year period was accomplished in five days and by only seven runoff events that constituted some 20% of the mean annual precipitation (31.6mm). The mean annual sediment yield was estimated to be 338 metric tons km^{-2}, of which only 1% was in dissolved load, 65% was suspended load and 34% was bedload. The discharge of sediment on to the alluvial fan (station 02) should equal the sediment removed from the fan (station 01) *less* the sediment stored on the fan. As Table 9.8 shows, the budget calculated for the period 2 April 1967 to 14 November 1973 does not balance to the extent that total deduced fan aggradation is five times greater than the measured aggradation on the fan, a discrepancy that remains unexplained.

Table 9.7 Sediment budget for areas near Santa Fé, New Mexico.

	Estimated average rates (t k^{-2} yr^{-1})	%
TOTAL EROSION	5452.9	100
Surface erosion (slope wash)	5335.2	97.8
Gully erosion	78.5	1.4
Mass movement	38.4	0.7
TOTAL DEPOSITION	1205.5	22
Deposition in channels	564.9	10
Trapped in reservoir	640.6	12

Source: Leopold et al. (1966).

Table 9.8 Tentative sediment budget for the Nahel Yael alluvial fan.

Net aggradation, alluvial fan, 27 April 1967 to 14 November 1973, as determined by cross-section surveys	10.57m³	21 tons
Sediment inflow, station 02, same period (events 7A, 7B, 10)		
(i) suspended	390 tons	
(ii) bedload (½ added)	195 tons	585 tons
Sediment outflow, station 01, as above:		
(i) suspended	176 tons	
(ii) bedload (½ added)	88 tons	264 tons
Deduced aggradation on fan, total load		321 tons
bedload only		107 tons

Source: Schick (1977).

CHAPTER 10

Slopes: forms and processes

10.1 Introduction

Whereas the literature on desert slopes is replete with general observations, it is only in recent decades that precise measurements of slope processes and forms have been undertaken systematically. In this context "slopes" are broadly taken to be those surfaces characterized by runoff in the form of overland flow, where water is not concentrated into channels. This section examines recent slope process–form studies. Cognate work on mountain slopes, hillslopes, pediments and alluvial surfaces is reviewed in other sections.

10.2 Runoff processes on slopes: general models

Hydraulic processes on desert hillslopes include rainsplash erosion and overland flow. Many studies of the former have been widely reported (e.g. Cooke & Doornkamp 1990). On bare surfaces it appears that rainsplash transports in the order of one-twentieth of the sediment carried; the remainder is moved by overland flow. In this sense, therefore, it may appear to be relatively unimportant in deserts, but it is vital in the context of particle *detachment*, especially on inter-rill areas, where it may be dominant (Abrahams 1990, pers. comm.; see also Ch. 7.2).

Runoff on desert slopes is commonly believed to be characteristically Hortonian flow (Horton 1945, Yair & Lavee 1985); that is to say, runoff is ultimately produced from circumstances in which the rate of supply of rainfall is greater than the infiltration capacity of the soil or surface debris. Infiltration capacity will be controlled by a range of soil characteristics such as soil structure, texture, vegetational cover, biological activity, soil-moisture content and surface condition (such as stoniness). Initially rainfall will

infiltrate into the surface. As the supply rate begins to exceed infiltration capacity, puddle storage will commence. Eventually the puddle storage will be exceeded and overland flow will begin, usually as unconcentrated flow (i.e. inter-rill or sheet flow). Such flow will at first be unable to detach loose particles, although it will carry particles detached by rainsplash. However, as its velocity and volume increase, its force will exceed the resistance of surface material, and flow detachment will begin. The force per unit area exerted by flow parallel with the soil surface, F, can be defined by the DuBoys formula. The description of Hortonian flow, first detailed by Horton in 1945, has been repeated many times and the argument will not be developed fully here. However, a crucial feature of it is the distance down slope (x_c) at which flow detachment begins, the distance at which the force of flowing water (as defined by the DuBoys formula) is greater than the resistance (R) of surface material. Horton summarized this relationship as follows:

$$x_c = \frac{65}{q_s n} \left[\frac{R_i \tan^{0.3}\theta}{\sin\theta} \right]^{5/3} \qquad (10.1)$$

where q_s is the runoff intensity (in/hr), n is the surface roughness, R_i is the threshold value of resistance of the surface to erosion, and θ is the angle of slope.

This synthesis of the overland-flow erosion system effectively identifies the main variables that need to be examined if slope runoff and forms are to be effectively interpreted in terms of Horton's theory.

General field observations of slope runoff in deserts tend to confirm the Hortonian interpretation (e.g. Doornkamp et al. 1980) For example, Abrahams & Parsons (1990, pers. comm.) found in Arizona that overland flow at an experimental site travelled 20–25m before it had sufficient shear stress to cause flow detachment. However, important qualifications are required. First, while the Hortonian model appears to be more appropriate in arid and semi-arid areas than in many well vegetated humid areas, it may not be the only acceptable explanation of slope runoff. The "saturation overland flow model" associated with Hewlett (e.g. Kirkby & Chorley 1967), in which water moves down slope through the soil before emerging in saturated zones at the surface, and in which rainfall duration and soil-water storage available for infiltration are crucial considerations, may also be relevant. In eastern Australia and in Chad, where throughflow may occur in solodized-solonetz soils, the slopes are usually long pediments or alluvial fan slopes (e.g. Hallsworth & Waring 1964, Bocquier 1968); but in Western Australia, throughflow has been observed on steeper slopes (e.g. Bettenay & Hingston 1964).

Thus, Scoging (1988) found that in southern Spain a model combining features of *both* Hortonian and throughflow models was necessary to explain observed runoff. She suggested that a fixed potential soil moisture storage capacity is depleted at a rate dependent on the relationship between rainfall intensity and infiltration capacity. She recognized a fundamental distinction between pre- and post-saturation phases of infiltration, so that infiltration only becomes controlled by the profile when the soil moisture storage volume is full. This relationship is called the infiltration envelope (Smith 1972; Fig. 10.1A) in which,

for a given rainfall intensity, there is a period of constant infiltration and no runoff while the soil is unsaturated. When the profile is saturated, infiltration declines according to profile conditions and especially saturated conductivity (K_s = final infiltrability). Similarly, for given antecedent conditions, it will take longer for a lower rainfall intensity to saturate the soil reservoirs (e.g. Scoging 1988). This infiltration model is summarized in Figure 10.1B. Crucial to its operation is the nature of the soil or regolith, and especially its water conductivity characteristics and available storage volume.

Infiltration rates vary greatly with soil type. For example, Reid & Frostick (1987) showed that final infiltration rates on sandy soils at experimental sites in semi-arid Kenya were 78mm hr^{-1}, whereas on silty soils that developed crusting the rates were less than 10mm hr^{-1}.

Scoging & Thornes (1979) demonstrated an unexpected paradox at runoff sites in semi-arid Spain. Here, the times from initiation of rainfall to the initiation of runoff were shorter than those reported in the literature *despite* the fact that the soils had high "final infiltrabilities". Indeed, the higher the infiltration rate, the shorter the time to runoff. The explanation of this relationship appears to lie in the fact that the soils have limited available storage volume. Such volume might have been limited, even in porous soils, by near-surface water accumulation in fine-grained materials, the effect of surface crusting,

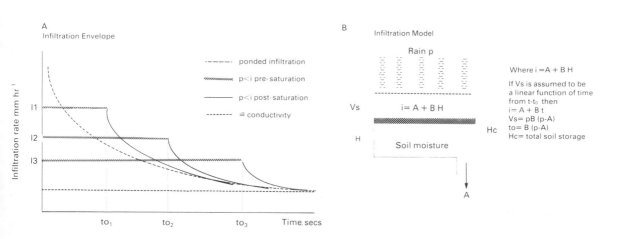

Figure 10.1 (A) The infiltration envelope and (B) the infiltration model, developed by Scoging (1988) in southern Spain. Key: i = infiltration rate; A = parameter related to wetting conductivity at residual air saturation (related to final infiltration rate); B = parameter relating to rate of fill of moisture storage; t = time; v_s = depth of soil moisture storage capacity that must be filled before saturation (Scoging 1988).

or by the pressure effects of trapped air on downward water movement.

These and other models can be used to predict runoff. All depend in some way on the relationships between rainfall and infiltration, and the prediction of the latter poses the greatest difficulties (e.g. Swartzendruber & Hillel 1975, Scoging & Thornes 1979).

10.3 Classification

Desert slopes can be classified conveniently into three types, according to the nature of their controlling or dominant processes. **Gravity-controlled slopes** comprise steeper slopes where debris is removed by the effect of gravity, and slope failure may occur as rockfalls or landslides. Such slopes are usually of bare rock and their development is *detachment- (or weathering-) limited*. The forms of such slopes strongly reflect the geological characteristics of the rock outcrops. On gentler slopes debris produced by weathering passes down slope by mass movement processes:

these are **debris-covered slopes**, whose inclination is often related to the angle of repose of thin debris mantles. They may be *transport limited*. Such mantles may be stabilized by cementation, so that the slopes may be effectively fossilized and inherited from previous conditions. On lower and gentler slopes, the debris is finer and is moved by slope wash: these are **wash-controlled slopes**. Slope wash is not confined to such slopes, because runoff can also remove debris from the other slope types, but it dominates morphological change here.

On many desert hillslopes, the slope profile from the drainage divide to a stream channel is composed of slope units of each of these types. For example, the upper slope may be gravity controlled, the middle slopes debris covered, and the lower slopes wash-controlled. It is also common for such a pattern to be repeated on a profile where complex sequences of sedimentary deposits produce alternating resistant and less resistant rocks (Fig. 10.2). Often the slope units may be separated by distinct breaks of slope. The boundaries between slope types are often morpho-

Figure 10.2 Sequences of slope units in gently dipping sedimentary rocks in Bahrain.

logically distinct. Moreover, there may be discontinuities of processes between the types. For example, there is commonly an important runoff discontinuity between rocky slopes and colluvial or debris-controlled slopes: runoff generated in the former often disappears or is greatly reduced by infiltration on reaching the latter (e.g. De Ploey & Yair 1985).

Are such desert slopes distinctively different from those in more humid areas? Qualitatively, desert slopes appear to be different, although lack of vegetation and long vistas may create a misleading impression. Few studies have tried to analyze the differences quantitatively. An exception is the study by Toy (1977) of 29 south-facing profiles in Kentucky and Nevada, developed on shales dipping at less than 5°. Although Dunkerley (1978) expressed some reservations about the approach, Toy concluded that arid hillslopes tend (i) to have a *smaller radius of curvature* at their crests than those developed in humid areas; (ii) to be *steeper* in their straight segments; and (iii) to be *shorter*. These characteristics *might* reflect generally higher drainage densities and stream-channel frequencies in arid regions. Similarly, in the Northern Hemisphere, north-facing slopes tend to be more humid than south-facing slopes, and tend to be gentler. This contrast is indeed quite common in deserts (Yair & Lavée 1976, and below).

10.4 Monitoring desert slope processes

10.4.1 Introduction

In recent years there have been significant empirical studies of slope processes at sites in deserts. Most are concerned with examining surface wash and creep processes. Some of these studies are now reviewed as a preamble to the discussion of slope types and slope evolution.

10.4.2 Runoff and erosion on rocky and debris-mantled slopes

(a) Sde Boqer

Yair and others have monitored soil and slope conditions, and rainfall and runoff events, on a number of clearly defined catchments on rocky and debris-mantled slopes at Sde Boqer and other sites in the Negev Desert since 1972 (Figs 10.3–5). One of the most interesting and potentially useful results of this important work is the recognition of a marked contrast between runoff on the two slope types.

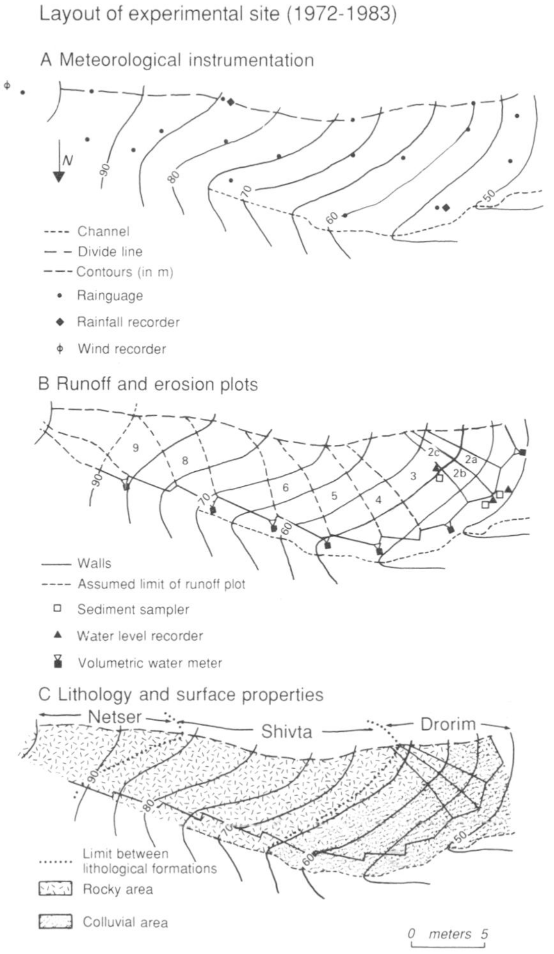

Figure 10.3 Layout of experimental site at Sde Boqer, Israel, 1972–83 (after Yair et al. 1987).

Runoff is both higher and more frequent on the rocky slopes, where the rock:soil ratio is high (Fig. 10.4, Table 10.1).

There is a pronounced discontinuity at the boundaries between the two slope types, because runoff from the rocky slopes can disappear down slope in the debris mantles as a result of the high infiltration capacity of the latter (e.g. Yair & Lavée 1985).

Yair & Lavée (1985) also showed that for both convective and frontal storms, rainfall intensities often exceed infiltration capacities in some locations, and slope runoff responds rapidly, producing characteristically short-duration, steeply rising and falling hydrographs (Fig. 10.4). Equally important, there are substantial spatial and temporal variability and discontinuities of runoff that reflect both the very varied

Table 10.1 Frequency and magnitude of runoff events at plot 2, Sde Boqer, Negev, 1975/6–1978/9.

Subplot	Planimetric area (m^2)	Slope length (m)	Slope angle (%)	Number of flows	Total runoff yield (litres)	Runoff per unit area (l m^{-2})
(i) Whole slope	439	63	27.4	40	5,149	11.7
(ii) Colluvial slope	307	34	28.3	36	4,084	13.3
(iii) Rocky slope	161	33	27.0	65	5,348	33.2

Source: Yair & Lavee (1985).

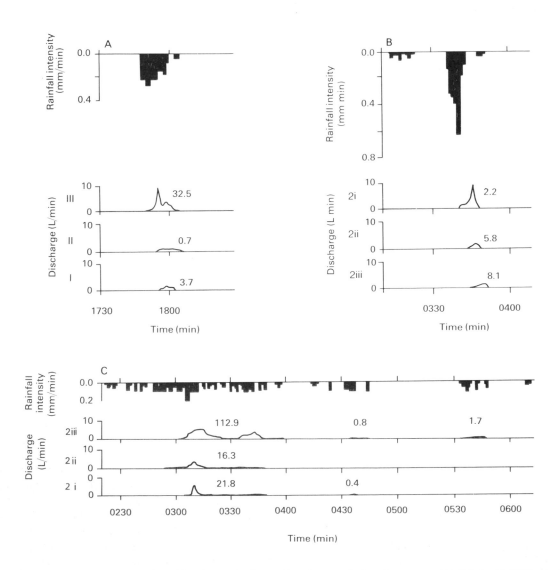

Figure 10.4 Rainfall and runoff at Plot 2 at Sde Boqer for characteristic rainstorms; (A) 24 April 1978, a single shower, short-duration, low-intensity rainfall; (B) 15 October 1978, a single rainshower of short duration and relatively high intensity; (C) 24 April 1977, a prolonged rainstorm (frontal) with multiple showers within it. (i) = whole slope; (ii) = colluvial slope; (iii) = rocky slope (after Yair & Lavee 1985).

Figure 10.5 A sediment sampler, a water-level recorder and a runoff collector at an erosion plot, Sde Boqer, Israel.

nature of climatic events, and the varied nature of unconsolidated sediments and bedrock types.

As indicated previously, the Sde Boqer sites also revealed the possibly unexpected fact that runoff coefficients vary inversely with slope inclination (Yair & Klein 1973). Where scree slopes are dominated by coarse material, for example, infiltration is high and runoff low. Nevertheless, where there are distinct flow paths or gullies in such screes, comprising large blocks set in dense and compact fine material, runoff can occur because the impermeable blocks shed water rapidly onto the underlying relatively more permeable, compact and fine-grained material at a rate exceeding its infiltration capacity (Yair & Lavée 1976, 1985).

It was also perhaps surprising to discover that *organic activity* plays an important rôle in sediment transfer rates (e.g. Yair & Rutin 1981, Yair & Enzel 1987). For example, it was shown that porcupines in search of bulbs break the otherwise surface-protecting soil crust, and burrowing isopods produce faeces that disintegrate easily under raindrop impact, thus pro-

ducing loose soil, which is easily removed in runoff. Broadly, from studies at several Negev sites, Yair & Rutin (1981) showed that the amount of sediment produced declines with rainfall (and would be negligible at *c.* 50mm yr^{-1}), and was higher on rockier and north-facing slopes where soil moisture is higher.

Do the larger rock blocks on steeper slopes move? In a laboratory experiment designed to simulate the Sde Boqer conditions, Yair & De Ploey (1979) showed that creep may be a relevant process. A boulder set in a loam pedestal moved slowly under dry conditions by "dry block creep". Movement increased when relative humidity rose to 100%, under wet—frost—thaw sequences, and when the rôle of burrowing animals was simulated.

Slope aspect provides a major influence on infiltration capacities under certain conditions. For example, on shale slopes in the Zin valley of Israel, differences in solar radiation and soil moisture led to the formation of a thick, dense crust on wet, north-facing slopes, and to a thin friable and relatively porous crust on south-facing slopes. Runoff can occur on the north-facing slopes more easily than on the south-facing ones (Yair & Lavée 1985).

Furthermore, it seems likely that the chemical properties of dryland soils may significantly affect infiltration and thus runoff. For example, the formation of a soil crust in clay soils resulting from clay dispersion may be promoted where the exchangeable sodium percentage is high and electrolyte concentrations are low. An increase in the latter may lead to flocculation and thus to increased infiltration (Yair et al. 1980).

(b) Walnut Gulch, Arizona

Runoff erosion studies have been undertaken at Walnut Gulch for many years at standard erosion plots. Abrahams, Parsons and Luk have initiated a new series of experiments using rainfall simulators and trickle systems to reproduce overland flow (Fig. 10.6; Luk et al. 1986, Abrahams & Parsons 1991, Abrahams et al. 1991, Parsons et al. 1991). In one recent study (Abrahams et al. 1988a,b) they monitored overland flow and sediment concentration on six lightly vegetated plots on slopes ranging from 5° to 33° (Fig. 10.7). Their analysis provides a very useful comparison with the work of Yair and others in the Negev. In Arizona, infiltration was so high on the steeper, coarse-textured slopes that little runoff occurred and equilibrium was not reached (e.g. sites

Figure 10.6 Monitoring runoff and erosion from simulated rainfall at the Walnut Gulch Experimental Watershed site, 1988.

5 and 6, Fig. 10.7), whereas on gentler, fine-textured plots (1–4, Fig. 10.7), not only was equilibrium discharge reached much more quickly, but it was reached in unusual steps. Such stepped hydrographs probably reflect the decrease of permeability at the top of the B horizon: discharge first levels off as surface infiltration approaches the saturated hydraulic conductivity of the A horizon; then it rises to a higher equilibrium as water penetrates and eventually saturates the B horizon. Figure 10.7 also shows that sediment concentration (on all except one plot) declines with time, and is negatively correlated with discharge, suggesting that the available supply of fine sediments is rapidly exhausted, and perhaps implying a *supply limitation* and limited erosional detachment.

Even more instructive is the comparison between surface runoff and sediment yield in Arizona and the Negev. In the Negev, sediment yield was directly related to the runoff coefficient, and was inversely related to hillslope gradient on slopes steeper than 15° (see above), the latter being positively correlated with particle size (Fig. 10.8B). In Arizona, however, there is no significant simple correlation between sediment yield and any plot characteristics; rather, sediment yield is curvilinearly related to slope (Fig. 10.9A) with a maximum at 12°; on lower slopes, runoff changes little with gradient so that yield increases with gradient; on steeper slopes, runoff decreases with gradient causing sediment yield to decline (Abrahams & Parsons 1991). This conclusion leads to a modified runoff–sediment yield model (Fig. 10.8A). Where the

gradient exceeds 12°, the Yair & Klein (1973) model operates. Where the gradient is less than 12° (and none of the Negev sites had a gradient of less than 12°), the runoff coefficient is insensitive to gradient, particle size and surface roughness, and sediment yield is positively correlated with these variables. It remains to be seen, from further studies, how significant the 12° threshold is: quite probably it will vary spatially with ground-surface conditions; and it may be significant that the 12° is near the upper limit of pediment slopes (Ch. 13).

(c) South Africa

In studies by Le Roux & Roos (1986a,b) wash traps ("Gerlach troughs") were used to monitor wash processes on low-angle, debris-covered hillslopes composed mainly of shale and siltstone and capped by dolerite and covered with 27–29% perennial grasses and shrubs, near Bloemfontein and Reddersburg in South Africa. They showed that erosion of sand-sized particles relative to clay-sized particles is inversely correlated with rainfall (probably because storm duration is inversely related to intensity); that the amount of clay fraction in the entrained slope wash is positively correlated with rainfall intensity (probably because higher intensity rainfall and drop size can disrupt the sand-armoured surface); and that the product of rainfall and its intensity is positively correlated to the clay fraction being removed (probably because of high rainsplash effects in conditions of high rainfall intensity). Overall, the rate of wash erosion measured over seven years on the Reddersburg slope was somewhat higher in wet years than in drier years, and gave values of *c.* 17.8mm 10^{-3} yr^{-1} for soil, and 9.6mm 10^{-3} yr^{-1} for rock, relatively low figures that reflect high vegetation cover and low human interference.

10.4.3 Particle monitoring on debris-mantled slopes

Downslope movement of surface particles on debris-mantled slopes has been monitored in several deserts as a means of determining rate and inferring process. In the Mojave Desert, for example, the movement of painted stones particles greater than 8mm in diameter on mountain/pediment slopes over 16 years was

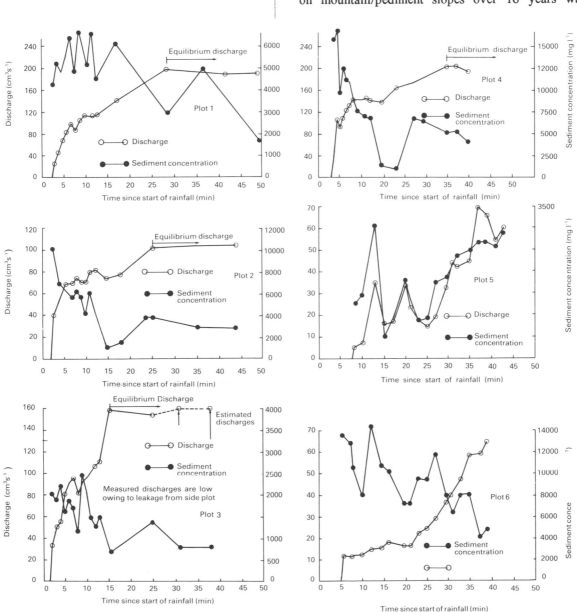

Figure 10.7 Graphs of surface runoff discharge and sediment concentration against time since the start of rainfall at six runoff sites at Tombstone, Arizona. Plots 1–4, slope <21°; plot 5, slope = 22°; plot 6, slope = 33° (after Abrahams et al. 1988).

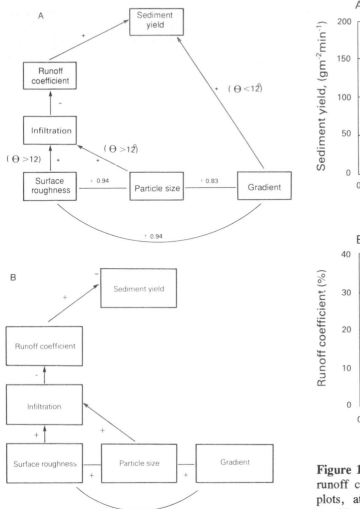

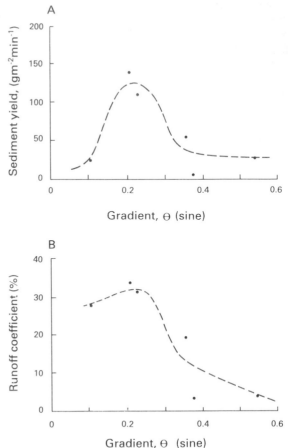

Figure 10.9 Graphs of (A) sediment yield and (B) the runoff coefficient against gradient for the six runoff plots, at Walnut Gulch, Tombstone, Arizona. The curvilinear relations were fitted to the data by eye (after Abrahams et al. 1988).

Figure 10.8 (A) Causal diagram showing the factors controlling the runoff coefficient Q_c and sediment yield G on the hillslopes studies at Tombstone, Arizona. The numbers shown on the diagram are Spearman rank correlation coefficients. (B) Causal diagram showing the factors controlling the runoff coefficient and sediment yield on desert hillslopes in the Nahel Yael Watershed, according to Yair & Klein (1973) (after Abrahams et al. 1988).

recorded by Abrahams et al. (1984). This study showed that the distance moved by particles was directly related to the length of overland flow and the hillslope gradient, and inversely related to particle size, leading to the inference that hydraulic action rather than surface creep was the dominant process of the 7–24° slopes (Fig. 10.10). In nearby southern Arizona, Kirkby & Kirkby (1974) analyzed the movement of painted stones of greater than 1mm in diameter across 12 hillslope/pediment profiles after each rainfall in a two-month period. They showed *inter alia*, that particles greater than 1mm moved at a rate between 4.2 and 5.4cm³ cm⁻¹ yr⁻¹, and that the major transport process was unchannelled and channelled surface wash, together with rainsplash. The distance of the particle movement was directly related to the hillslope gradient, the amount of unvegetated surface and the storm rainfall; and it was inversely related to particle size and surface roughness; and unrelated to distance from the divide. They also drew important conclusions on the mountain/pediment "break of slope" (see Ch. 13).

These observations, confirming the importance of overland flow, contrast with "seven years" of sandstone-fragment monitoring by Schumm (1967) on Mancos shale hillslopes near Montrose, Colorado, where the rate of movement was found to be directly proportional to the sine of the hillslope gradient, and thus to the gravity component acting parallel to the hillslope. Here, at higher altitudes than the previous

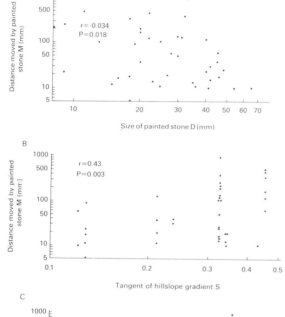

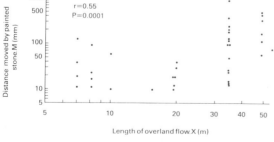

Figure 10.10 Bivariate plots of the distance moved by painted stones against the size of the stone, the tangent of hillslope gradient and the length of overland flow, at a site in the Mojave Desert, California. The sample size is 39 (after Abrahams et al. 1984).

studies, creep related to frost action seems to be the dominant process.

On a rectilinear, 30.7° debris-mantled valley slope in an expanding erosion system in the Turkana Desert of Kenya, Frostick & Reid (1982, 1983) used distinctively painted stones greater than 10mm in diameter to monitor debris movement over a two-year period (Fig. 10.11). Their results showed an exponential pattern of mass movement at each site (summit, mid-slope and basal site). Mean rates, in an area of 300mm yr^{-1}, were between 0.59mm yr^{-1} (summit) and 1.38mm yr^{-1} (mid-slope), with a median value of 0.82mm yr^{-1}, and they were directly related to particle size. Particle shape is a significant factor influencing rates, because spherical and rod-like fragments move more rapidly than blades and discs (Frostick & Reid 1983). Where shallow chutes are formed, runoff may undermine particles so that median rates of movement rise to 3.50mm yr^{-1}. Overall, they showed the valley slope to be steepening as the erosion system of which it is a part expands.

10.5 Gravity-controlled and bare-rock slopes

Bare rock slopes are more common in deserts than in most other environments, chiefly because the rate of debris production by weathering is commonly lower than the rate of removal. This is especially true in "watershed" locations, where resistant rocks outcrop and there is high available relief. Such surfaces ("free faces") occur near to the tops of composite hillsides or where there has been deep entrenchment. Their form depends greatly on the lithology and bedding characteristics of the rocks. For example, relatively unresistant rocks tend to form rounded upper slopes, as a result of weathering and creep processes. More resistant units form more rectilinear faces, and locally form may be controlled by joints and bedding planes (Fig. 10.12). Vertical and near-vertical free faces are not uncommon. Such slopes can develop by several processes, of which the most important are basal sapping, slope failure and "unloading".

10.5.1 Basal sapping
Basal sapping is a groundwater-controlled process involving localized weathering and erosion through seepage and "spring erosion" in relatively impermeable strata underlying a massive and highly transmissive rock formation. The free faces in the latter are undermined and collapse, usually by **rockfall** (e.g.

123

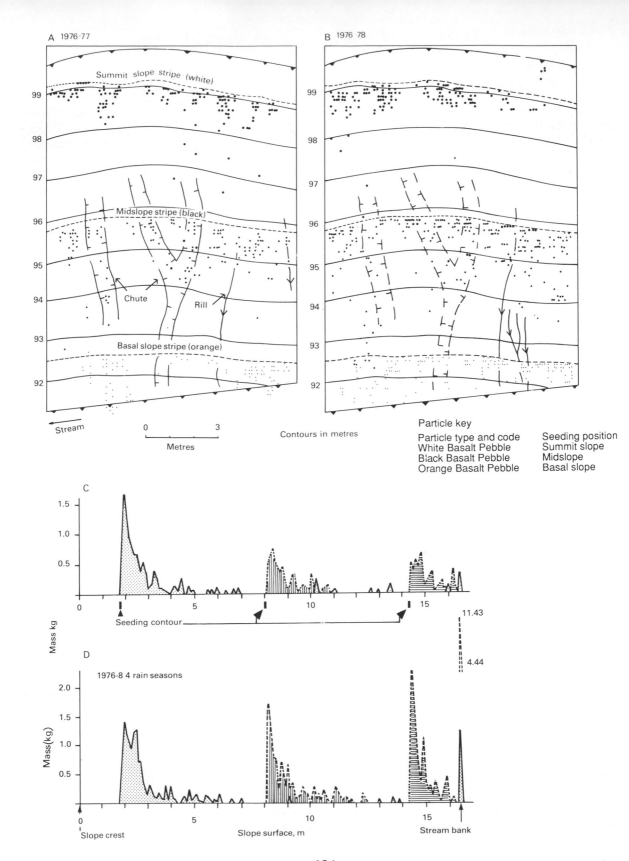

Figure 10.12 Sunrise over Mount Sinai: the bare surfaces of an exposed granite intrusion are differentiated by jointing.

Peel 1941, Ahnert 1960, Laity & Malin 1985, Howard et al. 1988). Schumm & Chorley (1964, 1966) identified the process in the fall of Threatening Rock in Utah in 1941. Here, rock movement in sandstone increased as moisture accumulated from precipitation, and it was particularly rapid in winter months when **frost action** and **wetting** by snowmelt of the shale beneath the resistant beds were active.

Laity & Malin (1985) recognized the importance of sapping in the development of free faces within the massive Navajo sandstone outcrops in theatre-headed first-order, west-trending valley networks of the Colorado Plateau (Fig. 10.13). The base of the free faces (headwalls) are marked by cavities and alcoves, soft, crumbly surfaces and evidence of salt weathering and frost action. Where sapping does not occur, as in the east-trending tributaries (because groundwater flow is westwards and down dip) the headwalls are absent, although steep sandstone slopes modified

mainly by overland flow do occur. Basal sapping is also likely to be particularly significant in limestone terrain where subterranean drainage emerges at a scarp foot.

The free face is only one of several slope units of bare rock associated with massive rock formations in canyon-side and scarp situations. Commonly, the scarps are characterized by an upper convexity, followed successively down slope by a free face, a rectilinear lower segment, and a basal concavity. Such scarp slopes, often initiated by the development of canyons or by faulting, adopt a form that is determined by rock resistance, the orientation and spacing of joints, the direction of dip, and the thickness of the dominant resistant member or caprock (e.g. Schumm & Chorley 1966).

In the Colorado River Canyon country, Nicholas & Dixon (1986) divided such scarps into **embayments**, **headlands** and **outliers**. This division reflected mean

Figure 10.11 Particle monitoring at sites in Kenya. (A, B) Plan of marked particle displacement over one- and two-year periods: note position of chutes. (C, D) two-dimensional distributions of mass movement from original seeding contours for one- and two-year periods on a 30.7° slope (after Frostick & Reid 1982).

fracture spacing within the marker rock unit underlying the main massive scarp rock unit. It is much higher in the headlands than in embayments. Also,

recession was more rapid in embayments because undercutting by "ravelling" (in which blocks drop out of the cliff one by one) was confined largely to the

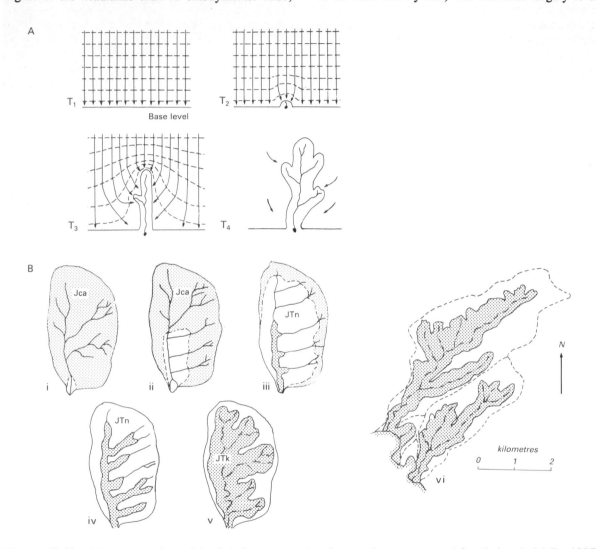

Figure 10.13 (A) A general model of drainage extension by sapping processes (after Laity & Malin 1985). Groundwater flowing through the Navajo Sandstone towards the base level (T1) encounters an irregularity (T2) (an outcropping joint or erosion notch) that causes groundwater to converge at this point. The increased flow enhances local weathering and initiates the sapping process, causing headward migration of the valley (T3). Tributaries grow as groundwater, emerging along the valley sidewalls, exploits zones of susceptibility, usually joint planes. The rate of lateral weathering approaches that of headward retreat as the drainage area of the spring head declines (T4). (B) Schematic illustration of canyon development in the Navajo Sandstone. As the Carmel Foundation (Jca) is eroded from the surface of the plateau, a network of surface runoff is initiated on the Navajo Sandstone (JTn). Runoff is commonly concentrated along joints (i) that enlarge to form canyons as groundwater converges at the valley head (ii). A network of valleys is formed by the process of basal groundwater sapping (iii and iv). (v) shows the network in an advanced state of development. The canyons occupy a very large proportion of the drainage basin, and may continue to grow until adjacent systems merge. (vi) illustrates two tributary networks that occupy most of their drainage area. A new drainage system of surface erosion is developing on the canyon floor.

embayments (Fig. 10.14).

Where such slope profiles are not progressively buried by their own debris, they might be thought of as being in a state of equilibrium, or at least as being supply-limited. The slopes should thus develop by parallel retreat of the scarps (e.g. Schipull 1980) and the relative dominance of free faces would be a function of the thickness and dip of the caprock (e.g. Schumm & Chorley 1966; Fig. 10.15A).

Oberlander (1977) developed variations on this theme to explain the evolution of massive sandstone cliffs in Utah (Fig. 10.15B), where joint-controlled slab collapse associated with spring sapping and granular disintegration are the main processes. He drew attention to major scarp variations *within* the apparently homogeneous massive sandstone, which he attributed to very minor intra-formational discon-

tinuities that lead to the production of "slick rock" smooth footslopes and secondary free faces ("slab walls") within the main scarps, and which owe little to the physical properties of the sandstone itself. Oberlander indicated the development may *not* be so much an *equilibrium* form, but rather an **allometric** one in which each sandstone outcrop produces a similar suite of diverse forms *as and when it is exposed*; only when the effect of basal sapping along thin-bedded underlying strata becomes dominant is regular cliff retreat established.

10.5.2 Slope failure

Slope failure, other than by rockfall, is also important on some desert bare-rock slopes, although landslides and related phenomena are often the product of antecedent wetter conditions in the Pleistocene: many

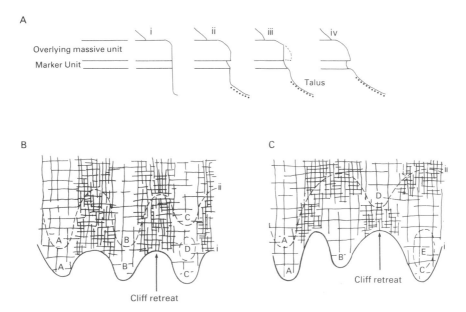

Figure 10.14 (A) Stages of embayment scarp retreat. (i) The initial cliff face is cut by spring sapping or streams. Weak top units of the Organ Rock Formation have begun to recede, leaving the overlying massive unit as a resistant caprock. Cliff rounding begins. (ii) Undercutting of the marker unit occurs along zones of high fracture intensity. Ravelling, spalling and granular disintegration are the likely failure mechanisms. Talus begins to accumulate and cliff rounding continues. (iii) Continued undercutting results in the collapse of the overlying massive unit. (iv) Undercutting resumes, and cliff rounding and talus accumulation continue. Once the overlying massive unit has failed, the underlying beds are exposed to greater surface erosion. Sloping embayment faces result. This completes the cycle which then returns to stage (ii). A steady state of embayment retreat will continue until fracture intensity changes relative to the cliff face. (B) Schematic diagram of "uniform" cliff retreat as the result of joint compartments in the marker unit. During retreat from time 1 to time 2, headlands A, B and C persist. Outlier D remains in direct line with headland C. (C) Schematic diagram of "varied" cliff retreat as the result of joint compartments in the marker unit. During retreat from time 1 to time 2, headland A persists, headland B disappears, and headland D appears where formerly there was an embayment. Headland C remains only as an outlier. Vertical walls and headlands may become detached from the cliff face (after Nicholas & Dixon 1986).

landslides on desert slopes are actually relict. For example, Grünert & Busche (1980) have described large fossil landslides along escarpments on the west side of the Murzuk Basin in the central Sahara (Algeria/Libya). Within these predominantly Nubian sandstone escarpments, **rotational** slides are found mainly in the north where the underlying Triassic clays are thin. In the south, **mudflows** dominate areas where the clays are thicker (Fig. 10.16). Some of the features are large: for example, over a kilometre long, about 200m wide and about 100m high. The latest evidence of movement, and only very little movement compared with the massive earlier failures, took place

c. 8,000BP The features have been modified by both fluvial and aeolian activity, the latter mainly north of 23°50′N (Grünert & Hagedorn 1976). Other examples of such fossil failures have been found elsewhere in the central Sahara (e.g. Grünert & Busche 1980), on the edge of the Tadmait Plateau in Algeria, in Iran (Watson & Wright 1969), by Cooke in the volcanic ashfields of the Atacama Desert of northern Chile (Fig. 10.17), and by DeGraff (1978) in Utah. In the semi-arid zones where geological conditions and locally seasonal precipitation are appropriate, both fossil and active landsliding is commonplace (e.g. in Los Angeles County, Cooke 1984; and in Lebanon,

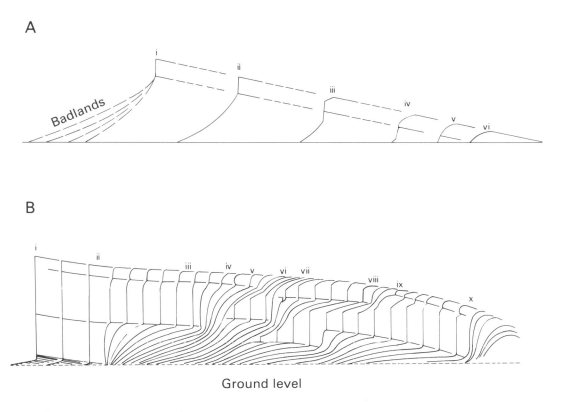

Figure 10.15 (A) Schematic illustration of the change in scarp form as the downdip retreat causes the caprock to become an increasingly important component of the scarp face (after Schumm & Chorley 1966). (B) Development of scarp forms in massive sandstone through time and space. For simplicity, a constant ground level (broken line) is assumed during scarp retreat, along with equal thicknesses of removal from major slab walls in each unit of time. At i a cliff is present due to sapping above a thin-bedded substrate. At ii the thin-bedded substrate passes below ground level, scarp retreat slows, and effective intraformational partings assume control of the scarp form. Effective partings close at iii, iv, v, vi, vii and viii, leading to local slab wall stagnation and rounding into slick rock. Partings that open at iv, v, vi and ix initiate growth of new slab walls. Note that the effect of former partings in rocks that have been removed continues to be expressed in the form of slick rock ramps and concave slope breaks. Lowering of the ground level during backwearing would cause cliff extension upwards from the major contact (after Oberlander 1977).

Khawlie & Hassanain 1984).

10.5.3 Unloading and exfoliation
Unloading is the process whereby cracks develop in bedrock parallel to the surface and open up as the overlying load is progressively removed. It is common in igneous rocks and sandstones, and has been described in many deserts (e.g. Bradley 1963, Kesel 1977, Twidale 1982a). Crucial to slope form is the frequency and spacing of joints and the weathering history of the rocks. Where joints develop by unloading, or "pressure release", they tend to be parallel to the surface and more frequent near to it, as in the sandstones of the Colorado Plateau (Fig. 10.18; Bradley 1963). Superimposed on such joints are those related to the cooling history of magma or post-depositional tectonic stresses. These joints are often

fundamental in determining the basic geometry of rock exposures: they too may be large-scale curved joints, due perhaps to compressional stress (e.g. Twidale 1981b), or they may provide reticulate patterns that define blocks of rock on many different scales.

Characteristic features resulting from unloading, and associated weathering processes, are **mountain fronts** and **inselbergs**. The latter are distinctive areas of isolated hills developed on massive rocks that often appear to rise abruptly from the surrounding plains (e.g. Osborn 1985, Selby 1982b). They are by no means always associated with unloading, but they are often striking features: for example, the Olgas and Ayer's Rock are exceptional examples, and they are so distinctive that they are almost synonymous with the Australian Desert. Inselbergs are not distinctively

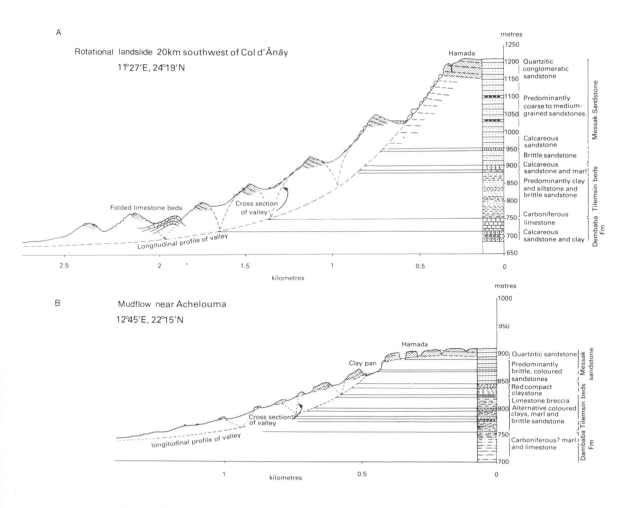

Figure 10.16 Slope failures in the Sahara (after Grünert and Busche 1980).

Figure 10.17 Massive complex slope failures in ignimbrites along the Rio Grande Valley in the Atacama Desert, Chile.

desertic, however: they are typical also of savanna landscapes, and it is possible that some desert inselbergs are inherited palaeoforms in deserts.

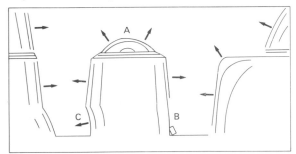

Figure 10.18 Diagrammatic sketch of exfoliation joints in massive Colorado Plateau sandstones. Arrows show inferred directions of expansion: A is an exfoliation zone, B an exfoliation cave, and C an overhanging exfoliation plate in a meander scar (after Bradley 1963).

Inselbergs adopt two major forms: **domed inselbergs** (**bornhardts**), and smaller **boulder inselbergs** (**koppies**, **tors** and **nubbins**). These forms often co-exist, but relations between them are a matter of speculation. Certainly one possibility is that domed inselbergs *evolve* into boulder inselbergs (e.g. King 1951, Selby 1977a, Kesel 1977); and many observers agree that the contrast lies in part in the *pattern of jointing* or faulting, for domed inselbergs occur where jointing is scarce, and it is in closely jointed areas of an intrusion that weathering produces regolith and corestones ultimately form boulder inselbergs (Fig. 10.19). Jointing also often controls bornhardt plan (Fig. 10.20), but the causes of the varied joint patterns vary. Twidale (1981b) suggested that domed inselbergs develop in deep, compressive zones of antiformal structures where jointing is limited, whereas boulder inselbergs originate in the tensional areas

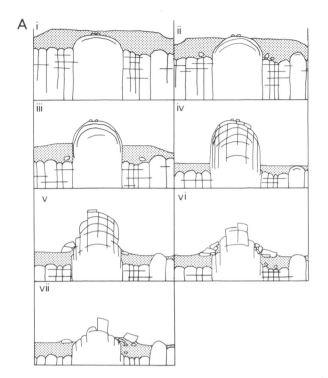

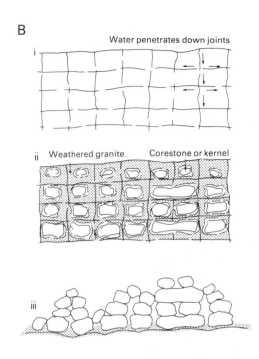

Figure 10.19 (A) A schematic diagram to show one mode of development and decay for domed inselbergs (after Thomas 1974). In diagrams i–iv both the land surface and the basal surface of weathering are differentially lowered around the developing domical summit which is stripped of regolith at an early stage. Stages iv–vii show the disintegration of the dome into a kopje as a result of tensional jointing and peripheral collapse. (B) Two-stage development of boulders by differential fracture-controlled subsurface weathering and subsequent exposure of corestones by evacuation of friable weathered debris (after Twidale 1982b).

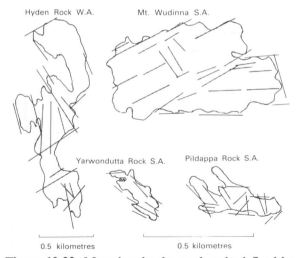

Figure 10.20 Many bornhardts are largely defined by fracture sets which are either orthogonal or rhomboidal in pattern, as shown by these examples from South and Western Australia (after Twidale 1982b).

in the upper part of an intrusion.

In contrast, Pye et al. (1984), who studied these co-existing features on the Matopos batholith in Zimbabwe, found the domed inselbergs to be associated with poorly jointed, slowly cooled porphyritic granite emplaced late in the evolution of the batholith. They were bounded by areas of late-stage foliation and delimited by major fracture systems created by post emplacement tectonism. Selby (1982b,c) similarly concluded from his examination of bornhardts in Namibia that the domal form arises from structures created during granite emplacement: here, the features are, as elsewhere, fundamentally controlled by structure which serves both to block out inselberg areas and to control slope angles through the influence of joint patterns on rock mass strength.

The evolution of bornhardts *as a whole* is also a matter of controversy. In many different areas, although not everywhere (Selby 1977a), a two-stage mechanism appears appropriate, in which differential

131

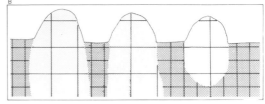
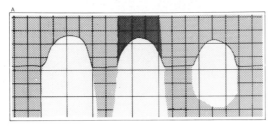

Figure 10.21 A two-stage mechanism to explain the evolution of bornhardts. (A) In the first stage, differential subsurface weathering takes place, controlled by rock type or by the density of fractures. Thus the weathering front, or lower limit of significant weathering, is irregular. Protuberances of fresh rock representing incipient bornhardts may project into the mantle of weathered rock. These may be of slightly different, harder rock, as in the centre; or they may be more massive, as shown at the right and left, with fewer open fractures than adjacent compartments. (B) In the second stage the weathered rock is partly or wholly eroded, depending on the depth of river erosion, and the resistant masses emerge as bornhardts (after Twidale 1982b).

Figure 10.22 Deeply weathered granodiorite, with relatively unweathered corestones in section and on the surface, in the southern Atacama Desert, Chile.

subsurface weathering is followed by exhumation often associated with base-level fall and drainage incision (Figs 10.19, 21, 22). The weathering of the first phase is often accomplished under surfaces of low relief, many of which are ancient (e.g. Twidale 1982b). If this general evolutionary model is correct, an individual inselberg needs to be seen as being at some stage or other in the second phase.

Goudie & Bull (1984) illustrated one possible way in which the weathered mantle may be stripped from granite slopes in Swaziland (Fig. 10.23). This model is based on the observation that *only* the uppermost particles in colluvium down slope show marked edge effects of abrasion. These "younger "particles, it is suggested, are buffeted more in transit because at a later stage of exposure the surface roughness and flow concentration are greater. Once the bedrock surface is exposed it may continue to develop.

There are some who argue that evolution of inselbergs may continue by parallel retreat of slopes, whereby the concentration of moisture at the free face/plain junction promotes weathering and causes collapse and retreat of the higher, drier slope. Twidale (e.g. 1982b) and others found little evidence of such retreat on a major scale.

Certainly, however, the inselberg surfaces are embellished with many weathering features, such as tafoni, basins, gutters, flared slopes, rills, spalling, crumbling, pitting, flaking and crusts and varnishes (see Part 2). In igneous rocks such features may be developing in the present conditions, but at least some of them are likely to have been initiated in association with earlier subsurface weathering.

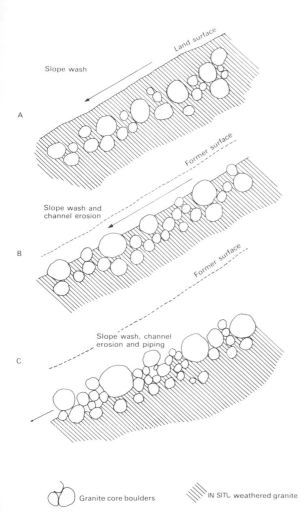

Figure 10.23 Slope evolution on tors and grus above colluvial sites (after Goudie & Bull 1984).

Figure 10.24 Parallel retreat of a free face (I, II, III . . . H) with scree accumulation (I′, II′, III′, . . . H′) and the formation of a buried rock core (F, A, B, C, H). β is the angle of the cliff, α is the angle of the scree (after Lehmann 1933, modified by Young 1972).

10.6 Debris-covered slopes

The lower limit of gravity-controlled slopes is usually marked by a fairly sharp boundary, often between mountain and plain, and it may be entirely in bedrock. The reasons for such "clean" boundaries are many, but fundamental to them is the fact that debris supply from above is less than that which can be removed: the supply is *weathering-limited* and in deserts is often relatively low. In places, falling debris may simply disintegrate, so that, for example, a sandstone block becomes more easily transported sand.

If the rate of debris production exceeds the rate at which it can be removed, debris will accumulate at the base of the gravity-controlled as a scree or talus slope (Statham & Francis 1986). Theoretically, as Fisher first pointed out in 1866, the scree may build up in front of the free face, progressively burying it and protecting a convex rock-cut slope beneath it. The buried slope is theoretically convex because as the free face retreats the supply of debris is reduced and spread over an increasingly wide area (e.g. Fig. 10.24). In fact, such convex slope elements rarely seem to have been identified, although it is not impossible that some of the "slick rock" slopes in sandstone have this origin (see Sec. 10.5). More commonly, relatively shallow talus accumulations over near-rectilinear bedrock slopes are found.

The debris mantles are often relict in deserts (e.g. Fig. 10.25). They are shallow, and more-or-less uniform in thickness. They are commonly stable, armoured by surface boulders, cemented (with gypsum, for example), and varnished. Such features are common throughout the central Sahara (Busche & Hagedorn 1980) the Middle East and the deserts of the Americas. Gerson (1982) saw palaeoclimatic

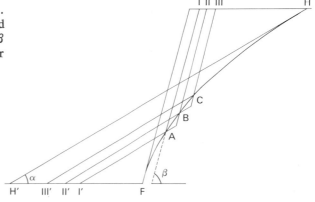

Figure 10.25 Relict debris mantles on bedrock slopes in Bahrain.

significance in these features (Fig. 10.26). He suggested that continuous talus aprons below gravel-producing free faces are formed mainly under mildly arid to semi-arid conditions, and that they are maintained and recede by debris flow and wash (Fig. 10.27). Under arid to extremely arid conditions, talus accumulation is limited, and talus stripping occurs by erosion, wash and gullying. The result is a sequence of flatirons comprising talus aprons formed under more pluvial conditions, separated and dissected by erosion and gullying in interpluvial conditions. The profiles of such alternating sequences merge down slope. Gerson stated that, in general, talus slopes are relatively insensitive to climatic changes and therefore

they only respond to major changes. Rates of scarp retreat through repeated cycles and range from 10–60 m/10^5 years. This intriguing idea has been developed

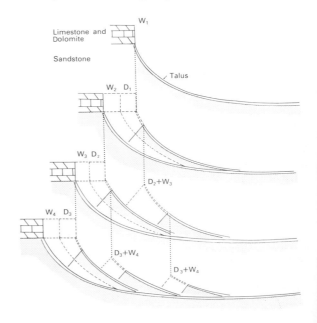

Figure 10.26 The evolution of a sequence of talus "flatirons": W = pluvial regime; D = interpluvial regime. During pluvial periods, the evolution and maintenance of talus aprons through relatively slow retreat prevail. During interpluvial periods, there is relatively fast retreat through erosion and gullying. Talus "flat-irons" mark the end of past pluvial periods (after Gerson 1982).

in Sinai (e.g. Bowman et al. 1986) and it is potentially applicable to many areas such as Arizona, Cyprus, Bahrain, the central Sahara and northern Chile, to name but a few areas where flatiron sequences have been observed.

There is, however, an important distinction between relict accumulations and stable accumulations. The latter may be achieved under almost any conditions, although the relations between angle of repose and slope inclination are not always clear, as Melton (1965b) and Carson (1971) discovered in Arizona and Wyoming respectively.

In Arizona, Melton found that the angle of the straight portions of desert debris-covered hillslopes varies from 12° to 37.5° on granitic rocks and from 17.5° to 34° on volcanic rocks (Melton 1965b, Fig. 10.28). In both cases the mean is about 26° and there is a very distinct change in the character of the slope at angles just above this (at 28.5°): steeper slopes are unstable (a block of a size common on the slope rolls down it), while gentler slopes are stable. Above this

angle the slopes become increasingly unstable until an angle of between 34° and 38° is reached. This last figure is a limiting angle for debris-covered slopes, since slopes steeper than this have no debris cover. Carson (1971) found a similar distribution of slope angles in Wyoming with a modal group between 25° and 28°.

These observations alone contradict earlier studies of desert hillslopes, some of which were actually made in the same area as Melton's study. They make it clear that an explanation involving the angle of repose of the debris alone is untenable, since most slopes are below this angle. Two kinds of explanation have been offered. Melton (1965b) suggested that the steeper angle (34°–37°) was indeed the angle of static friction of the debris (i.e. the angle of repose), but that the limiting angle for stable slopes (about 26°) was the angle of sliding friction of the debris. He found that experimentally produced values for these angles were very similar to the angles that he measured in the field. However, Melton's experi-

Figure 10.27 Debris fans in the high-altitude arid environment of the Hunza Valley of the Karakorams, Pakistan. In the centre, a recent debris flow has cut through fields to the river.

Figure 10.28 Boulder-covered slopes in granite rocks, north of Phoenix, Arizona. How safe is the house (left) from boulder fall?

mental results only concerned the movement of single cobbles over a rocky surface and, as Carson (1971) pointed out, this is not representative of the field situation where the blocks interlock in a mantle. Carson maintained that the real relationships should be sought in the angle of internal friction *of the debris mantle*. However, there is a contradiction here: the angles of internal friction for gravel–soil mixtures are known to lie in the 42°–45° range, which is considerably above the angles found in the field. Carson explained this anomaly by invoking positive porewater pressures in the mantle, and showed that if one postulates saturated mantles, the stability angle is indeed close to 26°. He maintained that although storms which could produce such saturation are rare in Wyoming (return periods in the order of 1,000 years), they may still be important in controlling slope angles. It is here, of course, that palaeoclimatic interpretations may begin to be invoked.

There is a considerable spread of slope angles about the modal slope value. One might suppose that the

size of the debris would have an effect on slope angle, but Melton (1965b) found that the relationship between angle and boulder size was very weak in Arizona. He suggested that the association (if any) would be due to weathering rather than frictional considerations: on steeper slopes smaller blocks roll away more easily, and finer material is washed away more quickly to leave only the coarser blocks. Carson (1971) related the angle of slopes to the state of weathering of the debris mantle. The steepest slopes (near the angle of repose of the coarse weathered blocks) have the least weathered mantles; the modal group of angles are in mantles that are partly weathered; and the low-angle slopes are in very weathered material. Locally, steeper angles are often maintained near to stream channels, where these undercut the slopes (Melton 1965b).

The processes taking place on debris-covered slopes are probably dominated by **mass movement** in the form of debris flows, by flowing water and runoff creep, and by movement of individual blocks under gravity at angles between 37° and 28°. Some large blocks roll down slope to the foot of the steep, straight unit. As the slope angles decline, wash becomes more and more important. In many deserts, most debris-covered slopes are below the angle of repose and runoff is the main process of movement. Large blocks detached from the rock by weathering are disintegrated *in situ* and the resulting material is removed by the wash. Melton (1965b) found that there were linear block-fields on some slopes running normal to the contours. He attributed these to local washing away of the finer debris to leave behind only the larger blocks.

In some circumstances, the material in the debris mantle may be formed *in situ*, and the size of the debris may be unrelated to slope inclination. For example, granitic boulder-covered slopes may be derived by the removal of a weathering mantle to reveal weathered corestones, and the slopes on which the boulders rest may approximate a former "weathering front". In short, the slopes may represent a two-phase evolution similar to that described for bornhardts (Fig. 10.19). Thus, in the Mojave Desert, California, Oberlander (1972) suggested that deep weathering of granite occurred some 8 million years BP, and subsequent stripping of weathered debris and exposure of corestones accompanied the onset of aridity since the late Tertiary. On this interpretation, the slopes are essentially palaeoforms that are only being embellished at present. This contrasts with the alternative but less likely interpretation which envisages parallel retreat of the slopes under present conditions of aridity, and the production of the next crop of boulders beneath the surface ready to be revealed as the slope retreats.

10.7 Wash-controlled slopes

The gentler slopes below gravity-controlled slopes are often dominated by fine-grained and incohesive materials whose movement is dominated by wash processes. Such slopes include alluvial fans and pediments, the characteristics of which are reviewed below, in Chapters 12 and 13. The preceding discussion of monitoring studies also focused on wash-controlled slopes (Sec. 10.4). This section will therefore be brief.

In general, it might be expected that particle size would decrease with distance and gradient across wash-controlled surfaces. Indeed, this is often the

case. For example, Dury (1966b) concluded that particle size greater than 2mm on the *b*-axis tended to decrease in an orderly fashion down slope across a concave pediment in Australia. Also, Cooke & Reeves

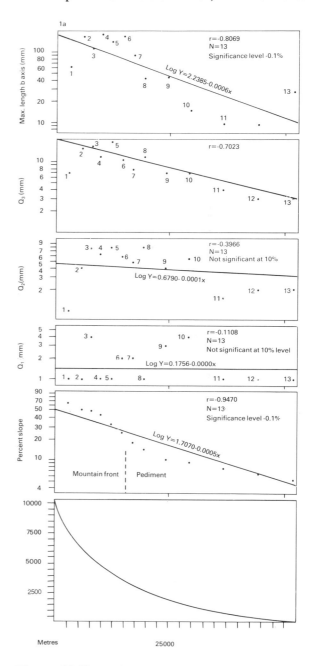

Figure 10.29 Relations between particle size and distance along a mountain–pediment profile in Stoddard Valley, Mojave Desert, California (after Cooke & Reeves 1972).

(1972) found that the size of the largest particles of debris-covered mountain front and pediment slopes in the Mojave Desert decreased most rapidly and consistently down slope, and that the rate of decline matched decline in gradient (e.g. Fig. 10.29). However, the decline in particle size with slope and distance is not simple. For example, the relationship with the larger particles is probably strongest because these relatively large particles are the most resistant to downslope movement by flow (and other processes such as rain, creep and rodents) and they trap finer particles: slopes may thus become adjusted to these larger particles, or vice versa. In addition, debris on wash slopes may be supplemented by locally derived weathered material from the underlying bedrock, so that the rate of decline depends on the relative proportions of transported and *in situ* debris.

Two subsequent studies in the same region have elaborated these relationships. Abrahams et al. (1985) analyzed particle-slope relations on debris-covered slopes in the Mojave Desert developed on closely jointed and mechanically weak rocks that were without free faces above them or basal stream channels down slope of them. They showed that hillslope gradient and mean particle size are related, *independently* of distance from the divide. Furthermore, the relationship, as defined by the regression coefficient, varied with the *plan form* of the slope, giving a value of about 1 for plan-planar slopes, less than 1 for plan-convex slopes, and greater than 1 for plan-concave slopes (Fig. 10.30); and, where plan form is held constant, the regression intercept varied with rock type (Fig. 10.30A). Hydraulic processes will vary with plan form (e.g. overland flow converges down slope down plan-concave slopes) and the observed relationships are consistent with the view that the slopes are formed by hydraulic processes and are adjusted to them. Abrahams et al. (1985) saw such slopes as being transport-limited, adjusted to contemporary processes and having a strong slope–particle size relationship. They represent one extreme of a spectrum of slopes which is marked at the other end by widely jointed, mechanically strong rocks that are weathering-limited, poorly adjusted to present processes, and having a weak relationship between slope and particle size. Such slopes include the boulder-clad granitic slopes discussed elsewhere.

The relationships between particle size and slope, while generally true, can vary from place to place, as the previous studies have shown. In the Sonora Desert of Arizona, Akagi (1980) also showed that there are differences in the relationships on profiles of similar inclination but on different rock types. On non-

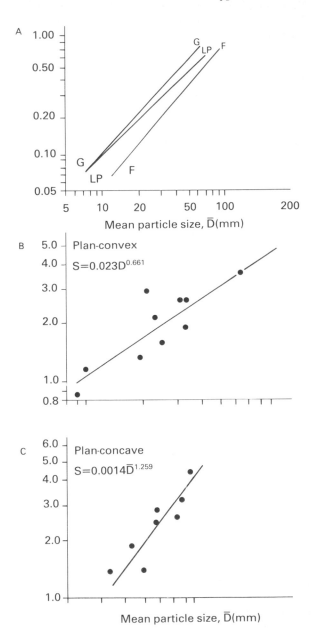

Figure 10.30 (A) Graphs of hillslope gradient against mean particle size for plan-planar debris slopes underlain by gneiss (G), latitic porphyry (LP) and fanglomerate (F). (B) Graphs of hillslope gradient against mean particle size for a plan-convex debris slope, and (C) for plan-concave debris slope. All examples are from the Mojave Desert, California (after Abrahams et al. 1985).

granitic rocks capped by resistant rocks, particle size varied inversely with slope on parts of some traverses. The general relationship may also be interrupted by the local accumulation of coarser material, for example in sieve deposits on alluvial fans (see Ch. 12) or by long-distance gravity-controlled rolling.

The wash processes on slopes have been studied by Emmett (1970), who examined the effects of wash by assuming that it obeyed laws similar to those followed by water flow in channels. In his field experiments on wash slopes in Wyoming, Emmett found that when rain fell on a slope, the upper part of the slope transmitted runoff as a thin sheet in which the flow was laminar. Lower down the slope the flow still had properties of laminar flow, but rain falling on to the surface added considerably to the efficacy of the water as an eroding agent, and here the flow could be described as "disturbed". Near the foot of the slope the flow became truly turbulent.

This flow can be characterized by the same equations as characterize the flow of water in channels, namely:

$$d \propto Q \qquad (10.2)$$
$$v \propto Q^m \qquad (10.3)$$
$$s \propto Q^z \qquad (10.4)$$

where d is depth of flow, v is velocity of flow, s is slope, Q is discharge and f, m and z are exponents. (The equation $w \propto Q^b$ in which w is the width and b is an exponent, does not apply to slopes, since width can be considered constant; increase in discharge is therefore assumed to be absorbed by the other variables.)

Emmett's observations of wash processes on slopes in Wyoming suggest that sheetflow without rills is the dominant process on slopes there. These processes are in dynamic equilibrium with the rest of the erosional system and with the prevailing conditions of vegetation (*Artemesia* scrub) and rainfall. Rills do not develop in this part of Wyoming, and Emmett suggests that they only occur when there is a change of the controlling conditions; and even then the system might adjust as a whole to eliminate the need for the high erosion rates associated with rills. However, where vegetation is inhibited, rills may well be the dominant form.

Rill erosion of wash slopes is also fundamental in the sense that the concentration that it represents provides greater erosive power. Thus, for example, Walker (1964) found that wash on a sandstone talus slope and pediment in arid Australia was only capable

of moving particles smaller than 5mm in diameter; but when the sediment entered small rills, the material finer that 100mm was washed out.

10.8 Some applications of slope studies

10.8.1 Water harvesting and soil erosion in the Negev

Undoubtedly, slope runoff and erosion processes have important practical applications to problems of soil conservation and water harvesting in deserts. These can be well illustrated by the problem of the Nabatean settlements in the Negev, settlements created in the Byzantine period some 2,500 years ago. It has been known for many years that these settlements were sustained by sophisticated techniques for collecting runoff from hillslopes and concentrating it in valley floors to sustain irrigation (Fig. 10.31). One view was that the runoff from slopes was increased (by over 24%) by piling surface stones into heaps so that the underlying loess could be crusted by rainfall to reduce infiltration capacity (Evenari et al. 1971). A second hypothesis states that the stone clearing was designed to promote soil erosion of loess so as to encourage soil deposition in the valley floors and improve the agricultural potential there (Kedar 1967), but this process was shown to be too slow.

Yair (1983) provided an alternative interpretation based on his field monitoring studies at Sde Boqer (Sec. 10.4). He showed that the irrigated farms were not only confined to valleys cut in the bedrock plateau of the Negev Highlands (and were absent from loessial plains in the north, partly discounting the other hypotheses), but that they also depended on canals drawing water not from the loess surfaces but from the steeper, stony surfaces up slope of them. The reason for this was that the frequency and magnitude of runoff are positively related to the extent of the bare bedrock outcrops, or the bare bedrock:soil ratio: thus runoff is greater from the higher, steeper, stony slopes, where porosity and absorption capacity are low; and, if captured *before* it disappears into the colluvium down slope, it provides a relatively reliable source of water, especially from frequent showers of short duration.

The contrast between runoff characteristics on rocky and debris-masked slopes in Sinai has led to suggestions for controlling soil erosion and developing water harvesting through controlled colluviation on lower slopes (De Ploey & Yair 1985). Such

Figure 10.31 Avdat in the Negev Desert: the site of a Nabatean settlement and recent runoff studies.

colluviation is in general promoted by heavy soil erosion up slope which increases sediment load concentrations in runoff. Colluviation can be promoted up slope by reducing discharge through, for example, infiltration or spreading. On the lower slopes, the judicious use of vegetation or storage structures may be possible. The preceding discussion, and the work at Sde Boqer, led De Ploey & Yair (1985) to focus on two possibilities: first, to increase water yield by *increasing* soil erosion on rocky slopes and thereby reducing infiltration and promoting runoff; and, secondly, to *exploit the wet zone at the discontinuity between the rock slopes and the debris-mantled slopes* by, for example, creating "mini-catchments" that collect runoff and concentrate it artificially at points where plants or trees can be grown and where infiltration depth and soil-water storage can be increased, salts can be evacuated to depths below the root zone, and the collection of organic matter from upslope might improve the water-holding nutrient capacity of the soil (De Ploey & Yair 1985).

10.8.2 Archaeological prediction

Kirkby & Kirkby (1976) demonstrated the value of monitoring slope debris movement in the context of archaeological debris-mound degradation by slope and rainsplash processes. They predicted that mound profiles should lower and spread out over time in a form approximating a normal curve (Fig. 10.32), and exemplified the trends at semi-arid sites in Mexico and Iran. Such curves can help to determine the relative ages of small debris mounds in an area, to interpret sherd concentration, and to inform archaeological evaluation of density and its related population implications.

10.8.3 Accelerated erosion by vehicles

Off-road vehicles (ORVs) have many uses in deserts: for recreation, racing, military manoeuvres, exploration, movement between settlements and wells, and animal herding. Where such vehicular activity is intense and surface disturbance is great, runoff and erosion can be seriously accelerated. In the Mojave Desert, recreational ORVs, especially motorcycles,

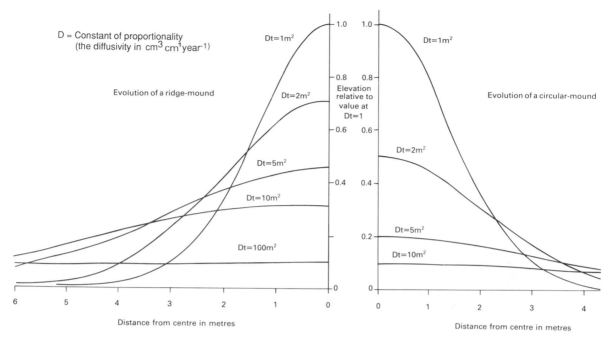

Figure 10.32 Mathematical model for the evolution of the house-mound by erosion over time for (A) a ridge-shaped mound and (B) a circular mound (after Kirkby & Kirkby 1976).

remove vegetation and disrupt soils over considerable areas (see Ch. 8). Iverson (1980) simulated rainfall and runoff at several 1m² sites on poorly sorted, stony soils, some of which had been affected by vehicular traffic, and some of which had not. He showed that vehicles had the principal effects of reducing infiltration capacity; loosening material and removing surface stabilizers and thus increasing the availability of sediment for removal; channelling runoff into wheel tracks and thus increasing the erosive power of flows; and smoothing micro-topography to the slopes, and so reducing friction. Iverson demonstrated that ORVs decreased the Darcy–Weisbach friction factors by c. 13 times and increased runoff Reynolds numbers by c. 5.5 times; sediment transport capacity and gully formation were enhanced; and runoff power (the equivalent to mean boundary shear stress × mean flow velocity) and its concentration were increased. For the same runoff power the amount of sediment yield was much higher on plots subjected to vehicle use (Fig. 10.33). The experiments also showed that ORV effects were limited on silty sands; but they were highest on poorly sorted gravelly soils, which are particularly susceptible to reduction of infiltration

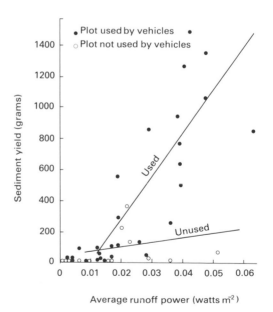

Figure 10.33 Computed regression lines for runoff power versus sediment yield from metre-square erosion plots used and unused by off-road vehicles (after Iverson 1980).

capacity by vehicle movement. Interestingly, gravel concentration correlated *negatively* with runoff on *natural* slopes and *positively* with runoff on vehicle-modified slopes. Iverson suggested that the rôle of surface gravel may change from detaining runoff and increasing permeability by promoting an open soil structure under natural conditions to that of making the surface more impervious after vehicular disturbance.

CHAPTER 11

Channels: processes and forms

11.1 Ephemeral flow in channels

The nature of flow in ephemeral-stream channels reflects the full range of climatic and drainage-basin characteristics. Especially important are the nature and distribution of rainfall events, the infiltration capacities of surface materials (which may vary seasonally), antecedent moisture conditions (which also vary seasonally), irrigation abstraction, and local topography. In many deserts, particularly where precipitation falls as short sharp showers, the runoff hydrograph characteristically rises very steeply, has a sharp peak, and the recession limb is initially steep followed by a relatively long, slow fall (Fig. 11.1; e.g. Schick 1970, Thornes 1976, Griffiths 1978). The steep initial rise is typical of the so-called "flash-flood": it is often almost vertical, indicating the passage of a bore (a rapidly advancing frothy wave of water; Fig. 11.2). But the steep-fronted bore is not always present. In monitoring floods in Turkana (Kenya), for example, Reid & Frostick (1987) recorded flood hydrographs with gentler rises, which they attributed to rainfall over the whole catchment reducing water-storage capacity in channel beds before the flood bores arrived. The falling limb of the hydrograph commonly reflects not only the passage of the ephemeral pulse of water, but also the loss of water by infiltration into the channel bed and possibly the return of groundwater back into the channel from bed and banks, especially where they are composed of alluvium. A clear example of flow reduction through infiltration or "transmission losses" is shown in Figure 11.3, where hydrographs derived from two flumes 10.9km apart along Walnut Gulch (Tombstone, Arizona) reveal a water loss of some 57% (or 1425.42 m³/km) during a single storm.

In a study of transmission losses within ephemeral-flow alluvial channels in southern Spain, Thornes (1977) likened the channel bed material to a reservoir overlain by a perforated lid (Fig. 11.4). The crucial consideration is the balance between inflow discharge (Q_i, Q_{si}), on the one hand, and total transmission loss, channel and subchannel storage (S_c, S_s), and outflow on the other (Q_o, Q_{so}). Where total transmission loss exceeds inflow, flow in the reach will cease. This point is likely to be reached relatively quickly in alluvial channels in deserts, and flow in channels is likely to be rapidly reduced, compared with wetter regions, for several reasons. First, the inputs are relatively low. Secondly, the transmission loss is relatively high, especially where the alluvial deposit is thick. Thirdly, the water table is relatively deep, the sediments have a high hydraulic conductivity, and high outflow from "reservoir" storage is possible because of relatively long intervals between runoff events. Fourthly, if channels have a high width:depth ratio, as is often the case in deserts, small increases in inflow are reflected in rapid increases in width, which encourages rapid infiltration through the bed.

The balance between inflow and transmission loss will change along a channel. In general, total discharge will decline exponentially as transmission loss increases. In addition, the survival of a flow will depend on the magnitude of the input, and the spacing between tributary inputs, if any (Fig. 11.5). Flow may, for example, be either continuous or intermittent during a flow event. Thornes (1977) developed Burkham's (1970b) equation for predicting discharge and an equation relating discharge to drainage area to form an expression which gives the **survival length** of a flow:

$$L = \frac{g^m}{mac\,C_1} \cdot \left[\frac{l}{p}\right]^{1.66m} \qquad (11.1)$$

where L is the survival length of flow, g is the specific yield (or gain) in volume per unit area per second, m, a, c are coefficients and exponents in the hydraulic geometry equations, and C_1 is the relationship between infiltration and depth. l/p is derived from the equation, $l = pA^{0.6}$, where l is the area increment function; and p is a coefficient. If L is greater than the average spacing between tributary inputs, axial flow will survive. The distance L reflects the size of the runoff event, so that in a large catchment there may be a *spatial* manifestation of magnitude in the sense that larger flows extend further down stream. This phenomenon is well illustrated in the Gur-Saoura-Messaoud catchment in the Sahara (Fig. 11.6).

Ephemeral flow hydrographs may reflect significant differences between small and large channels and

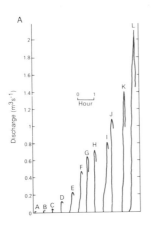

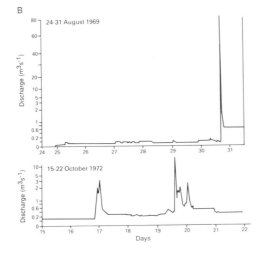

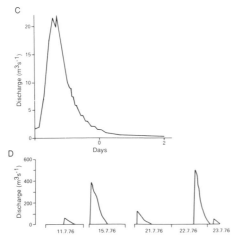

Figure 11.2 An advancing flood bore in a small channel during a rainstorm in central Bahrain.

Figure 11.1 (A) Hydrograph rises for various stations and watersheds within the Nahel Yael research watershed, southern Israel (after Schick 1970). (B) Selected trace of stage records from the Las Tosquillas station, Rio Ugijar, Spain (after Thornes 1976). (C) Recorded hydrographs in Mithawan Nallah, Pakistan (after Griffiths 1978).

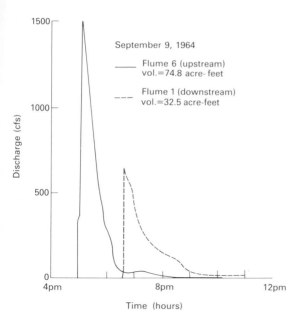

Figure 11.3 Transmission losses for a flood in Walnut Gulch, Arizona, represented by the hydrographs for two flumes 10.9km apart in the ephemeral-stream channel (after Renard & Keppel 1966).

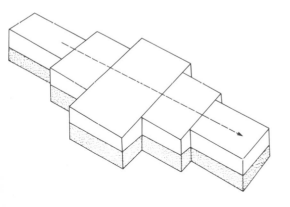

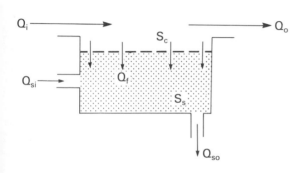

catchments. For example, small headwater catchments normally show a relatively rapid flow concentration, and short lag times to flood peaks. This is because slopes are short and steep; valley heads tend to be zones of flow concentration; soil cover is relatively low and time to saturation short; and smaller catchments are more commonly covered by a single storm. Larger catchments and channels, in contrast, show the opposite tendencies and additionally have greater potential for storage, which may delay flow concentrations: thus, their hydrographs tend to be flatter and to have lower peaks (e.g. Cooke et al. 1982).

Wolman & Gerson (1978) also emphasized the importance of catchment size. For example, they suggested that large catchments, lacking vegetation

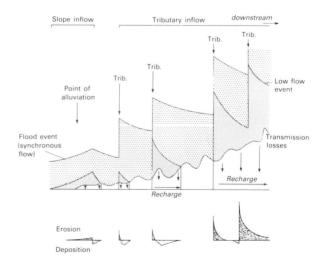

Figure 11.5 Schematic representation of transmission losses and their relation to channel flow and erosional processes for storms of two different magnitudes in an ephemeral drainage system. The shaded area represents discharge (after Thornes 1977).

Figure 11.4 Schematic representation of channel structure with depth in terms of water distribution. The shaded area represents the bed material. S_c and S_s are subsurface and channel storage. Q_i, and Q_{si} are inflows to this storage, and Q_o and Q_{so} are the outflows. Q_f is infiltration. The arrow denotes downstream (after Thornes 1977).

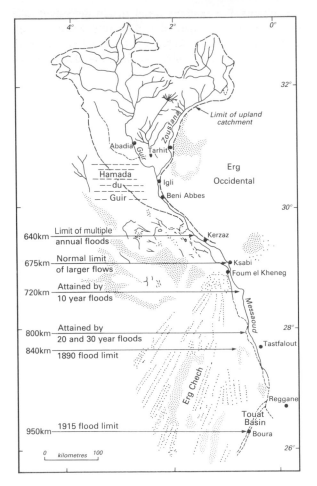

Figure 11.6 The limits of flooding in the Gur–Saoura–Messaoud catchment of the Sahara (after Vanney 1960, from Mabbutt 1977).

short-period (15 minutes) rainfall and peak total discharge (e.g. Renard 1970).

The determination of magnitude–frequency relations in ephemeral systems is also often impeded by the lack of hydrological data, but it is now possible in some areas (e.g. Ben-Zvi & Cohen 1975). Baker and others have developed a new approach to determining magnitude–frequency relations through the careful analysis and dating of sequences of **slack-water sediments** that are deposited in protected areas of low-flow velocity (such as at the mouths of tributaries to major canyons (e.g. Baker et al. 1983a). Individual flood sedimentation units are often capped by drapes of silt and organic detritus, and individual flows can be identified in section by such features as stratigraphic breaks, buried palaeosols, changes in induration, and colour changes. Organic detritus is often suitable for radiocarbon dating. From such information, flood units need to be correlated; for example, on the basis of stratigraphic position, thickness and structure, induration, soil horizonation and by dating. Then the geometry of flows can be determined, assuming that the top of the slack-water sediment corresponds to the height of the flood, and by calculating cross-section area, down-channel gradient and roughness. Using this approach, Baker et al. (1983b) estimated that the discharge of the highest recent flood in the ephemeral Finke River gorge (North/South Australia) had a discharge of 3,360m^3 s^{-1} and a mean flow velocity of 2.05–2.5 ms^{-1}, and they identified this as being one of the major floods in 1967, 1972 and 1974. Other areas where this technique has been applied include the Salt River (Partridge & Baker 1987) and the Verde River (Ely & Baker 1985) in Arizona.

11.2 Sediment movement in ephemeral flows

Ephemeral flows are undoubtedly effective agents of erosion and sedimentation along the channel stretches in which they occur. Commonly, but not universally, such flows show a rapid rise in suspended sediment concentration (Figs 11.7 & 8), possibly to the point of creating debris flows (see Ch. 12.2). There are often several reasons for this: there is an abundant supply of loose dry materials on the channel bed; relatively unstable, unvegetated alluvial banks are easily eroded during the passage of a flood wave; relatively high drainage densities facilitate the chances of water and sediment from slopes reaching the channels; and trans-

and frequent, low-magnitude events, retain the evidence of large-magnitude flows better than more humid catchments of a similar size. In addition, many large catchments may be influenced by snowmelt or frontal rainfall from time to time, so that they will have a second type of hydrograph, one characterized by higher peaks and longer duration than the hydrograph of short storms.

The prediction of ephemeral flow in channels is difficult in deserts, for the variety of reasons discussed above, but it is slightly easier for smaller than for larger catchments. For example, studies in Arizona suggested that the mean annual flood peak correlates strongly with catchment area for catchments less than 200km^2, and there is a good correlation between peak mission losses reduce the water available for transport. Reid & Frostick (1987) showed that such flows in the Il Kimere basin (Turkana, Kenya) are highly turbulent (with Reynolds number > 10^4) so that, with no limitation on sediment supply, suspended sediment concentrations by grade size are hydraul-

ically controlled; and the mean size of suspended load varies systematically with flow parameters (Fig. 11.7). This pattern of behaviour contrasts sharply with that of perennial streams. As the flood falls, depos-

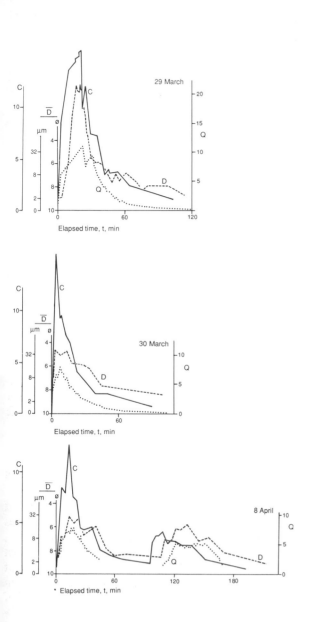

Figure 11.7 Suspended sediment concentration (C) by mean size-fraction distribution (D) through three flash floods (Q) in the Kimere Basin, Kenya (after Reid & Frostick 1987; Frostick et al. 1983).

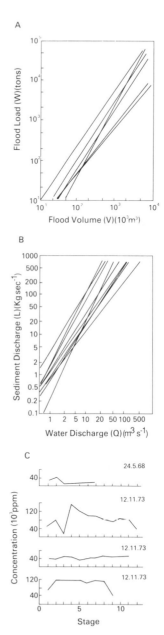

Figure 11.8 (A) the relationship between flood volume and flood load for various "streams" in Sinai; (B) relationship between water discharge and sediment discharge for various streams in the Negev (after Schick 1977).

ition begins to prevail, possibly in two phases: first as a plane bed of relatively coarse material; and secondly, by a well defined course that may be meandering (Thornes 1980). In general, the sediment load and sediment discharge appear to be related to flood volume and water discharge (Fig. 11.8).

The erosion and sedimentation accompanying a flood pulse (e.g. Leopold et al. 1966) reflect, *inter alia, local flow conditions* and the *rôle of tributaries* in the drainage net. The crucial rôle of tributaries and the drainage net is exemplified by Frostick & Reid's (1979) study of sand-bed channel deposits in Turkana (Kenya). They showed that here, in contrast to what

might be expected and to what is normal in perennial streams, mean particle size *increased* locally downstream (Fig. 11.9A). They attributed this trend to changes of stream power: small headwater flows spread quickly and have little power; downstream, increments from tributaries increase the flood wave and power, ensuring the entrainment of larger particles. At each point down channel, the stream becomes more competent to transport progressively coarser sediment, and so leaves behind in the bed material that becomes progressively coarser down stream. The frequency of alternating sequences of *paired* horizontal, parallel laminae in stream-bed sediments is also attributed to the number of tributaries above the sample point (Fig. 11.9B), reflecting the passage of flow pulses that in turn are controlled by the drainage net (similarly, each peak in the sediment concentration curve of Figure 11.7 reflects tributary inputs; Frostick et al. 1983). This pattern of repeated sediment activation led Thornes (1977) to suggest that the pool and riffle sequence typical of perennial streams *may* be replaced by horizontal sequences of coarse and fine material.

Bedload is often a significant (even a dominant) component of sediment movement in ephemeral streams (Fig. 11.10). Bedload movement may even be assisted by high suspended-sediment concentrations (Gerson 1977). For example, Leopold & Miller (1956) monitored bedload movement in an ephemeral-stream channel in New Mexico, and noticed that flows

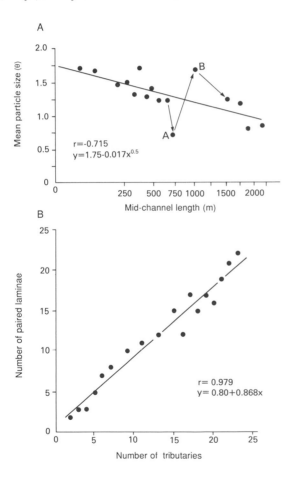

Figure 11.9 (A) Channel-fill mean particle size (phi units) as a function of distance from headwaters for a catchment near Lake Turkana, Kenya. (B) Relationship of the number of paired laminae and the number of upstream tributary confluences in the ephemeral stream-channel fill of a catchment near Lake Turkana, Kenya. (After Frostick & Reid 1979.)

Figure 11.10 Bedload trap in an ephemeral-flow channel, southern Israel.

no deeper than half the diameter of the cobbles moved them. Schick et al. (1987) noticed that desert stream beds often have a less well developed armour of bedload particles than perennial streams, and such particles tend to be better distributed through the bed sediments. Their monitoring of marked bedload samples in two small catchments in the Negev Desert (Israel) revealed that there is a substantial vertical *exchange* of bedload particles after several floods, a phenomenon of quasi-equilibrium reflecting the effect of medium-size "dominant" floods and a scour depth sufficient to encourage the exchange.

The relative proportions of suspended, bed and solution loads in ephemeral flows are not yet fully understood. In general, the ratio of clastic sediment to dissolved load is probably higher in drier areas, partly because the limited precipitation and its rapid runoff reduces solution load.

The sediment within ephemeral channels is often associated with several distinctive facies and bedforms (Fig. 11.11). The facies types and associated micro-features, such as lags, chutes, bars and levees, have been a matter of extensive research (e.g. Glennie 1970, Picard & High 1973, Rust 1981, Zwolinski 1985). Figure 11.11 provides a reasonable summary

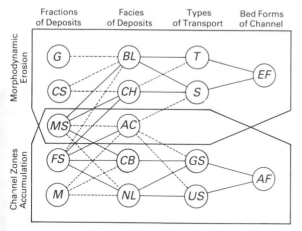

Figure 11.11 Observed linkages within a depositional model for a desert creek channel in central Australia. Key: G, gravel; CS, coarse sand; MS, medium sand; FS, fine sand; M, silt; BL, bottom lag facies; CH, chute facies; AC, abandoned channel facies; CB, channel bar facies; NL, natural levée facies; T, traction; S, saltation; GS, graded suspension; US, uniform suspension; EF, channel bed erosion forms; AF, channel bed accumulation forms. Continuous lines show first-order linkages, and dotted lines show second-order linkages (after Zwolinski 1985).

for one channel in central Australia, but it is also probably a model applicable elsewhere. Picard & High (1973) also provided a thorough description of bedforms and sedimentary structures along the beds of ephemeral channels. Very often, the bedforms seen in a dry ephemeral stream channel are those left after the cessation of the last flood. Thus, the bars that may move as kinematic waves under conditions of flood are left to circumscribe anastomosing lower-magnitude flows and to be eroded by them.

One lesson that is emerging from instrumental and historical studies of sediment movement in channels is that dryland channels may be *poorly coupled with slopes* (Warren 1985). Thus, for example, the supply of sediment from slopes may occur at different times from periods when flows occur in channels, and it may simply be stored on floodplains rather than delivered to the channels. A major channel-erosion episode then has first to remove accumulated flood-plain sediment, so that if supply from slopes exceeds that which channel flows can remove, almost all of the sediment will come from the floodplain store, excavated mainly from the *highest*-order channels, so that the erosional runoff event will not significantly affect the slopes (e.g. Graf 1983c). This decoupling appears to be destructive to drylands, and quite distinct from perennial-flow channels.

11.3 Hydraulic geometry

Leopold & Maddock (1953) and Leopold & Miller (1956) introduced the concept of hydraulic geometry, in which hydraulic characteristics of streams are related to water discharge. The most common of these relations are:

$$w = aQ^b \qquad (11.2)$$

$$d = cQ^f \qquad (11.3)$$

$$v = kQ^m \qquad (11.4)$$

$$L = pQ^j \qquad (11.5)$$

where w is the mean channel width, d is the mean channel depth, v is the mean flow velocity, Q is the water discharge, L is the suspended-sediment load, and a, b, c, f, k, m, j and p are constants. Because

$$Q = wdv, \quad Q = aQ^b \times cQ^f \times kQ^m,$$

it follows that

$$a \times c \times k \times 1 = 1 \text{ and } b+f+m = 1.$$

These hydraulic geometry relations may pertain to a given cross section (at-a-station hydraulic geometry) or to a number of cross sections at a specified flow duration or flood recurrence interval (downstream hydraulic geometry).

Interest here centres on the questions of whether desert streams have hydraulic geometries different from humid streams, and whether ephemeral streams have hydraulic geometries distinct from perennial ones. Schumm (1960) showed that the width:depth ratio varies inversely with the percentage of silt–clay in the bed and banks. The composition of the banks is especially important. Desert alluvial channels tend to have little silt–clay in their banks, and often to have wide, shallow channels with poorly defined banks. Such streams may also be braided (see Sec. 11.4). Desert streams thus *may* tend to have higher b values and lower m values than their counterparts elsewhere (e.g. Park 1977). There appears to be no significant difference between ephemeral and perennial streams in this respect. *Down stream*, Park (1977) showed that ephemeral streams tend to have lower b and higher m values than perennial streams.

An important contrast between ephemeral stream channels in a semi-arid area and "average river channels" arising from Table (11.1) is that in the case of the former, suspended-sediment concentration increases down stream more rapidly than discharge (e.g. $j = 1.3$), whereas in the case of the latter the reverse

is usually true (e.g. $j = 0.8$). The increase in sediment concentration down stream in ephemeral-stream channels has been widely observed (e.g. Thornes 1976, Schick 1970; see Sec. 11.2). It is also possible that the higher downstream rate of velocity increase that has been observed in ephemeral-stream channels may be a response to the increase in sediment concentration. These *general* hydraulic relationships may have superimposed upon them regular changes such as oscillations of width (e.g. Thornes 1976).

Bull (1979) conceptualized the relationships between some of the controlling variables in the fluvial system in terms of the **threshold of critical power**, defined as a ratio:

$$\frac{\text{stream power}(w)}{\text{critical power}} = 1.0 \qquad (11.6)$$

in which

$$w = \gamma QS/\text{width} = \gamma dsu = Tu \qquad (11.7)$$

where γ is the specific weight of the fluid (and is assumed to be roughly constant), Q is the discharge, S is the gravity gradient, d is the mean depth of flow, u is the mean flow velocity, and T is the mean boundary shear stress; critical power is the power *needed* to transport the average sediment to a slope or channel, and it reflects channel width, depth and flow velocity. This universally applicable concept is specifically exemplified by Bull in semi-arid and arid areas. Figure 11.12 depicts the relations between stream power and critical power for a barren granitic hill-

Table 11.1 Values for exponents b, f, m and j.

	Average downstream relations					Average at-a-station relations			
	Ephemeral streams			Perennial streams		Ephemeral streams		Perennial streams	
	Southeast Spain[5]	New Mexico[1]	Nebraska[2]	Average[3]	Nebraska[2]	New Mexico[1]	Nebraska[2]	Average[3]	Nebraska[2]
b	0.63	0.5	0.03	0.5	0.69	0.26[4]	0.35	0.26	0.24
f	0.20	0.3	0.48	0.4	0.12	0.33[4]	0.43	0.40	0.56
m	0.17	0.2	0.45	0.1	0.19	0.32[4]	0.22	0.34	0.20
j		1.3		0.8		1.30[4]		1.5–2.0	

Sources:
1. Leopold & Miller (1956). The authors express reservations about the small number and accuracy of their observations, but the data probably give a reasonable indication of reality.
2. Brice (1966). Data from the Medicine Creek basin. At-a-station data may be considered reliable; downstream data only indicate the correct order of magnitude.
3. Leopold & Maddock (1953). 4. Unadjusted median values. 5. Harvey (1984a).

slope in the headwaters of a large basin, subjected to a local thunderstorm. In reach A, where the ratio is higher than 1, additional sediment load is taken up, largely by vertical erosion. The ratio is approximately 1 in reach B, and stream power in reduced because of decreased hydraulic roughness, so that lateral cutting prevails; while in reach C, power decreases further so that the ratio is less than 1, sediment load decreases, and alluviation occurs. Changes in the controlling variables, especially climate and land use, are reflected in changes in the power ratio, and thus in the balance of erosion and deposition.

The following discussion of channel processes and forms focuses on the distinctiveness of channel forms in deserts. Other relationships are discussed further in the sections on pediments (Ch. 13) and alluvial fans (Ch. 12).

11.4 Channel forms and ephemeral flows

The range of alluvial channel types is very extensive, but Schumm (1985) conveniently classified them into five major groups according to cause-and-effect relations reflecting sediment load, flow velocity and

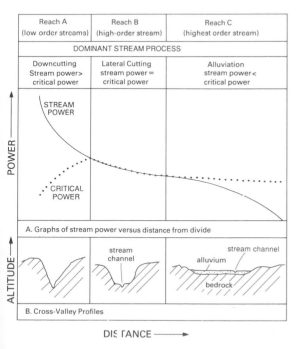

Figure 11.12 Diagrammatic sketches and graphs of stream power and critical power for an arid rock drainage basin (after Bull 1979).

stream power (Fig. 11.13). In general, at least for perennial flows, channel dimensions are due mainly to water discharge, whereas shape and pattern are related to the type and quantity of sediment load. As the channel pattern changes from 1 to 5, so too do peak discharge, sediment size and sediment load, and hydraulic characteristics such as flow velocity, tractive force and stream power (Schumm 1985). Within a system, change reflects intrinsic or extrinsic thresholds. For example, the difference between braided and meandering channels is in part defined by the definable threshold relating channel slope and discharge; and in deserts, progressive accumulation of sediment may increase channel slope until a slope threshold is crossed and entrenchment occurs (Schumm 1985; Fig. 11.14). Within such generalizations, which are now fully explored in textbooks (e.g. Schumm 1977, Richards 1982, Graf 1988a), the question arises: What are the *distinctive* features of process–channel form characteristics in drylands?

First, there are *networks*. Here *drainage density*, considered by some (e.g. Gregory & Walling 1973) to be an important reflection of process, is higher in drylands than in otherwise comparable, more humid areas. Drainage density is controlled by topography, lithology, soil and vegetation, and human influence, as well as climate, so it is difficult to segregate the climatic effect from other variables. Nevertheless, Gregory & Walling (1973) demonstrated, using a graph of total stream-length to basin area, that drainage densities were highest in semi-arid areas. Similarly, Melton's (1957) analysis of channel density for basins in Colorado, Utah, Arizona and New Mexico, showed that drainage density varied *inversely* with the $(P–E)$ index, being highest in those locations where the index was lowest (Fig. 11.15). The probable reasons for this peak in drainage density in semi-arid regions are that in even drier climates there is insufficient runoff to provide high drainage densities, whereas in more humid climates there is more vegetation to inhibit linear erosion and channel development.

Secondly, *channel networks* in deserts are not always as fully integrated as those in more humid areas, for the reasons already provided: runoff is ephemeral, and often it only affects part of a system at any one time. Morphologically, this has at least two important consequences. First, while accordant channel junctions are normal in deserts as elsewhere, they are by no means universal: a flow in a higher-order tributary can cause incision that leaves lower-order tributaries "hanging". Equally, a tributary valley flood may deposit sediment in a high-order valley, perhaps blocking it or deflecting flow (e.g. Schick & Lekach 1987). Secondly, because erosion and sedimentation may be segmented within a channel system at any one time, segmentation may lead to discontinuous gullying.

Thirdly, there is an impression that *channel networks adopt rather distinctive patterns* in deserts, for two broad reasons. Where they are evolved in hilly or mountainous bedrock areas, the innate geological structure is directly imposed on the pattern: and on low-angle, vegetation-free alluvial surfaces the drainage patterns are commonly pinnate or dendritic, depending *inter alia* upon the inclination of the surfaces. Such patterns are largely independent of bedrock structure, even where developed on pediments.

Fourthly, ephemeral stream-channel *long profiles* may be distinctive in drylands. In perennial streams a concave upward profile is a characteristic reflection of increasing downstream discharge. In ephemeral stream channels, this tendency may be counterbalanced by declining discharge, or increasing particle size, or by the restricted channel length down stream (e.g. Renard 1972, Brown 1983). A result of this balance may, in some circumstances, be reflected in relatively *constant* long profiles over considerable distances; alternatively they may even be *convex*.

Fifthly, *channel adjustment* appears to be *relatively rapid* in response to ephemeral flows. In part this is due to the sparsity of riparian vegetation along many channels. Whereas many perennial-stream channels may approach a steady state (and return to it following disturbance by external changes), ephemeral-

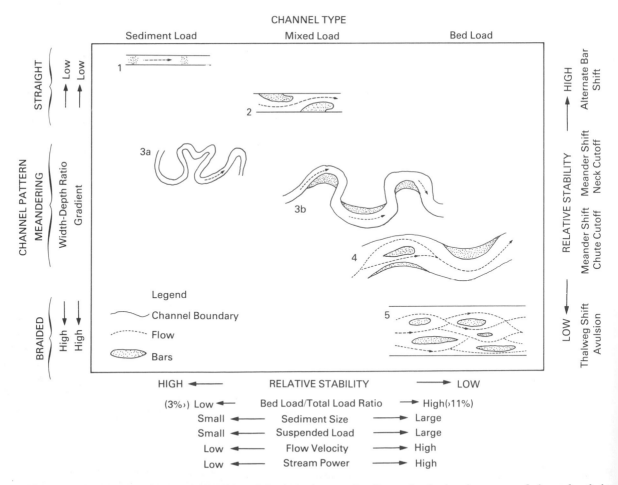

Figure 11.13 Channel classification based on pattern and type of sediment load, showing types of channels, their relative stability, and some associated variables (after Schumm 1985).

stream channels in drylands often appear to be adjusting continuously to changing circumstances, and approaching new different states, as Rendell & Alex-

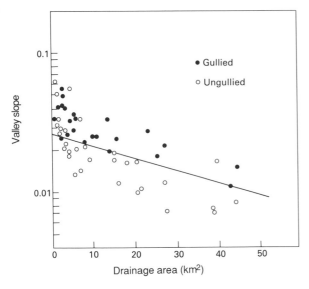

Figure 11.14 Relation between valley-floor slope and drainage area for small drainage catchments in the Piceance Creek area, Colorado (after Schumm 1977).

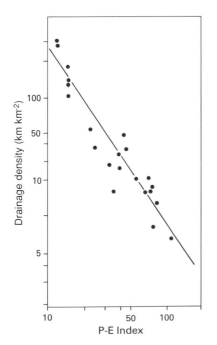

Figure 11.15 Relations between drainage density and the *P–E* index for the southwestern USA (after Melton 1957).

ander (1979) argued for clay-cut channels in the Basento Valley in southern Italy, and Rhoads (1988) argued for bedrock and alluvial mountain channels in southern Arizona. Continuous rapid adjustment may be reflected in seasonal contrasts, as described by Brown (1983) in a study of channels in the badlands of Borrego Springs, California. Here, light winter rains produced negligible flow and transport, and minimal change, but hillslope mass movements altered channel widths and caused local aggradation; intense summer storms, in contrast, caused rapid water and debris transfer, and local scour and fill. These continuously alternating seasonal adjustments are presumably reflected over longer periods in the continuous adjustment of channel form.

Perhaps this possibility of continuous adjustment is better expressed in terms of channel *recovery period*: the time taken for a channel to revert to its "equilibrium" following the disruption of a major flood. Wolman & Gerson (1978) showed that such recovery is a function of the rate of vegetation recovery and that, because vegetation is "rare or minimal" in deserts, flood damage would be nearly irreparable, so that *progressive* change occurs, rather than reversion. This is illustrated by channel width adjustment following large storms in rivers from different climates in Figure 11.16. Wolman & Gerson further demonstrated that, for drainage areas up to 100km^2 in arid regions, the rate of channel width enlargement with increase of drainage area is higher than in humid areas; and for larger drainage areas, the increase in width is less than for humid and subhumid regions. They attributed this contrast to the following: the small size of effective storms in deserts; the fact that the channels may continue to be widened by extreme flows until they can accommodate the largest flows; lack of "recovery" because of little vegetation; the wider channels become, the less likely small flows are to widen them; and transmission losses mean that discharge diminishes, or increases at a slower rate. They conclude: "In terms of channel form, the effectiveness of an event of given absolute magnitude and recurrence interval becomes progressively less as one moves from the humid to the arid region. The effect is the same as moving from large to small watersheds" (Wolman & Gerson 1978: 200).

One rider to these generalizations is essential. It is a mistake to assume that because **riparian vegetation** is generally less dense in deserts than elsewhere, it is of no consequence. In fact, phreatophytes such as tamarisk (e.g. *Tamarix pentandra*), cottonwood (*Popu-*

lus fremontii) and willow (*Salix gooddingii*) – which draw their sustenance from groundwater at depth – can significantly influence channel geometry by increasing bank resistance to erosion, inducing deposition and increasing roughness. In both ephemeral

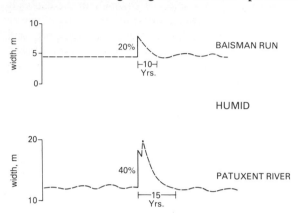

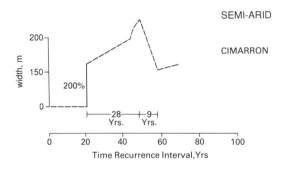

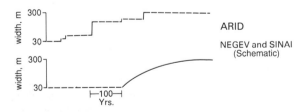

Figure 11.16 Sequence of changes in channel width showing recovery periods following large storm events and climatic variations for selected rivers in different climate regions. The initial increase in the Cimarron River width is during a major flood, followed by continued widening during a dry period. The second jump in width is due to a smaller flood in a dry period. Declining width occurs during a moist period with vegetation regrowth (from various sources, after Wolman & Gerson 1978; Cimarron River curve after Schumm & Lichty 1965).

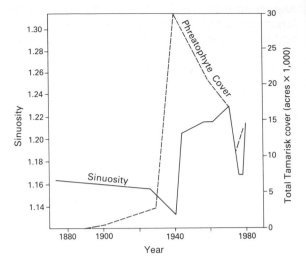

Figure 11.17 Relationship between density and coverage of phreatophyte growth and channel sinuosity on the Gila River, central Arizona. The channel changes occurred mainly during flood periods, so that there is a lag between changes in vegetation and corresponding changes in channel sinuosity (after Graf 1981).

and perennial stream channels, such vegetation can lead to significantly reduced channel width and, as a result, to increased overbank discharge (flooding). Thus, Graf (1978) showed that major river channels of the Colorado River and the plateau country had an average width reduction of 27% when tamarisk was established after 1930. Hadley (1961) demonstrated a similar effect on an arroyo in northern Arizona; and Graf (1981) demonstrated that in the ephemeral Gila River channel, sinuosity was increased from 1.13 to 1.23 as phreatophyte density increased in the 1950s (Fig. 11.17).

11.5 Floods and composite flood surfaces

The term "flood" is used ambiguously in deserts. It often merely means the passage of an ephemeral flow within the confines of an existing channel, with the characteristics and effects described above. Graf (1983a), in commenting on the passage of flows through historical gully systems (**arroyos**, see Sec. 11.7), came to the interesting conclusion that during high-magnitude, low-frequency floods, it is the process that controls the forms, but the forms created by such flows actually control processes during subsequent smaller events. This notion, that it is the larger event that controls channel development in drylands in contrast to the dominance of medium-frequency and -magnitude events in humid areas, is widely held and evidence for it is growing. For example, Baker (1977) indicated that in dry regions (Central Texas) it may be the rarer, catastrophic, event that is more important in shaping stream channels, because the frequency distribution of flow discharges is more strongly skewed to the right than is the case in humid regions.

This concept is relevant in several fluvial contexts, including channel width, terrace formation and the creation of palaeoforms (Sec. 11.7). There may be a distinctive general principle here: sometimes forms determine process effects, and sometimes the reverse is true; either way, forms and processes are never in equilibrium, as sometimes seems likely in temperate areas (Warren 1985).

Many desert floods leave their established channels and inundate the surfaces adjacent to them. The immediate area is not usually a floodplain, in the sense of a humid area floodplain which characteristically has a clearly delimited zone with morphological boundaries. It is often a less clearly defined zone, a composite mixture of alluvial features, and is found on most alluvial surfaces, including alluvial fans (Ch. 12). The flood itself is the equivalent of what is traditionally known in the desert literature as a **sheetflood** since McGee first described it in 1897. Although the term-

inology is somewhat confused, a sheetflood is best considered as a low-frequency, high-magnitude event, a sheet of unconfined turbulent floodwater moving down an alluvial slope; it contrasts with high-frequency, low magnitude and non-turbulent **sheet-flow**, and it may be erosive and highly charged with debris (Fig. 11.18; Hogg 1982).

The record of such floods in deserts is now extensive. To give a few examples: an extreme flood recurred in southeast Spain in 1980 (Harvey 1984a); huge areas of southern Tunisia were inundated in 1969 (e.g. Mensching et al. 1970); massive floods affected the Oued Saoura and the southern Atlas in 1959 (Vanney 1960; Fig. 11.6); Lake Eyre, a normally dry salt lake in central Australia was flooded in 1950, 1974 and 1984 (Kotwicki 1986); a major flood extended beyond the banks of the Santa Cruz arroyo in southern Arizona in 1983 (Saarinen et al. 1984); there have been major floods in 1976 and 1971 in southern Israel (e.g. Schick 1971, 1987); and a serious flood occurred in central Oman in 1988. In places, the record has considerable historical depth (e.g. Burkham 1970a), which occasionally allows the creation of magnitude–frequency curves.

Large, rare floods can accomplish significant and durable geomorphological changes that are only embroidered by the many low flows in the long periods between the floods. Graf (1988b) cogently argued that, in effect, channel systems in drylands are often compound and may have entirely different behaviour patterns depending on discharge. In addition, he found that flood-discharge features within the flood zone may be in equilibrium with high flows but not with subsequent low flows. He also showed that the change from gradual infilling in the low flows (causing a change from braiding to meandering) and rapid erosion during high flows (causing change from meandering to braiding) reflects the interplay between stream power and channel slope. Thus, the coexistence of different channel patterns in dryland flood zones can be explained in terms of this hypothesis. For example, in Cooper Creek, Australia, Nanson et al. (1986) attributed braids to high flows and co-existing anastomosing channels to moderate flows. Further examples of the effects of such flow variations on channel development on alluvial plains are provided by the Gila and Salt rivers in Arizona, where changes are evident from historical sources for over a hundred years (Graf 1983b, 1988b; Fig. 11.19). Here, successive floods have caused channel *relocation* and *rearrangement* with substantial lateral

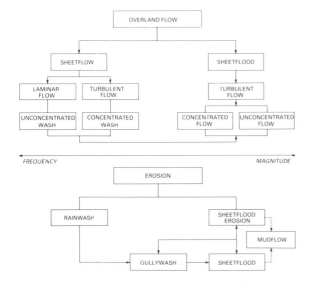

Figure 11.18 Hydrological and geomorphological classifications of sheetflood processes (after Hogg 1982).

migration, and some recent floods have caused deepening. Furthermore, the changes are spatially variable along the system, with stable and relatively unstable reaches.

The evidence of such flow variations is often to be found on aerial photographs and it can be used to classify potential flood hazard within the compound valley-floor features.

11.6 Perennial rivers

The major perennial rivers of deserts are exoreic, drawing their water and hydrological characteristics mainly from more humid watersheds. Such rivers commonly have complex and long histories that reflect the tectonic and climatic histories of the Quaternary and beyond – for example, the Nile (Harvey & Grove 1982, Said 1982), the Niger (Beaudet et al. 1977b) and the Colorado (Hunt 1983). This history is reflected in complex morphologies and sedimentary sequences along their courses.

In general, however, the perennial rivers' channel morphology appears to conform to that established for perennial streams in humid regions, with the important qualification that local processes may impose upon the forms established by the perennial flow. For example, **debris flows** generated locally in tributaries of the Colorado River play a major rôle in locally influencing the geomorphology of the main river. They provide a major source of sediment; they provide massive, largely immovable, boulders that create rapids which act as important hydraulic controls and are largely relict features (Graf 1979c, Howard & Dolan 1981); they create fans that can constrict flow or deflect it to one side of the canyon; the combination of debris fans and higher flow velocities over rapids causes flow-separation zones conducive to the formation of beaches up stream and down stream of rapids; and reworking of debris-flow material down stream of rapids creates debris bars (Webb et al. 1987; Fig. 11.20).

11.7 Channel changes and the generation of fluvial palaeoforms

The recognition of mutual adjustment between variables in response to environmental conditions is of great value in the study of channel changes. In principle, a change in any one variable is likely to cause an appropriate adjustment in the channel system. In what is patently a complex multivariate situation, similar adjustments – in terms of visible landforms – may result from changes to different variables or combinations of variables within the system.

The speed and timing of such changes is fundamental. The fact that many fluvial features in deserts are formed by large events of low frequency, and that

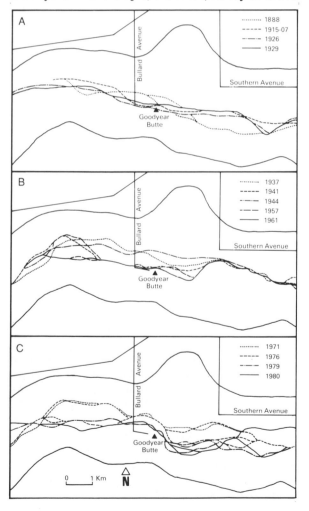

Figure 11.19 Locations of the thalweg of the channel of the Gila River 20km west of Phoenix, Arizona, 1868–1980. In the period 1868–1929 the channel was relatively stable in its location. In the period 1937–61, dense phreatophyte growth reduced the channel capacity and destabilized its location. In the period 1971–80, several major floods caused further locational changes (after Graf 1988).

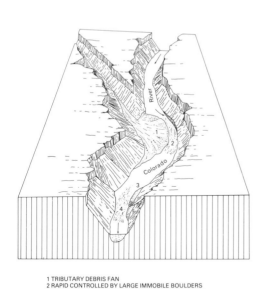

1 TRIBUTARY DEBRIS FAN
2 RAPID CONTROLLED BY LARGE IMMOBILE BOULDERS
3 DEBRIS BAR (synonymous with 'island' or 'rock garden')
4 RIFFLE OR RAPID CAUSED BY DEBRIS BAR

Figure 11.20 Geomorphological features of a typical rapid controlled by debris flows on the Colorado River (modified from Hamblin & Rigby, after Webb et al. 1987).

it is often the last such event that produced the contemporary forms, create a significant problem for the evaluation of desert landforms. When does an event become historical? When do the forms it creates become **palaeoforms**? When are such forms merely *dormant* and awaiting reactivation by the next major event? There are probably no simple answers to these questions, but a number of observations are pertinent at different timescales.

Over the *short term*, process monitoring provides some clues. Within historical times, it may be possible to recognize *secular trends* and to disentangle the complex webs of cause and effect, for example through the study of arroyos. Over longer periods, through the Quaternary, several different types of channel changes and methods of "palaeoform formation" can be recognized, including *channel metamorphosis*, *cut-and-fill* and *terrace development*; *drainage abandonment*; and *drainage replacement*. Each of these topics is considered below.

11.7.1 Short-term changes
In the short term, there is a small body of evidence

Figure 11.21 An arroyo, initiated in the late 19th century, in the Aravaipa Valley, southern Arizona.

that erosion and deposition in ephemeral systems responds to climatic change as well as, of course, to human interference. For example, Leopold's (1976) monitoring of change in ephemeral systems in New Mexico over a 15-year period led him to conclude that arroyos which were actively eroding in the early part of the century have been actively alluviating since 1961, and the rates of both headcut erosion and sheet erosion have been reduced between 1959 and 1975. He attributed this consistent set of changes to the worldwide cooling trend since *c.* 1940, a trend that may be reduced in the future as the CO_2 content of the atmosphere increases. A similar story is provided for the Little Colorado River, Arizona (Hereford 1984).

11.7.2 Secular trends: entrenchment and arroyo initiation
A particularly important reflection of changes to fluvial systems is the initiation of channel entrenchment in alluvial sediments, and its subsequent development in the form of gullies or vertical-walled arroyos (Fig. 11.21). Such changes have been widespread in many semi-arid areas, especially those that have been settled in colonial times, such as the southwestern USA, Southern Africa, India and Australia. This phenomenon has attracted much attention, and the successful analysis of it often requires both field and archival work. The crucial problem, to some, is the cause or causes of entrenchment – was it the result of climatic change, overgrazing, or artificial flow concentration, for example? Certainly, prehistorically

there can be no doubt that arroyo cutting and filling was driven by climatic change (e.g. Antevs 1952), but the historical possibilities are much more complex because increased erosiveness of flows and/or increased erodibility of valley-floor materials can be achieved in so many different ways (Fig. 11.22).

Increase in flow velocity could have resulted from increase of discharge or slope, reduction of surface roughness, or increase in depth of flow. Increased erodibility of valley-floor materials might have arisen through the reduction of riparian vegetation. Depth of flow might have been increased by concentrating a given discharge through the construction of embankments or the cutting of irrigation canals. A good example is Greene's Canal on the Santa Cruz River,

Arizona, which inadvertently collected and concentrated floodwater to such an extent that the original canal has now been transformed into an arroyo (Cooke & Reeves 1976). Surface roughness could have been reduced by removal of vegetation – especially the grasses – on the floodplain by cattle or by road traffic. An illustration of this change occurs in the Altar Valley, Arizona, where an arroyo has been formed in part along the line of an early waggon route. Increase of slope could locally have been effected by deposition arising from increased sediment concentration as discharge declined down stream. The slope could have been increased to the point at which gullying was initiated (e.g. in Wyoming and New Mexico: Schumm & Hadley 1957). Along a single

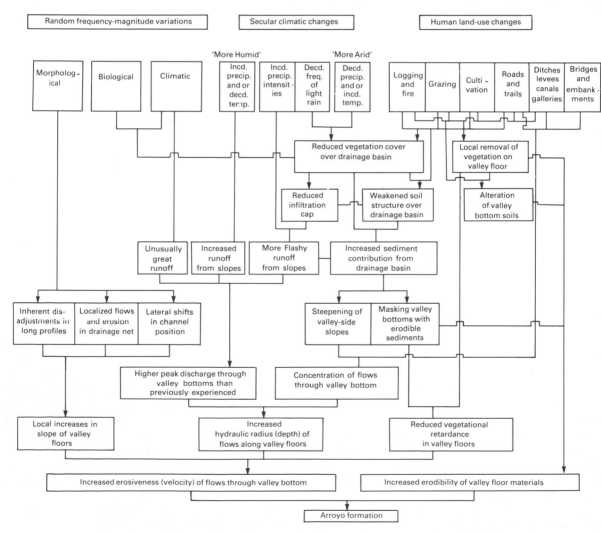

Figure 11.22 A model of arroyo formation (after Cooke & Reeves 1976).

valley there could have been several gullies which ultimately became integrated into a single arroyo. This mechanism could mean that entrenchment is *not* necessarily contemporaneous within or between valleys (e.g. Patton & Schumm 1981). Such slope changes would have been essentially random, and need not have been determined by specific environmental changes in the basin.

However, the most commonly advocated cause of arroyo formation is that of increased discharge, achieved either through a change of climate or through vegetational changes from other causes, leading to an increase of surface runoff and valley-floor discharge.

Several different secular climatic changes have been proposed. Some have argued that, during the critical period of trenching towards the end of the nineteenth century, the climate may have become drier (e.g. Antevs 1952). Others have suggested that it may have become wetter (e.g. in Arizona: Huntington 1914). Finally, there are those who argue for a period of relatively low frequency of small rains and high frequency of large, heavy rains, especially in summer (e.g. in New Mexico: Leopold 1951; Arizona: Cooke & Reeves 1976; Fig. 9.5). Each of these changes, of course, would have been accompanied by vegetation changes. Vegetation removal and alteration could also have been achieved by grazing of cattle and sheep, and perhaps by fire and deforestation, and such changes would probably have led also to greater surface runoff and increased discharge along the valley floors; and a further increase might have resulted from reduction of infiltration capacity by animal trampling.

In any given area there are likely to be several changes, some perhaps independent of one another, while others are consequent upon a single change. Melton (1965a), for example, argued that in some Arizonan **cienegas** (marshy, grass-covered valley-floor areas) reduction of grass cover in the valley led (i) to deposition of sediment on the cienegas and to increased inclination of their *transverse* slopes (slopes normal to line of flow) and thus to concentration of flow; and (ii) to reduction of critical erosion velocities of superficial sediments and reduction of hydraulic roughness. The net effect of these changes would have been to increase both flow velocities and the erodibility of the cienega deposits, so that entrenchment occurred. Similarly, Graf (e.g. 1979a,b) showed that there is a threshold of entrenchment relating the tractive force of flows to the resistance of valley

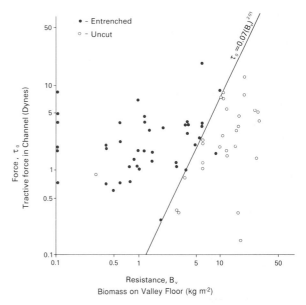

Figure 11.23 Force versus resistance and a discriminating function separating trenched and untrenched sites for the 10-year flow event in valleys of the Front Range, Colorado (after Graf 1979b).

floors to erosion (the latter measured in terms of biomass): at any one time whether a valley floor is entrenched or unentrenched depends on which side of the threshold it resides (Fig. 11.23). Either an increase of tractive force or a reduction of biomass can lead to entrenchment if the threshold is crossed. Mining was the main cause of increased tractive force and reduced biomass in the Front Range of Colorado, shown in Figure 11.23 (Graf 1979b).

The entrenchment–sedimentation sequence may be part of what Schumm & Hadley (Hadley 1977) called a "cycle of semi-arid erosion". They observed in entrenched drainage systems in eastern Wyoming and New Mexico the coexistence in the same environment of both aggrading and entrenched channels (Fig. 11.24). From this they hypothesized that an initial gully or arroyo (Fig. 11.24) migrates headwards (B) so that tributaries become entrenched and the rate of sediment removal is increased. When the abundant supply of sediment from the retreating headcuts (C) exceeds the capacity of flows to remove it, then deposition occurs in the lower reaches of the arroyo, leading to slope steepening and ultimately to new entrenchment. Such changes are essentially independent of external environmental changes.

This very brief review and sampling of the

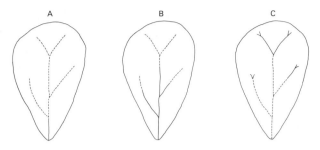

Figure 11.24 The semi-arid cycle of erosion. Dashed lines indicate stable or aggrading channels, and solid lines indicate trenched channels (after Schumm & Hadley 1957).

extensive "arroyo" literature serves to emphasize the variety and possible complexity of arroyo origins. Some explanations – climatic change for instance – are regional in character; others – such as concentration of flow in irrigation channels – are of only local significance. For each hypothesis there have often been both advocates and adversaries. It appears that much of the controversy has arisen because of the complexity of the problem, because there has been inadequate recognition of the fact that different changes in different areas can lead to similar results, and because precisely documented local explanations are rather scarce. Once entrenchment is initiated, the

arroyo development depends on the relationship between bank processes, such as slumping, and valley-floor fluvial processes. Thus, Graf (1979b) showed that width and depth changed at different rates. He also showed, in a study of arroyo development in the Henry Mountains, Utah, that changes in stream power may accompany such entrenchment. Thus, he found that in the 1980s, stream power declined down stream as deposition prevailed and overbank flows were common; today, stream power increases down stream with discharge and because flows are contained within arroyos (Graf 1983a).

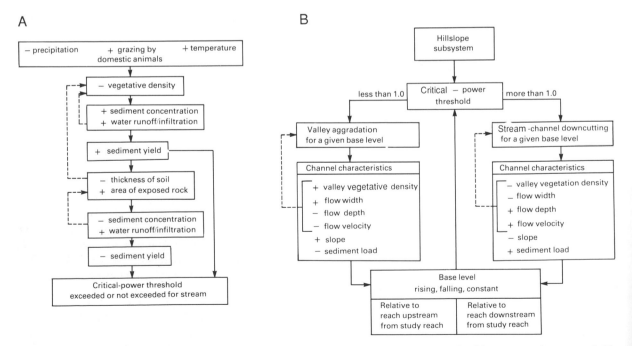

Figure 11.25 Increases (+) and decreases (−) in elements of a hypothetical semi-arid stream subsystem. Self-enhancing feedback mechanisms are shown by dashed lines with arrows (after Bull 1979).

11.7.3 Quaternary changes: channel metamorphosis

Bull (1979) elaborated his concept of threshold of critical power in streams (see Sec. 11.3 above) in the context of hillslope and channel systems in drylands (Fig. 11.12), and their changes during the Quaternary. His initial assumption was that during the Holocene in the deserts of North America and the Middle East (at least), precipitation decreased and/or temperature increased. On hillslopes, such changes would have reduced vegetation density, increased infiltration rates and increased the area of bare ground, thus increasing sediment concentration and water runoff for a given rainfall event, and increasing *critical* power, so that the critical threshold was not exceeded. Subsequent decrease in soil thickness and increase in bare ground led to a decrease in sediment yield, and a *reduction* in critical power, but to an *increase* in stream power as a result of valley fill deposition: the critical power threshold was crossed and erosion began in the fill (Bull 1979). In the semi-arid channel system, the critical power threshold separates two kinds of responses (especially in easily eroded materials) to changes in vegetative cover caused by climatic change or human activity. Where the threshold is greater than 1.0, downcutting is perpetuated; where it is less than 1.0, aggradation prevails (Fig. 11.25). Both changes are reflected in changing channel characteristics.

When climatic or similar changes are of sufficient magnitude and duration, whole channel systems may be metamorphosed. Schumm (1968, 1969) explored such metamorphosis both theoretically and in the context of the semi-arid Riverine Plain of the Murrumbidgee River in New South Wales and elsewhere.

Schumm first described the empirical relationships between channel characteristics and the discharge of water and sediment, and developed equations which not only summarize these relationships but also allow the prediction of changes likely to occur to the former with changes of the latter. These equations are summarized below, without the extensive data and argument on which they are founded, as a basis for discussion.

First, water discharge (Qw) is directly related to channel width (w), depth (d), and meander wavelength (L) and is inversely related to channel gradient (S). Thus:

$$Qw \simeq \frac{w,d,L}{S} \qquad (11.8)$$

$$Qs \simeq \frac{w,L,S}{d,P} \qquad (11.9)$$

In addition, at a constant discharge of water, where Qs is bedload sediment discharge, and P is sinuosity (ratio of channel length to valley length).

Four further equations describe the effect of changing, separately, water and sediment discharge. A plus or minus exponent is used to indicate how the various aspects of channel morphology will change.

$$Qw^+ \simeq \frac{w^+ d^+ L^+}{S^-} \qquad (11.10)$$

$$Qw^- \simeq \frac{w^- d^- L^-}{S^+} \qquad (11.11)$$

$$Qs^+ \simeq \frac{w^+ L^+ S^+}{d^- P^-} \qquad (11.12)$$

$$Qs^- \simeq \frac{w^- L^- S^-}{d^+ P^+} \qquad (11.13)$$

In practice, of course, environmental changes often lead to contemporaneous changes in both water and sediment discharge. In the following equations, which describe such changes, Qt represents the *proportion* of bedload in the total sediment load, F is the width:depth ratio, and the plus and minus signs indicate an increase or decrease in a variable.

$$Qw^+ Qt^+ \simeq \frac{w^+ L^+ F^+}{P^-} S^{\pm} d^{\pm} \qquad (11.14)$$

$$Qw^- Qt^- \simeq \frac{w^- L^- F^-}{P^+} S^{\pm} d^{\pm} \qquad (11.15)$$

$$Qw^+ Qt^- \simeq \frac{d^+ P^+}{S^- F^-} w^{\pm} L^{\pm} \qquad (11.16)$$

$$Qw^- Qt^+ \simeq \frac{d^- P^-}{S^+ F^+} w^{\pm} L^{\pm} \qquad (11.17)$$

Long-term channel changes are ultimately associated with climatic changes, except where tectonic activity is significant. Often, however, evidence of former channels and channel changes in the Quaternary has been eliminated, and the nature of former channels and their changes is a matter of speculation. But the Murrumbidgee river system provides a notable exception: here there is preserved evidence of successive channel systems, and the problem is to describe differences between them and to seek an explanation of the differences. Certainly the explanation lies ultimately in climatic changes, but the nature of these changes has been a matter of controversy (e.g Butler 1960, Langford-Smith 1962).

Although the local and regional sources of water and sediment in the basin of the Murrumbidgee River have remained largely unchanged, there is evidence on the Riverine Plain of three distinct channel systems (Schumm 1968). The *modern river* has a low width: depth ratio, a low gradient and a moderately low sinuosity, and it carries only a small, fine-grained (suspended) load. An older channel, that of the "ancestral river" has a similar width:depth ratio, gradient and sinuosity, but it is larger, and it has a higher meander wavelength; its sediment load is also fine-grained (suspended) and small. A still-older channel system, that of the "prior streams", contains relatively wide and shallow channels which have high gradients and width:depth ratios, low sinuosity and relatively high sand (bed)loads.

The prior stream channels were probably produced during a period drier than at present, during which there were high flood peaks (estimated from calculations based on measurements of channel cross sections, slope and mean velocity) but less total runoff than at present. Evidence for a drier period includes the correspondence between soil salinity and the distribution of prior stream channels, weathering characteristics of palaeosols, aeolian deposits and hydrological information (Schumm 1968). Transformation of the prior stream channels into those of the ancestral river could have been accomplished by an increase of water discharge and a decrease in the proportion of bedload, as described in Equation (11.16). Such changes may have accompanied a change of climate to more humid conditions on the plain and to higher precipitation in the headwaters of the Murrumbidgee, as occurred during the Australian "Little Ice Age" 3,000 years ago. The change from prior channel to ancestral channel as deduced from Equation (11.16) requires a significant reduction of channel gradient, and Schumm (1968) showed that this was achieved largely through lengthening of the channel by an increase in sinuosity.

Figure 11.26 (A) Diagrammatic cross section of an arroyo bank in the southwestern USA, showing the idealized relationship of depositional units: vertical lines indicate weathering (not to scale). (B) Depositional units arranged according to ^{14}C dates. Number of ^{14}C dates in parentheses (after Haynes 1968).

To create the present channel of the Murrumbidgee a further change was necessary: as the present channel is similar in most respects to that of the ancestral river, except that it is smaller and has a lower meander wavelength, a reduction of water discharge is required (see Equation 11.11). This deduction leads to the conclusion that the present climate is "drier" than that prevailing at the time of the ancestral river.

11.7.4 Cut-and-fill and terrace development

Short-term erosion and deposition trends can be matched by longer-term alternations. The commonest reflection of these alternatives is the creation of an often complex sequence of **cut-and-fill** within an ephemeral stream system, in which the different units are not necessarily continuous or sequential. An excellent example is Haynes' interpretation of arroyo development in the American Southwest which shows

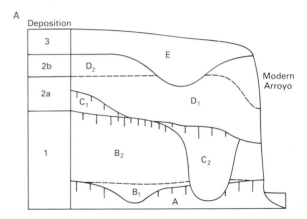

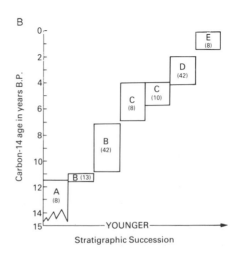

significant temporal overlap of some units – a fact that reveals that the different phases of deposition did not occur at the same time everywhere (Fig. 11.26, Haynes 1968). A contrasted example is provided by Talbot & Williams (1979). They demonstrated that alluvial fan sedimentation sequences in the alluvial fans on the flanks of fixed dunes in central Niger (Fig. 11.27), reflecting bioclimatic changes, comprise several cycles. Each is characterized by an aggradation phase (reflecting at least a greater incidence of intense storms than at present); an hiatus, marked by surface stability, no sediment accumulation and soil formation (reflecting a 115–130% higher rainfall and vegetation cover than today); and a phase of erosion (reflecting an interval of aridity and low frequency of intense rainstorms). The climatic interpretation of sedimentary sequences here appears to be substantiated. However, a general word of caution is in order: care must be taken in interpreting sedimentary sequences in terms of climatic change even if no obvious alternative explanations are evident, for erosion–deposition changes can take place endogenetically (e.g. Andres 1980; see Sec. 11.7.2).

Cut-and-fill on a grand scale, which almost certainly reflects climatic changes, was chronicled by Graf (1989) in Lake Canyon (a tributary of the Colorado River) in Utah. He showed that the canyon had been blocked on at least three occasions over the past 5,000 years by aeolian sediments in drier periods. Behind these barriers lakes were formed in subsequent wetter periods, and eventually they overflowed to cut away the aeolian barriers, leaving only remnants of the lake sediments.

In small desert catchments, the development of many terraces is common along the major floodways.

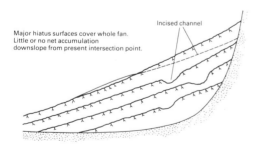

Figure 11.27 Diagrammatic cross sections through an alluvial fan in the Janjari area, central Niger. Hiatus surfaces completely cover the fan, indicating the episodic occurrence of intervals when little or no sediment accumulated on the fans (after Talbot & Williams 1979).

As indicated above, the main effect of major floods is to excavate sediment, leaving behind a broader flood zone that is only modified slightly until the next major flood. Such entrenchment may also create terraces out of earlier flood zones. Schick (1974a) developed a conceptual model of these changes for the Nahal Yael Research Watershed in southern Israel. His empirical observations led him to the view that quasi-continuous valley-floor deposition by relatively frequent minor floods is occasionally interrupted at long intervals by "superfloods" that clear out much of the earlier sediment and leave terraces in their wake (Fig. 11.28).

Commenting on this model, Schick emphasizes the following. First, terrace sequences do not *necessarily* correlate within and between desert drainage basins, chiefly because the flow events vary in space and time. Secondly, terraces do not *necessarily* reflect climatic changes, unless the occurrence of superfloods is a response to such changes. Thirdly, the *number* of terraces is a reflection of both the magnitude and frequency of superfloods *and* the pattern of events between them. Terraces may be obliterated by burial, lateral erosion or by gully and slope processes. Schick thought that in small catchments the number of superfloods sufficient to obliterate a terrace by erosion is between 2 and 3, and that the probabilities of finding one terrace are roughly 1 in 2; of two terraces, 1 in 3; and no terraces, 1 in 6.

Schick's model, which appears to apply reasonably well to small catchments in the Sinai, needs further testing. It does not necessarily apply to larger catchments with extensive terraces that may reflect the influences of base-level and climatic changes (such as the Huasco Valley, in Chile). Similarly, even in Sinai, it has to be compared with the apparently climatically controlled erosion–sedimentation sequences of slope deposits (e.g. Gerson 1982).

11.7.5 Drainage abandonment

Drainage abandonment occurs where a system formed under more pluvial conditions is merely abandoned, in whole or in part, with the advent of aridity. The extent of abandonment may vary greatly: for example, from perennial, to ephemeral or intermittent; or ephemeral drainage may become more sporadic and each event may affect a smaller portion of the system. In such circumstances, the system may superficially appear to retain its form, although the effective area of drainage is reduced and there may be changes in channel geometry.

There is extensive evidence of fluvial systems that

are on a scale too large to have been created or sustained under the present climatic conditions. The Chad Basin and the northeastern part of the Niger Basin are a remarkable example in West Africa (Fig. 11.29). The former Lake "Mega-Chad" and its predecessors in the Chad and Bodélé depressions were sustained by drainage from the surrounding massifs. Today the lakes are represented only by fluvio-lacustrine deposits and strandline features. The great drainage ways are generally dry; and aeolian land-forms have been created throughout much of the basin (Grove & Warren 1968). Similarly, in arid Western Australia (Van der Graaff et al. 1977) there is a very extensive palaeodrainage system that is preserved in a relict duricrusted landscape and was probably formed in the late Cretaceous to early Eocene (Fig. 11.30). The system functions only in part today.

11.7.6 Drainage replacement

Drainage replacement involves the abandonment of one drainage system or network and its replacement by another. There are not many examples of this change in the literature, although the river metamorphosis described in the Murrumbidgee system (above) is one form. Another is the so-called "suspendritic drainage lines" of parts of the Arabian Peninsula, in Nubia, Texas and New Mexico. In the Western Shar-qiya of Oman, for instance, Maizels (1987) has described a complex series of palaeochannels that form a series of superimposed gravel *ridges* (Fig. 11.31). In this area a thick sequence of Plio-Pleistocene alluvial fans has suffered both entrench-ment (creating terraces) and deflation of fine-grained, poorly cemented interfluvial sediments (exhuming cemented gravels in channels buried in the older

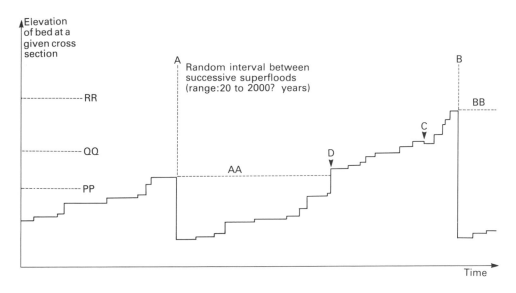

Figure 11.28 The elevation of a stream-channel bed as it develops over long periods of time, as envisaged by Schick (1974a) and simplified considerably. Superflood A removed from the given cross section all the alluvium aggraded since the previous clean-out. An intermittent period of aggradation followed for, perhaps, 700 years. Occasionally, a storm centre might occur in the immediate vicinity of the site of the cross section, which would then degrade for this single event designated C. Later, superflood B would again remove most of the previously aggraded alluvium. Remnants of the floodplain level at the time of event B would remain on the flanks of the valley at level BB. The terraces produced by event A, at level AA, were in existence during the first part of the period between events A and B. The quasi-continuous rise of the aggrading bed during this period progressively filled the valley until event D caused "over-terrace" deposition and obliterated terrace AA. If the superflood that occurred prior to the event A took place when the level of the channel bed was lower than AA, such as at level PP, then terrace AA would be the only one in existence during the period from A to D. If the level at that time was higher, say QQ, then the valley would have, during the period from A to D, two terraces, QQ and AA. Because of steep valley slopes and lack of cementation due to climate, preservation of terraces for more than a few removal cycles is unlikely. Remnants of a level such as RR, if subjected to the effects of a series of normal events plus superstorms such as A and B, would not survive as a definable terrace, even if they were to stay above aggradation levels.

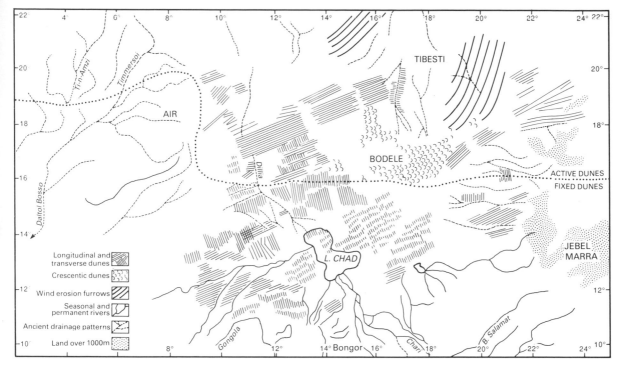

Figure 11.29 The central portion of the southern Sahara, showing a variety of active aeolian features, and palaeoforms such as ancient drainage systems and fixed dunes (after Grove & Warren 1968, and others).

deposits), to produce sinuous ridges up to 5m high (Fig. 11.31). The exhumed ridges vary in their sinuosity, gradient, width and roughness, so that the discharge conditions responsible for their formation can be predicted.

11.8 Some applications

Most of the processes and changes described above have practical applications. The nature of water and sediment movement is crucial to the operation of irrigation works and flood farming (e.g. Sundborg 1986), and to determining the location of placer deposits and geochemical exploration (Beeson 1984). The shifts between erosion and sedimentation, and arroyo cutting, are in many places both a *reflection* of human management of catchment areas and a major *influence* on human use of drylands in the ways they affect, for instance, water-table levels and rangeland quality (e.g. Rapp et al. 1972, Cooke & Reeves 1976). The study of sediment yield is fundamental to predicting the longevity of reservoirs (e.g. Sundborg 1986).

The analysis of sediments in ephemeral stream channels, through judicious sampling and geochemical analyses, is an important way of assessing mineral resources. For example, Soliman (1982) analyzed the −1mm fraction of channel sediments in the search for copper and other deposits in Egypt; Hinkle (1988) found that analyses of the non-magnetic fraction of heavy-mineral concentrates panned from stream sediments are an effective way of locating such metals as copper, lead, zinc, tin and tungsten in arid environments; and Beeson (1984) used the fine fractions of stream sediments in South African, Iranian and Australian drylands in the search for mineralization zones.

The occurrence of ephemeral floods, and floods in perennial rivers, is a major hazard to settlements (as was shown by the unanticipated Nile floods in the Sudan in 1988) and it requires special responses that reflect the catchment environments (e.g. Cooke 1984, US Dept of Commerce 1978).

Hazards of erosion, sedimentation and flooding often precipitate human responses, which in turn can cause major adjustments to dryland drainage systems. Mesopotamian rivers have been modified continuously

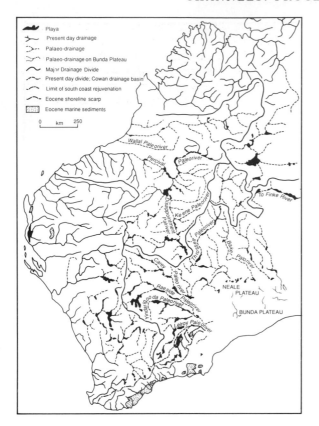

Figure 11.30 The palaeodrainage of Western Australia and adjoining areas (Van de Graaf et al. 1977).

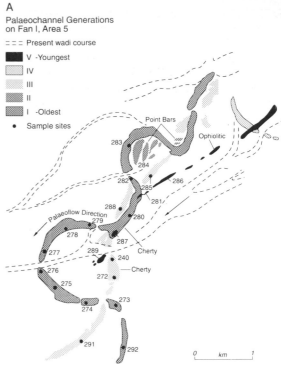

Figure 11.31 (A) Channel pattern and lithology changes associated with successive generations of palaeochannels on a fan west of the Wahiba Sands, Oman. (B) A simplified model of landscape evolution of the alluvial fans and raised channel systems in the Wahiba sands region, Oman: five major phases of landform evolution are identified (after Maizels 1987).

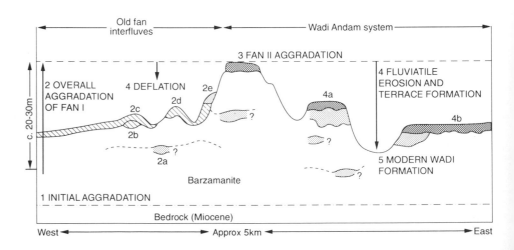

over many centuries in response to these hazards (e.g. Adams 1981). The creation of dams on the Colorado River has significantly increased the stability and the danger of downstream rapids (e.g. Graf 1980). On the other hand, the Aswan High Dam on the Nile has not only regulated water supply but down stream has reduced essential siltation, promoted bank and delta erosion, reduced seasonal fluctuations of water-table level and thereby increased the salinization hazard, and allowed sand encroachment to continue without inhibition in some places (e.g. Ibrahim 1982). Similarly, the drainage of lakes and swamps may profoundly affect the geomorphology of the related catchment and rivers, as in the Jordan catchment (e.g. Inbar 1982).

CHAPTER 12

Alluvial fans
and their environments

Within desert landscapes, slopes and channels are combined into several distinctive assemblages. As indicated in Part 1, the two basic types are shield or platform deserts, and basin–range topography. Within these types, the recurrent pattern is of mountain and piedmont plain (Fig. 4.2), and within these, the main assemblages are alluvial fans and their catchments, pediments and their related hillslopes, and base-level plains (playas etc.). In the following sections, attention is focussed successively on each of these assemblages and their evolution.

Figure 12.1 An alluvial fan derived from a small catchment in the Domeyko Hills, southern Atacama Desert, Chile.

12.1 Definitions and occurrence of alluvial fans

Alluvial fans are cone-shaped deposits of fluvial sediments. They usually radiate down slope from points which are commonly where drainage leaves a mountain area. They may also radiate from points within a mountain valley, or within the piedmont plain. They form part of fluvial systems that include the catchment areas supplying them, and commonly the base-level plains to which they are tributary (Fig. 12.1). Of the several types of fan, a common distinction is between "wet" and "dry" fans: the former have perennial flow, the latter only ephemeral flow. In deserts, alluvial fans are almost entirely of the "dry" type.

"Dry" alluvial fans occur in many deserts, especially those that have considerable local relief. They are widespread in the basin–range province of the American West, and they have been described in the Sahara (e.g. Busche 1972), throughout the Middle East (e.g Maizels 1987), in Iran (e.g. Beaumont 1972), Pakistan (e.g. Anstey 1965), southern Africa (e.g. Booysen 1974) and Spain (e.g. Harvey 1978). However, they are by no means confined to deserts. As Nilsen & Moore's (1984) bibliography revealed, alluvial fans occur in humid and temperate regions, such as central Virginia, USA (e.g. Kochel & Johnson 1984) and Tasmania, Australia (e.g. Wasson 1977a, b); and in glaciated mountain and periglacial regions (e.g. Alberta: Kostaschuk et al. 1986; Brooks' Range, Alaska: Nilsen & Moore 1984; Norway: Theakstone 1982). In addition, alluvial fan sediments occur throughout the geological column (e.g. Galloway & Hobday 1983, Flint 1985).

Alluvial fans have attracted substantial research interest, especially in recent years, for several reasons. They are intrinsically fascinating because they are often *clearly delimited systems*, in which the interplay of environmental variables and distinctive processes and their changes over time can be analyzed. More than this, however, alluvial fans pose serious problems for settlement and development in drylands (e.g. Cooke et al. 1982, French 1987) and their deposits are of interest because they may provide loci for waste disposal, and a valuable aggregate resource. Also, they may contain placer deposits, and they can be important sources of both floodwater and groundwater (Nilsen & Moore 1984). Fans have not only been studied in the field; they have also been the basis for several significant experimental studies (e.g. Schumm 1977, Hooke & Rohrer 1979, Rachocki 1981).

The occurrence of alluvial fans in deserts is broadly influenced by several factors.

(i) Lack of vegetation means that the positions of drainage channels are *relatively* unfixed. Nevertheless, the presence of vegetation may serve to dissipate flow on fan surfaces and to sustain sheetflow (e.g. Bull 1977) or even the reverse, to concentrate flow.

(ii) There may be long periods between floods when debris may accumulate in the catchment areas, and occasional heavy rainfalls and floods may lead to mass evacuation of material (Beaty 1963). (In fact, the evidence for the production of debris in large quantities at present in deserts

is limited, and in Arizona (Melton 1965a) and elsewhere there is evidence from weathering features that debris is *not* actually being generated rapidly at present.)

(iii) Fans commonly occur where there is a fall of base level in the depositional area relative to the catchment area, due either to erosion or tectonic movement (Bull 1977).

(iv) Fans occur most commonly where the ratio of depositional area to catchment area is small: that is, where large highlands border small lowlands. Such is the case in areas of faulting or where drainage is deeply incised. Thus, for instance, in the northern part of the Basin–Range Province in California, the ratio is small and fans are common; but in the Mojave Desert to the south, the ratio is large (mainly because of erosional modification of tectonic relief) and fans are rarer (e.g. Lustig 1969a). The mountain:piedmont ratio is commonly influenced by tectonic movements. Bull (1977) suggested that conditions favouring the accumulation of untrenched deposits adjacent to mountains reflect the relations between rate of uplift (Δu), rate of fan deposition (Δs) and rate of channel downcutting (Δw) as follows:

$$\Delta u/\Delta t \geq \Delta w/\Delta t + \Delta s/\Delta t \qquad (12.1)$$

(v) The occurrence of dissected mountain areas with channel sytems that focus on the mountain/plain boundary is a pre-requisite for fan formation. The geometry and distribution of fans owes much to the location, number, size and spacing of drainage channels at the mountain/plain boundary, and to the plan-shape of the boundary.

In their simplest form alluvial fans are "unsegmented". More commonly, fans comprise several segments arising from erosional and depositional changes over time. Whatever the overall form, fan morphology includes three basic elements (Fig. 12.2): channels; "abandoned" old or elevated fan surfaces between channels; and contemporary depositional surfaces down slope of channels. Channels are crucial to the functioning of the systems. The **feeder channel** is the essential link between the catchment and the fan. Beyond it there are commonly two channel networks. The first comprises **distributaries** from the feeder channel, which form a broadly **radial** pattern. They are often braided if entrenchment is slight, and they

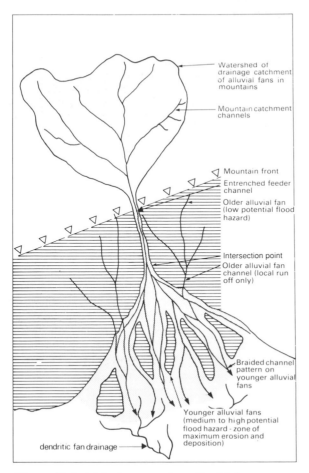

Figure 12.2 Main features of an alluvial fan system.

Watershed of drainage catchment of alluvial fans in mountains
Mountain catchment channels
Mountain front
Entrenched feeder channel
Older alluvial fan (low potential flood hazard)
Intersection point
Older alluvial fan channel (local run off only)
Braided channel pattern on younger alluvial fans
Younger alluvial fans (medium to high potential flood hazard - zone of maximum erosion and deposition)
dendritic fan drainage

portant features: the presence of a point at which drainage leaves a confined valley, usually near the mountain/piedmont junction; and divergent flow paths below that point. The initiation of deposition normally is not a simple function of a change of slope between mountain and plain, as was once commonly supposed. In fact, the longitudinal profiles of fans rarely show a significant change at the mountain/plain boundary. Fans are fundamentally formed by deposition beyond the confines of mountains or escarpments as the result of changes in the hydraulic geometry of flows. As Bull (1977) explained, stream discharge is by definition a product of width, depth and velocity of flow. When a stream reaches the end of a confining channel, it spreads out, so that increase in width is accompanied by reduction in depth and/or velocity, which causes sedimentation. In addition, discharge often decreases as the stream travels over permeable deposits, thereby increasing sediment concentration, and leading to further deposition (e.g. Bull 1977). Furthermore, it is important to emphasize that, while *individual flows* may each be slightly fan-shaped at their distal ends, not all flows increase in width, most will cover only a small percentage of fan area, and certainly none of them is likely to flood the *whole fan* (Fig. 12.4).

An inherent feature of fan development is the continually changing pattern of channels and loci of deposition. These shifts may occur within a single flood, or with successive floods, and they arise largely from the progressive filling and overflow of channels, often as the result of blockage by boulders or mudflows, or by flow diversion around coarse deposits. Over a long period of time these changes ensure the maintenance of fan form by distributing material widely over the surface. **Channel piracy** is a common feature of alluvial fans, and it arises in large measure because of the juxtaposition of many channels of differing bedloads and gradients at different levels. Channels originating in mountains, for example, may have coarser loads and steeper gradients than gullies confined to the fans (Denny 1967). Also, channels might change position by lateral migration and as a result of channel-bed aggradation at the intersection point (Hooke 1987).

merge with the fan surface at the **intersection point**. The second, not often mentioned in the literature, arises from local drainage *on the fan itself*, occurs down slope of the distributary network and may interdigitate with it, usually forming a **dendritic** pattern (Fig. 12.3).

Hooke (1987) has argued that channels on alluvial fans are characteristically wide and shallow and, as a result, are typically unstable and **braided**. The width: depth ratio is probably high because channel banks are easily eroded, and because there is only a low transverse slope in channels from their banks towards the centreline, so that transverse sediment flux is low. Instability is revealed through the formation of alternate bars along the banks *and* the centre of the channel, the number increasing with the channel width: depth ratio (Hooke 1987).

The fan shape is determined by at least two im-

12.2 Processes on alluvial fans and their associated deposits

The nature of alluvial fan deposits reflects in large

Figure 12.3 An alluvial fan in Panamint Valley, California, showing distributory drainage channels and a dendritic system of channels rising on the fan itself.

measure the type of bedrock in the catchment basin, the nature and rate of its destruction, its alteration during transport and its post-depositional modification; crucial too, are the relationships between debris supply and rainfall–runoff. Most flows on alluvial fans in deserts are ephemeral (see Ch. 11.1).

Although alluvial fans all result from the deposition of material carried from the mountains, personal observations of fan floods (Beaty 1963), laboratory experiments (Hooke 1967) and analyses of fan sediments (Bull 1962, 1963) suggest that the processes of debris movement are varied, and may range from viscous *debris flows* to *streamflows*. Not all forms include both debris forms and water-laid sediments. Some have none of the former; a few may be without the latter. Debris flows are rarer, mainly because they require an abundant supply of fine sediment, they may require an initiating mechanism (such as slope failures in saturated sediment) and they seem to be more important in dry areas where vegetation cover is low (Hooke 1987).

All the processes tend initially to follow established channels. Sheetflood material appears to be limited on fans, although sediments may locally overtop the established channels, and widening of deposits occurs, especially down slope of the channel ends.

The various processes may often be related *through time during a single flood*. Eyewitness reports of alluvial fan flooding in the White Mountains of California and Nevada (Beaty 1963) provide an illustration. Two-and-a-half hours after a heavy thunderstorm in the mountains, masses of debris advanced down channel in a series of waves along a low front of boulders and mud; the lower ends of this debris-flow consisted of silts and fine sands with cobbles and small pebbles which constituted a rapidly moving mudflow; after the debris-flows had halted, streamflows continued for up to 48 hours and dissected the newly laid sediments and older deposits.

Similarly, the New Year's Day flood on alluvial fans of the La Cañada Valley, in Los Angeles County (e.g. Cooke 1984) consisted initially of a series of pulses which led to debris from valley slopes accumulating in the mountain channel *near to* the fan apex, but only caused water dominated flows on the fan itself. The mass of sediment material was then struck

171

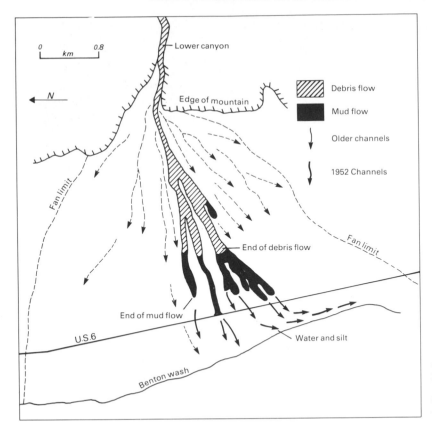

Figure 12.4 Milner Creek fan, White Mountains, California, showing deposits of the 1952 flood (after Kesseli & Beaty 1959).

by intense rainfall and high velocity discharge, creating a major debris flow on the fan that effectively evacuated available sediment from the catchment. The debris flow had a front reported to be up to 7m high in places and it transported boulders weighing as much as 70 tonnes. Subsequent water flows were underloaded and highly erosive, cutting into the debris flow deposits.

Debris flows differ substantially from stream flows on alluvial fans as elsewhere (Fig. 12.5). As Hooke (1987) observed, natural debris flows have, for example, higher viscosities and densities; they behave thixotropically; a semi-rigid surface layer may overlie an underlying viscously deforming mass; the weight of a rock in the flow is reduced relative to its weight in water (a "buoyancy effect"); bed shear forces and drag forces are higher than in a stream of comparable depth; and debris flows cannot selectively deposit finer sediment, unlike water flows, so that they cannot convert into a stream by deposition. Such flows often have steep fronts, and include differential rates of movement reflecting thicknesses of flow, so that surges are common. The causes of debris flows are

considered by Hooke (1987).

Debris-flow and streamflow deposits on fans are also distinctively different. For example Bull, (1962, 1963, 1964a) and Wasson (1977b) showed that, on alluvial fans in California and in Tasmania, they vary in their grain-size curves (Fig. 12.6A) and can be distinguished in terms of their CM patterns (Fig. 12.6B), with mudflows showing poorer sorting than streamflow deposits (a sample plotting on the CM line would be perfectly sorted in its coarser half).

Debris-flow deposits are derived from flows with over 50% sediment concentration, and with velocities from 0.5–16m sec^{-1}. The mean velocity, $\bar{V}$, of a debris flow can be approximately calculated after the event by the expression:

$$\bar{V} = \sqrt{r_m g \tan\beta} \qquad (12.2)$$

where r_m is the channel bend radius of curvature, g is the gravitational acceleration and $\beta = \Delta h/w$, where h/w is the elevation difference between the flow surface on the inside and outside of the bend and w is the width of the flow (when slope is less than c. 5°) (Wasson 1977b). The deposits are typically poorly

Figure 12.5 A mudflow deposit in and adjacent to an alluvial fan channel, in the central Atacama Desert, Chile. The figures are standing on an older debris flow.

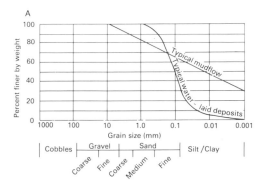

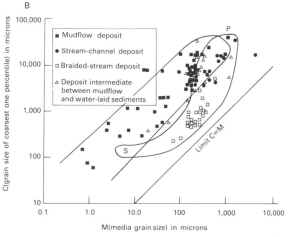

sorted, weakly bedded, and dominated by subangular clasts, a high proportion of fine sediment and a high void ratio. They often comprise long, narrow strips that follow channels. Runout can be over many kilometres (depending mainly on slope viscosity and transmission losses), and they spread out only at the distal ends of fans in lobate planforms. In cross-section they may be slightly concave, and bordered by levees, including well defined lateral ridges of coarse debris (Fig. 12.7). Their lateral boundaries are usually sharp, and they are often associated with **sieve deposits and lobes** (Fig. 12.8). Sieve deposits occur where the fan material is so coarse and permeable that surface flow infiltrates into it before reaching the toe of the fan, and a lobe of coarse debris is deposited. The debris acts as a sieve by allowing water to pass through while retaining sediment (Hooke 1967). The flows may develop *transverse* slopes, so that overlapping of channel banks may be asymmetric, and overtopping of obstacles is common.

Clumps or trains of boulders and armoured mud balls occur on some of the fans described by Bull (1964a), and they were probably transported in mudflows or slightly more fluid flows. Coarse material

Figure 12.6 (A) Typical grain-sized curves for mudflow and water-laid deposits on alluvial fans (after Bull 1963). (B) CM patterns of superficial alluvial-fan deposits in western Fresno County, California (after Bull 1962).

may be "rafted" or pushed along at the front of flows. The rapid deposition of such material can lead to the formation of sieve deposits. The horizontal orientation of flat, coarse fragments and graded bedding distinguish fluid *mudflows* from the more viscous flows which have no graded bedding and some larger rock fragments orientated and imbricated relative to the flow direction. Bubble cavities are common and they result from air being trapped beneath the mudflow or being entrained in it – a feature which may actually decrease the viscosity of the flow.

Water-laid sediments are derived from flows containing much less debris: a higher proportion of water allows much better sediment sorting. Bull (1963) distinguished two types of water-laid sediment. The first are those from shallow flows which continually filled small channels and changed position frequently. These deposits are generally well sorted and often show cross bedding or lamination. The second are water-laid sediments in major channels which are usually coarser and more poorly sorted. Water-laid deposits often have no clearly discernible margins. Mudflow deposits and water-laid sediments represent the two extremes; there are many possible mixtures between them. In general, debris flow deposits are likely to be dominant nearer to the fan apex; and water-laid sediments predominate in the distal zone (e.g. Wasson 1977a, b, Hooke 1987).

12.3 Morphometric characteristics of alluvial fans

Data for the analysis of several morphometric properties of alluvial fans are often recovered from topographic maps, although high-quality maps and field checking (or a least the study of matching remote-sensing imagery) are essential. It may be that alluvial fans do have distinctive representation on topographic maps. Doehring (1970), for example, considered that they can be distinguished on contour maps from pediments because their drainage texture does not become finer in a headward direction. However, maps do not normally provide conclusive evidence, and they certainly do not reveal the complexity of fan geography, such as the nature of segmentation. However, contour crenulations may be useful in identifying intersection points (Bowman 1978).

Morphometric properties that have received most attention are *shape*, *area* and *slope*.

12.3.1 Shape
Several authors have attempted to describe fan shape precisely, using simple equations (Troeh 1965, Murata 1966, Bull 1968, Lustig 1969b). For instance, the surface of a fan may be compared to that of a segment of a cone characterized by concavity of the longitudinal profile and convexity of the transverse profile, and an appropriate equation, suggested by Troeh and adopted by Bull and Lustig, is:

$$Z = P + SR + LR \qquad (12.3)$$

where Z is the elevation of any point on the fan, P the elevation at the apex of the fan, S is the slope of the fan at P, R is radial distance from P to Z, and L is the half the rate of change of slope along a radial line. $P + SR$ defines a right circular cone; R provides the curvature; S is negative and L is positive, giving

Figure 12.7 A debris flow, with well developed levées, in the Mojave Desert, California.

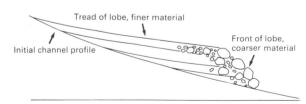

Figure 12.8 A schematic sketch of the growth of a sieve lobe (after Hooke 1967).

a concave fan with convex contour curvature. This equation, of course, only defines a single, simple fan. Application of it by Bull (1977) to a single fan demonstrated that it provided a reasonable approximation of the fan form. In the future, as Lustig (1969b) observed, it should be possible to find a general solution of the equation on a regional basis and thereby provide an index for regional comparison.

12.3.2 Area

It has frequently been demonstrated that fan area, A_f, is related to catchment drainage area, A_d, by the general equation:

$$A_f = cA_d^n \qquad (12.4)$$

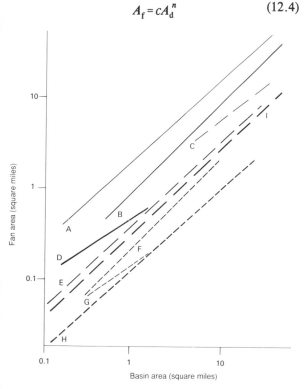

where the constant c is the area of fan with a drainage area of one square kilometre, and the exponent n is the slope of the regression line (Fig. 12.9; Bull 1964b).

The intercept c is affected by four important factors: (i) the *erodibility* of rocks in the catchment area, with which c will normally be positively correlated (e.g. Bull 1968, Hooke 1972), so that it may be possible to use c as an index of relative erodibility (Hooke & Rohrer 1977); (ii) *tectonic* tilting may influence c by altering basin area and relief (e.g. Rockwell et al. 1985), as in Death Valley, California, where fans associated with the dip slope of upfaulted blocks tend to be larger than those adjacent to the fault-scarp mountain fronts (e.g. Denny 1965); (iii) *precipitation*, with which c will normally be positively correlated; and (iv) the *amount* of space available locally for deposition, with which c is positively correlated.

The value of the regression coefficient, n, is normally less than one, implying negative allometry (Church & Mark 1980): that is to say, large drainage basins yield proportionally less sediment onto fans than smaller ones. The value of n is rather consistently c. 0.9 (although, as Church & Mark (1980) pointed out, the variation from 1.0 may not always be great) and there are several probable reasons for this (e.g. Hooke & Rohrer 1977, Lecce 1988, 1991): (i) large catchments may be less frequently covered by a single storm; (ii) more sediment may be stored on valley-side slopes or in the channels of large basins; and (iii) valley-side slopes tend to be lower in larger basins, and therefore yield less sediment (e.g. Hooke 1968); (iv) larger catchments may be more likely to generate floods that can carry sediment *beyond* the fans (e.g. Bull 1972); (v) larger systems may take longer to adjust to available space than small ones where transfer distances are short and larger fans hem them in (e.g. Church & Mark 1980); and (vi) a limited available space will more seriously inhibit the development of large fans (e.g. Kostaschuk et al. 1986). The value of n may also reflect the nature of processes forming the fans. For example, Kostaschuk et al. (1986), in discussing the A_f/A_d relationship for debris flow fans in a deglaciated area near Banff, Canada, suggested that smaller, more steeply sloping basins could produce debris flows of lower strength and thus proportionally larger areas than larger basins with lower slopes, thereby reducing n to below a value of one.

There are variations in the value of n in response to specific second-order effects such as variations in

Figure 12.9 The relationship between fan area and basin area for fans from various regions (based on Lecce 1988). (A) $A_f = 2.1A_d^{0.81}$; shale source San Joaquin Valley (Bull 1964b). (B) $A_f = 0.96\ A_d^{0.98}$; sandstone source, San Joaquin Valley (Bull 1964b). (C) $A_f = 1.05\ A_d^{0.78}$; Death Valley, west side (Hooke 1968). (D) $A_f = 0.44\ A_d^{0.82}$; dolomite and quartzite source, Deep Springs Valley (Hooke 1968). (E) $A_f = 0.42\ A_d^{0.94}$; Owens Valley (Hooke 1968). (F) $A_f = 0.24\ A_d^{1.01}$; Cactus Flats (Hooke 1968). (G) $A_f = 0.16\ A_d^{0.75}$; quartzite source, Deep Springs Valley (Hooke 1968). (H) $A_f = 0.15\ A_d^{0.90}$; Death Valley east side (Hooke 1968). (I) $A_f = 0.39\ A_d^{0.94}$; White Mountains (Lecce 1988).

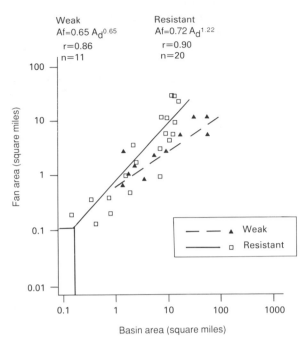

Weak
$Af = 0.65\, A_d^{0.65}$
$r = 0.86$
$n = 11$

Resistant
$Af = 0.72\, A_d^{1.22}$
$r = 0.90$
$n = 20$

Fan area (square miles)

- - - - ▲ Weak
——— □ Resistant

Basin area (square miles)

Figure 12.10 Fan area versus basin area by drainage-basin erodibility; White Mountains, east, California (after Lecce 1988).

slope, vegetation, climate and lithology. For example, Lecce (1988, 1991) showed that along the west side of the White Mountains (California, Nevada), relatively erodible catchments produced unusually small fans, and a low value of n (= 0.65) indicated greater sediment storage in the larger erodible catchments. On the other hand, catchments in relatively resistant rocks had a value of $n = 1.22$, suggesting that their sediment storage does *not* increase with catchment size (Fig. 12.10).

Hooke (1968) argued that the general A_f/A_d relationship reflects the fact that fans are *space-sharing systems*. He recognized a tendency towards equilibrium in which all fans increase in thickness at approximately the same rate: if a fan is too small for the volume of material it receives it will increase in thickness more quickly than, and will increase in area at the expense of, adjacent fans, until an equilibrium is established. Thus, fan area is proportional to the volume of sediment supplied to it, and that volume is in turn a reflection of catchment area, so that:

$$V = c\frac{dh}{dt}A_d^{\,n} \qquad (12.5)$$

where V is the volume of debris supply per unit time,

dh/dt is the rate of increase of fan thickness, and c and $A_d^{\,n}$ are as in Equation 12.4 (Hooke & Rohrer 1977).

In addition, Hooke (1968) argued that, in an enclosed basin containing an aggrading playa surrounded by alluvial fans, the rate of playa deposition beyond the fan should equal the rate of fan deposition. For instance, if the playa is too large for the amount of debris being supplied, it would increase in thickness more slowly than the fans, which would therefore encroach on it, reduce its area and thereby increase its rate of thickening. Because of the ephemeral nature of change within the system, a steady state may not be achieved at any one time, and the tendency towards it may continue for a protracted period. Over longer time periods, fans in enclosed basins will continue to grow, and they will progressively bury the mountain front.

The type of steady-state relationship envisaged by Hooke contrasts with that proposed by Denny (1965, 1967) who suggested that the rate of fan deposition equalled the rate of fan erosion. There is no doubt that both erosion and deposition occur on fans. To take a simple example, new deposition onto fans is associated with the main channel in the mountains; but downslope there may be quite separate channel systems, signifying erosion confined to the fan itself. Denny's hypothesis has not been established – indeed it would be almost impossible to demonstrate empirically – and it seems rather unlikely, unless material eroded from the fan is *removed entirely from the system*. If eroded material merely accumulated on the adjacent playa, the playa would influence the form of the fan. For example, if the playa encroached on the fan, the fan deposition rate would increase, but erosional processes on the fan would operate over a smaller area and the erosion rate would decrease until the deposition rate was positive on fans and equalled that on the playa (Hooke 1968).

Church & Mark (1980), in discussing Hooke's model in terms of allometry, suggested that *systems* of fans adjusting towards a form of equilibrium will approach a limiting A_f/A_d relationship, and will do so in one of two ways (Fig. 12.11): (i) as long as there is room for fans to increase in absolute area, the value of n will be roughly constant, but the relationship will move steadily towards the limiting condition; or (ii) small fans may approach equilibrium much more rapidly than large ones, as described above.

12.3.3 Slope

In much the same way as a perennial stream profile, the longitudinal profile of an alluvial fan, is usually a smooth exponential curve, an hydraulic profile determined by such variables as debris-size, water-sediment discharge, catchment morphology and sediment production rates. It is usually less than 10°.

Alluvial-fan slope, S_f, is inversely related to *drainage catchment area* A_d (e.g. Bull 1964b):

$$S_f = CA_d^{-b} \qquad (12.6)$$

Fan slope is also directly related to *basin relative relief* (Melton 1965a) as follows:

$$S_f = C(H/\sqrt{a_d})^d \qquad (12.7)$$

in which S_f is the upper-fan gradient, and H is the vertical relief above the fan apex. In the case of Equation 12.6, the exponent b might be expected to be *c.* 1.0, a simple linear relationship (Church & Mark 1980). However, in a study of fans in Alberta, b was greater than 1, indicating positive allometry (i.e. fan slope increases more rapidly than the ruggedness of the basin), probably as a result of non-linear sediment size and discharge effects (Kostaschuk et al. 1986).

Catchment area lithology is a crucial influence on fan slope, because it affects the nature of debris, sediment production, depositional processes and sediment concentration in flows reaching the fans – all of which, in turn, influence fan slope. Thus, for example, Bull (1968) found in the Central Valley of California that the slopes of fans with source areas having high rates of sediment production were steeper than those from areas with low rates of sediment production. Hooke's (1968, Hooke & Rohrer 1979) laboratory experiments demonstrated the following:

(i) Fan slope is generally inversely proportional to fan area, fan radius and to discharge (Figs 12.12 & 13). Discharge is especially important, because large discharges can transport debris on a lower slope than small discharges as they have higher flow velocities and higher bed-shear stresses (Hooke 1968). One consequence of this situation is that increase of discharge across a single fan may cause erosion or even trenching near the fan head, and deposition towards the fan toe, thus decreasing slope overall. Smaller discharges will tend to deposit higher up the slope. Actual fan slope will thus reflect a balance between these tendencies.

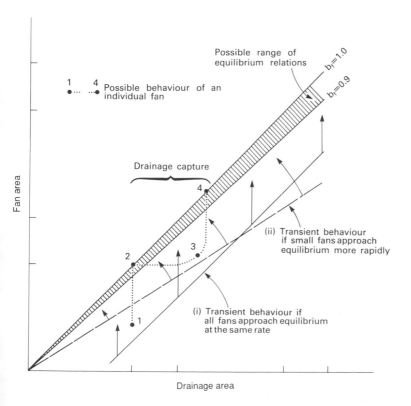

Figure 12.11 Schematic diagram of possible variations in the areal relationship of the alluvial fan to the contributing drainage basin. Note that the behaviour of an individual fan is quite different from the static allometric relation (after Church & Mark 1980).

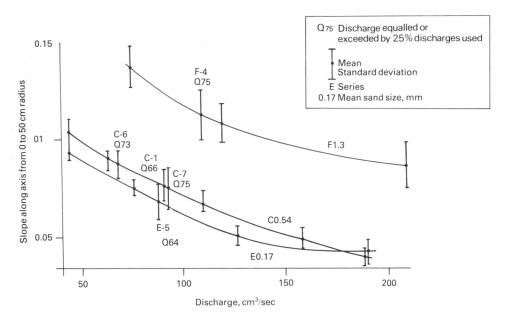

Figure 12.12
Relations between
slope and discharge
for three different
sediment sizes in
laboratory studies
of alluvial fans.
The open bars (for
C-6, etc.) were
not used to draw
the curves (after
Hooke & Rohrer
1979).

(ii) Fan slopes normally vary directly with particle size, so that the slope of fan profiles usually declines with particle size in a downstream direction. This view is supported by some field evidence (e.g. Bluck 1964, Denny 1965), but it

is also contradicted in some locations (e.g. Beaumont 1972, in Iran). Thus, in contrast to what is usual for "wet" fans, the relationship does not always hold.

(iii) Fans with flows of higher sediment concentration generally have steeper slopes.

(iv) Steeper slopes occur where depositional processes are associated with predominantly coarse sediments, and gentler slopes occur where fluvial processes alone are dominant. In particular, debris flows are often associated with steeper slopes, as are sieve deposits.

The previous discussion refers specifically to the general longitudinal profiles of fans. Within any one fan, the axial slope is gentlest, and slopes on the flanks are steeper (Fig. 12.14), a difference which Hook & Rohrer (1979) attributed to variations in average discharge. They considered the contrast to be greater on steep fans composed of relatively non-cohesive materials because higher discharges would tend to flow down the axes of such fans, whereas lower discharges can be turned to move down the flanks. On gentler fans, and/or those composed of more cohesive sediment, the higher discharges can also be turned down the flanks and thus the slope contrast is less.

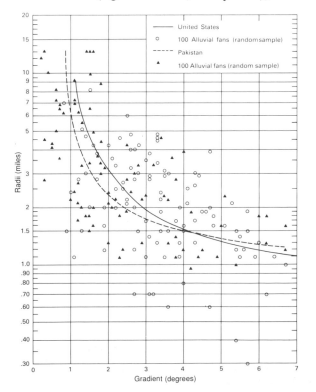

Figure 12.13 Radii and gradient relationships of alluvial fans for random samples of fans in the USA and Pakistan (after Anstey 1965).

12.4 The development of segmented fans

12.4.1 Patterns of segmentation

Unsegmented fans are unusual in areas where fans coalesce and where they have undergone prolonged development. More commonly, the piedmont zone is composed of often complex segmented fans (Fig. 12.15). Criteria for distinguishing between fan segments of different ages, apart from their relative topographic positions, include drainage pattern and depth of incision, morpho-stratigraphic relations, the nature of the soil profiles (including the presence/absence of **caliche** and the nature of desert varnish) and the weathering of stone pavements, vegetation patterns, and degree of dissection (e.g. Table 12.1). Recently, these means of creating relative chronologies have been supplemented by the new, and still experimental, technique of cation-ratio dating (e.g. Dorn 1983; Ch. 6.2.1). In this method the elemental ratio (K+Ca)/Ti in rock varnish is determined using particle-induced X-ray emission (PIXE). The premise is that the K+Ca:Ti ratio decreases with time because

K and Ca are more mobile than Ti. The cation leaching curve (CLC) needs to be calibrated in each region using established chronologies, such as radiometric ages of underlying surfaces or of organic material in varnish, or the K–Ar ages of volcanic rocks. Once this is done, the method can be applied to give the minimum age of surfaces that have not been dated absolutely, but do have a stone varnish on surface particles (e.g. Dorn 1988).

One commonly observed sequence of segmented-fan development was summarized by Denny (1967; Fig. 12.16). Many unsegmented fans (Fig. 12.16A) become entrenched near their apices with the result that the loci of deposition are moved down slope; secondary fans are built on or beyond the former segments, and the upper parts of the initial fans are no longer zones of deposition (Fig. 12.16B). The initial segments may become dissected into gullies by local runoff. In Figure 12.16B, the head of the gully at X may be below the level of the main channel at Y, so that, if the left bank of the main gully is eroded, discharge in it may eventually flow down the fan

Table 12.1 General characteristics of young, intermediate and old alluvial-fan deposits.

Characteristic	Young	Intermediate	Old
Drainage pattern	distributary: anastomosing or braided	tributary: dendritic	tributary: dendritic or parallel
Depth of incision	less than 1m	variable (1–10m)	greater than 10m
Fan surface morphology	bar-and-channel	variable, generally smooth and flat	ridge and valley, most of surface slopes
Preservation of fan surface	presently active	incised, but well preserved wide flat divides	basically destroyed, locally preserved on narrow divides
Desert pavement	none, to weakly developed	none to strongly developed	none (surface destroyed), to strongly developed (surface preserved)
Desert varnish	none, to weakly developed (most varnished clasts reworked from older surfaces or bedrock)	none to strongly developed	none (surface destroyed), to strongly developed (surface preserved)
B-horizon	none, to weakly developed	weakly to strongly developed	none (surface destroyed), to strongly developed (surface preserved)
Calcic horizon	none, to weakly developed, CaCO$_3$ disseminated throughout	weakly to strongly developed	none, carbonate rubble on surface (surface destroyed) to strongly developed petrocalcic horizon (surface preserved)

Source: Christenson & Purcell (1985).

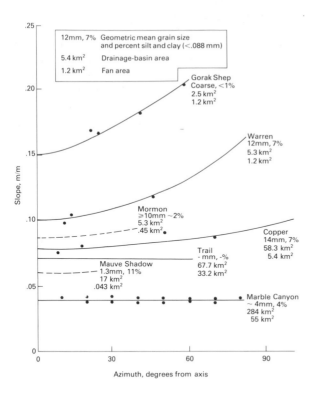

Figure 12.14 Relation between slope and azimuth on natural fans with different grain-size characteristics, drainage basin areas and fan areas in Death Valley, California (except for Gorak Shep fan, which is in Eureka Valley, California; after Hooke & Rohrer 1979).

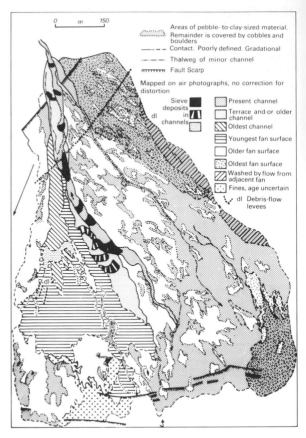

Figure 12.15 Geomorphological map of Shadow Fan in Death Valley, California (after Hooke 1967).

gully, leading to the preferential development of a third fan segment (Fig. 12.16C). During these changes, the overall area of the fan has increased, and more and more of its surface has become abandoned by depositional processes and dominated by erosion.

The sequence shown in Figure 12.16 is a common but not universal one. Figure 12.17 shows a contrasting example, from Death Valley, California, which reflects a threefold sequence of climatic and tectonic changes (Dorn 1988). Here, successively younger fans are to be found set within the broad boundaries of the initial fan, producing a "telescopic" pattern. One reflection of this evolution is the migration of the intersection point. The intersection point may migrate down fan (as in Fig. 12.17) or up fan, depending on the rates of entrenchment and the profiles involved (Bowman 1978). The specific patterns of segmented fans vary greatly from region to region, and the causes of this variability are a matter of major debate.

The following discussion focuses on the *mechanisms* responsible for entrenchment and the movement of loci of deposition.

12.4.2 Entrenchment and moving loci of deposition

Several different hypotheses have been put forward to explain fan segmentation, entrenchment and the movement of loci of deposition.

(a) Hypotheses related to temporary changes at present

Denny (1967: 85) maintained that entrenchment is "the normal consequence of large variations in flood discharge". This inherently reasonable hypothesis implies that modification to fans during the most frequently recurring floods consists of aggradation, but the exceptional, large event with greater competence causes fan-head entrenchment. Beaty (1974a) also inclined to the view that entrenchment is a contemporary phenomenon associated with extreme, but normal and expectable events. Entrenchment may occur during the later stages of a flood, when high waterflows continuing after initial debris-flow deposition dissect the older deposits. Hooke (1967) maintained that the feeder channel is often incised into the fan head area because, in this zone, water is able to transport on a lower slope the material deposited by earlier debris-flows, i.e. the hydraulic gradient required to move debris-flows is higher than that for water transport because the former have higher viscosity and finite yield strength. However, Hooke recognized that some fan-head entrenchment is too deep to have been produced by this process alone.

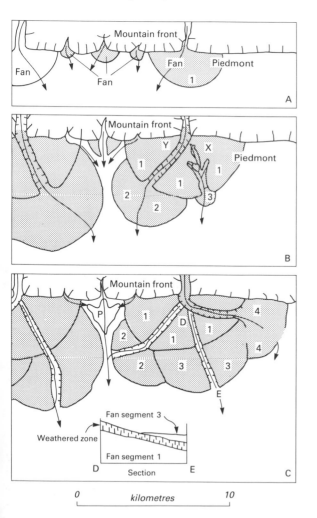

Figure 12.16 The development of segmented fans on a piedmont plain (after Denny 1967).

(b) Hypothesis of intrinsic slope threshold (complex response)

Schumm (1977) recognized that the feeder channels of some fans are entrenched, and others are not, *often in the same region*. In reproducing fan evolution in laboratory experiments, Schumm was able to create the same diversity. First, water and sediment issued from the catchment dividing into shifting, braided channels. Then this pattern was interrupted *periodically* by fan-head trenching. Following entrenchment the catchment area was *rejuvenated*, sediment delivery was *increased* and the trench became filled. Schumm (1977) explained this sequence in terms of *intrinsic slope thresholds* within the system, which require no change in *external* circumstances. Velocity decrease across the fan causes deposition at the fan head until the fan slope has risen to its threshold for the given discharge, whereupon entrenchment is initiated, with the consequences described above.

In one sense, this view of trenching is similar to that of Eckis (1928), who maintained that trenching was a function of an overall decline of sediment yield from the catchment as a long-term erosion cycle progressed: the main difference is that Eckis envisaged permanent, late-stage entrenchment.

(c) Hypothesis of catchment evolution

In a survey of catchment and piedmont characteristics in the Tunisian Southern Atlas, White (1988, unpublished) concluded that stream capture, and the progressive exploitation of structural and lithological weaknesses in the upland catchment, led to entrenchment, movement of the intersection point and the creation of new fan segments (Fig. 12.18).

(d) Tectonic hypotheses

Without doubt, alluvial fan development in some areas is influenced by tectonic activity. Bull (1977) argued that in such areas, the forms produced reflect the relations between rates of uplift (Δu), fan deposition (Δs), and channel downcutting (Δw). Conditions favouring the accumulation of untrenched deposits are described in Equation 12.1. Entrenchment, and the downslope migration of deposition, are favoured where

$$\Delta u/ \Delta t < \Delta w/\Delta t > \Delta e \Delta t \qquad (12.8)$$

(where e is the erosion of the alluvial fan deposit adjacent to the mountain). Thus, for alluvial fans in the San Joaqin Valley, California, Bull (1964b) recognized two situations, in which the profiles of fans

181

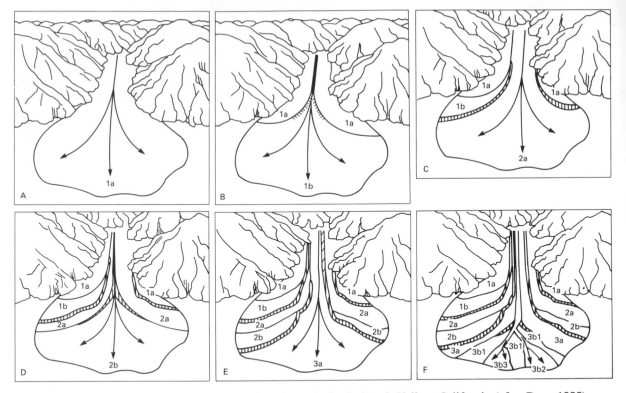

Figure 12.17 Stages in the evolution of a telescopic fan in Death Valley, California (after Dorn 1988).

became segmented: one where the rate of downcutting exceeded the rate of uplift, giving rise to progressively lower stream-channel gradients; and the other where the rate of uplift exceeded the rate of downcutting, giving rise to successively steeper gradients (Fig. 12.19).

(e) Hypotheses of climatic change
An alternative explanation of fan entrenchment is that it results from a long-term change in discharge conditions which, in turn, is a consequence of climatic change or, perhaps, removal of vegetation by overgrazing. It is possible to conceive of a variety of climatic changes which could lead to the desired change from deposition to entrenchment, such as increasing storm frequency, increasing storm intensity, increasing total precipitation, and decline of total precipitation with increased storm intensity.

That climatic change has occurred during fan formation in many areas is beyond dispute. Lustig (1965) cited several features associated with fans in Deep Springs Valley, California, which might be

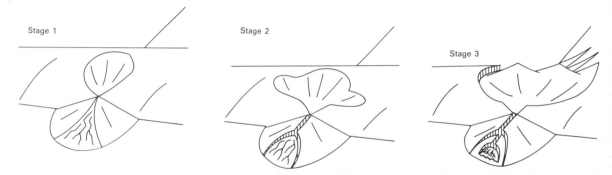

Figure 12.18 The development of the Oeud el Tfal alluvial fan and catchment in Tunisia (after White 1988).

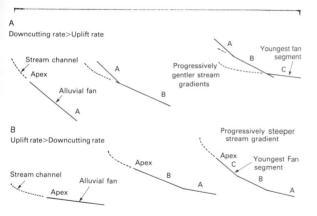

Figure 12.19 (A) An alluvial fan profile formed when the stream downcutting rate exceeds the mountain uplift rate. (B) A segmented fan profile resulting when the uplift rate exceeds the downcutting rate (after Bull 1964a).

indicative of climatic change. Amongst these were great movement of the loci of fan deposition; misfit trenches near fan apices; paired terraces continuous with former fan surfaces; desert varnish on abandoned fan surfaces; and greater estimated tractive forces within present active channels than on the fan surfaces. In discussing fan-head trenching in Deep Springs Valley, Lustig pointed out that the phenomenon is so widespread that a regional explanation is required. He indicated that in most instances the estimated tractive force within the trench is greater than on adjacent fan surfaces: "Because modern floods are confined within channels, the tractive force associated with the modern process must be greater than that which accompanied transport in the past". If this is true, then one or more of the three component variables that comprise tractive force must be greater today than in the past (Lustig 1965: 175). Of the possibilities, increased density of flow seems most likely; and this may result from a change in climate.

Lustig went on to outline a climatic cycle of fan development. In a fan-building period, aggradation is general, and is associated with heavier precipitation than at present, higher water:sediment ratios and lesser tractive forces. In the following fan-trenching period, precipitation is more localized and less frequent, water:sediment ratios are low, mudflows are more frequent and tractive forces are greater, so that the former aggradation surface is dissected.

Whether predominant aggradation coincides with more humid or more arid conditions, or the transitions

between them, or vice versa, is a matter of considerable controversy – even within one area – and is probably a rather simplistic notion. For example, on the alluvial fans of Death Valley, California, Dorn et al. (1987) envisaged fan aggradation during relatively humid periods of the Pleistocene, whereas this view was contested by Wells et al. (1987), who regarded deposition as coinciding with time-transgressive changes in climate, and related several depositional units to the arid Holocene. In a later paper, Dorn (1988) argued that aggradation occurred from head to toe of the fans in semi-arid (more humid) climates, and that fan-head entrenchment coincided with the glacial-to-interglacial transition, when climate locally was becoming more arid. Dorn recognized that the sequence was repeated at least three times, and that the resulting pattern was at times influenced by tectonic activity.

A simple humid/arid contrast is also blurred elsewhere. In southeast Spain, Harvey (1984b) recognized pre-Würm phases of predominant aggradation followed by complex cut-and-fill in predominantly dissection phases post-Würm. He envisaged a long-term Quaternary trend of progressive sediment diminution that was punctuated by local responses to climatic changes and trenching thresholds. He also observed that some Spanish fans, unlike those in the American southwest for example, showed trenching not only at fan-head but also at *mid-fan intersection points* (Harvey 1987). Such mid-fan trenching occurred where there was a calcrete crust, the discrepancy between fan and channel slope was high, and channel width was limited.

In central Niger, Talbot & Williams (1979) argued that periods of fan aggradation occur when there is a relatively high incidence of intense rainstorms, degradation is promoted in rather arid intervals (such as the present day) and surface stability characterizes relatively humid intervals of higher vegetation cover. In Iran, Beaumont (1972) concluded that lower mean temperatures during glacial phases, when frosts and snow might occur in winter, provided considerably better conditions for widespread alluviation, and that the fans are today largely fossil features. These varied views can be compared to those of Gerson (1982, Fig. 10.26) on talus aprons, in which the formation of the aprons is attributed to arid/semi-arid conditions, and their dissection to arid/extremely arid conditions. On the Lake Torrens piedmont plain in South Australia, Williams (1973) recognized a fan-building phase in mid-Würm times (from *c.* 38,000 to *c.* 30,000yr BP) under cold, arid conditions with brief periods of high rainfall or rapid snow-thaw, a fan dissection phase in cold, arid and windy conditions in Late Würm (*c.* 24,000–16,000yr BP), and fan aggradation in the middle Holocene (75,000yr BP) and late Holocene (*c.* 3,500–1,800yr BP) when there was relatively high stream discharge in arid to semi-arid conditions. Williams suggested that the deposition/erosion balance crucially depends on rainfall intensity: deposition is dominant when intensity is high in brief periods of water abundance, while dissection coincides with low stream discharge. A further conclusion, derived from these observations, is that fan aggradation did not coincide with glacial periods, and that sandy deserts were not formed in interglacials in Australia, as has sometimes been supposed.

These varied views reveal at least two important facts: alternations of entrenchment and deposition are common on alluvial fans throughout the arid realm; but their explanation in climatic or other terms varies from one location to another, and possibly even from one time to another. Often the alternations are prompted by quite subtle climatic shifts; and certainly the simple equation of alluviation = pluvial = glacial is not universally tenable. To evaluate all such interpretations, absolute dating of events is now essential.

(f) Chronologies of fan evolution

Whatever the cause or causes of fan entrenchment, chronologies of fan evolution are beginning to emerge from studies in different deserts. Fans can vary in age from only a few years to much of Quaternary time. Their surface characteristics, as discussed above, sometimes reveal evidence of relative age, and cation-ratio dating, radiocarbon dating and K–Ar dating have all been used in attempts to date fan surfaces and deposits. It is unrealistic at present, however, to think in terms of successful global correlations of fan development: the data are too few, and the dating is too controversial. In terms of fan evolution within a specific area, it may be realistic locally and in the short term to think of tendencies towards equilibria, but, in most areas, fans will continue to develop over very long periods of time as the volume of sediment in them rises. Entrenchment may be temporary, permanent or cyclical within this broad timespan of evolution.

12.5 Applied studies of fans

Alluvial fans pose serious hazards to human activities on them (Fig. 12.20; French 1987). There are several

Figure 12.20 An alluvial fan in Sinai, Egypt, with a settlement vulnerable to flooding.

reasons for this. Flows are relatively *rare* and are therefore often unexpected. The flows are difficult to monitor (Bull 1977). When they do occur, flows may arrive without much warning, because of the rapid hydrograph rise and because they may be generated by mountain storms not experienced on the fans themselves. *Within* an individual flow, flow type can vary from water-dominated to debris-dominated movement, a variety that can pose a wide range of different erosion, transport and depositional problems. Flow type can also vary, often unpredictably, *between* flows. Finally, the location of flows can vary unpredictably, within and between flows, because of flow diversion and capture. In short, the nature, timing and location of flows is characteristically unpredictable.

Geomorphological evidence, including that which allows the relative age of fan segments to be determined (Sec. 12.4), can be collated and translated to produce flood-hazard zoning maps, such as that of the Suez region, Egypt (e.g. Cooke et al. 1982). Simi-larly, Rhoads (1986) used hydrogeomorphic analysis to identify five hazard zones on the rapidly urbanizing piedmont plain near Phoenix, Arizona, and to devise appropriate management strategies (Table 12.2). Often process data are absent, and process has to be inferred from the evidence of form and deposit. For example, bedform evidence of old sheetfloods and debris flows can be discerned (e.g. Wells et al. 1985), even if their age cannot, and conclusions may be possible on matters such as their velocity and viscosity. The morphometry of the fan systems can help: for example, catchment area may be a surrogate for discharge (e.g. Cooke et al. 1982). In the USA, the Federal Insurance Administration, as part of the National Flood Insurance Program, has developed a new methodology for defining flood-hazard zones on alluvial fans that recognizes their distinctive properties (Magura & Wood 1980). The method is based on predicting the 100-year flood discharge at the fan apex (using a rainfall–runoff regression model) and then determining

Table 12.2 Flood-hazard zoning on alluvial fans, Phoenix region, Arizona.

Flood-hazard zone	Hydrogeomorphological features	General flood-hazard management strategy for north Scottsdale, Arizona
I	Floodplains along major washes and active portions of alluvial fans; flow confined by valley or banks; entrenched or unentrenched feeders, debris flow and water-dominant flows.	Reserve for flow conveyance as permanent natural open space. Prohibit development until a detailed hydrological analysis of flood characteristics is performed.
II	Alluvial slopes located down slope from fans and pediments, plus some inactive fan segments. Sheetfloods and channel flows.	Maintain low development densities. Require structures to be elevated or flood-proofed. Require structures to be positioned on the highest elevation of a lot if this location does not conflict with other lot requirements. Prohibit the positioning of structures in the path of any clearly identifiable channel, regardless of size. Require stormwater retention basins in subdivision designs.
III	Alluvial plains, runoff not highly channelized because of very low gradients; sheetfloods.	Design street patterns as runoff collectors to improve drainage.
IV	Pediments and inactive alluvial fans. Channels more deeply incised than in II and III. Over-bank flooding.	Maintain moderate development densities. Require structures to be positioned on higher areas separating drainages wherever possible. Require stormwater retention basins in subdivision designs. Restrict development from swales containing large washes.
V	Abandoned alluvial fans; no major fluvial hazards.	High development densities allowable. Require stormwater retention basins in subdivision designs to prevent exacerbation of downslope flooding.

Source: Rhoads (1986).

the depth, average velocity and area of flooding for each type of "reach" or surface (such as "entrenched" and "unentrenched" zones) by taking into account such factors as channel depth, distance from channel and, of course, geomorphological context.

Once the hazard is perceived, a variety of responses may follow (e.g. French 1987). In Los Angeles County, for example, several strategies have been adopted (Cooke 1984). The problem of sediment supply is attacked through debris dams and dredgeable debris basins in feeder channels and catchments. On the fans themselves, efforts are made to make determinate the indeterminacies of flows, especially by concentrating and canalizing drainage. In addition, flood-warning devices may be installed (e.g. Porath & Schick 1974), emergency management plans prepared

and insurance policies purchased (e.g. Cooke 1984).

In many non-urbanized areas of alluvial fans, transport routes (roads, pipelines etc.) commonly are obliged to traverse the relatively gentle fans rather than cross the mountains. Which path should they follow? This question often has to be answered by engineers without runoff data, and with only geomorphological data to guide them. If they cross fan apices, large bridges may be necessary and the flood-erosion hazard is greatest, but at least entrenched channels are relatively stable; in mid-fan areas, the amplitude of relief is lower, but the frequency of channels is higher than in the apical zones; and in the distal areas, many shallow channels commonly require many bridges or culverts.

CHAPTER 13

Pediments and glacis

13.1 Definitions and difficulties

A **pediment** may be simply defined, for the purposes of this review, as a gently sloping surface developed across bedrock that may or may not be thinly veneered with alluvium and/or an *in situ* weathering mantle. Coalescent pediments form **pediplains**.

However, the problem of definition is serious, fraught with difficulties of ambiguity and translation. Even the etymology of the word, with architectural overtones, is confused: the *Oxford English dictionary* suggests that "pediment" may be a corruption of "pyramid", or may be derived from the Latin *pes* (*ped*-) meaning "foot". It is quite clear that only the most general qualitative definition can command wide acceptance; and it is equally clear that the reason for this is that the morphology of pediments is enormously varied. Definitions are almost as numerous as authors, and much controversy arises from discussion in common terms of different phenomena by different investigators. There are even discrepancies between thoughtful, general definitions. Compare, for example, "pediments are erosional surfaces of low relief, partly covered by a veneer of alluvium, that slope away from the base of mountain masses or escarpments in arid and semi-arid environments" (Hadley 1967: p.83); "*les pédiments sont des surfaces rocheuses ou glacis rocheux, possédant une faible pente (inférieure à 5°), généralement localisés au pied de massifs montagneux*" (Mammerickx 1964a: 359); and "A pediment is a terrestrial erosional footslope surface inclined at a low angle and lacking significant relief in all three dimensions. It usually meets the hillslope at an angular nickline, and may be covered by transported material" (Whitaker 1979: 432). Discrepancies between these three definitions relate to "erosional", "veneer of alluvium", the presence of a mountain mass (with "footslope" and "nickline"), slope, relief and climatic environment. The huge range of definitional subtlety is conveniently reviewed by Whitaker (1973, 1979). In French, *pédiments* are surfaces usually developed on resistant rocks (such as granites) that also dominate the mountain catchments (Fig. 13.1); *glacis d'érosion* are also pediments, but *glacis d'érosion en roches tendres* refer to surfaces discordantly eroded across less resistant sedimentary rocks that often differ from more resistant rocks in the mountains (e.g. Dresch 1957, 1982, Tricart et al. 1972, Twidale 1983b; Fig. 13.2). In German, *Pediment* is common, and synonyms include *Fußfläche* and *Verebnungsfläche* (Whitaker 1979); and *glacis d'érosion* is a phrase used in the German literature increasingly to describe "soft rock" surfaces (e.g. Busche & Hagedorn 1980).

For many years, pediments attracted more study and controversy, and sparked the imagination of more geomorphologists, than most other desert landforms. One reason is clear: they are unexpected features, a surprise to the scientist from temperate lands. Witness McGee's (1897) astonishment when he first encountered them through his horses' hooves in the American West. The surprise arises mainly because the pediment is usually cut in the same rock as the adjacent mountain, and there is thus a *poor* relationship between form and structure. Surprise is an excellent spur to speculation. The relevant body of literature is enormous (e.g. Whitaker 1973). Views of pediments vary widely. To some, pediments pose problems; to others, there is nothing magical about them. Birot (1968) regarded the development of pediments as the central problem of arid-land geomorphology; whereas to Lustig (1969b, p D67) "the only real 'pediment problem' is how the reduction or elimination of mountain mass occurs".

A review of the pediment literature reveals several weaknesses. First, few studies present precise data which are in a form that is both verifiable and of use to subsequent investigators: most data are general and,

Figure 13.1 A pediment surface developed across quartz monzonite in the Mojave Desert, California.

Figure 13.2 Glacis d'érosion, a gravel-mantled pediment surface cut across steeply dipping sedimentary rocks in the Dead Sea Basin, Israel.

at times, ambiguous. For example: "pediments in area x slope at between six and one degrees" – do these figures represent the range of slopes for single pediments or the range within which all pediment slopes occur?

Secondly, the explanations of pediments have, until recently, made gross assumptions about processes. For example, sheetflooding, a common *deus ex machina*, has been observed only rarely on pediment surfaces; backweathering of mountain fronts is more honoured in the deduction than in the observation; and lateral stream planation is not a phenomenon com-

189

monly seen on pediments. Coupled with deductions concerning process is the common cause-and-effect error arising from relations between a deduced process and the visible landform. Clearly, for instance, sheetflooding cannot produce a planar surface, because a planar surface is necessary for sheetflooding to occur. Again, widespread weathering and occasional removal of weathered debris is unlikely to produce a pediment surface; it is more likely to maintain, probably at a lower level, a pre-existing form. In short, there is frequently confusion between *pediment-forming* and *pediment-modifying* processes. And yet a more justifiable assumption concerning processes is frequently ignored: processes have almost certainly changed in many areas – in nature, magnitude, frequency, etc. – as a result of climatic changes during the course of pediment evolution. A related assumption is that pediments are the product of the arid environment in which they occur: evidence is accumulating in some areas that they may be ancient surfaces inherited from very different conditions. The lack of any satisfactory dating of pediment evolution is a major difficulty.

Thirdly, in the past, pediments – unlike alluvial fans – have been studied in isolation from their catchments. This is much less true today, especially because of monitoring studies of processes that cross the mountain/piedmont divide (e.g. Sec. 13.4).

13.2 Distribution of pediments

As gently sloping surfaces developed on bedrock, pediments are clearly a worldwide, azonal phenomenon; but they certainly exist throughout the arid realm (Fig. 13.3). Indeed, King (1953) maintained that most of the world's landforms are transformed into either pediments or their collective counterparts, pediplains; and the innocent might be forgiven for recognizing similarities between pediplains and peneplains (Dury & Langford-Smith 1964). Pediments have also been recognized within the stratigraphic column (e.g. Williams 1967). Some have argued that the occurrence of pediments is related to particular climatic conditions. In the southwestern USA, for instance, Corbel (1963) recognized three geomorphological zones: (i) where mechanical weathering is predominant, (ii) where chemical weathering is most important, and (iii) an intermediate zone where both occur. It is in the latter zone, he argued, where climate is "regularly irregular" between the zones of winter snow and summer floods, that pediments are

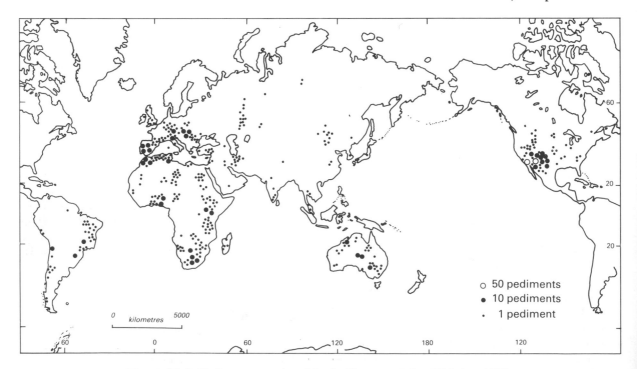

Figure 13.3 Pediments mentioned in the literature (after Whitaker 1973).

formed. Such assertions are interesting but unproven.

The distribution of pediments *within* deserts is a matter of considerable interest. Three locations are characteristic: (i) at an intermediate location between watershed and base level, and usually between a mountain and an alluvial plain (such pediments are characteristic of basin–range country, and are called, *inter alia*, **apron pediments**); (ii) where apron pediments coalesce without being surmounted by mountains or inselbergs, a **pediment dome** is found; and (iii) adjacent to major base levels, such as exoreic streams or lakes, **terrace pediments** may occur.

Twidale (e.g. 1983b), drawing upon observations of pediments throughout the world, has offered a morphogenetic classification into three main types: *mantled* and *rock* pediments on crystalline outcrops, and *covered* pediments on sedimentary strata. Mantled pediments, he argued, carry a veneer of weathered debris on bedrock, and their form is shaped by weathering of bedrock and wash removal of the debris (c.f. Mabbutt 1977: 83). Rock pediments are exposed weathering fronts, developed by mantle-controlled planation (see below) and are therefore often the successors to mantled pediments. Covered pediments are those surfaces developed discordantly across sedimentary strata, having a veneer of coarse debris (e.g. Fig. 13.2). These strata may be less resistant than those in the catchments, but certainly some of them arise from scarp retreat. (They approximately equate with the glacis d'érosion en roches tendres and *glacis d'épandage* of some French scientists.)

However, pediments by no means occur throughout all desert basins. For example, Lustig (1969a) identified a singular distinction between the distribution of pediments and alluvial fans in basin–range country. He argued that if pediments originate by mountain-mass reduction, "the average mountain mass should be significantly smaller in certain areas that are characterized by the existence of pediments" (Lustig 1969a: 67). His morphometric analysis of basin-range parameters in the western USA showed that areas occupied by smaller ranges (in terms of width, height, relief, area and volume), such as southwestern Arizona and southeastern California, are indeed those characterized by pediments. By contrast, in areas with larger ranges, alluvial fans are common and pediments are rare. In short, it could be that the distribution of pediments and alluvial fans is in part determined in the basin–range context by morphometric relationships which, in turn, may be a reflection of erosional and tectonic history.

Indeed, pediments and alluvial fans, far from being mutually exclusive, are commonly closely related, the major difference being that the former reflect predominantly *erosional* environments on the piedmont plains, whereas the latter reflect predominantly *depositional* environments. Alluvial fans often rest on bedrock surfaces, and pediments may have alluvial veneers. When does a veneer become a fan? Bull (1977) suggested that fans may be distinguished from pediments where the alluvium's thickness is more than one hundredth of the *length* of the landform. And it is quite possible for an alluvial fan to be planated, perhaps as a result of lowered base level, to create a pediment (or at least a glacis), and for the pediment to be buried by a subsequent fan deposit.

Although there is little doubt that many pediments are erosional features, because they are cut discordantly across structures and rocks of different lithologies, some believe that pediments are preferentially developed on particular rock types. Coarsely crystalline rocks, and notably granite, are certainly commonly associated with pediments (e.g. Warnke 1969, Cooke 1970a, Akagi 1972, Mensching 1978, Twidale 1983b). In a detailed study of the Sherman erosion surface of Colorado and Wyoming, Eggler et al. (1969) convincingly demonstrated that the surface is associated exclusively with a granite facies which disintegrates rapidly and totally to gruss, as the result of the expansive effects of altering biotites; other granites and crystalline rocks give rise to parkland–tor and more rugged topography. That is to say, the surface is closely related to bedrock lithology, and in this case to a particular type of granite. However, there is no doubt that pediments occur on rocks of other lithologies. In southern Arizona, for instance, they are found on alluvium, indurated sedimentary deposits, metamorphic, and several contrasted types of intrusive and extrusive igneous rocks; and elsewhere (especially in the Sahara) they have been described on alternating "hard" and "soft" rocks or generally weaker rocks (the *glacis d'érosion* of Birot & Dresch 1966), and on sandstone, schist, mudstone, shale and limestone (e.g. Mensching 1970, Barth 1976, Weise 1978, Kar 1984).

13.3 The forms of pediments

Unlike alluvial fans, pediments are not features that can easily be considered as parts of clearly circumscribed, functioning systems. Often the surface occurs in more than one drainage basin. In addition, it may be impossible to assume that present drainage networks on a pediment were associated with its formation. In other instances, there may be several isolated pediments in what is essentially a uniform plain within a single drainage basin. Elsewhere, the pediment may legitimately be considered as a functional component of a drainage basin. Nevertheless, for the purposes of precise description and exploring explanations, it is desirable and useful to distinguish a pediment system. In the context of the basin–range topography of the Mojave Desert, Cooke (1970a) suggested that a beginning might be made by using a unit (called a pediment association) which includes the pediment, the mountain area tributary to it, and the area of alluvial plain to which it is tributary. The unit is not entirely satisfactory, even in this context, but it is useful, especially for evaluating *profile* development and where the surface is a product of the drainage system in which it lies.

13.3.1 Gross morphometry of pediments

Analysis of quantitative pediment data is confined to a few articles (e.g. Corbel 1963, Mammerickx 1964b, Dury et al. 1967, Cooke 1970a). Such analyses depend in the first instance on the definition of variables; in the following account the definitions used are those of the authors concerned. One scheme of definition which may be of general value is shown in Figure 4.2. Most of the working definitions are explained by Cooke (1970a), but it should be emphasized that a critical point of definition, which can affect most calculations, is the lower limit of the pediment on the piedmont plain. Cooke defined it as the line at which the alluvial cover becomes continuous.

(a) Descriptive data

Pediments vary enormously in area. Cooke found in the western Mojave Desert, California, a range of values from 0.78 to 35.1km^2, and a mean value of 5.98km^2. Pediments in the neighbouring Sonoran Desert, Arizona, are generally larger: Corbel (1963) cited a range from 2.5 to 620km^2, and a mean value of 100km^2. Similar variation is apparent in other deserts: in the Atacama Desert, Chile, for instance, both very small and very large pediments have been described (e.g. Cooke & Mortimer 1971, Clark et al. 1967a); and in Africa and Australia, pediments may be very extensive indeed (e.g. King 1953).

Also of importance is the fact that the proportion of an area occupied by pediments varies greatly. To

Table 13.1a Comparison of landform areas in part of the Mojave Desert and parts of Arizona, USA.

	Average percentages of landform areas	
	Western Mojave Desert	Parts of Arizona
Mountain	28.9	10
Piedmont	71.1	90
Pediment	6.7	30
Pédiment zone couverte	–	10
Alluvial plain	64.4	50

Source: Cooke (1970a), Corbel (1963).

contrast the Mojave and Sonoran deserts again (Table 13.1a), it is clear that the proportion of the area in pediment is greater in Arizona than in California. The ratio of mountain area to plain area also varies. In the western Mojave Desert, the ratio is 0.36, and 9.4% of the plains' area is composed of pediments; McGee (1897) suggested that the ratio in southern Arizona is 0.25, and that 40–50% of the plains' area is composed of pediment.

Slope is an important feature of pediments because it is more susceptible to analytical evaluation than most pediment properties (see below), and it is also variable, usually within the range 0°–11°. Table 13.1b shows a sample of pediment-slope data from southwestern North America. The data are derived from study of several pediments in each of the areas mentioned. Slope also usually varies along the profile of a single pediment: generally, the profile is concave upwards and slope varies within the range 0°–11°; sometimes, however, profiles may be rectilinear (e.g. Cooke & Mason 1973) or even convex.

(b) Analysis of morphometric data

Analysis of morphometric data is only in an inchoative stage, and here commentary is confined to a consideration of pediment-slope data through the examination of several simple explanatory hypotheses. Unless otherwise stated, pediment slope is defined as the slope of the line joining the highest and lowest points of a pediment profile along the line of pediment association length (Cooke 1970a).

It can be hypothesized that pediment slopes in an area are positively correlated with the relief and length dimensions of the pediment associations in which they occur. This hypothesis is substantiated in the western Mojave Desert, where the correlation of the pediment

Table 13.1b Pediment slope data from arid and semi-arid areas of southwestern North America.

Area	Range of longitudinal pediment slope (°)	Mean pediment slope (°)	Reference
Western Mojave Desert	30′–11°	2°35′	Cooke (1970a)
Papago Country, Arizona	30′–2°12′	–	Bryan (1925: 95)
Ajo region, Arizona	48′–3°18′	1°–1°40′	Gilluly (1937: 332)
Southeast Arizona	1°30′–4°(?)	–	Tuan (1959)
10 pediments in Arizona	30′–4°	2°	Corbel (1963: 53–4)
Mojave and Sonoran deserts	1°30′–5°40′ (means range) 30′–7°	–	Mammerickx (1964b: 421), Balchin & Pye (1955: 171)
Ruby–East Humboldt Mountains, Nevada	west flank: 48′–3°48′ east flank: 4°54′–5°25′	–	Sharp (1940: 357)
Southwest USA	30′–7°	2°30′	Blackwelder (1931b: 137)
	6′–2°42′	–	Bryan (1936: 770)
	7′–8°5′	2°11′–3°15′	Leopold et al. (1964: 494)
	15′–10°	–	Tator (1953)
Cape region, Baja, California	1°–5°	–	Hammond (1954: 79)

slope with the pediment association relief:length ratio gave an r value of 0.4416 ($P = 0.001$). In this region, pediment association relief results largely from tectonic movements and, as relief and length of pediment association are positively correlated ($r = 0.6728; 0.001 > P$), it seems reasonable that the pediment association relief:length ratio may be considered as an indirect expression of the effects of tectonic movements.

To explore this interpretation further, a second hypothesis is proposed: the slope of pediments associated with faults is significantly steeper than the slope of pediments not associated with faults. This hypothesis is confirmed by data from the western Mojave Desert (Table 13.1c). Thus, although the evidence is not conclusive, it does appear that, in one area at least, tectonic activity may be a significant factor in determining the general slope of pediments.

Table 13.1c Relations of pediment slope and faults.

Pediments	No.	Mean slope	Range	Standard deviation
1. Associated with faults	31	2°55′	39′–5°23′	1°12′
2. Not associated with faults	22	2°10′	31′–3°50′	48′

$t = 2.7$; significance level ≈ 0.01.
Source: Cooke (1970a).

On the assumption that pediment slope is a function of hydraulic variables in a fluvial system (an assumption open to debate; see below), additional hypotheses may be formulated. First, it might be hypothesized that pediment slope is inversely related to catchment basin area, on the deductive grounds that pediment slope should be inversely related to water discharge, and that discharge should be proportional to catchment basin area. This hypothesis was not substantiated in the western Mojave Desert by Cooke (1970a; $r = -0.2335; P > 0.1$), or in Arizona and California by Mammerickx (1964b). (For an explanation of this, see Ch. 15).

Secondly, it might seem likely that pediment slope will vary along a single profile according to the nature, and especially the size, of debris transported across it. At Middle Pinnacle, near Broken Hill, Australia, Dury (1966b) demonstrated that two surveyed pediment profiles were adequately described by logarithmic curves, that particle size decreased in an orderly fashion down slope, and that rates of particle-size decrease were less than the rates of slope decline. Cooke & Reeves (1972) came to similar and additional conclusions from pediment profile studies in California (Ch. 10.7). They described differences in the nature of slope changes between mountain front and pediment on different lithologies. In quartz monzonite areas there is often a marked contrast between debris sizes on mountain fronts and on pediments, which may account for the distinct break of slope between the two landforms in these areas. On other rock types, where the particle-size contrast is less marked, the change of slope is less abrupt. Akagi (1980), in a similar study in Arizona, also showed a strong relationship between particle size and slope angle, but he demonstrated that rolling boulders confused the relationships and that the relationships varied significantly with rock type.

If pediment slope is in part related to debris-size, the broader possibility exists that pediment slope is significantly related to rock type. Despite some assertions to this effect, the idea remains to be rigorously tested, and Mammerickx (1964b) concluded that bedrock lithology is not a decisive factor in determining pediment slope.

Finally, it might be hypothesized that pediment slope decreases with pediment length, in the deductive belief that, as a pediment extends simultaneously in the horizontal and vertical planes, its relief:length ratio will decline. At present this hypothesis does not appear to be justified: both Mammerickx (1964b) and Cooke (1970a) identified only weak relationships between pediment slope and pediment length in the southwestern USA. Dury et al. (1967), in a short study on pediments in New South Wales, demonstrated that pediment steepness fails to correlate significantly with pediment length.

Mabbutt (1977) suggested that pediment gradient is lower (and the piedmont angle is accentuated) in areas of greater aridity. This seems an unlikely possibility, despite some support for it from southern Africa and elsewhere, because very sharp junctions are to be found in semi-arid areas (e.g. southern California) and modest slopes are to be found in extreme aridity (e.g. the Atacama Desert): lithology is certainly a fundamental control; climate may be.

13.3.2 The piedmont angle

The upslope margin of pediments is frequently clearly demarcated as the boundary between mountain and plain. In plan it may be linear, or crenulate with **embayments** into the mountains where major lines of drainage emerge; where two embayments meet across a divide, breaking up the continuity of a mountain mass, a **pediment pass** is formed (e.g. Warnke 1969). The transverse profile of the boundary may be relatively flat and horizontal or, where there are embayments, undulating.

The boundary often separates slopes of very different inclinations; the angle produced by the two slopes is called the **piedmont angle** (Fig. 13.4). (Kesel 1973, defined the angle as the maximum difference between two consecutive slope readings from the piedmont to the mountain slope). Two comments are appropriate on the form of the boundary and the piedmont angle. First, the piedmont angle may be as little as 90°, but it is usually much greater.

Figure 13.4 Granite inselbergs and pediments, central Namibia, with distinct piedmont angles.

The abruptness of the junction has been exaggerated, often because it appears to be more abrupt than it actually is when viewed from a distance across the piedmont plain. In many cases, the longitudinal profile forms a continuous logarithmic curve, uninterrupted by a break of slope (e.g. Cooke & Reeves 1972). Secondly, there is definitely no simple single, universally applicable explanation of the feature, as some authors have suggested. The explanation of the origin and development of piedmont angle (or "nickline" of Whitaker) is central to discussions of topographic evolution and surface processes in deserts.

13.3.3 Details of pediment topography
The contention that pediment surfaces are enormously variable can be substantiated by a brief discussion of

Figure 13.5 A trench profile in a quartz monzonite pediment in the Mojave Desert, California. The profile includes an alluvial veneer on deeply weathered bedrock that is laced with calcium carbonate. A and reddened B soil horizons, common in the region, are absent; they may have been eroded from the pediment surface.

several features, some of them of diagnostic value. Although many published accounts may give a contrary impression, a pediment which is a clean, smooth bedrock surface is rare indeed. In most cases, the pediment is a complex surface, comprising patches of bedrock and alluvium, in places capped by weathering and soil profiles, punctuated by inselbergs, and scored by a network of drainage channels.

The patchiness of bedrock and alluvium may result partly because some material is temporarily in transit across the surface and partly from the incomplete removal of formerly more extensive alluvial covers. Inselbergs vary in number and dimensions: often they appear to become larger and more numerous towards the upper limit of the surface; and frequently they occur either on more resistant bedrock or on major interfluves.

Rather surprisingly, little attention has been given to the occurrence of weathering and soil profiles on pediments, although their evidence may be critical to the interpretation of pediment development. On the Apple Valley pediment, California, Cooke & Mason (1973) have described a soil profile which has been eroded in places and buried in others, and the weathering profile in quartz monzonite beneath the bedrock surface (Fig. 13.5). Oberlander (1974) has considered the extensive evidence of Tertiary saprolite formation in the same region. Mabbutt (1966a) sketched the weathering features on granite and schist pediments in central Australia and elsewhere; and Twidale (e.g. 1964) has recorded evidence of deep weathering at the upslope margins of pediments in South Australia. The relationships between pediments and zones of supergene enrichment of copper minerals in the southern Atacama Desert have also been studied (e.g. Clark et al. 1967a).

Another important yet neglected feature is the presence of cut-and-fill features on pediments. Channels 1–3m deep and now filled with alluvium have been described by Cooke & Mason (1973) in the Apple Valley pediment, California; Dury (1966b) recorded a buried channel at Middle Pinnacle, New South Wales; and there are many examples in southern Arizona and northern Saudi Arabia (e.g. Fig. 13.6). The presence of buried channels indicates that the relations between erosion and sedimentation in the pediment zone have changed during the period of pediment development. The filling of channels and other depressions in bedrock by alluvium is commonly responsible for the general smoothness of many pediments. In some cases, it seems likely that what

Figure 13.6 A channel cut into a pediment surface and subsequently filled, near Tabuk, Saudi Arabia.

may appear to be **dissected pediments** are actually undulating surfaces that have always been so – that is, they were "born dissected" (Gilluly 1937; Fig. 13.7). In places, the pediment surface may be fan form (a so-called **rock fan**), the erosional equivalent, perhaps, of an alluvial fan.

Closely related to buried channels are pediment drainage nets. These, too, have rarely been con-

sidered. There are three common types:

(i) Channels occurring in the upper part of the piedmont plain, which commonly form a distributary system and die out lower down the surface. Such channels often straddle the piedmont angle, and they are deepest at intermediate positions on their longitudinal profiles.

(ii) Channels occurring on the lower part of the piedmont plain, which are generally deepest at the lowest point in their longitudinal profiles, and usually form part of a drainage system that has been rejuvenated on one or more occasions by lowering of base level. Such systems may cover the whole pediment. When drainage in this type of net is rejuvenated it often leads to the destruction of the pediment surface. For instance, as Figure 13.8 shows, drainage channels on a pediment near Hodge, California, are in general accord with the slope of the pediment, but a number of low-order channels follow strike directions normal to the pediment slope and thus disrupt its longitudinal continuity.

(iii) On relatively undissected surfaces, often between areas characterized by types (i) and (ii), drainage nets may consist of complex and frequently changing patterns of shallow rills.

These drainage nets are similar in pattern and location to those on alluvial fans, and they may

Figure 13.7 A pediplain surface in the Norte Grande, Chile: this topography has the characteristics of a "born dissected" pediment.

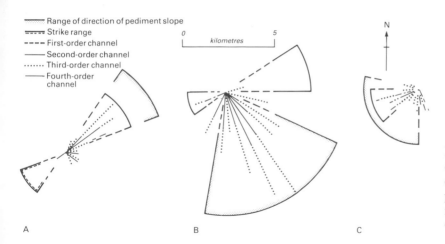

Range of direction of pediment slope
Strike range
First-order channel
Second-order channel
Third-order channel
Fourth-order channel

0   kilometres   5

N

A   B   C

Figure 13.8 The orientation of drainage channels, the range of orientation of rock strike, and pediment slope on three pediments near Hodge, Mojave Desert, California.

perhaps be explained in similar terms. Type (i) is probably generated by drainage in the catchment area behind the pediment, type (ii) may result from runoff on the pediment surface itself, and type (iii) probably arises from rillflow, perhaps characteristic of declining sheetfloods, in the intermediate zone. Drainage incision may reflect adjustment to climatic or tectonic changes, or changes in the nature of water flow within the system. Such changes could have accompanied pediment formation, or they could be younger and lead to pediment destruction.

Where bedrock is exposed at the surface, the latter is often discordant to the structure of the former. For example, Mabbutt (1966a) described the longitudinal profile of a schist pediment in which the surface *generally* slopes at right angles to the strike; *in detail*, the surface comprises discontinuous ridges of schist parallel to the strike and up to 0.6m high, separated by flatter, debris-mantled zones. In contrast, surface and structure occasionally accord where, for example, the surface coincides with semi-horizontal joint planes in quartz monzonite. Down slope a pediment often disappears beneath an alluvial cover, where it is known as a *suballuvial bench*. There has been much speculation on the nature of this feature. One persistent deduction is that it ought to be convex (e.g. Lawson 1915, Davis 1933; c.f. Fig. 10.24). The limited available evidence does not confirm this idea. The suballuvial bench of the Apple Valley pediment, California, is a continuation of the rectilinear exposed surface, and its slope is only 2′ steeper that of the overlying alluvium. In a gravimetric and seismic survey of the suballuvial bench of Cima Dome, California, Sharp (1957) was unable to identify any convincing convexity of the profile; and a seismic

refraction survey of suballuvial profiles at Middle Pinnacle, Australia, did not reveal any convexity in the suballuvial bench (Dury 1966b).

13.4   Pediment processes

The occurrence of pediments in so many different environments is testimony enough that they are not exclusively desert features and that the range of processes working on them, even in deserts, is likely to be large. Without doubt it is essential to recognize that essentially similar pediment forms may be produced by different processes in different areas. The study of pediment processes falls into three groups: a small body of monitoring studies; a larger group of general observations; and a very large body of inferences drawn from the evidence of landforms and deposits.

13.4.1   Monitoring studies
Empirical observations of pediment processes comprise, first, studies of erosion and sedimentation on small-scale analogues of the mountain–plain system, and secondly, several attempts to record processes operating in full-scale systems.

(a) Small-scale analogues
The Badlands National Monument, South Dakota, is an outstanding area of rapidly changing badland-slope and miniature-pediment topography, developed on Oligocene clays, shales, sandstones and volcanic ashes of the Chadron and Brule formations. Smith (1958) surveyed landforms on the Brule Formation in detail and concluded that the badland slopes retreat by dis-

aggregation, slumping and especially sheet and rill erosion; that the pediments are extended by spreading sheetwash close to rill mouths at the badland-slope/pediment junction, and that sheetwash is the principal process of transporting debris across the pediments. As sheetwash declines following a rainstorm, it changes from continuous layer flow into subdivided rillwash. Detailed though Smith's observations are, his conclusions are mainly inferences drawn from inspection of form and deposit.

Fortunately, Schumm (1956, 1962) has monitored changes in the same area by means of 15 reinforcing rods (8.4mm in diameter, 46cm long) inserted into the ground until they were flush with the surface. He observed exposure of these rods and the retreat of badland slopes between 1954 and 1961. Two of the

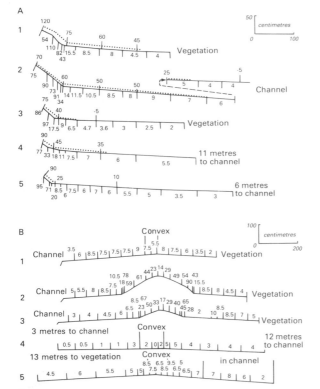

Figure 13.9 Slope profiles in Badlands National Monument, Dakota. (A) Profiles in May 1961. Numbers below profiles are percentage slopes; those above profiles are erosion depths (in mm) at stake positions; dashed lines represent the profile in July 1953. (B) Profiles illustrating headward extension and coalescence of a miniature pediment; the profiles were measured at the same time in different locations. Numbers above profiles give percentage slopes (after Schumm 1962).

stakes were buried; the remainder were exposed by 10–75mm. Erosion was greatest on the badland slopes; and on the pediment it decreased away from the badland-slope/pediment junction (Fig. 13.9). Two important facts are clear. First, the badland slopes have retreated by 6–12cm, leaving behind extensions of the pediments which are steeper than the older parts of the pediments. Secondly, the pediment surfaces are regraded as they are extended. Smith's view that sheetwash is important on the pediment is confirmed by Schumm, who showed that it both regrades and transports debris across the surface. Schumm elaborated on one important point: *the transition from steep badland slope to gently sloping pediment is not accompanied by deposition at the boundary*. Sediment is removed across the pediment to a zone of deposition farther down slope; and at the same time the pediment is regraded. The reason for this is almost certainly that the velocity of flow is of the same order of magnitude on both badland slope and pediment, despite their different inclinations. This argument may be illustrated by reference to the Manning equation for mean velocity of flow which states

$$V = \frac{1.49D^{\frac{2}{3}}S^{\frac{1}{2}}}{n} \qquad (13.1)$$

where V is the mean velocity of flow in feet per second, D is the mean depth of flow (or hydraulic radius), S is the slope, and n is a roughness coefficient. If it is assumed that the hydraulic radius is equal to the mean depth of flow, and that the depth of flow is the same on hillslope and pediment, D can be dropped from the equation. Mean relative velocity (V_r) is defined as

$$V_r = \frac{1.49S^{\frac{1}{2}}}{n} \qquad (13.2)$$

Then, if S is the 0.8 on the hillslope, and 0.1 on the pediment, and $n = 0.06$ on the hillslope and 0.02 on the pediment, V_r would be 24 on the pediment and 23 on the steep, rough hillslope, i.e. a decrease in roughness on the pediment apparently compensates for the decrease in slope. Furthermore, D is likely to be greater on the pediment, and thus pediment flow velocities would be proportionately greater. This slope-surface roughness contrast is one that is common in full-scale pediment systems, especially those developed on coarse-grained igneous rocks, and Schumm's argument may apply to them. For example, in the western Mojave Desert (using Cooke's data), quartz monzonite mountain fronts have a mean slope of 28°40′ and a roughness of, say, 0.06; quartz monzonite pediments have a slope of 2°31′, and a roughness of, say, 0.02. Given these facts, V_r on mountain fronts is approximately 18, and that on pediments 15; the relative velocities of flow on the two landforms are probably very similar.

(b) Monitoring full-scale systems
Pediment surfaces are involved in several of the monitoring studies described in Chapter 10.20. That by Kirkby & Kirkby (1976; Ch. 10.8.2) is particularly important in the context of the piedmont angle and the contemporary evolution of pediments. They viewed the piedmont angle in semi-arid southern Arizona not as a break, but as the local concentration of what is a widespread slope concavity that ranges in width from 20–80m for granite and 250–750m for schists and basalts. It is, they suggested, a **transition zone of process** between mountain slopes dominated by gravity-controlled talus processes that require no hydraulic force to initiate movement, and pediment slopes where wash processes are dominant and the gravitational component is insignificant. On the mountain slopes, release of material is determined by weathering rates; on the wash slope, removal is limited by the transportation capacity of the wash. In general, the slope is concave and related to particle size. It is the *combination* of gravitational and wash processes that determines the transitional slope. Where only coarse and fine debris are available, as on granite slopes, the change of slope is most abrupt and the transition shortest, because there is no need for intermediate slopes to accommodate intermediate-size debris. The basic concavity can be embellished by other processes, such as subsurface weathering of the boundary. Monitoring of pediment runoff at the Sacaton Mountain sites showed that surface wash in shallow braided channels was much more effective than unchannelled wash, and that the pediment surface was being *lowered* overall in the 1.5yr period of measurement.

13.4.2 General observations of processes
Since the oft-quoted and vivid description by McGee (1897) of a sheetflood (or of something which has come to be interpreted as a sheetflood), the phenomenon has been considered as an important process in developing pediments. Several comments are appropriate here. First, there certainly are occasions when water may flow across relatively undissected desert surfaces in a sheet. Secondly, sheetflow of flood proportions is extremely unusual, and interpretation of early travellers' graphic descriptions of runoff in terms of sheetflooding has rightly been questioned. Thirdly, whatever the effect of a sheetflow, it cannot create the surface, for the surface is a prerequisite for its establishment. It can at best merely modify the surface by removing debris and perhaps by undercutting steep slopes. Fourthly, because records of sheetflow are rare, knowledge of their effects is limited. One of the few informative studies was made by Rahn (1967), who demonstrated from observations of four floods in southwestern Arizona that runoff from heavy storms on the bajada may be of a sheetflood character, whereas runoff from mountain storms is more commonly in the form of a streamflood, and that the flow is usually supercritical.

That streamflow occurs on pediments is a fact more commonly verified by observation, although measurements of discharge are nevertheless scarce. Anastomosing drainage patterns on undissected surfaces suggest streamflow, although it is possible that many of these ephemeral patterns result from the translation of sheetflow to streamflow as a runoff episode closes. A problem exists in the interpretation of channel data with respect to pediment formation, even if such data were available, because it is not clear in most instances whether the channel system created, or is creating, the pediment, or whether the system was established on the pediment after its formation.

The consensus view is certainly that the channel system is at least related to pediment development. In particular, many authors place emphasis on the possibility that a stream flowing across a pediment may swing from side to side and progressively erode the surface by lateral planation. Proponents of this process are many: Gilbert (1875, 1877) is credited with its discovery; Johnson (e.g. 1932) with its promotion; and Sharp (1940), Rahn (1966) and Mackin (1970) with its reinvocation. (These examples are taken from the American literature; the idea is encountered in other languages.) Evidence used to identify lateral planation on pediment margins is circumstantial and it includes that of rock fans, and that of inselberg slopes which are steeper where channels impinge against mountain/piedmont junctions (Rahn 1966). In addition, Sharp (1940) suggested that evidence for lateral planation in the Ruby–East-Humboldt Range, Nevada includes

"The facts that terraces cut largely by lateral planation along adjacent streams coalesce to form partial pediments; that the largest and smoothest areas of pediment are adjacent to the streams and are restricted in areas of hard rocks . . . Of the several features cited, the formation of partial pediments by coalescing terraces, obviously cut by lateral planation as shown by meander scars and related features, is considered the most convincing argument favouring the lateral planation theory." (Sharp 1940: 363).

It is well known that lateral planation often occurs in the floodplains of perennial rivers, and it could be that the process is most often responsible for fashioning pediments where perennial flows cross piedmont plains from mountain areas. Or perhaps the process is most effective where relatively unresistant rocks are involved (e.g. Denny 1967, Sharp 1940). On the other hand, the effectiveness, and indeed the presence, of the process have been denied by some (e.g. Lustig 1969a); empirical observations have not confirmed its efficacy (e.g. Schumm 1962); and some evidence cited in its favour in basin–range situations is, to say the least, unusual. As Lustig (1969a: D65–D66) observed: "No doubt, channels will migrate with time on these (pediment) surfaces, and some erosion will therefore be accomplished, but the part of any pediment that abuts the mountain front cannot be explained in this manner. The hypothesis virtually requires that streams emerge from a given mountain range and, on occasion, turn sharply to one side or the other to "trim back" the mountain front in interfluvial areas. Such stream paths, nearly perpendicular to a sloping surface, would defy the laws of gravity and have not been observed except in those areas where drastic tilting has occurred". The requirement that streams should "trim back the mountain front" for lateral planation to be important is removed by Parsons & Abrahams' (1984) notion of **divide removal**, whereby divides are eliminated and the mountain front made to retreat by processes (including lateral planation) in embayments – that is along the valley sides of streams before they emerge from the embayment onto the piedmont plain. This notion may be crucial to understanding the evolution of some piedmonts.

In some areas, and notably in Australia, emphasis has been placed on the rôle of weathering processes in fashioning pediments (e.g. Mabbutt 1966a, Twidale 1967). Mabbutt (1966a), in describing pediments of central Australia, demonstrated that granite pediments are characterized by weathering mantles up to 1.7m thick beneath thin and mobile alluvial covers, and that

these mantles are more stable on middle-pediment slopes, where the degrees of clay weathering and horizon differentiation are greater. The rock face in contact with the mantle is more uneven than the surface, indicating differential weathering at depth and, perhaps, reduction of surface irregularities by alluviation. Cooke & Mason (1973) and Moss (1977) identified a similar weathering of monzonite beneath a thin alluvial mantle on pediments in California and Arizona. In addition, subsoil weathering is not only more effective than surface weathering but may also be more important in trimming back the hillslope and extending the pediment. In places on schist pediments, surface irregularities reflecting lithological differences are associated with an uneven, chemically weathered suballuvial floor, but in contrast to granite pediments, rock breakdown is more rapid at the surface. "Ground-level trimming" is the name given by Bryan (1925) and adopted by Mabbutt for this process in which the mantle imparts a general levelling to an otherwise uneven and heterogeneous bedrock surface, directly through its base level control of ground-level sapping, and indirectly through its provision of an even surface of attack upon weathering outcrops by rainwash and sheetflow (Mabbutt 1966a: 89).

Subsurface weathering is likely to be particularly pronounced at the piedmont junction because of the natural concentration of water there. Reasons for this concentration include the facts that runoff tends to disperse as it reaches the junction, and percolation may thus be increased (especially if a debris veneer is present), and that more runoff-producing rains occur in the mountains, and the runoff often travels only a short distance on to the plains (Twidale 1967). Twidale (e.g. 1987) identified zones of especially intense weathering at piedmont junctions and related these to a sequence of pediment-surface development and to the development of flares on adjacent mountain fronts (Fig. 13.10). However, such zones are by no means a universal phenomenon. Surface weathering pits, created without any mantle, may be significant in lowering pediment surfaces, especially if they are sufficiently abundant and the mountain front is relatively stable, as appears to be the case in granite pediment areas of the semi-arid Mongolian Ikh Naart Plateau (e.g. Dzulynski & Kotarba 1979).

Two further comments arise from a recognition of the presence of near-surface weathering phenomena. First, the relative rates of weathering and fluvial erosion processes require careful examination. Does the presence of a soil profile indicate a stable surface,

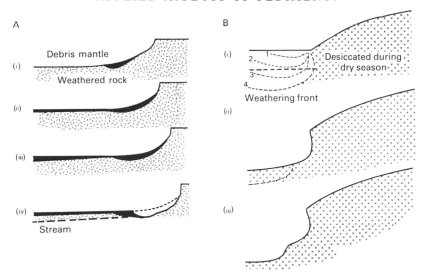

Figure 13.10 (A) Pedimentation, scarp retreat and preferential weathering at the junction of a pediment and a scarp (after Twidale 1968). (B) Stages in the development of flared slopes at the piedmont junction (after Twidale 1982b).

or a balance between rates of weathering and rates of surface lowering? Secondly, it is important to know the relationship between soil-profile development and pediment extension. For example, if a pediment grows at the expense of a mountain mass, the surface is clearly diachronous, being youngest at the mountain front. Is this fact reflected in the pattern of pediment soils, or are the weathering features much younger than the pediment?

Corbel (1963) concluded from a study of pediments in Arizona that chemical processes are relatively, and often absolutely, more important in the piedmont zone than in other areas, especially where a pediment is thinly covered with alluvium. He estimated, on the basis of rather sparse data, that in Arizona the rate of pediment erosion is $36m^3$ km^{-2} yr^{-1} (of which 70% is chemical), and in California it is said to be $6m^3$ km^2 yr^{-1} (of which 85% is chemical). These estimates emphasize the potential importance of chemical processes in pediment development. Taken together with the comments of others mentioned above, it would seem that such processes deserve greater attention.

13.5 Applied aspects of pediments

If pediments surprise geomorphologists, they also surprise developers who expect to excavate in alluvium on the piedmont plain. The surprise can become expensive, if trenching, for example, requires extensive blasting. In the Apple Valley Ranchos development in the Mojave Desert (Cooke & Mason 1973), most of the service trenches required blasting; and the same is true of several subdivisions in Scottsdale, Arizona. In Al-Ain (UAE/Oman) limestone pediments underlay recent urban projects and also required expensive excavation. On the other hand, pediments underlain by sound bedrock provide strong foundations for development. In addition, as Graf (1988a) pointed out, pediments, like fans, suffer problems of flooding, sedimentation and channel mobility.

CHAPTER 14

Base-level plains: playas and sabkhas

14.1 Introduction

14.1.1 Definitions and terminology

The lowest areas within enclosed desert drainage basins are often marked by almost horizontal, largely vegetation-free surfaces of fine-grained sediments. Similar surfaces may also occur at intermediate locations in a desert drainage network and some may once have been connected to the sea (e.g. in the Wahiba region, Oman). Such base-level plains are common and distinctive features. They have acquired many different local names. General terms include *nor* (Mongolia), **pan** (South Africa), *sebkha* (North Africa and the Middle East), and **dry lake** and **playa** (North America). The variety of local terminology is evident in the Sahara, where terms such as *sebkra* (*sebjet*, *sebchet*, *sebkha*, *sabkhah*, *sebcha*), *zahrez*, *chott* and *garaet* are used for similar features in different areas (Coque 1962, Smith 1969). Here the term **playa** is used as a general term for base-level plains in desert drainage basins (Fig. 14.1). Different types of playa have acquired specific names. A clay–silt playa, for instance, is known as a **clay pan** in Australia, a *takyr* in Russia, a *khabra* in Saudi Arabia, and a *qu* in Jordan; and a saline playa is called a *kavir* in Iran, a *salar* in Chile, and a *tsaka* in Mongolia (Neal 1969).

In many ways, coastal flats (Fig. 5.12), known in Arabic as *sabkhas*, are very similar to inland playas: they, too, are almost horizontal, vegetation-free surfaces of fine-grained sediments; but they are associated with areas of coastal aggradation, especially around the Arabian Peninsula and along the western littorals of several deserts.

The literature on playas and sabkhas is more extensive, with the compilation edited by Neal (1975) usefully bringing together material in English, and volume 54 (1986) of *Palaeogeography, Palaeo-*

climatology, Palaeoecology providing some excellent Australian perspectives arising from the SLEADS (salt lakes, evaporites and aeolian deposits) Project.

Playas have received much attention, for five main reasons. First, they are sites of mineral wealth – notably salts and especially chlorides, sulphates, carbonates, nitrates and borates – and their economic potential has engendered a large literature (e.g. Gale 1914, Goudharzi 1970). Secondly, the record of playa deposits has been valuable in determining the nature of climatic and ecological change in desert basins during the Quaternary (e.g. Morrison 1966, Chivas et

Figure 14.1 The playa surface of former Lake Chew Bahir, in southern Ethiopia. The spit was formed at a time of high lake levels.

al. 1986). Thirdly, horizontal smooth surfaces are attractive for high-speed vehicular movement and military activity (e.g. Neal 1965a, b, 1968). Fourthly, playas form unusual, distinct and varied geomorphological and hydrological environments (e.g. Coque 1962, Motts 1965). Finally, they are the loci of some serious engineering problems in deserts (e.g. Cooke et al. 1982).

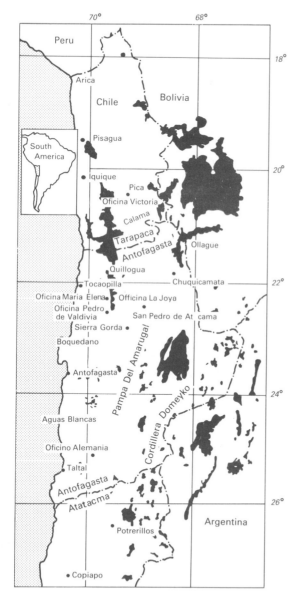

Figure 14.2 Salars and lakes (shaded black) of the Atacama region in Chile (after Stoertz & Ericksen 1974).

The distribution of playas has now been mapped in many deserts. For instance, there are over 1,000 playas recorded in North Africa; of some 300 playas in the USA, at least 120 are the relics of Pleistocene lakes (e.g. Morrison 1968); Stoertz & Erickson (1974) and others described the pattern of predominantly moist salars in the Atacama Desert and the Andes (Fig. 14.2); and there are hundreds of silt–clay pans in Western Australia.

14.1.2 Origin: drainage blockage
The occurrence of playas is principally controlled by climate, but as most playas are located within enclosed drainage basins the causes of enclosure are clearly of some interest. Enclosure of drainage basins most commonly results from tectonic movements, for example by the creation of basins or *blocking of drainage* by faulting, folding or subsidence. Death Valley, California, is an example of the first, and the Chott Djerid, Tunisia, of the last. Drainage may be blocked by alluvial or aeolian deposition (e.g. interdune playas, in the Empty Quarter of Saudi Arabia) or erosion (e.g. Qattara Depression, Egypt), or by volcanic activity (e.g. on the Puna surface, northern Chile).

14.1.3 Origin of smaller hollows
The desert surface is occasionally pock-marked by many, small, enclosed hollows that are a major component of the pattern of playas. They, too, have a variety of different names and they occur throughout North Africa and the Middle East, the High Plains of Texas, southern Africa, the fringes of the Indus Plain in Pakistan, and in southern and western Australia. Although superficially similar, these hollows vary greatly in form and origin.

On limestone or calcrete areas, small depressions arising from *solution* or *collapse* occur in deserts as elsewhere. They are usually only a few metres deep and tens of metres wide, and they have flat floors that may occasionally be flooded. The *dayas* of north Africa (Mitchell & Willimott 1974), Qatar and Bahrain (Doornkamp et al. 1980, Babikir 1986; Fig. 14.3), the *balte* of Libya, the *dongas* of Australia, the *vleis* of southern Africa, the *chor* or *sor* of central Asia, and the *bas-fonde humide* of French literature, appear often to be of this type.

On the southern High Plains of Texas and New Mexico, there are over 30,000 small ephemeral lakes, known as playa lakes, that are not dissimilar to dayas. Osterkamp & Wood (1987) showed that these occur

Figure 14.3 A shallow vegetated daya in the limestone plain of southern Bahrain.

where water is periodically stored at the surface and is transmitted to the subsurface groundwater table through an unsaturated zone. Suitable locations occur along, for example, ephemeral stream channels, along fractures where infiltration is higher and at low points in aeolian deposition. The process of water recharge through the unsaturated zone is crucial to the further development of the playas. It leads, for example, to *dissolution* of the underlying limestone (oxidation of organic material generates CO_2 which in turn becomes weak carbonic acid); to *piping* and to *eluviation* of fine-grained sediments; deflation is also important. Some of these changes can in turn cause slow subsidence and playa enlargement.

Goudie & Thomas (1985) located many thousands

of small playas or pans in South Africa, Botswana, Zimbabwe and Zambia. The features have densities in some areas of up to 1.4 pans km^{-2} and are up to tens of metres deep and several kilometres wide. In places they are elongated parallel to prevailing winds. Their origin differs from that of the dayas. Goudie & Thomas suggested a model for their development (Fig. 14.4), in which they are formed under dry conditions, when surface materials (loosened perhaps by salt weathering and animal scuffing) are removed by deflation. Their preservation depends on their *not* being infilled by fluvial deposition, perhaps because drainage is blocked by dunes, or is disorganized by tectonic deformation. Deflation is also a common cause of small lake basins and playas in other areas, such as Australia (Fig. 14.5).

14.1.4 Playa and basin geometry

Very few data are available for relating the size and shape of playas to drainage-basin conditions. As mentioned in the discussion of alluvial fans (Ch. 12), mutual adjustment of the playa area and the fan area might be envisaged. Similarly, the playa area might be related to other morphometric properties of drainage basins. Using approximate map data, a preliminary analysis by Cooke showed that for 38 playas in the Mojave Desert, California, playa area is positively correlated with drainage basin area ($r = 0.6645$, significance level 0.1%), and negatively correlated with the ratio of drainage-basin relief to basin area ($r = -0.4055$, significance level 1.0%). Such an analysis does tend to confirm the notion that playa area is related to drainage-basin conditions, and

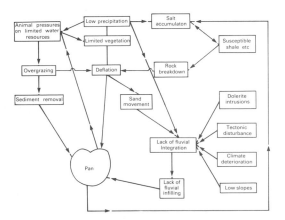

Figure 14.4 A model of pan development (after Goudie & Thomas 1985).

Figure 14.5 A string of salt lakes and playas (pans) formed by deflation of an ancient drainage line near Lake Grace, Western Australia.

this line of enquiry could profitably be pursued, especially with reference to changing playa dimensions and hydrological circumstances.

The actual areas of playas vary greatly, from a few square metres to over 9,000km². Lake Eyre in Australia is one of the largest, with an area of approximately 9,300km². The shape of playas, despite short-term variability, may reflect the origins of, and the contemporary processes at work within, playa basins. For instance, a straight margin may be imposed by faulting, or a jagged, irregular boundary may be determined by the pattern of deposition at the termini of drainage lines in a complex network.

14.2 Distinctiveness and classification

14.2.1 Playa environments
Playas are pre-eminently receptacles for sediment and water, and their nature is in large measure determined by their sedimentary and hydrological properties. As erosion–depositional systems their character is a function of many, interrelated variables (Fig. 14.6), the most important of which are groundwater, runoff, surface water, pore water, sediments, salts, aeolian processes, and chemical and biological reactions.

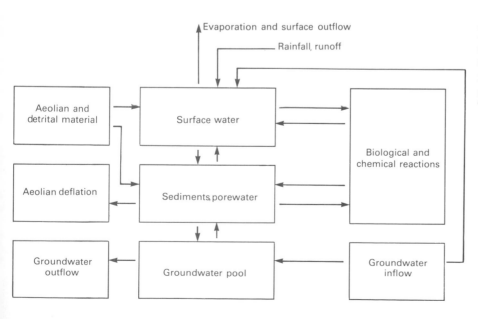

Figure 14.6 Diagrammatic representation of the possible exchange pathways for water and salts in a salt-lake system (after Torgerson et al. 1986).

These variables affect, *inter alia*, the nature of surface morphology, deposition and diagenesis, and the input and output of material (e.g. Torgersen et al. 1986).

Playas are characterized by fine-grained clastic and non-clastic sediments. The material is derived by fluvial and aeolian processes from within and outside the catchment (as suggested for Bahrainian playas in Fig. 14.7). Clastic sediments are mainly derived from surface runoff by deposition, either from flows of water or from standing water bodies. The clastic material is usually fine-grained, consisting of clay, silt and granular particles. Non-clastic sediments, com-

posed mainly of saline deposits, come largely from groundwater, although it may be difficult to determine precisely the proportions resulting from groundwater and surface runoff. Playas are distinctively areas of evaporite formation, as discussed in Chapter 6.2, and in the discussion of solonchak soils (Ch. 7): salts are fundamental controls on playa morphology.

The accumulation of fine-grained sediments has three main geomorphological consequences. First, many playas – especially those with smooth, hard surfaces – tend to be relatively impermeable, and thus the accumulation of surface runoff is encouraged. Secondly, because of the shrinkage properties of the sediments, desiccation phenomena are common (Ch. 2.5 Fig. 14.8). Thirdly, aeolian activity is often effective in these areas of fine-grained sediments.

Water on playas results from surface runoff, direct precipitation or groundwater discharge. Groundwater discharge at playa surfaces may be from capillary rise (see below) or, in areas where the water table intersects the surface, directly from the groundwater zone. Surface runoff is removed by infiltration and evaporation; groundwater is lost by evaporation and perhaps some transpiration. All playas exist in areas where annual evaporation is considerably greater than annual precipitation. However, as the ratio is reduced, and the supply of surface water and groundwater is increased, so the period in which water is at the surface is extended, and eventually perennial lakes may be formed (Langbein 1961). Motts (1970) suggested that the term playa should be used where the surface is flooded for 25% of the time. The duration of flooding is very variable. Where runoff is the cause, a single flow may last for only a season or, as in the case of large catchments such as the Lake Eyre Basin, for up to three or so years. Where groundwater is the main source, surface water may be permanent or less variable. Lakes in closed basins can exist only where precipitation on the lake and inflow to it are equal to or greater than the rate of evaporation.

The local relations between present climate and basin-floor conditions, and the hydrological balance, are influenced by many variables. The geometric, soil and vegetation characteristics of the drainage basin are all important. The rate of evaporation in any one area depends not only on climatic conditions but also on other factors such as water salinity and the geometry of the water body. Further complications arise if, for example, part of the water supply is from outside the arid region, or from fossil groundwater which has

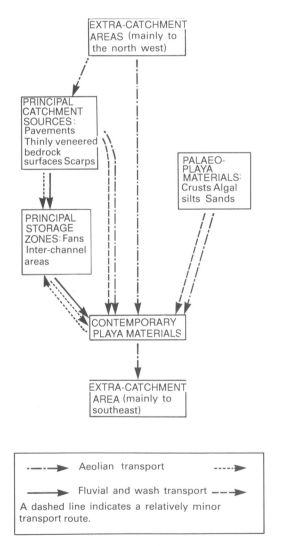

Figure 14.7 The contemporary transport system of sediments in the playas of Bahrain (after Doornkamp et al. 1980).

Figure 14.8 A giant desiccation crack on a dry, hard, playa surface, Panamint Playa, California.

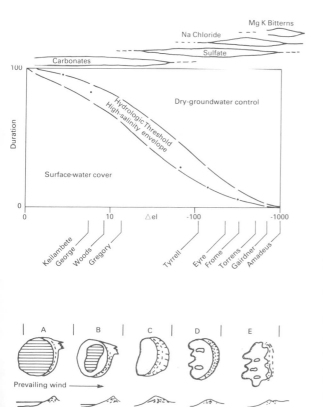

moved long distances; or if, as is often the case, the playa is inherited from prior pluvial conditions and is now inundated only by infrequent floods. Nevertheless, dry playas and perennial closed lakes may be viewed as the two extremes of a continuum representing the conditions of enclosed-basin floors in arid lands.

Bowler (1986) classified playa basins in Australia, taking many of these considerations into account (Fig. 14.9). His classification is based on the percentage of time the surface is flooded, a **disequilibrium index** (a measure of how far a basin's climate today differs from that necessary to maintain a steady-state surface water cover) and the extent to which the basin is dominated by surface water or groundwater.

Figure 14.9 A diagrammatic classification of some Australian terminal basins across a climatic gradient from humid (Keilambete) to arid (Amadeus) regions. The classification is based on an equilibrium index (Δel). The index is a measure of how far the basin's modern climatic setting deviates from that necessary to maintain a steady-state surface-water cover. Different morphological and sedimentological features, reflecting successive stages in the hydrologic series are used to divide the basins into five types, A–E, from surface-water- to groundwater-dominated systems (after Bowler 1986).

Motts (1965) also suggested a simple, qualitative, hydrological classification of playas based on the relative amounts of groundwater and surface inflow (Table 14.1). Examples of the different types occur in many deserts. For example, the hard dry playas of the western USA and the *garaets* of Tunisia are supplied largely by surface runoff (Table 14.1, 1), whereas the self-rising moist playas in the former area and the *chotts* of the latter are fed mainly by groundwater discharge (Table 14.1, 5). The relations between groundwater discharge within groundwater divides were criteria used by Motts (1965) to classify further the groundwater discharging playas.

Table 14.1 Hydrological classification of playas.

1	2	3	4	5
		increase of GWD		
Increase of SWD		increase of SWD		All GWD
All SWD	Relatively large SWD and small GWD	Approximately equal SWD and GWD	Relatively large GWD and small SWD	
1A 1B				
Small Large SWD SWD				

GWD: groundwater discharge
SWD: surface-water discharge
Source: Motts (1965).

14.2.2 Surface features

Adjectives commonly used to describe playa surfaces are hard, soft, wet, dry, rough, smooth, flaky, puffy, salty or non-saline (Neal 1969). Several combinations of adjectives frequently recur. For example: hard, dry, smooth surfaces; and soft, friable, puffy surfaces. Often a particular surface type is associated with a specific group of environmental conditions. Thus, hard, dry, smooth surfaces commonly occur on playas dominated by surface runoff and silt- and clay-size sediments. Many of the surface properties can be precisely measured, but quantitative data are few. Consequently, the following discussion relies on the use of rather general qualitative terms. Such terms do not make for precision, but they do facilitate comparison and generalization and they permit simple hypotheses to be outlined.

In a detailed survey of playas in Iran, Krinsley (1970) recognized seven major playa surface types and determined their extent and frequency (Table 14.2). This analysis emphasizes the relative predominance of

Table 14.2 Extent and frequency of playa-surface types in Iran.

Surface type	Area (km^2)	% of total playa area	Number of playas (partly or wholly of particular surface type)	% of all playas
Salt crust	27,624	41	50	83
Clay flat	23,724	35	24	40
Wet zone	6,702	10	14	23
Fan delta	3,229	5	6	10
Swamp	2,889	4	3	5
Intermittent lake	1,912	3	2	3
Lake	1,172	2	2	3
TOTAL	67,252	100	60	100

Source: Krinsley (1970).

salt-crust and clay-flat surfaces in Iran, a common feature elsewhere.

Motts (1965) further classified playa surfaces according to the position of the water table and the nature of groundwater discharge, as these hydrological factors also influence the sedimentology and ultimately the geomorphology of playa surfaces (Table 14.3).

Table 14.3 Classification of playa surfaces.

(i) Total surface-water discharging playas where the water table is so deep that no groundwater discharge occurs at the playa surface.

(ii) Where discharge occurs at the surface by capillary movement.

(iii) Where discharge occurs directly at the water table.

(iv) Where discharge occurs by phreatophytes and other plants.

(v) Where discharge occurs by springs.

(vi) Combinations of all of these.

Source: Motts (1965).

This classification provides a suitable basis for describing playa surfaces, although each category is not necessarily exclusive because different hydrological processes may produce similar surface features. Each of the six categories is examined below, the general comments coming mainly from descriptions of USA playas by Motts (1965), Neal(1969), and Langer & Kerr (1966).

(a) Total surface-water discharging playas

The surfaces of these playas, which by definition lie above the capillary limit, are characteristically smooth, hard, commonly dry, and composed of fine-grained clastic sediments. Over half of their material may be of clay, and there may be some calcium carbonate accumulation, but the proportion of salines is often low. They are relatively impermeable and have relatively high bearing strength. Because there is a downward component to water movement when the surface is flooded, a soil profile may be developed below the surface and this may have a distinctive zone of caliche development. The surface of some playas in this category, such as that of Sabzever Kavir, Iran (Krinsley 1970), may show traces of a distributary drainage network. Aggradation is by deposition from surface floods or by aeolian deposition. The quantity and quality of incoming sediment will depend largely on the geological and climatic environment within the basin. Motts (1965) suggested that in many playas of the western USA, for example, more clay and less silt were deposited in times of more humid climate than is the case at present, because of greater chemical weathering. The actual surface type which develops, however, is in part related to the conditions of sedimentation. Different surface types could develop from essentially similar sediments entering the basin, depending on the nature of the major salines at the time of deposition. For example, waters with high carbonate contents may have deflocculated clay particles, leading ultimately to the formation of homogeneous, fine-grained, compact crusts.

High shrinkage coefficients for surface and subsurface material in playas of this type mean that desiccation phenomena are common (Ch. 8). Neal (1969) noted that the surface glaze on these playas is perhaps related to a high degree of fine-particle orientation.

(b) Capillary-movement playas

Capillary movement of water to the playa surface will vary mainly according to the relations between the opposing forces of evaporation and capillary pressure. Evaporation may vary seasonally and over long periods of time. Capillary pressure increases as the radii of particles in the sediment decrease. Capillary rise of water above the water table, H_c, is theoretically between 1m and 10m, with field values of up to about 4m being quite common (e.g. Fookes et al. 1985). Terzaghi & Peck (1948) defined it as follows:

$$H_c + \frac{C}{eD_{10}} \qquad (14.1)$$

where C is an empirical constant that depends on grain shape and surface impurities, and ranges from 0.1 to 0.5, e is the void ratio and D_{10} is the effective size, in which 10% of particles in grain-size analysis are finer, and 90% are coarser than this value. The limit of the capillary fringe may be within or above the ground surface (Fig. 5.10); in the latter case, evaporites may occur on the surface. It is a line of considerable practical importance because of its relation to salt weathering (Ch. 6.2). In places, artesian pressure may significantly supplement capillary pressure (e.g. Neal & Motts 1967). Rapid capillary discharge is often associated with soft, friable, puffy ground – the self-rising ground of some authors – and with salt-thrust polygons. Slow capillary discharge may yield a thin salt crust. In either case, the surface tends to be saline, soft, fairly permeable and loosely compacted, and it may have a micro-relief of up to about 15cm. In comparison to the hard, dry crusts of the type in (a) (Table 14.3) the surfaces of capillary-movement playas in the USA appear to have a lower clay content, and a higher proportion of salines and larger-size particles. The moisture content of the surfaces, and hence the surface character, may vary greatly during the year.

(c) Playas with direct groundwater discharge

Salt crystallization characterizes playa surfaces where groundwater discharges seasonally or perennially onto the surfaces. There may be thin salt crusts, commonly of halite or gypsum, or thicker salt pavements up to a metre or more in thickness. Pressures of salt crystallization tend to cause surface disruption, creating salt ridges, salt thrust polygons, and extremely irregular micro-topography (Ch. 2.5). The surfaces are generally wet, soft and sticky; and they may be puffy in places. Solution phenomena – such as small pits and sinkholes – may also occur.

(d) Playas with phreatophyte discharge

Although many playas are without vegetation, loss of water by evapotranspiration through phreatophytes and other plants can occur where the water table is sufficiently close to the surface for it to be tapped by plant roots and where the water is of low enough mineralization to allow plant survival. Features associated with playa plants are ring fissures (Ch. 2.5), slight surface subsidence, and phreatophyte mounds. Evapotranspiration may lead to local lowering of the water table beneath plants and, as a result, there might be slight subsidence of the ground. **Phreatophyte** mounds arise from the accumulation of fine-grained windblown material and from the precipitation of salts near to the base of plants that tap the groundwater (such as *Sarcobatus* sp. in the USA). They may be up to 5m high. A sequence of phreatophyte mound development is described by Neal & Motts (1967). Initially the plant grows to maturity near to the playa surface. Aeolian debris begins to accumulate around it, and salts may form a surface crust on the mound. As the mound becomes larger, the plant continues to survive at the top of the mound by extending its root system. There will come a time when the plant is so high that its roots can no longer draw water from the water table, and it will die. The mound may resist erosion because of the salt crust, but, once the crust has been dissolved by rainwater, the aeolian deposits beneath will soon be degraded by sand, rainwater, streams, or even by lacustrine erosion.

(e) Playas with spring discharge

Springs occur where the water table or piezometric surface is higher than the playa surface. If the piezometric surface should rise, the springs may become foci of evaporite deposition, producing **spring mounds**. Spring mounds can theoretically grow to the level of the piezometric surface. They have been described in the Lake Eyre region, Australia (e.g.

Watts 1975), the southwestern USA (e.g. Hardie et al. 1978), in Djibouti (Fontes & Pouchan 1975), Ethiopia (Englebert 1970) and Tunisia (Coque 1962). Roberts & Mitchell (1987) described belts of spring mounds associated with artesian groundwater in the chotts Djerid and Fedjadj (northern Sahara, Tunisia; Fig. 14.10). They are up to 30m high, and their morphology is produced by the interaction of several processes, including aquifer pressure-head, chemical content of the water, net evaporation, plant growth and aeolian activity. The material in the mounds comprises sand- and silt-size particles, mainly of gypsum and calcium carbonate, organic material and aeolian deposits. Some mounds have central pools or marshes (Fig. 14.11). Where artesian water emerges on the seasonally flooded chott surface, springs (**aioun** in Arabic) occur without significant mounds, and they can be identified on satellite imagery (Rebillard & Ballais 1984). They are up to 4m deep and 5m wide and have nearly vertical sides. The aioun are of un-

Figure 14.10 A spring mound at Chott Fedjadj, near Seftimi, Tunisia.

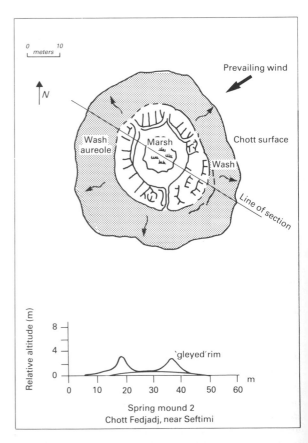

Figure 14.11 A diagrammatic view of the spring mound shown in Figure 14.10 (Roberts & Mitchell 1987).

certain origin, but they may arise from surface solution and subsidence (Coque 1962).

(f) Combination of surface types

Any single playa surface may include the characteristics of more than one of the above groups. For example, a hard, smooth clay surface may grade into a puffy crust, and a puffy crust may merge into a salt pavement. Such a spatial gradation would probably be accompanied by a decrease in the argillic component and an increase in salt content (Langer & Kerr 1966), and by an increase in surface moisture. Similar changes may affect a single part of the playa surface as conditions, especially hydrological conditions, change through time. Such spatial sequences may include the annular arrangement of different salts around the playa, reflecting their relative solubilities (Ch. 6.2).

14.2.3 Wind activity

The accumulation of material in playas is related mainly to evaporation processes and to surface runoff. The latter may also be a locally significant process of erosion. However, as most playas occur in the lowest areas of enclosed basins, solid material can be removed from them only by the wind. The mechanisms of wind erosion and dune formation are fully discussed in Part 4. Here it is only necessary to state a few points specifically pertinent to playas.

First, not all playa sediment is equally susceptible to deflation. For example, hard dry crusts with a high clay content are very much more resistant to deflation than surface material rich in salts (Coque 1962). In addition, the susceptibility of playa sediment to deflation varies with its moisture content, which is itself often quite variable in area, in depth, and through time. The effective lower limit of deflation in playas is the water table. This is a particularly important deflation datum in sabkha.

Secondly, removal of playa sediment by the wind does not necessarily mean that there is net erosion from the drainage basin. Some material may be carried in suspension out of the basin, but most will be deposited either on the playa (e.g. in phreatophyte mounds) or around its margins, perhaps in the form of dunes (such as lunettes, Ch. 18). Some of the marginal deposits may be returned to the playa. Indeed, a cycle within a closed basin may be envisaged in which clastic and saline materials are supplied to the playa by rainfall, runoff and groundwater, are removed by deflation, and are subsequently returned to the playa by runoff or from recharged groundwater. This pattern of sediment movement associated with playas has rarely been monitored. An example is the 5yr study by Young & Evans (1986) of dust movement on and around the dry, hard Grass Valley Playa (Nevada). Sampling transects on the lee (eastern) side of the playa and adjacent slopes showed that deposition was greatest beyond the playa margin (e.g. 2,930g m^{-2} yr^{-1}), was about 1,920gm^{-2} yr^{-1} on the playa, and only 6% of total deposition occurred beyond 1.9km from the playa itself. It is doubtful if the dust collected on the playa in this study would have stayed there if it were not for the artificial traps. The deposition in the zone sampled, however, varied greatly from year to year (inversely with annual precipitation) and was highest in winter (the time of strongest winds).

Thirdly, wind action has been invoked to explain rather unusual stone tracks on some playa surfaces in the USA. The tracks, which often change their direction, are up to about 3cm deep, several centimetres wide, and vary in length from a few metres to a recorded maximum of 270m. At one end of each track there is usually a stone or some other object, embedded in the surface and weighing up to 200kg. Clearly, the tracks have been created by the scraping of the stones across the playa. Three principal mechanisms of stone movement have been suggested.

Stanley (1955) proposed and tested in detail on Racetrack Playa, California, the hypothesis that the stone tracks are produced by windblown ice floes dragging protruding stones across the wet playa surfaces. This hypothesis is strongly supported by the identical signatures of groups of tracks, which imply that the scraping stones were held in fixed positions relative to one another during movement. Only ice could provide the planar body required to maintain those positions. A significant feature of Stanley's analysis is his demonstration that mathematical prediction of tracks produced in a rotating floe closely corresponds to observed patterns. The occurrence of ice-ramparts on playa shores and reports of ice floes on playas in winter lend support to this hypothesis. As Stanley emphasized, stone tracks occur on only a few playas; it could be that they are absent from many because there are no suitable stones or because there are no ice floes. Certainly in some playas, such as North Panamint Playa, California (Cooke 1986), stones are scattered across the surface but are not associated with tracks. In such cases, tracks may not have been formed, or they may have been obliterated by subsequent changes in the distribution of fine surface sediments.

Although the ice-floe hypothesis is convincing, a second, more likely hypothesis is that stones are blown by strong winds across wet playa surfaces (Kirk 1952, Sharp & Carey 1976). The theoretical discussion and experimental work of W. E. Sharp (1960) cast doubt on the idea, because he showed that the force produced by the drag of the wind must be sufficient to overcome the frictional resistance of the mud, and that the wind velocities required to move scrapers over wet mud are higher than those which commonly occur naturally. These observations, and the fact that there is no evidence at present of stone tracks in playa surfaces where ice floes never form, suggested that the wind- and wet-surface mechanism may be relatively unimportant. Sharp & Carey (1976) monitored stone movement on Racetrack Playa, and came to different conclusions. While they recognized that some stone tracks *may* have been made by windblown ice floes, they noted that some could *not* have been moved in this way because, *inter alia*, some stones had moved out of an encirclement of stakes, some tracks had crossed in ways that would have been impossible if the ice floes had been involved, distances between some stones had changed during movement and some matching tracks were of different lengths. They concluded that wind-driving of individual stones is the basic process, and that it probably occurs at a crucial time when the surface is characterized by a thin, slippery, superficial layer of water-saturated mud.

A third hypothesis proposes that the sliding of playa scrapers owes little to wind activity. Wehmeier (1986) argued that stone tracks arise on Alkali Flat (Nevada) from subaquatic or inaquatic sliding in which the driving force is the **hydraulic energy of runoff**. Arguments for this process include the fact that tracks follow flow directions, no ice floes were present at the time of the track formation, and maximum wind velocities were probably inadequate to *initiate* sliding. Wehmeier demonstrated that the shearing stress required to initiate movement was easily attained with water flow velocities of 5.5km h^{-1} or more (com-

pared with > 190km h^{-1} for wind, given assumptions of stone size, buoyancy etc.).

14.3 Playa change and evolution

Short-term changes in the character of playas are mainly associated with seasonal and annual variations in the availability of surface and subsurface water. Such changes are evident, for instance, in the chotts of southern Tunisia. Coque (1962) showed in the Chott el Djerid that there is surface water in most, but not all winters, and that the area inundated varies from year to year. Similarly, Millington et al. (1987), using difference and ratio images from LANDSAT thematic mapper and multispectral scanner data backed up by field observations, were able to chronicle seasonal and annual changes on the Chott el Guettar, especially in terms of water fluxes and surface wetness (Fig. 14.12A), the redistribution of water, salt efflorescences and crusts, and sediment, changes in vegetation cover, aeolian activity and

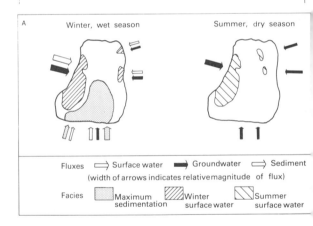

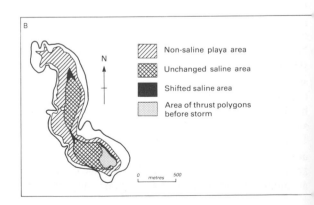

Figure 14.12 Playa dynamics. (A) A provisional model of seasonal water, salt and sediment dynamics at Chott el Guettar, Tunisia (after Millington et al. 1987). (B) Pago Playa, Estancia Valley, New Mexico, showing shift of saline contacts after a storm in 1964 (after Neal & Motts 1967).

fluvial processes. Such changes may be significant even as a result of a single storm, as Neal & Motts (1967) showed for the Pago Playa, New Mexico (Fig. 14.12B). In a later study, Millington et al. (1988) and Townshend et al. (1989) used LANDSAT thematic mapper data to formulate a dynamic process-based model of, and to delimit process domains in, the Chott el Djerid, in which the contributions of groundwater, dissolved salts, surface runoff and eolian processes, and their changes over time, are evaluated. They found, *inter alia*, that Band 7 reflectance data are related to total soluble salts, and to sodium, manganese and calcium carbonate content, such that radiance increases as total soluble salts decline; that difference images from Band 7 help location of runoff zones; and that mixture modelling aids the monitoring of salt and sediment distribution.

The commonest short-term playa changes include alterations in the distribution of saline and non-saline areas of desiccation and salt phenomena, and especially of crust types. For example, at Harper Playa, California, Neal (1968) attributed the conversion of soft, dry, friable surfaces to hard, dry crust, over four years, to the dissolution of surface evaporite minerals during flooding. An additional type of change may occur where a playa is underlain by water-soluble evaporites, and the water table falls. Downward-percolating water may here dissolve salts and cause surface subsidence in forms ranging from vertical-sided collapse sinks to smoothly rounded dolines (Neal & Motts 1967). In the Aghda Playa of Iran, sink holes and collapse depressions are due to groundwater discharge reaching the surface and leaching the salt from the clayey-silt deposits (Krinsley 1970).

Progressive changes over a period of years may lead to the creation of some distinctive surface features. Changes of vegetation, surface runoff and water-table levels in the western USA during the nineteenth century have had significant effects on playas (Neal & Motts 1967). The main consequences of a falling water table, for example, appear to have been the formation of giant desiccation polygons and stripes, and the creation of relict spring mounds; the solution and subsidence of surfaces underlain by evaporites; the local replacement of soft, puffy surfaces by hard, dry, compact playa surfaces; and playa expansion.

Some have seen particular significance in the alternation of wet and dry conditions on playa surfaces. Wet periods are generally times of sediment accumulation. In dry phases, erosion, largely in the form of deflation (but also by running water around playa margins), is ascendant. It is therefore possible to envisage a regular alternation of erosional and depositional episodes. The net effect of such an alternation would clearly depend on the relative rates of the erosional and depositional processes, and the frequency and duration of the episodes. This is an attractive hypothesis for, although evidence on rates of processes and frequency and duration of episodes is very limited, it is clear that some playas are characterized by erosional forms such as deflation hollows, striae and dissected lacustrine sediments, whereas others are without such phenomena and seem only to have undergone a net accumulation of sediment.

Over the much longer timespan of the Quaternary there is ample evidence of change in some playas. The two main categories of evidence are erosional and depositional features marginal to the playa, and the stratigraphic record of the deposits beneath it. Marginal features include strandlines, tufa deposits, spits, bars, deltas, abrasion terraces, cliffs and overflow channels (Figs 2.4 & 14.13). Such phenomena clearly represent formerly more extensive lakes, but they only have chronological value if they contain or can be related to datable material, such as artefacts, mollusca, wood, etc. Often more illuminating is the strati-

Figure 14.13 Tufa pinnacles developed at spring orifices at Searles Lake, California, during high lake and groundwater levels during the Pleistocene.

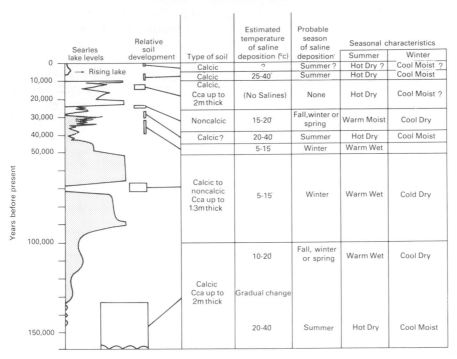

Figure 14.14 Relationships between lake history, degree of development of fossil soils crystallization temperatures and season of correlative salines, and inferred climatic characteristics of selected interpluvial intervals in Searles Basin, California (after Smith 1968).

graphic record which may contain valuable evidence of environmental conditions during the evolution of the playa, together with datable materials. The stratigraphic interpretations are today increasingly based on the analysis of many different features of the sediments, such as their organic content (including faecal pellets, diatoms, pollen or shells), their chemical and sedimentary composition (including ionic and carbonate mineralogy analysis) palaeomagnetic and amino-stratigraphic characteristics, and soil horizons.

The detailed work of Smith (1968) in Searles Basin, California (Fig. 14.14) provides an outstanding early illustration of studies of playa evolution. Searles Basin is extremely sensitive to regional climatic changes. When inflow of water from the Sierra Nevada increased, a lake was formed and mud layers were deposited. At times the lake overflowed into Panamint Valley. When the flow diminished, the lake waters were evaporated, and a salt lake or salt pan was created. Fortunately, in this area the sensitivity of the basin to climatic changes is associated with a wealth of evidence of such changes in the past. Lacustrine sediments and shoreline information allow reconstruction of high lake levels. Marginal sediments can be related to subsurface sediments. Subsurface saline deposits indicate the nature of low lake levels, and some of these deposits can be correlated with inter-

pluvial soils in the basin. Gaps in the depositional sequence are identified by unconformities and fossil soils. The succession has been dated by the ^{14}C method for ages less than 40,500yr BP, and by estimation of ages over 45,000yr BP from rates of mud sedimentation determined within the ^{14}C dated beds. Analysis of the mineral assemblages in the saline deposits allows estimation of the temperatures of deposition, and permits inferences concerning the season of saline deposition and, finally, the temperature/precipitation characteristics of summer and winter. The changing lake level in the Quaternary and associated information on playa conditions are summarized in Figure 14.14. As the details derived from lake cores and the sophistication of relative and absolute dating techniques have improved, so interpretations of the climatic evolution of playa basins have evolved. For example, the interpretation of Pleistocene Lake Bonneville (USA) is now very much more complex than that envisaged by G. K. Gilbert over a hundred years ago (e.g. Fig. 14.15; Morrison 1966).

Many other studies have now been made of the Quaternary history of playas in this and other deserts (e.g. Coque 1962, Morrison 1966, Chivas et al. 1986, Kelts & Shahrabi 1986, Kezao & Bowler 1986). Although the evidence varies from place to place, it is beyond doubt that many playas were formerly

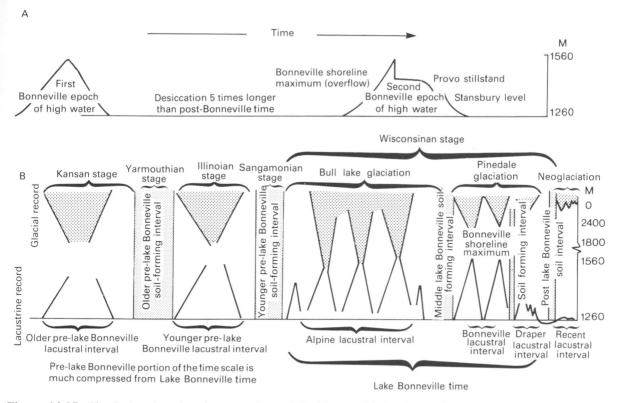

Figure 14.15 Classical and modern interpretations of the history of Lake Bonneville. (A) Classical interpretation (Gilbert 1890). (B) Modern interpretation and correlations with glacial fluctuations in the Wasatch Range, Utah (after Morrison 1966).

occupied by more extensive and more permanent lakes in one or more periods during the Quaternary, and that the present forms are inherited from climates which were wetter or colder, or both. In some areas, the removal of large water bodies has resulted in isostatic recovery which may be reflected in warped shorelines and other phenomena. In the area of former Lake Bonneville, which covered some 50,000km², the Bonneville shorelines have been raised in a broad domical uplift of about 64m (Crittenden 1963). However, not all playas have related evidence which points towards more extensive lakes. In Western and South Australia (Neal 1969) and in parts of the Sahara, for instance, no marginal features indicating more extensive ancient lakes have yet been identified.

An example of a more recent interpretation of playa cores is that by Kezao & Bowler (1986) of the Late Pleistocene evolution of lakes in the Qaidam Basin, China (Fig. 14.16). They showed that from about 40,000 to 25,000yr BP the climate was humid and the lake fresh to slightly saline; after 25,000yr BP the climate became drier and the lakes were evaporated;

they finally dried out between 15,000yr BP and 9,000yr BP, when the highest halite layer was deposited. A contrasting example comes from the sediments of small playas that represent the relics of interdune lakes in the Rub Al-Khali (McClure 1978; Fig. 14.17) which date from c. 36,000 to 17,000 years BP, when the climate became hyper-arid and the lakes dried up, except for a subpluvial interruption between 9,000 and 6,000yr BP.

Bowler (1986) offered a general model of Quaternary evolution of playa-lake basins in Australia, arguing that his spatial sequence (Fig. 14.9) could be considered as an approximate *temporal* analogue within a basin as it was modified by climatic fluctuations (Fig, 14.18).

Throughout this section, diversity of playas has been noted, in terms of their origin, hydrology, mineralogy, processes and surface forms. Diversity also characterizes the age and evolution of playas, while local conditions determine the forms they take.

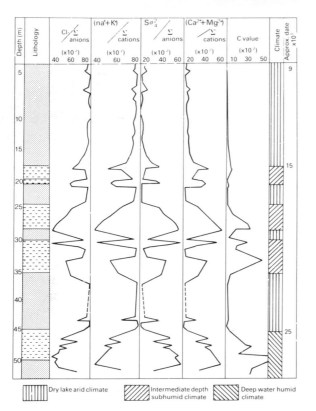

Figure 14.16 The distribution of soluble ions through a core from the Qaidam Basin, China, with an interpretation of palaeohydrological conditions (after Kezao & Bowler 1986).

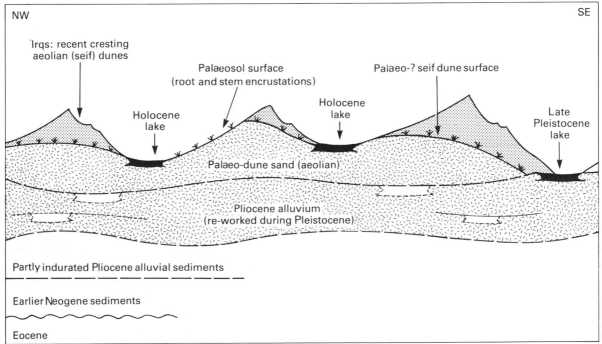

Figure 14.17 A stratigraphic cross section of Quaternary sediments of the southwest Rub'Al Khali (not to scale; after McClure 1978).

216

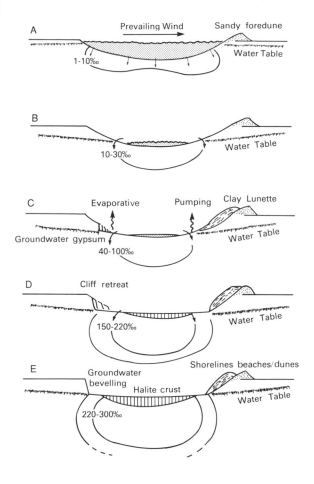

Figure 14.18 Sequential diagrams illustrating the nature of changes experienced by basins in different states in the surface-water- to groundwater-controlled hydrological series in Australia. Stages A–E correspond in a general way to regional types A–E (Fig. 14.9), or to successive changes a single basin may undergo during a period of major climatic change from a humid (lacustal phase, surface-water-dominated) to an arid (playa phase, groundwater-dominated) environment (after Bowler 1986).

14.4 Sabkhas

Sabkhas – low coastal flats just above high-tide level along desert coasts (Figs 5.12, 14.19 & 20) – are similar to such features of coastal aggradation as the Chenier plain of Louisiana and the Wash in England. As Evans et al. (1964) and Purser & Evans (1973) explained in their classic description of the sabkha along the Trucial Coast in the Arabian Gulf, sabkhas differ from these features mainly because carbonate sediments and evaporite minerals are a major component. Typically, as Evans et al. (1964) explained, a sabkha is composed of a variety of surface forms, including marine features (such as beach ridges, strand lines, creek systems, and abrasion surfaces cut across bedrock), aeolian features of deflation and deposition and many evaporite salt-crust phenomena. The level of the surface may in part be determined by groundwater level, which is the lower limit of deflation. Sabkha sediments, dominated usually by carbonates, evaporites, fluviatile, aeolian and marine debris, and

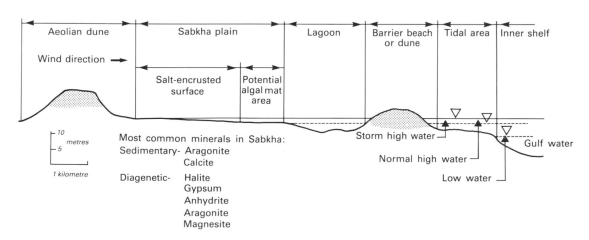

Figure 14.19 A generalized cross section across a typical coastal sabkha with typical surface features (after Akili & Torrance 1981).

Figure 14.20 Saline efflorescence on a sabkha surface in Bahrain.

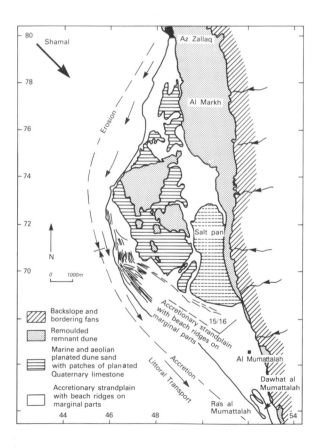

Figure 14.21 The southwest sabkha in Bahrain (after Doornkamp et al. 1980).

sometimes cemented (e.g. with carbonate or gypsum), are developed by intertidal sedimentation, by chemical deposition, and by deposits derived from calcareous organisms. On the seaward side of the sabkha, there are characteristically intertidal flats, lagoons, and off-shore bars and islands; and beach ridges may separate the two zones. When tides are exceptionally high, or augmented by strong winds, the sabkha may be flooded, possibly through channels in the beach ridges. Flooding by terrestrial runoff may also occur. The evolution of the sabkha is reflected in the patterns of beach ridges and strandlines, and it is often strongly influenced by aeolian processes between floods.

Many of the features described above are to be found in the Southwest Sabkha of Bahrain (Fig. 14.21; Doornkamp et al. 1980). The *c.* 60km² area is being eroded in the northern sector and is suffering littoral transport southwards towards a spit in the south. It has extensive sand and gravelly-sand beach ridges, an accretionary sand plain, flats dominated by loose, homogeneous, laminated and poorly graded medium-fine gypseous calcareous quartzose sand. In addition, there are some algal peat layers, patches of wave-trimmed bedrock, remoulded and planated dunes, blistered salt-crust surfaces with polygonal features (comprising chiefly halite, gypsum and anhydrite), and an inner salt pan. This complex morphology arises from the drowning of a sand-dune topography during the Holocene transgression (*c.* 6,000–7,000 BP) and the fashioning of the coast under the influence of the prevailing **shamal** wind from the

218

northwest, which also causes extensive deflation.

14.5 Engineering and geotechnical problems

The two major engineering problems associated with playas and sabkhas are *foundation protection* and *trafficability*. Akili & Torrance's (1981) recognition of eight major geotechnical problems on sabkhas is equally applicable to playas. These problems are: (i) a decrease in strength of surface crust as a result of inundation or the absorption of water from the atmosphere; (ii) variation in compressibility, which could lead to differential settlement; (iii) volume changes due to the hydration/dehydration of salts, especially gypsum; (iv) the corrosive effect of chlorides and sulphates; (v) evaporative pumping, which moves soluble salts to the surface or into foundations; (vi) densification of material at the surface, which can break off cementation bonds in material at depth, reducing load bearing capacity; (vii) inflow of *fresh* water (e.g. from water-supply pipes), which could dissolve salts and reduce material strength; and (viii) seasonal inaccessibility when areas become saturated. Salts are fundamental to many of these problems, for not only do they influence the nature of weathering processes, but they also modify (with other factors) a range of material properties, such as shear strength, consolidation properties, cohesion, plasticity index, permeability and compressibility (e.g. Khan & Hasnain 1981, Stipho 1985, Hossain & Ali 1988).

The problem of foundation protection has been extensively reviewed in Cooke et al. (1982), and also in the context of salt weathering (Ch. 6.2). The main aspects of the problem are evident in the list above, but most serious is the possibility of saline solutions penetrating the foundations of buildings and roads as groundwater or by capillary pumping. The crucial considerations are the shallowness of groundwater and its salinity, to which the hazard can be said to be directly related (e.g. Fig. 5.20), but of course the type of salts and the materials being attacked are also

important. In this context, Fookes et al. (1985) emphasized problems of surface settlement due to solution by flowing groundwater, heave of the surface due to salt crystallization, and the creation of chemically aggressive environments.

These problems are very common in sabkhas, where groundwater is almost always saline, and in those playas where groundwater or the capillary fringe is within the foundation zone. The problem of trafficability (the ability of vehicles to drive over the natural surface) has two principal dimensions. First, the creation of desiccation phenomena such as giant desiccation cracks (Ch. 2.5) poses a major hazard to vehicles. The giant cracks, for example, can break open roads and runways, and if they are unexpected they can cause serious traffic accidents.

Secondly, the nature of playa and sabkha surfaces is spatially and temporally very variable, and their conditions can range from hard, dry surfaces that make excellent racetracks and airfields (as in Lake Eyre, Australia, and Lake Bonneville, USA) to groundwater discharge surfaces that can be extremely irregular, perilously soft and almost impassable. Sometimes a hard crust (especially if it has a carbonate content of greater than 50% by weight, or a lower carbonate content but with a sulphate content ≥5% by weight, or a high sodium chloride content with high proportions of carbonate and/or sulphate, and the water table is more than 1m deep) provides a good base for roads (e.g. Ellis & Russell 1974). But such roads can become quagmires if they are wetted because, for example, the soluble salts (such as halite) in the stabilizing crusts are removed.

A first step towards solving both of these problems is normally to prepare a detailed geomorphological and geological map, to analyze systematically the depth to groundwater, the thickness of the capillary fringe, and the quality of groundwater, and to determine the geotechnical properties of the materials (e.g. Stipho 1985). Remote sensing imagery can be invaluable, revealing not only the nature of surface morphology but also the spatial relations between wet and dry surfaces.

CHAPTER 15

Models of mountain–plain evolution

It is difficult to seek a single comprehensive hypothesis of mountain–plain development in deserts, because the form of the land varies enormously from place to place, and the effective processes are variable in both space and time. Furthermore, detailed diagnostic evidence of evolution is rare, and much interpretation rests on impression, intuition and imagination. Some essential evidence may never be reclaimed. In this chapter some of the more convincing explanatory hypotheses are outlined.

15.1 A general model

In most localities, the pediment forms one area within a system generally comprising a mountain area, a pediment, an alluvial plain and a base-level plain. The pediment is often intermediate between zones of bedrock and alluvium, erosion and sedimentation and, perhaps, uplift and subsidence. The critical boundary in this system is that between areas where erosion is dominant and areas where deposition is preponderant. More often than not, this boundary coincides with the piedmont junction and there is no pediment; elsewhere, the boundary may be on the piedmont plain, and there is a pediment. The location of the boundary will be determined in large measure by the relationships between the rates of debris supply and removal. An equilibrium condition can be envisaged where the boundary is stable and the two rates are approximately equal. In these circumstances, the actual position of the boundary will be determined by local circumstances – it could be anywhere along the profile. If the equilibrium is upset, so that supply changes with respect to removal, the boundary will migrate in the appropriate direction until a new equilibrium is established. The movement of the boundary may or

may not be accompanied by a change in form (such as dissection). In areas with pediments it appears that erosional forces are adequate to remove available debris efficiently from at least part of the piedmont plain. This general model accommodates an important notion: that the landforms may not be formed under aridity, but merely developed under it, and the basic form persists through successions of climatic change. This is the **traditionelle Weiterbildung** concept of Büdel (e.g. 1970), in which *persistence* of form is fundamental.

Within this general framework, several evolutionary sequences may be envisaged. First, the pediment or piedmont plain may be extended at the expense of the mountain mass within the erosional zone. Here the major explanatory hypotheses include the *slope retreat hypothesis*, and *drainage basin hypotheses* (including the *lateral planation hypothesis*). Secondly, erosional processes may become ascendant over depositional processes in the piedmont zone, leading to the stripping of the alluvial cover to expose a pediment – the *exhumation hypothesis*. Thirdly, where the processes of debris supply equal or exceed those of removal, the piedmont surface may be subject to extensive weathering or etching beneath an alluvial cover – the *mantle-controlled planation hypothesis*. Beyond this, there is evidence in some areas that present-day mountain-pediment features are inherited from past conditions with or without substantial modification – the *inheritance hypothesis*. Finally, *sequences* of piedmont surfaces may be formed, reflecting changes externally or internally to drainage systems.

In addition, there are four *phases* of piedmont evolution common to most explanatory hypotheses: *initiation, extension, modification* and *destruction*. Initiation is usually associated with the creation of relief, and this can only be achieved either directly by tectonic movements or indirectly by drainage incision consequent upon changes in climate or base level. Tectonic uplift is an important cause of relief initiation especially, for example, in basin–range country. Once initiated, the mountain–piedmont relations will depend on the relations between *subsequent* tectonic activity and subaerial processes.

Bull (1984) summarized his view of the tectonic possibilities as follows (Fig. 15.1): immediately up stream and down stream of a fault, the key relationships are the relative rates of uplift ($\Delta u/\Delta t$), stream-channel downcutting in the mountains ($\Delta cd/\Delta t$), piedmont aggradation ($\Delta pa/\Delta t$) and piedmont degradation ($\Delta pd/\Delta t$). In Figure 15.1, high, moderate and low tectonic activity are associated, respectively, near to the mountain front, with (A) thick accumulations of piedmont sediments, (B) entrenchment and (C) undissected pediment. Within category (C), Bull recognized three typical landform assemblages (Table 15.1). Weise (1978), working in Iran, argued that a fault scarp is not the only way of tectonically creating the mountain front: even a relatively gentle folding is capable of leading to incision that leads to a similar profile.

15.2 Slope retreat hypotheses

The idea that pediments develop as a result of the weathering and erosion of a retreating steep slope – in effect that the pediment is a surface of transportation – is an old one. Emphasis in the desert literature has been on the nature of the retreat, whether by parallel recession or downwearing.

15.2.2 Parallel retreat hypothesis

The notion of parallel retreat of both gravity-controlled and debris-covered slopes has been explored in Chapter 3.2. It is a notion which is easily extended to the development of pediments, especially in circumstances where debris is constantly removed from the base of the steeper slopes, the surface debris of the steeper slopes can be removed by prevailing forces of erosion, the piedmont angle can retreat, and a "caprock" helps to "hold up" the mountain front.

Lawson (1915), for example, viewed the geometry of mountain and plain development in basin–range country as follows. A mountain front, however it may have been initiated, undergoes weathering and erosion, achieves a characteristic slope (determined by lithology) and retreats roughly parallel to itself (Fig.

Table 15.1 Classification of Quaternary relative tectonic activity on mountain fronts (*t*, time; *u*, uplift; *cd*, channel downcutting; *pa*, piedmont aggradation; *pd*, piedmont degradation).

Relative tectonic activity	Relative rate of activity	Typical landforms at mountain fronts	
		Piedmont	Mountain
Active			
Class 1, highly	$\Delta u/\Delta t \geq \Delta cd/\Delta t + \Delta pa/\Delta t$	untrenched fan	V-shape cross-valley profile in bedrock
a			
b	same as above	untrenched fan	U-shape cross-valley profile in alluvium
Class 2, moderately	$\Delta u/\Delta t < \Delta cd/\Delta t > \Delta pd/\Delta t$	entrenched fan	V-shape valley
3	same	same	U-shape valley
4	same	same	embayed mountain front
Inactive			
Class 5	$\Delta u/\Delta t \lll \Delta cd/\Delta t > \Delta pd/\Delta t$	dissected pediment	dissected pediment embayment
a			
b	$\Delta u/\Delta t \lll \Delta cd/\Delta t = \Delta pd/\Delta t$	undissected pediment	pediment embayment
c	$\Delta u/\Delta t \lll \Delta cd/\Delta t < \Delta pd/\Delta t$	undissected pediment	may have characteristics of active mountain front

Source: Bull (1984).

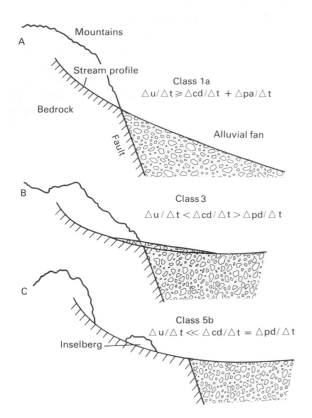

Figure 15.1 Sketches along longitudinal profiles of fluvial systems, showing the landform associations that are typical of different uplift rates; and inter-relations of local base-level processes that are conducive to (A) the accumulation of thick alluvial-fan deposits next to a mountain front, (B) the development of an entrenched alluvial fan, and (C) the development of an undissected pediment. See Table 9.15 for further explanation (after Bull 1984).

15.2A). As the mountain front retreats, alluvium accumulates at its base, and a bedrock bench is produced beneath it. The nature of the mountain-front-piedmont plain profile is determined by many features which include the initial geometric framework; processes and rates of denudation; structure and lithology; and tectonic conditions. Lawson says little about the *plan* characteristics of the situation. This omission is important because mountain fronts are not always straight in plan, and embayments may be crucial loci of evolution (see Sec. 15.3).

In the "ordinary case" envisaged by Lawson, of a piedmont plain with a rising base level, the mountain front retreats at a uniform rate, but the alluvium

increases in thickness at its foot at a decreasing rate because successive alluvial increments are spread over a progressively larger alluvial plain (Fig. 15.2A). As the piedmont plain continues to develop, however, the smaller quantities of alluvium derived from the mountain may be insufficient to keep the extending piedmont plain covered, and a bedrock bench may be formed subaerially (a pediment). Ultimately, the mountains may be completely eliminated, and only piedmont slopes will remain.

Since it was introduced, Lawson's imaginative hypothesis has been critically examined, modified, embraced and rejected (e.g. Davis 1930, Tuan 1959, King 1962). In general, if not in detail, its assumptions accord well with observations of pediments in many basin-range deserts (although, as discussed above, a convex suballuvial bench has rarely, if ever, been observed). Its principal failing appears to be that

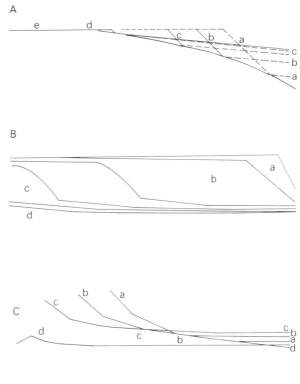

Figure 15.2 Hypothetical sequences of mountain and piedmont plain development in the desert. (A) Lawson's scheme in an area with a continuously rising base level (after Johnson 1932). (B) Stages in the evolution of a granite mountain area in the Sudan (after Ruxton & Berry 1961). (C) Successive profiles of regrading allied to lateral plantation of pediments (after Johnson 1932).

the rôle of major drainage courses from the mountains is underplayed. Ruxton & Berry (1961) formulated an interesting extension of the ideas of Lawson and others in explaining the evolution of granite landforms in arid, semi-arid and savanna areas of the Sudan (Fig. 15.2B). They suggested that once an equilibrium profile is established, it continues to develop by parallel retreat of both the hillslope and the "plinth", until eventually downwearing becomes more important than backwearing. No progressive accumulation of basin sediments is involved in the area, but central to the hypothesis is the presence of a regolith zone along most profiles, which comprises a superficial layer of migratory debris and beneath it a sedentary weathering zone in which material is prepared for removal. Both of these models imply that the pediment slope declines over time as the surface is regraded.

Rates of mountain-front retreat and pediment extension are largely a matter of speculation. Melton (1965a) reckoned that pedimentation in southern Arizona may have proceeded, assuming a uniform rate, at about $0.1cm$ yr^{-1}. This compares with King's (1953) general estimate of $0.1-0.2cm$ yr^{-1}, with Schumm & Chorley's (1966) estimate of scarp recession rates in Colorado of $0.06cm$ yr^{-1}, and with Twidale's (1976) estimate of $0.05mm$ yr^{-1} in the southern Mount Lofty Ranges in Australia.

A problem closely allied to the parallel retreat hypothesis is that of the origin and development of the **piedmont angle**. As noted previously, the piedmont angle varies greatly from place to place, and the boundary between mountain and plain may be anything from a gradual change of slope to an abrupt break. In places the mountain front is extremely steep, and occasionally it appears to have been oversteepened. There appears to be general agreement – but little proof – that the piedmont angle is maintained as the landforms evolve. A corollary of this view is that the mountain front retreats parallel to itself. This appears to be established in certain places where change is rapid (e.g. Schumm 1956), and is eminently probable in arid and semi-arid areas, where there is *unimpeded removal of weathered debris from the base of mountain fronts*, and slope inclination is adjusted to environmental conditions. A further corollary is that as the mountain front retreats it *automatically* creates a pediment surface which can later be modified (e.g. Büdel 1970, King 1953).

Explanations of the origin and maintenance of the piedmont angle include (a) lateral planation by streams (Johnson 1932, Rahn 1966); (b) uplift of a mountain block along a fault or fold; (c) change from one type of bedrock lithology or structure to another (Twidale 1964, 1967, Denny 1967) and, allied to this, maintenance of steep mountain fronts by resistant caprocks (e.g. Everard 1963); (d) differences between processes operating on mountain front and pediment (e.g. change from rillwash to sheetwash; from turbulent to laminar flow; from gravity-controlled slopes to fluvially controlled slopes); (e) change from one type of debris to another, such as the bimodal debris sizes associated with granite weathering (e.g. Cooke & Reeves 1972); (f) maintenance or even oversteepening of the mountain front by ground-level trimming (Mabbutt 1966a) or locally intense weathering followed by exhumation (Twidale 1987, 1982a); and (h) concentrated subsurface erosion at the hillfoot due to lateral flow or water in the weathering profile (Ruxton & Berry 1961, Bocquier 1968).

One of the weaknesses of the parallel retreat hypothesis is that it seems to assume that the mountain front is *linear*. Often, however, this is not so. Many, perhaps most, mountain fronts in deserts have **embayments** where slope development may be strongly influenced by lateral planation and other channel processes (see Ch. 15.3).

15.2.2 Retreat by downwearing

Retreat may not be parallel, especially where debris from the mountain front accumulates near to the slope base (i.e. covering the pediment) and there is no steep-slope sustaining caprock. An excellent example of the rapid downwearing development of initially steep scarps is provided by Wallace (1977; Fig. 15.3A & B) who showed that alluvial fault scarps in north-central Nevada decline with age (from about 60° to 8–25°) as *processes* on the initial tectonic surfaces alter with time from mass-washing control to wash control (Fig. 15.3C). In a similar study, Sterr (1985) not only described retreat, rounding and lowering of fault scarp crests, but also used morphometric indices to estimate rate of degradation. In a similar way, precise ^{14}C and K–Ar dating of cinder cones and their development in the Cima volcanic field in California led Dohrenwend et al. (1986) to conclude that about one million years of development was characterized by general *downwasting*, progressive decrease in cone height and *decline* in mean maximum slope, trends that were accomplished through a sequence of Quaternary environmental changes (Wells et al. 1985).

There is little doubt that some slopes have re-

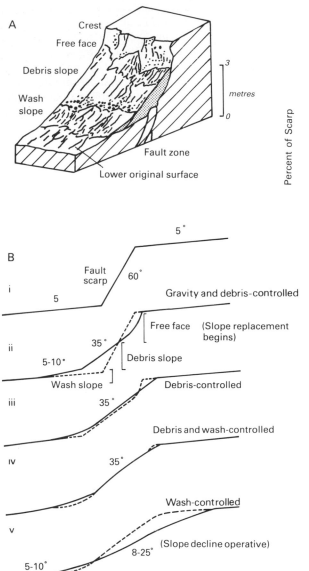

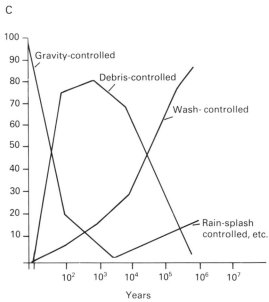

Figure 15.3 (A) A block diagram of a fault scarp. (B) The sequence of fault-scarp degradation; to show incremental change, the dotted line represents the solid line of the previous profile. (C) A speculative interpretation of the relations between the dominant erosional processes acting on scarps in unconsolidated alluvium and colluvium as they age. The shapes of the curves were derived largely by deduction, but the time element is estimated from the data presented in Wallace (1977).

treated, either by parallel retreat or by downwearing, to leave pediments at their bases, and this hypothesis has been adopted, in different forms, to explain many desert landscapes. For example, Weise (1978) recognized an evolutionary sequence in Iran in which an initial, tectonically generated mountain–bahada landscape is successively transformed into pediments, pediplains and desert dunes, by both parallel retreat of steeper slopes and regrading of pediments. Here, as elsewhere, the major controversy is over *how* the

slopes retreated. As explained below (Sec. 15.5) in some areas there is also a controversy over *whether* they retreated.

15.3 Drainage-basin hypothesis

The drainage-basin hypothesis represents a logical extension of Lawson's view in the context of modern drainage-basin studies. It has been succinctly propounded and encapsulated by Lustig (1969a: D66–7):

It has been proposed that pediments result from parallel slope retreat of the mountain front . . . The writer's observations in several deserts of the world suggest that few, if any, escarpments are not dissected by prominent drainage systems . . . The existence of drainage basins in the mountain ranges is of central importance to the various pediment arguments. These basins are the loci of the most effective erosional processes that operate on mountain ranges . . . The mountain fronts may well retain some characteristic slope angle that reflects rock strength, structure, weathering characteristics, and other variables, and they may retreat at this angle. This does not prove, however, that ranges are primarily reduced by parallel retreat of escarpments. . . .

. . . There is no question that processes of subaerial and suballuvial weathering occur on pediments today, nor that fluvial erosion also occurs. A pediment must exist prior to the onset of these processes, however, and in this sense the origin of pediments resides in the adjacent mountain mass and its reduction through time. . . .

To say that pediments result from the reduction of adjacent mountains through time is to give voice to a seemingly obvious and intuitive argument . . . Each range has some local base level in an adjacent basin, and it cannot be eroded below this level. Given stability for a sufficient period of time, the consequences of mountain reduction must inevitably include the production of a pediment . . . The only real "pediment problem" is how the reduction or elimination occurs.

Observation shows that the course of any master stream channel from a given drainage basin from the mountains onto the pediment surface and thence to the basin floor has no sharp change of slope . . . The interfluvial areas, however, generally do exhibit a marked change in slope, at least within a narrow zone parallel to the mountain front. The reason for the existence of such a zone is precisely that it is an interfluvial area; the dominant process that operates on a mountain front is not fluvial.

Qualitatively, it can be argued that two basic processes are operative on a given mountain mass. The steep slopes of the mountain front are interfluvial areas that are subject to weathering. Runoff on these steeply sloping surfaces is of short duration and is not concentrated. The runoff serves largely

to remove the finer weathered debris that is transportable. Larger particles generally remain in place until they are reduced in size by weathering. The rates of mountain front retreat are basically unknown, but by any reasonable assessment they are slow in relation to rates of processes that are operative in drainage basins. . .

A version of this hypothesis has been formulated by Parsons & Abrahams (1984). Working in the Sonoran and Mojave deserts, they showed that a high proportion of mountain fronts are in fact embayments, and that this proportion is directly related to the area of the mountain mass and is inversely related to its relief. The embayments may in places make only slight indentations into the mountain front. Elsewhere the embayments are effectively valley floors that may extend many kilometres into the mountain range. In embayments, mountain-front retreat (Fig. 15.4A) is replaced by divide removal (Fig. 15.4B). Thus, where a mountain front consists largely of embayment, mountain mass reduction occurs mainly by divide removal. Field measurements confirmed that both hillslope processes and lateral erosion by channelled streams play a part in piedmont development inside and outside embayments. This is a view that accords with Sharp's (1940) analysis of the Ruby–East–Humboldt ranges in Nevada, where it was estimated that about 40% of the mountain front was subject to "lateral planation". Similarly, Weise (1978) attributed a major rôle to lateral planation in the development of pediments in Iran.

The **lateral planation hypothesis**, so denigrated by Lustig (1969a) is thus not without its supporters. In Douglas Johnson's (1932: 656–8) words:

The essence of the theory is that rock planes of arid regions are the product . . . of normal stream erosion. A peculiarity of the stream erosion theory is the relative importance attached to lateral corrosion . . .

Every stream is, in all its parts, engaged in the three processes of (a) vertical downcutting . . . (b) . . . aggrading, and (c) lateral cutting . . . If we imagine an isolated mountain mass in an arid region, it will be evident that the gathering ground of streams in the mountains . . . will normally be the region where vertical cutting is at its maximum . . .

Far out from the mountain mass conditions are reversed . . . Aggradation is at its maximum . . .

Between the mountainous region of apparently dominant degradation and the distant region of

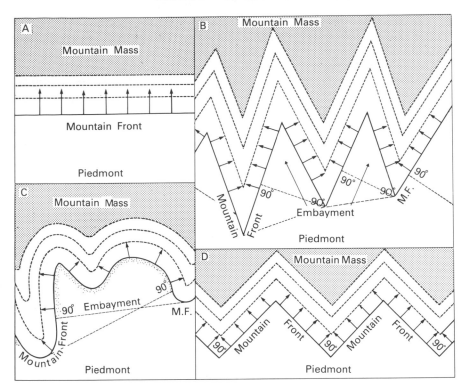

Figure 15.4 Piedmont extension by (A) mountain-front retreat along a straight mountain front, (B) by divide removal in embayments, (C) at embayment heads and promontories with embayments, (D) by mountain-front retreat along a sinous mountain front (after Parsons & Abrahams 1984).

apparently dominant aggradation there must be a belt or zone where the streams are essentially at grade . . . Lateral corrasion is again dominant not only in fact but in appearance . . . Heavily laden streams issuing from the mountainous zone of degradation are from time to time deflected against the mountain front. This action, combined with the removal of peripheral portions of interstream divides by lateral corrasion just within the valley mouths, insures a gradual recession of the face or faces of the range.

As the processes described by Johnson proceed, the pediment profile is progressively regraded (Fig. 15.2C). Abrahams & Parsons (pers. comm., 1990) feel that lateral planation is the major process of divide removal and that the process can operate in embayments that extend far into the mountains and along streams that create them.

15.4 Mantle-controlled planation hypothesis

This hypothesis may most appropriately be reviewed by quoting Mabbutt (1966a: 90–91). The remarks apply only to central Australia, but they may well be relevant in other areas:

. . . it is concluded that grading on these granitic and schist pediments has largely acted through the mantles, whereby the smoothness of depositional profiles has been transmitted to suballuvial and part-subaerial pediments. The action of the mantle is most direct on granitic pediments, in that suballuvial notching and levelling proceed through weathering in the moist subsurface of the mantle. On schist pediments control of levelling by the mantle is less direct in that its upper surface is the plane of activity of ground-level sapping and of erosion by rainwash and sheetflow. In this way, continuity of grade is established across a succession of separate schist outcrops. On schist pediments, where trimming is more rapid, a single depositional cycle may leave a widespread erosional imprint; on massive granitic rock, where weathering advances on a broad front and much more slowly, each cycle of mantling may produce only minor effects. The processes described are most active near the hill foot, where mantling and stripping alternate more frequently; the lower parts of both schist and granitic pediments in central Australia appear to be largely and more permanently suballuvial . . .

It might be objected that the processes described are tantamount to insignificant secondary modification, under stable conditions, of a pedimented landscape left by other primary agents. However, this would lose sight of landscape evolution in these areas since the Tertiary. Relics of a weathered Tertiary land surface survive as laterite-capped low platforms on plains adjacent to all study areas and occur close enough to the hills in the Aileron and Bond Springs areas for the former piedmont profiles to be reconstructed. The profiles show that the present plains were already the sites of lowlands on the Tertiary surface, topped by the hills of today, and that subsequent evolution has involved mainly the breaching of the duricrust and the etching away of a layer of soft weathered rock . . . In most sites the hill base was set back to a structural boundary from which there can have been little subsequent retreat . . . What has been involved in pedimentation has been slight back-trimming at the hill base, partial weathering of an irregular, exhumed weathering front, and the imposition of continuous, concave profiles . . .

This scheme is genetically allied to that of Thomas (1966), who found in Northern Nigeria that the main processes working in the inselberg landscape were etching and stripping, and to that of Ruxton & Berry (1961) noted above. It has been widely acknowledged to be of importance, especially in Australia (e.g. Twidale 1983b). Surface weathering *without* a mantle may also be a relevant process of pediment lowering, as Dzulynski & Kotarba (1979) argued for pediments in Mongolia.

15.5 An exhumation hypothesis

An exhumation hypothesis differs from the others in that it assumes a mountain–piedmont-plain relationship to be established; indeed, it could have been established in terms of any of the hypotheses described above. All that is required is a transfer of the alluvial boundary across the piedmont plain to expose the bedrock pediment. Evidence for such movement of the boundary may include truncation of soil and weathering profiles, and the presence of alluvial outliers. The exhumation will normally be accomplished by running water, which may be concentrated in channels, or may merely be unconcentrated wash. General causes of such changes may be: changes in the geometric relations of mountains and plains in the

system; climatic or tectonic changes; or base-level changes. Paige (1912) and Tuan (1962) are amongst the several proponents of this hypothesis.

15.6 Inheritance hypotheses

In the few localities where pediment surfaces have been dated absolutely, some surprising results have emerged. For example, many pediment surfaces may be of great antiquity and be inherited from quite different morphogenetic conditions.

Thus, K–Ar dating of successive late Cenozoic basaltic lava flows resting on pediments and pediment domes in the Mojave Desert, California (Dohrenwend et al. 1987) not only established the early Tertiary *origin* of the surfaces. It also revealed that the lavas erupted onto a *downwasting* erosional environment, because progressively younger flows bury progressively lower and younger surfaces, so that each lava-protected remnant stands above the present surface to an extent that is proportionate to its age (Fig. 15.3). Average rates of lowering were determined by dividing the average height of each remnant above the modern surface by the K–Ar age of its basaltic cap. The average downwasting rates against distance from dome crest are shown in Table 15.2; they lead to a simple model of dome evolution in which pediment surfaces are progressively lowered (Fig. 15.5) down to the mid-dome zone. It is a model which Dohrenwend et al. (1987) thought could well apply to many

Table 15.2 Average downwasting rates versus distance from dome crests.

Distance from dome crest (km)	Downwasting rates (cm/10³yr)			
	Profile B 3.88myr	Profile E 0.99 myr	Profile F 0.85myr	Average
2.0	2.8	–	–	2.8
2.5	2.6	2.8	–	2.7
3.0	1.5	2.1	–	1.8
3.5	1.3	1.7	1.8	1.6
4.0	1.2	1.2	1.2	1.2
5.0	0.5	0.4	0.7	0.5
6.0	0.2	–	0.3	0.2
7.0	0.0	–	0.0	0.0
12.0	–	–	0.0	0.0

Source: Dohrenwend et al. (1987).

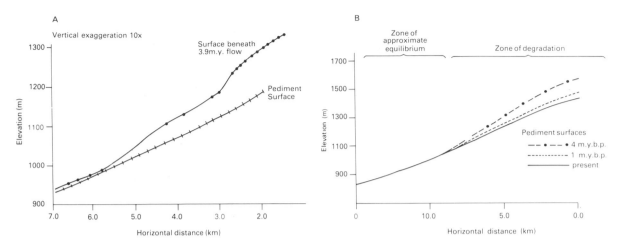

Figure 15.5 (A) Longitudinal profiles comparing middle Pliocene and modern pediment surfaces in the Mojave Desert, California: the horizontal distance is measured from the dome summit. (B) An empirical model of pediment dome evolution synthesized from the longitudinal profiles and data in Dohrenwend et al. (1986). The present surface is a generalized composite of two representative profiles of the modern surface of Cimacita dome. The Pliocene and Pleistocene surfaces were reconstructed using a smoothed plot of average downwasting rates versus distances from dome summits. Crestal and upper flank areas have downwasted, mid-flank areas have remained in a state of approximate equilibrium (i.e. little or no downwasting or aggradation over time), and lower flank areas (not shown) have probably aggraded. The horizontal distance is measured from the dome summit (after Dohrenwend et al. 1986).

upland areas flanking closed, arid basins in the American southwest. Of course, given that these pediments are not attached to mountains at their highest points, downwearing is certainly the most likely trend of development, especially if upwarping is involved.

This model still begs two questions: How old are the *original* surfaces, and how were they *formed*? Oberlander (1974) expertly addressed these questions in the Mojave Desert. He concluded, using the evidence of dated volcanic deposits on pediment surfaces and of the weathered profiles beneath them, that the pediments were inherited features entirely formed in the Tertiary *prior* to the establishment of full desert conditions. The pediments were formed under what he calls "denudational equilibrium", whereby "parallel" hillslope erosion was balanced by regolith renewal through rapid chemical breakdown along a subsurface weathering front in semi-arid conditions over eight million years ago. During this phase of pediment evolution, *backwearing* of soil-covered slopes, in circumstances similar to those described above (Ch. 13; Oberlander 1972, Moss 1977), was the norm and the surface may have been "born dissected". Subsequently, during the late Pliocene and Quaternary, the pediments developed by stripping the weathering

mantle, and backwearing of no more than 5m, an evolution quite compatible with that envisaged by Dohrenwend et al. for the pediment domes. Present-day weathering is slow, as revealed by the varnish cover on boulders.

This interpretation does more than illustrate the antiquity of pediments and their subsequent evolution. It throws into severe doubt the models of parallel retreat developed by Lawson (1915) and others for the same region. It means that pediment evolution, at least in this area, is not a peculiar product of aridity; and it implies that morphometric analyses (such as those of Cooke 1970a), because they assume that the pediment is an integral part of the present system, may be rather fanciful. The failure of morphometric analysis to explain the origin of pediments may actually reveal the *lack* of adjustment between present forms and contemporary processes (Oberlander 1974).

The idea that pediments and their associated mountains are palaeoforms is not confined to the American Southwest. It has been explored in detail, for example, by Twidale in Australia (e.g. 1976). He demonstrated that there are substantial remnants of ancient pediplains ("palaeo-surfaces"), surfaces that have not been exhumed, throughout the continent. For example, he described a laterized palaeoplain of early-

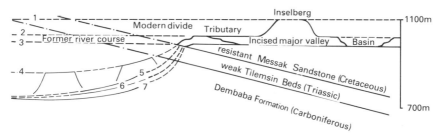

Figure 15.6 A simplified profile showing major stages in the development of the Msak/Manghini escarpment, central Sahara (not to scale). Landslides, Quaternary slope debris mantles and their dissection are not shown. 1, Former plateau level; 2, present plateau level; 3, floor of large hanging valleys and of large intra-plateau basins – river beds and the plateau surface actually converge towards the right; 4, one of several surfaces formed above the present foreland level during differential lowering; 5, 6, slope profiles of increasing concavity merging at the present foreland level – younger stages of escarpment growth; 7, youngest surface and slope (pediment) probably already dissected towards the end of the Tertiary (after Busche 1979).

to mid-Tertiary age in Northern Australia, pre-Cenozoic (possibly even Cretaceous or older) surfaces in central and southern Australia, and surfaces that may be inherited from the middle Jurassic, or even the Triassic, in the Mount Lofty Ranges of South Australia. Although not exhumed, some of these surfaces may be etchplains – surfaces from which a weathering mantle has been removed to expose the base of weathering. The "New Plateau" in western Australia is such a surface, formed by stripping a laterite mantle. In addition, Twidale & Bourne (1975) argued that some inselberg slopes associated with palaeoplains have not retreated significantly over exceptionally long periods of time, and that they do not therefore conform easily to the notion of parallel retreat under arid conditions.

A comparable view is expressed in the central Sahara, where German workers such as Hagedorn and Busche (e.g. Busche 1979; Fig. 15.6) have shown that the major period of pediment formation preceded Quaternary landsliding in the scarps surrounding the Murzuk Basin, that there has been scarcely any scarp retreat since, and that it is unlikely that the pediments were formed under arid conditions. Indeed, some of the surfaces may be stripped etchplains, initially produced under savanna conditions (e.g. Busche & Hagedorn 1980).

Twidale's (1976) views are summarized in a useful model of evolution (Fig. 15.7), in which initial structural contrasts are accentuated by weathering and erosion of weaker zones, in which relief amplitude is increased, and in which ancient surfaces survive or develop only slowly over millions of years. This

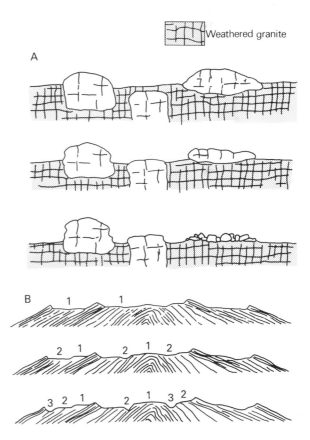

Figure 15.7 Two models of landscape development proposed by Twidale. (A) Granite: stippled areas represent weathered granite. (B) Fold mountain range: 1, oldest palaeoplain; 2, second palaeoplain; 3, present surface of low relief (after Twidale 1976).

229

Figure 15.8 A sequence of piedmont surfaces in the southern Atacama Desert, Chile.

model of palaeoform persistence, he suggested, may be most appropriate in arid and semi-arid environments.

15.7 Sequences of piedmonts

Most of the previous discussion has been concerned with the nature and evolution of a single piedmont in an active system. In many areas, however, several piedmonts may coexist as a "geomorphological staircase" and, although they may have been initiated at different times, it is possible that they are all continuing to be extended (Fig. 15.8). Sequences of

pediments have been described in many areas: for example, by Twidale (1982a) in South Australia, by Butzer (1965) at Kurkur in Egypt, by Mensching (1970) in the northern Sahara, by Rich (1935) at Book Cliffs, Utah, by Sharp (1940) in the Ruby–East-Humboldt Range, Nevada, by several authors in Arizona and New Mexico, and by Weise (1978) in Iran.

There are, for instance, magnificent flights of terrace pediments along the valleys of the San Pedro, Rio Grande and other major rivers in Arizona and New Mexico. The principal explanatory hypothesis states that the succession of surfaces results from

Figure 15.9 A longitudinal profile of the lower San Pedro Valley, Arizona, showing the elevation of piedmont surfaces on both valley flanks (after Tuan 1962).

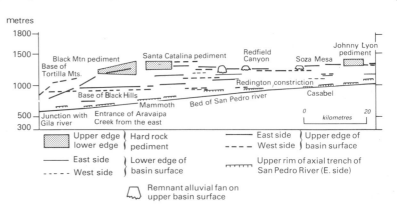

valley-side planation graded to successively lower base levels of the downcutting rivers in late Tertiary and Quaternary times (e.g. Bryan 1936). Tuan (1962) considered the causes of the suite of surfaces in the San Pedro Valley (Fig. 15.9) in some detail. He showed that the simple "base-level hypothesis" was not strictly applicable in this area. More probably, he claimed, it was tectonic movements and their effects on the fluvial system which led to the succession of surfaces. Evidence for tectonic activity and against base-level change includes the recognition of structural deformation during the period of landform development, the irregular arrangement of basin surfaces, their differential dissection and exhumation, and the straightness of scarps between upper and lower surfaces.

Rich (1935) put forward an alternative explanation of multiple pediment surfaces where the pediments are developed on weak rocks adjacent to mountain catchments comprising more resistant rocks, in which no recourse to *external* base-level change is required. He envisaged (Fig. 15.10) streamflows from the mountains removing coarse sediment on a steeply graded surface in the piedmont zone, whereas streamflows arising from within the piedmont cone itself on the soft rocks require a gentler gradient but cut down more deeply than the mountain streams near the piedmont angle. As a result the local piedmont streams (C, D) will *capture* the higher streams, leaving an abandoned pediment, (E) (Rich 1935, Schumm 1977). If repeated, a sequence of pediments may be formed. These pediments, as Chorley et al. (1984) pointed out, are irregular bedrock surfaces mantled with sediment that makes them appear smooth. However, they differ fundamentally in origin from pediments formed by scarp retreat or lateral planation.

Finally, the multiplication of piedmont surfaces may be attributed to climatic changes. The alternation of pluvial and arid conditions, or the local modifications of aridity, may alter fluvial conditions sufficiently to provoke dissection or planation, especially where there is no evidence of tectonic or base-level changes. This is a view advocated, for example, for the southern Atlas Mountains (Morocco to Tunisia) in the Sahara and in Egypt by French workers (e.g. Coque 1960, 1969, 1979a). Here, three to five *glacis d'érosion* (Fig. 15.11) are said to reflect the alternation of *périodes d'aridité attenuée* (contemporaneous with pluvial periods, when diffused runoff formed the glacis) and *périodes d'aridité accentuée*

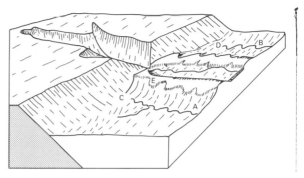

Figure 15.10 The development of multiple pediments by stream capture (after Rich 1935). For explanation see text.

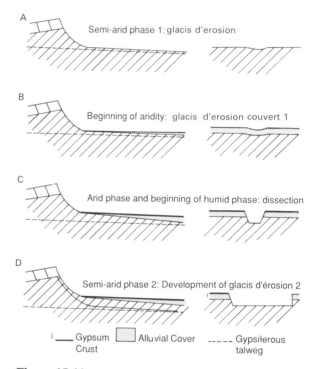

A   Semi-arid phase 1: glacis d'erosion

B   Beginning of aridity: glacis d'erosion couvert 1

C   Arid phase and beginning of humid phase: dissection

D   Semi-arid phase 2: Development of glacis d'érosion 2

i ⸻ Gypsum Crust   ▨ Alluvial Cover   ----- Gypsiferous talweg

Figure 15.11 A morphoclimatic sequence in a piedmont zone of the Tunisian Sahara (after Coque 1969).

(interpluvial periods, when linear erosion caused their dissection). Mabbutt (1966b) recognized similar pluvial/humid alternations in the McDonnell ranges in central Australia.

Thus, the explanation of piedmont sequences ultimately depends on the local recognition of climatic base-level, tectonic or "internal drainage basin" changes and their relationships. Such recognition is often still a matter of controversy and conjecture.

PART 4

WIND IN DESERT GEOMORPHOLOGY

CHAPTER 16

The rôle of wind

16.1 Introduction

16.1.1 The competence of wind

Wind blankets a quarter of the deserts in sand (Fig. 16.1), and strips hundreds of million tons of dust from them each year. It was even more effective in the past than it is today. Mesozoic winds assembled massive sandstones in Europe and North America, and nearer in time, dust preserved on ocean beds and in ice sheets demonstrates that they were even more vigorous 20,000 years ago than they are today. Farther in space, wind is by far the most powerful surface agent in the deserts of Mars.

On Earth, desert winds can carry more sediment, measured in m^3 (km width)$^{-1}$ yr^{-1}, than any other geomorphological agent. Hundreds of square kilometres of dunes, most over 3m high, each progressing at 15m yr^{-1}, are not an unusual occurrence. Sven Hedin's yardangs, which cover some 10,000km^2 of central China, are alleged to have been excavated by wind to depths of 2–3m in historical time.

16.1.2 Wind versus water

Observations like these provoked early geomorphologists to declare that wind dominated desert erosion (Penck 1905b, Tolman 1909, Keyes 1912, Paige 1912, Walther 1924, Passarge 1930). Of these the most audacious was Keyes, who claimed that wind had repeatedly levelled landscapes, a process he termed "eolation". Keyes's recklessness, and the evidence of effective water erosion in the more accessible deserts, converted the majority of later geomorphologists to fluvial models of desert denudation. The most influential writers dismissed wind erosion as little more than cosmetic (Bourcart 1928, Davis 1930, 1954).

For all these rebuffs, the aeolianist argument is not dead. The Sahara probably produces 60 to 200 million tonnes yr^{-1} of dust, while the Niger, the only major river to drain the same area, carries a mere 15 million tonnes yr^{-1} of sediment (Grove 1972, Junge 1979). The imbalance was even greater in windier, drier periods in the Quaternary. But the comparison of rates is a crude form of debate. More than one modern geomorphologist, moreover, has proposed candidates as plains of wind erosion.

16.1.3 Collaboration of wind and water

The evidence of dust transport and aeolian planation may not please obdurate fluvialists, but neither does it vindicate Keyesian aeolianists. However, the polarized debate ignores the diversity of deserts, and, more important, the co-operation between the two erosional and sedimentary processes.

The debate, at its crudest, does not admit that wind and water dominate different ends of the humidity range in deserts (Fig. 16.2). With little rain, as in the central Sahara or central China, wind indisputably achieves more than water, but the rate of fluvial transport overtakes wind at modest rainfalls, as in most deserts. Although the data on which these estimates of erosion rates rest are not entirely without problems, as Walling & Webb (1983), acknowledge, and as L. Wilson (1973) and Schmidt (1985) have shown, the essential truth of the present assertion (that wind erosion dominates very dry deserts and water erosion increases towards the desert margins) remains.

A second, more profound mistake is to assume that wind and water work in isolation, for quite on the contrary, they are in close partnership, in most of the desert (which is neither very wet nor very dry). Little windblown sediment has not first been released from rock by weathering processes dependent on the presence of water and then been transported by water. More surprisingly, almost all of the contemporary fluvial sediment in some (perhaps most) semi-arid areas was recently aeolian. In the Quaternary, the alluvium of wet phases became aeolium of the dry, and vice versa. Because fluvial groundwork was more dynamic in semi-arid than very arid regions, producing more alluvium, the paradox is that there is now

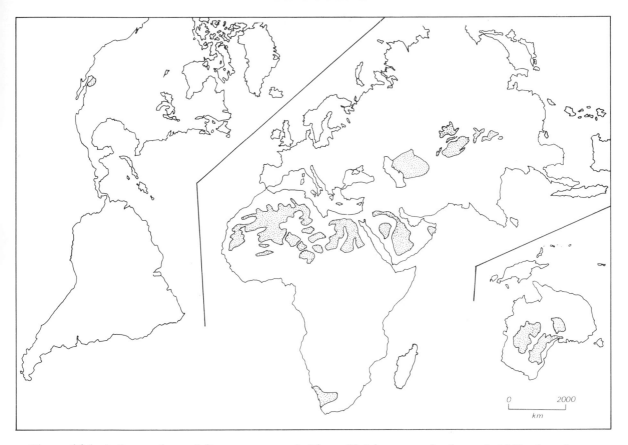

Figure 16.1 Active sand seas (after many sources). Figure 29.1 is a map of active and stabilized sand seas.

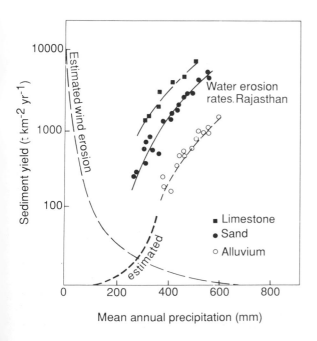

more windblown sediment (loess and sand) in wetter than in drier deserts. Though marked enough for sand, the semi-arid distribution of loess is even more pronounced, for loess is trapped in large quantities only on vegetated surfaces and is more effectively brought down by rain than by simple fall-out. Co-operation can be seen in erosional as well as depositional forms, for wind is impotent even to abrade without an input of fluvial sand, and the highest rate of fluvial erosion on Earth occurs in the Chinese loess, an aeolian sediment. Depressions of the order of hundreds of square kilometres in many deserts were probably created in a combined attack by both water and wind.

Even the contrast in the behaviour of wind and

Figure 16.2 Measured and estimated rates of erosion by wind and water in different climatic conditions (data after Walling & Webb 1983; estimate after Marshall 1973).

water seems more to augment than restrict collaboration. Water erosion begins on the broad surfaces of slopes, but is then rapidly narrowed into channels. Wind works on the broad front at all scales. Moreover, a comparison of the great streams of sand in sand seas with those in stream channels shows stark contrasts in rhythm and spatial pattern, for the magnitude–frequency relations of sediment-moving winds are quite unlike those of sediment-moving desert streams. However, although dunes usually over-ride low-order stream channels, they seldom bury high-order rivers. Each process therefore has its domain, one taking over where the other leaves off, ensuring that desert erosion as a whole follows neither a purely aeolianist nor a purely fluvialist agenda.

16.1.4 The distinctiveness of aeolian geomorphology

Aeolian geomorphology has always drawn less attention than its fluvial, slope, coastal and glacial counterparts. Remoteness, lack of data and fewer practical applications are doubtless responsible. Persistent low priority has permitted slow-moving debates, of which the controversy over the relative rôles of wind and water was but one. The ballistic hypothesis of ripple formation, the roundness of wind-blown sand, the rôle of eddies in the formation of the lee slopes of dunes, the "wind-rift" hypothesis of linear-dune formation, the "response diagram", and the size of roll vortices in the planetary boundary layer have been others. The only major contribution of the aeolian branch to geomorphology as a whole, albeit a massive one, was Bagnold's classic, *The physics of blown sand and desert dunes* (1941).

Aeolian geomorphology has, however, been brought to life by discoveries on Mars; by the realization of the value of modern and ancient dune sands as reservoirs for water and oil; and by the increasing accessibility and economic value of terrestrial deserts. It is profiting from the same kinds of advance in theory and technology that have revived the rest of the discipline, but it still labours under two peculiar constraints:

(i) Above all, the great three-dimensionality of processes in most aeolian environments: winds can come from many directions during an annual cycle, whereas all stream and slope movements are unidirectional (down slope) and most coastal movements are narrowly longshore.

(ii) Because of the intensification of atmospheric circulation in some Pleistocene periods, many extant aeolian landforms were shaped by substantially stronger winds than those of today. As in glacial geomorphology, whole classes of feature are essentially inactive, with few, if any, modern analogues (for example, large vegetated parabolic dunes, lunettes, pans and megayardangs).

These constraints mean that great challenges remain, as this part of the book will show. The structure of the discussion is based, in part, on the belief that understanding at a small scale underlies understanding at larger scales, but also on the geomorphological maxim that the nature, relationships and number of factors increases with increasing scale (Schumm & Lichty 1965). The approach, therefore, is through a series of broadening horizons: from grain entrainment and movement to ripples; from desert winds to wind abrasion and deflation; from the character of the resulting forms to aeolian sediments; from processes on individual dunes to patterns of dunes influenced by complex wind regimes; and, finally, from dune fields and sand seas to the influence of past climates on aeolian landforms.

16.2 Mobility in wind

Wind is more effective in deserts not because it is stronger than in wetter environments, for in general it is not, but for two other reasons: dry surfaces and sparse vegetation. Within the deserts themselves, where moisture and plant cover diminish, surface characteristics and wind velocity join them as factors determining the timing and location of aeolian activity.

16.2.1 Moisture and mobility

When the pores between them are saturated with water, loose particles on a surface are only partly exposed, and can experience only a small part of the drag and lift forces of the wind. As the sediment dries, more grains are uncovered and become more vulnerable, but they are still held tightly together by the surface tension in the meniscuses between them. This form of cohesion is at a maximum when meniscuses are tightly curved and when they extend over a maximum grain surface. Further drying contracts and ultimately breaks the meniscuses, cohesion dissipates, and the surface becomes progressively more vulnerable (Svasek & Terwindt 1974, de Ploey 1980, Sherman 1990).

The most significant effect of moisture is to increase the threshold of movement (defined in Ch. 17.3). Rule-of-thumb estimates have often chosen 4% water-content as the limit to wind erodibility (the pores then being about 15% full) (Azizov et al. 1979, Logie 1982, Zhu Zhenda 1984), though Goldsmith et al. (1988) found no movement, even in high winds, when soil moisture was as low as 1.8% by weight.

More rigorous models and observations show the process to be much more complex. Chepil's (1956) pioneering study found that soil moisture content at 15 atmosphere suction (the wilting point) marked the threshold between erodible and non-erodible soil. Johnson (1965, quoted by Goldsmith 1985) found a direct relationship between threshold shear velocity and the log of the water content. More recent work has been reviewed by Hotta et al. (1984) and Sherman (1990) (Fig. 16.3). They noted many problems in comparing the results. Discrepancies arise because of differences in the content of organic matter, fine particles or algae and bacteria, which, even in small numbers, have a conspicuous effect. Various research designs have accounted for evaporation in different ways. When winds are strong, evaporation is also high, and the threshold on a wet surface may be very little different from that on a dry one (Hotta et al. 1984).

Disagreement, therefore, is not surprising. Hotta et al. (1984) believed the most reliable experiments to show that the threshold increased by 7.5cm s^{-1} for each 1% increase in soil moisture content up to 8%. Grain-size had no appreciable effect on Hotta's relationship, but other workers have noted that the actual curve of mobility against moisture content in any one sediment must depend on the pore-size distribution and the pF curve (the curve of moisture content against the pressure needed to extract it), which itself depends on grain-size distribution. If the pores have a narrow size-distribution, a high proportion of them may suddenly empty of water as the soil dries, and sediment may then abruptly become available for transport by wind (Nickling 1988).

Moisture contents must reach much higher levels before they inhibit saltation, once it has begun, for then the chief method of entrainment is by bombardment (Sarre 1990; Ch. 17.3c). Sarre suggested that saltation was not inhibited until moisture contents reached above 14%.

It follows that if rain saturates a surface, transport is considerably reduced (Seppälä 1974). Winter rainfall of about 1570mm on the Oregon coast reduces seasonal sand transport to 36% of its potential, as calculated from wind data (Hunter et al. 1983). In northern Sinai, with winter rainfall at only about 97mm, the reduction is 14% (Tsoar 1974). In hot, dry conditions, showers penetrate quickly into sands, the

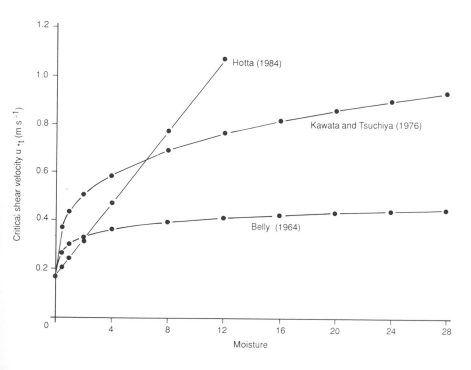

Figure 16.3 Different measurements of the shear velocity threshold of sand movement by wind as a function of moisture content of surface soil (after Sherman 1990).

surface dries rapidly, and movement is restrained only for an hour or two (Agnew 1988). The extent of restraint depends on the pattern of rainfall with respect to sand-moving winds, and must vary greatly from place to place, from season to season and from year to year. For example, wind erosion increases noticeably in drought years (Zingg 1953). The effects of moisture on dune form are discussed in Chapter 23.6.

If accompanied by a driving wind, rain may increase the transport rate by splashing particles into the path of the wind (de Ploey 1977, Logie 1982, Draga 1983, Rutin 1983, Sarre 1988b). Rain may also slake and pelletize fine soils and render them more susceptible to deflation when they dry.

Meniscuses are more gently curving between larger particles, such as sands, than between silt and clay particles, so that the same amount of moisture holds sands less firmly than finer particles (Bisal & Hsieh 1966). Sands have two further properties that render them more erodible by wind: they drain quickly; and because water is held loosely between sand particles, it is more readily evaporated from the upper horizons.

The effects of atmospheric humidity are more demanding to measure than those of soil moisture. Belly's (1964) relationship for the threshold velocity in humid conditions was: $u_{*th} = (1 + 0.5 \, h/100) \, u_{*t}$; where u_{*t} is the threshold shear velocity for dry conditions; and h is percentage atmospheric humidity (u_{*t} is defined in Ch. 17.3). Another experiment showed that the threshold velocity was 8.5m sec^{-1} (10m above-ground) when atmospheric relative humidity was 50%, and 12.5m s^{-1} when it was 80% (Knottnerus 1980). In the field, Nechaec (1969) found little wind erosion at relative humidities above 50%.

16.2.2 Plants and mobility

Because plants are more persistent than soil moisture, they are more effective at regulating sediment movement by wind. They play some increasingly complex rôles in this respect.

(a) Roughness height

When wind blows over a immobile surface, there is a shallow layer above the surface where the velocity is zero, as shown in Figure 16.4. The thickness of this layer, z_0, is known as the roughness height (or length). Because it is so small, z_0 is not an easy property to estimate. Its value can only be inferred from projections of velocity profiles towards the surface, and velocity measurement near the surface is very difficult, especially in blowing sand. Yet the

value of z_0 turns out to be critical to many models of sand transport and dune formation.

These problems notwithstanding, there have been some useful studies on the controls on the value of z_0, most based on laboratory studies rather than field work. It has been found to depend on the height, size, shape and the arrangement of roughness elements at different scales, but the relative rôle of each is difficult to discover (review by Rasmussen 1989, Greeley et al. 1990). On the level surfaces of wind tunnels, the value of z_0 has been found to rely mainly on the size of the particles. Bagnold (1941) proposed that z_0 was about 1/30 of the median diameter of the surface grains (a similar formulation was made by Zingg 1952).

In wind tunnels, z_0 has been found also to depend on the arrangement of grains. When coarse grains are close or far apart, the 1/30 rule (or other approximations) apply, but higher values of z_0 occur when the largest particles are spaced at about twice their diameter (Greeley & Iversen 1985, 42–3). Values of z_0 over sandy, level terrain, but without sand in saltation, are of the order of 0.005m. Gillette et al.

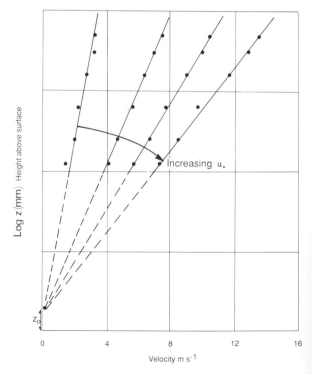

Figure 16.4 Definition of the roughness height, z_0, and shear velocity, u_*.

(1980) provided a list of z_0 values over a number of desert surfaces.

On near-uniformly vegetated surfaces, z_0 is displaced considerably upwards from its position over bare surfaces (Fig. 16.5; Olson 1958a, Bréssolier & Thomas 1977, Hesp 1983, Sarre 1988a). In situations such as these, the value of z_0 is affected by four characteristics of the vegetation.

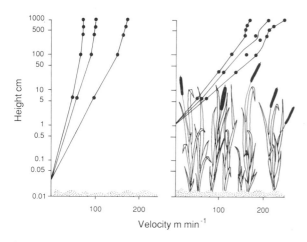

Figure 16.5 The effect of plant height on the roughness height, z_0 (after Chepil & Woodruff 1963).

(i) *The height of plants* Thomas's (1975) review showed values between 0.1cm (for 1cm-high grass) to 9.0cm (for 60cm-high grass). Thomas verified Tanner & Pelton's (1960) empirical equation for vegetation heights between 10 and 100cm: $\log z_0 = 0.997 \log_{10} H - 0.883$ (all lengths in cm), where H is the height of the vegetation. Tanner & Pelton themselves maintained that the relation held for vegetation up to 10m high.

(ii) *Vegetative characteristics* Thomas (1975) found different z_0 values for different species on a temperate coastal dune. One relevant plant characteristic is flexibility: Thomas's review showed z_0 values of 9.0cm for rigid 50cm-high "grass", and 5.0cm for flexible grass of the same height (also Hesp 1983, 1989, Morgan & Finney 1987, Sarre 1988a). The shape of individual plants is also important. Raupach et al. (1980) found a relationship in which the height of the zone in which winds were very considerably modified, $z_w = h + 5l$, where h is the height of the obstacle and l is its breadth across-wind. With many plants there may be a windspeed minimum at mid-height where the leaf biomass is greatest (Hesp 1989).

(iii) *Density of plant cover* Hesp (1989) studied this relationship experimentally on coastal dunes. Dunes formed round low-density vegetation were longer and lower than those round high-density plantings. Plants 1m apart acted independently; closer than 60cm, they encouraged accretion.

(iv) *Plant litter* In agricultural fields, and presumably also in some semi-arid areas, the depth of plant litter on the surface also controls the rate at which sediment is released to wind (Chepil & Woodruff 1954, Bisal & Ferguson 1970, Lyles 1976, Gillette 1979).

Some of these characteristics have been brought together to produce an approximate estimate:

$$z_0 = 0.5 h s/S \qquad (16.1)$$

where h is the plant height, s is the silhouette area, and S is the specific area (Lettau 1969). The shelter afforded by plants has been extensively studied in the context of shelter-belts for agricultural fields (bibliography by Busche et al. 1984). Adjustment is rapid at boundaries between patches of plants with different heights, flexibilities and densities (Maiti & Thomas 1975, Thomas 1975, Hesp 1983). The form of the dunes created in the presence of grassy vegetation is discussed in Chapter 25.2.2a.

(b) Effect of isolated plants on sand-carrying capacity
On the edges of the African and Middle Eastern deserts, the common grass *Panicum turgidum*, which favours sandy soils, may act in much the same way as grasses on the temperate coastal dunes (Barbey & Carbonel 1972, Williams & Farias 1972), but continuous grass-covered surfaces are rare in deserts themselves, where bunched vegetation is more characteristic (Waisel 1971, Evenari et al. 1986). In clumped vegetation, the displacement of z_0 is less serviceable a measure of reduction in the effectiveness of the wind than the wind's sand-carrying capacity.

Laboratory and field measurements show that common shapes and arrangements of natural semi-desert vegetation effectively curtail the erosive force of wind (Marshall 1971, Raupach et al. 1980). In the field, there is a sharp drop in sand movement as shrub cover thickens from 10–25%, and there is little movement below 30% – with shrubs about as high as wide (Marshall 1970, 1971, Ash & Wasson 1983, Wasson

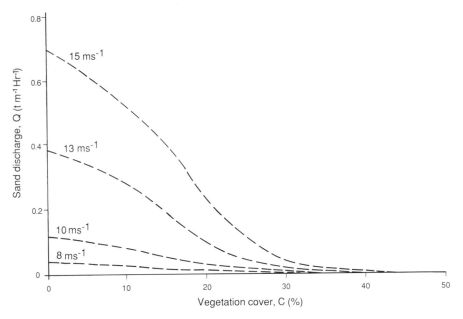

Figure 16.6 The effect of shrub cover density on sand movement, for various windspeeds (after Wasson & Nanninga 1986).

& Nanninga 1986; Fig. 16.6). Wind-permeable, gramminiform and erect and spreading plants are less effective in this respect than squatter, dense plants (Buckley 1987). At a given wind velocity, the amount of sand in movement is said to be a cubic polynomial function of the percent projected cover of plants. For example, the velocity of a wind which blows at 15m s^{-1} over bare sand is reduced by 50% when it blows over a surface where the projected cover of plants is 9%. Velocities are reduced to zero on surfaces with 43% cover (Buckley 1987).

(c) Suitable rooting media
The relationship between live plants and moving sand is, however, more complex and dynamic than this, for plants must survive if they are to control sand movement. The two main limiting ecological constraints on plant growth in desert sands are soil moisture and the stability of the rooting medium. Much further down the list come things like soil nitrogen (Buckley et al. 1986).

In arid and semi-arid areas, sands have some properties that render them better storers of water than finer soils. Pre-eminent among these are permeability, allowing a large water store, and the slowness or absence of water movement to the surface through capillaries, so that water that sinks below a shallow surface layer is protected from evaporation for many months (Hack 1941, Tolstead 1942, Bagnold

1954a, Hillel & Tadmor 1962, Sorey & Matlock 1969, Alizai & Hulbert 1970, Gupta 1979, Agnew 1988). This contrasts with the behaviour of soil moisture in wetter climates, where sands are poor seed-beds because they drain more quickly than finer sediments. The mean annual rainfall threshold at which sands gain the advantage in moisture supply over finer soils is about 300mm (Noy-Meir 1973). In addition, sandy soils allow root-penetration to deeper water tables than do finer soils, and rooting depths of 10m have been reported in sandy soils (Baitulin 1972).

Where, as in deserts, water is the prime ecologically limiting factor, sandy soils are therefore potentially better media for plant growth than other soils, and it is a common observation that they support more vegetation than nearby finer or coarser soils (e.g. Smith 1949, Obeid & Seif el Din 1971, Twidale 1981a, Le Houérou 1984).

However, against their attractive water-storage capability, sandy soils are liable to be too mobile in the wind to allow growth to maturity. Hence there is a delicate balance between the stability and mobility of loose sands. In semi-arid areas, this delicate balance is reflected in the sharp boundaries of dune fields (Ohmori et al. 1983; Ch. 28.3a). In semi-arid climates and even some very humid ones, high enough sand mobility can keep sand free of vegetation. Nearby zones with the same rainfall and essentially the same sandy soil, but where there is slightly less sand in movement by reason of sand supply or wind velocity, are covered with vegetation. Sites where the sand supply has been reduced, for one reason or another, are rapidly recolonized by plants.

Some plants are able to compete well on sandy soils, and some, such as *Ammophila* spp. on northern-hemisphere coastal dunes (Ranwell 1972), or *Spinifex hirsutus* and *S. sericeus* in Australian coastal dunes (Hesp 1983), may even thrive in moderate disturbance, perhaps using mycorrhizas to help them fix nitrogen and phosphorus (Read 1989). Some desert plants are also adapted to sand encroachment. In Tunisia *Rhanterium sauveolens* and *Aristida pungens* are able to extend adventitious roots and to develop new above-ground biomass when swamped in sand, when most neighbouring plants are destroyed (Bendali et al. 1990). In the Middle East, the low plant *Cyperus aucheri* appears to need to collect sand as a moisture reservoir. Bushes of the genus *Calligonum*, which are common in parts of southwest Asia and the Indian subcontinent, require large areas of bare sand from which to extract moisture, and therefore compete well on bare sand (Warren 1988c). But even for these plants there are limits to the mobility they can withstand, especially when frequent water stress is the other limiting factor.

(d) Survival in droughts

Wasson (1984) showed that the relationship between plants, moisture and sand movement was still more complex when the high variability of desert rainfall was taken into account. Figure 16.7 shows that periods of high rainfall variability (even without a change in the mean annual figure) can lead to increases in sand activity.

In parts of the world where a change in climate since the late Pleistocene has encouraged revegetation of dunes, as in many now semi-arid areas, many small dune-fields have been reactivated, by one process or another, but most commonly by intensive grazing (Grandet, 1957 in Mali, Tricart 1959 in Niger, Suslov 1961 in central Asia, Hurault 1966 in Chad, Smith et al. 1975 in South Australia, Melhorn & Trexler 1976 in Nevada, Ehlers 1982 in Iran, Hellden 1984 in the Sudan, Tsoar & Møller 1986 in southern Israel, Ohmori et al. 1983 and Wasson 1986 in Australia, Mainguet 1986 in Niger, Mortimore 1989 in Nigeria).

This kind of reactivation involves more thresholds than those in wind velocity and plant growth, for there must also be thresholds of variability in soil moisture, drought intensity and frequency, rainfall, trampling and feeding behaviour of grazing animals, and the intensity and frequency of fires (Ohmori et al. 1983). Very few of these thresholds are understood.

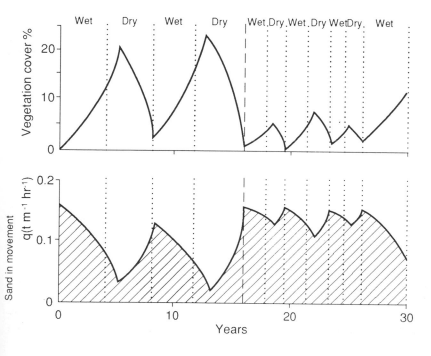

Figure 16.7 The effect of inter-annual rainfall variation on shrub cover and hence sand movement (after Wasson 1984).

16.2.3 The limits of sand deserts

The nature of the boundary between stabilized and moving sand on the edge of the deserts (which is a good, if rough, measure of where that edge is) may have practical significance, for example in defining desertification, and has received some attention.

The desert edge is in reality a wide zone where sand mobility interdigitates with immobility, spatially and over time. As mean annual rainfall increases across a desert boundary, the lower slopes of dunes are the first to be vegetated, while dry dune crests remain mobile into higher rainfall areas. Years of drought can bring moving sands into areas formerly considered stable, and years of rain can take them back again. Despite these difficulties in defining the edge of the desert, many authorities have considered it useful to attempt a definition.

The easiest way of examining the boundary of sand movement might seem to be on satellite imagery, but there are problems here of relating the image interpretation to the ground truth. The results of one study which aimed to delimit the boundary of sand movement using Meteosat and NOAA images are shown in Figure 16.8 (Mainguet et al. 1980a, Mainguet 1986).

In view of the importance of plants in controlling sand mobility, one approach to defining the edge of mobile sand deserts is to estimate plant cover. The Chinese sand deserts, for example, are commonly divided into "shifting" (< 10% cover); "semi-fixed" (10-50% cover); and "fixed" (> 50% cover) (Chao Sung-chiao 1984). The 35% cover of Wasson & Nanninga (Fig. 16.6), referred to above, seems an excellent approximation of a measure of the desert edge. Capot-Rey (1957b), more whimsically, defined the edge of the desert in West Africa as the line beyond which one's clothes became tangled with burs of the *cram-cram* grass (*Cenchrus biflorus*) (Fig. 16.8).

However, plant cover, let alone species composition, is difficult to record over extensive areas, and surrogate measures are thought necessary by some authorities. Because the primary limiting factor on plant growth is available moisture, and because this is more readily estimated from standard meteorological data, some attempts to define the regional limits of sand movement have been based on annual rainfall

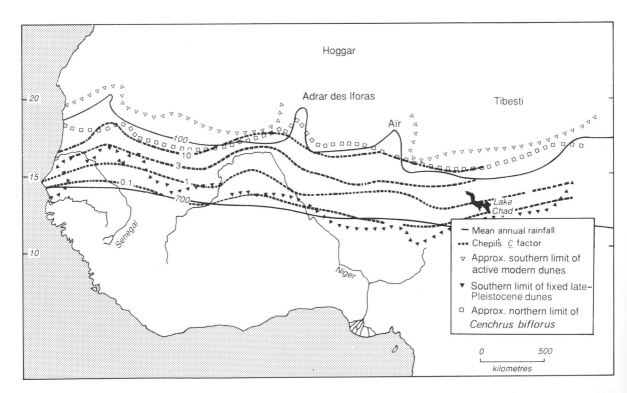

Figure 16.8 The southern boundary of sand movement in the Sahara according to various criteria (after Talbot 1980, using Chepil's *C* factor – for which see text – and Mainguet et al. 1980b).

statistics. However, these attempts have produced contradictory results. Estimates of the position of rainfall limits of sand mobility range from 25mm mean annual rainfall in the western Sahara (Sarnthein 1978), through 150mm in the southern Sahara (Mainguet et al. 1980a) to anywhere between 100mm (Lancaster 1979, quoted by Thomas 1986a) and 300mm (Flint & Bond 1968, Goudie 1983b) in southern Africa.

One of the problems with this approach is doubtless the difficulty of defining the level of sand activity across a broad frontier in a variable climate, but a more serious one is the interference of other factors, particularly evaporation and windspeed. More elaborate indices of sand mobility take these into account. The oldest and best tried of these is Chepil et al.'s (1962) index:

$$C = U_{ma}^3 / (P-E)^2 \qquad (16.2)$$

where C is a wind erosion factor; U_{ma} is mean annual wind speed (mph); and $(P-E)$ is Thornthwaite's climatic moisture index (precipitation minus actual evapotranspiration).

The Chepil equation seems to be widely applicable. It has proved to be a good measure of soil erosion by wind in Britain (Briggs & France 1982); dust-blowing in Israel (Yaalon & Ganor 1966); and of the limit of active dunes in the Sahel (Talbot 1984). Talbot found the limit to be defined by $C = 10$; nabkhas and isolated active dunes occurred between $C = 5$–10 (Talbot's limit is compared to that of Mainguet et al. 1980b in Fig. 16.8). Chepil's climatic factor has been revised by Skidmore (1986a), but this revision has yet to be tested.

Wasson (1984) derived another measure:

$$M = 0.21(0.13W + \ln E_p/P) \qquad (16.3)$$

where M is the mobility index ($M \geq 1$ means mobility), W is the percentage of days with sand-moving winds, E_p is annual potential evapotranspiration, and P is mean annual precipitation.

Wasson also took into account the acceleration of wind up the flanks of dunes, so that he could predict movement for either crests or swales.

Lancaster developed two indices (1987b, 1988a). In his second:

$$M = W/(P/PE) \qquad (16.4)$$

where W is the percentage of time wind blows above the threshold velocity for sand movement (taken at 4.5m s^{-1}), P is mean annual rainfall in millimetres, and PE mean annual potential evapotranspiration (also in mm).

In southern Africa, Lancaster claimed, dunes were fully active where $M > 200$; plinths and interdunes were stabilized with $M = 200$ to 100; only dune crests were active with $M = 50$ to 100; and where $M < 50$, all dunes were stabilized.

16.2.4 The controls on sediment movement by wind within deserts

(a) Moisture, plants, evaporation

Moisture and plants regulate sediment movement by wind in deserts much as they regulate it on their margins (Secs 16.2.1 & 2). Mobility is reduced in patches of moisture or vegetation in river valleys, around springs or after rain. However, in the common absence of these controls, four additional factors come into play.

(b) Surface roughness

If pebbles are close enough, they reduce the mobility of finer sediment between, partly by absorbing a proportion of wind shear (Fig. 16.9, Lyles 1977). Marshall (1971) found that, if the ratio of the silhouette area of pebbles to the area of surface was >0.1, erosion was severely restricted. This represents quite a sparse cover. Below this "inversion point", with pebbles far apart, mobility may increase, as wind scours round the stones. Scouring is more active beneath angular than round pebbles, and the inversion point is reached at lower cover values for small than for large pebbles (Logie 1982). If the density of pebbles is above the inversion point, deflation removes the first few of the finer grains quickly, but, as erosion uncovers further large particles, they enhance the protection, and erosion decelerates (Chepil 1950, Carter 1976, 1977, Gillette 1977). Deflation is thus one of the many mechanisms that produce desert pavements (Ch. 7). If roughness is produced by features like plough-ridges, then optimum protection occurs with ridges 10cm high and 43cm apart (Hagen & Armbrust 1985). Part of the effect is due to the maintenance of humidity in deep furrows (Verma & Cermak 1974). The same effect must occur on naturally rough surfaces, as on recently channelled alluvial plains.

(c) Salts

Particles are aggregated by salts, if present, and even

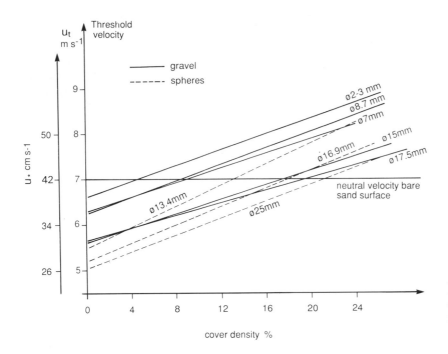

Figure 16.9 The effect of cover density of pebbles on the velocity necessary for movement of the sand in between (after Logie 1982)

small quantities of salt can greatly increase the threshold of movement (Lyles & Schrant 1972, Nickling & Ecclestone 1980, Pye 1980); the type of salt makes little difference (Nickling 1984). However, while they may help to reduce movement in the short term, salts also slowly break up the surface and prepare it for transport, either in single-particle or pellet form (Ch. 25.3.1b). Moreover, salt crusts are fragile and can be easily disturbed by bombarding sand (Pye 1985).

(d) Crusts
Crusts may develop after rain, as clay is deposited by muddy suspensions in the voids between coarse particles close to the surface (Fletcher & Martin 1948), and as the clay minerals dry out and knit together. Even a 0.3cm crust can protect the soil almost completely from erosion. The strength of the crusts is related to clay type, but not significantly to clay content above a low minimum (Gillette 1978, Gillette et al. 1980, 1982). A crust increases the velocity needed to dislodge particles, but bombardment steadily breaks it up (Chepil 1945a, Gillette et al. 1974, Hagen 1984). Some sandy surfaces in humid and semi-arid climates are aggregated by fungi, bacteria, actimomycetes and algae. In semi-stabilized dunes and on temperate coastal dunes algal filaments bind the sand, helping to keep it moist, and can markedly reduce sand movement (Fletcher & Martin 1948, Barbey & Couté 1976, Forster & Nicholson 1981, Van Den Anker et al. 1985, Yair 1990); mycorrhizas may act in the same way (Read 1989).

Finally, and most significantly, sediment movement is controlled by wind velocity. This factor is examined in depth in the next chapter.

CHAPTER 17

Entrainment by wind

On a loose, dry surface, free of plants and other obstructions, a wind of sufficient strength dislodges and entrains particles of clay, silt and sand.

17.1 The forces in entrainment

17.1.1 Shear at the surface: drag velocity or u_*

The energy available for dislodging, entraining and transporting particles comes from wind. High wind velocities near the bed provide high shear forces. Shear is best expressed in terms of the **shear velocity** (or drag velocity, or friction velocity), u_*:

$$u_* = \sqrt{\tau_0/\rho_a} \qquad (17.1)$$

where τ_0 is the shear force per unit area on the bed (g cm^{-1} sec^{-2}), ρ_a is the density of the air (in g cm^{-3}). u_*, which therefore has the dimensions of velocity (m s^{-1}), is a very important concept in aeolian geomorphology, and will be referred to repeatedly in what follows.

In a neutral, stably stratified atmosphere, with high Reynolds number (fully turbulent flow), velocity usually increases logarithmically with height above the surface, and u_* is related to the slope of the of the velocity/log.height curve in the following way (Fig. 16.4):

$$u_z/u_* = 1/K\ln z/z_0 \qquad (17.2)$$

u_z is wind velocity at height z; u_* is as defined immediately above; z_0 is defined in Chapter 16.2.2a; and K is von Kármán's constant, usually taken as 0.4. This is one form of the Kármán/Prandtl equation.

If the surface is rough, u_* is higher in the same ambient windspeed than over smoother surfaces. This is seen in Figure 17.1, where wind is shown passing from a smooth to a rough surface, as from an alluvial plain to a lava flow (Greeley & Iversen 1986). u_* has a higher value over the rough than over the smooth surface.

When there is strong surface heating or cooling, or movement over topographic obstacles, u_* does not bear such a simple relationship to the velocity profile (Ch. 20.4.2 & 7). Fuller descriptions of u_* are given in Bagnold (1941: 50–56) and Vanoni (1975: 231–2).

17.1.2 The effects of turbulence

The rôle of turbulence in entrainment is less well understood in air than in water, although "flurries" of grain movement indicate its importance (Willetts et al. 1990). The most obvious effect of turbulence is to vary lift and shear forces on the bed (Chepil & Siddoway 1959; summary in Vanoni 1975: 252–3).

Figure 17.1 A streak pattern and small ephemeral dunes in sand transport by wind (photograph by T. Linsey).

Rasmussen et al. (1985) found that turbulence over periods of 240 seconds caused u_* to vary by 10%. These fluctuations cause particles to vibrate (Bisal & Nielsen 1962, Lyles & Krauss 1971), and this must make grains more vulnerable to entrainment. It may be that the non-stationary wind velocities produced by turbulence make little difference except near the threshold of movement because of a lag between wind velocity changes and transport rate changes (Fan & Disrud 1977), although the reality and importance of this lag are still matters of discussion. For dust entrainment, on the other hand, turbulence may be much more significant, because Nickling (1978) found that the rate of dust transport is better correlated with levels of turbulence than with u_*.

Turbulence, however, must play other rôles in entrainment. In water, there are organized bursts of turbulence, known as "kolks" (Jackson 1976). Grass (1971, 1982) made hydraulic turbulence visible with hydrogen bubbles, and showed that parcels of higher-velocity flow were carried suddenly towards the bed in "sweeps", while parcels of low-velocity flow left the bed as "bursts". Similar bursts undoubtedly take place in wind, and must have an effect on entrainment, but little is known about them. Evidence for organized secondary motion in air lies in the almost universal pattern of streaks of moving sand or dust on dry surfaces, sometimes called "sand-snakes" (Fig. 17.2). Similar streaks occur close to hydrodynamically smooth beds under water and, like the burst–sweep mechanism, these are better understood than those in the wind (Kline et al. 1967, Sumer 1985, Weedman & Slingerland 1985). They must be zones of higher and lower velocity, associated with small, low-level, wind-parallel vortices, which lift sand in some zones and not others. Under water, and probably also in the wind, the spacing of these low-level streaks is proportional to u_*.

There are likely to be great complications in the relationship of turbulence and entrainment, although there has been little research in this area. Turbulent erosion of the bed must create an uneven surface, which must then itself help to create irregularities in the flow (Barndorff-Nielsen 1989). Another feedback process must be the dampening, or at least modification, of turbulence by saltating particles (Anderson et al. 1990).

17.2 Modes of entrainment

Loose particles on a surface over which a wind is blowing experience a horizontal drag force and a vertical lift force. These forces are proportional to $\rho_a d^2 u_*^2$, where ρ_a is the density of the air, and d is the particle diameter (Iversen et al. 1987). The particles are held to the surface by the gravitational force $(g(\rho_p - \rho_a)d^3$, where ρ_p is the particle density) and by interparticle cohesion.

17.2.1 Lift

Wind is accelerated over any protrusion from the bed. Since higher velocity is accompanied by lower pressure (by Bernoulli's equation), a loose particle over which there is acceleration must experience an upward suction or lift. The force should be significant only very close to the bed where the velocity gradients are steepest (Fig. 17.3). Even then, lift can only be important to the first ejections, and can raise grains only slightly. It may well be more effective on rough beds in the field than in smooth beds in wind tunnels, and on particles already set in motion by drag than on static ones (Chepil 1945a, 1959, Sumer 1985), but it is of little significance once grains begin to leap, descend and eject others (Anderson 1989). Inflows of air into a permeable bed seem also to make little

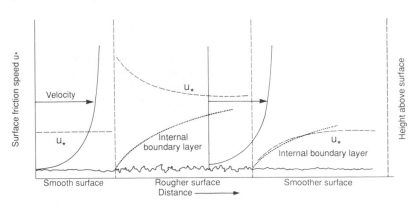

Figure 17.2 The response of shear velocity, u_*, to changing roughness (after Greeley & Iversen 1985).

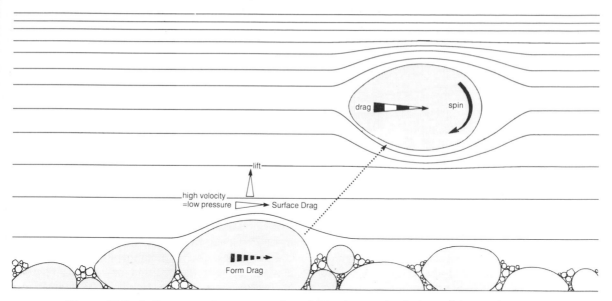

Figure 17.3 A diagrammatic representation of lift, shear and spin in particle entrainment.

difference to entrainment (Williams et al. 1990). Lift is more important in dust devils than in normal entrainment by wind (Greeley & Iversen 1985: 85–9; Iversen 1986b; Ch. 20.4.3).

17.2.2 Drag

Drag includes **surface drag**, which is skin friction between the particle and the air; and **form drag**, produced by the difference in pressure between the windward and lee sides of the particle (especially if there is flow separation, as over angular particles and/or at high Reynolds numbers). Because surface drag is greatest on top of the particle, it induces rolling, while form drag induces rolling and sliding. Drag is the first observed process down wind from the

leading edge of a sand patch (Willetts et al. 1990). It may even induce ejection when moving particles collide or are dragged over small projections (Nalpanis 1985). But as a mechanism of entrainment, drag, like lift, is negligible once bombardment begins.

17.2.3 Bombardment

When the wind is carrying a large number of grains, the near-ground velocity declines, and almost all further entrainment is achieved by the momentum acquired by sand grains descending from higher levels in the air stream. When the wind meets a sand patch, the transition to this condition happens almost immediately (Bagnold 1956). High-speed films show that a descending grain usually produces one high-energy

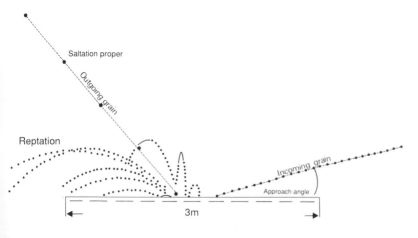

Figure 17.4 The impact and ejection process during sand transport by wind. The diagram is a tracing of a strobe-lit photograph in Mitha et al. (1986), on which a greater distance between successive positions of grains denotes faster travel (after Anderson 1987).

ejection (Fig. 17.4), taking some 50% of its impact energy (the ejectum is often the impacting grain itself) and about ten lower-energy, short-distance movements from a zone around the original impact (Nalpanis 1985, Willetts & Rice 1985a, 1986, Mitha et al. 1986, Werner 1990).

17.3 Thresholds of entrainment

One of the most important concepts in aeolian geomorphology is the size-limit of particles that can be carried by a wind of a given velocity: the "threshold of movement". Research into thresholds in aeolian geomorphology dates back to Sokolow (1894; quoted by Högbom 1923 and Bourcart 1928). The **threshold velocity**, or better the **threshold shear velocity** (u_{*t}), for any size of particle is the velocity (or shear velocity) at which those grains begin to move. Thresholds have been defined by rule-of-thumb, by theory and by experiment. Capot-Rey's (1953) whimsical definition was that a sand-wind was one that made eating distasteful. Examined in detail, the threshold is a complex transition in which wind-

velocity fluctuations, induced by turbulence, make a large difference to particle behaviour (Grass 1970; Fan & Disrud 1977). Given this transition, the best definition of u_{*t} is probably the value of u_* at which movement of the whole bed begins (Nickling 1988). The theory of the threshold of aeolian movement is reviewed by Greeley & Iversen (1985: 74–6) Iversen (1986b), and is not discussed further here.

Experimental work reveals a distinctive relationship between particle size and u_{*t} (Fig. 17.5). In the experiments, beds of particles, each of grains in a narrow size range, are exposed to a range of u_*. The u_* value at which the particles begin to move is plotted against the size of particle. A minimum u_{*t} for particles has always been found, with threshold values increasing away from the minimum for both finer and coarser grains. The minimum u_{*t} decreases with particle density. For quartz sand in common terrestrial atmospheric pressures, it is about 0.1mm; for lead particles it is about 0.05mm (Iversen 1986b).

Experiments always reveal a difference between the **static or fluid** threshold, which marks the point of first movement, and the **dynamic or impact threshold**, which is the threshold under bombardment, and

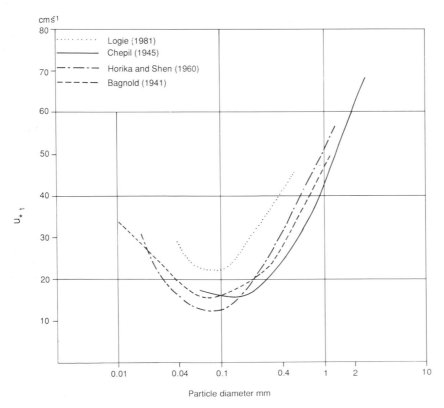

Figure 17.5 A comparison of different curves of the relationship between particle size and threshold velocity (after Savat 1982). Note the minimum, with an increasing threshold speed for finer and coarser particles.

which, because of the energy carried down by the descending particles, is about 80% of the static threshold (Bagnold 1941, Anderson & Haff 1988). Once bombardment starts, movement can continue at a lower overall wind velocity than before. The two thresholds are compared in Figure 17.6.

17.3.1 Fine particles

The need for increasing shear to mobilize finer and finer particles less than about 0.1mm in size has several explanations, as follows.

(a) Surface smoothness

A bed of only fine particles provides a very smooth surface, with few projections subject to lift, and totally within a layer near the bed where wind velocity is very low (the "viscous sublayer", equivalent in thickness to z_0). Drag is also more evenly shared on a bed of only fine particles than on a bed with coarse ones. But smooth surfaces are not common in the field, for even dry lake surfaces are broken up by cracks and boils, and yet, in the absence of bombardment, something still prevents fine particles from rising from rough surfaces. Moreover, the steepness of the curve of u_{*t} against the decreasing size of fine particles (Fig. 17.5); suggests that surface roughness effects are not very important.

(b) Cohesion

Cohesion is a more general and important control on the behaviour of fine sediments in wind, for cohesive forces begin to equal gravitational ones when particles are less than about 0.10mm. Fine sediments cohere better than coarse ones for three reasons: packing density; co-ordination number of the particles (the number of other particles with which each is in contact); and interparticle bond strength. These mean that cohesion is a function of $(1/d)^3$ (Smalley 1970). However, other relationships with particle diameter have also been suggested (Iversen et al. 1987). Fine particles have four properties that induce cohesion:

 (i) They are generally more platy than coarse ones, and so have more contact surfaces.
 (ii) They have greater specific surface (greater surface area per unit volume), again giving a greater likelihood of contact, and greater surface activity per unit volume. This encourages weak chemical bonds, such as the Van der Waals forces. The relative and cumulative value of these forces also relies on such properties as organic content and mineralogy (Belly 1964, Sagan &

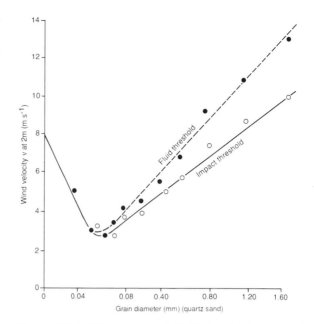

Figure 17.6 The impact and fluid threshold compared (after Chepil 1945, Hsu 1971).

Bagnold 1975, Fletcher 1976, Iversen et al. 1976a, Greeley & Leach 1978).

(iii) In some circumstances, electrostatic charges may be important for particles with high specific surface area (Iversen & White 1982; Nickling 1988).
(iv) Finally, moisture, held more firmly between fine than coarse particles by capillarity, is probably the most important factor in the resistance of fine particles to movement in the field (Ch. 16.2a).

None of these forces is yet well understood, because they are difficult to control and measure in the laboratory, and this explains some the range of results for thresholds in fine particles (Fig. 17.5). The difficulty is compounded in the field. Pye's (1987: 30–43) review of field results showed that threshold values (u_{*t}) for dust movement lay in the 0.2–0.6m s^{-1} range (approximating to windspeeds at 10m of between 4 and 12m s^{-1}). Actual values depend on all the factors we have mentioned, as well as the nature of the grain-size mix of the particles, the crusting of the surface (Ch. 16.2.4).

Many clay soils cohere into large clods which prevent erosion, not only by cohesion but by supplying a rough, pebble-like protection (Ch. 16.2d; Bisal & Ferguson 1970, Gillette 1986b). In contrast, small enough aggregates require a lower u_{*t} than their individual clay particles, for they can then move as sand-size particles or "pellets" (Ch. 25.3.2b). These may bombard the surface and further increase entrainment, although they break up more readily than mono-mineralic sands.

(c) Lack of bombardment
Smoothness and cohesion explain why dusts are not readily mobilized in wind tunnels, and are the controls on the entrainment of dust on great expanses of bare clayey or silty sediment. However, a third factor is probably more important in practice, for, even on such surfaces, dust is entrained if there is bombardment by saltating sand particles. If there are few sand-size particles, very little dust is entrained. When there is bombardment, the inverse relation of particle size and u_{*t} disappears (Gillette & Walker 1977).

17.3.2 Coarse particles
Gravity is much more important than cohesive forces in controlling the entrainment of particles coarser than 0.1mm. Increasing size and therefore mass (if density is constant) increases the gravitational force that must be overcome before movement can occur.

Bagnold's (1941: 86) relationship holds better for particles coarser than about 0.1mm diameter than for fine ones (Greeley & Iversen 1985: 77–82; Iversen 1986b):

$$U_{*t} = A \left[\frac{\rho_a}{\rho_p} gd \right]^{1/2} \qquad (17.3)$$

where A is a constant. Bagnold believed A to be related to the grain Reynolds number ($R_g = \rho_a u_0 d/\nu$); d is grain diameter and ρ_p and ρ_a are the density of the particle and the air; u_0 is the ambient windspeed and ν is the kinematic viscosity.

At the initiation of movement, Bagnold found $A = 0.1$ (agreeing fairly well with Iversen et al. 1987); when saltation was in progress, Bagnold found $A = 0.08$. Vanoni (1975: 234–7) and Greeley & Iversen (1985: 76–83) and Iversen et al. (1987) provided further analysis of the constant A. The dependency of u_{*t} on air and particle density means that its value will differ from the "normal" at high altitudes where ρ_a is small, or in very cold conditions, as in the Antarctic, where it is very high (Selby et al. 1973), or where sand densities differ from those of quartz (for example in pelletized clay, gypsum or calcareous sands). It is also different on other planets, such as Mars or Venus, where there are different values for gravity and atmospheric density (Greeley & Iversen 1985:

89–92; Iversen et al. 1987).

The static threshold is rather a theoretical notion in the field, since bombardment usually begins very soon after windspeed passes the static threshold. But even the impact threshold is difficult to establish in the field, because of the variety of grain sizes and surface conditions. Nevertheless, Lettau & Lettau (1978) found that the threshold velocity was rather well defined in the field for sands predominantly between 125μm and 177μm. Below 4m s^{-1} (measured at 63cm above ground) there was rarely movement; between 4 and 5m s^{-1} there was some creep; above 5m s^{-1}, there was some leaping movement; and above 5.5m s^{-1} movement was "uniformly strong". This last velocity value was equivalent to a u_* of 25cm s^{-1}, and it agreed with Belly's wind-tunnel experiments on sand of an equivalent diameter. Rasmussen et al. (1985) found values of u_{*t} to be about 20m s^{-1} on beaches in Denmark. Using a portable wind tunnel, Gillette (1986b) found that sandy soils had a range of u_{*t} values of 25–40m s^{-1}.

17.3.3. The effects of different grain-size mixes
Most sediments in deserts are mixtures of sizes of particle, and these mixtures behave rather differently from single-size sediments. Entrainment from mixtures is poorly understood in aeolian as in fluvial entrainment. A few examples will illustrate the complexities.

(a) Mixtures of sand with clay and silt
In this kind of sediment, fine sands are the first to move, having the lowest fluid threshold and, once moving, they bombard the clays and silts and raise them regardless of their individual fluid thresholds. Thus, the dynamic threshold of mixtures with sand and finer particles is considerably lower than the fluid threshold of the mean particle size (Chepil & Woodruff 1963).

(b) Mixtures of different sizes of sand
There is likely to be a wide range in types of behaviour with mixed sands, depending on the nature of the size mixture and wind velocity. Only a few examples can be illustrated here. The simplest case is for mixtures with grains that are too coarse to be entrained under any conditions, for these will accumulate and eventually prevent further entrainment: in other words, they will provide a "lag deposit" (Belly 1964). When mixtures are more complex, a range of surprising processes may take place. One of these is

the pivoting angle effect: the angle through which a grain has to be rotated before it will start to move is smaller for a coarse grain resting on a bed of finer ones than on a bed of grains of its own size (Li & Komar 1986). The coarse grains should therefore be entrained more easily in this condition. If coarse grains are also brought to the surface by shaking (Ch. 18.1), their preferential removal would leave the bed progressively finer, as there would be fewer coarse grains remaining to stick their heads over the parapet. Another strange effect may occur in mixtures with fine sands that can be entrained at the fluid threshold, and coarse ones that cannot: the coarse ones might be mobilized by bombardment with the finer sands (Nickling 1988). In poorly sorted sands, the coarsest grains may protect the finest, leaving the medium-size sands to be preferentially removed. The results of this kind of process are further discussed in Chapter 18.8.1b.

(c) Mixtures with pebbles

If the density of the pebbles is below the "inversion point" (Ch. 16.2.4b), turbulence is increased, and the threshold of movement for the sands is lowered.

17.2.4 Other factors influencing thresholds of movement

(a) Packing density

In theory, increasing the packing density should increase the threshold velocity, but Logie (1981) found no evidence for this in her wind-tunnel experiments. u_{*t} varied little as bulk density was changed from about 1.54 to 1.73g cm^{-3}. Bulk densities in barchan sands in Peru range from 1.12 to 1.48gm cm^{-3} (Hastenrath 1967, 1978), so that on dunes, packing density should make little difference to entrainment.

(b) Disturbance

Surfaces disturbed by water erosion, vehicular traffic or trampling by animals are more vulnerable to wind erosion than are undisturbed surfaces. Disturbance makes surfaces more erodible in three ways: (a) by roughening the surface and stimulating turbulence and

lift; (b) by loosening interparticle bonds; and (d) by turbulence surrounding the disturbance itself. Alluvial surfaces may yield large amounts of dust just after disturbance by a flood, before a crust forms, or before it comes to be protected by a "lag" of gravel. Many observers have now documented the way in which dust is raised by vehicles. The first large-scale example on record was when dust-storm frequencies at Maryut in Egypt rose from 3 or 4 a year to 50 during the British, German and Italian campaigns of the 1940s (Oliver 1945). Other examples have been recorded in Saudi Arabia (Jones et al. 1986), and in the Mojave Desert (Clements et al. 1963, Nakata et al. 1976, Wilshire 1980).

(c) Surface slope

Howard (1977) showed that, in theory, the threshold should be raised on upflow slopes and lowered on downflow slopes. By adopting some simplifying assumptions Allen (1982b) was able to produce a curve of the effect of slope on threshold velocity (Fig. 17.7). The curve shows that the slope effect should be virtually undetectable on the low angles of most windward slopes of desert dunes, a hypothesis confirmed in Howard et al.'s (1978) model and by Hardisty & Whitehouse's (1988) experimental results, also shown

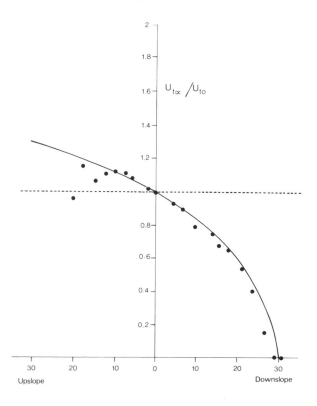

Figure 17.7 The effect of surface slope on the threshold of movement by wind. $u_{t\alpha}/u_{t0}$ is the ratio of the threshold on a sloping surface to that on a level surface (prediction after Allen 1982b; experimental results after Hardisty & Whitehouse 1988).

on Figure 17.7. However, the slope effect is likely to be significant in the short periods after changes in wind direction; especially during reversals, when threshold velocities will be noticeably lowered on steep, erstwhile slipfaces. This should have effects on the process of dune reversal (Ch. 26.4.3).

(d) Particle shape
Willetts et al. (1982) followed Williams (1964) in showing that the shape of particles had a considerable effect on u_{*t}, but Nickling (1988) found little effect. It is probable that shape differences have to be very great, as between the shelly and quartzose sands in Willetts et al.'s experiment, before shape has an appreciable effect.

(e) Mulches, micro-topography, crusts
These effects are discussed in Chapter 16.2.4.

CHAPTER 18

Grains in motion

Once entrained by wind, grains move in four ways: creep (and related near-surface activity), reptation, saltation and suspension.

18.1 Near-surface processes, including creep

When sand begins to move over a loose surface, the particles of bed are activated in a number of ways. These are very difficult to observe or model and, despite some theorizing and a little observation, little is known about them. Very recent work is showing that the processes are very complex.

18.1.1 Creep

Surface movements are often categorized simply as "creep". In practice, creep is difficult to distinguish from other near-surface processes (described immediately below) and even from reptation. Although these distinctions may be difficult in practice, it is possible in theory to regard creep as being two types of motion: first, rolling induced by saltation impacts and, secondly, rolling induced by gravity into craters created by saltation impacts. These kinds of movement involve particles that are generally much coarser than saltating or reptating ones.

By his definition, Bagnold (1941: 35) found creep to include particles up to six times the size of saltating grains. Willetts & Rice (1989) placed the division between the sizes of particle in creep and saltation at about 0.150mm diameter, but the exact distinction must be a function of the grain-size mix and of u_*. Sand in creep is less well sorted than saltating sand (Stapor et al. 1983). Willetts & Rice (1985b, 1986) filmed what they regarded as creep at 3,000 frames sec^{-1}, and then observed it in slow motion. With u_* at 48cm s^{-1}, they found that 0.355–0.6mm diameter grains moved at about 0.5cm s^{-1}. Groups of similar-size grains started to move as a wave over the surface, but rapidly dispersed as some grains moved faster than others. The wave disappeared completely in three minutes. Some creeping grains are buried for long periods of time, often in ripples (Barndorff-Nielsen et al. 1982).

The proportion of the total load travelling in these kinds of ways must vary with grain-size mix, although it may be independent of u_* (Horikawa & Shen 1960). The creep:saltation ratio has been found to be 1:3 (Bagnold 1941: 34, Willetts & Rice 1985b); between 1:1 and 1:3 (Chepil 1945a); or 1:7 or 1:8 (Borówka 1980). But these observations need to be treated with great caution, because of the difficulty of isolating "creep" from the other near-surface processes.

18.1.2 Deflation and deposition

When a mobile surface is examined more closely, it can be seen that there are processes at work other than simply creep. First among these is the differential loss or accumulation of different size fractions, depending on whether the bed is eroding or accumulating. By definition, grain-size does not change on an equilibrium surface, but there are few if any such surfaces in nature, because dunes are dynamic bodies composed of erosional and depositional portions. On an eroding dune surface, the probability that a grain will be removed is a logarithmic function of the logarithm of its size; thus, small grains are much more likely to be removed. On an accumulating surface, by the same argument, the probability of deposition is likely to be a logarithmic function of the logarithm of size; here fine grains are much more likely to be deposited (Bagnold & Barndorff-Nielsen 1980). Thus, erosional surfaces are overloaded with coarse grains and depositional ones with fine grains. When these two probability distributions are combined, a log-hyperbolic function is produced (Ch. 22.2.1a).

18.1.3 Shaking

There are yet other processes at work. The uppermost few grains are always coarser and better sorted than those immediately beneath, suggesting a variety of sorting processes (Sarre & Chancey 1990). These undoubtedly include the winnowing of finer particles, but must also involve other phenomena, for they occur even when all the grains are fine enough to be entrained. Sarre & Chancey identified one of these processes to be the preferential upward movement of coarse particles to the surface when a size mixture is shaken (in this case by bombardment). The process may extend less than five grain diameters beneath the surface, and is probably very fast. It is probably the result of a compressional–dilatational wave radiating from the point of impact of a saltating grain. Models of the shaking of size-mixtures show that the amplitude of shaking is directly related to the speed of sorting, and that large particles rise by rotating and ratcheting themselves against the smaller ones, their roughness therefore being an important control on the rate of rise (Haff & Werner 1986, Rosato et al. 1987). If the shaking process in bombardment is simulated by dropping trays of sand, it appears that there is the necessary amount of energy to produce this kind of sorting (Sarre & Chancey 1990).

Sarre & Chancey (1990) believed that the shaking mechanism was relatively independent of reptation, but this is debatable, for others have observed how, after saltation impact, a grain may travel through the surface layers and jettison another particle many grain-diameters down wind (Willetts & Rice 1985a), and the sorting to the surface of coarse grains on beds of finer ones is likely to make them more vulnerable to entrainment.

There are still other processes at work in the surface, some working in one way and some in the opposite way. For example, bombardment both consolidates the surface, rendering it harder to mobilize (Haff 1990), and may elevate some grains to positions where they are more vulnerable to later dislodgement (Iversen et al. 1987). There is clearly much more to learn about what happens in the few layers of grains on moving beds of sand.

18.2 Reptation

A distinction between saltation and reptation has been made in recent research. While saltation is the long-leaping, rebounding motion of high-energy grains,

reptation (from the Latin "to move slowly") is the less vigorous "splashing" or low-hopping of grains dislodged by the descending high-energy particles (Anderson & Haff 1988). The term "reptation" was used by early geomorphologists, such as Bourcart (1928), apparently to denote what Bagnold termed creep, but, much later, Haff rediscovered the term independently for use in the present sense. In view of the difficulties in defining what Bourcart meant, or indeed just what are the processes at on a mobile sand surface, it is better to use Haff's definition.

Each saltating grain dislodges about ten reptating ones (Werner & Haff 1988). Reptation therefore accounts for a very high proportion of the sand in transport at any one time, and may also account for a high proportion of the total transport rate, although measurement of reptation is very difficult (Anderson et al. 1990). Unlike those in creep, grains in reptation continually pass between the reptation and saltation modes or through immobilization by burial (Anderson 1987b).

Unlike saltation, the distribution of the velocities of reptating grains is strongly exponential, and heavily weighted towards small velocities (Fig. 18.1; Anderson 1987b). As in saltation, the number of ejecta is distinctly related to the impact speed of the incoming grain, and so to u_*, but the ejection speed has no such clear relationships. The mean launch angle in reptation is between 60° and 70°, and many grains

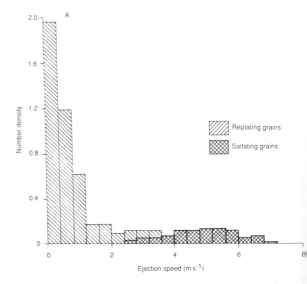

Figure 18.1 The distribution of ejection speeds for saltating and reptating particles (after Anderson & Haff 1988).

bounce back against the flow. These angles have little relation to the impact speed, and so presumably nor to u_* (Willetts & Rice 1985a, Anderson & Hallet 1986, Anderson & Haff 1988).

18.3 Saltation

18.3.1 Introduction

The word "saltation" was coined by Gilbert (1914) from the Latin *saltare* ("to leap"). In the saltation process an ejected grain is hurled into the air and it gathers momentum from the wind before it descends back to the surface. Saltating grains, unlike reptating ones, are able to splash up others. This is because they have higher impact velocities, and these are directly related to u_*. Unlike reptation velocities, saltation velocities have a wide Gaussian distribution (Fig. 18.1; Willetts & Rice 1985a, Anderson & Hallet 1986, Anderson 1987b, Anderson & Haff 1988, Nalpanis et al. 1990). Saltation can occasionally be violent. A severe windstorm in the San Joaquin Valley of California, with windspeeds at 10m above ground up to 190km hr^{-1}, imbedded particles of 23mm diameter into a wooden telegraph pole at 4.9m above ground (Sakamoto-Arnold 1981), and individual saltating grains may travel a few metres horizontally (Willetts 1989). Although usually less violent, saltation is still very significant, for it is the mode of travel of most windblown sediment.

The understanding of saltation is fundamental to aeolian geomorphology. It has a direct bearing on: sediment transport rates, sediment sorting, the pattern of aeolian abrasion, the formation of ventifacts, the character of yardangs, dust entrainment, and processes on the windward slopes, crests, brinks and the lee slopes of dunes. Each of these processes in turn affects processes and forms at still larger scales.

18.3.2 Ejection

An ejected particle may be either a rebounding or a newly dislodged grain. Rebounding grains lose some of their energy to reptation, and change some of their horizontal to vertical momentum. Dislodged grains are those few that are flung high enough by a "splash" to feel the forward impulse of wind and so enter saltation. Mean launch angles are between 20° and 40° down wind (not 90° as thought by Chepil & Milne 1939 and Bagnold 1941: 17) (Fig. 18.2; White & Schulz 1977, Nalpanis 1985, Sørensen 1985, Willetts & Rice 1985a, Anderson & Haff 1988,

Anderson 1989, Nalpanis et al. 1990). As in reptation, some grains bounce back against the flow. Ejection speeds are related to u_* and are 50–60% of impact speeds. Ejection speed may be independent of particle size (Nalpanis et al. 1990). Ejection velocities are greater from slopes inclined into the wind, such as the windward sides of ripples (Willetts & Rice 1989). This should mean that saltation is more vigorous on rippled than plane beds.

18.3.3 Flight

The ejected grain is hurled into the wind and it descends along a path determined by gravity and the drag imparted by the wind. Sørensen (1985) and Nalpanis et al. (1990) maintained that the height of

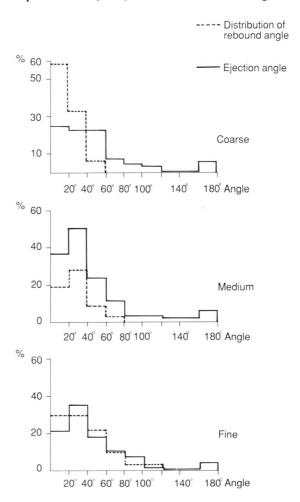

Figure 18.2 Ejection (launch) angles for rebounding and newly ejected particles with different sizes of sand (after Willetts & Rice 1986).

bounce was about $h = (1.5 \pm 1)u_*^2/g$. Flight paths are not much affected by collisions with other saltating grains, even though there may be a number of these (Bagnold 1985). Spin (the "magnus effect") is important, if at all, only for particles larger than 0.1mm, where spin may give some lift and downwind acceleration on the upper part of the upward path, and downwind acceleration on the downward path (White & Schulz 1977, White 1985, Anderson & Hallet 1986).

The gravitational force is greater on large, heavy particles, some of which therefore move sharply up and sharply down (White & Shulz 1977), the notion of the sharp ejection angle being better supported by experimental results than that of the sharp impact angle (Nalpanis et al. 1990). Smaller particles, where gravity and drag are more equal, appear to have gentler rises and descents. Some large particles rise high into the saltation layer, perhaps because they have lower specific surface areas and therefore suffer proportionally less vertical drag (White & Shulz 1977, de Ploey 1980, Draga 1983, Nalpanis 1985, Jensen & Sørensen 1986). In a sandstorm, fine particles of the order of 0.1mm rise only a few centimetres, while a few with diameters of 1mm may reach 1.5–2m (Sharp 1964). Sharp found that some large particles bounced higher than finer ones in strong winds (Fig. 18.3). Despite these findings, most authorities have found, both in the field and the wind tunnel, that there is a rapid fall-off in the mean size of particles and also a rapid improvement in sorting with height above ground (Williams 1964). Nickling (1983) found a consistent decline in the size of the suspended load

with height. u_* apparently makes little difference to the size grading with height (Chepil & Milne 1939, Williams 1964).

Saltation hop-lengths are about 12 to 15 times the height of bounce, but have a very wide distribution. There is disagreement about the relationship between grain size and distance of travel in saltation (White & Shulz 1977, Barndorff-Nielsen et al. 1982, Greeley et al. 1983, Sørensen 1985, Anderson & Hallet 1986, Nalpanis et al. 1990). Trajectory lengths may be directly related to u_*, but perhaps only for high-travelling grains (Fig. 18.4; Hunt & Nalpanis 1985, Sørensen 1985, Werner 1990). Nalpanis et al. (1990) found hop-length to be proportional to about $16\,u_*^2/g$. These findings have a direct bearing on hypotheses of ripple formation (Ch. 19).

18.3.4 Descent and impact

The great majority of grains descend at angles of between 10° and 15°, and here models and experiments are in good agreement (Anderson 1989, Nalpanis et al. 1990). Most experiments show that the mean angle decreases with grain size and u_* (Sørensen 1985, Willetts & Rice 1985a, 1989, Jensen & Sørensen 1986). Coarser grains have a wider spread of descent angles (Willetts & Rice 1989). Ungar & Haff's (1987) model showed little variation in impact velocity over a wide range of u_*. The horizontal velocity of large grains is close to that at the top of their trajectory, but small ones move at nearer the windspeed close to the bed (Nalpanis 1985). In one experiment, sand with a median diameter of 0.3mm had a mean impact velocity of about 360cm s^{-1} when

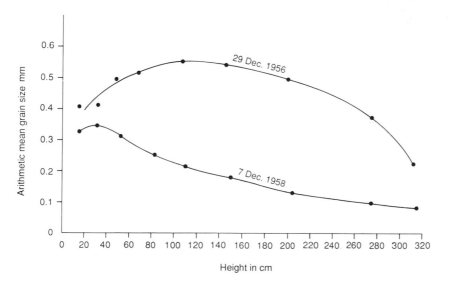

Figure 18.3 The variation of particle size with height in a saltation curtain (after Sharp 1964).

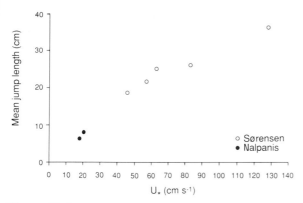

Figure 18.4 The relationship between u_* and trajectory lengths (after Sørensen 1985).

u_* was 39cm s^{-1} (Willetts & Rice 1985a). Modelled impact velocities for grains of 0.25mm diameter in u_* of 0.5m s^{-1} are of the order of 1.5m s^{-1}, showing that grains gather about ten times their lift-off energy in flight (Anderson & Hallet 1986).

18.3.5 The wind-velocity profile near to the bed; steady-state saltation

Because saltating sand is abstracting energy from the wind, the wind-velocity profile in the saltation curtain adopts a form that is quite different from that of a wind without sand. Most recent modelling more or less confirms Bagnold's (1941: 57–61) finding that wind-velocity profiles in saltation come to a focus at a roughness height, z'_0, which is higher than z_0 over a fixed surface, although the pattern is probably complex (Fig. 18.5; Werner 1990). Werner's model showed that, as u_* increased, wind velocity near the ground decreased, a result he believed might be because the greater number of grains ejected above the focus at higher u_* values, which carried more momentum high into the flow. However, these findings must still be regarded as provisional (Anderson et al. 1990).

The roughness height with sand in saltation increases with windspeed. Owen (1964) predicted $z'_0 \approx cu_*^2/2g$, where $c = 0.02$, and Anderson & Haff's (1988) model confirmed the increase of z'_0 with u_*. Chamberlain (1983) found Owen's $c = 0.016$. In the field Rasmussen et al. (1985) found higher values (between 0.07 and 0.09); they were unable fully to explain the discrepancy with theoretical and laboratory results, suggesting that it might have been due in part to sloping surfaces. They used $z'_0 \approx 5$mm in saltating sand on a beach.

These changes in the wind-velocity profile must be manifestations of a feedback mechanism, because the retardation of wind by the saltating grains cannot continue indefinitely. A steady state is achieved when wind velocity near the bed is slowed to a value at which as many grains are leaving the bed as are falling onto it (the mean reproduction rate is zero, Anderson et al. 1990). Shear on the bed is then lowered to a value at which aerodynamic entrainment is not possible, and entrainment is by bombardment only (Owen 1964, Anderson & Haff 1988, 1990).

18.4 Suspension

In suspension, grains follow turbulent motion; in saltation they do not. A simple measure of the distinction is the ratio of u_* to u_f, the fall velocity of a particle in air, $u_f/u_* < 1$. u_f is defined by Stokes's law: it is a function of the balance between the weight of the particle and the drag of the air upon it. The vertical velocity in turbulence near the ground is approximately equal to u_*, so that, with $u_f < u_*$, particles stay aloft. Intermediate sizes of grain take

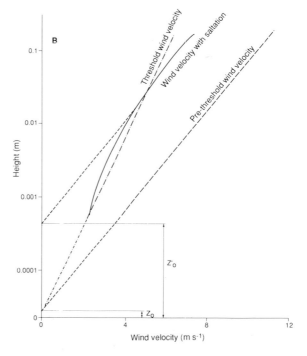

Figure 18.5 The modelled wind velocity/height profile within the saltation layer under various circumstances (after Anderson & Haff 1988).

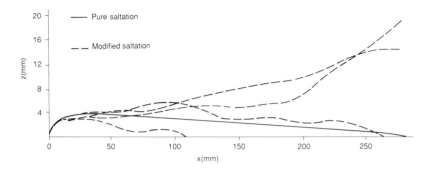

Figure 18.6 Paths of particles in "pure" and "modified" saltation (after Anderson 1987a).

part in "modified saltation", where saltation paths are made irregular by turbulence (Fig. 18.6; Hunt & Nalpanis 1985, Anderson 1987b). The distinctions between suspension, modified saltation and saltation are shown in Figure 18.7.

Observations over level surfaces show a sharp tran-

sition to suspension or modified saltation when grain size falls below about 0.1mm (Nalpanis 1985). There is a fairly clear distinction between silts and clays, which can be held aloft in many winds, and sands which rarely go into suspension (Pye 1987: 46–7). However, there may be complications near the bed. Saltation may dampen turbulence, and this may help to retain dust-size particles within the saltation layer (De Ploey 1977, 1980, Jensen et al. 1984). There are further complications in strong vortical flow, as in the lee of dunes with slipfaces. Here quite large particles can be held in suspension (Hunt & Nalpanis 1985). Strong turbulence in heated conditions may also hold quite large grains in suspension (Hunt & Barrett 1989).

Some distinctions can be made between particles in suspension. Particles larger than $70\mu m$ cut through some of the turbulence. They may descend near the bed occasionally, but when they do, they experience enhanced lift forces and may again be swept aloft. Particles smaller than about $70\mu m$ virtually follow the turbulent paths of parcels of air. Those finer than about $20\mu m$ can remain aloft indefinitely (Pye 1987: 49). These particles stay up, once they have been lifted, until the wind velocity falls, turbulence declines, or until they are precipitated by rain or snow.

Tsoar & Pye (1987) calculated the maximum distance that particles could follow in a moderate neutral-atmosphere windstorm (Fig. 18.8). Another estimate is found in Vanoni (1975: 238). These workers believe that particles larger than $20\mu m$ rarely travel more than 30km from their point of entrainment, and that loess (commonly 20–30μm particles) has travelled well under 300km. There are many assumptions in these calculations, notably about atmospheric stability and the lack of resuspension, but Tsoar & Pye point out that the figures tally with what is known about the behaviour of dust.

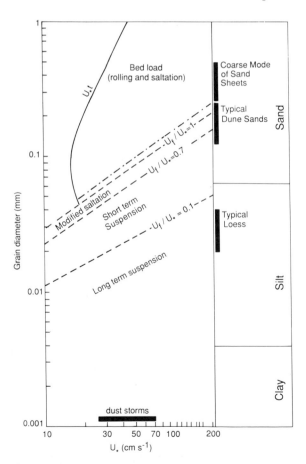

Figure 18.7 The distinction between dust and sand (after Tsoar & Pye 1987). u_f is the fall velocity.

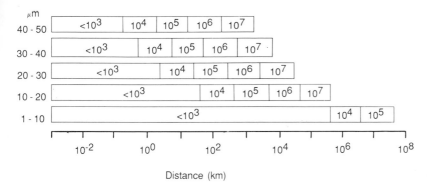

Figure 18.8 Distances normally travelled by dust particles according to size. Figures in the boxes refer to different coefficients of turbulent exchange (after Pye & Tsoar 1987).

18.5 Other modes of entrainment (and transport)

Wind may be able to move boulders on the wet slippery surfaces of playas. These "playa scrapers" are discussed in Chapter 14.3.

Pebble-size particles can occasionally move *up* wind. In the field, Sharp (1964) suspected that the undermining of windward edges of some pebbles resting on sand led to this form of movement. His observations have been confirmed: (a) in a wind tunnel, where Logie (1981) observed marbles being moved in this way, and moving slowly up wind, at the same time as digging themselves into the sandy bed; and (b) in the field by Mattsson (1976), who found differently shaped pebbles behaving in different ways, as they retreated up wind.

18.6 Transport rates

18.6.1 Relationship of horizontal mass flux and kinetic energy flux to height above the bed

There is an exponential decrease in the mass flux of particles with height (Fig. 18.9), and this has now been adequately modelled (Chepil 1945a, Jensen et al. 1984, Rasmussen et al. 1985, Nalpanis 1985, Anderson & Hallet 1986, Gillette & Stockton 1986, Anderson & Haff 1988, Werner 1990). In one wind-tunnel experiment it was found that the transport rate was 100 times greater at 1cm than at 10cm (Sørensen 1985). In another, 57% of saltating grains travelled less than about 5cm (Chepil & Milne 1939), and in yet another 79% travelled less than 1.8cm from the bed (Butterfield 1990). The height of travel varies as the quality of the rebound surface (Sec. 18.6.3).

However, the exponential decrease of mass flux with height does not mean that kinetic energy follows

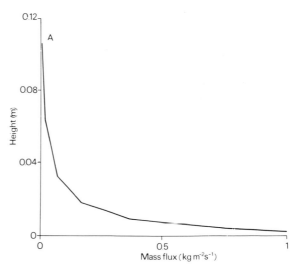

Figure 18.9 Modelled variation of mass flux of sand in transport by the wind with height above the bed (after Anderson & Haff 1988).

the same pattern, and the curved profiles of surfaces eroded by wind show that it does not (Ch. 21.2.3c). When kinetic energy flux for saltating sand is modelled for a realistic (exponential) distribution of particle trajectories, it peaks above ground (Fig. 18.10). If the model is set using $u_* = 1.0$m s^{-1} and particles of 0.25mm diameter, the maximum comes at about 0.8m above ground; with a mixture of grain-sizes the peak is lower. The main reason for the above-ground peak is that particles in saltation spend most of their time near the peak of their trajectories (Anderson 1986). By adjusting for different surface conditions (Sec. 18.6.3), Anderson was able to bring his modelled maxima for kinetic flux in saltation close to the maxima measured in the field. Anderson found that the maximum of kinetic energy flux for suspended

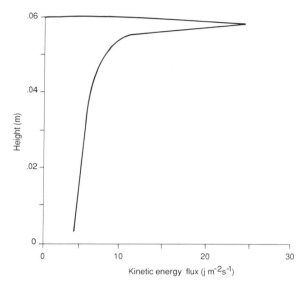

Figure 18.10 Modelled distribution of kinetic energy due to sand in transport with height above the bed (after Anderson 1986).

grains came at a much greater height, and was not as peaked as for the saltating particles. For both types of movement, kinetic energy was related to u_*^5 (Anderson 1986).

18.6.2 Horizontal transport rates

The relationship between discharge rate of sediment and wind velocity is a central concern in aeolian geomorphology. The sand discharge rate is defined as the quantity passing through a plane of unit width (usually 1m across the wind) and infinite height above ground, perpendicular to the wind, per unit time.

Because of the difficulties in understanding saltation, difficulties in measuring u_*, the highly turbulent nature of flow near the surface, and the difficulty of observing what happens close to the ground, let alone complications such as moisture, there is still no definitive relation of sediment discharge to wind velocity. The relationship has been investigated by many workers since Sokolow in 1894 (quoted by Högbom 1923 and Bourcart 1928). Greeley & Iversen (1985: 100); Swart (1986) and Sarre (1987) reviewed different formulae, noting how difficult it is derive and confirm them. It is still uncertain to what power of u_* the discharge rate is proportional (Anderson & Hallet 1986, Jensen & Sørensen 1986, Ungar & Haff 1987, Willetts & Rice 1988, Butterfield 1990, Nalpanis et al. 1990, Werner 1990). The power may vary with

the nature of the sand (its shape and density; Willetts et al. 1982). The rôle of turbulence is also unclear; it undoubtedly plays a part in controlling the rate of dust transport (Nickling 1978), and may play a rôle in sand transport in certain situations, such as the wake region down wind of dunes and perhaps even the windward toe of dunes.

One of the most widely used formulations is that of Lettau & Lettau (1978):

$$q = C(\rho_a/g)u_*^3 \,(1 - u_{th}/u_z) \qquad (17.4)$$

where q is the discharge rate of sand in g (m width)$^{-1}$ t^{-1} (t is a specified time period); C is an empirical constant related to grain size, commonly of the order of 6.5; ρ_a is the density of the air; u_{th} is the impact threshold velocity; and u_z is the wind velocity at height z. u_{th} here should be taken as the impact rather than the fluid threshold (Werner 1990).

The predictions of some formulae are compared with some field measurements in Figure 18.11. Most formulae converge to a u_*^3 relationship.

18.6.3 The effects of sorting and surface

(a) Sorting
Williams's (1964) experiment showed that sorting

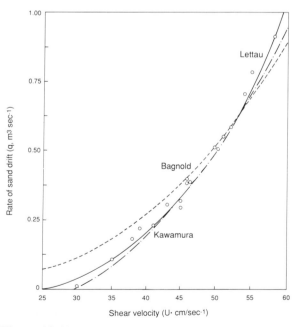

Figure 18.11 Predicted and actual sand movement rates in relation to shear velocity (after Fryberger & Dean 1979).

made little difference to transport rates. This was confirmed for fairly spherical particles by Willetts et al. (1982), but they found that, when a bed contained a significant proportion of dust, the transport rate was considerably depressed.

(b) Surface

An increase in the proportion of large grains on the surface increases the likelihood that saltating grains will rebound, and so increases the transport rate (Bagnold 1941: 71–2). The larger the grains, the more effective the rebound. Bagnold (1941: 67) found that the coefficient C (similar to the C in formula (1)) was 1.5 for nearly uniform sand, 1.8 for "naturally graded" sand, 2.8 for a poorly sorted sand, and 3.5 over a pebbly surface. Chepil (1945a) found values of C between 1.0 and 3.1. Rebound from hard rock should give even higher values, and McKenna-Neuman (1989) has found good rebound off frozen surfaces as well. Bagnold believed that at some point increasing pebble size and number generated a qualitative change in the nature of saltation. There was no longer be a steady-state in which sand flux to and from the surface was in balance, and large quantities of sand could be carried. The importance of this observation in relation to the formation of dunes is discussed in Chapter 23.2.3.

18.6.4 The effect of distance and time (lag)

There is a lag between the arrival of a wind at the edge of a sandsheet and the achievement of equilibrium saltation. The particles first have to gather momentum on the surface and then in saltation. Because of the difficulties of observing and modelling this process, however, the lag distance, its character and significance are still much in discussion. Bagnold (1936) found that sand-flux fluctuated wildly in the lag zone. He believed that the lag existed because the drag exerted on wind by the grains took time to propagate through the whole saltating flow. The fluctuations were the outcome of feedback between wind and moving particles. As the first ones accelerated, they abstracted momentum from wind, and decelerated it near the bed. This increased the flux of momentum from above, which in turn increased particle entrainment (Fig. 18.12). Positive feedbacks eventually dampened the oscillations. Bagnold found the lag to be extend over 4–7m. The distance seemed to be independent of u_*. He related this kind of finding to the establishment and minimum size of dunes (Ch. 23.2.1).

Other workers have also found lags, but none has reported the complex oscillation. Chepil & Milne's (1939) lag distances were between 2m and 10m. Anderson & Haff's (1988) model shows a somewhat different pattern to Bagnold's (Fig. 18.13). They found a period of about a second in which lift and drag were the only processes of entrainment, and then a rapid cut in as bombardment started. Each newly lifted grain at first ejected many others, but in one or

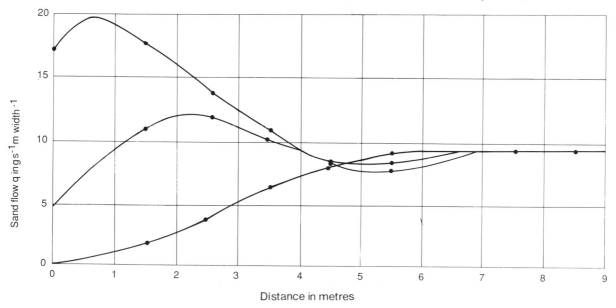

Figure 18.12 The oscillating transport rate of sand after the contact of wind with a sand patch (after Bagnold 1941).

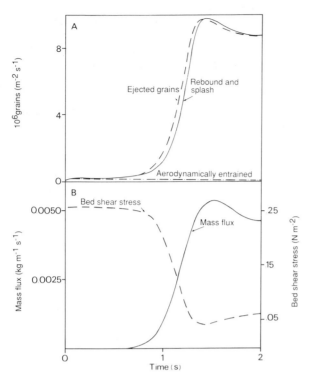

Figure 18.13 The behaviour of sand transport at the start of sand movement (after Anderson & Haff 1988).

two seconds the number of grains in transport reached a peak and then declined slightly to a stable number. Total adjustment took place over "several long trajectory times", the distance being a weak function of u_*. The modelled lag distance, therefore, was of the order of 10m. Willetts (1989) found that the removal rate of sand rapidly adjusted to changing values of u_* (within 20cm), but that the full rate of sand transport appropriate to a particular value of u_* was not achieved until somewhere between 2m and 25m. Howard et al. (1977, 1978) found that the incorporation of a small lag of between 1.6m and 6m made significant differences to their dune model, but that longer lags decreased its predictive power. The part of a dune most sensitive to lag was the brink.

Another "lag" phenomenon was termed, somewhat unfortunately, "avalanching" by Chepil (1957b). A better term might be **"cascading"**. It is the process in which deflation increases down wind from the start of a patch of erodible sediment which includes silts, clays and sands (such as a bare field) in response to the increasing bombardment provided by the growing volume of saltating grains. On agricultural fields with a cloddy soil structure and plough-ridges, cascading is accelerated by the progressive breakdown of the clods, and the smoothing of the surface as the plough-furrows are filled with sediment. The distance to full load is independent of u_*, but does depend (inversely) on the availability of erodible material. Chepil found that the distance varied from about 60m on very erodible to over 1.5km for very slightly erodible surfaces.

18.6.5 The effect of bed slope

The effect of surface slope on thresholds of motion has an indirect effect on the rate of transport (Ch. 17.3.4c). There are also direct effects, for transport must be against gravity on an upward slope. Bagnold (1956: 294) introduced a theoretical expression for the transport rate on slopes. This expression (Fig. 18.14) showed that on gentle up-slopes, as on the windward slopes of most dunes, the effect should be negligible. Howard et al. (1978) employed the following relationship:

$$q = Q/\cos\theta \, (\tan\alpha + \tan\theta) \qquad (17.5)$$

where q is the sand discharge on a slope; Q is the sand discharge on the flat up wind; θ is the slope angle in the transport direction and α is the dynamic friction angle (about 33° for most sands).

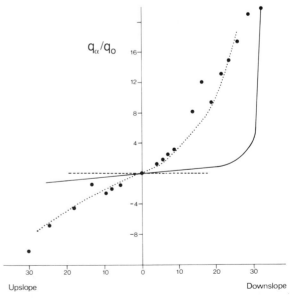

Figure 18.14 The effect of surface slope on the transport rate of sand in the wind. q_α is transport on the slope and q_0 is sand transport on the flat (curve after Bagnold 1956, data of Hardisty & Whitehouse 1988).

In their modelling of dunes Howard et al. (1978) found that slope angles on dunes made little difference to transport rates, as implied in Figure 18.14.

18.6.6 Measured and estimated rates

There are few reliable measurements of rates of sand transport in the field. Table 18.1 gives some figures estimated from various relationships between conventional meteorological measurements of windspeed and sand movement.

Table 18.1 Some reported rates of yearly sand movement (equivalent to Fryberger & Dean's (1979) drift potential), calculated from conventional meteorological observations. Many more estimates are found in Fryberger & Dean (1979).

Location	Rate [m^3 (m width)$^{-1}$ yr^{-1}]	Source
Northeast Sahara	5–100	Dubief (1952)
Kuwait	20	Khalaf et al. (1990)
Nouadhibou, Mauritania	62–162	Sarnthein & Walger (1974)
Pomona, Namibia	235	Fryberger & Dean (1979) (sample figures estimated from curve A in Fig. 94:148)
Nouadhibou, Mauritania	186	
Bilma, Niger	80	Fryberger & Dean (1979)
Ghadames, Libya	55	Fryberger & Dean (1979)
So Ch'e, China	3	Fryberger & Dean (1979)
Oregon Coast	34	Hunter et al. (1983)
Jafurah Sand Sea	18	Fryberger et al. (1984)
South African coast	30	Illenberger & Rust (1988)

18.7 Magnitude–frequency relationships

Sediment transport by wind has magnitude–frequency relationships quite different from those of transport in streams. The contrast is seen in the relationship of dunes to drainage networks. Low-order channels are overwhelmed by dunes; only when sand seas reach major channels do they meet their match in terms of long-term transport rate. Only the high-magnitude–

low-frequency transport in high-order channels can match the low-magnitude–high-frequency transport of sand by wind.

A more precise picture is given by measurements of wind velocity. High-magnitude–low-frequency winds in the Australian desert make little difference to the resultant of sand movement, most of the transport being achieved by moderate-magnitude–moderate-frequency winds (Brookfield 1970). A similar pattern was observed by Illenberger & Rust (1988) on the South African coast. Data from the Namib sand sea (Fig. 18.15) show that in a high-energy, unimodal wind regime, a high proportion of sand is moved by strong winds with low frequency (80% of sand by winds blowing for only 6% of the time); in a moderate-energy, directionally bimodal wind regime, 69% of sand is moved by winds blowing about 5% of the time; in low-energy complex regimes 89% of sand is carried by winds blowing for 6% of the time (Lancaster 1985b).

Sand transport by wind in deserts apparently occurs at a much lower magnitude and higher frequency than sediment transport by desert streams.

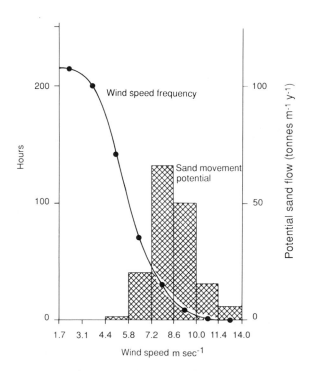

Figure 18.15 The magnitude–frequency relationship of sand transport to windspeed in the Namib Sand Sea (after Lancaster 1985).

18.8 Size and shape sorting processes

18.8.1 Size sorting

This section focuses the discussion of sand movement by wind onto the question of the sorting of particles by size and shape. The measurement of sorting in windblown sands and the grain-size characteristics of aeolian sands are discussed in Chapter 22.2.1. The sorting of dusts is discussed in Chapter 22.1.4.

(a) Basic processes

Sorting happens both in entrainment and in transport. In entrainment, sorting is a function of the distribution of probabilities that grains of one size range will or will not be dislodged, rolled or suspended by descending grains of the same or a different size. In transport, sorting is a function of the aerodynamic properties and rollability of various shapes. In practice, it is very difficult to distinguish these two processes, so that the physics of the processes is still obscure. The discussion of entrainment and transport shows that the sorting of sands is a complex process. In most cases, aeolian transport, as any other form of sedimentary transport, produces better sorting (McLaren & Bowles 1985). For example, in the zone below 1.5m over a bed of mixed silt and sand, the sorting and positive skewness of the transported load is greater than on the bed (Fig. 18.16).

(b) Sorting from mixtures of sizes

Aeolian sorting is much more complex when the source sediment contains a wider range of particle sizes. When the bed is composed of a mixture of clay and sand, saltation abrades clay from the surface itself, and from the surfaces of sand grains lifted into wind. Even more fine material can be produced as sand-size clods are lifted and then disintegrate when they again hit the bed. The grain-size mix of the saltation and suspension load above an eroding fine soil varies with the availability of different size fractions in the parent soil, the availability of sand with which to blast the surface, and on u_* (Gillette 1979).

(c) The production of bimodal sands

Bimodal (or tri- or multi-modal) size mixtures of windblown sand (two or more modes in the grain-size distribution) are common (Ch. 22.2.1). Four models have been produced to explain them.

(1) A model for the production of a bimodal sand from a poorly sorted parent sand was developed for the zibars of the Ténéré Desert by Warren (1972). In this model the medium sand (the saltation load) is taken first, and saltates away quickly. The coarse sand, travels slowly as creep, while the fine sand is protected by the coarse, and remains behind to form the second mode in a bimodal mixture (Fig. 18.17; Folk 1968, Warren 1972, Lancaster 1986a). There is some support for this model in the experiment of Willetts & Rice (1988). Sands less than 0.25mm in diameter lodged behind coarser ones, and were entrained rather erratically. Sharp's (1963) explanation of ripple formation was that fine sands, perhaps retained in this way, could subsequently be shaken downwards into the body of a ripple as it moved forwards (Ch. 19.2). An alternative explanation, in

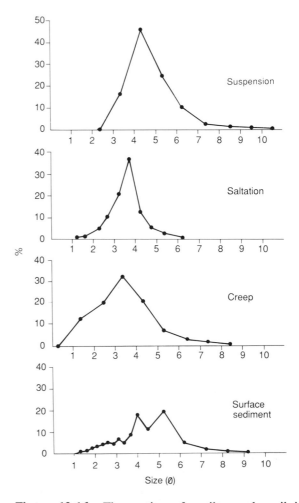

Figure 18.16 The sorting of a silty–sandy soil in transport close to the surface (after Nickling 1983).

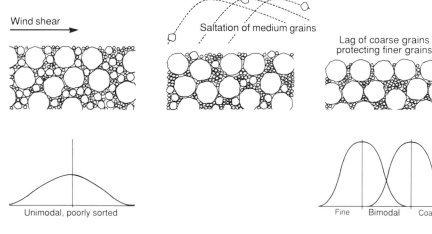

Figure 18.17 The hypothesized sorting mechanism for producing a bimodal-sized sand from a unimodal, poorly sorted sand. The resulting grain-size distribution is shown in Figure 22.7 and sample sites in Figure 23.25.

which fine material remains because it is difficult to mobilize (Folk 1971a), is untenable where there is a large saltating (bombarding) fraction, as in most desert dune areas (Warren 1972).

(2) Wood (1970) compared aeolian transport on the desert surface with the movement of particles in a separating distillation column, for which mathematical models had been developed. The model assumed a distinct separation of creep and saltation transport, now a dated concept (Ch. 18). Wood's model produced bimodal size distributions from unimodal sands in quite a short time.

(3) The third model is that the fine mode has been added subsequent to initial aeolian deposition. The source is either dust (Skocek & Saadallah 1972) or weathering (Haynes 1989).

(4) There is some doubt about the effects of particle density on the transport rate, and therefore about density sorting. Some experiments show that size may be more important than density. Gerety & Slingerland (1983) found that Bagnold's definition of the threshold did not apply to heterogeneous sands. It under-predicted the threshold for the heavier grains and over-predicted it for lighter ones. In sands of the same size, but different density, heavy and light grains travelled at the same rate. Density seems to have some effect on sorting in the field, where sorting is by both size and mineral composition (Willetts 1983).

18.8.2 Shape sorting

The shape of real windblown sands is discussed in Chapter 22.2.3. This section examines the meaning of shape and some of the physical processes in the selection of different shapes in entrainment and transport.

(a) Definition of shape

Barrett (1980) differentiated between the meaning of shape at three scales: **sphericity** was shape at a grain scale; **roundness** was shape at the scale of an edge; and **surface texture** was shape at a microscopic scale. Surface texture is discussed in Chapter 22.2.5.

"Sphericity" is the nearness of a shape of the grain to that of a perfect sphere. It is not easy either to define or to measure, because it is a three-dimensional property (Willetts & Rice 1983). The available graphical methods have dubious dynamic meaning, and are tedious to operate. Of the graphical methods, Willetts & Rice found Snead & Folk's (1958) sphericity criterion to be the most satisfactory.

"Roundness" in sedimentology refers merely to the edges of grains (their "overall smoothness" according to Barrett), so that even rather cubic grains can be "rounded". Roundness is often measured visually by comparing grains with the standard charts of Powers (1953).

The dynamic meanings of sphericity and roundness are obscure. One proposal for a dynamically meaningful measure of shape is that it should be based on settling velocities in a liquid (Briggs et al. 1962), but Willetts & Rice (1983) rejected this approach as inappropriate for transport by wind. They favoured Winkelmolen's (1971) "rollability" criterion. This is based on the timing of a large population of grains as it rolls down a cylinder inclined at 2.5° to the vertical. Willetts & Rice found that the results of this test correlated closely with the graphic sphericity criterion of Snead & Folk. Although "rollability" is easy to measure, and may have more dynamic meaning than the other measures of shape, its exact dynamic significance for reptation and saltation is still not clear. Rollability may be more appropriate as a measure of the effect of shape on the "creep" load, whatever that is, than on the saltation load.

(b) The theory of shape-sorting

In theory, the shape of a grain should affect the angle through which it must be rotated to move. This angle is commonly equated with the angle of initial yield (Ch. 23.3.7d). A long, thin particle, if upright, needs to be rotated through a smaller angle before it is dislodged, than a spherical particle, and through a smaller angle than if the long particle were recumbent (Li & Komar 1986). On the other hand, a long thin particle wedges more firmly between others, and a more spherical (or rounder) grain rolls more easily, while a long grain has to be dragged. Irregularly shaped particles have more contact points with others than smooth ones, and so are harder to dislodge. Theory also suggests that shape affects the value of drag when particles are larger than about 0.3mm (Willetts & Rice 1983).

(c) Experimental evidence

In movement, low-sphericity particles depress the transport rate, especially at high windspeeds, and are therefore presumably selected against. Shelly or shaly sands are less effective at dislodging others, less effective at maintaining themselves in saltation, and have lower, longer trajectories (Fig. 18.18; Williams 1964, Willetts 1983, Willetts & Rice 1986). In one experiment, the exponent, n, in the transport equation, $q = cu_*^n$, fell from 2.65 for fairly spherical sand to 2.15 for shelly sand (with sorting held roughly constant) (Willetts et al. 1982). Willetts et al. confirmed Williams's (1964) results to the extent that shelly sands did move at marginally higher rates in

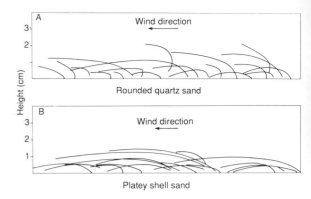

Figure 18.18 Saltation paths of particles with different shapes (after Willetts 1983).

low u_*, but at lower rates in high u_*. Thus, although there is doubt about shape selection, there is less doubt that more spherical particles are selected in wind transport (Ch. 22.2.3).

18.9 Electrical effects

Dust- and sandstorms can generate considerable electrical potential close to the ground (Stow 1969, Kamra 1972, 1977), but whether this has any effect on sand transport and dune formation is less certain. Gill (1948) interpreted experimental results to show that moving sand became positively charged by friction, while dust was negatively charged. Ball (1927), Walter (1951) and Gabriel (1964) believed that electric effects were responsible for controlling the arrangement of sand into dunes. Few others subscribe to these views.

CHAPTER 19

Ripples

19.1 Description

19.1.1 Common aeolian ripples

Ripples (Fig. 19.1) cover most dry, bare sandy surfaces in deserts. They are absent in only four situations which, although common, cover very small parts of dune landscapes: where there is very coarse, usually bimodally sized sand, as on zibar and sand sheets; at high u_*; where there is grainfall into local areas of low wind velocity; and on actively avalanching slipfaces. Most ripples develop and travel rapidly and have short lives. In a wind tunnel, ripples in well sorted sand (mean size $c.$ 0.15mm) take less than 10min to reach equilibrium with a new wind condition (Seppälä & Lindé 1978), although in the field some mega-ripples may take years to develop and may last for centuries.

Ripple wavelengths range from a few centimetres to tens of metres, with heights from less than 1cm to about 30cm. The wavelength:height ratio (the "ripple index") is said commonly to be larger than for subaqueous ripples at 17 or 18 (Cornish 1914, Tanner 1967), although it varies widely (at least 12 to 50), and is probably not diagnostic of aeolian ripples (Sharp 1963, Seppälä & Lindé 1978). Mean wavelength is strongly related to u_*, but as u_* increases, so does the range of wavelengths, and at high velocities many small, secondary ripples cover the large ones (Fig. 19.2). Field observations confirm the relationship between u_* and wavelength: ripples are farthest apart on dune crests and at times of the day when winds are strongest (Hunter & Richmond 1988). Windward slope angles lie between 8° and 13°, and in the extreme 25° (Werner et al. 1986), although close inspection shows windward slopes to be covered with very small bumps, perhaps grain concentrations,

perhaps "mini-ripples" (Fig. 19.3). Summits are broadly convex; lee faces have angles up to 30°. Some ripples have sharp brinks, and some smooth (Sharp 1963).

Most ripples are flow-transverse, although on sloping surfaces, where the downwind component of grain movement is supplemented by gravity, they are slightly flow-oblique (Howard 1977). Most ripples are sinuous in plan, although sinuosity varies greatly. At the same windspeed, ripples of coarse sand are more curved than those of fine; with the same sand size,

Figure 19.1 Wind ripples, showing Y-junctions (photograph by R. Turpin).

267

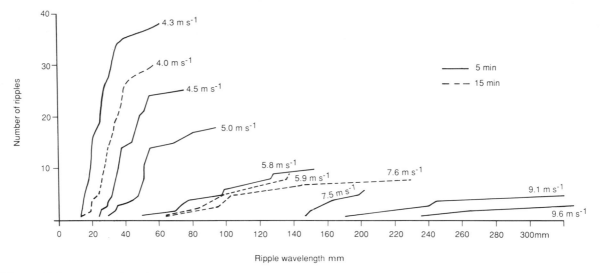

Figure 19.2 The relationship of ripple wavelength statistics to standardized wind velocity in a wind tunnel (after Seppälä & Lindé 1978).

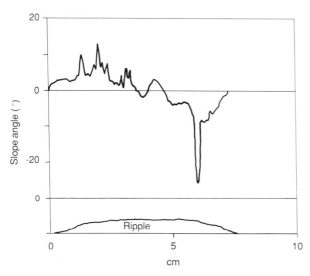

Figure 19.3 A detailed profile of a wind ripple produced by a shadowing technique (after Werner et al. 1986).

ripples in strong winds are more sinuous than those in gentle. Wavelength and height are also related to grain size: ripples in coarse sand are more widely spaced than those in fine (Seppälä & Lindé 1978). Wavelength is also related to the slope of the underlying surface: if sand size is constant, ripples have the longest wavelengths, and the gentlest slopes on the steepest part of the windward slope of a dune, and

some ripple lee slopes may even have upward inclinations in this position. Ripple-spacing is smallest on the steepest downwind slopes (Werner et al. 1986).

19.1.2 Mega-ripples

Mega-ripples (Fig. 19.4) are ripples with wavelengths between about 1m and 25m. They have also been called "ridges", "residual ripples", "grit waves" (Bagnold 1941: 153–7), "residue ridges" (Cornish 1897), "granule ripples" (Sharp 1963) and "giant ripples" (Wilson 1972b). They are found where there is coarse sand, as at the bases of windward dune slopes (Tsoar 1990), and where there are high winds, as in mountain passes. Mega-ripples and ripples are usually two distinct populations, the mega-ripples having a greater contrast in grain-size between crest and trough (Tsoar 1990). Some mega-ripples are nearly symmetrical in cross section, perhaps because longevity means that they are attacked by winds from different quarters (Greeley & Iversen 1985: 154). Bagnold believed that his Egyptian mega-ripples had taken centuries to form, although Sharp (1963) and Sakamoto-Arnold (1981) found that similar ridges had formed "in hours" during severe windstorms in California.

19.1.3 Chiflones

Simons & Eriksen (1953) termed ripple-scale flow-parallel bands of coarse and fine sand "chiflones" or "sand streams". They were 100–150m long, of relatively uniform but variable width and had a 1.0–1.5m wavelength at right angles to the flow. The finer sand formed low, linear ridges. Pits dug across the chiflones revealed alternating dark and light bands, indicating that they were subject to lateral migration. Figure 19.5. shows another more stable (and more common) type of flow-parallel ripple-scale feature, formed where coarse sands are channelled between obstacles such as bushes.

19.1.4 Adhesion and rain-impact ripples

Adhesion ripples are a distinctive form of aeolian ripple, found only on wet surfaces. They are probably initiated in much the same way as in other ripples, but, because water rises into them by capillarity, saltating sand and suspended dust adhere to upwind surfaces, and the ripples grow up wind (Hunter 1980). They may be broken into smaller "adhesion warts"

(Kocurek & Fielder 1982). On slightly dryer surfaces, sand may accumulate as a plane bed (Hunter 1980). **Rain-impact ripples** are formed in driving rain; they have the normal asymmetry of aeolian ripples (Clifton 1977).

19.2 Processes

Saltation drives creep and reptating sand up the windward slope of a ripple, the rate increasing from the trough towards the crest, where it then declines somewhat as the slope levels out. The increasing discharge rate stimulates the departure of finer grains, so that the surface becomes covered with coarser sand (Sharp 1963, Tsoar 1990). The coarseness of the ripple crests may be due to a decline in the creep rate, so that incoming coarse grains cannot leave (Willetts & Rice 1989); or to the inability of coarse grains to leave the crest, because they cannot be propelled down the lee face, where saltation impacts have less energy (Bagnold 1941: 155).

Coarse sand eventually rolls over and accumulates on the windward face of the ripple. This process, although never avalanching, creates foreset bedding, as on dunes. The lee eddy is never strong enough to move sand (Sharp 1963). Ripple movement is achieved by the erosion of the windward slope and the creation of foreset beds in the lee. Movement incor-

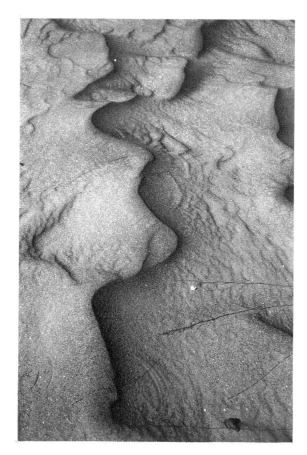

Figure 19.4 Strongly curved mega-ripples with concentrations of coarse sand (in this case dark) at the crests.

Figure 19.5 Coarse (dark) sand channelled between nabkha, forming a parallel-sided streak parallel to the wind.

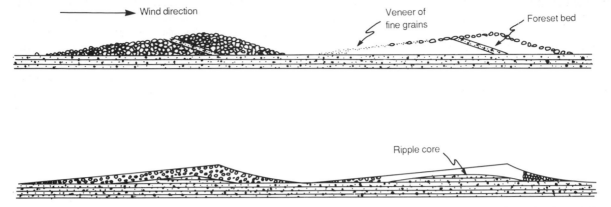

Figure 19.6 The internal structure of wind ripples, showing fine sand shaken into the core of the ripple and flat-bedded strata produced by the migration of successive, climbing ripples. Produced by the sectioning of resin-impregnated ripples from the field (after Sharp 1963).

porates the foreset beds into the body of the ripple, where they form its bulk. Ripple deposits, however, seldom preserve foreset strata. Forward movement shakes down fine sand, which accumulates in the core (Fig. 19.6; Sharp 1963, Hunter 1977a). These processes generate coarse-skewed, coarse sand on crests, and symmetrical to fine-skewed sand in troughs (Binda 1983). Ripples in uniform sand adjust quickly to new conditions; those of coarse sand adjust to stronger winds by a flattening of the crest, perhaps as the coarse grains on the crest go into saltation (Jensen & Sørensen 1986).

Like dunes, large ripples move more slowly than small ones (Seppälä & Lindé 1978). Sharp found a good fit of data to the formula $U_{rm} = (U_4 - 15.5)/7$; where U_{rm} is the ripple celerity, and U_4 is the wind velocity 4ft above ground. This was more or less confirmed by Hunter & Richmond (1988), who found ripples migrating about 5m a day in a moderate sea breeze.

19.3 Hypotheses of ripple formation

There are three main hypotheses as to the origin of ripples.

19.3.1 The ballistic hypothesis

(a) The ballistic model
The ballistic model was Bagnold's (1941: 62–4, 146–53). It was based on his belief that there was a correspondence between ripple wavelength and salta-

tion jump length. The model needed a chance distortion of the bed for ripple initiation, of which there are plenty in the desert. A chance strong turbulent eddy, which deformed the bed, might be even more likely (Williams & Kemp 1972). More sand was then ejected from the windward slope of this bump, and more landed one saltation path down wind to form the first ripple. Saltation from this second site landed down wind to form the next and so on.

The ballistic hypothesis apparently explained the relation between windspeed and ripple wavelength, following the relationship between windspeed and saltation length. It also apparently explained the greater wavelength achieved by ripples with coarse sand at the crest: as coarse sand accumulated, in this model, it provided a better rebound surface for saltation; this increased the length of saltation paths, and ripple wavelength therefore also increased. This last process lead eventually, in this model, to the formation of mega-ripples (Ellwood et al. 1975).

(b) Problems with the ballistic hypothesis
The ballistic hypothesis is now in serious doubt, for several reasons. The doubts began with Bagnold himself (1941: 165–6), who with Greeley & Iversen (1985: 155), described a class of ripple ("aerodynamic ripples") in fine sand under high winds with wavelengths much longer than the saltation path-length. Early doubts were also raised by Sharp (1963), who noticed that, in a constant wind and with the same sand, ripples started small and increased in size, a sequence he believed to be incompatible with Bagnold's version of the hypothesis. Sharp believed that

the parallel growth of height and wavelength showed that spacing was controlled more by the height of the ripple and the angle of the descending grains. Folk (1977a) later pointed to the common habit of transverse ripples of meeting in Y-junctions (Fig. 19.1), a phenomenon he believed to be hard to explain with the ballistic hypothesis.

Much more fundamentally damaging to the hypothesis, however, is the lack of evidence that ripple wavelength corresponds to saltation length (Belly 1964, Willetts & Rice 1989, Walker & Southard 1982, Walker 1981, quoted by Anderson 1987a). Folk (1977a) pointed out that a remarkably narrow peak in the distribution of saltation lengths would be needed if the ballistic process were to operate, and the Anderson & Hallet (1986) model of saltation produces no such peaks (Fig. 18.1). Anderson (1987a) showed that saltation trajectories of the length of ripple wavelengths did not have sufficient energy to drive reptation or creep, and therefore could not play a significant rôle in ripple formation.

It appears that Bagnold's picture of a "rhythmic barrage of grains travelling trajectories equivalent to the ripple wavelength . . . is not a correct image of the process" (Anderson 1987a: 954). Although it did have some attractive explanatory powers, the surprising thing is that, with so little empirical evidence, the ballistic hypothesis should have been so little questioned for so long.

19.3.2 Wave hypotheses

A set of alternative hypotheses, which explain some of the features left unexplained by the ballistic hypothesis, particularly the three-dimensional shape, are the **"wave" hypothesis**, which come in four forms (not necessarily mutually exclusive).

(a) The bed as a fluid
This hypothesis is now usually dismissed (Kennedy 1969). In the hypothesis, the sandy bed itself was seen as a very dense fluid which deformed beneath the moving fluid above, as predicted by the Helmholtz equation.

(b) The saltation curtain as a fluid
This form of the hypothesis was resurrected by Brugmans (1983). The saltation curtain certainly has some of the properties of a dense fluid (Bagnold 1956, Williams 1964, Raupach 1990). It has been calculated that the volume concentration of sand in the saltating

layer is about 10^{-3} particles per m³, giving an atmospheric density of 2.5kg m⁻³ (Hunt & Barrett 1989). In Brugmans' model, the discontinuity between the two fluids (air and the saltation curtain) was deformed into gravity waves, and when these are anchored by chance nuclei, ripples developed. With suitable values for density, the Helmholtz gravity-wave formula gave Brugmans wavelengths consistent with field observations.

(c) Wave-like instability in the boundary-layer
This hypothesis has been popular (Matchinski 1955, de Félice 1956, Wilson 1972b, Folk 1976a, 1977b), although there seems to be little empirical support. Some kind of secondary air motion does seem to be necessary to explain the three-dimensional waviness of ripples (Fig. 19.4), perhaps associated with flow-parallel regularities at the ripple scale. Flow-parallel ripples were first remarked upon by Cornish (1897). Simons & Eriksen believed that chiflones (Sec. 19.1.3c) were formed by local jets, and they compared them to Bagnold's (1941: 178–9) bands which were 40–60m apart and 500m long. Wilson (1972b) suggested that one explanation of flow-parallel "ripples" was regular ripple-scale longitudinal turbulence in the saltation curtain. The band structure of sand and dust movement (Ch. 17.1.2; Fig. 17.2) is further evidence for this kind of processes. Similar explanations have been developed for subaqueous banding (Weedman & Slingerland 1985).

(d) Secondary motion in the lee of transverse ripples
It is notable that flow-parallel ripples seldom occur in the absence of transverse elements (Wilson 1972b, Ellwood et al. 1975). Wilson therefore suggested that they might also arise because of centrifugal instability of flow in the lee of the transverse ripple ridges. This follows Allen's (1969a) explanation of such forms in subaqueous ripples, and is very similar to Wilson's model for the three-dimensional shape of transverse dunes (Ch. 26.1.2).

19.3.3 The reptation ripple hypothesis
The best developed hypothesis for ripple formation at present is that of Anderson (1987a). The model begins, like Bagnold's, with an initial irregularity in the bed. This irregularity generates perturbations in the population of reptating grains. Given reasonable assumptions, and an exponential (or gamma) distribution of reptation lengths, the model gives repeated ripples after about 5,000 saltation impacts. The mod-

elled ripples have a strong peak in wavelength in the order of six mean reptation lengths, which is realistic and has a reasonable relationship to u_*. The reptation hypothesis is supported by the experiments of Willetts & Rice (1989). As their ripples grew, reptation lengths increased on upwind slopes and decreased on the downwind ones, producing rhythmic fluctuations and asymmetric ripple shapes. They found that the vertical component of velocity was greater from the windward slopes of ripples, and the creep rate may increase on these slopes. This suggests that ripples could even increase the total transport rate above that on a flat sand bed. Thus, ripples, like dunes, may be an "equilibrium" response to transport processes.

The reptation ripple model has yet to explain the coarsening of sand towards the crests of ripples, mega-ripples, Folk's Y-junctions or flow-parallel features, but it is the best available.

CHAPTER 20

Desert winds

Four wind systems at the macro-scale ($>10^6$m), three at the meso-scale (10^6 to 10^3m) and a number of structures at the micro-scale ($<10^3$m) are responsible for wind erosion, the movement of dust and sand, and dune formation.

20.1 Macro-scale wind systems

20.1.1 Trade winds

The most extensive wind systems over tropical deserts are those generated by high-pressure cells between the Equator and about 30°N and S (Reiter 1969; Fig. 20.1). These "Trade Winds" (or simply "Trades") produce some high wind-energy environments, as in parts of Libya and Mauritania, but also some low-energy ones, as in the Kalahari (Breed et al. 1979b). Trades blow clockwise around the anticyclonic cells in the northern hemisphere and anti-clockwise in the southern. In both hemispheres the cells dip east, producing stronger winds on the eastern than western arcs.

The oceanic anticyclones are more stable than the continental ones, and bring some constant, and often strong, winds to western tropical desert coasts. They touch the Peruvian and Chilean coasts between 25° and 35°S (Prohaska 1973, Lettau 1978, McCauley et al. 1977a), coastal Namibia (Lancaster 1985b), and the northwestern shores of Western Australia (Wasson 1984). The Azores anticyclone generates strong winds on the Atlantic coast of the Sahara, from Morocco to Dakar (Fryberger & Dean 1979, Fryberger 1980) and on the Canary Islands (Chamley et al. 1987). These North Atlantic winds, like the others, have slight seasonal variations as the parent anticyclone fluctuates in position. The same anticyclone drives the winds

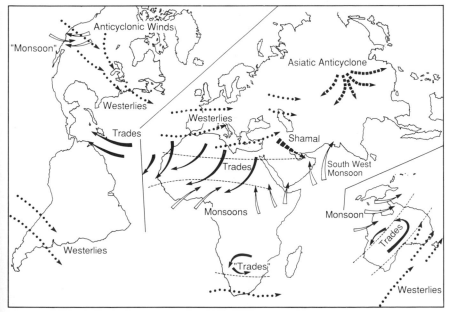

Figure 20.1 The major wind systems in the world's deserts.

that blow over the northern deserts of Venezuela and Colombia (Goddard & Picard 1973).

On land, the Trades are strongest in winter when high-pressure cells are best developed over the cool continents, and it is then that they reach nearest the Equator. In summer, when terrestrial high-pressure systems are replaced by low pressure, terrestrial Trades significantly moderate.

In the Sahara, two wind systems are driven by these terrestrial anticyclones (Dubief 1979, Fig. 20.2). Most powerful and extensive are the winds that reach West Africa in winter as the **Harmattan**. This wind moves massive amounts of sand, and is the main determinant of dune patterns. Aufrère (1935) saw the sand dunes of the western Sahara as "immense anemograms written by the Trades on the surface of the desert". They also carry very large quantities of dust (McTainsh 1985). When the harmattan reaches the Gulf of Guinea, it rises over the intertropical front (Sec. 20.1.2), and takes dust out over the Atlantic (Prospero et al. 1981). In the west, it also rises over the Trade wind system of the Mauritanian coast. In addition, the Saharan high-pressure cell drives the weaker Etesian winds of summer which blow onto the eastern Mediterranean coast of Libya and Egypt (Dubief 1959, El-Baz 1986). The Saharan Trade wind system has an eastern echo over Arabia, where the winter winds swing around in an arc over the eastern subcontinent becoming northeasterly over the Rub' al Khali.

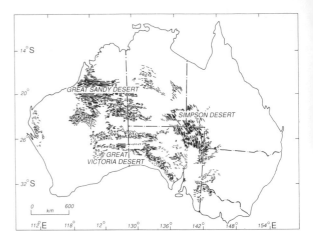

Figure 20.3 The Australian continental dune pattern, apparently a result of flow of winds round a high-pressure cell (but see text; after Brookfield 1970).

In the southern hemisphere, the Australian continental dune pattern forms a great gyre around the eastern edge of the desert (Fig. 20.3). The dune pattern apparently testifies to a symmetrical high-pressure cell, though it is difficult to fit dune and wind trends (Brookfield 1970). This may be because the dunes are relics of a time when the anticyclone was displaced or compressed, but a comparison of the dune and wind trends suggested to Brookfield that the dunes could be aligned to contemporary winds blowing during the seasonal migration of the anticyclone southwards in the spring.

In southern Africa, subtropical high-pressure cells are centred about 32°S and fluctuate in position seasonally by about 4° (Van Zinderen Bakker 1980). The strongest (winter) winds blow in an anticlockwise arc (Thomas & Goudie 1984). In summer the winds are lighter and more variable, as elsewhere. As in Australia, the Kalahari dune patterns form a great swirl, and these patterns have also been linked to the anticyclone (Goudie 1970, Lancaster 1980a, 1981a).

At the height of the last glaciation, Pole-to-Equator temperature gradients intensified and the Trade wind systems were much stronger (Parkin & Shackleton 1973, Flohn & Nicolson 1980, Van Zinderen Bakker 1982). The continents cooled more than the oceans, and winds from the continents therefore intensified, encouraging aridity (Manabe & Hahn 1977). Newell et al. (1981) calculated that wind velocities at this time were 24% greater in the northern hemisphere than at present in winter, and 124% greater in

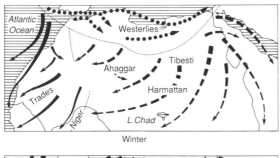

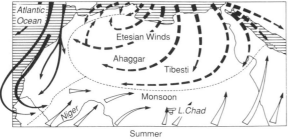

Figure 20.2 Saharan wind systems.

summer; in the southern hemisphere they were 17% greater in both seasons. Ash & Wasson (1983a) deduced that wind velocities could have been 30% higher in the Australian deserts at this time. Much more dust was then being deposited in the oceans than it is today (Shackleton & Opdyke 1977).

In the southern Sahara 18,000yr ago, strengthened harmattan winds were able to move sand hundreds of kilometres farther south than they do today (Grove & Warren 1968). Although a change of wind direction may also have been involved (Warren 1970), present opinion is that the winds followed much the same track as do today. The evidence lies in the alignment of ancient yardangs and mega-yardangs in the Borkou area (Ch. 21.2c; Mainguet et al. 1980a); offshore deposits of sand and dust (Sarnthein et al. 1981); and dune alignments in the central Sahel (Talbot 1980).

In Australia, the intensified Trades may alone have been enough to reactivate dunes, without any change in rainfall (Wasson 1984), but the hydrological equations are complicated by the fact that higher winds must have encouraged greater desiccation. Westerlies were forced northwards, so that there is a divergence of modern winds and the trends of ancient dunes in the Mallee country and in the Strzelecki Desert (Wasson 1986). Wind directions farther north were much as today (Brookfield 1970).

Quaternary climatic patterns of southern and central Africa are even more a matter of controversy than they are elsewhere (Ch. 5.3; Lancaster 1984a, Thomas & Goudie 1984). Unlike Australia or the Sahara, wind patterns may have shifted. In the Kalahari, as in Australia, Trades should have been more intense in the last glaciation (Heine 1981, 1982, Thomas 1984a), and it was probably then that the main dune patterns were formed, for modern drift potentials are too low for dune activity. Even if, as some authorities believe, the Kalahari was wetter then than it is today (Heine 1981), the dunes may have been activated and the pans excavated by stronger and more unidirectional winds.

20.1.2 Monsoons

Low pressure over heated summer continents draws in monsoons over some deserts. These are shallow, 1–3km-thick wedges of air behind the intertropical front (ITF) (Fig. 20.1).

Monsoons are strongest over the huge Asian landmass. Strong southwesterly winds cross the dry eastern tip of Arabia and arid southern Pakistan and western India (Findlater 1969), where they are the main determinant of contemporary and ancient dune patterns (Wasson et al. 1983, Warren 1988a). They also bring large amounts of dust from East Africa (Middleton et al. 1986). Although strong on the southern Arabian coast, they have less wind-energy in the Thar. The *Shamal* ("north wind"), or *tog* of the Arabian Gulf, is also drawn in by the Asian monsoon (Membery 1983). This northwesterly wind, confined on the north by the Iranian mountains, may blow at over 10m s^{-1} at 10m above ground. It carries great quantities of dust from Iraq to the Gulf states, and moves dunes along both coasts. In the Chinese deserts, nearer the centre of the low pressure, the Asian monsoon winds are light and variable.

The rather weaker monsoons of the Sahel of northern Africa, reach to about 20°N. Behind the ITF these monsoons bring rather variable, generally weak, winds, but also some strong, short-lived squalls. These *haboobs* raise massive quantities of dust, some of which, reaching the easterly winds about 3km above ground, is carried over the Atlantic as far as the northern Caribbean (Carlston & Prospero 1972, Tetzalff & Wolter 1980). On the surface the Sahelian monsoon winds are generally southwesterly, and they create increasing variability in the annual mix of winds as one goes south away from the desert (Breed et al. 1979b, Fryberger 1980; Fig. 20.4).

Southern California experiences a similar, though

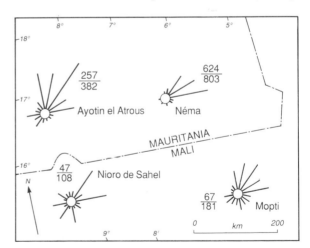

Figure 20.4 Sand movement roses in a part of the West African Sahel, showing the dominance of the harmattan in the north and the increasingly strong effect of the West African monsoon towards the south. The fraction is the ratio of the resultant to the total drift potential, and sand-movement roses are explained in Figure 20.6. (After Breed et al. 1979.)

much feebler monsoon in summer. Heating of the continental interior creates low pressure, towards which winds blow from the Pacific high-pressure system; these are the strongest winds of the year in parts of the Californian desert (Laity 1987).

The southern hemisphere monsoon systems achieve only about 50 kg m^{-1} s^{-1} mass transport compared to 100–180 in the north (Flohn 1984). In Australia the ITF reaches between 20° and 27°S, with centres of cyclonicity in the northwestern and northeastern desert (Brookfield 1970). As in the Sahel, the cyclones produce a pattern of increasing wind variability as one moves equatorwards (Wasson 1984). Also as in the Sahel, winds associated with the ITF may raise dust and take it great distances.

In the last glaciation, some authorities believe that monsoons did not reach as far into the Sahara as they do today (Flohn & Nicolson 1980), whereas others believe that the ITF was in roughly the same position, though bringing much less rain (Sarnthein et al. 1981). The monsoons were re-established after the last glaciation and, during the climatic optimum 5,000–6,000yr ago, they may have been stronger and have penetrated farther north than today (Van Zinderen Bakker 1980, Kutzbach 1981). Thus, in the Sahel, the strong northeast Trades moderated as rainfall increased, and the dune pattern of the previous period of intense Trade-wind activity was stabilized with little change in pattern.

In southern Asia the post-glacial readjustment of climate produced rather different dune patterns. The southwest monsoon was light in the glacial periods (the last one being about 18,000yr ago), and strong in the interglacials (as at present) (Manabe & Hahn 1977). As in the Sahel, monsoons reached their post-glacial maximum some 9,000yr ago (Prell 1984, Prell & Van Campo 1986). In the Thar, increasing rainfall (reaching perhaps 50% more than at present), coincided with stronger winds than today's, and these created dunes at the same time as vegetation was being re-established. Dunes remained active for longer into the Holocene than they did in the southern Sahara. This unique climatic situation may explain the unusually extensive development of vegetated parabolic dunes in the Thar (Wasson et al. 1983).

In Australia at the glacial maximum, monsoons seem to have been less vigorous, and to have kept farther north than today (Wasson 1986). The climatic optimum (3,000–1,000yr ago), unlike its counterpart in the Sahara, was a period of renewed dune formation (Wasson 1984). About 18,000yr ago the

northern Kalahari, unlike the Northern Hemisphere deserts, received more rain that it does now, although the Namib remained hyper-arid (Van Zinderen Bakker 1982).

20.1.3 Continental and polar anticyclones

The immensity of the Asian continent generates an almost unique wind system over its deserts. In winter, an intense, shallow anticyclone forms, and is forced northwards to about 55°N over Mongolia by the Tibetan Plateau. Winds blow clockwise around this high-pressure zone, but, when the northeast winds on the eastern side of the anticyclone meet the Tibetan Plateau, they split. One set of winds blows west into the Tarim basin and on into the (formerly) Soviet deserts. The other blows east over the Ordos Plateau and into central China (Chao Sung-chiao 1984, Mainguet & Chemin 1988; Fig. 20.1). These winds are very dusty. Their ancestors took dust from central Asia to form the massive loesses of the Hwang He valley.

In general, these winds today have little energy (Breed et al. 1979b, Fryberger & Ahlbrandt 1979); they are strongest in the spring and early summer when they are northerly in the Ala Shan, northeasterly in the Taklimakan, and easterly in the Kara Kum (Zhu Zhenda 1984). Southwards, weak winter winds from the Asian anticyclone blow across India and the tip of the Arabian peninsula as the "northeast monsoon", but make little impact on dunes in the Thar or the Wahiba Sands (Wasson et al. 1983, Warren 1988a). They mix with Westerlies to produce a complex wind pattern.

A pale echo of the Asian winter anticyclone sometimes occurs in the western USA, in the area of Nevada. It may send strong winds southwestwards over southern California and occasional dusty winds, known as the "Santa Ana", southwestwards to the coast (Bowden et al. 1974, Havholm & Kocurek 1988). Further winter anticyclones cover the Arctic and Antarctic, driving strong winds that deflate, abrade and build dunes in the small ice-free areas (Selby et al. 1973, McKenna-Neuman & Gilbert 1986).

The central Asian anticyclone and its associated aridity intensified during the late Pleistocene as rapid uplift of the Tibetan Plateau shut off moist winds from the south (Chao Sung-chiao & Xing Jiaming 1982, Zhu Zhenda et al. 1987). During glacial periods, anticyclones also intensified over the continental ice sheets of central Asia and northern Europe, but, like those in North America, the strong northeasterly katabatic winds which circulated off them affected only a narrow zone near the glacial margin (David 1981, Filion 1987).

20.1.4 Westerlies

In winter, the poleward margins of the tropical deserts are brushed by the outer edges of mid-latitude cyclones as they track eastwards, bringing generally westerly winds (Fig. 20.1). The Westerlies extend their zone of influence towards the Equator in winter and retreat poleward in summer. Because the Pole-to-Equator temperature gradient is steeper in the southern hemisphere, the southern Westerlies are stronger than those of the north. The southern hemisphere depressions brush the tip of Africa in the winter, drawing in hot, dry, strong "berg winds" off the interior plateau into the Namib desert. In Australia, Westerlies dominate south of 30°–33°, especially in the Mallee country (Wasson 1984). In summer, the southern depressions are displaced far to the south (Tyson 1964, van Zinderen Bakker 1980, Flohn 1984).

In the northern hemisphere, the upper Westerlies move in great meanders around Earth, and occasional strong waves may draw in depressions right across the Sahara (Dubief 1979). The more common depressions of the main zone of surface Westerlies between 45° and 60°N track across the Mediterranean and the Middle East and draw in strong surface southwesterly winds across the northern edges of the deserts. These winds affect the Sahara and Arabia south to about 30°N in winter, and dominate their annual sand-movement roses (Fig. 20.5; Breed et al. 1979b, Jones et al. 1986, El-Baz 1986). The interaction between Westerlies in winter and Trades in summer brings complex wind regimes to the northern edges of the Sahara and Arabian deserts (Dubief 1952, Wilson 1971, Fryberger 1980).

The cold-winter, continental deserts of North America also feel the effects of the eastward-tracking depressions, especially in winter (Sweet et al. 1988). They even penetrate the edges of the central Asian anticyclone from time to time, where they have an appreciable effect in the southern Tarim basin and in the Dzungarian basins of westernmost China, and even penetrate the Badain Jaran Desert in north-central China (Walker et al. 1987). At the margins between these Westerlies and the anticyclonic easterlies, there are complex wind regimes and dune patterns (Chao Sung-chiao 1984, Zhu Zhenda 1984).

Westerlies occasionally bring strong winds to bear on the desert surface, where they are very effective at raising dust. In the Mediterranean, the notorious Sirocco, Gibli and Sharav/Khamsin winds are most active in May and June, when they carry dust from the Sahara deep into Europe and the Middle East (Dan & Yaalon 1966, Dubief 1979, Yaalon & Ganor 1979, Ganor & Mamane 1982, Jones et al. 1986). Winter fronts also bring the dustiest conditions to Arizona (Brazel & Nickling 1986). There are other strong dusty winds in central Asia and in the southern hemisphere. Some of this dust is swept into the mid-latitude jet streams, and moved thousands of

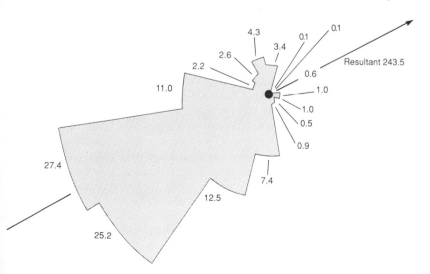

Figure 20.5 A sand-movement rose for northern Arabia, showing the dominance of southwesterly winds associated with the passage of depressions in the zone of Westerlies (after Jones et al. 1986).

kilometres in matters of hours, as between Australia and New Zealand, or between China and Hawaii or Alaska (Pye 1987: 111–14, Pye & Tsoar 1987).

During glaciations, Westerlies intensified more than any other global wind system; in interglacials, they waned. In the cold phases, powerful westerly and northwesterly winds blew across the Great Plains North America (David 1981), building mega-barchans in the Nebraska Sand Hills (Warren 1976a, Ahlbrandt & Fryberger 1980). In California, winds were stronger, although near to their present directions (Christian 1970, Kutzbach and Wright 1985, Orme & Tchakerian 1986, Johnson 1977, Laity 1987), and in the Pacific, where winds left deposits of dust, and where windspeeds must have increased dramatically, the position of maximum wind activity stayed between 38° and 40°N throughout the late Pleistocene (Rea & Leinen 1988).

In Europe, close to the expanded ice sheets, a persistent anticyclone blocked the entry of Westerlies for most of the time, and created its own strong westerly circulation over the North European Plain and southwards into Hungary (Williams 1975, Gozdzik 1981, Koster 1988). Farther south in the Old World, Westerlies appear to have moved closer to the Equator, reaching more often to the Hoggar and Tibesti massifs in the central Sahara (Van Zinderen Bakker 1982, Flohn & Nicolson 1980), and strengthening considerably in the Canaries and the coast of Morocco (Weisrock 1982, Rognon & Coudé-Gaussen 1987). However, the Westerlies may have been more confined than they are today, squashed as they were between the glacial anticyclone and the tropical high-pressure cells, and this may have concentrated the deposition of loess (Coudé-Gaussen & Rognon 1988). There may even have been a change in direction at least of the wind resultants in the northern Arabian desert (Whitney et al. 1983).

Southern-hemisphere Westerlies also intensified in the cool Pleistocene phases (Harrison et al. 1984). In Australia they moved farther north and may have acquired a slightly more northerly trend (Sprigg 1979, Bowden 1983, Wasson 1986).

20.2 Meso-scale (local) winds

At the meso- or local scale ($< 10^6$m), sediment-carrying winds are affected by three kinds of process.

20.2.1 Large topographic effects

Deflection of wind around high ground is particularly marked where there is stable stratification, and where hill massifs project through the surface air masses. When the Froude number (with the height of the hill as the depth factor) is less than 1, there is strong horizontal flow around a hill. One of the most effective of these deflections at a large scale has been referred to in Sec. 20.1.3: the splitting by the Tibetan Plateau of the winter northeasterlies associated with the shallow winter anticyclone in China. There are many other examples. In the Peruvian coastal desert, maritime air lies below 600–800m in winter and 200–500m in summer. Hill massifs which reach near to or above these heights force winds to blow around them. Huge masses of sand accumulate on their upwind slopes, and there is intense wind erosion on their flanks (Howard 1985).

Deflection creates strong and constant winds in mountain passes. The flow is strongest in mid-pass, and retardation by the walls of the pass cause a fanning pattern away from the central zone (Dokka 1978). There are hydraulic jumps as a wind comes out of narrow passes and in the lee of large massifs (Queiroz et al. 1982, Gaylord & Dawson 1987). In many places the funnelling produces a markedly reversing wind pattern, all winds from one quarter being funnelled in a narrow range of directions, while those from the opposite quarter are funnelled in the diametrically opposite one (Laity 1987). Even quite major variations in climatic patterns do not produce variations in wind directions through these passes, which can therefore experience strong winds from a constant direction for many thousands of years.

Intense funnelling and its geomorphological effects have been noted by Sharp (1964) in the Coachella valley in southern California; Laity (1987) and Sakamoto-Arnold (1981) elsewhere in California; and Gaylord (1982), Kolm (1985) and Gaylord & Dawson (1987) in Wyoming. In the Sahara, Trades followed the same curving path around the eastern flank of the Tibesti massif, perhaps since the Miocene, and have created deep mega-yardangs in Borkou (Ch. 21.3.2c). Areas with funnelling are characterized by yardangs, ventifacts, mega-ripples, barchans and transverse dunes (McCauley et al. 1977a, Breed et al. 1979b).

Deflection is less obvious over long, gentle rises, but can nonetheless be detected in the central Australian dune fields (Brookfield 1970).

Large topographic obstacles also generate lee effects such as lee waves and "Kármán" vortex trails. Kármán trails in the Trade Winds down wind of the Canary Islands reach up to 320km in length (Chopra & Hubert 1964, Zimmerman 1969). Turbulence is intense in these trails, creating long sand-free "shadows" down wind of large hill massifs on land (Ch. 25.2.2d).

20.2.2 Anabatic and katabatic winds

Anabatic (upslope) and katabatic (downslope) winds characterize the climates of many areas close to mountain masses over about 10km diameter. Katabatic winds are generally the stronger. A model of katabatic winds can be developed if assumptions are made about mountain slope, depth of the flow layer and surface drag (Thorson & Bender 1985).

Anabatic/katabatic wind systems have diurnal and seasonal cycles. In the Namib Sand Sea, "thermo-topographic winds" are upslope by day and downslope by night. The strongest downslope winds are in winter; the strongest upslope ones in summer (Tyson & Seely 1980). Anabatic/katabatic winds are felt around many desert mountain massifs, as in Morocco (Burrell 1971). They may intensify pressure-gradient winds, such as the notoriously hot and dusty "Santa Ana" in southern California (Bowden et al. 1974), the "berg" winds off the scarp east of the Namib Sand Sea (Tyson & Seely 1980, Lancaster 1983b), and the dry dusty winds that blow over Patagonia (Middleton et al. 1986). Katabatic dust storms are frequent in Soviet central Asia, Afghanistan, Pakistan and the Tibetan Plateau (Middleton et al. 1986). The strongest katabatic winds are felt in the polar deserts, intensified by the low temperatures over snow and ice (Mc-Kenna-Neuman & Gilbert 1986). Violent winds blow off the high ice plateau of Antarctica (Mather & Miller 1966), and are the chief determinants of dune patterns in the few ice-free areas (Calkin & Rutford 1974). Katabatic winds may also be generated by the North Polar ice cap on Mars (Tsoar et al. 1979). Some large accumulations of sand may themselves create anabatic and katabatic winds which may help to create even higher dunes.

20.2.3 Land and sea breezes

Land/sea breeze systems are the meso-scale equivalents of the macro-scale monsoons. They are strongest and most regular on hot desert coasts in summer, where land/sea temperature contrasts are greatest (Flohn 1969). On the coast of Israel they occur on 80% of summer days (Halevy & Steinberger 1974). Sea breezes are also common, and are effective sand-movers on many temperate-sea and even large lake coasts in summer (Hunter & Richmond 1988).

Onshore sea breezes begin in mid-morning at the coast and work their way a few tens of kilometres inland during the afternoon, sometimes behind a gusty "sea-breeze front" (Johnson & O'Brien 1973). They can reach 300km inland if coupled with other wind systems. The active layer is about 1,000–1,400m deep on tropical coasts, thinning inland. Windspeeds may reach storm intensity (Defant 1951, Tyson & Seely 1980), but are more usually in the 3–6m s^{-1} range. Surface drag usually reduces windspeeds inland (Illenberger & Rust 1988), so that sand transport and dune celerity decline. Sea breezes are intensified where the coast is concave, and dissipated where it is convex to the sea (Flohn 1969). They are subject to the Coriolis force, such that they swing rightwards or clockwise as they strengthen in the northern hemisphere. The contrast between the smoothness of the sea and the roughness of the land means that the inland breeze may be deflected by up to 35° from its angle of incidence at the coast. Land breezes (the nocturnal complements of sea breezes) can carry dust to the coast, but are generally less powerful.

Sea breezes are well developed on the Peruvian and Namibian coasts where land/sea temperature contrasts are marked. In both deserts, the southerly Trades are intensified and swung to a more westerly direction by the sea breeze (Tyson & Seely 1980, Lettau 1978, Prohaska 1973, McCauley et al. 1977a). In northern Sinai in summer a strong northwesterly sea breeze blows off the Mediterranean almost every afternoon (Harris 1957, Tsoar 1974). Another system exists off the coast of Mauritania (Jänicke 1979). Sea breezes on many temperate and wet tropical coasts are the principal force in keeping dunes free of vegetation, as in southern California (Orme & Tchakerian 1986).

de Félice (1968) believed that a system rather like land-/sea-breeze systems occurred between sand seas and hammadas in Algeria.

20.3 Individual wind climates

The combination of seasonally different global winds and locally forced winds creates a great variety of individual wind climates. In terms of sand- and dust-movement, the character of individual wind climates is best summarized in the form of sand (or dust) movement roses.

20.3.1 Sand-movement roses

A knowledge of the relationship between wind velocity and sand movement, and some data on the wind regime (velocity and direction) at a particular station, allow the construction of "sand-movement roses".

There are several ways in which a sand-movement rose could be constructed from wind data, depending, among other things, on the choice of formula for linking wind velocity to sand movement, and the choice of a threshold velocity for sand movement (Bagnold 1951, Brookfield 1970). The choice of formula may not be critical to the results of the exercise, since many of them yield similar results at the velocities at which most sand is moved. However, as will be apparent in some of the argument that follows, the choice of threshold may be critical. This was noted by Wilson (1972b), when he discovered that bedforms in fine sand had quite different orientation to the resultant as calculated for coarse sand.

Table 20.1 gives a simple system for constructing sand-movement roses, using the method of Fryberger & Dean (1979). Fryberger & Dean used the roses to produce two further parameters: resultant drift direction (RDD); and the resultant drift potential (RDP), being the potential relative amount of sand moved in this direction. Resultants can be calculated mathematically or graphically. Any resultant must be seen as relating to a specified grain-size of sand, a specified time period, and a specified area (Wilson 1972b).

20.3.2 Classification of wind regimes

The classification of wind regimes with respect to dune types dates from Aufrère (1931a, 1932). Aufrère talked of "dunes of conjunction" when two or more sand-carrying winds in an annual cycle were at a narrow angle; "dunes of opposition" when two dominant winds blew at 180° to each other; "dunes of incidence" or "quadrature" when these two winds were divided up by topographic obstacles; and "multidirectional dunes" when at least two dominant winds created a complex regime. Other systems of classification were discussed by Brookfield (1970) and Wilson (1971).

Fryberger & Dean (1979) classified sand roses in the following way (Fig. 20.6). The ratio of the resultant drift potential (RDP) to the total drift potential from all directions (DP) is first computed (terms as defined above). This produces an index of directionality: when wind comes from one direction

Table 20.1 The construction of sand-movement roses (after Fryberger & Dean 1979).

Data Wind data must include records of velocity and direction throughout the year. Fryberger & Dean used several sources, but principally "N Summaries" prepared by the World Meteorological Organization (WMO). Like most standard measurements, these record wind velocity at 10m above ground. The N Summaries record winds in 5–11 velocity categories in knots at 3- to 6-hour intervals to the nearest 10° compass direction. Fryberger & Dean chose average velocities for each category and converted the velocities to cm s^{-1}. The quality of a sand rose depends on the quality of the wind data.

Conversion to sand movement Only potential sand movement can be estimated from this data, since they do not include data on sand availability and surface characteristics. Fryberger & Dean chose Lettau & Lettau's formula for converting the wind data to sand-movement estimates (Ch. 18.6.2), and a "universal" threshold velocity of sand movement (measured at 10m above ground) of 12 knots. This can be varied according to local sand sizes. They simplified Lettau & Lettau's formula to:

$$\text{sand movement, } q \propto V^2(V-V_t)T$$

where V is the wind velocity measured at 10m, T is the percentage of the total time during which a wind of a particular velocity (V) blew from any one direction, and V_t is the sand-movement threshold.

The q values are summed for all velocity classes from each direction to provide the relative potential sand movement ("drift potential", DP, in vector units) from that direction. Drift potential figures for each compass direction allow sand-movement roses to be constructed as in Figure 20.7.

throughout the year the *RDP:DP* ratio approaches 1; where it comes from many directions the ratio approaches 0. Using this approach, Fryberger & Dean (1979) recognized five types of wind regime (Fig. 20.6). They also characterized wind regimes into:

high energy (>400 total annual drift potential, or about 33m^3 per m width yr^{-1});
intermediate energy (399–200, or about 33–17m^3 per m width yr^{-1}); and
low energy (>200, or less than about 17m^3 per m width yr^{-1}).

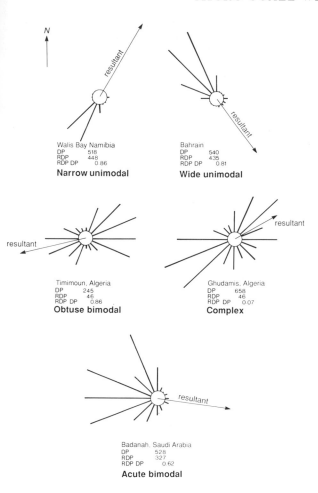

Figure 20.6 Fryberger and Dean's (1979) classification of sand-moving wind systems. The terms are explained in the text.

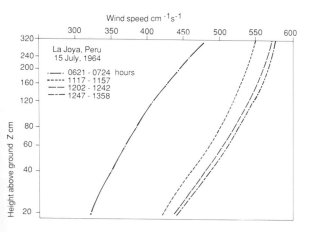

20.4 Micro-scale wind structure

20.4.1 General characteristics of desert winds

Most deserts experience some seasonality. Non-desert conditions intrude in either winter (on the poleward margins) or summer (on the equatorial margins). In the remaining months in these seasonal deserts, and continuously in a few hyper-arid tropical deserts, there are "truly desert" conditions in which there are intense diurnal variations in temperature and in lapse rates, and in consequence in winds.

At night, rapid cooling in clear skies creates a stable surface layer in which winds are usually too light to move sand or dust. A strong inversion can develop early in the morning, with cool air near the ground and warmer air above (Dubief 1979). Thermal contrasts between different types of surface and different elevations intensify as the Sun rises, and these drive strengthening winds, but at first only above the inversion, where "decoupling" from the surface allows them to reach high velocities. These upper winds may create gravity waves (Sec. 20.4.5) at the surface of the inversion, and as the ground warms up, the waves, if present, grow and eventually break up the stratification, in the process sending strong "ground jets" to rake the surface (Knott 1979, Warren & Knott 1983). The possible significance of ground jets in dune formation is discussed in Chapter 23.2.2c. Windspeeds usually continue to build up into the afternoon, driven largely by convection. Regular fluctuations in windspeed and direction may be associated with convection cells, but, like the convection itself, winds are usually erratic (Reid 1985). The heat of the middle of the day is the time of greatest dust-devil activity. The windspeeds decline again after sunset.

20.4.2 The effects of surface heating and cooling on wind velocity profiles near the ground

Surface heating affects the velocity profile near the surface by introducing buoyancy (Stearns 1978, Olson 1958a, Jensen et al. 1984; Fig. 20.7). Velocity profiles are flattened (velocity increases away from the ground at a lower rate than it would in unheated conditions). If, on the other hand, there is an inver-

Figure 20.7 The effects of temperature on wind profiles above the surface. The early-morning wind profiles were recorded in cooler conditions than those at mid-day (after Stearns 1978).

sion, say at night, with cool air near the ground, the velocity profile sharpens (velocity increases away from the ground more steeply than it would otherwise). These effects can distort the calculation of u_* and z_0 values from anemometers placed above a few tens of centimetres above ground, leading to errors in the estimation of u_*. Even with gentle heating on a Danish beach, Jensen et al.(1984) found substantial errors in predicting u_* and z_0 from anemometers at only 4.5m above ground, although they noted that a good estimate of z_0 would allow u_* to be predicted fairly accurately.

20.4.3 Dust devils

Dust devils (*djins* in Arabic; "willy-willies" in Australia) are familiar features of desert landscapes (Fig. 20.8). These thermal vortices, made visible by dust and other debris, arise when and where there has been intense surface heating, and where some local process encourages vorticity (Sinclair 1969). In one set of field measurements, the steepness of the lapse rate in the 0.3–10m layer above ground correlated quite well with the velocity, size and frequency of dust devils (Ryan & Carrol 1970, Ryan 1972). There have been various suggestions as to the source of vorticity (including flow around small hills and even passing rodents; Sinclair 1969, Hallet & Hoffer 1971, Idso 1974), but Carrol & Ryan (1970) believed that it arose from shear in horizontal flows in convective, unstable atmospheres. As many dust devils revolve clockwise as do anticlockwise, although at any one time most rotate in the same sense (Carrol & Ryan 1970).

Once a dust-devil vortex has formed, pressure drops in the calm centre, and windspeed reaches a maximum in a tight ring around it. Secondary vortices may rotate around the primary ring, and at a larger scale, several dust devils may rotate around a slowly moving disturbance centre of greater diameter (Hallet & Hoffer 1971). As the vortex intensifies, it draws in hot surface air, which is carried aloft on the edge, and this movement is balanced by a downdraught in the centre. The axis is frequently off-vertical and twisted (Fig. 20.9; Sinclair 1973). Most dust devils migrate at random in light winds, but down wind in stronger ones, at a speed rather greater than the ambient wind (Kaimal & Businger 1979). Some are stationary for long periods and may then lift large amounts of dust.

Tangential velocities in dust devils reach 22m s^{-1},

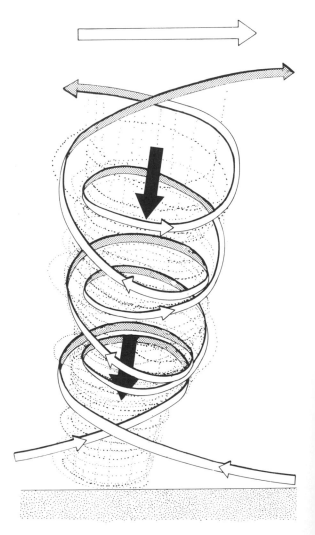

Figure 20.9 Circulation in a dust devil (after Idso 1974).

Figure 20.8 Dust devils as seen by a mediaeval artist.

and vertical velocities 3.5m s^{-1}, and they may raise 250kg km^{-2} of dust (Pye 1987). Simulated dust devils in a wind-tunnel show that lift forces are independent of particle diameter (except for very large and very small particles), so that they can lift fine dust as readily as medium-size sand, unlike unidirectional winds. This is evidently because of the sudden pressure difference between the bottom and top of the surface sediment, rather than drag and small-scale lift as in unidirectional winds (Greeley & Iversen 1985: 85–9, Iversen 1986c). Dust devils may be the way in which most Martian dust is lifted, for it is difficult to see how else it could be entrained in such a low-density atmosphere (Sagan et al. 1971, Iversen 1986c).

Dust-devil frequency can be of the order of 50–80 per day. Heights range from 75m to 100m, occasionally reaching 1km, and diameters from a few to hundreds of metres (Sinclair 1969, Hallet & Hoffer 1971, Idso 1974). Dust devils are more frequently seen where there are sources of dust, as in river valleys or recently burnt areas.

20.4.4 Other intense micro-scale systems

Thunderstorm systems often generate intense cool downdraughts as they advance, and these entrain dust in a spectacular fashion. These are termed "downdraught haboobs". The dust storm appears as billowing wall of dust up to 3km high, indented at the base (Lawson 1971), travelling at about 50km h^{-1} for up to about an hour. The Khartoum area of Sudan experiences about 24 of these during the monsoon season; southern Arizona only has two to three per year, also in summer (Middleton et al. 1986).

20.4.5 Secondary flow patterns over flat terrain

Two types of secondary flow relevant to dune formation have been reported by meteorologists: gravity waves and roll vortices.

(a) Gravity waves

Gravity waves (Kelvin–Helmholz instability) occur when there is layered flow in the atmosphere, as at a temperature inversion. Wavelengths are related to the depth of the lower layer, density contrasts and wind velocity (Emmanuel & Bean 1972). The development of gravity waves in the boundary layer has been extensively studied by Scorer (1978: 154–223) and Jackson (1977b). The spacing of dunes in the lee of hills has been associated with these lee waves by Dawson & Marwitz (1982) and Kolm (1982 and

1985). Some examples are shown in Figure 20.10. Wippermann (1969) believed that transverse vortex systems would be generated in linear vortex flow, and that transverse dunes on the backs of linear dunes were evidence of these transverse vortices.

(b) Cellular convection and roll vortices

Horizontal-axis, or roll, vortices are widely observed in the atmosphere. They are seen in the behaviour of birds, gliders, balloons and insects and the paths of pollution, moisture and "wind-rows" on the sea (Hanna 1969, Kuettner 1971, Otterman 1975, Knott 1979, Brown 1983). Dust storms are frequently arranged in parallel lines, presumably related to vortex flow (Hastings 1971, Swift et al. 1978, Middleton et al. 1986). The most visible signs are "cloud streets" or lines of cloud created by the updraughts of air between pairs of roll vortices (Fig. 20.11). The commonness and regularity of cloud streets, particularly in the Trade Wind belt, first became apparent on satellite images. They can be followed continuously for hundreds of kilometres.

Two mechanisms are said to be involved. First is

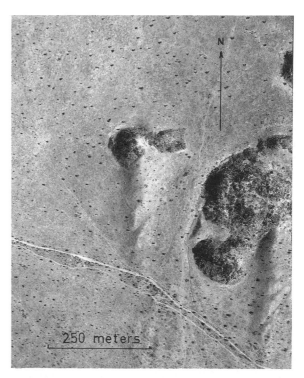

Figure 20.10 A dune pattern apparently created by lee-waves down wind of a small inselberg in stabilized dunes in central Sudan (after Warren 1968).

Figure 20.11 The structure of roll vortices, and their possible relationship to linear dunes (redrawn from various sources).

the heating of the atmosphere from below. In still air this creates regular cells of rising air, which have been linked to star-dune formation by some workers. If there is a horizontal wind, the vertical cells are stretched to become roll vortices (Brown 1983). This is vividly demonstrated during forest fires (Haines & Smith 1983). Another way in which roll vortices can develop may be curvature introduced by the Earth's rotation (Wippermann 1969).

Empirical observation shows that roll vortices occur when there is an strong inflection in velocity profile (LeMone 1973, Wippermann 1969). Kuettner's survey (1971) concluded that cloud streets occurred "in strong flows heated from below with a curved velocity profile of rather uniform direction". Other conditions include flat terrain, little variation of wind direction with height, and maximum wind velocity at about 1km above the surface, all conditions that are common in deserts. The pattern develops over about 20km down wind of the start of the appropriate conditions (Hanna 1969). Roll vortices are aligned between 13° and 16° to the geostrophic wind (Wippermann 1969). They have about 10% of the energy of the mean flow, and have more energy in highly convective conditions (Brown 1983). Kelly's (1984) compilation of evidence on the wavelengths of roll vortices (wavelength = **two** vortices) is shown in Figure 20.12.

The size of roll vortices is said to be related to the depth of the planetary boundary layer (Brown 1983).

This relationship was explored by Swift et al. (1978) for streaks of dusty air separated by clear bands of air, spaced between 25m and 30m. They hypothesized that the streaks should have a wavelength 2 to 4 times the depth of the boundary layer, putting this then at

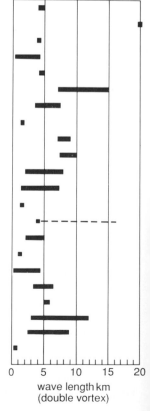

Figure 20.12 Dimensions of roll vortices recorded in the literature (after Kelly 1984).

15–30m, a figure that was consistent with its growth, and the expansion of the vortices. In the Trade Wind belt and in the monsoons, where the inversion is commonly at 1–2km, this relationship suggests that pairs of roll vortices would be spaced 2–8km apart. Some dune spacings may be determined by vortices beneath shallower, shorter-duration early-morning inversions (Knott 1979). The association of roll vortices and linear dunes is discussed in Chapter 26.2.4.

20.4.6 Roughness effects

Wind direction varies with height according to the "Ekman Spiral" (Scorer 1978: 150–53). At altitude the geostrophic wind direction is determined by the pressure gradient and the Coriolis force, and so flows parallel to the isobars. Nearer the ground, roughness causes the wind to swing to the left of the geostrophic direction (in the northern hemisphere) or to the right (in the southern). The most obvious geomorphological effect of this is a swing as winds leave the smooth sea and meet the rougher land (Fig. 20.13). Similar effects can be seen in the desert where winds pass from smoother linear-dune landscapes to rougher transverse dunes (Warren 1976b, Lancaster 1982a). This kind of swing was measured on the South Australian coast by Hails & Bennet (1980).

20.4.7 Flow over hills

(a) Gentle windward slopes

In the past two decades, meteorologists have been developing and testing models for flow over low hills, such as dunes (Taylor et al. 1987). It is now generally accepted that, on a hill that is much wider (in the wind direction) than it is high, the principal effects are felt in an "inner region", whose height (l) is given by $(l/L) \ln^2 (l/z_0) = 0.32$. L is the length of the dune in the wind direction at half its height. l is typically about $0.05\ L$ (Jackson & Hunt 1975, Taylor et al. 1987; Fig. 20.14). For a 10m-high sand dune, the Jackson & Hunt formula gives a depth of the inner region of the order of 1.5m. On a dune on the Danish coast which was about 8.5m high and 140m wide (L = 30m) in the wind direction, with z_0 = 0.01m (over vegetation), the inner region was found to be about l = 1.8m high (Rasmussen 1989). The greatest change in the velocity profile from the upwind condition occurs within the inner region at a height of about $l/3$ (Fig. 20.14). On Rasmussen's dune, referred to above, the maximum windspeed (and divergence from the upwind velocity profile) occurred at the crest at 0.5m above ground.

At the base of windward slopes of gently sloping hills, most models and measurements show a distinct fall in velocity and u_* (Fig. 20.15; Jackson & Hunt 1975, Taylor et al. 1987). On Kettles Hill in Alberta,

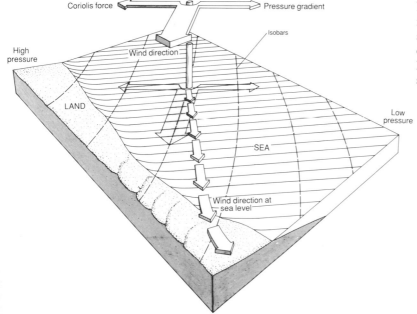

Figure 20.13 The effect of changing surface roughness on the near-surface wind direction. Wind coming off the smooth sea is turned to the left as it moves over the rough land surface.

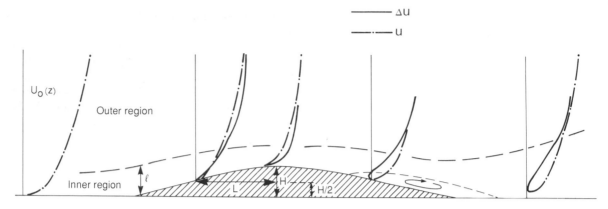

Figure 20.14 Wind flow over a low hill, showing the inner region and changes in the velocity profile above the surface. Δu is the difference between the windspeed at a site and the equivalent upwind value (after Carruthers & Hunt 1989).

which was about 100m high and about 2,000m wide in the windward direction, this zone of decreased velocity extended over about 200m (Walmsley et al. 1982). It may be related to an upwind zone of flow separation described below for flow over steeper bluffs. Its dynamic significance on dunes is discussed in Ch. 23.3.3).

On the main part of the windward slope, the models show a steady increase in velocity and shear. The maximum fractional speed-up (the factor by which the upwind velocity must be multiplied) is predicted by the Jackson & Hunt model to occur near the top of the hill and to be $\approx 1.6 H/L$, for three-dimensional, axially symmetric hills.

Most meteorological studies find that the highest velocities at the summit occur in a very narrow region, just short of the crest (Fig. 20.16; Jackson & Hunt 1975, Pearse et al. 1981, Lai & Wu 1978,

Mason & Sykes 1979, Bradley 1980, Taylor 1981, Walmsley et al. 1982, Jensen & Zeman 1985, Mason 1986, Rasmussen 1989). However, Jensen et al. (1984) found that the point of maximum speed-up on a dune was up to four times lower than predicted by Jackson & Hunt, and Jensen & Zeman (1990) also found u_* peaking just before the crest. It is likely that the position of the point of maximum shear is very sensitive to the shape of the windward slope; Richards (1980, 1986) found the maximum to be nearer the crest on dune-shaped hills than on sine-wave hills. Moreover, if there is flow separation in the lee, the maximum speed-up is reduced. The significance of this for dunes is discussed in Ch. 23.3.5).

The pattern of flow described in Figure 20.14 shows that measurements of flow over dunes that rely on measures of windspeed taken at only one height above ground at any site, or on more than one, but

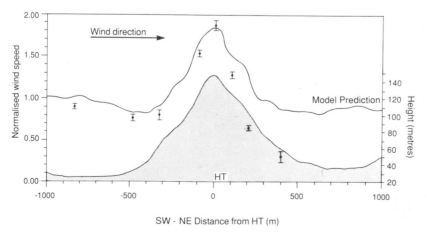

Figure 20.15 The decrease in wind velocity and u_* at the base of the windward slope of Askervein Hill in South Uist. Dots and error bars refer to actual measurements of windspeed (data from Taylor et al. 1987).

286

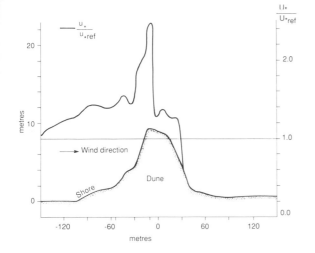

Figure 20.16 The sharp peak in wind velocity at the summit of a low hill (a coastal dune ridge). The ratio u_{*0}/u_{*ref} is the ratio of the windspeed at the site to the windspeed at a reference site on the beach (after Rasmussen 1989).

with one or more above the height of maximum velocity, are of very limited value for deducing shear on the surface and its effects on sediment transport. Such measurements seriously underestimate shear and sand transport on windward slopes. This has been demonstrated on dunes, by Lancaster (1987c) and Mulligan (1987). Lancaster (1987c) found highest discrepancies at the crest.

Atmospheric inversions introduce different patterns. In shallow inversions the speed-up may be less than if there is no inversion, but as the air above becomes more stable, speed-up on the windward slope increases, especially at the summit, and there is a lull up wind ("blocking") where there may even be separation and the development of a rotor. Speed-up is greatly increased in very stable conditions (Carruthers & Choularton 1982). Inversions tend to intensify flow around rather than over the hill. In moderate, stable stratification, gravity waves appear in the downwind wake (Taylor et al. 1987).

If wind approaches at a horizontal angle to a slope, the flow near the surface is diverted to flow more at right angles to the windward slope (Rasmussen 1989).

(b) Steep windward slopes

If the windward slope of an obstruction is steep enough (for example a scarp), and wind turbulent enough, an upwind vortex develops. The simplest pattern of separation is the two-dimensional case: when wind blows over an infinitely long ridge at right angles to the flow. There is then a vortex in front of the obstacle, developed from vorticity inherent in the approaching wind by virtue of the velocity gradient away from the ground. Another vortex with intense velocities may occur immediately down wind of the crest of the obstacle if it has a sharp edge, as in Figure 20.17 (Bowen & Lindley 1977). If the obstacle's shape varies in three dimensions (as is commonly the case), it presents facets oblique to the flow. There is then a three-dimensional pattern of separation. The 2-D upwind vortex is now bent around the obstacle and it feeds two horizontal axis "horseshoe" vortices, spinning off down wind (Fig. 20.18). The extent of development and character of these 3-D patterns depend on the size of the obstacle and the strength of the wind.

(c) The lee: flow separation and wakes

When wind with enough strength to carry sand flows over a steep downward step, such as a scarp, or brink of a sand dune, it enters an area of very high turbulence, which has two major zones: the separation bubble and the wake.

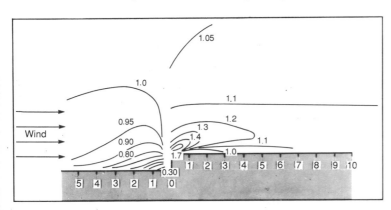

Figure 20.17 Contours of wind velocity in the vicinity of a steep, wind-facing scarp. A zone of low velocity, and probably separated flow occurs on the downwind (upper) side of the scarp face. In some conditions, a dune might accumulate just downwind of the crest of the scarp (after Bowen & Lindley 1977).

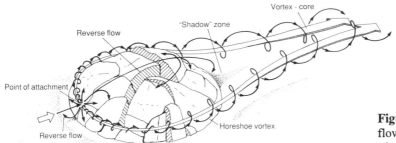

Figure 20.18 The three-dimensional flow pattern around a hill (redrawn after Greeley & Iversen 1985).

The **separation bubble** or "lee vortex" or "lee eddy" is a zone where wind flows back towards the hill. The wind overshoots the crest, reattaching to the ground some distance down wind (the reattachment point; Fig. 20.19). For a hill which is very broad across the flow, the separation bubble is between five and ten times the height of the hill; where the hill is as broad (across wind) as long (in the wind direction), the length of the bubble is only about as long as the height of the hill (Hunt & Barrett 1989), although the position of the reattachment point fluctuates wildly (Kiya & Sasaki 1983). Ambient u_*, turbulence and roughness height, z_0, are controls on the size and character of the separation bubble.

The lee vortex takes on a different character if wind approaches a ridge at a horizontal angle. Seginer (1975) found that the width of the lee vortex was greatly reduced as the wind flowed more and more parallel to a barrier. Mean flow in the lee is diverted to flow parallel to the ridge (Tsoar 1978, Rasmussen 1989, Tsoar et al. 1985). At angles in the 30°–50° range, velocities in this flow can reach higher values

than in the upwind unaffected flow (Seginer 1975, Tsoar 1978, Tsoar et al. 1985). The significance of these findings for linear dune formation is discussed in Chapter 26.2.3c.

Above the separation bubble, and extending down wind of it, is the **wake**. This is a zone of much less organized turbulence than the separation bubble in which there can be three types of turbulent motion. First, there are small, very irregular vortices with horizontal axes aligned transverse to the flow, which

Figure 20.19 A separation bubble in the lee of a dune, visualized with smoke.

Figure 20.20 "Wakes" or sand-free "courts" in the lee of barchans.

arc shed from the crest of the hill and travel down wind, growing and dispersing as they do so (Kiya & Sasaki 1983). Secondly, when the hill is narrow, its flanks shed vertical-axis "Kármán" vortices into the wake, whose pattern is a complex function of the height, orientation to the flow and width of the obstacle (Mason & Morton 1987). Thirdly, the flanks of narrow obstacles may also shed horizontal-axis "corkscrew" vortices, parallel to the flow, from either flank (Greeley & Iversen 1985: 208). Down wind of the reattachment point, and beneath the wake, an internal boundary layer re-forms.

Wind-tunnel studies show that bed shear in the wake is a function of the dimensions of the obstacle and the upwind value of u_* (Greeley & Iversen 1985, 220–22). The wake may extend many times the length of the hill down wind (Fig. 20.20; Hunt & Snyder 1980, Greeley & Iversen 1985: 211–32). The character of the wakes of angular bodies is not yet fully understood (Iversen et al. 1990), but it is apparent that the intensity of turbulence increases up to a point as the relative height (ratio of height:breadth across the flow) of obstacles increases (Fig. 20.21). Mean velocities are reduced in these wakes, but turbulence increased (Hunt & Barrett 1989), and this helps to keep sand in movement, and even to increase the transport rate above ambient levels.

Wakes occur behind obstacles at a great range of scales from mountains to bushes. Large "Kármán" vortices trailing down wind of some major mountain massifs have been described in Sec. 20.2.1. At a smaller scale, the wake of a 75m-high by 460m-

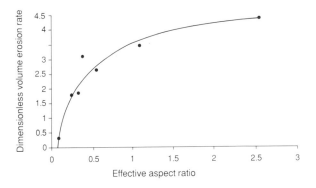

Figure 20.21 The intensity of erosion related to the aspect ratio of obstacles. Taller obstacles are to the right on the horizontal axis. The plot shows that, up to a certain height, the erosion rate increases with the height of the obstacle (after Iversen et al. 1990).

diameter crater in the Californian desert can be about 4km long, and u_* values in this zone can be higher than in the unobstructed flow up wind (Greeley & Iversen 1986). Smaller still, wakes are experienced down wind of isolated barchan dunes, whose length may be ten times the height of the dune and more (Lancaster 1987b; Fig. 20.20). Wakes in the lee of sharp, continuous scarps are much longer than those in the lee of smooth, isolated obstacles (Engel 1981).

The geomorphological significance of the lee eddy and the wake are discussed in Chapters 23.2.3b and 23.3.7a.

CHAPTER 21

Wind erosion

Wind erodes by:

deflation, whereby incoherent sediment is entrained; or
abrasion, in which consolidated, cohesive materials are worn down.

21.1 Deflation

21.1.1 Controls

The controls on deflation have already been described. They include: moisture content of the surface soil (Ch. 16.2.1); surface roughness (Ch. 16.2.4); and grain size (Ch. 17.3).

In the field, Gillette et al. (1980) measured "deflatability" with a portable wind tunnel. They found a descending order of vulnerability:

disturbed soils > sand dunes > "alluvial/aeolian sands" > disturbed playas > skirts of playas > playa centres > desert pavements.

Although even a light wind can deflate incohesive sand at $32m \ yr^{-1}$ (Wilson 1971b), such a rate is seldom sustained, even over short periods of time or small areas, because deflation is almost always severely supply-limited. Deflation can persist only in three situations: where there are deep deposits of loose sediment of sizes appropriate to the threshold velocities of the wind (as in a sand sea); where there is continuous fluvial input of loose particles of the appropriate size (as on active alluvial plains); or where appropriately sized particles are released by weathering (as from a soil or sabkha). Collection of dust with a vacuum cleaner from desert ground surfaces in Mali revealed that they held $78-659kg$ m^{-2} of deflatable material (a large proportion of which was of dust size). This is a small amount relative to the power of local winds to remove it. Weathering (supply) processes appear to release deflatable material on a slow and regular basis (Nickling 1990).

21.1.2 Rates

Time- and space-dependence is illustrated by the very few measurements or estimates of deflation rates that have been made. On the small scale, comparisons of surface level on staked plots in a devegetated area in New Mexico between 1933 and 1980 discovered very great variability in erosion rates over short distances, with both losses and gains, and a maximum loss of 64.6cm (Gibbens et al. 1983, Hennessey et al. 1986). Estimates of longer-term rates are much lower, and although one might expect them to be very variable, depending as they must on u_* and soil erodibility, they show some consistency. Using the size of a lunette as a measure of deflation in Algeria, Boulaine (1954) estimated a rate of $1mm \ yr^{-1}$. Wilson (1971b) estimated rates of $0.5-2mm \ yr^{-1}$ over 2,000yr near Biskra in Algeria. In Peru, Lettau & Lettau (1969) used data from barchan dunes, which were apparently gaining sand from deflation, to arrive at a rate of $0.15-0.22mm \ yr^{-1}$. Extrapolating to northern Peruvian hollows, they estimated rates of $1.6-5.4mm$ yr^{-1}.

The estimation of the rate of deflation from agricultural fields has received an enormous quantity of research. The most widely used empirical formula is that given by Woodruff & Siddoway (1965)

$$E = f(I, K, C, L, V) \qquad (21.1)$$

where E is the potential erosion loss in tons per acre per annum; I is a soil erodibility index; K is a ridge-roughness factor; C is a local climatic index (Ch. 16.2.3); L is a factor relating to field shape in the prevailing wind direction; and V is a vegetation cover index.

These factors are given empirical weightings in the formula, and the erodibility of a specific field is then estimated (Wilson & Cooke 1980). The equation has served well, and is about to be updated.

21.2 Abrasion

21.2.1 Processes

Abrasion occurs when wind hurls grains at consolidated or cohesive rocks. On hitting a hard target, roughly spherical grains create circular "Hertzian fractures", with a diameter of 10–30% that of the grain, depending on the impact velocity (Fig. 21.1). Repeated impacts increase fracturing until the surface stabilizes with a blocky pattern. Fracturing then ceases, although small particles may still be chipped off. The rate of abrasion thus reaches a maximum and then declines. Angular grains chip away and produce a steadier rate of loss (Marshall 1979).

The mass of target removed per impact, A, can be expressed as:

$$A = S_a \rho_p (V\sin\alpha - V_0)^n \ (D - D_0)^m \qquad (21.2)$$

where S_a is the susceptibility of the material (depending on hardness and other properties), ρ_p is the particle density, α is the impact angle, V is the grain

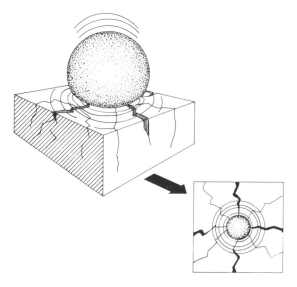

Figure 21.1 A Hertzian fracture pattern produced by the impact of a sand grain on a hard surface (redrawn after Greeley & Iversen 1985). An electron-microscope photograph of a Hertzian fracture appears in Figure 22.9.

speed, and V_0 and D_0 are the threshold particle speed and the threshold diameter of impacting particle needed to initiate abrasion. Experiments give the following values: $n = 2.6$, $m = 3.6$, $D_0 = 0.002$mm and $V_0 \approx 0$ (Scattergood & Routbort 1983).

Scaling up to a whole rock surface, the rate of abrasion is directly related to the supply rate of the abrasive and the kinetic energy of the impacting particles, and indirectly to the compressive strength of the abrading material (Suzuki & Takahashi 1981). Where compressive strength is uniform, this suggested to Anderson (1987) that the erosion rate L could be written:

$$L = S_a (q_{ke}/\rho_t) \qquad (21.3)$$

where q_{ke} is the kinetic energy flux to the eroding surface, ρ_t is the density of the target; and S_a, as above, is the susceptibility of the abrading material.

Most authorities maintain that abrasion is accomplished mostly by sand (as implied by the equations above), but some believe that dust and even snow can abrade. It is usually assumed that dust can only polish rock (Bourcart 1928, Maxson 1940), but Higgins (1956) found Pleistocene ventifacts imbedded in a position where abrasion by sand seemed to be impossible, and Teichert (1939), Maxson (1940), Whitney & Dietrich (1973) and Whitney & Splettstoesser (1982) produced evidence that dust and snow had eroded ventifacts. Whitney & Brewer (1967) even claimed that airborne ions could abrade. Dietrich (1977b) demonstrated in a wind tunnel that soft dust or snow were able to abrade hard rock, given time, but field evidence shows that saltating snow has little abrasive power compared to saltating sand (McKenna-Neuman & Gilbert 1986), and Scattergood & Routbort (1983) dismissed as "unreasonable" the evidence of their own experiments when they suggested that smaller particles would erode more effectively. Most authorities agree: it does seem reasonable that fine particles abrade more slowly than larger ones (Greeley & Iversen 1985: 116–17), for, if nothing else, they are more susceptible to being carried by turbulence away from the impacting surface (Anderson 1986).

21.2.2 Rates

The kinetic energy of impact depends on the diameter of the particle and the impact velocity, itself dependent on u_*^5 (Anderson 1986) – hence the importance of high winds in abrasion. Another vital control is a supply of loose grains. In the Coachella Valley in southern California, which experiences high winds, there is more abrasion after fresh supplies of sand have been brought by streams (Sharp 1964, 1980). The rate of abrasion has also some finer controls. If the incoming grains are at too high a concentration, they collide with each other, and abrasion is checked (Wood & Espenschade 1965, Suzuki & Takahashi 1981). There is a critical angle of the abraded surface to the incoming grains. If it is at right-angles to the wind, it abrades more slowly than if it is at an angle slightly inclined to the wind (Whitney 1979, Sharp 1980). The optimal angle varies with rock type, probably depending on crystal structure (Greeley et al. 1982). As a new facet is abraded, it alters its angle, and this may accelerate or decelerate further abrasion, depending on the original orientation.

Observations in Nova Scotia convinced Hickox (1959) that ventifacts could form in ten years and, in southwest Egypt, Grolier et al. (1980) noted that Neolithic artifacts had been abraded. If the material is soft, erosion is undoubtedly fast: it has been observed that massive mud-brick walls around ancient Egyptian cities have been completely eroded, and although some of the erosion was undoubtedly by water, most must have been by wind. Petrie (1889) calculated long-term erosion rates of over about 1mm yr^{-1} at one of these sites (with presumably much faster rates in the early years). However, while Sharp (1964, 1980) found about 6.6mm yr^{-1} maximum abrasion on bricks, Sharp & Saunders (1978) found only 0.07mm

Figure 21.2 A three-faceted ventifact from the coastal desert of Oman. The facets have been smoothened and cut by shallow flutes by wind abrasion.

yr^{-1} on gneiss. Rates undoubtedly vary widely, depending on rock type, supply of abrasives and windspeed. Rates were everywhere greater in windy late-Pleistocene times.

21.2.3 Height distribution

After a severe windstorm in central California, average maximum abrasion on telegraph poles occurred at 0.24m above ground (Sakamoto-Arnold 1981), and a survey by Clements et al. (1963) concluded that maximum abrasion occurred at between about 0.07m and 0.5m. These observations accord with other field measurements by Sharp (1964), Wilshire et al. (1981) and Gillette & Stockton (1986), and with laboratory experiments (Suzuki & Takahashi 1981). Anderson's (1986) theoretical models add further confirmation: there should be a distinct maximum of abrasion between 0.1m and 0.4m above ground (depending on u_*), and little abrasion above 2m (Ch. 18.6.1; Fig. 18.10). Anderson believed that the upper part of an abrasion profile was controlled by high winds, the lower by gentler winds.

21.3 Wind-erosion forms

21.3.1 Ventifacts (of the order of < 1m)

Ventifacts (literally "made by wind") are rocks shaped by aeolian abrasion. They may be loose pebbles or the country rock. Ventifacts at any one site are commonly strongly aligned in one direction (Sharp 1949, Lindsay 1973a, Lancaster 1984b, Laity 1987), perhaps because abrasion is accomplished more by strong than gentle winds, and because these winds tend to blow from one quarter. Ventifacts present some interesting problems, which can be divided into those on ventifacts above and below 10cm from the ground.

(a) Small ventifacts (< 10cm)

Small ventifacts have a number of diagnostic attributes, some of which are shown in Figure 21.2. The "type" ventifact is a "brazil-nut" shape, with two or three **facets** at between 45° and 70° to the ground surface on which they rest (Needham 1937). Smaller facets are straight, and larger ones slightly curved. Facets are generally abraded at right angles to the formative wind (though see below), and their angle with the horizontal appears to reflect the lower part of the profile of kinetic-energy flux in a sand-charged wind (Sharp 1949; Sec. 21.2.3). However, many small ventifacts exhibit **multiple faceting**, a phenomenon that has been explained in several ways; the explanations may apply in different situations, and are not mutually exclusive.

(i) *Original shape*   Many early authorities maintained that ventifacts inherited their shapes, and hence their pattern of faceting, from their original shapes (for example, King 1936, Higgins 1956). This view seems to be confirmed by some of Kuenen's (1960a) laboratory experiments, and by Sugden's (1964) field observations. However, other experiments show little relationship between the original and abraded shapes of small pebbles, except in the early stages of abrasion (Schoewe 1932, Greeley & Iversen 1985: 128).

(ii) *Multi-modal wind regimes*   A second ready explanation is that multiply faceted ventifacts are formed by sandblasting in multidirectional wind regimes (Kuenen 1960a). The passage of mobile dunes, or the appearance of such ephemeral features as bushes, could produce changes in the ground level wind regime near a ventifact even in an ambient wind regime that was unidirectional. Although this kind of explanation must explain many faceted ventifacts, there are still some that are said to occur in unidirectional wind regimes, as in mountain passes (Sharp 1949), and multi-faceting has been created in laboratory experiments in unidirectional winds (Schoewe 1932). Schoewe concluded that while multi-modal wind regimes were likely in many cases, their effects might be difficult to distinguish from those of unidirectional regimes.

(iii) *Overturning and swivelling*   A more acceptable explanation is that pebbles have been repeatedly swivelled or overturned. Overturning is strongly suggested by the fact that, where large and small ventifacts occur together, large, less mobile ones are single-faceted, while most smaller, movable ones are multi-faceted (Sharp 1949, Lindsay 1973a). Sharp (1949) and Whitney & Splettstoesser (1982) found overlapping sets of flutes and grooves (below) of different orientation on the same ventifact, which are yet another indication of rotation. Further evidence for overturning comes from dated sequences of ventifaction. In an Antarctic sequence, small ventifacts on the youngest exposures have many facets, presumably because abrasion has not yet eliminated the shapes of the original pebbles. Where exposure has been slightly longer, ventifacts have fewer facets, as the original shapes are eliminated. On the oldest exposures, the number of facets on pebbles increases again, and there is a decreasing relationship between facet orientation and wind direction, presumably as overturning increases (Lindsay 1973a). Overturning can be achieved by wind erosion itself, if the pebble is resting on loose sand or soil (Sharp 1964, Mattson 1976), or where winds are strong, as in Antarctica (Selby et al. 1973) and in ancient glacier-margin environments. Considering the discussion in Chapter 16.2.4, these kinds of overturning would be more frequent where the pebble density was low. Solifluction, frost heaving, swelling and shrinking clays, floods, or kicking and tunnelling by animals could also be responsible for rotation.

(iv) *Dust abrasion*   Rotation may be the most satisfactory and widely applicable explanation of multiple faceting, but it is difficult to reconcile it with consistent multi-faceting (with systems of faceting facing similar directions on many pebbles), nor can it explain instances where multiply faceted ventifacts are imbedded in cemented rock where overturning is impossible (Higgins 1956). These problems lead Whitney to propose the dust-erosion of ventifacts (as explained in Sec. 21.2.1). Whitney's wind-tunnel experiments showed her that erosion occurred on downwind facets that could not be reached by saltating sand. Dust was carried to these facets by complicated secondary flows. Breed et al. (1989) believed that Anderson's simulations (Sec. 21.2.3) – which showed that the maximum flux of suspended particles was metres above ground, and that abrasion by particles in suspension was likely to be slight – did not take account of secondary flows around obstacles. If dust blasting does occur, it would be effective only on those areas inaccessible to sand blasting. Where the sand blasting creates the dominant, upwind facet, dust blasting, acting more slowly, would only be able to create the downwind facets, and some of the fine detail, discussed next.

Apart from facets, small ventifacts are also fluted and scarred with pits, re-entrants, salients and sills. Some authorities distinguish **flutes**, which have closed ends, from grooves, which are longer and have open

Figure 21.3 Fluting on a small ventifact of soft volcanic rock in northern Saudi Arabia.

ends. They are not separated here. Flutes usually occur in parallel groups on the edges and corners of ventifacts (Fig. 21.3). They are better developed on large than small rocks, and are said to develop in the early stages of ventifaction. Flutes usually originate at pits and become shallower down wind. In the Namib they have mean widths of 5.5mm and mean depths of 2mm (Sweeting & Lancaster 1982); in southern California they are up to 58cm long and 4cm deep (Steiner 1976). Flutes may splay out on the upwind ends of large hardrock ventifacts (in effect yardangs) (Fig. 21.5). Fluting is generally confined to within about 2m above ground (Steiner 1976, Sweeting & Lancaster 1982), and is aligned to the strongest winds (Tremblay 1961, Sweeting & Lancaster 1982). Fluting that picks out lithological variations in the rock can be called **etching** (Lancaster 1984b).

There can be little doubt that fluting is the result of vortical movements over the ventifact (Maxson 1940, Richardson 1968), but whether it is sand or dust that is involved in the abrasion in the paths of the vortices is less certain. Whitney's experiments, referred to above, showed that fluting could develop in air with no saltating particles, and that it rarely occurred on the windward face which was most subject to sand blasting (Whitney 1979). Dust was said to have been the abrasive.

Early geomorphologists believed that small dendritic rills, known as "*rillensteine*" (Fig. 21.4) had also been produced by wind abrasion, but the absence

of rillensteine on all except soluble rocks (typically limestones) and the seeming impossibility of producing dendritic patterns by wind abrasion in experiments, led Lowdermilk & Woodruff (1932) to conclude that rillensteine were solution features. They did suggest that wind-driven films might on occasion produce rillensteine with preferential orientations parallel to prevailing winds.

Pits, **re-entrants** and **salients** result from inhomogeneities in the rock, such as cracks, mineral grains of different hardness, or volcanic vesicles (Higgins 1956). Pits are commonest on facets inclined between 55° and 90° to the wind (Greeley & Iversen 1985: 125). Whitney (1979) believed that pits were formed by vortical lift created by secondary flow patterns rather than by sandblasting.

Sills are formed when the upper part of a ventifact is abraded, leaving the lower portion protected by the surrounding soil (Sharp 1964, 1960). In some areas pebbles are completely worn down level with the soil surface (Delo 1930), although effects of this kind could easily be confused with those of salt weathering (Whitney & Splettstoesser 1982).

The spatial distribution of small ventifacts reflects several controls, Babiker & Jackson (1985) found ventifact distribution in Qatar to be uneven, the highest concentrations being up to 30 m^{-2} or 25×10^3 ha^{-1}. There are four main controls on uneven distribution:

(i) *Wind velocity* There are more ventifacts where u_* is high, as where wind is funnelled in passes

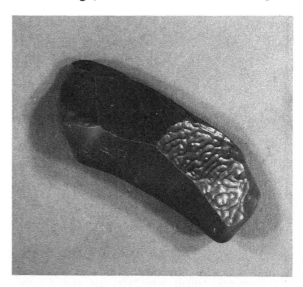

Figure 21.4 Rillensteine on a small limestone ventifact from Oman (photograph by A. Spratt).

or where there is speed-up over mountains (Laity 1987).

(ii) *Lithology* Ventifacts occur only in rocks which do not disintegrate by weathering faster than they can be abraded (Selby 1977b). Limestones, dolerites and hard sandstones are particularly likely to retain facets and other features (Sweeting & Lancaster 1982). Quartz is not often abraded, probably because it is too hard (Steiner 1976, Selby 1977b).

(iii) *Pebble spacing* There are few ventifacts where pebbles are close together; where they thin out, concentration increases (Babiker & Jackson 1985).

(iv) *Ground character* Ventifacts on loose sand, which absorbs the energy of saltating grains, are not as well developed as those resting on hard rock or calcrete (Babiker & Jackson 1985, Laity 1987). Rebound is also more effective off frozen, than off loose, unfrozen ground (McKenna-Neuman 1989). This accords with Anderson's modelling of abrasion (Sec. 21.2c), where it was shown to be up to four times more effective on hard ground.

(b) Large ventifacts (> 10cm above ground)
When wind abrasion reaches above about 10cm, ventifact formation is affected by the profile of kinetic energy flux in the wind (Ch. 18.6.1) and profiles are concave to the wind, with a maximum depth of erosion somewhere about 0.5m above ground. Large ventifacts are also often heavily fluted (Fig. 21.5).

(c) Age
Some ventifacts in soft rocks and in windy environ-

Figure 21.5 Fluting on the upwind side of the hard-rock yardang in central Egypt.

ments (such as Heard Island in the South Atlantic) may be active today (Staal 1958, Rude 1959), but many ventifacts are inherited from the Pleistocene. This is true of most periglacial ventifacts (e.g. Sharp 1949, Higgins 1956), and possibly of many desert ones. Some ventifacts in the Californian desert are covered by desert varnish, others are spalling or show weathering crusts, while many split ventifacts show no signs of abrasion on newly exposed surfaces, all indicators of inactivity (Smith 1967, Steiner 1976, Laity 1987). Steiner (1976) and Dorn (1986) estimated that abrasion had ceased in California about 5,000 years ago.

The primary explanation of intense ventifaction in the late Pleistocene must be that winds were stronger, and possibly because there was more hard, frozen ground, but the story is certainly more complex, for the drier climates of that period might have experienced a reduced supply of sand for abrasion (Laity 1987). Periods of high climatic variability in which there was both intense water erosion and strong winds would have provided the best ventifact environment. Glacial outwash plains with large quantities of loose sand, poorly colonized by plants, a cool dry climate, hard frozen soil and intense katabatic winds are clearly the most favourable environment for ventifacts.

21.3.2 Yardangs
"Yardang" is a Turkmen word (Hedin 1903), now used in geomorphology for a wind-abraded ridge of cohesive material (Fig. 21.6; McCauley et al. 1977, 1980). The classic form has a length:width ratio of about 4:1, smoothed sides, and a sharp "keel" along the ridge, but shapes vary widely. Many, if not most, yardangs have no keel, and in some large yardangs the length ratio can reach well over 10:1. It is convenient to class yardangs, like dunes, as micro-,

Figure 21.6 A meso-yardang.

meso- or mega-forms (of the order of 1m, 10–10^2m and 10^3m respectively).

(a) Micro-yardangs

Micro-yardangs include the small centimetre-scale ridges seen behind pebbles or shells on a wet sandy beach (Allen 1965) and the "nubbins" of McCauley et al. (1977). Allen's ridges, which are very common, are as elongated as meso-yardangs, but the nubbins are much less so, and seem to relate more to pre-existing inhomogeneities in the parent rock than to aerodynamic forms. Allen noted a variety of erosional features at this scale, including crescentic hollows on the upwind sides of immovable boulders.

McCauley was anxious not to confuse his nubbins with pedestal rocks (Figs 5.19, 21.7), which were thought by some early geomorphologists to be wind-abraded forms because their lower, indented profile appeared to conform to the abrasion profile. This interpretation is now generally rejected for most pedestal rocks, for the indentation seldom conforms either to the orientation of prevailing winds or to the ideal profile of wind abrasion. Pedestal rocks have been convincingly shown to be produced mainly by weathering, in many cases salt weathering (Ch. 5.2), although wind has deflated much of the weathered debris, for many pedestal rocks are surrounded by moats from which fluvial transport is impossible (Bryan 1923c, Leonard 1927).

Striations, which might be classed as micro-yardangs, have been described from Antarctic and high mountain areas. These are spaced at about 0.2m, occur on level ground, and are expressed in sorted stripes of (alternately) pebbles and finer material. They are reputed to have been formed largely by wind erosion (in intense katabatic winds), perhaps with the aid of needle ice (Beaty 1974b, Hall 1979).

(b) Meso-yardangs

Hedin's classic yardangs are the "type" meso-yardang. These were between 1m and 4m high and up to hundreds of metres long, although most were only 5–10m long. Many meso-yardangs in Egypt, Chad, Mali and Algeria are of the same order of size, although some reach 10m high (Beadnell 1909b, Capot-Rey 1957a, Grolier et al. 1980). In coastal Peru meso-yardangs are up to 6m high (McCauley et al. 1977, 1980). Length:width ratios vary from field to field, a characteristic that McCauley et al. (1977) attributed to differences in age. The profile of meso-yardangs is strongly asymmetric with a steep windward slope or "prow", and a gentle lee slope (Fig. 21.6). Meso-yardangs can occur singly or in groups, and groups may cover many square kilometres; the Kharga yardang field in Egypt covers over 1,500km² (Grolier et al. 1980).

(c) Mega-yardangs

Mega-yardangs are best developed in Borkou on the southeastern and southwestern flanks of the Tibesti massif in the central Sahara, where they are up to 200m high, many kilometres long and spaced between 0.5km and 2km (Fig. 21.8), and cover an area greater than 650,000km² (Capot-Rey 1957a, Durand de Corbiac 1958, Grove 1960a, Mainguet 1968, Hagedorn 1968, Peel 1968a). Others have been reported from an area 150km long and "tens of kilometres wide" in central Egypt (Breed et al. 1979a), and in the Lut in southeastern Iran, where they reach 70m high, many hundreds of metres in length, and are spaced between 0.5km and 1km (Gabriel 1938, Krinsley 1970,

Figure 21.7 A pedestal rock (photograph K. Gardner).

N

→ Axis of barchans

0   20
kilometres

Figure 21.8 Mega-yardang patterns in Borkou, Chad. The dark streaks are the summits of mega-yardangs. The trend of the barchan axes indicates the direction of an apparently strongly directionally unimodal wind regime (after Mainguet 1968).

McCauley et al. 1977, 1980, Selivanov 1982, Bobek 1969). There are also mega-yardangs in Peru, although they are usually single. They are kilometres long and 100m high (McCauley et al. 1977).

Mega-yardangs are of the same order of size and have the same parallelism as linear features covering large tracts of loess terrain in southeastern Europe, where they are known as "*gredas*" or "*gorbiszcza*" (Rozycki 1968); the central (former) Soviet Union (Bartowski 1969); China, where they are called "*liagns*" (Rozycki 1968, Bartowski 1969); the American Great Plains, where they extend over 300,000km² (Russell 1927, Flint 1955, Crandell 1958, Zakrewska 1963); and over 65,000km² of the Colorado Plateau (Shawe 1963, Stokes 1964). Like the Borkou mega-yardangs, some of this alignment may be ancient, for in Colorado and Nebraska some parallel troughs are filled with Peorian (last glacial) loess (Hill & Tompkin 1953).

Although generally thought to have been formed by

wind erosion, an alternative explanation for aligned topography in the Middle West, proposed by White (1961), involves collaborative wind and stream action: wind blows sand into valleys cutting transversely across its path, leaving wind-parallel valleys to be eroded preferentially by streams. Some of this ridging, however, may be more depositional than erosional, and as such is comparable to the formation of linear dunes discussed in Chapter 26.2.5.

(d) Materials

Most small yardangs are composed of sediments which cohere moderately, but abrade readily (McCauley et al. 1977). Quaternary lacustrine silts, clays, chalks or diatomites are ideal. Micro-yardangs can even be eroded in wet sand, although they are quickly destroyed when the sand dries (Cooper 1958, Allen 1965, Hunter et al. 1983). Some mega-yardangs, such as those of the Lut desert of Iran, are also eroded in soft lacustrine marls (Bobek 1969), but mega-yardangs in Borkou and central Egypt have been incised into hard sandstones and limestones, some as old as the Cambrian (Mainguet 1968, McCauley et al. 1977).

Dip, strike and faulting in the parent rocks considerably affects the shape of yardangs. Incision is faster where faulting or bedding run parallel to the wind. Hard basements act as a base level for wind erosion. Where hard sediments overlie soft, abrasion may undermine the slope, and encourage slope retreat by collapse.

(e) Processes

Wind is unquestionably the main yardang-forming agent. Although some yardangs have much the same dimensions as drumlins and hard-rock bedforms caused by massive flooding, as in the scablands of Washington State (Ward & Greeley 1984, Baker 1978), and some mega-yardangs are of the same size as "giant grooves" said to have been excavated by ice-sheets in northern Canada (Smith 1945, Lucchitta 1982), the possibilities of confusion are small (McCauley et al. 1977). The Borkou and neighbouring mega-yardangs have been assigned to "basement fracture trends" by Pesce (1968), but with little evidence.

Most yardangs occur in unidirectional wind regimes, as in Peru (Bosworth 1922) or central China (Hedin 1903), and most are aligned parallel to these winds, as recorded in meteorological observations and by nearby dunes (McCauley et al. 1977). The form of meso-yardangs, with their widest portions up wind and their hull-like shapes, is similar to other aero-dynamic shapes (Fig. 21.6). Ward & Greeley (1984) hypothesized that skin-friction would decline as the relative width of the yardang increased, because more of the surface was protected from the oncoming wind; on the other hand, pressure drag would increase as relative width increased, because of flow-separation effects (Fig. 21.9). The balance of these two effects produced minimum total drag when the width:length ratio was in the region of 1:4. The width:length ratios of mega-yardangs reach well over 1:10 in Borkou and the Lut, and this may point to some other aero-dynamic process, as yet unknown. Even more myster-ious are the details of the form of mega-yardangs, particularly in the Lut, where air photographs show they are formed by the intersection of many boat- or eye-shaped erosional hollows at a number of scales (Bobek 1969).

Many yardang corridors are sprinkled with coarse sand which is hurled at the yardangs in high winds, as described by Hedin himself, and by Peel (1968a) in Borkou, where there are ample signs of abrasion, and where little debris has accumulated from the clearly active water erosion and weathering of the surround-ing mega-yardangs (Mainguet 1968). The upwind faces of yardangs are abraded, and in hard-rock yardangs, the "prows" are pitted and grooved, and eroded in the curved profile of kinetic energy flux in the manner of ventifacts (Fig. 21.5; Breed et al.

1979a). In soft materials, abrasive undercutting may lead to collapse of blocks which are then rapidly abraded. On many steep upwind faces there is a "moat" (McCauley et al. 1977). Abrasion removes incohesive sandy beds more quickly than cohesive clayey ones (following the threshold behaviour of sediments described in Ch. 17.3; Ward & Greeley 1984).

Although yardangs are found only where fluvial action and weathering are slow, most have been initiated from forms created by, or have been later modified by non-aeolian processes. Many have formed out of gullied terrain: gullies may first dissect the ground, as on a scarp edge, after which they are exploited by wind erosion (Blackwelder 1934, Ward & Greeley 1984). Gullies parallel to the wind are the most quickly enlarged.

On most mega- and many meso-yardangs there is little abrasion above the height of saltation in the corridors, as shown by the delicacy of the forms above that height, such as weathering pits and cem-ented root channels. The soft rocks of many mega- and meso-yardangs can retain delicate rills on their upper surfaces, while lower slopes are abraded (Fig. 21.10). These are generated by a number of non-aeolian processes such as: salt weathering, desiccation cracking, slumping (caused by abrasive undercutting or during heavy rainfall), piping and rilling. The signs of these processes are best preserved down wind of the abraded "prow" (Ward & Greeley 1984). Some Egyptian yardangs retain a relic red soil on their summits (Breed et al. 1979a). The crests of the Borkou mega-yardangs are covered by a thick desert varnish (Mainguet 1970, Peel 1968a).

Above the abrasion zone on some yardangs, nevertheless, there are smooth aerodynamic forms, including sharp "keels". These may simply be formed by the virtual deflation of soft, actively weathering materials (McCauley et al. 1977). Clouds of dust are raised when walking on these slopes, indicating very weak cohesion. Wind-tunnel experiments show that there are complex flow-fields around yardangs with some reverse secondary flows, and zones of low pres-sure where deflation of this loose material is encouraged, leaving aerodynamic shapes (Ward & Greeley 1984). Abrasion by dust has also been suggested as an explanation of these aerodynamic forms. It might be particularly intense where the meeting of the divided flow creates high turbulence and over the keel (Whitney 1978, 1983).

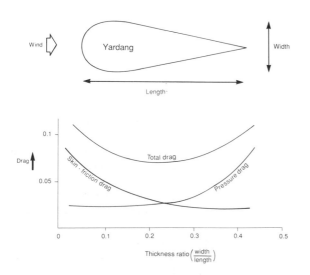

Figure 21.9 Yardang aerodynamics, showing that the total drag is at a minimum when the width:length ratio is at about 2.5 (after Ward & Greeley 1984).

Figure 21.10 The delicate pattern of rills preserved unabraded on the summit of a small meso-yardang in Oman. The flanks of the yardang are strongly abraded, wiping out evidence of the rilling episode.

(f) Size and spacing

The size and spacing of yardangs have still not been adequately explained. Wind velocity, lithology and age may go a long way to explain size, but none of these explains regularity. There is a surprising absence of data about spacing, even though regularity does appear to occur in cursory examination. There are four possible hypotheses:

(i) *Lithology* The Borkou mega-yardangs have sudden changes in spacing (Fig. 21.8), which may be related partly to pre-existing breaks in the topography or to accompanying changes in lithology (Mainguet et al. 1980a). The problem is to understand the process that links lithology to spacing. Mainguet also maintained that there may be some form of structural control on the spacing of the Borkou features.

(ii) *Inheritance* from a quasi-regular gullied landscape ((d), above).

(iii) *Bedform hierarchy* Size and spacing could relate to a bedform hierarchy (Ch. 24.7.3). The size groupings and range, particularly of mega-yardangs, correspond closely to those of dunes. Nonetheless, there is even less evidence to support the bedform hierarchy hypothesis for yardangs than for dunes. The boat-shaped hollows of the Lut mega-yardang field (referred to above) appear to be at several scales.

(iv) *Bush-spacing* For smaller meso-yardangs, it may be that the spacing of yardang cores is determined by the spacing of bushes (Capot-Rey 1957b, Peel 1968a). Desert vegetation is typically bunched and discrete, and commonly regularly spaced. Many desert bushes accumulate silty or clayey nabkhas, and many nabkhas go through a cycle of accumulation and erosion, as wind is channelled around the growing mound (Ch. 25.2.3a). If erosion continues into the cohesive sediments beneath the nabkhas, then yardangs form, and inherit the regular spacing of the plants. Capot-Rey (1957b) and Mahmoudi (1977) noted that many yardangs had originated as nabkhas, and many contained concretions around plant roots (rhizoconcretions).

(g) Rates of formation, age

Experimental yardangs in wind tunnels have shown that there is first rapid incision, with little alteration of form. When the base (such as a hard basement rock) has been reached, the yardangs themselves begin to change. The upwind corners are the first to be abraded, followed in turn by the windward slope; leeward areas and flanks; and finally the upper faces down wind of the crest. The optimal shape minimizes flow separation, and this shape is maintained during further erosion (Ward & Greeley 1984).

The incision of meso-yardangs into Holocene lake deposits shows that they can be formed quickly. Many Saharan meso-yardangs post-date the Neolithic pluvial, the last time that many lakes filled, giving rates of incision of between 4mm and 0.4mm yr^{-1} (Beadnell 1909b, Peel 1968a, Hagedorn & Pachur 1971). If Hörner (1932) and Norrin (1941) were right that the Lop Nor yardangs in China had been eroded since the fourth century AD, then their rate of incision would be about 0.2cm yr^{-1} (McCauley et al. 1977). Forty-three years of wind erosion of Blackwelder's (1934) yardangs in California produced 1m of undercutting on one, and obliterated rills on others, giving a maximum rate of 2.3cm yr^{-1} (McCauley et al. 1977). The present-day frequency of meso-yardangs may be no more than a function of the creation of pluvial lakes and the period since their desiccation. If the rate of incision is as great as the observations imply, there will be few meso-yardangs left within a few more hundred years, if present conditions continue.

The Lut mega-yardangs are also recent, having been cut in Pleistocene lake deposits (Bobek 1969). In Borkou, on the other hand, where corridors between the mega-yardangs exhibit deposits from three different lacustrine phases, and where the upper yardangs are covered by thick desert varnish, the yardangs may be "hundreds of thousands of years" old (Hagedorn 1968, Peel 1968a, Mainguet 1970). The period of greatest activity was almost certainly during the dry, windy glacial periods of the Pleistocene, but they could be millions of years old, for the Sahara has been arid since the Miocene (Mainguet 1984d). The mega-yardangs in Egypt may also date back to Pliocene or even middle-Miocene times (Breed et al. 1979a).

21.3.3 Pans and other wind-eroded basins (1 to c. 10km²)

(a) Morphology and distribution

Pans are shallow, elliptical or kidney-shaped depressions with smooth, rounded outlines (Fig. 21.11). They are abundant in some semi-arid parts of the world, as in parts of the High Plains in North America, the Argentine Pampas, the Pantanal in Brazil, the Senegal Delta, southeastern and southwestern Australia and southern Africa from Zaire and Mozambique to Cape Province, Manchuria and the western Siberian Plain (Goudie & Wells, in prep.).

Not all shallow wind-eroded lake basins are smoothly rounded. In Australia, wind-eroded lake basins in semi-arid areas are rounded, while those in arid areas have irregular outlines (Fig. 14.7; Bowler 1986). This correlation does not extend to North African wind-eroded basins, which have irregular outlines even in quite humid conditions (Fig. 21.12; Coque 1979b). However, with the exception of wave-adjusted beach processes, most of the processes of formation of smooth and irregular pans seem to be similar. For some reason, the North African basins were either not as frequently flooded as those in other parts of the world, or were not subject to such uniform wave action.

Pans would be difficult to detect were they not

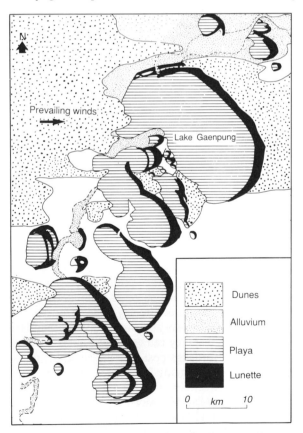

Figure 21.11 Large pans (playas) and lunettes in central New South Wales (data from Bowler & McGee 1978).

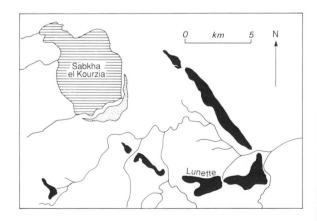

Figure 21.12 Angular, uncurved pan/lunette shapes in northern Tunisia (after Coque 1979).

filled with water or salt. Because small wind-eroded hollows are seldom detected in this way, the lower limit of pan size is not well known. For example, it needed rain to show Bryan (1923b) that there were many pan-like hollows of just a few metres diameter in Arizona. There may be many more of these, as yet unrecorded, features. There are also problems with the upper size-limit of pans, for some large wind-eroded basins, such as the Richat depression, have a pan-like shapes, and at this size-limit irregular wind-eroded hollows merge with major desert depressions (Sec. 21.3.4). These problems aside, the sizes of features recorded as pans in southwestern Australia are between 0.004km^2 and 100km^2 and they reach a fairly well defined peak at about 0.05km^2 (Killigrew & Gilkes 1974); in South Africa the range is 0.05–30km^2, with a mean of 0.2km^2 (Goudie & Thomas 1985). The largest feature recognizable as a pan in eastern Australia is about 45km across (Bowler & McGee 1978).

In parts of southern Africa and the High Plains of the USA, the density of pans reaches 1 km^{-2}, and over large areas it is greater than 0.15 km^{-2} (Fig. 21.13; Frye 1950, Reeves 1966, Lancaster 1978a, Goudie & Thomas 1985). Pans may cover 12% of some landscapes. Density and size-distribution vary widely, there being isolated clusters and zones with larger and smaller pans. Size seems to relate to the space available: those formed between linear dunes, with little room for expansion, are smaller than ones developed on open plains (Suslov 1961, Goudie & Thomas 1985). Size may also be related to stream order (below).

(b) Processes

Pans are the outcome of a number of processes, mixed in varying degrees.

(i) *Surface-water hydrology* Semi-arid climates favour closed lakes. The boundary of the area in which closed lakes can occur is a function of precipitation and temperature (Langbein 1961). In the USA, closed lakes exist only in Thornthwaite's "dry" climate, west of a line from the eastern Dakotas south to central Texas. In some areas, tectonic doming may also have encouraged ponding on rivers in semi-arid climates, and thus initiated pan formation (Marshall 1987).

(ii) *Alternation of humid and arid geomorphic processes* Closed lakes are also favoured by the short-term climate rhythms and longer-term

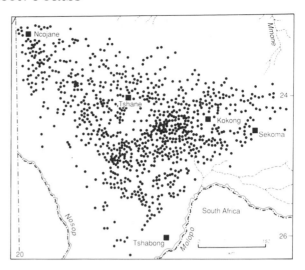

Figure 21.13 Pan distribution in southeastern Botswana (after Goudie & Thomas 1985).

climatic history of semi-arid environments, with their alternations between fluvial and aeolian activity. Whether quasi-annual or secular, desiccation dislocates drainage systems by permitting floods to reach only short distances down drainage nets, and to inundate depressions which then act as potential pans (Bettenay 1962, Osterkamp & Wood 1987). Dunes migrate over channels, block them, and create other potential sites, as in Western Australia (Gregory 1914). Many pans are aligned along ancient channels (Wellington 1943, Bettenay 1962, Osterkamp & Wood 1987, review by Goudie & Wells 1990), and thus larger pans should occur in higher-order drainage channels. This may account for the difference in size between the large pans of the lower plains of the Murray–Darling system in southeastern Australia and the smaller ones in the upper Orange River basin in South Africa.

(iii) *Solution* Solution may have played a rôle in initiating pans in some areas, notably those in the High Plains of the USA (Reeves 1966). In some of these, solution and wind excavation may both have played a rôle, for there are usually fringing lunette dunes (Gilbert 1895, Judson 1950, Reeves 1966, Osterkamp & Wood 1987). More purely solutional hollows, for example the "*dayas*" of the semi-arid parts of North Africa, are less smooth in outline (Mitchell & Willimott 1974). Animal activity, as in buffalo or elephant "wallows", seems to play only a small part in pan formation (Thomas 1988c). Wallows, it should be noted, are also associated with semi-arid climates where water is sparsely distributed.

(iv) *Salt weathering, cracking clays* In semi-arid climates salts can be mobilized, but not completely removed from the landscape. They are therefore concentrated in dry lakes, and they help to disintegrate the lake-bed sediments. Pans are specially frequent where the underlying rock is saline, because saline conditions encourage the pelletization of clays. Smectite clays, also susceptible to pelletization are also associated with pan formation (Goudie & Wells 1990). Alternating flooding and desiccation encourage these processes (Price 1963).

(v) *Deflation* Deflation follows salt weathering and pelletization. Deflation has been the dominant process in the excavation of pans. In many areas, substrates that are easily eroded by wind, particularly saline ones, have the highest densities of pans. The most convincing evidence of deflation is the existence of sandy, clayey or saline lunette dunes on many downwind shores. Deflation by rotating dust devils (Shand 1946) can be discounted.

(vi) *Wave and current action* The leeward sides of many pans, including their smooth curves, beach ridges and cliffs, strongly suggest wave activity as a major factor in their creation. In southwestern Australia, pans show a fairly sharp peak of ellipticity at 0.33, as measured by the length: width ratio, and this strongly suggests wave and current activity (Killigrew & Gilkes 1974, Bowler 1986). The smooth downwind edge of many pans conforms closely to the log-spiral form of many shorelines (Yasso, quoted by King 1972, 373–4). Where both shores are smooth, there have apparently been wave-forming winds from opposing directions (Killigrew & Gilkes 1974).

The long axis of most pans is transverse to the prevailing wind (Bowler 1986). In southwestern Australia they are aligned transverse to winter winds, for it is then that the pans are full of water (Killigrew & Gilkes 1974). Depressions excavated by wind alone (without the aid of waves) resemble blow-outs in being longer than broad in the windward direction (Frye 1950, Lancaster 1986c). Aligned lakes with more angular shapes, as in northern Alaska and Bolivia, are thought to be tectonic in origin (Plafker 1964, Allenby 1988).

Pans bear a close similarity to smooth-sided, elliptical, aligned lakes in active or former periglacial areas, the best known and most extensive areas of which are on the Arctic slope of Alaska (Carson & Hussey 1962), the Liverpool Bay area of the North West Territories in Canada (McKay 1956b), the "Carolina Bays" (rounded lakes) of the southern Atlantic coastal plain of the USA (MacCarthy 1937, Johnson 1942, Odum 1952, Prouty 1952, LeGrand 1953, Rex 1962, Thom 1970, Stolt & Rabenhorst 1987), and the Paris Basin (Cailleux 1960, Pissart 1960, Matchinski 1962a,b). Others occur in the northern (former) USSR (Black & Barksdale 1949).

These periglacial aligned lakes are similar to pans in their size, shape and concentration in the landscape. The half-million or so Carolina Bays are spattered over a huge swathe of the Atlantic coastal plain, from northern Florida to southern New Jersey. As with the South African pans, the Bays cover up to 50% of some areas.

Although there have been advocates of subsurface solutional "swirling" (LeGrand 1953), direct aeolian deflation (Livingstone 1954), spit development in coastal lagoons (Robertson 1962) and meteorite impact (Melton & Schriever 1933, MacCarthy 1937, Prouty 1952) as the origin of some or all of these features, and although many of the European ones are undoubtedly marl-pits (Prince 1961, 1964) or dolines (Sperling et al. 1977), many authorities now agree with a model something like the following for the smoothly outlined depressions: lakes collect in irregular wind-eroded hollows (perhaps blowouts); wind-induced currents in the lakes then concentrate erosion at the tighter ends of the ellipses, elongating the lakes transverse to the wind (Cooke 1943, Livingstone 1954, Carson & Hussey 1960, Stolt & Rabenhorst 1987; Fig. 21.14). The Carolina Bays would have formed in periods of higher winds in the last and/or penultimate glaciation.

Although pans may form by erosion in this way, there is probably still a lot to learn about the hydraulics and geomorphology of shallow lakes. Lees (1989) proposed an alternative sequence in which wave action and return flow created bars that segmented lakes of the order of only a metre deep. Segmentation-converted lakes were originally elongated with the wind into ones that were longer transverse to the wind. The bars developed into incipient lunette dunes.

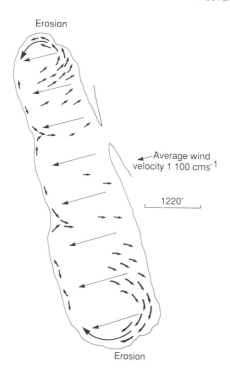

Erosion

Average wind
velocity 1 100 cms⁻¹

1220'

Erosion

Figure 21.14 Currents in curved Alaskan lakes, suggesting erosion at the apices and enlargement transverse to the wind (after Carson & Hussey 1962).

(c) Climatic associations
This necessary combination of processes means that pans are features of semi-arid rather than arid landscapes. In South Africa, pan density drops off sharply to the wetter side of the 500mm isohyet, and gently on the drier side, although there are some notable exceptions in wetter areas (Wellington 1943, Goudie & Thomas 1985).

(d) Summary model of pan formation
A simplified model of pan formation would therefore be:
(1) Impeded drainage allows a temporary lake to form.
(2) On drying, salt accumulates and crystallizes, breaking up the bed of the lake.
(3) The broken, pelletized sediment is blown out of the depression while it is still dry.
(4) Renewed flooding in the deepened depression produces a lake in which wave action shapes the shore, and elongates the lake transverse to the wind.

However, most pans are palaeoforms. Pans in southwestern Australia are aligned to contemporary winds (Killigrew & Gilkes 1974), and some North African, Mauritanian and Zimbabwean pans may still be active (Tricart 1955e, Flint & Bond 1968, Benazzouz 1987), but some of the pan–lunette systems in southeastern Australia have been inundated by rising sea levels, and the nested nature of many pans and the multiplicity of lunettes testifies, here and elsewhere, to a long history of stabilization and reactivation (Sprigg 1979). In the Kalahari, a complicated climatic history is indicated by the pattern of lunettes: the outer one is sandy, indicating wind erosion of the Kalahari Sand; while the inner one is more clayey, indicating deflation of a lacustrine deposit from the already- formed pan (Lancaster 1978b). The pans of the southern High Plains of the USA also seem to date from a former dry period (Reeves 1966).

21.3.4   Large enclosed basins (of the order of 10³–10⁴m size)

There are five kinds of evidence for wind erosion of large basins:
(i) Pans (Sec. 21.3.3).
(ii) Wind-eroded features such as yardangs in desert depressions (Secs 21.3.2 & 21.3..5); for example, protection of basin sediments by Pleistocene lavas in California shows that nearby unprotected sediments have been deflated many metres in quite a short time (Blackwelder 1931a).
(iii) Direct observations of dust being removed from playas (Young & Evans 1986).
(iv) Palygorskite, a clay mineral found in arid-zone soils, but in much larger quantities in saline lakes, is found in far-travelled dust, showing that large quantities of sediment must be leaving such lakes (Ch. 22.1.3).
(v) The paucity of alluvium and salt in many desert basins relative to expected inputs (Blackwelder 1923a, Langbein 1961).
The problem is to relate all this circumstantial evidence to the equally undoubted existence of many large enclosed basins in deserts, for some enclosed basins in deserts are unquestionably tectonic, and the simplest explanation for their survival would be that the desert environment had produced neither the water nor the sediment to fill them.

Although direct evidence is hard to come by, many large closed depressions have been ascribed to wind

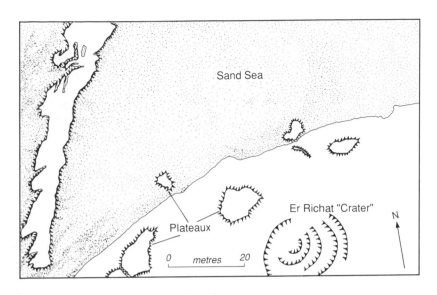

Sand Sea

Plateaux

Er Richat "Crater"

N

0   metres   20

Figure 21.15 The Richat Crater, Mauritania, a large apparently wind-eroded basin (redrawn from a satellite image in Breed et al. 1979).

erosion. Examples are found in the Turgai Plateau, parts of Khazakhstan, the Ust-Urt Plateau and the Bel-Pak-Dala in the (former) USSR (Suslov 1961). Others occur in northern Peru (Lettau & Lettau 1969). The Richat "crater" in Mauritania (Fig. 21.15) is an almost circular depression, about 40km in diameter, eroded out of an anticlinal structure (Daveau et al. 1967). Although they could find no conclusive proof, Daveau and her colleagues believed that the Richat and nearby Chemchane depressions were wind-eroded features.

The most diligently debated depressions have been the accessible basins of North Africa (e.g. Passarge 1911, Walther 1924). The largest and most thoroughly researched of all is the Qattara depression in northern Egypt, 192,000km² of it below sea level (bsl), with the lowest point at 134m bsl (Said 1960, 1962). There are others at Farafra, Dhakla and Kharga in the Western Desert of Egypt and in southern Tunisia and Algeria. Opinion has been divided over these for more than 90 years. Beadnell (1909b) believed in wind erosion, Ball (1900) and Bourcart (1928) did not.

If they were wind eroded, the process must have been complex and very extended, and always constrained by the water table (Beadnell 1909b, Pogue 1911). Hobbs (1917) and Anderson (1947) postulated recurrent rises and falls of the water table over long periods of the Pleistocene, in phase with fluctuating sea level. High water tables provoked weathering, the products being removed by deflation. Kentsch & Yallouze (1955) agreed, adding structural and lithological controls to the picture, and this model was

developed further by Said (1960), who believed that solution of an overlying limestone cap had initiated the sequence. Said still considered deflation to be the major agent, but scarp retreat by water erosion to be an important subsidiary process. Since uplift in the post-middle Miocene, some 20,000km³ of debris had been deflated to form sand dunes down wind. Another sequence that began with the solution of limestone may have occurred in the Ust-Urt Plateau in the (former) USSR (Suslov 1961).

21.3.5 Plains of deflation
The most derided of the aeolianist proposals of the early part of this century was that wind had produced peneplains equivalent to those then being suggested as the products of water erosion (Keyes 1909b). Davis (1930, 1954) dismissed the notion almost out of hand. Were he alive, Davis might be dismayed to find that the notion lives, and that its champions have found evidence that is hard to refute. The most persistent candidate for aeolian planation is in southwestern Egypt and the northwestern Sudan (Sandford 1933b). Breed et al. (1982, 1987) developed a model of the slow destruction of the Tertiary fluvial landscape of this area as it dried up in the Quaternary. The wind abraded the scarps, leaving conical hills, which were themselves eventually levelled. It was not clear what controlled the level of the resulting plain.

Other candidates for "eolation" are the dry coastal deserts of Peru, where the end result of aeolian erosion is said to be a gently undulating plain cut across the slightly folded Tertiary sediments (Mc-

Figure 21.16 An apparently wind-eroded plain cut across aeolianite strata in the southwestern Wahiba Sands in the Sultanate of Oman.

Cauley et al. 1977); the "Serir" north of Tibesti in the east-central Sahara; and the region between the Ahnet and Adrar des Ifoghas in the south-central Sahara (Mainguet et al. 1980a). In the southwestern Wahiba Sands in Oman, a nearly level, or very gently sloping, smooth plain is cut across aeolianite strata, the only explanation for which yet again appears to be wind erosion (Fig. 21.16; Gardner 1988, Warren 1988a).

Less level plains, but with rather more convincing evidence of deflation, are found in many deserts. The evidence lies in ridges, whose meanders and braids are unequivocal proof of their origins as river channels, but which are now raised up to 20m above the intervening plain (Fig. 21.17; Maizels 1990). They have been called "suspendritic drainage", "gravel trains", "perched wadis", "pseudo-eskers", "wadi-ridges", "drainage in bas-relief" and "suspenparallel drainage" (Miller 1937, King 1939, Butzer & Hansen 1968, Glennie 1970, Hotzl et al. 1978, Beydoun 1980, Reeves 1983, Anton 1984). The generally accepted interpretation is that raised channels are preserved either by coarse channel-bed deposits or by

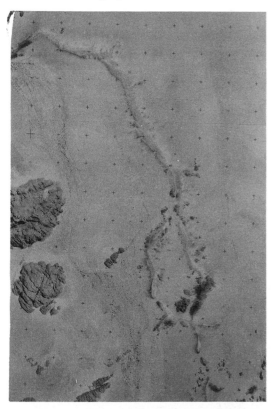
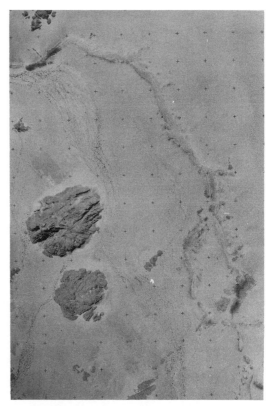

Figure 21.17 A stereopair of raised channels in southeastern Libya. For explanation, see the text.

the preferential cementation of material beneath the channel. Finer or less cemented sediments in the interchannel areas have been deflated, with minor fluvial erosion in places. This explanation follows the pioneering observations of Hörner (1932) around Lop Nor in central China and Miller (1937) in the Nejd in northern Saudi Arabia. In both places yardangs between the raised channels are irrefutable evidence of wind erosion, although elsewhere yardangs are not so evident. The channel systems have been given various ages from Holocene to Pliocene, creating therefore a range of rates of wind erosion (Maizels 1990).

CHAPTER 22

Aeolian sediments

This chapter is devoted to the nature and origin of aeolian sedimentary particles. Sedimentary structures are described in Chapter 27.

22.1 Dust

22.1.1 Sizes

A distinction between sand and dust has been made in Chapter 17.3, and some distinctions between different sizes of dust in Chapter 18.4 (Fig. 18.7). Coarse dust, with the potential to travel short distances in suspension, has diameters of 20–70μm. Particles smaller than 20μm are defined as fine dust, with the potential to travel right around the globe (Gillette 1981).

22.1.2 Quantities and distribution

Dust is hard to quantify (Pye 1987) but, measurement notwithstanding, large amounts of dust do appear to leave deserts. Measured by dust-storm days, the two dustiest places on Earth are the Siestan Basin in Iran and the floodplain of the Amu Darya in the (former) Soviet Union. The Sahara spews vast quantities northwards into Europe, northeastwards into southwest Asia, southwards into West Africa, and across the Atlantic to the Caribbean. Some estimated figures are shown on Figure 22.1. The Chinese desert is another major zone of dust generation (Middleton et al. 1986, Middleton 1989). Even more dust left the deserts in the recent geological past. Ice cores show that 18,000yr ago conditions were a hundred times more dusty than the mean Holocene level (Goudie 1983a).

Despite the undoubted pre-eminence of deserts as sources of dust, Goudie (1983a) found no simple relation between quantities of dust and mean annual rainfall. The dustiest areas were areas with rainfalls of the order of 100–200mm. There are three possible, not mutually exclusive, explanations:

(i) little dust is generated in very dry areas by weathering or fluvial action;

(ii) most dust has been removed from very dry areas, particularly during the windy conditions of the last glaciation, which may have cleaned the deserts out of dust for millennia; or

(iii) there is less wind disturbance in deserts.

The most arid parts surveyed by Goudie, in the Sahara, have been arid since the Miocene at least, so that there can be little dust left and little more can be generated. These observations relate to the opening remarks of Chapter 16.1.3 about the inter-relations of wind and water erosion.

22.1.3 Sources and generation

Most dust has had a complicated history. After its initial weathering from an igneous or metamorphic rock (the "P1" phase of Smalley & Smalley 1983), it has gone through repeated transportational (T1, T2, T3, . . .) and depositional (D1, D2, D3, . . .) episodes before reaching its present resting place.

In any one place in arid and semi-arid areas, the immediate source of most of the dust in transit, or being deposited, is very local (Yaalon & Ginzbourg 1966, Coudé-Gaussen et al. 1984b). Among high-yielding local sources of dust, very young, poorly sorted alluvium, as yet uncolonized by vegetation and

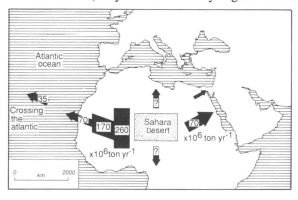

Figure 22.1 Annual dust export from the Sahara (after Ganor & Mamane 1978).

not yet protected by a crust, is thought to be the most important (Dan & Yaalon 1966, Péwé 1981, Goudie 1983a, Coudé-Gaussen 1984b, Gerson & Amit 1987, Pye 1987: 58–62). Large quantities of fresh alluvium are produced in actively eroding areas, for example those subject to recent tectonic uplift. Playa lakes and coastal sabkhas are other important dust generators, which are also most productive in actively eroding landscapes (Jänicke 1979, Coudé-Gaussen & Blanc 1985, Kolm 1985). Unweathered rock debris is mixed in the playas with re-formed clay minerals, biogenic deposits such as diatomite, and salt, and salt weathering prepares and loosens these sediments. Palygorskite is a mineral of the high-pH and silica-mobility environment of sabkhas (Velde 1977), and has been found to be common in Saharan dusts blown over the Atlantic (Coudé-Gaussen 1982, 1984a).

Ancient dry lakes store more potential dust than active ones do (Coudé-Gaussen & Blanc 1985). Much of the harmattan dust comes from ancient lake beds near Bilma in Niger and Faya-Largeau in Chad (Kalu 1979, McTainsh & Walker 1982). Some of this and many other dusts is diatomite, some is salt, some gypsum – all deposits in these lakes. The incision of yardangs into the ancient lake deposits indicates intense erosion; in parts of coastal Peru most dust appears to come from eroding yardang fields (McCauley et al. 1977). Nowadays large quantities of dust may come from tilled soils in semi-arid areas. Another source is volcanic ash.

Nevertheless, much dust is quartzose and, since Smalley's provocative assertion (restated in Smalley & Smalley 1983) that most quartz dust is produced by glacial grinding, the production of this kind of dust has attracted considerable research. It now appears, *pace* Smalley, that quartz dust can be produced in several ways in the desert (see summary by McTainsh 1987): its generation, like that of sand, is a function of a range of crystallization, breakage and selection processes (Dacey & Krumbein 1979).

Weaknesses in quartz grains, which appear as igneous rocks cool from their molten state, yield very small quantities of dust-size quartz (Krinsley & Smalley 1973, Smalley 1974, Krinsley 1978). Simple heating and cooling, and wetting and drying, do not appear to break up quartz grains sufficiently (Smalley & Vita-Finzi 1968, Pye & Sperling 1983), and neither does grinding in desert rivers, although dust produced in this way could have been inherited from a wetter past. Another inherited supply could come from ancient rock-weathering (Little et al. 1978, Nahon &

Trompette 1982, Pye 1983a). A seasonally dry climate, as on the present desert margins, might be the most favourable for such a process, because in these conditions silica is not as liable to solution and leaching as in a wetter climate. Frost weathering could be important in cold deserts, particularly if combined with salt-weathering, but Pye (1987), although cautious, believed that frost weathering produced little quartz dust.

The two most probable causes of dust generation in deserts are salt weathering and aeolian abrasion. Salt weathering has been discussed in Chapter 5.2. The effectiveness of aeolian abrasion in saltation, once thought of as a minor generator of dust (Kuenen 1969), is gaining credibility. Whalley et al. (1982b, 1987) and Coudé-Gaussen (1990) found that considerable amounts of dust could be produced from angular grains as they knocked together in wind machines. Veblen et al. (1981) examined patterns on quartz produced in a similar way, and concluded that it could produce particles of the same size as desert dust. Although the conditions in these experiments are very different from those in the field, their products are similar to natural dust particles: both show signs of considerable abrasion under the electron microscope (Coudé-Gaussen & Balescu 1987), and Coudé-Gaussen (1982, Coudé-Gaussen & Rognon 1983) maintained, with this kind of evidence, that the southern Tunisian loesses were derived by abrasion of sand in the nearby Great Eastern Sand Sea. She also showed how fine and soft material was winnowed out of dunes as they moved down wind in a sand sea (Coudé-Gaussen et al. 1982). Many desert-margin loesses are indeed found down wind of sand seas or dune fields. Other examples are the Negev of southern Israel, Nebraska and northern Nigeria. There seems to be an abundance of ways in which dust-size quartz particles can be produced in deserts.

22.1.4 Entrainment and transport

Of the transport-limiting controls on entrainment, the supply of sand grains for bombardment is the most important (little sand means little bombardment). As to wind power, studies with portable wind tunnels show that the rate of dust entrainment is proportional to about u_*^2 (Nickling 1990). Other transport-limited controls include the surface-roughness (increasing roughness, from pebbles or vegetation, reduces entrainment), aggregation, and the moisture content of the soil. At any one site, however, there may be little correlation between u_* and dust concentration, largely because much of the dust is far-travelled (Offer & Goosens 1990).

In transit, coarse dust travels near the ground, and fine dust moves higher up, concentrations and grain-sizes following power-law patterns of decline with height (Nickling 1988). Nickling found that 20–60% of dust travelled below 5m. The height–decay curve, however, is likely to depend on a number of factors, in particular whether the dust is local or far-travelled. Nickling showed that the rate of transport was more strongly related to measures of turbulence than to u_*, and if so, more dust would travel in unstable than in stable atmospheric conditions.

Most dust storms start as plumes, or narrow concentrations of dusty air streaming down wind from a source. The plumes act as jets of dense dust-laden air in less dense clear air. Like jets, they are narrow at first, but they suddenly expand down wind, both vertically and horizontally. Plumes carry the highest concentrations of dust, up to $10^5 \mu g\ m^{-3}$ (Goudie 1978a). Large, well defined plumes sometimes appear over Iraq in the Shamal season, spewing large amounts of dust over the Arabian Gulf. One extended over 400km, widening from about 10km near its source to 60km at dispersal (Middleton 1986a). Others originate in dry lakes, as in Patagonia (Middleton et al. 1986), and from the tracks of off-road vehicles, as in the Mojave (Bowden et al. 1974). These last reached 30km, and covered an area of 300km². Plumes leave recognizable deposits, with distinctive soils. In Wyoming the deposits of Pleistocene plumes are easily distinguished 5km from their origin (Fig. 22.2; Kolm 1985, Pillans 1989).

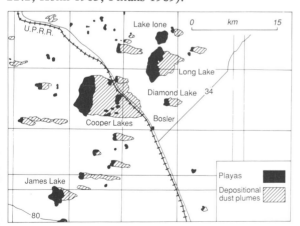

Figure 22.2 Dust deposited by plumes emanating from wind-eroded basins in central Wyoming (after Kolm 1982).

If there is a wide dust source, several plumes may form, perhaps driven by roll-vortex type motion in the planetary boundary layer (Ch. 20.4.5b; Swift et al. 1978). The plumes merge down wind into a less concentrated but more extensive dust storm. This kind of dust storm can carry very great quantities of dust. In southern Israel, some 1 to 2×10^7 tonnes of dust were at one time in transport within an area of $10^3 km^2$ (Dan & Yaalon 1966). A Saharan dust storm measuring $2,200 \times 2,200 km$ has been observed on METEOSAT images (Kästner & Kästner 1978).

Dust is sorted as it travels. Pye's (1987) review of dust sorting showed that particle-size decreased down wind from a source. The harmattan dust which blows across Nigeria each winter is refined from a median grain size of 0.0743mm at Maiduguri in the northeast to 0.0089mm at Sokoto in the west (McTainsh & Walker 1982). Most "proximal" dust (near its source) is poorly sorted, often being mixed with sand (as in Sinai – Coudé-Gaussen et al. 1984b), but after a few hundred kilometres of transport, most dust is less than $30 \mu m$. In the Antarctic and in Hawaii, far from sources, most dust is less than $5 \mu m$. Many dust deposits are bimodal in size, either because they are mixtures of far-travelled and local material, or because they are mixtures of particles, some having disaggregated in transport and some not (Pye 1987).

22.1.5 Deposition

The deposition of desert dust in the oceans and in the humid parts of the globe (in soils and as sheets of loess – Pye 1987: 198–265) is beyond the scope of this book. The emphasis here is on dust deposition in deserts and their semi-arid margins.

(a) Size distribution

In general, desert loess has a wider size-range, and has moved further than glacial loess (McTainsh 1987). A distinction can be made between "proximal" coarser deposits, generally near to sources, and "distal", finer deposits further down wind (Coudé-Gaussen & Rognon 1988). Coarser dust settles back to Earth by gravity in places where windspeed declines, as around topographic obstacles or where vegetation increases the surface roughness. Moderately fine dust is deposited on vegetated ground and, at the far end of the spectrum, particles smaller than $15 \mu m$ are deposited only if they are washed out by rain; if they are aggregated by electrostatic charges; or if they are brought down by adhering to coarser grains (Yaalon & Ganor 1975).

(b) Very arid areas

Little fine dust settles permanently in very dry deserts. There is little rain to bring it down, and few traps to hold it. Where there are traps in these areas, it can be seen that there is a large amount of dust in transit (Gerson & Amit 1987). Artificial traps in Arizona have collected dust at up to 54tkm^{-2} yr^{-1} (Goudie 1983a), but long-term accumulation in the open desert is much slower: 0.02–0.05mm yr^{-1} during the Holocene in the arid Negev of Israel. In these arid areas most dust is trapped in wadis or playas (Yaalon & Dan 1974). Much of the dust which does settle may be washed off higher areas and may accumulate in basins and at the base of slopes, as in lava fields in the Mojave (Wells et al. 1985, McFadden et al. 1986), or in Saudi Arabia, where the reworked dust of the dried lakes sharply contrasts in colour with the dark lavas.

There are three types of primary dust-trap in very arid terrain: rock fissures (Coudé-Gaussen et al. 1984); damp or heavily cracked playas; and coarse alluvium and desert pavements. The trapped dust is eventually washed into coarse pores and protected from further entrainment. Silty and clayey upper soil horizons develop, and over thousands of years these soil layers become plugged with dust (Wells et al. 1985, Gerson & Amit 1987).

(c) Semi-arid areas

The controversy of the 1960s and 1970s (Smalley & Vita-Finzi 1968) over the very existence of loess on desert margins seems now to have been resolved: most authorities accept that there are large quantities of desert-margin loess (McTainsh 1987, Tsoar & Pye 1987, Coudé-Gaussen & Rognon 1988). Desert loess is found on the margins of the subtropical deserts, as in Tunisia (Coudé-Gaussen et al. 1982); in the Levant (Yaalon & Dan 1974); in northern Nigeria, where some 3m (on average) has accumulated over about 40,000yr (Smith & Whalley 1981, McTainsh 1984, 1987); other parts of West Africa (Coudé-Gaussen 1987); and in the Murray–Darling Lowlands in southeastern Australia (Dare-Edwards 1983).

Dust is more effectively trapped in sparsely vegetated semi-arid areas, than in the bare, arid deserts (Yaalon & Dan 1974). Indeed, many present-day desert-margin loesses may have accumulated in periods rather wetter than at present, when there was more vegetation (Coudé-Gaussen 1987, Tsoar & Pye 1987), and when much of the dust was almost immediately redistributed by surface wash (Coudé-Gaussen

& Rognon 1988). Present-day rates of deposition are between 0.1 and 0.4mm yr^{-1} in the semi-arid northern Negev (Rim 1952, Gerson & Amit 1987), and up to 0.7mm yr^{-1} or 92t km^{-2} in parts of the semi-arid Sahel (McTainsh 1984). However, most present-day dust is much finer than Pleistocene desert-derived loess and the rate of deposition is probably much less (Tsoar & Pye 1987).

Some desert loess is silty, and some clayey. Loessic clay has been described in Israel (Bruins 1976) and in southeastern Australia where the aeolian clay deposits were earlier termed "**parna**" (Butler 1956, 1974, Dare-Edwards 1983). Because weathering in semi-arid tropical areas takes place at much the same rate as dust accumulation, desert dust is more commonly incorporated into soils on the margins of deserts, than forming a distinct horizon (Vine 1987). This may explain why it was not detected for so long.

(d) Local spatial patterns

Where topography is subdued, dust can accumulate as a blanket-like covering of the landscape. This is particularly so with fine loessic clays, such as the Australian "sheet-parna" of Butler (1956), perhaps better termed "sheet loessic clay" (Dare-Edwards 1983). These "clay sheets" are often restricted to the dust plume zones down wind of ephemeral lakes (Sec. 22.1.4; Kolm 1985, Pillans 1989). These dust mantles are formed both of pelletized and single-particle clay (Bowler & McGee 1978, Wasson 1986). The particle size of sheet loessic clay in Australia is bimodal, with peaks in the clay and silt or sand fractions.

Most workers consider the source of sheet clays to be the erosion of fluvial or palaeolacustrine areas, although Sprigg (1979) maintains that many of them came from a deflated sea floor, exposed when sea level fell. Most sheet clays in Australia are palaeo-features, being leached of carbonate to several centimetres; there may have been several phases of clay sheet accumulation (Butler & Hutton 1956).

When loesses are coarser, as in the silt range, there are more marked depositional forms. One of these is a deposit very close to source, as in parts of the Mid-West of the USA and the Danube lands where loess hills (sometimes called "loess lips") rim the great floodplains, such as the Missouri, Mississippi and Danube (Fehrenbacher et al. 1986). A recent study of dust movement from a Nevada playa showed that most was deposited within a short distance down wind, with a maximum intensity of deposition of 2,930g m^{-3} yr^{-1} (Young & Evans 1986).

It has often been assumed that, apart from these features and dune-like features in sandy loess (Bariss 1967, Rozyki 1968), loess is deposited fairly evenly over the ground. However, where loess has been deposited over hilly terrain, Goosens (1988a–c; Goosens & Offer 1990) believed that loess tended to accumulate on the upwind sides of isolated hills and more markedly in a narrow zone in their lee. These findings contradict studies in the Negev of Israel, where loess was found to accumulate preferentially on lee slopes (Yaalon & Dan 1974). If the rate of loess deposition is as affected by vegetation cover as has been claimed, then ecological controls, such as aspect and soil moisture, may also play rôles in determining the local patterns of loess thickness (Reed 1930).

22.2 Sand

22.2.1 Size parameters

The conventional approach to the analysis of the size-distribution of sediments has been through the Gaussian distribution, and the usual measures have been mean size, sorting (standard deviation), skewness and kurtosis (see in particular Folk & Ward 1957). The discussion below must therefore depend on this approach for most of its data. However, Bagnold's (1937b) use of the log-hyperbolic function has been

gaining popularity (e.g. Bagnold & Barndorff-Nielsen 1980, McArthur 1987, Hartmann & Christiansen 1988). There is no need in this section to describe the conventional measures, but the log-hyperbolic function is not as familiar and is introduced below.

(a) The log-hyperbolic distribution

The log-hyperbolic function is based on a plot of the log of size on the x-axis (the log of the weight in a particular size-category) against the log of frequency density (the weight in a size category divided by the width of that size category) on the y-axis. Most grain-size analyses of aeolian sands produce plots like that shown in Figure 22.3. The best-fit curve for this kind of data is the log-hyperbolic function, as shown in the Figure. The log-hyperbolic curve is adequately described with the following parameters (Christiansen & Hartmann 1988):

c and s, which are the slopes respectively of the asymptotes on the coarse and fine sides of the hyperbola; μ, which describes the peak size reached by the asymptotes of the distribution, which is closely correlated with the modal size in conventional sedimentary analyses; and μ^* is the peak size of the distribution itself; $\xi = [1+\delta(sc)]^{\frac{1}{2}}$, which describes the peakedness of the distribution; and π, the asymmetry, where

$$\pi = ((c-s)/2)(cs)^{1/2} \qquad (22.1)$$

Another parameter, τ, describes sorting near the modal point (McArthur 1987, Christiansen 1984). c and s can be plotted on a triangular diagram to give a range in types of log-hyperbolic distribution, as in Figure 22.4. Most aeolian sands fall in the upper left-hand side of the triangular diagram.

Bagnold & Barndorff-Nielsen (1988) and Barndorff-Nielsen & Christiansen (1986) believed that the two slopes of the hyperbolic distribution could be given physical interpretation: vigorous entrainment, by removing fine material, would increase s, the fine-side slope, and decrease c the coarse-side slope, in the remaining sediment; vigorous deposition would do the opposite, as fine material settled out. Thus, on the "shape triangle" in Figure 22.4, entrainment would produce a shift leftwards and upwards; and deposition would produce a shift rightwards and downwards.

(b) Conventional descriptions of aeolian sand-size parameters

Some evidence for the physical sorting mechanism has been discussed in Chapter 18.8.1; the discussion here

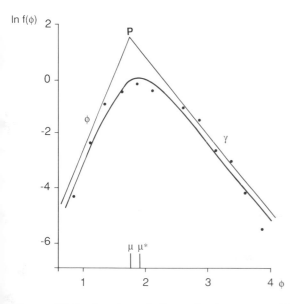

Figure 22.3 Log-hyperbolic grain-size curves (data points for an aeolian sand after McArthur 1987).

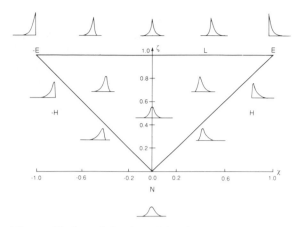

Figure 22.4 A 2-D plot of log-hyperbolic parameters showing conventional grain-size curves for comparison (after Hartmann & Christiansen 1988).

examines the character of actual aeolian sands. Most are fine (means from 0.125mm to 0.25mm) and moderately well sorted, and particles larger than 1.6mm and smaller than 0.1mm are rare (Ahlbrandt 1979). They are usually unimodal (but see below), leptokurtic and fine-skewed (Folk 1971a, Chaudhuri & Khan 1981, Barndorff-Neilsen et al. 1982, McArthur 1987). Where there are violent winds or dense cold air, particles up to 2cm in diameter have been observed in saltation (Hedin, quoted by Bourcart 1928, reviewed by Folk 1968; Lindsay 1973a, Selby et al. 1973), but in more normal conditions the coarsest desert aeolian sands are in sand sheets and zibar dunes, such as the Selima sand sheet where mean sizes are about 2mm (Breed et al. 1987).

In situations where they have been traced to their alluvial or marine sources, or where they have been observed to change down wind in a dune field, aeolian sands are generally seen to become finer, a process known as "aeolian fining" (Finkel 1959, McKee 1966a, Hastenrath 1967, Ahlbrandt 1974, Lancaster & Ollier 1983, Pye 1985b, Sweet et al. 1988). McLaren & Bowles (1985) modelled the effects of transport on the size characteristics of sand grains, basing the model on the probabilities of selection (Fig. 22.5). Their model fitted reality quite well. In transport, sands became finer, better sorted and more negatively skewed. In higher-energy environments there was less negative skewing. McLaren & Bowles adapted the model to situations in which energy changed in the downstream direction. A positively skewed lag was left behind.

McLaren & Bowles's model was for any kind of sedimentary transport (be it aeolian, fluvial or marine), so that aeolian sands may differ little from their source sediments. There are three specific reasons for expecting similarity: short distances of travel in the wind; sources with substantially similar characteristics; or complicated sedimentary histories, including more than one phase of aeolian, fluvial or marine transport (as explained in Sec. 22.2.6, below). The difficulty of discovering diagnostically aeolian characteristics has become increasingly clear as more and more sands have been analyzed. Some authorities despair of ever finding a reliable measure of the distinction (McLaren 1981). Others persist, their methods becoming more and more elaborate (Friedman 1979, Flenley et al. 1987, Bui et al. 1989).

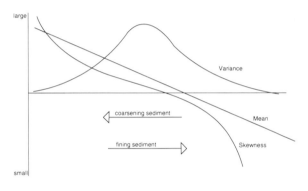

Figure 22.5 A model of sand-sorting in sedimentary transport (reformulation of a diagram by McLaren & Bowles 1985).

(c) Bimodal and multi-modal aeolian sands
When examined with care, most aeolian sands can be seen to contain at least two modes (Fig. 22.6; Vincent 1984), but most sands from interdunes and sand sheets are bimodal or multi-modal, even when examined more cursorily (Fig. 22.7; Bagnold 1941: 125–43, Verlaque 1958, Behiery 1967, Folk 1971a, Skocek & Saadallah 1972, Warren 1972, Lindsay 1973a, Ahlbrandt 1979, Chaudri & Khan 1981, Jawad Ali & Al-Ani 1983, Khalaf & Al Hashash 1983, Vincent 1984, Anton & Vincent 1986, Breed et al. 1987, Thomas 1987a). Folk (1968) maintained that the size of the coarse mode in these bimodal sands was very similar worldwide, and that this mode was usually very well sorted. The ways in which bimodal sands might be produced are discussed in Chapter 18.8.1.

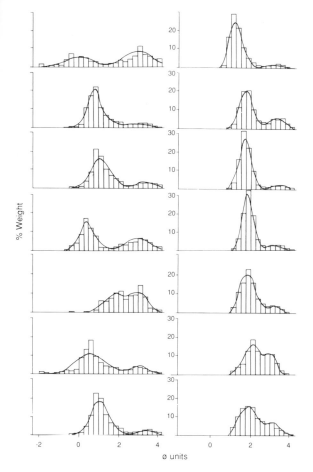

Figure 22.6 Multiple modes in aeolian sands revealed by very fine size-analysis (after Vincent 1984).

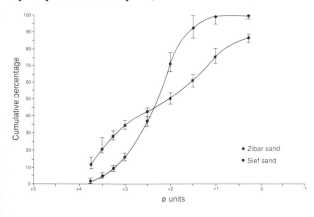

Figure 22.7 Analysis of bimodal zibar dune sands and unimodal sayf dune sands from the Ténéré Desert, Niger. An hypothesis to explain this distribution is shown in Figure 18.17. Sample sites are shown in Figure 23.25.

(d) The "response diagram"

Besler (1983) believed that active and inactive dunes could be distinguished using Friedman's (1961) "response diagram". This is a plot of the mean size against the sorting of sands on which, she claimed, active and inactive sands were found in different zones. Besler has been supported by Harmse & Swanevelder (1988), but criticized by Vincent (1985, 1988), Thomas (1986) and Livingstone (1987, 1989). At best, the response diagram is of unproven worth; at worst, it confuses the study of aeolian sands.

22.2.2 Acoustic sand and fulgurites

(a) Acoustic sand

One of the greatest oddities of dune sands is that some make a sound when disturbed. The effect has been called "musical", "sonorous", "vocal", "roaring", "sound-producing", "barking", or most commonly "singing" or "booming". The Al Murra bedouin in eastern Saudi Arabia call it *al Hiyal*. The phenomenon is quite common, with a history in literature back to early Chinese accounts and to Marco Polo, who likened the sound to that of a drum. Later travellers heard successively a steamship's siren or a piston aircraft. Lord Curzon (1923: 261–339) collected together a host of early accounts. Victorian tourists used to go out of their way to Jebel Nakus ("the Mountain of the Bell") in Sinai and to some Hawaiian coastal dunes that boomed, and this stimulated some early, rather inconclusive speculation (Bolton 1889, Goldsmid 1897). There has been no shortage of later publications, but still no definitive explanation (van der Walt 1940, Gibson 1946, Yarham 1958, Twidale 1978, van Rooyen & Verster 1983).

Unlike the less common "squeaking" or "whistling" beach sands, which emit a high pitch (>500Hz; Takahara 1973), most desert dune sands issue a deep, complex sound whose main notes are between 80Hz and 264Hz (Lewis 1936, Lindsay et al. 1976), or even lower, between 50Hz and 80 Hz (Criswell et al. 1975), best called "booming", although Haff (1986) described a complex mixture of "yips", "yaps", "wheezes" and "oomphs" when acoustic sand was disturbed in the laboratory. Criswell et al. (1975) noted an initial "roar" that lapsed to a deeper drumming. They and others also noted reports of a lower beating frequency at between 1Hz and 10Hz, and a geophysical shockwave (at 50–60Hz) that could be felt as one handled the sand (likened to an electric shock).

The sound can happen spontaneously during avalanching, or when a slipface is pounced upon, or ridden over by a squadron of Marco Polo's Chinese horses. It starts almost as soon as the sand is disturbed and can last for up to 15 minutes. It stops immediately the sand stops moving; it can be heard 8–10km away. Lewis (1936) and Haff (1986) found that sound varied with the speed of disturbance. Lewis found that the sound could be made to vary according to the speed with which a plank was drawn through the sand. The sounds can be produced in the same sand in the laboratory.

Most workers follow Bagnold (1941, 1966b) in the belief that uniform size-grading is important to sound production. Twidale's review (1978) showed that mean sizes of acoustic dune sands were generally below 0.5mm, that freedom from dust and complete dryness were essential, and that pitch was possibly related partly to grain size. Haff (1986) found that the 0.10–0.25 fraction made the loudest noise. The need for complete dryness explains why some dunes boom only intermittently (Haff 1986). Particle shape, packing and temperature appear to be of little consequence. Sound can be produced in quartz or carbonate sands.

Bagnold's early conjecture (1941), that it was a quality of their surfaces that made sands whistle or boom, was not confirmed for squeaking beach sands by Ridgway & Rupp (1970) and Ridgway & Scotton (1972), but was corroborated by Takahara (1973) for beach sands and by Lindsay et al. (1976) for dune sands, all of whose acoustic aeolian sands were highly polished. This was also corroborated by Takahara (1973) and Haff's (1986) ingenious experiments. Takahara destroyed the sound-making properties of squeaking sand by repeated pounding, but then resuscitated them by washing with acid. Haff found that even very small quantities of dust adhering to the sand prevented booming.

There are two schools of thought on the acoustic mechanism in natural sands. Bagnold's (1966b) model of sound-production on slipfaces seems to fall into the "air-cushion" school, which maintains that the sound is produced by the vibration of air between the particles. The alternative is that sound is produced by friction between smooth particles (Takahara 1973). The argument is unresolved.

(b) Fulgurites (or fulgarites)
Another, but much rarer, peculiarity of aeolian sands is long root-like features of fused silica known as fulgurites. In the Kalahari, Lewis (1936c) dug out fulgurites that were about 1cm in diameter and over 2m long, some with many branches. Explanations have included adax urination, lightning strikes (these two perhaps in unison) and root nodulation. Lewis believed that burnt sand around the fulgurites, charcoal near them in the sand, and a mineral composition similar to meteorites, showed that they must have been caused by lighting strikes, and SEM examination confirms this as a high probability (Pye 1982e). Lumps of **silica glass** up to about 7kg in weight have also been reported from some dune areas in North Africa (Clayton & Spencer 1933, Oakley 1952). These lumps appear to be tektites formed by the fusion of dune sand in meteorite impacts.

22.2.3 Shape
In discussing the shape of aeolian sand, distinctions between roundness and sphericity must be recalled (Ch. 18.8.3). The shape of a particular aeolian sand is affected by three processes: selection in entrainment, selection in transport, and abrasion in transport. The earlier discussion showed that neither theory nor experiment produced clear conclusions about shape selection in entrainment or transport. Comparisons of natural aeolian sands with their reputed parent non-aeolian sands do not clarify the issue: when aeolian sands are compared to non-aeolian ones, it appears that in some cases low-sphericity grains have been selected (Free 1911, Mattox 1955, Morris 1957, Winkelmolen 1971, Stapor 1973, Veenstra 1982, Stapor et al. 1983), while in others that more spherical grains have been taken (MacCarthy 1935, MacCarthy & Huddle 1938, Shepard & Young 1961). It seems that selection depends on the range of shapes available and on ambient wind conditions.

The effects of abrasion are more certain, for rounding can only increase as grains collide in transport, and this in turn must lead to greater sphericity. This has been shown in "wind machines", where grains are quickly rounded at first but then more slowly (Whalley et al. 1987). Rounding results from contacts between rapidly moving and rotating particles. Rounding does appear to happen in the field, and quickly, when the particles are soft aggregates of clay (Gillette 1986a), but the speed of rounding of quartz sand in natural aeolian transport is less clear, for the dynamic similarity of wind machines to real aeolian transport has yet to be proven.

The nineteenth-century wisdom was that aeolian sands were "perfectly rounded" like "millet-seed"

(Phillips 1882, Mackie 1897), and when real aeolian sands are examined, some are indeed found to be rounded or moderately rounded (Chaudri & Khan 1981, Goudie & Watson 1981, Ashour 1985, Mazullo & Magenheimer 1987). But many more are poorly rounded (Cayeux 1928, Cailleux & Wuttke 1964, Folk 1978, Goudie & Watson 1981, Khalaf et al. 1984, Thomas 1987a, d). The discrepancy doubtless arises because: (a) of rounding in other sedimentary environments (Barrett 1980); and (b) of short distances of transport in by wind. Rounding cannot therefore be taken as diagnostic of aeolian transport.

Where rounding does exist in aeolian sands, it increases with grain size (Goudie & Watson 1981, Thomas 1987a), but not to the coarsest grains of the creep load; although not as rounded, these may be more spherical than the saltation load (Thomas 1987d).

22.2.4 Mineral composition

Most aeolian sand is quartzose, which in Folk's (1968) terms, is "mature" or even "super-mature". ("Maturity" refers to the extent of weathering and removal of less stable minerals, and so to the preferential accumulation of quartz.) There is no mystery about the origin of the quartz. Quartz is common in igneous, metamorphic and sedimentary rocks, and most of it breaks down preferentially to grains of sand size (Krinsley & Smalley 1972). Quartz is also substantially more resistant to chemical disintegration than other common rock minerals.

Other rock minerals, particularly feldspars and dark minerals such as magnetite, are common secondary constituents in dune sands. A few are dominated by other minerals such as: calcium carbonate in the southern Wahiba Sands of Oman and western India (Goudie & Sperling 1977, Goudie et al. 1987); gypsum in White Sands, New Mexico (McKee 1966a; Fig. 22.8), around the Great Salt Lake in Utah and in parts of southern Tunisia (Coque 1962, Watson 1983); salt in other parts of southern Tunisia (Besler 1977b); and clay minerals.

22.2.5 Grain surfaces

(a) Colour

Redness (hues redder than 5YR in the Munsell system) is a property of many free-draining sediments and soils in arid and semi-arid areas (Pye 1983h). It is especially characteristic of dune sands. Red sediments and soils are not, however, confined to deserts, being better developed in seasonally humid areas, but because many ancient dune sands (both in deserts and their margins) are very red, the red colouration of desert sediments is discussed here.

In bulk, dune sands range from brilliant white to deep red, with yellows, reddish-yellows and browns between. Individual grains may be black, and they form a high-enough proportion of a few sands to make them also dark in bulk. Dark sands are composed of various heavy minerals such as magnetite; light-coloured sands are composed of halite, gypsum or carbonate. Redness is a more common colour, has a more complex explanation, and has attracted more attention. Redness is associated more with aeolian than with marine or fluvial sands, probably because of the greater likelihood that they will provide oxidizing environments (Gardner & Pye 1981).

Quite marked redness can derive from as little as 0.4% ferric oxide by weight (Phillips 1882). The oxides may cover grain surfaces, but are best preserved in indentations; the redness may also derive from red clays held between the sands (Pye 1983h). Because fine grains are more irregularly shaped and suffer less violent collisions, they are usually the red-

Figure 22.8 Brilliant white gypsum sand at White Sands National Monument, New Mexico.

der. A range of colours accompanies a range of poorly crystalline iron minerals, from yellow or brown hydrated ferric oxides, including goethite, to the much redder haematite. Although haematite is deeper red, the colour varies, even at the same haematite content, depending probably on the clustering pattern of the crystals (Torrent & Schewertmann 1987). More rarely, redness may result from fragments of red rocks, such as chert. Redness seldom extends through a whole column of sand, being most commonly confined to the upper layers of the soil profile. When this redness is associated with stabilized dunes, the red soils are known as "oxisols" in the USDA Soil Classification System (Dregne 1976a).

Six factors control redness in dune sands:

(i) *Age* It is widely maintained that dune sands redden with age (Wopfner & Twidale 1967, Folk 1976b, Breed & Breed 1979, Walker 1979, Gardner 1981, Gardner & Pye 1981, El-Baz 1984b). This could be due to the slow rate of production of haematite, although there is disagreement on this (Gardner & Pye 1981). It could also be due to a slow accumulation of iron-rich dust (Bowman 1982). Ageing is thought by some authorities to occur along the direction of aeolian transport (Norris 1969), but Gardner & Pye noted first that the evidence for reddening in the transport direction was equivocal; and secondly that reddening in the transport direction seemed unlikely, because redness was likely to be removed by abrasion. Redness is only preserved in dusts which suffer few damaging impacts. Anton & Ince (1986) could find no evidence of reddening in the transport direction of Saudi Arabian dune sands. The same arguments apply to fluvial sediments, which also do not preserve redness in transport; only colluvium seems to preserve the colour of the parent sediment (Pye 1983h). In favourable conditions (humid but with a dry season) deep redness may develop within 5,000yr, but because of variations in other controls, the rate of reddening must vary widely from place to place (Pye 1983h). The time requirement means that only fairly stable land surfaces can redden thoroughly.

(ii) *Climate* The weathering of iron and its crystallization to haematite needs moisture. Thus, although the reddening of sands in low-rainfall areas may be possible in brief periods of wetting by occasional rain or dew, reddening is likely to be faster in more humid climates, and indeed sands in these climates are generally the redder (Norris 1969, Folk 1969, White 1971, Walker 1979, El-Baz 1984b). Coastal dune sands in tropical climates are some of the reddest of all sediments (Gardner 1981a). A phase of desiccation may be necessary to precipitate the haematite (Pye 1983h).

Redness can be retained by sediments when the climate changes, particularly when the change is to drier conditions, and this may explain the redness of dune sands in many deserts. Deserts with long histories of hyper-aridity have fewer red sands than semi-arid ones, or than ones that have had a fluctuating climatic history (although there are those, such as Walker 1979, who disagree with this interpretation). In high-rainfall areas, dune sands quickly acquire deep bleached podzolic A_2 horizons. Temperature may also play a part, for few contemporary cool-climate sands are red, even in dry environments (Gardner & Pye 1981).

(iii) *Source* The source of the redness in dune sands puzzled early workers, who noted that it could not have come from within the sands themselves, which were largely quartz (Phillips 1882). The redness in an aeolian sand may derive from the parent sand (an already red alluvial sand or, more rarely, one with minerals which supply iron when weathered), from infiltrating waters, or perhaps most commonly, from dust added to the soil surface (Walker 1979, Wasson 1983a, Pye 1983h). Some authorities have found that source is the most important control on colour (Anton & Ince 1986).

(iv) *Mechanical stability* More active sands are constantly abraded and lose redness.

(v) *Chemical environment* Weathering to redness requires high Eh (oxidizing) and high pH (alkaline) conditions. Both these conditions are common in semi-arid aeolian sands where there is both aeration and little leaching. They also occur in some alluvium, and much alluvium is red. The breakdown of organic matter, which is scarce in deserts, provides powerful reducing agents that remove redness. Local concentrations of ions, such as calcium or magnesium, accelerate reddening (Gardner & Pye 1981). Bacteria may play a rôle in oxidation.

(vi) *Preservation* If conditions change, redness may be lost, although once haematite has formed it may take quite considerable changes to reduce it (Gardner & Pye 1981), as is shown by the preserved red colour of Permian aeolian sands in cool temperate climates such as that of the English Midlands.

Although red sands do occur in many semi-arid environments, they are not very precise indicators of climatic conditions for two reasons: first, because reddening depends on local as well as ambient conditions; and, secondly, because many red colours have been inherited from past, unknown environments.

(b) Surface texture

Most dune sands are "frosted" (Fig. 22.9), a feature that appears to be the result of two processes, commonly acting together (Margolis & Krinsley 1971, Al-Saleh & Khalaf 1982, Wilson 1979): *chemical solution and precipitation* of silica; and *mechanical abrasion*. The acquisition of these characteristics is not necessarily independent, for solution is more effective on newly abraded grain surfaces (Krinsley &

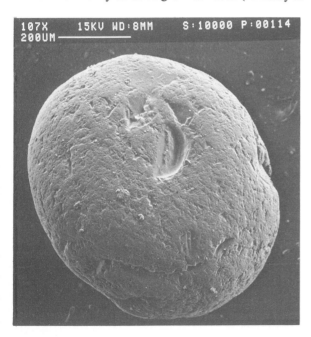

Figure 22.9 A photomicrograph of an aeolian sand grain from a dune near the airport at Derby, northwestern Australia. The small angular indentations indicate grain-to-grain collisions by subangular grains at low energies (5–8m s^{-1}). The central horseshoe-shaped Hertzian fracture indicates collision of subrounded grains at higher energies, possibly in a storm (>18m s^{-1}). (Photograph and interpretation by P. A. Bull.)

Smalley 1972). The solubility of silica is considerably raised at high pH. pH can be raised by the presence of salts, and a form of frosting can be produced experimentally by subjecting quartz grains to saline solutions (McGee et al. 1988). The necessary moisture for the solutional process could come from occasional rain or more frequent dew. Reprecipitation could follow evaporation or a change in pH.

Mechanical abrasion occurs as the grains clash in transport. Mechanically produced microscopic surface textures are typical products of aeolian transport. "Upturned plates" are said to be virtually diagnostic (Kaldi et al. 1978, Krinsley et al. 1976). Le Ribault (1978) also identified "crescent-shaped impact features" with sharp edges as diagnostic. Although the complex history of most aeolian sands (Sec. 22.2.6), means that most also exhibit microscopic features inherited from marine, glacial or fluvial transport (e.g. Baker 1976, Al-Saleh 1982), both Krinsley & Le Ribault believed that a microscopic aeolian character is stamped quickly onto sand grains.

22.2.6 Sources

As sediments move away from their origins, they "mature", in Folk's terms (Sec. 22.2.4), and it is a measure of their varied and lengthy history that most dune sands are mature or even super-mature. Like aeolian dust, most aeolian sand has moved far from its P1 event (release from igneous or metamorphic rock), and experienced repeated cycles of transport (T1, T2, . . .) and deposition (D1, D2, . . .), in Smalley's terms.

(a) Direct release from rock

Very little aeolian sand comes directly from P1 events. There may be instances in which aeolian sand is derived from a P1 weathering of igneous or metamorphic rock, as on the Pampa La Joya in Peru (Lettau & Lettau 1969), but these seem to be rare. Another instance in which sand for dunes comes from the directly subjacent floor occurs in southwestern Egypt, where Haynes (1989) describes the derivation of barchan sands by deflation of a Holocene soil; but the soil here was itself derived from ancient dunes.

(b) Alluvium

Alluvium, which has itself experienced many T–D cycles, is the immediate origin of most aeolian sand. Sandy material is carried as wash load in streams, and deposited in point- or braid-bars and deltas (Wilson 1971b, Mabbutt 1980). In low-relief deserts, such as the ancient shields of the Sahara, Kalahari and Australia, and in the central Asian plains, wash load is deposited on the floodplains of high-order streams. One of the questions that has exercised geomorphologists about many sand seas has been whether transfer from alluvium has been direct (contemporary with the deposition of the alluvium), or indirect (from ancient alluvial deposits).

Direct transfer, it can be hypothesized, is the more likely process, for alluvial sands would then have no time to acquire crusts, soils or a vegetation cover, or to be buried by layers of less erodible sediment such as pebbles or clay. Direct transport has been observed in Kansas (Simonett 1960); on the alluvial plains of the Jo Shui River, whence sand is blown into the Ala Shan desert, at a calculated rate of many tonnes of sand per day (Hörner 1936); and in Arctic Canada (McKenna-Neuman & Gilbert 1986). But these observations hardly amount to proof that direct transfer predominates over indirect.

Direct transfer can, however, be confidently inferred where dunes are close to alluvium, so-called "source-bordering dunes" (Fig. 22.10). Many small dune fields are situated directly down wind of large bends in river channels (Skocek & Saadallah 1972, Jawad Ali & Al-Ani 1983, Khalaf & Al-Hashash 1983 in reference to the Mesopotamian rivers, Denny & Owens 1979 in Delaware, Wasson 1983b in central Australia, Ma Chengyuan 1982 in China). In the Kalahari, alluvial sources are indicated by the greyness of sand which characterizes the downwind banks of many river channels (Thomas & Martin 1987). Madigan (1946) even suggested that the spacing of linear dunes might have resulted from the spacing of meanders of some of the larger Australian desert rivers.

Further, but less conclusive, evidence of direct transfer is the positioning of many large sand seas down wind of the alluvial plains of the greatest desert rivers. The Kara Kum is derived from the Oxus River among others (Suslov 1961); the Thar and Thal in the Indian subcontinent from the Indus and its modern and ancient tributaries and distributaries (Higgins et al. 1974a, Wasson et al. 1983, Gupta et al. 1979, Kar 1990); the Trarza and Cayor stabilized sand sea from the Senegal River (Tricart & Brochu 1955); the Gran Desierto in Mexico from the Colorado (Merriam 1969, Lancaster 1987d); and the Great Western Sand Sea in Algeria from the Saoura (Capot-Rey 1943, 1945).

In deserts with high relief, direct fluvial—aeolian transfer probably occurs on alluvial fans (Brown 1960, Holm 1960, Anton & Vincent 1986 in Saudi Arabia, Wilson 1973 in Algeria, Ochsenius 1982 in South America, Wells et al. 1982 in the western USA, Besler 1982b,d in the Namib). Alluvial fans are profuse sources of aeolian sand where there has been active tectonic activity and fluvial erosion throughout the Pleistocene, as in the Great Basin of the western USA and in Central Asia. In these deserts, high rates of glacial and fluvial activity coincided in the glacial periods with strong winds (Xu Shuying & Xu Defu 1983).

Although *delayed transfer* may be less effective than direct transfer, there is evidence that it has occurred, above all in the association of sand seas with major alluvial lowlands. The thickest aeolian

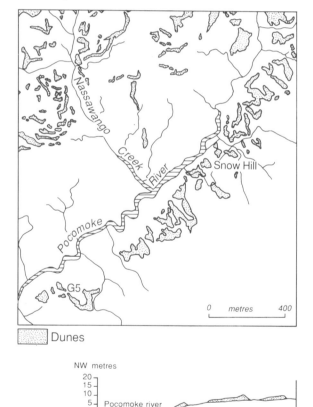

Dunes

Figure 22.10 "Source-bordering dunes" in Delaware. The sand in the dunes has clearly been derived from the nearby river. Pleistocene winds blew from the northwest (after Denny & Owens 1979).

sand in the Simpson sand sea in Australia occurs where alluvium would be expected to have been thickest (Mabbutt 1980, Wasson 1976, 1983a, 1986, Breed & Breed 1979), and analysis of heavy minerals indicates southeastward transport, in the alluvial rather than the aeolian direction (Carrol 1944). Similar conclusions have been reached in the northwestern Sahara (Capot-Rey 1970), the Selima sand sheet in northwestern Sudan (Breed et al. 1987), the Rub' al Khali (McClure 1978), other Arabian sand seas (Anton & Ince 1986), the Thar in India and Pakistan (Seth 1963, Hegde & Sychantharong 1982, Wasson et al. 1983), the Chinese sand deserts (Ma Chengyuan 1982, Chao Sung-chiao 1984a, Zhu Zhenda 1984, Zhu Zhenda et al. 1987), in the Asiatic (former) USSR (Suslov 1961), and the Trarza and Cayor sand sea in Senegal and Mauritania (Barbey et al. 1975).

The proposal that linear dunes have been excavated out of an alluvial cover (the "wind-rift" hypothesis) is discussed in Chapter 26.2.6.

(c) Marine and lacustrine sources

Oceans, seas and lakes have been vehicles for sorting and transporting large amounts of sediment to the windward borders of many dune fields and sand seas. On beaches near Port Elizabeth in South Africa, it has been estimated that about 40% of the longshore drift on a beach is blown inland to dunes (Illenberger & Rust 1988). A similar pattern of large transfers from a beach to dunes was found by So (1982). As with alluvium, transfer from marine sediments can be either direct or delayed. However, there is better evidence for direct marine, than for direct alluvial, transfer.

In *direct transfer* from a beach, as from alluvium, there is a close interdependence between aeolian and marine sedimentary processes. On beaches, longshore drift transports and sorts sediment, and only when it is fine enough is it blown onto dunes (Pye 1983b, 1985b). Longshore drift itself is fed from many sources, of which river mouths and deltas are the most important, and it is for this reason that marine and lacustrine sources are not always clearly distinguishable from fluvial ones. There is still uncertainty, for example, about whether some of the central Australian sand seas were derived from alluvium from shorelines of vast ancient lakes (Wasson 1983a).

There are many well authenticated examples of fluvio—marine—aeolian sedimentary sequences: in southern California and northern Mexico from the Colorado (Norris & Norris 1969, Sweet et al. 1988),

in southeastern Australia from the Murray–Darling (Hesp & Short 1980); northern Sinai and southern Israel from the Nile (Tsoar 1978); the Namib Sand Sea from the Orange River (McKee 1982, Lancaster & Ollier 1983); most of the dune fields in Peru (Douglass 1909, McCauley et al. 1977); and the Skeleton Coast in Namibia (Lancaster 1982a).

Longshore drift is also fed by coastal erosion. In the absence of significant river activity in the very arid eastern tip of Arabia, most of the sand in the latest dune phase in the Wahiba Sands must have been created by cliff erosion into soft sandy rocks, which were themselves created by the cementation of earlier sand seas (Fig. 22.11; Warren 1988a). Cliff erosion accelerated in many parts of the world as sea levels rose at the end of the last glacial period, and this

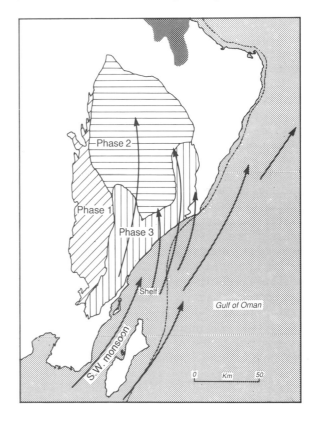

Figure 22.11 The Wahiba Sands related to source. Carbonate sands derived from the rich shelf environment (fed by upwelling currents) have been blown onshore by the southwest monsoon in at least three major phases. The sands of the third (Holocene) phase have been derived partly by marine cliff erosion of cemented dune sands of earlier phases, encouraged by rising sea levels.

encouraged cliff erosion to release large amounts of sediment, to be taken first as longshore drift and then inland by wind (Pye 1984b).

Longshore drift is also fed from offshore, in many cases with biogenic sediments: marine organisms, such as formanifers, oölids, meliolids and shells which are broken by wave movement and transported shorewards by wave action. The largest of the sand seas created in this way is the Wahiba Sands in Oman, where sands, even 50km from the sea, are one-third calcium carbonate (Fig. 22.11; Goudie et al. 1987). A more extensive, but thinner, layer of marine sediments has been spread over southwestern India by the same wind system (Goudie & Sperling 1977). Many smaller, but collectively extensive coastal dune fields are composed of sand that is largely marine-biogenic, as for example on the contemporary coasts of arid and semi-arid southeastern (Coulson 1940) and Western Australia (Bird 1968), and in the Canary Islands (Chamley et al. 1987). Most of these have been lithified to form aeolianite (Ch. 25.3.3).

Delayed transfer from inactive marine or lacustrine deposits may be easier than from ancient alluvium, for marine sediments are generally better sorted. Delayed transfer has been suggested in China, where dunes to the east of the Qinhai Lake are said to be derived from ancient deltas (Xu Shuying & Xu Defu 1983). The dunefield down wind of the field of mega-yardangs in the Lut Desert of Iran (Ch. 21.3.2), themselves eroded into Pleistocene lake deposits, is another example of delayed transfer from a lacustrine source (Krinsley 1970, McCauley et al. 1977). Many coastal dunes are thought to have been derived from marine deposits exposed by the fall of sea level at the height of the glaciations (see review by Pye 1984b).

Evaporites in sabkhas (inland and coastal) are much less important, although locally dominant, sources of aeolian sand. The distinctive products of these areas are sand-size pellets of salts (commonly halite and gypsum). Gypsum dominates the dune sand at the White Sands National Monument in New Mexico (McKee 1966a; Fig. 22.8), and around the large sabkhas or *chotts* in southern Tunisia and nearby Algeria (Coque 1962). The association of evaporite salts with clay produces pellets which are built into clay dunes (Ch. 25.3.2b).

(d) Complex sequences

The example quoted above, in which barchan sands in southwestern Egypt are derived from an underlying soil, itself weathered from more ancient aeolian sands, exhibits a very complex sedimentary history (Haynes 1989). The great maturity of windblown sands shows that complexity is the rule rather than the exception.

Most aeolian sands have experienced repeated cycles of fluvial, marine and aeolian activity. In ancient shield deserts, most aeolian sand has been through many of these phases: weathering, slope transport, fluvial transport, marine or lacustrine transport, ancient aeolian transport, recementation, and re-erosion by water and then by wind. This is said to be the case in the Kalahari (Jones 1982, Thomas & Goudie 1984, Thomas 1988a), the Namib (Lancaster 1989a), and in Arabia and the Sahara (Beadnell 1910, Brown 1960, Holm 1960, McKee 1962, Whitney et al. 1983, Breed et al. 1987). The white gypsum sands of New Mexico are derived from playa lakes, themselves drawing gypsum from Permian desert basins (Talmage 1932). In the Republic of Niger, fluvial sands were winnowed to dunes, which were themselves reworked by streams, not once but many times in the past 50,000 years (Durand et al. 1982). In some places, successive periods of dune building have fed on earlier dunes. In the Simpson and Strzelecki deserts in Australia, renewed wind activity has eroded the upwind ends of linear dunes and added the resulting sand to their crests or to their downwind ends (Wasson 1983a). Also in Australia, the Great Sandy, Great Victoria and Gibson dune fields are fed with sand from upwind sand seas (Wasson 1983b), and in the Sahara there are well established downwind transfers of sand from sand sea to sand sea (Ch. 28.4.1).

CHAPTER 23

Dune processes

23.1 Introduction and definition

This chapter examines the shape of individual dune slopes. It addresses, in order, the following properties and processes in dunes: minimum size, initiation, replication, the bases and main portions of windward slopes, crest-brink areas, lee slopes, the effects of moisture and grain-size, dune movement and dune-scale grain sorting. Later chapters examine the three-dimensional and larger-scale properties of different types of dune.

23.1.1 Dunes defined

A dune is a subaerial body of sand between about 30cm and 400m high and between about 1m and 1km wide, whose shape has been adjusted to ambient wind conditions by the piecemeal deflation of sand-size particles. The meanings of various terms for the parts of a dune are shown in Figure 23.1.

The size limits in this definition follow Glennie (1970), Wilson (1972c) and Mainguet & Callot (1978), the upper ones being those of the largest reported dune. The lower limits are more ambiguous, for size alone does not distinguish ripples and dunes in the 1–10m size range. However, the distinction is unproblematical if grain size is considered, for, in this bedform size-range, ripples are composed of coarse sand, and dunes of fine sand (Wilson 1972c). The dynamic significance of the lower size limit of dunes is discussed more fully below. Piecemeal deflation distinguishes dunes from yardangs, whose aerodynamic form is acquired by the abrasion of coherent material.

Dunes are built of mineral sand or sand-size aggregates of clays, salts or ice. Quartz sand (the commonest material) generally remains loose. Dunes of loose sand may be anchored to topographic obstacles, or plants, or may move freely. Carbonate sand, which is locally common, can also accumulate in loose anchored and mobile dunes, but may eventually cohere as "aeolianite". Most sand-size aggregates disperse shortly after emplacement in

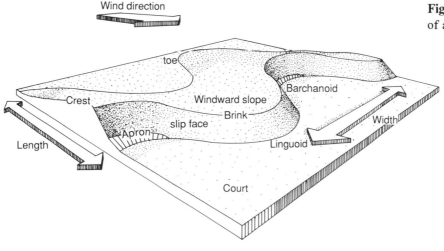

Figure 23.1 The parts of a dune.

dunes, and then rapidly cohere, preventing dune movement but preserving dune shape. Dunes of all these materials can be stabilized by plant colonization or soil formation following a change of climate or of sand supply; in this state dunes slowly lose their characteristic shapes.

23.2 The genesis of mobile dunes

There are three main issues in dune genesis: minimum size, initiation on firm desert surfaces, and replication on sandy surfaces.

23.2.1 Minimum size
Bagnold (1941: 180–3) believed that a dune could not be smaller than the zone of readjustment in sand-carrying capacity as a wind passed over the edge of a sand patch (Ch. 18.6d). In his model, patches less than 2.5–6m wide could not survive the violent fluctuations in sand-transport in this zone. In Bagnold's view the failure of attempts to create artificial dunes was due solely to their being smaller than this minimum.

Bagnold's hypothesis can be tested by observations on two dunes within 1km of each other, and within the same ambient wind regime (Fig. 23.2). The smaller of the two, about 2m long, was a mobile, natural barchan within a field of other natural barchans. The larger, about 3.5m long, was built with a dump shovel on an open, otherwise dune-free, plain. The small, natural dune outlived the larger, artificial one, which although metamorphosing into a recognizable barchan,

lasted only three weeks.

The natural dune in Figure 23.2 was smaller than the smallest of Bagnold's estimates of minimum size, yet survived. The larger, artificial dune was within his range, but did not. The lag distance of Bagnold's argument is apparently less important to the survival of a dune than an adequate supply of sand, for the natural dune was the better supplied. Sand supply does not affect Bagnold's explanation, for in his experiment supply made little difference to the lag distance. Besides, his model cannot apply to dunes in continuous sand, which are by far the most common form of dune. This is not to say that there is no minimum size, for no dunes less than about 1m across have been reported. What the case study does indicate is that the minimum size of dunes is poorly understood. The problem is related closely to dune replication, discussed below.

23.2.2 Initiation
In this context, "initiation" means the formation of a single dune on a more or less level firm surface, as opposed to the replication of dunes down wind of other dunes or on extensive sandy surfaces (discussed in Sec. 23.2.4). The appearance of dunes *ab initio* on nearly level desert surfaces is not difficult to explain. The three main models of initiation are not in conflict. They operate in different but common situations.

(a) Contrasts in surface roughness
The most common explanation for the initiation of dunes on apparently flat surfaces is that sand is trapped by small differences in the roughness of the

Figure 23.2 A small natural barchan and a barchan formed from an artificially created mound of sand in Oman, 1989. For explanation, see text.

surface, or in slight hollows. It is notable that dunes are seldom initiated on artificially smooth surfaces such as roads or airfields, for when dunes do appear on these surfaces, they have almost always been initiated elsewhere. When wind meets a contrast between a rough desert surface and a smooth one, there is a sudden drop in u_* (Ch. 17.1a, Fig. 17.1). This induces the deposition of the sand which the wind had been carrying over the desert surface. When wind meets a slight hollow, there is a decline in near-surface velocity, even in quite gentle dips, and this too induces deposition (Kocurek et al. 1990).

When a dune, initiated in either way, grows to a certain "equilibrium" size, in this explanation, it moves away from its template. Another dune starts to grow in the same spot and in its turn follows the first.

(b) Calving
Many mobile dunes are "calved" or separated from the ends of the anchored dunes, and they then migrate down wind as free dunes (Fig. 23.3; Howard 1985). The obstacles may be hills, scarps or bushes (Kocurek et al. 1990). The size of calved dunes reflects the size and characteristics of the obstacle only shortly after calving.

(c) Ground jets
Knott (1979) observed the break up of waves on an early-morning atmospheric inversion. The break-up released short-period "ground jets" (Fig. 23.4), driven towards the surface, which were capable of sweeping sand from limited areas and depositing it when they dissipated. These patches, Knott suggested, could be templates for dunes.

A recurring but never popular notion is that dunes are a response to electrical potentials produced in sand storms (Ch. 18.9).

Figure 23.3 Calving of barchan dunes from the lee dune in central Sinai (after Warren & Knott 1983).

23.2.3 Replication
The replication of mobile dunes from their first fixation is almost as mysterious as it was when Högbom (1923) and Bourcart (1928) collated early hypotheses. One early school declared that dunes were formed by gravity waves at the boundary between the less dense atmosphere and the denser cloud of

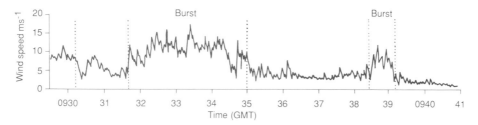

Figure 23.4 Windspeed variations with time at a site near In Salah in Algeria. The "bursts" of higher windspeed and turbulence are probably caused by "ground jets" resulting from the break- up of an early morning inversion. Data for July 1978 (after Knott 1979).

saltating particles, or even the denser bed itself (Baschin 1899, 1903, Walther 1924, Solger 1910a, Exner 1920). Both Högbom and Bourcart dismissed this model, and, although proposed again, it has been repeatedly dismissed by most authorities. Beds do not deform *en masse* like fluids, but by the piecemeal movement of particles (Kennedy 1969, Raudkivi 1976: 81). Chapter 19.3b describes some forms of the wave hypothesis which have been applied to ripples.

Bourcart and Högbom also discounted the notion that dunes were nucleated around obstacles, pointing out, very rightly, that no set of obstacles could be as regularly spaced as dunes (Fig. 23.5). Despite rejection, the proposal that large (mega-) dunes are merely veneers of sand over hidden hills has persisted (Foureaux 1903, Chudeau 1911, Cornish 1914, Gautier 1935, Hill & Tompkin 1953, Bakhit & Ibrahim 1982).

Three main groups of hypotheses endure to explain how an air stream saturated with sand might become unstable and create a persistent dune pattern. Each set of models is better developed for subaqueous than for aeolian dunes, though, as most authorities point out, the similarities between the two situations mean that basic principles should apply to both. The models have in common an initial disturbance or geometric discontinuity, although the character of the bedforms thereafter has little relationship to the character of the discontinuity (Yalin 1977: 212).

(a) Delayed response (or kinematic instability)
Kennedy (1969) introduced a model of dune formation

Figure 23.5 Regularly spaced transverse dunes in the Wahiba Sands, Sultanate of Oman.

in subaqueous flow that has received wide discussion, and it forms the basis of many other models (Raudkivi 1976: 82-7, Yalin 1977: 214-19, Engelund & Fredsoe 1982). The crux of the model is a delay factor that helps to magnify an initial disturbance until some equilibrium size and regular bed-configuration is reached, and then to propagate the pattern down stream.

Kennedy believed that this factor reflected two kinds of delay: first, the adjustment of flow properties following a change in the shape of the bed and, secondly, of sediment transport following a change in shear. He admitted (1969: 154) that it was "a factor of ignorance" and Yalin and Raudkivi saw it as little more than a mathematical device to produce dunes, to which Kennedy had given *post hoc* physical meaning. Moreover, Kennedy's model showed that suspended load played a part in determining dune pattern, which is very unlikely in subaerial dunes, let alone subaqueous ones (Yalin 1977: 218). Smith (1970) also used a lag factor in his analysis, and Engelund & Fredsoe (1982) continued to use Kennedy's factor, adding the effects of bed-slope to the model. The approach was further developed by Richards (1980), who later (1986) linked this model to the models of flow over hills described in Chapter 20.4.7.

Kennedy (1969) believed that his model could apply to desert dunes, given small modifications, but since it assumes high Froude numbers (usually associated with relatively shallow depths below an upper flow-boundary, as in a river), and a surface with waves, it may be of limited applicability in this context. Richards (1986) was uncertain as to whether aeolian dunes were analogous to subaqueous ripples (in which flow depth plays no part) or subaqueous dunes (where flow depth is significant). If the latter was the case, he speculated that aeolian dunes might form only where there were atmospheric inversions gave an upper surface analogous to the free surface in water (below), which seems an unlikely general explanation.

(b) Flow response
Beyond the brink of an existing dune, flow takes off and reattaches down wind, enclosing a "separation bubble" or "lee eddy" (Fig. 23.6). Turbulence is intense at the upper boundary of the separation bubble, and this causes intense erosion at the reattachment point, allowing no accumulation of sand. Down stream of the reattachment point, there is an expanding internal boundary layer, and above it a turbulent wake (McLean & Smith 1985). Near-ground velocities

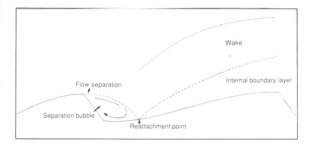

Figure 23.6 The response of flow over a "negative step", as in the lee of a dune (after McLean & Smith 1986).

are kept high where the wake dominates, but then rapidly stabilize as deceleration in the growing boundary layer comes into equilibrium with the wake. This pattern of flow first induces erosion, then deposition, thus producing a second dune down stream of the first. Much the same argument was followed by Fredsoe (1986). McLean & Smith found that their model fitted observations of flow over large dunes on the bed of the Columbia River.

(c) Organized turbulence

In this explanation, some pre-existing pattern in the flow, fixed by a discontinuity on the bed, determines the spacing of dunes. Yalin (1977: 222–32) was a strong champion of this explanation of subaqueous dunes. His model is based on the observation that wavelengths (although not heights) of dunes appear to be already determined before growth begins. Moreover, the wavelengths of subaqueous dunes, unlike ripples, are strongly related to the depth of flow, suggesting that they relate to the largest possible eddies, these being of the order of flow depth. Other "turbulence" models link subaqueous dunes to the turbulent burst–sweep or "kolk" process as described in Chapter 17.1.2 (Jackson 1976); and to surface waves (Allen 1982a, Hammond & Heathershaw 1981). Yalin (1977: 231–2) saw little problem in extending his model to aeolian dunes. The main problem with this type of model is to decide whether waviness in the flow precedes waviness of the dune surface, or vice versa.

(d) Evidence for organized turbulence in aeolian dunes

Both Högbom (1923) and Bourcart (1928) followed many early geomorphologists in the belief that this kind of explanation applied to aeolian dunes. The hypothesis as it relates to aeolian dunes has been adopted in one form or another by Wippermann

(1969), Wilson (1972b) and Folk (1977a).

Both meteorological and morphological evidence has been invoked in support of the model. The meteorological evidence is better at mega- than at meso-dune scale, and is best (but still weak) for linear mega-dunes (Chs 20.4.5, 26.2.4). Some meteorological evidence has also been found at the ephemeral dune scale, where the wavelength of snow dunes (3–15m) correlates with frequency spectra of turbulence in the formative wind (Kobayshi & Ishihara 1979).

Seven pieces of morphological evidence appear to support the model of regular turbulence:

(i) Ephemeral dunes, similar to those of Kobayshi & Ishihara (1979), referred to above, were reported by Bourman (1986); his dunes were 10cm high and 12–13m apart; Figure 23.7 shows yet more of these "spontaneous" ephemeral dunes in Oman; although of regular size, shape and spacing, they were only a few centimetres high. Others are shown in Figure 17.2.

(ii) The very puzzling "chevrons" reported by Maxwell & Haynes (1989). These are flow-transverse, parallel, sinuous, low-amplitude (10–30cm), long-wavelength (130–1,200m) belts of fine, pale sand migrating over darker coarse sand at rates of 500m yr^{-1}. Slope cannot play a part in the process of segregation or movement (as is required by delayed-response and flow-

Figure 23.7 Proto-ephemeral dunes in the Wahiba Sands, Sultanate of Oman. Vehicle tracks for scale. Similar features are shown in Figure 17.1.

response models of bedform development). However, it is also difficult to see how the chevrons relate to dunes with more relief.

(iii) Shortly after their appearance, ephemeral dunes begin to merge across the flow, suggesting the capture of regular flow-transverse patterns (Kocurek et al. 1990).

(iv) There is sometimes a good correlation between wavelength in transverse meso-dunes and the size of their sands (I. G. Wilson 1972b, 1973a, Lancaster 1982c, 1983b, 1985a). Wilson's argu-

ment was that the higher-velocity winds needed to move the coarse sand were associated with larger-scale turbulence which then generated longer wavelength dunes. A link between wind velocity and size of turbulence was also used by Havholm & Kocurek (1988) to explain the increase in the wavelength of superimposed transverse meso-dunes as wind velocity increased on the windward slope of mega-dunes; areas with high wind velocities in the Namib Sand Sea also correspond with areas of long-

Figure 23.8 Edge-enhanced LANDSAT TM image of part of the west-central Namib Sands, Sultanate of Oman, showing "proto-mega-linear" dunes in the lower left. Higher meso-linear dunes occur to the top right.

wavelength dunes (Lancaster 1988c); and the same relationship has also been noted in pyroclastic surge dunes (Sec. 23.9). However, the Wilson/Lancaster correlation was not found by Wasson & Hyde (1983b) in data from a large sample of dunes of many kinds in Australia, and may apply, if at all, only to transverse dunes.

(v) Mega-barchans cannot be explained as the result of a shortage of sand supply (the conventional explanation of barchans; Ch. 26.1.4); they hint at some kind of organized flow-parallel and flow-transverse pattern at the mega-scale.

(vi) Proto-mega-linear dunes (Fig. 23.8) and proto-mega-barchans (Fig. 23.9) are dunes at the mega-scale, but without mega-height. Proto-mega-barchans are collections of small barchans arranged in mega-barchanoid shapes, but with sand thicknesses of no more than a few metres. These are quite common: in southwestern Egypt (Haynes 1989); in the Thar Desert in India and Pakistan (Kar 1990); and in the Fachi–Bilma sand sea in Chad (Mainguet 1984b). Wing-to-wing distances are over 2km; they appear always to be much more elongated in the wind direction than full mega-barchans. They could be responses to pre-existing flow-parallel secondary flows in the Earth's boundary layer.

(vii) Wave-like motion in the lee of hills has been described in Chapter 20.4.5. Its association with dune forms is well documented, for example by Kolm (1982, 1985). Lee-wave patterns are evident in the patterns of vegetated dunes in the lee of small hills in the Sudan (Fig. 20.10).

This evidence is, however, insufficient to confirm the regular turbulence model of dune formation. This primary question about desert dunes is still unanswered.

23.3 Shape

This section discusses the shapes of the dunes that develop from sand patches, however accumulated. It is confined to two-dimensional shape in constant wind regimes. Three-dimensional shapes of mobile dunes in a number of wind regimes are discussed in Chapter 26. Separate discussion does not imply that two- and three-dimensional shapes are not connected; the connections are discussed in Chapter 26.

23.3.1 Equilibrium at the dune-slope scale

Dunes, both subaqueous and subaerial, ultimately appear to achieve and maintain a stable or "equilibrium" configuration. Equilibrium can be defined theoretically as the point at which the local sediment transport rate varies directly with bed elevation (Kennedy 1969). Yalin (1977: 227–8) believed that an equilibrium shape was the one that produced the smoothest flow and thus equalized energy loss in the streamwise direction. The claim is similar to the assertion that stable patterns in river meanders are those that equalize the loss of energy along the

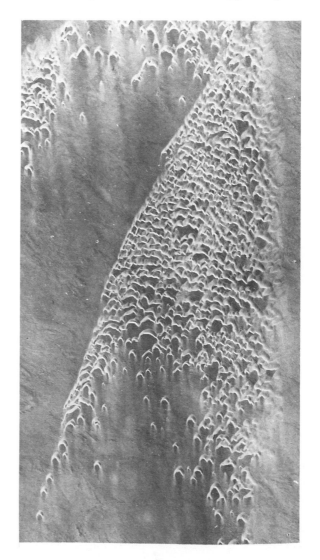

Figure 23.9 Proto-mega-barchans in the southern Jafurah Sand Sea, Saudi Arabia (after Al-Hinai 1988).

channel (Langbein & Leopold 1966), although in neither case do these claims yield much insight into the processes involved. Although hard to define, "meso-" scale dynamic equilibrium of this kind is nonetheless a vital concept in the understanding of dune sizes and patterns at a large scale, and it is discussed further in Chapter 24.7.

However, at a small spatial and temporal scale, which is the concern in this Section, meso-scale dynamic equilibrium is little more than a statistical abstraction, for real dune slopes are in constant change (Allen 1976). Readjustment takes time (the relaxation time), during which the slope is not fully in equilibrium with any wind. Readjustment to an increase in wind velocity or a new wind direction is rapid at first and then slows down, probably following some kind of exponential law (the "rate-law" of Graf 1977). Readjustment to a lowering of velocity is a lengthier process. An attempt is made in Figure 23.10 to plot a daily path of adjustment for some transverse/network dunes in Oman, following the system evolved for subaqueous dunes by Allen & Collinson (1974).

Winds that persist long enough to establish equilibrium forms are very rare, for the relaxation time probably exceeds the length of the usual semi-diurnal cycle of speed and direction in desert winds. Thus, a definitive field study of dune form is difficult, if not ultimately impossible. Laboratory modelling of mobile dunes is impractical because of the size of the wind tunnel that would be needed, and it has never been attempted. The only way forwards may be with wind-tunnel studies of dune shapes with fixed beds, and mathematical modelling, tested where possible. These problems, among others, mean that there are still many uncertainties about dune shapes.

23.3.2 Growth of sand patches to asymmetrical dune shapes

Sand travelling over a hard, pebbly desert pavement moves more rapidly than over a newly formed patch of sand (however initiated), because the wind is retarded by the increased volume of saltation. The sand carried towards the new sand patch is therefore deposited on its surface and a dune begins to grow (Bagnold 1941: 169–74).

The cross sections of all dunes in unidirectional flow are asymmetrical, even those of dome dunes which have no slip faces. An explanation of asymmetry was made by Exner (1927, quoted by Graf 1971; Fig. 23.11). As the sand patch grows upwards and becomes a dune, its upper parts are subject to faster winds than the lower parts and they move forwards to produce the asymmetry. Shear reaches a local maximum somewhere near the crest of the dune (Ch. 20.4.7), after which there must be deposition. In snow-dunes, where the snow can cohere, an overhanging cornice can develop as the snow is deposited

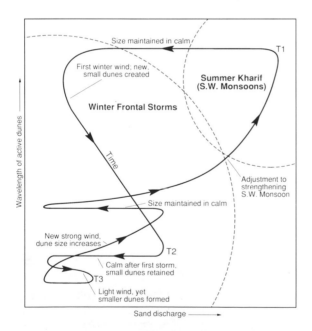

Figure 23.10 A hypothetical plot of dune-size variation with changing windspeeds in an annual cycle. Derived for the network system described by Warren (1988b) and shown in Figure 26.34. "T1", "T2" and "T3" refer to the dune systems in Figure 26.33 (after the system used by Allen & Collinson 1974).

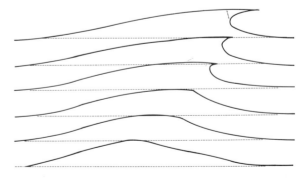

Figure 23.11 The modelled development of an asymmetrical dune form in flow from left to right (after Exner 1920).

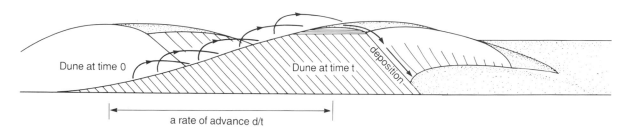

Figure 23.12 The mechanism of forward movement in sand dunes.

at the upper boundary of the separation bubble, but in the non-cohesive sands of most dunes, sand jettisoned over the steep lee slope avalanches down the slip face. Erosion of the upwind slope and deposition in the lee move the dune forwards and maintain the asymmetry (Fig. 23.12).

23.3.3 Bases of windward slopes
The windward base of an isolated dune is a point of considerable change, and potentially one of many feedback adjustments. Slope angle and roughness are sharply adjusted, and sand discharge may also suddenly change. The windward bases of continuous dunes

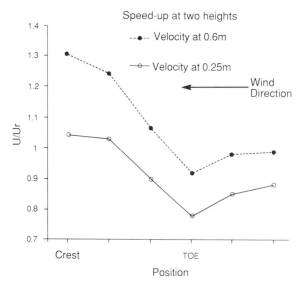

Figure 23.13 The decrease in windspeed at the base of the windward slope of dunes. Measurements made in 1990 in the Sultanate of Oman (courtesy of G. F. S. Wiggs).

are also zones of change, but of a different character, and they are discussed below (in Sec. 23.3.6).

It was shown in Chapter 20.4.7 that there is commonly a decrease in wind velocity at the base of the windward slopes of low, isolated hills. The phenomenon has also been repeatedly found at the bases of real and model dunes (Fig. 23.13; Tsoar 1985, 1986, Tsoar et al. 1985, Lai & Wu 1987, Jensen & Zeman 1985, Howard & Walmsley 1985, Livingstone 1986, Lancaster 1987c). If sand transport is related in some way to wind velocity, as in all the common models of sand transport, then this decrease should be accompanied by a fall in sand-carrying capacity, and an accumulation of sand. Mathematical models of dunes (Howard et al. 1978, Jensen & Zeman 1990; Sec. 23.3.4) develop such an accumulation. Yet this does not happen in nature.

The explanation of this paradox is elusive, but Wiggs (1992) believes that, despite falls in velocity near the base of the windward slopes, shear is maintained because of curvature effects on the flow. The implication is that the curve of the base of windward slopes is delicately adjusted to maintain sand transport.

23.3.4 Main portions of windward (or stoss) slopes
Above the toe regions), slope angle, roughness, windflow and sand transport are in continuous feedback relationships. The fundamental requirement is that shear (u_*) should increase up slope "such that the increasing volume of sand eroded from the . . . slope can remain in transport" (Lancaster 1987a: 519). In subaqueous dunes, where the same principles apply, it can be shown that, at equilibrium, sand transport must be proportional to the height above the base

(Fredsoe 1986). The form that delivers these conditions in subaerial dunes is apparently a straight slope at a low angle: below the crestal area, most natural windward slopes are nearly straight in cross section (e.g. Hastenrath 1967), with angles of between 5° and 10°.

Using the models and observations of flow over low hills described in Chapter 20.4.7, and relationships between wind velocity and sand transport, erosion and deposition on a windward slope can be modelled using a simple continuity equation. Most of the workers who have employed this approach have used simulations of flow over hills that have been superseded, or have based their calculations in the field on measurements of wind that do not adequately predict u_*, because they have been taken too far above the surface. Although few of them have produced stably migrating dune forms, they have nonetheless given significant insights into the nature of the adjustments.

The first model, that of Howard et al. (1977, 1978) was based on field measurements on a real barchan dune and wind-tunnel observations on a fixed-bed model of this dune. Howard & Walmsley's (1985) development of this model was based ultimately on the mathematical model of flow over hills introduced by Jackson & Hunt (1975). The Wippermann & Gross (1986) model used a different and simpler model of flow over hills, in which there was a logarithmic velocity profile over the whole slope (in spite of the flow pattern reported in Ch. 20.4.7), and a rather oddly shaped lee eddy. Jensen & Zeman's (1990) dune model used yet another (this time two-dimensional flow) model of air flow over hills. The Fisher & Galdies (1988) model was much simpler than these three, assuming, for example, uniform windspeed over the dune.

Howard & Walmsley (1985) and Walmsley & Howard (1985) showed that erosion and deposition were sensitive to minor variations in the shape of the slope, and especially to the value of roughness height, z_0, and its spatial variation. This is confirmed by field observations of rapid responses to changes in wind direction (Reid 1985, Warren 1988b). Wippermann & Gross (1986) produced a recognizable barchan from a conical pile of sand in 8 days with $u_* = 7m\ s^{-1}$ (Fig. 23.14), a situation not dissimilar to the field observations on the dune in Figure 23.2. Their model reached a steady-state slope of about 7°. In Jensen & Zeman's (1990) model, fluid forces were always trying to steepen the dune, a tendency they tried to counter by introducing a slope correction for the transport rate. Although Howard's (pers. comm. 1992) modelling also shows that a slope effect is important, it was not sufficient to bring stability to Jensen & Zeman the model. Jensen & Zeman therefore suggested that some kind of lag behaviour might also play a part, although Howard et al. (1978) had found that lag had little effect. However, Jensen & Zeman's model did produce a straight windward slope. Neither Howard & Walmsley's nor Jensen & Zeman's models achieved a dune which conserved its shape in migration, suggesting that there are still many basic questions to be answered. Other results from these models are discussed in appropriate sections below.

The controls on the angle of the windward slope are still obscure. Some reflections on the possible rôle of sand grain-size are given in Section 23.5, and the relationship between three-dimensional shape and slope angle in Chapter 26.1.2. Lancaster's (1985a) model predicted steeper slopes at higher windspeeds,

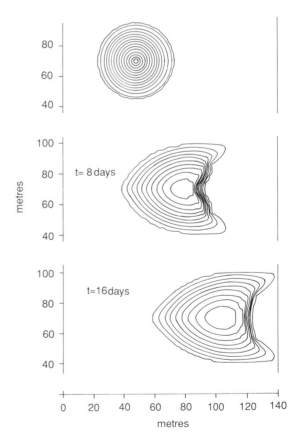

Figure 23.14 The evolution and migration of a dune in a numerical model (after Wippermann & Gross 1986).

in contrast to both the Howard et al. (1978) and Wippermann & Gross (1986) models, which suggest that high winds result in gentle slopes, and to observations in strong-wind environments, such as those experienced in pyroclastic flows (Sec. 23.9), in mountain passes (Gaylord & Dawson 1987), and in subaqueous dunes (Engelund & Fredsoe 1982, Fredsoe 1986).

23.3.5 Crests and brinks

Dune crests assume a spectrum of shapes with two end-members: straight to the brink or broad domed convexity (Fig. 23.15). The extreme of crest/brink separation, where there is no brink (because there is no slipface), occurs on dome dunes (Ch. 26.1.3). Explanation of this range of shapes is difficult, because the crest/brink area is apparently even more sensitive than the windward slope. It adjusts to small variations in wind velocity and direction, roughness height (z'_0), and the shape and intensity of the separation bubble in the lee (Bagnold 1941: 197–8, Walmsley & Howard 1985, Lancaster 1985a, Hesp et al. 1989). There are probably a number of feedback processes, and some complex aerodynamics. There are five sets of explanation for differences in crestal configuration.

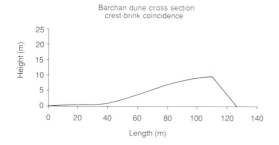

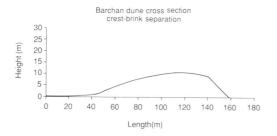

Figure 23.15 Neighbouring crest-brink separated and crest-brink coincident dunes in the Sultanate of Oman in July 1990 (courtesy G. F. S. Wiggs).

(a) Evolutionary change

Evolutionary change is the most common explanation. In most evolutionary schemes, dome dunes, and dunes with crest/brink separation are seen as precursors to dunes with straight slopes (King 1918, Högbom 1923, Chu Chen-Ta aka Zhu Zhenda 1963, quoted by Mainguet & Callot 1978, Hastenrath 1967, Embabi 1970/71, McKee & Moiola 1975, Jäkel 1980, Lancaster 1985a, 1987a, Kocurek et al. 1990). A problem with this sequence is that, although some dunes with crest/brink separation are undoubtedly immature, many, including some dome dunes, are apparently equilibrium meso-forms (Bourcart 1928, Aufrère 1931, Breed et al. 1980). Dome dunes closely juxtaposed to dunes with slipfaces may reach 6m high (King 1918, McKee 1966). What is more, exactly the reverse evolutionary sequence was observed on barchans by Capot-Rey (1963) and Verlaque (1958), and all sizes of Fryberger et al.'s (1984) dome dunes in the Jafurah Sand Sea in Saudi Arabia could seemingly evolve into barchans and vice versa. The most serious problem with evolutionary-change models is their aerodynamic explanation.

(b) Aerodynamics

If flow were to decelerate up wind of the crest, then deposition would also occur there and create a crest higher than the brink. Some such explanation is needed if dome dunes are accepted as equilibrium forms in unidirectional winds.

There is disagreement among meteorologists as to where the point of maximum windspeed occurs, and about its character (Ch. 20.4.7). Some see it peaking at the crest, some just before the crest, some as being a narrow zone of very high shear, some as more diffuse. In the field, Mulligan (1987) found u_* to peak well before the crest, but Lancaster (1987c) found the u_* peak at the crest. These differences could have many explanations, including measurement at a non-equilibrium stage of dune-slope development, for the character and position of the peak value are probably sensitive to the shape of the hill, to the ambient u_*, to wind direction and to the character of separation in the lee (Jensen & Zeman 1990).

An early version of Jensen & Zeman's model (Sec. 23.3.4) developed quite unrealistically marked crest/brink separation, perhaps because the velocity peaked before the crest, but it is hard to see how such a pattern of flow could do anything other than increase the size of the dune indefinitely, and this is what happened in the Jensen & Zeman model. A later version of the same model produced a straight slope to the brink, perhaps because the authors kept the maximum of u_* at the crest by manipulating the size and position of the separation bubble. But this was still not an equilibrium configuration.

Most subaqueous dunes also exhibit crest/brink separation, or at least broadly convex crests, in environments where flow is strongly unidirectional. McLean & Smith (1986) found crest/brink separation in dunes on the bed of the Columbia River, although they believed that a profile straight to the brink would be the more stable. Fredsoe (1986), however, used much the same model as McLean & Smith to conclude that broadly convex crests were more stable, but, because the degree of curvature in Fredsoe's model depended on the ratio of dune height to depth of flow, it may be of limited applicability to desert dunes.

Empirical studies in a wind tunnel suggested to Tsoar (1985) that dome shapes were more stable than dune shapes with straight slopes to the summit. In the field, Chu Chen-Ta (1963, quoted by Mainguet & Callot 1978), Lancaster (1987c) and Mulligan (1987) all found deceleration between crest and brink, but all these observations may have been undertaken in wind conditions that were not in equilibrium with the dune form (the usual problem in dune studies; Sec. 23.3.1). They cannot distinguish between pre-existing and *post hoc* crest/brink separation, or if the separation of the peaks of u_* and the dune itself was leading to a stable shape or one that was growing. These problems lead on to a third set of explanations.

(c) Variable wind velocities

This explanation also has its origin with Bagnold (1941: 197–8). In varying windspeeds, the point of maximum shear would be in constant migration (Lancaster 1987a, Hesp et al. 1989), and the point of maximum deposition would also migrate. Crest/brink separation would persist only until the new wind had achieved a new equilibrium crest/brink coincidence. If the crest is very sensitive to changes in wind characteristics, and if the period of adjustment is long enough, even minor diurnal variations could maintain the separation. This is the model of crest/brink separation adopted by Hunter & Richmond (1988). In this form, the variable-wind hypothesis combines the aerodynamic model of (b), above, with the evolutionary model of (a), above.

The crest is undoubtedly more mobile than the base of the windward slope. When winds are light, the dune crest may be mobile when the base of the slope is static. In a series of measurements, Lancaster (1985a) found that in light winds there was a ratio of 1:58 between the sand transport rate on the crest and that at the windward base; at higher-wind ambient velocities the ratio dropped as low as 1:3. This pattern means, in effect, that light winds can be lowering the crestal area when the base of the dune is inactive, and so creating a convex profile, if not crest/brink separation.

(d) Variable wind directions

Despite Hastenrath's (1967) finding that small barchans in Peru had more crest/brink separation than large barchans, even some transverse mega-dunes have crest/brink separation. In California, the wavelength of the meso-dunes superimposed on these mega-dunes decreases down wind of the crest in apparent response to a drop in wind velocity (Havholm & Kocurek 1988), perhaps a confirmation of the aerodynamic argument above. Nonetheless, Havholm & Kocurek suggested that crest/brink separation on their mega-dunes was a result of seasonal wind changes, the separation becoming less pronounced in strong transverse winds, and more pronounced when

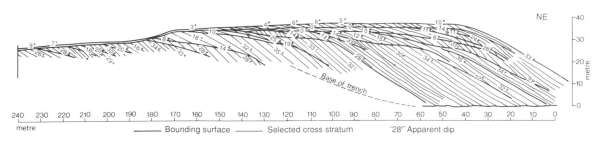

Figure 23.16 The common pattern of bedding on the upper part of a transverse dune, showing the shallow "cap" of ripple strata with many third-order truncation surfaces, overlying the main mass of the dune, which is built of slipface strata (after McKee 1979).

winds were variable in direction. A similar explanation has been applied to much lower dome dunes, many of which contain slipface bedding (McKee 1966, 1982). In McKee's explanation, dome dunes acted as normal transverse dunes in winds of some directions, but had their slipfaces occasionally destroyed by crosswinds.

The greater activity on crests, and their greater vulnerability to winds from different directions, is probably responsible for the cap of dense ripple strata which underlies the summits of most barchans and transverse dunes (Ch. 27.2.2; Fig. 23.16; McKee 1966, Embabi 1970/71). The cap is commonly about 0.5m deep and it consists of low-angle sets of ripple strata, separated by many truncation surfaces, dipping down wind (Ch. 27.2.2; Embabi 1970/71, McKee 1979c). One interpretation of the cap is that many small, dome-like dunes have formed and migrated over the crestal area. The sets of strata (between truncation surfaces) thin upwards, a pattern which may result from the migration of small, lower-memory dome dunes over slightly longer-memory features (Hunter & Richmond 1988).

(e) Lagged response
In addition to his other explanations, Bagnold (1941: 200) suggested that crest/brink separation might be a consequence of lag between changes in u_* and the rate of sand transport (Ch. 18.6.4). This would shift the point of maximum sand transport down wind of the point of maximum shear stress, which Bagnold believed to be at the crest. The point where deposition took over from erosion, the brink, would be even further down wind. Because the delay depended only on windspeed and sand size, and not on dune size, it would become progressively less marked as the dune grew. This prediction was corroborated by Hastenrath's (1978) observations.

The ultimate in delay occurs when sand completely overshoots the dune, although in normal circumstances this happens only on very low dune profiles, as on the wings of barchans. Overshooting was the explanation adopted for dome dunes by Breed et al. (1980), who saw them as expressions of high wind-energy environments, with long saltation paths, where there could only be slipfaces on very high dunes. The model may also apply to the dome-like shape of zibar cross sections, where there is seldom any saltation, and where, when saltation does occur, it must be in very long jumps.

More than one of these effects may be needed to explain crest/brink separation or dome dunes in the diversity of situations in which they occur. More than one may be in operation at any one site, and any one time. However, few of the models have been adequately tested.

23.3.6 The two-dimensional form of continuous transverse dunes

The "flow-response" model of dune replication explained above apparently produces windward (or upstream) slopes with a gentle upward convexity just below mid-slope (Fig. 23.6; McLean & Smith 1986). A more exaggerated form of this "lower convexity" was also produced by a computational model of flow over continuous dune terrain developed by McCullagh et al. (1972). The model was based on very simple assumptions, yet the triangular, peaked shapes at the start of the model were lowered and they adopted recognizably dune-like shapes.

Little is known about the shape of or windflow over continuous transverse aeolian dunes or their equilibrium shapes. Lancaster (1985a) found less speed-up over continuous than over isolated transverse dunes. Warren (1988b) found a lower-slope convexity on continuous dunes in Oman, and found that it intensified when windspeeds increased and became more unidirectional (Fig. 23.17); Hesp et al. (1989) found that flow at the base of the windward slope of a dune

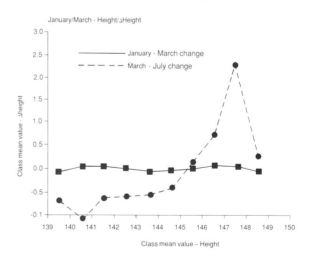

Figure 23.17 A comparison of dune form during the weak winter winds and strong summer winds, at 'Urayfat in the Wahiba Sands, Sultanate of Oman. Note the accentuation of a lower convexity on the windward slope in strong summer (*kharif*) winds (after Warren 1988b).

was strongly influenced by the presence of the dune up wind.

As to the spacing of continuous dunes, Fredsoe's (1986) model for subaqueous dunes suggests that an equilibrium spacing depends on u_* and grain size.

23.3.7 Lee slopes

There are two classes of transverse dune: those with no flow separation and no slipface (dome dunes and zibars) and those with both. As has been explained in the previous subsection, little is known about the aerodynamics or geomorphological dynamics of the first category. The discussion below is therefore confined to the second group.

(a) Flow separation and the lee eddy

As a dune grows upwards, there comes a point, depending on flow characteristics, at which the wind can no longer follow its form, and flow separates from the bed. The point is achieved at lower lee-slope angles in winds of higher turbulence. The separated flow reattaches at some point down wind, depending on the height of the dune and the velocity and turbulence of the wind. From the reattachment point, the wind flows both back towards the dune, forming a "separation bubble", and onwards down wind (Fig. 23.6). On dune-shaped hills, lee eddies are of the order of four times the height of the hill (Fig. 23.18).

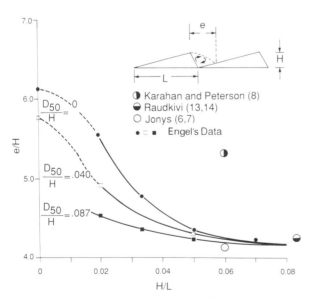

Figure 23.18 The size of the lee eddy or "separation-bubble" down-flow of dune-shaped obstructions in water (after Engel 1981).

There is little evidence for the claim that vortices form when the lee slopes of dunes are heated (Jäckel 1980).

There is no doubt that return eddies exist in the lee of dunes, both in air and under water (Fig. 20.19; Raudkivi 1976: 70, Engelund & Fredsoe 1982). In the lee of aeolian dunes, ripple-patterns and small shadow dunes show that wind velocities are great enough to move fine sand back towards the slipface, at least in high ambient windspeeds (Hoyt 1966, Wojciechowski 1979). They also bring large amounts of light plant debris to the slipface, and keep it there (Anderson & Büttiker 1988). In eastern Mauritania, Sevenet (1943) and Monod (1958) even noticed yardangs and other evidence of excavation into soft sands between transverse dunes. Other signs of wind erosion in this position have been reported from the Taklamakan (Aufrère 1928, quoting Hedin), from the Algodones Dunes in southern California (Sharp 1979, Sweet et al. 1988), from Oman (Gardner 1988), and from the Oregon coast, where strong winds erode the lee hollow, and gentle ones fill it up again on a daily cycle (Hunter & Richmond 1988).

Cornish's (1897, 1914) assertion that the "lee eddy" was responsible for the erosion of the lee face began an acrimonious debate in English (King 1918, Lansberg & Riley 1943, Cooper 1958, Inman et al. 1966, Hoyt 1966) and, in French, the notion had an independent life (Aufrère 1928a, Bourcart 1928, Verlaque 1958, Capot-Rey 1963, Coursin 1964, Mainguet 1984a). There is no evidence, however, that lee eddies affect the morphology of the slipface in any significant way. They cannot move sand up it, even at high ambient wind velocities. Their only accomplishment is low-angle accumulations ("aprons"; Fig. 23.1) of fine sand at the base of the slipface (Cooper 1958, 1967, Hoyt 1966, Reid 1985).

Grains projected well beyond the slipface may also contribute to the formation of aprons (below). A proportion of this projected sand travels well beyond the slipface, especially in high winds (Coursin 1964, Howard et al. 1977, Hunt & Nalpanis 1985). It may be taken over 60m beyond the brink (Hunter et al. 1983), but modelling and observation suggest that the great majority of grains, even in high winds, falls close to the brink.

(b) Grainfall

In some conditions grainfall may take place on the crestal area (above), but on most dunes most of the saltating sand is projected into the gentle winds of the separation zone. Grainfall begins at varying lee slopes, depending on ambient wind conditions. In one, common set of conditions, it has been observed to begin at 10° (Kocurek et al. 1990).

There should be an exponential decline in grainfall along a horizontal plane projected from the brink, and some authorities have assumed that this also means an exponential decline in the rate of deposition on the lee face (after about 10cm where the rate was constant) (Hunter 1985). However, Anderson (1988) found a bulge down wind of the brink, which he attributed to the fact that the slipface falls away at an angle greater than the mean return angle of grains, and to drag on the falling particles when they fall into the quieter conditions in the lee of the brink. In one case, Anderson's bulge was 0.2–0.4m from the brink of a dune, where it was 12mm above the expected position assuming a straight slope (Figs 23.19 & 20). The

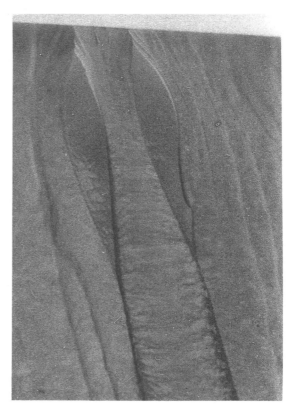

Figure 23.20 The upper part of an active slipface, showing the effects of intermittent slumping, including an actively retreating scarp below the brink, and "bottleneck" below the bulge created by the pattern of grainfall over the brink.

position must depend on u_* and grain size. Anderson's model predicted that coarse grains (with low lift-velocities) would fall near the top of the face, and finer ones further out; this contradicts an earlier model in which coarse grains were carried further (Chakrabarti & Lowe 1981).

Grainfall deposits are described in Chapter 27.2.1.

(c) Minimum sizes of slipfaces
Bagnold (1941: 211) predicted that slipfaces could not be smaller than a saltation jump length. He confirmed this in the desert, where there were no slipfaces below his predicted minimum size. The figure shows that, as a slipface declines in height, there comes a point below which height rapidly falls off. In Anderson's model ((b), above) the maximum projection of particles is about 1m, when u_* is 0.5m s^{-1} and grain size 0.25mm (common values), and this sets the minimum

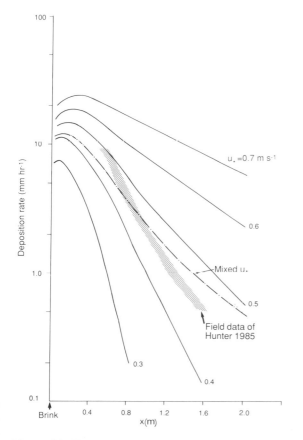

Figure 23.19 The modelled deposition rate down wind of the brink of a dune in sand-transport conditions. Note that the point of maximum deposition is down wind of the brink (after Anderson 1987a).

height of a 32° slipface at 0.6m. Anderson predicted that dunes over about 1m in height would be bound to have slipfaces in most conditions. This is not apparently confirmed in observations of dome dunes (Ch. 26.1.1), suggesting that there is something about the aerodynamics or geomorphological dynamics of these that has yet to be understood.

(d) Sandflow (avalanching)

When grainfall on the lee has built up the lee slope beyond a critical angle, sand flows down the slipface. There have been claims that there is more to the triggering of avalanching than just the overloading of the upper slipface. For example, Kaiser (1970, quoted by Yaalon 1974) claimed that changes in temperature played a rôle, for he had observed movement to begin as the slipface heated up in the morning, even before saltation began. However, whether this was an effect of heating or of the evaporation of cohesive moisture brought by dew is, however, debatable. These other effects are likely to be very minor, if they are at all important, because most slipface movement takes place in high winds in the middle of, or late in, the day.

The point at which failure due to overloading by grainfall occurs is known as the "pivot point" (Fig, 23.19). At the pivot point, failure reduces the slipface from the angle of initial yield to "the angle of repose" (more precisely, the "residual angle after shearing"). The difference between these two angles is on average about 2.5° (Bagnold 1966, Allen 1969a, 1970a, Carrigy 1970).

The residual angle (the angle of most of the slipface) alters little with grain size in the sand range, but is sensitive to grain angularity, moisture, salt content, and perhaps static electricity (Van Burkalow 1945, Land 1964, Allen 1969a, 1970a, b, Carrigy 1970, Li & Komar 1986). Mixtures of grain sizes behave differently from pure grain sizes (Li & Komar 1986), but this factor is small in well graded dune sand. These factors singly or together must explain discrepancies between results from the field and the laboratory. Laboratory measures vary from 30° to 32° (Carrigy 1970, Allen 1971), through 33° (Chepil 1959) and 34° (Fryberger & Schenk 1981) to 32°–35° (Van Burkalow 1945). Field values vary from 32° (Finkel 1959, Hastenrath 1967, 1978), through 33°±1° (Inman et al. 1966), to 31°–34° (Embabi 1970/71, McKee 1979c).

After failure, a 5–10mm scarp cuts back up slope from the pivot point to the brink (Hunter 1977b). The scarp remains active for many minutes, feeding a sandflow avalanche. The flow narrows or "bottlenecks" below the pivot point, and then rapidly expands down slope until it reaches a constant width (Fig. 23.19). The velocity profile within the flowing sand is parabolic, with a surficial low-velocity layer, a fast-flowing centre, and a rapidly declining velocity towards the base. These tongues cannot be more than a few centimetres thick (Lowe 1976). They move at about $0.2m\ s^{-1}$ (Hunter 1985). Sandflow tongues tend to be of roughly the same width across the slope, in the same ambient conditions, but vary from site to site and time to time, for unknown reasons. Widths are commonly of the order of 50cm. Subsidiary or tertiary sandflows may develop on either side of the main tongue (Hunter 1977b). When saltation is weak, sandflows do not reach the base of the slipface. Avalanching in aeolian dunes is almost always intermittent, for they experience far lower sand transport rates than subaqueous dunes, where avalanching can be nearly continuous (Hunter 1985).

Avalanche flow is turbulent, mixing the grains, and this in itself must dilate the moving mass of sand. Bagnold suggested that if some force, such as someone jumping onto the slipface, were to increase the descent velocity above the its normal speed, then an oscillating dilatation and compaction might take place, and this might be the mechanism of sound production in acoustic sand. His model of this process produced frequencies close to those of acoustic sands measured in the field (Ch. 22.2.2). He observed that the descending sandflow was sheared along near-horizontal planes as it came to rest on the gentler slope at the base. When the descending sand hits the base of the slope, a wave travels back up the sandflow at 5–10cm s^{-1}. Allen (1971) believed that this wave marked the limit between settled stationary sand below, and the more fluid active avalanche above.

The turbulent motion of the avalanche brings coarser and more platy grains (mica, shale, shell or bark) to the surface and carries them to the sides, and especially the toe (Bagnold 1954b, Fryberger & Schenk 1981, Sneh & Weissbrod 1983, Schenk 1983). Avalanche deposits are at almost the minimum possible bulk density (Allen 1971), and they include sand grains imbricated with axes pointing directly down slope (Ellwood & Howard 1981).

(e) Slumping

When there is cohesion in the slipface, as from early morning dew or in salty environments, whole blocks of sand may slowly slump without mixing (Fig. 23.21; McKee 1979c). When they are only weakly coherent, the slumps soon disintegrate, and movement on most lower slipfaces is almost wholly by sandflow (Schenk 1983). When the sand is very wet, other kinds of structure are found (Sec. 23.6).

(f) Lee slopes of mega-dunes
When a strong enough wind blows at right angles over the brink of a transverse mega-dune, a slipface sometimes 100m high may form, and avalanches may descend to the base. However, when winds blow at an angle to the brink, the lee slope is covered in low meso-dunes migrating across its surface (Havholm & Kocurek 1988, Sweet et al. 1988).

Slipface deposits are described in Chapter 27.2.3.

23.4 Minor surface forms

The smooth, expansive curves of dune slopes, which attract photographers, develop only when the wind has been blowing in one direction for some time, although they persist when it declines below the threshold, as at the end of a day. Winds from new directions, often in the early morning, can achieve their own smooth curves only after a period of adjustment, and during this time minor, low-wavelength features disturb the smooth outlines created by the previous winds. The most vulnerable parts of a dune to this kind of disruption are the steepest and highest, where slopes are most likely to be out of adjustment with the new winds and where winds are strongest. The features of early readjustment include "notches" or "flutes", of the order of 1–2m long and 50cm across (in effect micro-yardangs, Ch. 21.3.2a) which form when the wind blows diagonally across old brinks (Lettau & Lettau 1978, Hastenrath 1967, 1987); and reversed slipfaces where it blows in a direction diametrically opposite to the previous wind. Verlaque (1958) charted the development of one of these minor slipfaces.

23.5 The effects of grain size

Howard's model of dune form (Sec. 23.3.4) predicts that dunes of coarse material will have gentle slopes. This is consistent with Lancaster's (1985a) reasoning (although the assumptions are different) and with field observations, although it must be remembered that, in the field, coarse sands are almost invariably associated with zones of high wind velocity (Warren 1972, Wilson 1972c, Tsoar 1978, Lancaster 1982c, 1983b). A sample of dunes from the southwestern USA shows a simple, direct, almost linear relationship between the height:length ratio of dunes and the ratio of mean size to standard deviation of their sand: low, flat dunes

Figure 23.21 Slumping on an active slipface, shortly after dawn when the sand is slightly coherent with dew.

have coarse, poorly sorted sand (Tsoar 1986). Fredsoe (1986) concluded that subaqueous dunes in coarse material would also have gentler slopes than those in fine.

23.6   The effects of moisture

The effects of moisture on the entrainment of sand have been described in Chapter 16.2.1. This section examines the effects of moisture on dune shape.

Near-surface sand dries quickly after rain, and prevents evaporation from beneath, often for many months. A "wetting front" moves increasingly slowly down the column of sand until all the smaller pores are filled (Agnew 1988). Water moves down slipface strata more quickly than through ripple strata (Bagnold 1941: 246), and this concentrates moisture at the base of slipfaces.

As drying sand is eroded from the surface, the behaviour of the wet sand that is exposed depends on the ratio of the drying rate to the deflation rate. Only if evaporation is slower than deflation will the wet sand have an effect on sand movement and dune slope. Drying rates depend on solar radiation (a function of latitude and cloud cover), slope angle, albedo, windspeed, salt content and grain size. One would therefore expect varying effects. In western New South Wales at about 33°30'S (mean annual rainfall about 250mm), the north slope of an east-to-west dune receives much the same amount of radiation through the year, and therefore suffers little variation in potential evaporation; the south slope receives less radiation in winter and therefore has potentially a much more annual marked moisture cycle (Hyde & Wasson 1983). Thus, in winter, the wetter southern slope is not as readily eroded as the northern, and stays steeper. The actual pattern of wet and dry sand layers on a slope is a complex function of the recent history of erosion, burial, rainfall and drying.

Drying is quicker closer to the Equator. In Oman, at about 22°N, with less than 50mm mean annual rainfall, and a higher-energy wind environment than in New South Wales, Warren (1988b) observed that sand dried rapidly when it was exposed by deflation on the windward slopes of dunes, so that moisture had little effect on dune morphology. This conforms to some of the observations on surfaces subject to high evaporation rates by Hotta et al. (1984). On the Omani dunes, slopes passed smoothly from wet to dry sand.

The slowing of dune advance in wet conditions has been noted in northern Sinai by Tsoar (1974) and on the Oregon coast by Hunter et al. (1983). However, Norris & Norris (1961) and Sharp (1966) noted that rain made little difference to dune movement in southern California.

If rain is heavy, large coherent slumps can occur on slipfaces, leaving tension cracks above, and forming brecciated deposits at the base (Besler 1981, Hunter et al. 1983; Fig. 23.22). Where wet sand is exposed by deflation of the overlying dry sand, and where it does

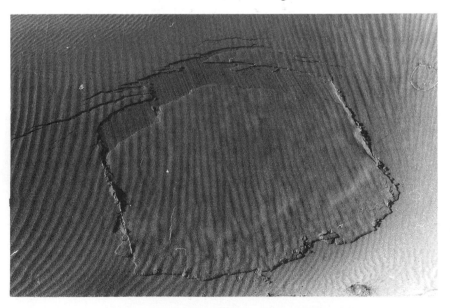

Figure 23.22 Coherent slumps on a water-saturated slipface with brecciation at the base.

not dry too quickly, it may be eroded into small yardangs and flutes (Hunter et al. 1983).

23.7 Movement

23.7.1 Individual dunes

Mobility is a very striking property of dunes, matched only by their sound-production. Sound and movement ostensibly bring dunes closer to life than anything else in the inorganic world. Movement is inexorable and can be exasperating. Haynes (1989) discovered that a barchan, at the base of which Bagnold had camped, had moved relentlessly at 7.5m yr^{-1} in the 57 years since. Capot-Rey (1957a) described how a dune which had covered the camel market at Faya-Largeau in 1935 had moved a kilometre by 1955. Movement of a metre in a day during windy seasons is a common phenomenon (for example, Hunter & Richmond 1988), and this can be a considerable nuisance: dunes repeatedly bury roads, gardens and buildings (Busche et al. 1984).

Movement is the natural condition of all un-anchored, unstabilized transverse dunes (Fig. 23.23). It is the unavoidable consequence of piecemeal defla-

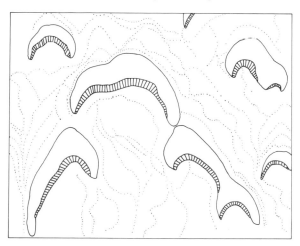

Figure 23.23 Former positions of dunes, as shown by traces on a sabkha up wind of actively migrating dunes. Salty solutions penetrate into the lower parts of the dunes, and preserve them when the rest of the dune moves forward. The trace from an air photograph also shows the enlargement of neighbouring wings of a barchan, but in opposing senses (redrawn from a photograph in Kerr & Nigra 1952). No scale given; the estimated distance between the wing-tips of the largest dune is 200m.

tion on the windward slope, and of deposition in the lee, as described above.

The migration rate of a dune, V_d, can be expressed as follows (Simons et al. 1965):

$$V_d = q_b / kH\rho_{pb} \qquad (23.1)$$

where q_b is the transport rate of the sand trapped by the slipface. $q_b + q_{th} = q$, the total transport rate, where q_{th} is the through-going (untrapped) transport rate. In dunes in near-unidirectional wind regimes, most of the sand arriving at a slipface is trapped, and $q_b \geq q_{th}$. Where winds are more variable in direction, q_{th} approaches q_b, and dune advance is slowed relative to the total drift potential, DP (Ch. 20.3.1; Embabi 1986/7). $k = A_c/LH$; A_c is the two-dimensional cross-sectional area of the dune, and L is its wavelength (distance between neighbouring dunes down wind). For a simplified, triangular, continuous dune, $k = 1/2$ (Rubin & Hunter 1982); H is the height of the dune; and ρ_{pb} is the bulk density of the sand in the dune.

Variations of this basic formula have been produced for different purposes (Bagnold 1941: 204, Wilson 1972b, Greeley & Iversen 1985: 185, Jensen & Zeman 1990).

An assumption of the above formula is that dunes maintain their shape in movement. This can only apply to mature dunes, supposedly in equilibrium with their wind and sand-supply environment, and, given the vicissitudes of individual dunes, it should only be applied to the mean values of dunes in bulk. If these conditions are met, however, the assumption seems to be justified by field observations (Long & Sharp 1964, Hastenrath 1967, Warren 1988b). When bulk changes in dune volume have been observed over long periods of time, or as they migrate, it has usually been asserted that wind or sand-supply conditions have changed (Lettau & Lettau 1969, Hastenrath 1987, Smith 1981).

The implication of the formula – that small dunes move more quickly than large ones – has been established beyond all reasonable doubt (Fig. 23.24). Doubts remain, however, about the exact form of the relationship, especially at either end of the size range. Finkel (1959) found that a linear formula, as above, was a good predictor of dune movement for dunes 2–7m high, but was wildly out for small dunes of the order of 1m high. This may have been due to the higher bulk densities of low dunes (Hastenrath 1978), or to a more variable wind environment for smaller dunes, sheltered by large ones. At the other end of the size range, Figure 23.24 shows that in many cases the relationship should be exponential rather than linear (Sarnthein & Walger 1974). Larger dunes (in the meso-size) seem to reach a plateau in their rate of movement beyond which size makes no difference. This may be due to greater wind speed-up on higher dunes (Ch. 20.4.7). Very large mega-dunes probably move very slowly indeed.

Both Wippermann & Gross's (1986) and Jensen & Zeman's (1990) models, reported in Section 23.3.4, simulated dune movement quite adequately.

23.7.2 Bulk transport

Bulk transport is the transport of sand by dune movement (the rolling over of sand in the dune), as opposed to transport across the desert floor by saltation (Lettau & Lettau 1969). Hunter et al. (1983) developed a formula for amount of sand in bulk transport (Q_b):

$$Q_b = kHV_d \qquad (23.2)$$

where k, the form factor $= A/LH$; A is the cross sectional area of the dune; L is dune spacing; H is

dune height and V_d is dune migration speed.

Bulk transport can be calculated from data on the shape (volume), bulk density, and rates of movement relative to size in a field of dunes. In Baja California Inman et al. (1966) found bulk transport in a field of low but continuous transverse dunes to be 23m³ m-width⁻¹ yr⁻¹. In a field of dispersed barchans in Peru, Lettau & Lettau (1969) found bulk transport to be 5×10^3 m³ m-width⁻¹ yr⁻¹ (Fig. 23.25). In another field of barchans, in Mauritania, the figure was calculated at only 1.1 m³ m-width⁻¹ yr⁻¹ (Sarnthein & Walger 1974).

Lettau & Lettau (1969) found that bulk transport was between 40% and 60% of the total discharge including saltation over the desert floor, but in Mauritania, saltation discharge was 50–100 times more than bulk transport (Sarnthein & Walger 1974). Lettau & Lettau (1969) noted that despite their slower progression, the larger dunes in a field of barchans transported much more sand than small, fast-moving dunes. They also found bulk transport to be sensitive to changes in wind speed: an increase of 20% in

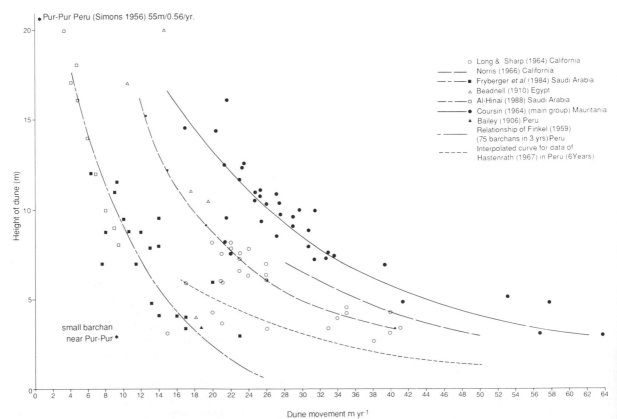

Figure 23.24 Dune movement data, from various sources.

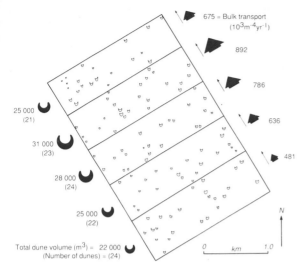

675 = Bulk transport
$(10^3 m^{-4} yr^{-1})$

892

786

636

481

25 000
(21)

31 000
(23)

28 000
(24)

25 000
(22)

N

Total dune volume (m^3) = 22 000
(Number of dunes) = (24)

0   km   1.0

Figure 23.25 Bulk transport of sand by barchans in Peru (after Lettau & Lettau 1969).

windspeed at 320cm could increase bulk transport rates by 50–100%.

23.8 Sorting on dunes

Sand is undoubtedly size-fractionated at the dune scale, but a mere list of four potential groups of sorting processes shows how complex this fractionation must be.

(i) *Sorting processes at the grain scale*: differential loss or gain on eroding and depositional surfaces and surface sorting under the shaking induced by bombardment (Ch. 18.1.3); sorting by entrainment and transport in saltation (Ch. 18.8.1); slope effects on thresholds (Ch. 17.3.4c); and sorting on ripples (Ch. 19.2).

(ii) *Sorting processes on single dune slopes* include: wind speed-up on the windward face, and the consequent differential rates of and periods of activity on that slope (Sec. 23.3.4); the probable existence of a short, peaked zone of high-velocity winds at the crest and the formation of a capping of coarse ripple-strata on the crest, and the variation of the area covered by this capping as winds fluctuate in velocity and direction (Sec. 23.3.5); the loss of some material in suspension from the brink (Sec. 23.3.7a); deposition of grain-fall on upper

slopes (Sec. 23.3.7b); and the accumulation of wind-born dust.

(iii) *Processes involved in dune migration*: the sorting of coarse grains to the foot of slipfaces (Sec. 23.3.7d); and the forward movement of the dune which brings these already-sorted slipface deposits to the windward slope (Sec. 23.7.1).

(iv) *Three-dimensional sorting around the dune* (Ch. 26.1.2).

Most of these processes are active and interactive when the dune is in motion, and they are constantly adjusting to different windspeeds and directions. At a position on the windward slope of a meso-dune, the surface may or may not be in equilibrium with the prevailing wind conditions. It may or may not be capped by ripple strata, for the upwind feather edge of the capping must migrate up and down the windward face. If slipface strata are outcropping, they may be relics of a very gentle or very severe episode on the slipface, of the order of five years previously. There may or may not have been side winds in the recent past. There may or may not have been upslope migration of coarse sands from the toe in the recent past.

Very few of the very large number of studies of sorting on dunes have explicitly recognized such complexity, and very few have employed sampling procedures fine enough to discriminate between the various processes. What is worse, many if not most of them have approached the question of sorting purely empirically, with little or no explicit model of the processes involved. The result is a great body of data that is contradictory and at worst, nearly worthless. Nonetheless, its findings are summarized below.

Some dune-scale sorting mechanisms are gross enough to be detected by coarse sampling methods, and for these there is broad agreement among workers. The first of these is the accumulation of coarse sand in a narrow zone at the base of the windward slope, where it forms mega-ripples (Sidwell & Tanner 1939, Amstutz & Chico 1958, Verlaque 1958, Finkel 1959, Simonett 1960, Hastenrath 1967, Tricart & Mainguet 1965, Lancaster 1981c, 1986a, Vincent 1984, Watson 1986, Livingstone 1987b). The reasons for this process are obscure. There may be some threshold-slope effect limiting the transport of coarse grains.

The second widely observed pattern may be a consequence of the first, and the accumulation of coarse sand on barchan wings and (less certainly) the linguoid sections of continuous transverse dunes. The coarse sand left at the base of the windward slope may be channelled around the curved plan-shape of the windward slope of the barchanoid section, unable to mount the slope even in the high winds (Hastenrath 1967). Hastenrath postulated a compensating mechanism whereby coarse sand was funnelled towards the apex of the barchanoid slipface, and so appeared preferentially again at the toe of the windward slope.

There is much more disagreement about what occurs on a two-dimensional cross section of a dune, for three reasons: first, all the processes mentioned above are in complex interaction here; secondly, many of them are in operation at scales well below the sample size of most workers; and, thirdly, windspeed and directional variation leave complex sequences of deposits. The most common finding is that crests contain finer, better sorted and less fine-skewed sand than lower windward slopes do (Alimen 1953, Verlaque 1958, Vaché-Grandet 1959, McKee & Tibbitts 1964, Glennie 1970, Embabi 1970/71, Lancaster 1981c, 1986a, Sneh & Weissbrod 1983, Binda 1983, Vincent 1984, Ashour 1985, Watson 1986b, Lancaster et al. 1987, Livingstone 1987b, Hartmann & Christiansen 1988, Tsoar 1990).

This pattern of upslope fining has been explained in two ways: first by a slope-related threshold effect (Vincent 1984, Watson 1986, Livingstone 1987b); and secondly, by sorting on the slipface, leaving only fine particles to be circulated in the upper dune as it turns over in movement (Sidwell & Tanner 1939, Sneh &

Weissbrod 1983). This process undoubtedly occurs, but it could only continue if coarse material were able to mount the dune for recirculation. It appears that these two explanations are mutually exclusive.

Fewer workers have found coarser sands on crests than in troughs (Bellair 1952, Alimen 1953, Alimen et al. 1958, Sharp 1966, Folk 1971a, Chaudri & Khan 1981, Barndorff-Nielsen et al. 1982, Lancaster 1986a, McArthur 1987). Many of the dunes found to have this kind of distribution are linear. One explanation may be to do with the source sand: if dune sand is derived from interdune corridors which are rich in silt and clay (low mean grain size), then the dune sands from which the silt and clay has been winnowed will exhibit coarser mean values (Folk 1971a). Another explanation could be that fine material has been winnowed out of upper slopes by winds with high velocities, leaving only coarse sand.

Sorting can also occur in dune landscapes. In the Ténéré Desert there is a repeating pattern of sorting between dunes of different type (Fig. 23.26). One hypothesis to explain this is that an alluvial sand with a broad mix of sizes is worked over by strong winds (the only ones able to move the coarsest of the grains). As explained in Chapter 18.8.1c, coarse and fine grains stay behind on zibars, while medium-size ones go on to form linear (sayf) dunes (Warren 1972). A similar explanation has been adopted to explain the bimodal sands in interdunes in the Namib Sand Sea by Lancaster (1986a).

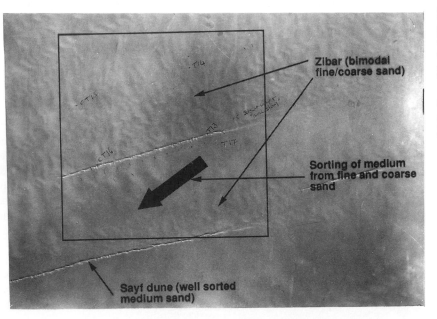

Figure 23.26 The spatial pattern of sorting of sands within a dune pattern in the Ténéré Desert, Republic of Niger (field data from Warren 1972). Sample characteristics are shown in Figure 22.7, and a hypothesis to explain them in Figure 18.17.

Zibar (bimodal fine/coarse sand)

Sorting of medium from fine and coarse sand

Sayf dune (well sorted medium sand)

23.9 Dunes in pyroclastic deposits

During volcanic eruptions, dunes are formed in pyroclastic surges whose velocities can reach 300m s^{-1} (Allen 1982a, Cas & Wright 1987). In many surges, dunes form only near the vent itself, and dune height and wavelength decrease away from the vent, presumably as velocity declines. Some pyroclastic dunes have steeper slopes facing the oncoming current, and these forms have been termed "anti-dunes". Many dunes exhibit bedding that does not approach the angle of repose, suggesting that they were formed in strong shear and by processes other than grainfall and avalanching. They share these characteristics with dunes formed in very high winds (Gaylord & Dawson 1987). The high velocities in pyroclastic surges create dunes with a wider range of grain sizes than normal dune-forming winds.

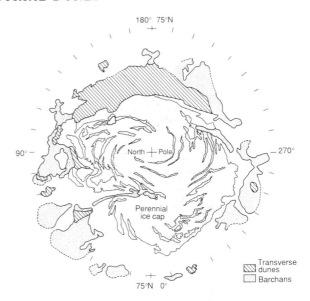

Figure 23.27 Sand seas surrounding the north-polar ice cap on Mars (redrawn after Tsoar et al. 1979).

23.10 Extraterrestrial dunes

Since 1971, Mariner and Viking orbiters and landers have transmitted to Earth images of sand dunes and aeolian ripples on Mars, and other probes have shown dune-like features on Venus and on the Saturnian moon, Titan (Greeley & Iversen 1985, Greeley et al. 1987). The north-polar sand seas on Mars are very large (Fig. 23.27; Lancaster & Greeley 1990). They and the many other Martian dune fields (Ward et al. 1985) exhibit dune forms very similar to those on Earth, including anchored and linear dunes, although the dominant dune-form is transverse, with some beautifully symmetric mega-barchans (Tsoar et al. 1979). Moreover, there are also yardangs, ventifacts and active dust storms on Mars (Greeley & Iversen 1985).

On Mars, gravity is only about 0.27 times and atmospheric density only 0.0007 times that of Earth.

These conditions produce a minimum value of the threshold/grain-size curve (Fig. 17.5) at 115μm grain size and 2m s^{-1} (u_*) on Mars (cf. Fig. 17.5 for Earth) (Sagan & Bagnold 1979, Smalley & Krinsley 1979, Iversen et al. 1987). Thus, Martian dust storms must either be raised by very strong pressure-gradient winds, or by other processes such as dust devils or slope winds. Dune sands may be moved by strong winds generated in these ways, the north-polar sand seas probably being moulded by katabatic winds off the ice cap (Tsoar et al. 1979). Other explanations for the existence of dunes in such a low-density atmosphere could be that they are composed of low-density sediment (Miller & Komar 1977, Greeley 1979), or that they are relics of a time when the atmosphere was denser (Greeley & Iversen 1985).

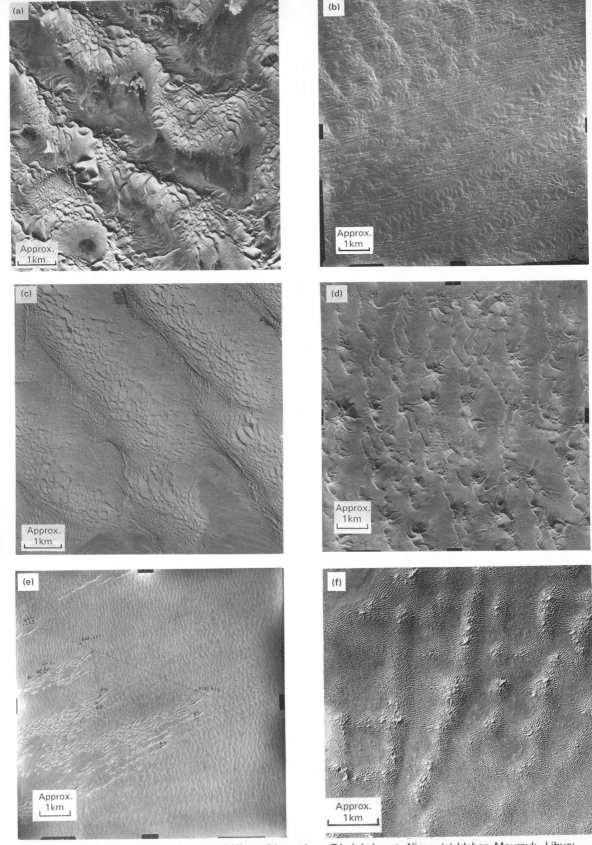

1 Varieties of dune form. (a) central Libya: (b) southern Ténéré desert, Niger; (c) Idehan Mourzuk, Libya; (d) southern Great Eastern Sand Sea, Algeria; (e) central Ténéré Desert, Niger; (f) northeastern Great Eastern Sand Sea, Algeria.

DUNE CLASSIFICATION

OBJECTIVES

Dunes have immense variety (Fig. 1). Diversity arises from infinite variation in the intensity and combination of dune-forming processes: diurnal and annual wind patterns, atmospheric stability and stratification, wind strength, sediment size, sediment supply, plant cover and plant shape, hard-rock topography, rainfall, and recent geological history. There may be other controls, such as secondary flow patterns, of which little is yet known.

With so little of this variety explained, a comprehensive genetic classification is impossible, but some form of classification must underlie any attempt at explanation. The simple interpretive classification used here defines a small number of apparently elemental types. Different purposes, such as the understanding of sedimentary structures or going-conditions for vehicles, would produce different classifications, as would different scales of enquiry (for example, a classification based on space observations, where the smallest dunes cannot be seen, or a classification evolved for a small area without the full range of dune variation).

A SIMPLE CLASSIFICATION

The classification

The classification is shown in Figure 2. The terms used are defined in later parts of the text.

The choice of terms

A classification for explaining dune forms cannot use terms that prejudge genesis, where it is still doubtful. For example, the term "longitudinal" must be avoided (as it is in many recent publications), because of its questionable genetic implication, while "transverse" can be retained because this form of dune is firmly established as flow-transverse. In general, the terms used here have been chosen to be unambiguous, short and readily imageable, and vernacular names are used if they meet these criteria. The Turkic word "*barchan*" is preserved because it is now widely understood in western European languages to mean a crescentic dune opening down wind. The Arabic words "*nabkha*" and "*zibar*" are also unambiguous shorthand for "a dune formed around a plant", and "low, regularly undulating coarse-sand sheet", whereas the terms "*draa*" (meaning an arm in Arabic and applied in North Africa to any low ridge) and "*oghrourd*" are ambiguous or awkward and have been replaced by the short and readily imageable "mega-dune" and "star dune".

2 A simple classification of dunes.

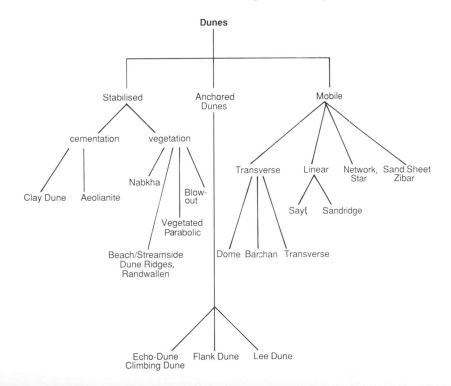

CHAPTER 24

Size and memory

Dunes span a width-range from a metre to well over a kilometre, and a height range from about 30cm to over 300m. An understanding of this variation must precede discussion of dune shapes, for, as in all geomorphology, explanation is closely connected to scale (Schumm & Lichty 1965). Eight processes control the size of dunes: sand supply, wind power, wind directionality, trap efficiency, upwind roughness, sand size, time and equilibrium form (hierarchy).

24.1 Size and sand supply

A dune cannot grow without sand. Dunes are large where there is copious supply, as in mountain basins fed by alluvial rivers, such as the basin–range deserts of Central Asia and the central and northern Sahara, or near biologically productive shelf seas as in the Wahiba Sands of Oman (Warren 1988a), or where sand has been generated by a long series of climatic alternations between phases dominated by either fluvial or aeolian activity. On coasts, the largest dunes occur down wind of the most rapidly accreting beaches (Short & Hesp 1982, Psuty 1989). Dunes are small or absent where there is little sand.

24.2 Size and wind power

No amount of sand will build a dune unless there is wind to move it. Other things being equal, dunes in high wind-energy environments are larger than those in low (Fryberger & Dean 1979). Some authorities assign the building of the largest dunes to the strong winds of the most recent glacial period. In general terms, bedform size is a function of the transport rate, itself related to sand supply and wind power (Rubin & Hunter 1982).

It appears, however, that the size–power relationship does not hold in two situations. First, the relationship seems to reverse at high windspeeds, where dune slopes are low, and where sand is in rapid movement. Secondly, high dunes in some sand seas accumulate down wind of zones of strong winds, where wind velocities decline. The highest dunes in both the Wahiba Sands in Oman and the Namib Sand Sea occur in this situation (Warren 1988a, Lancaster 1989c).

24.3 Size and wind directionality

Linear dunes in acute bimodal and wide unimodal wind regimes (as defined in Fig. 20.6) are bigger and more closely spaced than those in more unidirectional regimes (Fryberger & Dean 1979). A similar rule applies to star dunes, which are associated with multimodal regimes, and are the largest of mobile dunes (Lancaster 1989c). In the Great Sand Dunes in Colorado, it is believed that dunes have grown more in the more variable wind regime of the Holocene than they did in the less variable winds of the late Pleistocene when they began to form (Andrews 1981).

24.4 Size and trap efficiency

Some dune forms are better at trapping sand than others. Trap efficiency depends on three factors.

24.4.1 Dispersal
If sand supply is meagre, dunes are dispersed and they trap only a small fraction of the sand travelling across the desert floor between them. In a field of dispersed barchans in Peru (Fig. 23.24), streamers from the wings of barchans have a 50:50 chance of travelling 1,200m before they meet another dune (Lettau &

Lettau 1969). In another field of dispersed barchans, in Mauritania, the saltation discharge is 50–100 times greater than the bulk transport (Sarnthein & Walger 1974). Moreover, small, fast-travelling dunes in such dispersed dune fields have low probabilities of bumping into and merging with larger dunes. As sand supply increases, and dunes close up, they become more efficient at trapping sand, and intercepting small dunes, and are therefore able to grow to greater sizes. The transition from low to high trap-efficiency may be a sharp threshold.

24.4.2 Dune form
Continuous transverse dunes trap a high proportion of the sand transmitted over them. Very small quantities leave over linguoid sections and by suspension from brinks. In many transverse dune fields there is a downwind decrease in dune size, perhaps because upwind dunes are trapping large quantities of sand (Lancaster 1982a). Linear dunes in narrow bimodal wind regimes, in contrast, are efficient sand transmitters. In wider bimodal or more complex wind regimes, linear dunes trap more sand and grow to larger sizes. In effect, they become more like transverse dunes in their trap efficiency.

24.4.3 Obstacle size and the size of anchored dunes
The size of an anchored dune is related in part to the size of the obstacle to which it is anchored (Ch. 25.2.1). Dunes in the Isaouanne n'Irrarene in Algeria, where there is both a large supply of sand and large scarps, reach to over 350m high (I. G. Wilson 1973a). In the Taklamakan in China and in the Namib Sand Sea the highest dunes are near hills and scarps (Lancaster 1983b, Zhu Zhenda 1984). Other very large dunes occur in the Great Sand Dunes in Colorado near to the Sangre de Cristo Mountains (Andrews 1981). Near mountain massifs, katabatic/anabatic wind systems add to the complexity of wind regimes, and help to pile up sand (Lancaster 1983b).

24.5 Dune size, upwind roughness height and sand size

Theory suggests that, in a transverse dune system, the length of a barchanoid element (in the wind direction), and therefore its height, should be directly proportional to the upwind roughness height, z_0: greater roughness should produce higher dunes (Howard et al. 1977, 1978). Howard found few data to support this hypothesis, although he did note that barchan dunes at White Sands, New Mexico became larger down wind from the start of the dune field, perhaps in response to increasing upwind roughness (although perhaps also as a result of greater dune age). He noted that this relationship also implied that, if u_* were held constant, dunes with larger grain sizes would be smaller, since z_0 in saltation decreases at higher grain sizes. This is the opposite of Wilson's (1971) prediction.

24.6 Size and time

All other things being equal, the size of a dune is time-dependent.

24.6.1 Simple relationships
Dunes are frustrated from growth in many environments. A common example is shown in Figure 24.1, where the check is an ephemeral stream which periodically erases the dunes from its bed. By the same token, the largest dunes need thousands of years to accumulate. Estimates of the minimum time needed to build dunes over 100m high vary from 10,000yr in the Great Eastern Sand Sea in Algeria (Wilson 1972b) to 42,000yr in the Namib Sand Sea (Lancaster 1988c). Where growth takes so long, an additional requirement is continuing aridity, and this is met in the Namib Sand Sea (Lancaster 1982b, 1983a, McKee 1982). It may be significant that sand deserts that are more recent or have suffered more changes of climate, as in Australia, have smaller dunes (Wasson 1986).

Most large dunes have not grown continuously. Periods of higher windspeeds or of rapid input from

Figure 24.1 Small (1.5m high), low-memory transverse dunes on the bed of an ephemeral stream channel.

streams or beaches have alternated with hiatuses. There are signs of this kind of pulsed growth in the Strzelecki Desert in Australia (Wasson 1983a), in the Wahiba Sands in Oman (Warren 1988a).

24.6.2 Migration and dispersal

Dune size is also related to position in an evolving dunefield. A migrating dunefield is preceded by a vanguard of small, rapidly travelling dunes, followed by the main mass of large dunes (Porter 1986; Fig. 24.2). Moreover, if the input of sand to a dunefield is suppressed after the formation of a sand sea, the main mass of dunes moves down wind from the source, followed by a trailing edge of small dunes. This sequence has been proposed for the Algodones Dunes in California, which developed from sand on the ancient shoreline of an enlarged Salton Sea (Lake Cahuilla), but were abandoned as the lake retreated (McCoy et al. 1967). The McCoy model has been updated by Sweet et al. (1988) but, in respect of the present argument, its essentials remain unchanged. A similar model has been proposed for the dunes on the Ninety-Mile Plain in southeastern Australia (Coaldrake 1954), where there may have been a succession of pulses of dunes moving away from the shore, derived from pulses of sediment delivered to the beach. Yet other such sequences are in progress in Mauritania (Sarnthein & Walger 1974), and Oman (Warren 1988a).

24.7 Equilibrium form

24.7.1 Equilibrium and sand trapping

For all the generality of these first six explanations, they are not sufficient to explain the size of dunes. Dunes reach sizes above which they do not continue

to grow: viz. an "equilibrium" size.

The notion of equilibrium may be difficult to grasp at the scale of an individual duneslope (Ch. 23.3.1), but it is a vital notion in the study of dunes that adjust to longer-acting processes. At these longer scales, dunes are in "dynamic equilibrium" with their wind and sand-supply environment, which is to say that long-term and spatially averaged form is constant, with short-term fluctuations about this mean position. Empirical evidence of "dynamic equilibrium" is found in the peaked distribution of properties such as dune height and slope, the uniformity of morphometric parameters in any one dunefield (Fig. 24.3), and the maintenance of form as dunes migrate (Fig. 24.4; Finkel 1959, Hastenrath 1978, Lettau & Lettau 1978, Warren 1988b).

In the present context, the important implication of the notion of dynamic equilibrium is that dunes at equilibrium cease to be effective sand traps. Extra sand delivered to them is either transmitted down wind or goes towards the vertical accumulation of a sand body as dunes climb over each other. An equilibrium surface form is maintained during

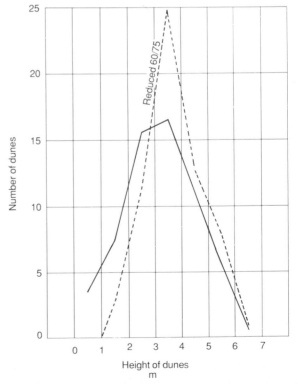

Figure 24.3 Gaussian distribution of dune height in a field of barchan dunes in Peru (after Hastenrath 1967).

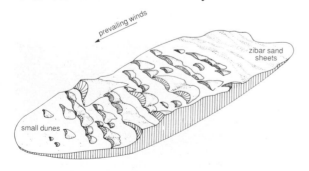

Figure 24.2 A model of dune form in a dune field, showing the variation of dune height in relation to position in the field (adapted from Porter 1986).

transmission or accumulation.

However, the argument needs to be taken further, for in many dunefields there is more than one "equilibrium" size: in other words there is a dune hierarchy.

N

——— 1949
- - - - 1983

Sabkha

0   0.5   1.0
|————————|
km

Figure 24.4 The maintenance of dune form during migration (after Al-Hinai & Moore 1987).

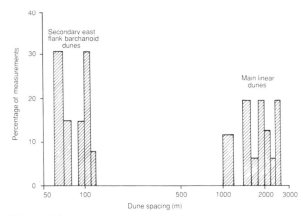

Figure 24.5 The coexistence of dune forms of different sizes in part of the Namib Sand Sea (after Lancaster 1981a).

24.7.2 Bedform hierarchies in aeolian dunes

A bedform hierarchy is the term applied to a bed on which bedforms of distinctly different size co-exist, with little or no overlap between the size groupings. Bedform hierarchies occur beneath water (in streams and on the bed of the sea), beneath air (including sand dunes and yardangs), and in exotic atmospheres such as those of Mars and Venus (Allen 1968b, Kennedy 1969, Yalin 1977, Greeley & Iversen 1985).

Four pieces of evidence suggest a hierarchy in aeolian dunes:

(i) Above all, the persistent coexistence of bedforms of distinctly different size (Figs 24.5 & 6).

(ii) The distinct upper size limit to dunes in any one area. The upper limit on size is very noticeable in the largest dunes, which, if mobile (not anchored), reach to about 100m in many sand seas. Wilson (1972b) maintained that large dunes in the Algerian sand seas had 1 million yr to grow, so

that they could have reached their actual sizes in about 10,000yr (as calculated from their size and the ambient sand transport rate); hence, he argued, some other mechanism was controlling their size.

(iii) The retention, on the surface of large accumulations of sand (such as Howard's "sand *sierras*" in Peru, 1985), of a hierarchy of smaller dunes.

(iv) The similarity of the size ranges of dunes and of mega- and meso-yardangs, themselves of remarkably uniform size in any one area. The sizes of yardangs can in no way be explained by sediment budget, nor do they seem to be related only to rock strength (Ch. 21.3.2).

The explanation of bedform hierarchies has taxed geomorphologists for decades, and is still obscure. Two hierarchy models for aeolian dunes are explored here.

24.7.3 Bedform hierarchies and atmospheric turbulence

The connection between subaqueous dunes and turbulence has been introduced in Chapter 23.2.3. It is not the only explanation of subaqueous bedform hierarchies, for some authorities see ripples as determined by bed roughness, but dunes by depth of flow (Richards 1980). However, this last form of explanation does not readily extend to aeolian bedforms (nor to subaqueous bedform hierarchies with more than two members), and Wilson (1972c) discarded it in favour of a turbulence model for the aeolian bedform hierarchy, which he saw as determined by different sizes of atmospheric eddy. His hierarchy had

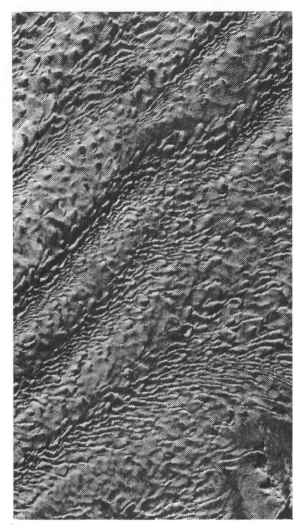

Figure 24.6 Coexisting dunes of different sizes (Yemen SPOT image © CNES).

three members: ripples, dunes and draa (sand bodies of the order of 100m high and 1km wavelength).

There is little argument about the distinctiveness of aeolian ripples, whether one accepts the ballistic, gravity-wave or the reptation hypothesis (Ch. 19.4), but both the existence and formative mechanisms of Wilson's dune and draa categories are more dubious. Although many sand bodies do contain only two size-groupings of dune, some contain more. In parts of the Wahiba Sands in Oman four groups coexist: ephemeral dunes; dunes with wavelengths of about 30m (most modern and active); dunes with wavelengths of about 500m (apparently stabilized); and dunes with wavelengths of about 2km (also stabilized, perhaps

relic). In other parts of the same dunefield, there are only two or three coexisting groups (Warren 1988a). The absence of Wilson's draa in the Australian sand seas has drawn remarks from many authorities.

Adherence to Wilson's bedform model would demand a variable number of nested sets of atmospheric turbulence to explain these anomalies. This requirement introduces the major problem with this explanation of the aeolian bedform hierarchy: the absence of appropriate atmospheric data, although some is emerging (Chs 20.4.5, 26.2.4).

These problems prompt the search for alternative or additional explanations of hierarchies.

24.7.4 Dune memory

Dunes of different sizes integrate the effects of different sizes of event; most dunes of the order of a metre high have experienced events only within the year, whereas 100m-high dunes have been subject to events for over tens of thousands of years. Small dunes can quickly be destroyed; the largest ones take thousands of years to destroy. Persistence allows different forces (such as more complex wind regimes) to come into play.

For subaqueous dunes this kind of behaviour has been described in terms of the reconstitution time, or the time it takes to form, re-form or to migrate one wavelength. Small dunes have short reconstitution times; large ones have long reconstitution times (Allen 1974). The idea has been extended to aeolian dunes (Fig. 24.7; Wilson 1971). Thomas (1975) believed there to be a logarithmic relation between the size of an aeolian bedform and its persistence: a meso-dune survived for more than 10 years; an embryonic dune survived between a day and a week; a "crest" could survive for less than one hour. The relationship was: $L = 36.47 \times T^{0.67}$, where L was the wavelength of the dune and T its minimum time of survival.

For aeolian dunes, with their potentially long reconstitution times, the concept is better expressed as "dune memory" (Warren & Kay 1987). Small dunes "memorize" the events of only a few hours, days or weeks; larger ones memorize the events in a series of annual cycles; the largest dunes memorize, or integrate into their form, the events of thousands of years, even of palaeowind regimes. Memory is not everywhere related in the same way to size: in an annual wind regime of gentle winds, quite small dunes have meso-memories, while in a variable regime of violent winds, even quite large dunes are re-orientated. Memory is therefore related to the ratio of the size of the

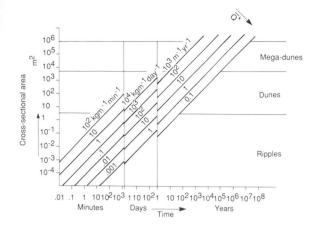

Figure 24.7 Reconstitution times (*x*-axis) of dunes of various sizes, at different sand discharge rates, Q^b (after Wilson 1971).

dune to the annual rate of sand movement (H/Q_{ann}). In other words, where wind regimes are constant over an area, size and memory are closely related.

The dune-memory concept raises many questions, for few of the relationships between dune size and wind regime are well understood. There are probably threshold sizes at which a dune can no longer be reformed by seasonal winds, and it begins to retain the imprint of the whole annual cycle. Another threshold apparently separates dunes that can or cannot survive as an annual cycle changes slowly under the influence of climatic shift.

24.7.5 Dune memory and dune hierarchy

Different climatic sequences, at different scales of operation, create suites of dunes with different hierarchical arrangements. The universal occurrence of low dunes, of the order of 1m high, is associated with short-period wind cycles, particularly those in daily cycles (Hunter & Richmond 1988). A second, larger (and also widespread) size group of dunes maintains its form for periods of over a year, adjusted to the almost universal annual cycle of climatic variation. However, the conformity of most dunes of this size to a distinct upper size limit suggests that there may be other controls, perhaps akin to the Wilson model.

For still larger dunes the interpretation of "memory" is more speculative. The largest mobile, unanchored dunes could be seen as responses to Milankovitchian climatic cycles, needing a long arid phase to reach their great size. The size of Wilson's mega-dunes in the Great Eastern Sand Sea would in this hypothesis be related merely to the length of an arid

phase, although it seems likely that some meteorological factor is also active in controlling the upper limits of these large dunes. Intermediate groupings, such as those in the Wahiba Sands might be related to smaller Milankovitchian cycles. A similar sequence, in which events of different magnitude and frequency determine size, has been proposed for subaqueous bedform hierarchies by Allen & Collinson (1974).

The dune-memory hierarchy is an additional rather than alternative explanation of the hierarchical arrangement of dune sizes. The closely defined upper limits of meso- and mega-dunes cannot be readily explained by an hypothesis, like the dune-memory hypothesis, that relies on rather irregular climatic cycles. Nor can a memory hierarchy explain regularities in spacing and form. Nevertheless, apart from its use in explaining some mega-dune hierarchies in desert sand seas, the dune memory hierarchy also usefully explains periodic beach-side dune ridges (Ch. 25.2.3a), and corresponds closely to the hierarchical scheme of nomenclature for strata and truncation surfaces devised by sedimentologists (Ch. 27.1.4).

24.8 Terms for dune size

24.8.1 The meaning of "size"

The size of dunes varies in three dimensions: height, width and length. Height:width (and/or spacing) ratios of many dune types (but perhaps not star dunes) fall within a narrow band (Breed & Grow 1979, Lancaster 1982c, 1983b, Twidale 1972, Mabbutt & Wooding 1983, Grove 1969, Lancaster & Greeley 1990). The relationships are specially consistent within individual dunefields (Lancaster 1988c). Dunes where height increases faster than spacing are ones where there is an abundance of sand and the dunes are piling up, as against escarpments; dunes where spacing increases faster than height (where there is a paucity of sand) exist on broad rocky pavements, as in parts of the western Sahara (Lancaster 1988c). These extremes are sufficiently rare to allow height and spacing to be discussed as one for most purposes. The length of a dune is less easy to define and is not often considered an important measurement. Size, therefore can most conveniently, and with only rare inaccuracy, be measured by spacing. This is fortunate, for height data taken from maps and remotely sensed imagery are rarer and less reliable than spacing data.

24.8.2 Size categories

The inadequacy of Wilson's terms for dune size

(ripples, dunes and draa) has been explained. Another approach to size-nomenclature is to term dunes "simple" where they have no superimposed dunes, and are therefore small; and "compound" where they are covered with simple dunes, and are therefore large (McKee 1979b, Breed & Grow 1979). This system has two problems as a system of size classification. First, it was developed for use on satellite images, where the smallest dunes cannot be seen. Compoundness therefore depends on the scale of enquiry, and is thus ambiguous. Secondly, as a description of size, the system is superfluous, for Breed & Grow themselves acknowledged a close correlation between the "compoundness" and size.

A system of terms for dune size can be based upon several criteria: first, a clear distinction in process is used to separate *ripples* from all other aeolian bedforms.

Dunes themselves can be divided according to two changes in behaviour within the "memory" continuum (Fig. 24.8). These breaks are similar to those identified between "geologic"; "graded" and "steady" in the system of Schumm & Lichty (1965).

Ephemeral dunes have memories of less than an annual wind cycle, and often of only a day or two. These are at the "steady" scale of Schumm & Lichty. They are the kinds of dune that reach 1m or more in height in a few hours if the wind is strong enough and the sand sufficient, and which may be destroyed as quickly if the wind direction changes (Coursin 1964, Verlaque 1958, Cooper 1958, 1967, Glennie 1970, Mainguet & Callot 1978, Reid 1985, Warren & Kay 1987, Warren 1988b). Ephemeral dunes are those on which slopes are in constant change. At any one site, dunes of several ephemeral scales may coexist. One group that is probably universal are those adjusted to daily cycles, as described by Hunter & Richmond (1988). Others may adjust to other local cyclic patterns, such as the passage of frontal storms, or longer-scale seasonal variation (dunes of summer; autumn or winter, or of the northeast or southwest monsoon, for example).

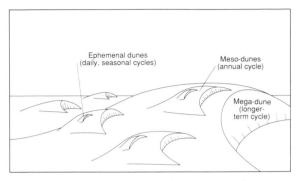

Figure 24.8 Terms used for "dune memory".

Meso-dunes "memorize" the ambient annual wind regime. To be able to do this, dunes must reach a size at which they can withstand minor seasonal or year-to-year changes in wind regime. Estimates for the memories of meso-dunes are fairly consistent: 25yr in Chad (Wilson 1972b) and 30yr in the Namib Sand Sea (Lancaster 1988c). Meso-dunes are the equivalent of Schumm & Lichty's "graded-scale" landforms.

Mega-dunes are those on which meso-dunes are superimposed; some are adjusted to contemporary wind conditions; some are "mega-memory" palaeofeatures. These are Wilson's (1972c) "draa". Estimates of the reconstitution time of mega-dunes have been given in Section 24.6.1. Mega-dunes are equivalent to Schumm & Lichty's "geologic-scale" landforms.

The actual sizes of dunes in these classes vary from site to site, depending on the nature of the wind regime, but it is a simple matter at any one site to classify dunes into these categories.

Finally, three further categories of feature can be distinguished:

Macro-dunes: massive accumulations of sand, for example where flow is impeded by mountain massifs (equivalent to Howard's (1985) "sand sierras"). These dunes create climbing sets of dune strata.

Dunefields and *sand seas*: these are defined and described in Chapter 28.

CHAPTER 25

Anchored and stabilized dunes

25.1 Definitions

Anchored dunes are dunes attached to objects which cannot be moved by wind (topography or vegetation); they do not move bodily, but they have mobile surfaces.

Stabilized dunes have been immobilized by cementation or vegetation after their formation; both surface and body are immobile.

The term "fixed dune" has not been employed here because it has been applied in the literature to both anchored and stabilized dunes. In the present senses, the term "anchored" was employed by Smith (1982), and the term "stabilized" by Howard (1985). Mobile, anchored and stabilized dunes are not everywhere unmistakable. Many anchored dunes experience changes in shape and size on seasonal or longer cycles, and the form of many apparently mobile dunes is affected by proximity to hills and scarps. Nor is there a sharp distinction between activity and immobility as dunes become progressively more stabilized.

25.2 Anchored dunes

25.2.1 Introduction

Two classes of anchored dunes are distinguished: those anchored by topographic obstacles such as hills, rocks or buildings, and those anchored by plants (phytogenetic dunes).

Size is controlled in the same way for both groups:
(i) The frontal area of the obstacle (height × width);
(ii) u_* (or better $(u_* - u_{*t})$, the excess of friction velocity over threshold friction velocity.

[Items (1) and (2) can be combined in a single formulation: the deposition rate around an object is a function of the Froude number ($Fr = u_*/(gh)^{1/2}$, where h is the height of the obstacle; Iversen 1983, 1986a).]
(iii) The volumetric concentration of sand in the wind.
(iv) Wind regime: because anchored dunes do not survive great changes in wind direction, they can grow to meso- or mega-size only in constant wind regimes, as in the Peruvian coastal desert (Howard 1985).

If they are continuously fed by sand, anchored dunes grow to an equilibrium size and shape at which any addition of sand increases the erosion rate, and any erosion decreases it. Variations in wind speeds and directions produce fluctuation about the mean size and shape (Howard 1985).

The form of an anchored dune is remarkably independent of the size of the parent obstacle. Dunes are anchored in two ways: (i) where flow around the obstacle forces sand streams to converge; and (ii) where sand is projected into zones of low wind velocity in the vicinity of the obstacle from where it cannot be re-entrained. Common types of anchored dunes are named in Figure 25.1.

25.2.2 Topographically anchored dunes

(a) Climbing dunes
Flow on the windward face of obstacles such as hills is described in Chapter 20.4.7. The shape of these patterns of flow and the supply of sand determine the shape of the resulting anchored dune (Fig. 25.1). If the slope of the upwind face of the hill is less than about 30°, there is no major blockage of flow, and no dune. With slopes of 30°–50°, a dune forms at an upwind distance of about 1.5–2 times the height of the hill, where the wind begins to rise over it. This "climbing dune" banks up against the hill as a sandy ramp. When the ramp reaches an equilibrium shape further sand is carried over it (Tsoar 1983). Climbing dunes characterize many hills and scarps in deserts

(Evans 1962) and on coasts where dunes encounter ancient marine cliffs (Brothers 1954).

(b) Echo dunes

If the upwind slope of the hill is over 50°, the leading edge of the dune is up wind of the scarp at about

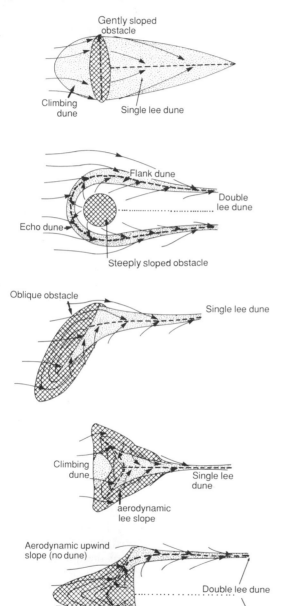

Figure 25.1 Types of anchored dune, related to the form of the obstacle (after Howard 1985).

three times the height of the scarp. A vortex forms immediately up wind of the scarp, and sweeps out a corridor in front of it. With large hills these corridors may be of the order of 1km wide. The oncoming wind and the return flow of the vortex between them create an "echo-dune" (Fig. 25.2). As the echo dune grows, both the vortex and flow over the crest of the dune are accelerated. A steady state is reached when the height of the dune is somewhere between 0.3 and 0.4 of the height of the obstacle. The vortex itself has then grown to a height near to that of the cliff (Tsoar 1983).

These simple two-dimensional models must be modified for scarps which are not at right angles to the oncoming wind. If, as is common, the scarp is sinuous in plan, then the form of the dune varies along its length. Crosswinds move sand laterally to positions where it can be carried as climbing dunes up gentle slopes, for example in gullies (Tsoar 1983). If the obstacle rests on an incoherent, erodible bed, the upwind eddy is strong enough to erode an annular hollow around an obstacle (Allen 1965).

(c) Flank dunes

Wind is accelerated around the flanks of obstacles, and can be very erosive in these positions. Commonly, a corridor is cleared between the obstacle and a flanking dune by the vortices shed down wind (Fig. 25.1). These effects are more marked in stratified flow.

(d) Sand shadows

The pattern of deposition and erosion in the lee of an obstacle depends on the flow patterns around it (Ch. 20.4.7), but dune topography around an obstacle is not a perfect reflection of flow around the uncluttered form, mainly because growing dunes themselves affect the flow (Iversen et al. 1990). For low obstacles, the horizontal-axis vortex in the lee dominates turbulence and patterns of erosion and deposition in the wake, but for higher obstacles the turbulence is dominated by vertical-axis "Kármán" vortices (Ch. 20.4.7), travelling down wind as zones of high turbulence, increasing u_* and sand transport above their ambient values, and encouraging deflation. Obstacles may also shed horizontal-axis vortices from either flank, and these add to the turbulence.

Because sand transport is a power function of u_*, small changes in u_* make great differences to the rate of erosion in the wake. In the lee of a small lava cone in southern California, the surface shear stress in the

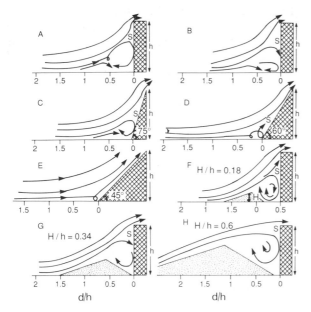

Figure 25.2 Types of climbing and echo dune, related to the upwind form of the parent obstacle (after experiments by Tsoar 1983b). A and B are two experimental runs in which the position of the stagnation point (S) was different. C, D and E show less recirculation in front of the cliff. In F, G and H an artificial dune shape was placed in front of the cliff, producing different circulation patterns.

ed out of a constriction, as between two hills or rocks. Speed-up through the gap increases the sand-carrying capacity of the wind, but expansion beyond decreases it, and a dune is deposited as a result. In extreme circumstances there may be an hydraulic jump in such a position, and, if there is sufficient sand in the wind, this is associated with an anchored dune.

(f) Lee dunes

Lee dunes are invariably linear in gross form, although they may consist of longitudinal groups of transverse dunes, and linear mega-lee-dunes are usually covered with transverse meso-dunes. Obstacles generate either one or two lee dunes, depending on the sand supply, ambient u_* and the geometry of the obstacle (Fig. 25.1). Obstacles that are narrow across wind generate only one lee dune. Obstacles with different geometries and orientations generate two lee dunes which enclose a sand-free lee hollow (Fig. 25.1). Hills of the same width might have one lee dune in a zone of strong sand movement, but two in one of gentler movement.

wake was 23% greater than its ambient level (Greeley 1986). Larger massifs reduce u_* values in their lee and prevent ingress of sand to lengthy sand-free zones or "sand shadows", for example in the West African Sahel (Urvoy 1933a, 1942). The sand shadow of the Koutous Massif in the southern Sahara extends for 270km (Fig. 25.3; Mainguet et al. 1980a).

(e) Sand drifts

"Sand drifts" was the term used by Bagnold (1941) for dunes formed where a sand-carrying wind expand-

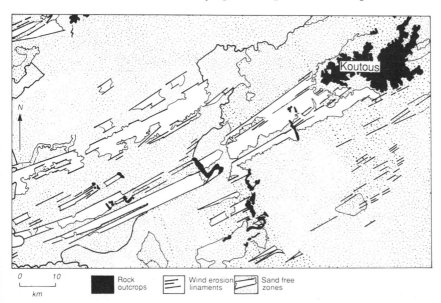

Figure 25.3 A large lee-shadow down wind of a hill massif, the Koutous massif in Mali (after Mainguet & Chemin 1981b).

355

Lee dunes can extend down wind at least to the point, several times the height of the obstacle, where the excess vorticity created in the wake of the obstacle declines to low values. This point is only up to 20 times the height of the hill (Howard 1985), but lee dunes have been observed which are much longer than this, implying that the dunes themselves generate flow patterns that elongate them. One such process may be an additional downwind component of sand transport imparted by the flow on the slopes of the dune (Howard 1985). Another process of self-regulation could happen on lee dunes that are large enough to induce temperature differences strong enough to produce sand-moving secondary flow. Lee dunes extending several tens of kilometres are not unusual. The longest seen by Breed & Grow (1979), the Draa Malichigdane in Mauritania, is 100km long. Large trailing dunes stretch up to 72km down wind of plateaux in the Western Desert of Egypt (El-Baz 1984a). When lee dunes reach an equilibrium size, smaller dunes may migrate over them.

Most lee dunes are straight linear features, but some, for reasons unknown, are sinuous (Howard 1985). Most are assumed to trail down wind parallel with the mean direction of flow, but obviously oblique lee dunes have been reported from several places (Fig. 23.3). In the confined flow beneath the coastal inversion in Peru, there may be an hydraulic jump in the lee of some hills, causing perturbations to the shape of the lee dunes (Howard 1985).

(g) Other lee features

Large and moderately sized hills generate lee waves if meteorological conditions permit. Kolm attributed the spacing of dunes down wind of hill massifs in Wyoming to lee waves, and Bowden (1983) believed that the cyclic pattern of mean grain sizes down wind of a hill in Tasmania might be linked to the wave-like variations in velocity in lee waves. Dunes apparently formed by wave-like wakes are shown in Figure 20.10. Others were recorded by Mainguet & Callot (1978).

Dunes may form immediately down wind of the crests of step-like, wind-facing scarps, but thereafter there is a zone of increased turbulence where there is no accumulation, extending about ten times the height of the scarp down wind (Fig. 20.17).

25.2.3 Dunes anchored by plants (phytogenetic dunes)

There are two types of dune anchored by plants:

dunes formed around plants (nabkhas and beach- and stream-side ridges), and those eroded into generally plant-covered surfaces (blowouts and vegetated parabolic dunes).

(a) Dunes formed around plants

NABKHAS Because there is a substantial scattering of bushes or clumps of grass in many, if not most, deserts, dunes associated with plants are very common indeed. Similar dunes develop on sandy beaches in many climates (Fig. 25.4). This kind of dune was described and analyzed from an early date (review in Hesp 1981), and has a rich ethno-geomorphological and neo-scientific terminology: "bush-mounds", "shrub-coppice dunes", "knob dunes", "phytogenetic hillocks", "dune tumuli", "rebdou", "nebbe" or "takouit". The concise, and now widely recognized North African word "nabkha" (or nebkha; Gautier & Chudeau 1908, Bourcart 1928, Dufour 1935) is preferred here. "Coppice dune" (Melton 1940) is an unfortunate misnomer (coppice is a wood-harvesting technique, or a small wood), and "bush-mound" is ambiguous as to the formative mechanism (as will be explained directly).

NABKHAS DISTINGUISHED FROM OTHER BUSH MOUNDS The ambiguity of the term "bush-mound" arises because some nabkhas are outwardly very similar to other mounds capped by bushes, but built by other processes. These processes include:
- the projection of rain-splashed particles into the areas beneath bushes; the accumulation is protected from further splash erosion, and a mound slowly accumulates;

Figure 25.4 A moderate sized nabkha, collected around a shrub, in this case silty material collected around a bush of *Salvadora persica* in southern Pakistan.

- wash-erosion between clumps of vegetation (Rostagno & del Valle 1988);
- frost heave, creating "palsa mounds" (Washburn 1979, Seppälä 1986);
- biological activity;
- and perhaps seismic activity.

The features most likely to be confused with nabkhas are the so-called **mima mounds**, named after Mima Prairie in Washington State (Dalquert & Sheffer 1942, Price 1949, Cox 1984, Cox & Gakahu 1986). Mima mounds are generally larger than nabkhas (up to 50m in diameter), and most contain gravel and pebbles (Cox & Gakahu 1986). They are best developed in damp hollows. The origin of mima mounds has been fiercely debated (Schwarz 1906, Krinitzky 1949). Although Krinitzky believed that their great regularity of size and spacing (which he illustrated beautifully) showed them to be aeolian, the most vociferous modern school believes them to formed by burrowing mammals, particularly rodents (commonly pocket gophers in North America) and/or termites (Cox 1984). Stones are sorted downwards within these mounds, as would be the case if termites brought finer material to the surface (Lovegrove & Siegfried 1986, Cox et al. 1987, Rostagno & del Valle 1988). Larger mounds, more certainly attributable to termites, but outwardly resembling large nabkha, are found only in savanna climates in Africa (Pullen 1979). The latest hypothesis is that mima mounds are formed by seismic disturbance (Berg 1990). Berg noted that mima mounds were found in alluvial lowlands liable to certain kinds of seismic disturbance in a wide range of climatic conditions.

Some mounds in semi-arid areas may have been created by seismic disturbance; and some in cold semi-arid climates by frost-heave. Splash and wash-erosional processes have certainly contributed to the formation of some predominantly wind-built mounds, and many are the setting for intense biological activity, for they attract subsidiary grasses and forbs, their soils contain more humus than nearby desert soils (Killian 1945, Coque 1962, Wood et al. 1978), and many are riddled with animal burrows. The mycelia and spores in some nabkha soils are strongly pigmented (Nicot 1955). Nabkhas, *sensu stricto*, must therefore be defined as mounds, most of whose sediment was windborn, whereas in mima (or other types of) mounds, most of the sediment has been carried by animals (or other non-aeolian processes).

NABKHA PROCESSES   The clumped vegetation of desert margins restricts but does not totally halt sand movement. Sand moving between the bushes feeds the dunes within them. Individual plants need to be higher than 10–15cm before they begin to trap sand effectively (Hesp 1979). If built of incohesive sands, nabkhas vary in volume as wind velocity and direction changes, being large when winds are light, and small when they are strong (Worral 1974, Warren 1988c, Hesp 1989). If built of finer, cohesive sediment, nabkhas are more permanent. The core of the nabkha grows to a height at which wind velocity is high enough for re-entrainment; this height is presumably greater in more cohesive sediments.

The form of a nabkha is a function of the size, density and growth habit of its parent plant (Capot-Rey 1957a, Monod 1961, Coque 1962, Barbey & Carbonel 1972, Mahmoudi 1977, Kosmowska-Suffczynska 1980). In the Saharo-Sindian belt from Mauritania to India, including central Asia, bushes of species *Calligonum, Calatropis, Zizyphus, Tamarix* and *Salvadora* build dunes up to 5m high and 10m in radius (Killian 1945, Coque 1962, Petrov 1962, Capot-Rey 1970, Saxena & Singh 1976, Warren 1988c). The largest nabkhas (**mega-nabkhas**) accumulate around clumps of trees. In the Wahiba Sands these can be 10m high and up to 1km long, and must have dune memories of hundreds of years (Warren 1988c). Other mega-nabkhas have been reported on the edges of the Oregon coastal dunes, where they have been termed "precipitation ridges" (Cooper 1958), and in Holland where they have been called "bordering dune ridges" or "*randwallen*" (Koster 1968).

As neighbouring nabkhas grow, flow is accelerated in the zones between, and these may suffer erosion (Bourcart 1928, Killian 1945, Coque 1962, Gross 1987). Hesp (1988, 1989) described a cycle which began with a nearly continuous vegetation cover and ended with a series of nabkhas isolated by erosion. When a plant on a silty, cohesive nabkha dies, erosion proceeds in earnest, and a form of yardang may be the result (Mahmoudi 1977, Quieroz et al. 1982).

Although anchored, most nabkhas are relatively ephemeral. When formed around annual plants they survive for only a season. Nabkhas around perennials may suffer changes in water table, rainfall, sediment supply or land-use (Gile 1966, 1975, Gibbens et al. 1983). Suslov (1961) conceived a cycle of nabkha activity in which sand accumulation raised the plant so far above the water table that it died, and the nabkha

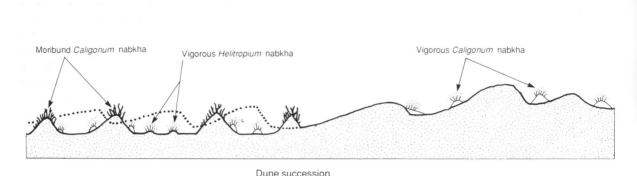

Former position of active dune field ← → ← Active dune field →

Moribund *Caligonum* nabkha

Vigorous *Helitropium* nabkha

Vigorous *Caligonum* nabkha

Dune succession

Figure 25.5 A nabkha succession, in which plants adapted to large amounts of sand movement are out-competed by those adapted to greater stability as a small dunefield moves down wind (after Warren 1988c).

was then eroded. Another form of "cycle" results from the onward march of dunefields: *Calligonum dumosum* in Oman can survive only on bare sand in active dunefields, since it extends its roots widely and shallowly beneath bare sand in order to glean water. As a dunefield moves on, the *Calligonum* is stranded on a more stable surface, and dies as it has to compete for water with invading *Heleotropium* bushes, which are adapted to more stable conditions. Large *Calligonum* nabkhas are replaced by small *Heleotropium* nabkhas (Fig. 25.5; Warren 1988c).

NABKHA MATERIALS Nabkhas can be built of sand, silt or pelletized clay. In many desert areas they are built of finer-textured material than nearby mobile dunes, perhaps because these sediments are more readily cemented by organic agents, and once trapped in nabkhas, are more difficult to re-entrain than sands (Suslov 1961, Snead 1966, Mahmoudi 1977, Khalaf et al. 1984).

UNVEGETATED ANCHORED DUNES ASSOCIATED WITH NABKHAS Climbing dunes, echo dunes, flank dunes and lee dunes commonly accumulate around clumps of vegetation. The lee dunes have been termed "shadow dunes", "elephant-head dunes", "pyramidal wind shadows", "sand drifts" and "tongue hills". With non-porous bushes, climbing-, echo-, flank- and lee dunes develop as around any impervious obstacle. If, as is common, the bush is porous to wind, the processes of echo- and lee dune formation are similar to those around porous fences (as built to protect installations from sand or snow). Iversen (1986a) showed that the form of such dunes was related to the

Froude number (the length dimension being the height of the fence, or bush). Higher winds created higher echo- and lee dunes. Although Willetts (1989) found that the crests of lee dunes created in this way stayed in much the same place irrespective of u_*, Iversen found that it moved down wind as $(u_* - u_{*t})$ increased (Fig. 25.6).

Lee dunes behind bushes can reach great lengths in constant winds, if sand supply is sufficient. They adopt a triangular shape, the base against the lee of the bush. The shadow dune usually has a sharp crest, created between the fields of action of alternating vertical vortices, shed off either flank of the bush (Hesp 1981, Clemmensen 1986, Hesp 1989). Plant width and lee-dune height have a direct relationship; the relationship with length is more complex, perhaps being related to u_* (Hesp 1981). Sandy echo- and lee dunes are more mobile than nabkhas themselves, and are readily re-orientated as the wind changes direc-

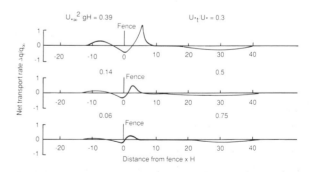

Figure 25.6 The position and size of lee and foredunes around a porous fence (or plant), as a function of the Froude number (after Iversen 1986a).

tion. Bedding in the lee dunes of nabkhas is discussed in Chapter 27.3.4.

BEACH AND STREAM-SIDE DUNE RIDGES On humid-climate coasts, sand blowing off the beach creates dune ridges parallel to the shore (Cooper 1967, Olson 1958a–c, Ranwell 1972, Hesp 1983, 1989, Sarre 1988a). Similar "stream-side dune ridges" are found beside ephemeral stream channels in arid and semi-arid areas (Wasson 1983a, Warren 1988a). Beach dune ridges have been called "vegetated foredunes", "frontal dunes" and "retention ridges".

Incipient dune ridges accumulate where the stems of grasses growing near the shore or stream channel slow the near-ground wind (Kuhlman 1958, Hesp 1983). They may be initiated by the extension of existing patches of vegetation, or where new plants germinate among litter on freshly exposed surfaces (Bird and Jones 1988). The cross section of incipient ridges is not unlike that of mobile dunes: sand transport rapidly falls off in the vegetation (within 3m), producing a gentle windward and a steep lee face (Hesp 1989). The rate of growth may be up to 1m yr^{-1} at first, declining over time (Stembridge 1978, Goldsmith 1985, 1989). The rate of growth and shape are related as much to such ecological factors as temperature, salinity, seasonality, rainfall and sandblasting, as to sand accretion (Ch. 16.2.2; Hesp 1983, 1989). Another factor is the rate of beach or channel progradation.

Incipient dunes are the locus of most sandtrapping on coastal and presumably streamside dune ridges (Sarre 1988a), but they are low-memory features, surviving only between storms (on beaches, Hesp 1989) or floods (on ephemeral streams). There is thus continual movement of sand between the beach or stream and the dune, some of it in cycles of only a few days, although more permanent movements occur in cycles of years or even decades. The interchange of sediment is a vital part of the maintenance of medium-term equilibrium on the coast (Leatherman 1979). On beaches, alongshore repeating patterns in incipient ridges may be related to cusp forms on the beach. On stream sides, the size of the dunes is related to variations in sediment supply along the channel.

Established dune ridges are of two sorts, both of them much longer-memory features than incipient ridges, integrating longer-term wind patterns. Both are commonly parallel to the coast, for the strongest winds are usually sea breezes. Along stream channels, they may reflect ancient courses.

The first type of established dune ridge occurs where there is strong sand supply. Massive, almost unvegetated, ridges may develop and move slowly inland, reaching well over 100m high and maintaining remarkably straight slip faces. An example is the Grand Dune of Pilat near Arcachon in southwest France, whose growth followed short-term littoral changes that released copious supplies of sand to the wind (Goldsmith 1985, Froidefond & Prud'homme 1990).

The second type is a more stable, vegetated ridge, with a more complex plant community than on incipient ridges, even including longer-memory plants such as trees. These usually have an extremely irregular topography. Successive subparallel ridges can form on prograding coasts or migrating stream channels. There may be some true slip faces, but most lee slopes are gentle, and quite commonly the windward slope is steeper than the lee slope. This may have two causes: (i) occasional undercutting by waves or ephemeral streams (Carter & Stone 1989); or (ii) a temporary or longer-term decline in sand supply from up wind, in which case the windward slope, although wind-eroded, maintains a steep slope angle because it is bound by plant roots. Established ridges can reach heights of 20m. They migrate slowly inland: at Braunton Burrows in western England, facing high winds off the Atlantic, established ridges have been moving at about 0.43m yr^{-1} (Kidson et al. 1989).

A pattern of established ridges may reflect a sequence of climatic events, being built of sediment released as the sea rose and released large quantities of sand by wave erosion (Ch. 22.2.6; Warren 1988a, Pye 1984b, Goldsmith 1985). However, Goldsmith (1985) pointed out that most were the products of only the past few hundred years, and that the formation of repeated ridges was more likely to be the product of periodic storm surges. As such, these ridges conform quite neatly to the dune-memory notions of Chapter 24.7.5.

As George Perkins Marsh appreciated even in 1864, beach-dune ridges in temperate climates, as in northwestern Europe, the Mediterranean, North America and eastern Australia, have been considerably interfered with by recreation, warfare (real and pretend) and aggressively protective management (Cooper 1958, Hewett 1970, Depuydt 1972, Ayyad 1973, Dolan et al. 1973, Gautier 1975, Kidson et al. 1989).

Rudimentary beach-dune ridges occur on some desert coasts, but the lack of moisture restricts the growth of the grasses and so of dunes (Warren 1988c). Sandy beach-dune ridges beside seasonal lakes in semi-arid areas behave much as temperate marine coastal dunes. Clayey beach-dune ridges adopt quite different forms, described below.

(b) Dunes eroded into generally plant-covered surfaces

BLOWOUTS   Blowouts, the prototype of the golf bunker, are elliptical sand-free hollows in otherwise vegetated dunes (Fig. 25.7). This is the sense in which the term is used by most authorities, despite its use by McKee (1982) and Lancaster (1989c) for hollows between bare dunes. Blowouts occur in continental and coastal vegetated dunes. They have also been called "amphitheatres".

On Dutch coastal dunes, blowouts are reported to be between about 10m and 30m long, with rare ones reaching 100m (Jungerius et al. 1981, Jungerius 1984). Modal size may vary slightly from year to year and is significantly different between sites, even over short distances (Jungerius & van der Meulen 1989). Continental blowouts are reportedly larger, reaching 100m in the High Plains of Texas (Hefley & Sidwell 1945) and 50ha in the Nebraska Sand Hills (Seevers et al. 1975). Blowouts everywhere have their longer elliptical axis parallel to the main wind direction. Blowouts are smaller on the upward-sloping portions of the original dune surfaces, and larger on dune crests and downwind slopes.

Blowouts are common on vegetated dunes on desert margins, as in the semi-arid High Plains of the USA (Hefley & Sidwell 1945, Smith 1965, Seevers et al. 1975), in semi-arid Australia (Ohmori et al. 1983), and in the eastern, vegetated part of the Namib Sand Sea (Lancaster 1983b).

Devegetation, which initiates blowouts, may happen in three ways:

(i) *Disturbance* Domestic, feral or wild animals, or people, may cause disturbance (Cooper 1967, Seevers et al. 1975, Oertal & Larsen 1976, Boorman 1977, Lancaster 1986b). Feral rabbits have been widely blamed for blowouts on temperate coastal dunes (Boorman 1989) and in inland dunes in Australia. There were certainly fewer blowouts on British coastal dunes after the introduction of the rabbit disease myxomatosis in the late 1950s (Ranwell 1960), but Watt (1937) found no evidence that rabbits had anything to do with inland blowouts in eastern England, and Rutin (1983) found it difficult to quantify their effects (especially of grazing and trampling). He did note that burrowing moved material down slope, exposed it to wind and water erosion, and caused slumping; the effects were uneven, being greatest on damp slopes. In the Cape Province of South Africa and in the Nebraska Sand Hills, blowouts have appeared since the introduction of grazing animals and cultivation (Seevers et al. 1975, Lancaster 1986b). Fire may be another mechanism for opening up vegetation to blowouts (Pye 1982a). On many coasts, blowouts are initiated or

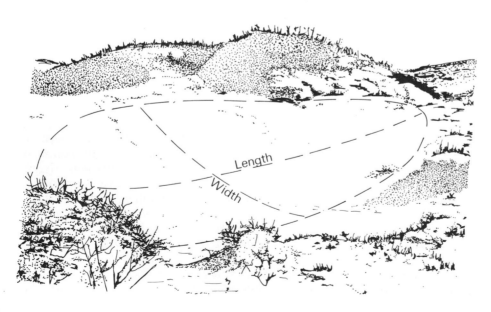

Figure 25.7 A typical blowout in Dutch coastal dunes (after Jungerius & Van der Meulen 1989).

enlarged by casual visitors.

(ii) *Localized aridity* Plants on dune crests are the first to suffer the effects of drought. Wind erosion occurs preferentially on dune crests kept dry by rapid drainage through the sandy soil (in moist climates), high evaporation rates in high winds, and the sweeping away of rain in high winds (Rutin 1983). Rutin found up to 30% less rainfall on the upper slopes of blowouts than in the bowls. This may explain why there are more blowouts in irregular terrain (more dry spots). Aridity may also explain why blowouts can appear even in completely protected plots (Jungerius & van der Meulen 1989).

(iii) *Rare high winds* Hurricanes and extreme storms can also remove vegetation and initiate blowouts, quite naturally (Bird 1974).

BLOWOUT PROCESSES Devegetation lowers the roughness height, z_0, and so increases surface drag and sediment entrainment. There is little evidence (or likelihood) that blowouts are initiated by "whirlwinds" or dust devils (Watt 1937), although they may be initiated by strong gusts (Jungerius et al. 1981). Most exposed patches revegetate naturally in a short period of time, only a minority becoming blowouts (Jungerius & Van der Meulen 1989). In those that do, wind is increasingly funnelled through a deepening devegetated area (Olson 1958b, Nordstrom et al. 1986, Rasmussen 1989). Windspeeds reach a maximum in the base of the hollow, and produce double vertical-axis vortices, sweeping sand from the base towards the sides (Hails & Bennet 1980). There is first a positive feedback process: erosion increases as the blowout deepens, but then negative feedback as the blowout expands, funnelling becomes less pronounced, and an "equilibrium" shape is reached. Enlargement may continue at a decelerating pace, but eventually sand cannot be carried up the steepened slopes, and there may be "spontaneous" stabilization. Erosion may also be halted if a hard or wet basement is uncovered by erosion (Hefley & Sidwell 1945).

Actively growing blowouts are net exporters of sand, for, once out of the blowout, sand cannot be as easily re-entrained for return from among the surrounding plants (Rutin 1983). The preferred direction of export is determined by the prevailing wind, or the dry-season wind. Erosion may be concentrated at the windward end, and the blowout may enlarge or migrate up wind (Jungerius et al. 1981, Hunter et al.

1983, Lancaster 1986a, Jungerius & van der Meulen 1989). If erosion is severe, a dune may form on the downwind side and advance over nearby vegetation, enlarging the area of bare sand down wind, and perhaps eventually advancing and expanding to become a parabolic dune (below). Blowouts in different vegetation types react in different ways (Watt 1937).

Blowouts have a distinctive life cycle. Most have a short life, revegetating quickly; some grow only to the modal size, and some continue to deflate to become large features, eroded to the basement. In a period of nine years, about half of a population of Dutch blowouts disappeared, quite naturally; artificially revegetated blowouts were, ironically, more vulnerable to reactivation than natural ones were (Jungerius & van der Meulen 1989).

Some blowouts originate when the wind exploits the breaking and burial of vegetation by surface water erosion and deposition (Jungerius & van der Meulen 1989). Where there is appreciable rainfall, water erosion may also play a part in the positive feedback enlargement of a blowout, helping to maintain bare sand. The processes include rain-splash, rilling and soil flowage (Watt 1937, de Ploey 1980, Rutin 1983, Yair 1990), all sometimes stimulated by the water-repellent qualities of some dune soils (Ch. 29.2.4d; Smith 1965, Jungerius & van der Meulen 1988, van der Meulen & Jungerius 1989). The lower slopes of most blowouts are adjusted more to water than to wind erosion (Rutin 1983).

VEGETATED PARABOLIC DUNES These dunes have been called "*garmada*", "U-shaped", "upsiloidal", "V-shaped", or "hairpin" or "dune plumes" (Fig. 25.8). The term "parabolic" was Steensrup's (1894). The parabola (if parabola it be) opens to the wind. When first described, this kind of dune was not linked to vegetation, and vegetation-free "garmada" were described, but the association with vegetation has now become axiomatic, even when there is little vegetation to be found (Anton & Vincent 1986). In the present discussion a distinction is made between vegetated parabolic dunes and unvegetated "zibar parabolic dunes", the latter being described in Chapter 26.5.2.

Vegetated parabolic dunes occur in three different environments:

(i) In cold climates, both ancient and modern (e.g. Nevyazhsky & Bidziev 1960, Galon 1969, Filion & Morisset 1983, Kolm 1985, Castel et al. 1989).

(ii) On temperate and wet-tropical coasts (Cooper

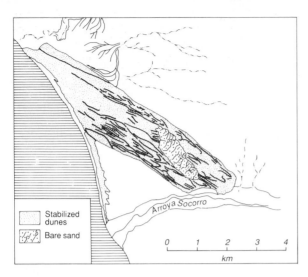

Figure 25.8 Vegetated parabolic dunes on the California coast (after Cooper 1967).

1958, Bigarella 1979, Story 1982, Pye 1982a, Aimé & Penven 1982).

(iii) On the edges of warm deserts, as both active and stabilized features (Grove 1958, McKee 1966, Verstappen 1968, Bourne et al. 1974, Bowler & McGee 1978, Lancaster 1983b). The largest body of vegetated parabolic dunes in the world occurs in the Thar Desert of India and Pakistan (Meddlicott & Blanford 1879, Wasson et al. 1983).

The restriction of this kind of parabolic dune to the semi-arid areas demonstrates that vegetation must play an important rôle in its formation, although the wetness of the sand, and springs from dune water tables could also contribute (Cornish 1908, David 1981). Other factors are: wind velocity, for it appears that strong winds are needed to overcome the increase in resistance offered by vegetated surfaces; and sand supply, for where there are large supplies of sand in lightly vegetated landscapes, vegetated parabolic dunes give way to mobile dunes (McKee 1966, Nevyazhsky & Bidzhiev 1960).

Vegetated parabolic dunes are thought by most authorities to be enlargements of blowouts (above; Fig. 25.9). Among trees, parabolic dunes have more lobes than in grassy areas, perhaps because windflow is more disturbed (Filion & Morisset 1983). The advancing dune grows in size as it feeds on sand from the erosion of the underlying sediments, and this slows it down (according to the laws of dune migration). It may also decelerate as erosion exhausts the supply of sand in the hollow of the dune. As the dune advances, trailing arms are left on either side, and the lower activity on these allows plants to recolonize and eventually to fix them. In almost all cases where vegetated parabolic dunes have been described, there are signs of the migration of a series of parabolic dunes down the same path, building up a series of concentric "nested" dunes (Fig. 25.10). These testify to alternating periods of activity and

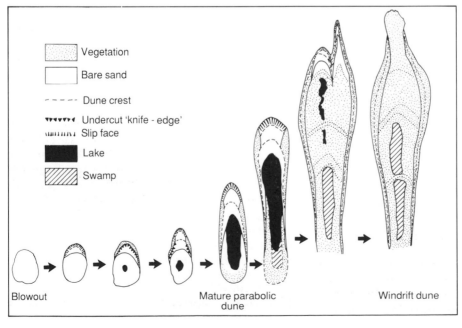

Figure 25.9 Development of a parabolic dune (after Pye 1982a).

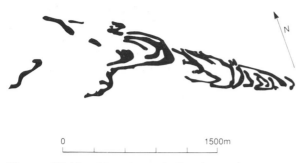

0   1500m

Figure 25.10 Nested parabolic dunes in northern Sweden (after Seppälä 1972).

inactivity. The fact that most vegetated parabolic dunes now appear to be inactive may indicate that we are in a period of generally low wind activity in semi-arid and coastal areas.

Few measurements have been made of the movement of vegetated parabolic dunes, but those that have show a wide range of celerities. Heavily vegetated parabolic dunes in northern Australia appear to move at only 0.05m yr^{-1} (Story 1982), and Cooper's (1958) and Pye's (1982a) direct measurements and McKee's (1979c) review also showed most to be slowly migrating (at about 2m yr^{-1}), but Hesp et al. (1989) measured maximum rates of 13m yr^{-1} in coastal Cape Province.

Story (1982) in northern Australia, and Kolm (1985) in Wyoming, could find no consistent relationships between dimensions and possible formative factors, but in places with sharp local contrasts in wind velocity, as in the vicinity of mountain passes, stronger, more unidirectional winds certainly generate longer ("long-walled" or "elongate") parabolic dunes (Brothers 1954, Short 1988, Pye 1982b, 1985, Filion & Morisset 1983, Gaylord & Dawson 1987). The formation of vegetated parabolic dunes may require a wind regime above a certain threshold of constancy, such as one with sea breezes. In Wyoming parabolic dunes (perhaps zibar parabolic dunes) are best developed in the divides between mountain ranges where wind directions are constant (Kolm 1985). In the Thar, vegetated parabolic dune axes are aligned to the strong and constant winds of the southwest monsoon (Wasson et al. 1983).

"Imbricate" or "multi-lobate" parabolic dunes with overlapping cusps, and shorter arms are said to form in more variable wind regimes (Filion & Morisset 1983, Hesp et al. 1989). Active vegetated parabolic dunes may have angular apices, as in the White Sands of New Mexico (McKee 1966), while less active ones,

as in the Thar, may have rounded apices (Wasson et al. 1983). In the Thar, where the sand comes from local alluvium, erosion in the parabolic hollows reaches down to harder, less erodible strata, and produces parabolic dunes that are clustered, with frequent superimposition, giving characteristic rake-like patterns (Fig. 26.24; Breed et al. 1979b). Melton (1940) believed that some vegetated parabolic dunes were constructional, but that longer, straighter ones were erosional. This notion has been incorporated into the wind-rift hypothesis of linear dune formation (Ch. 26.2.5).

The explanation of the uniquely large area of vegetated parabolic dunes in the Thar may lie in a unique postglacial climatic history (Ch. 21.1.2).

There are still some mysteries about vegetated parabolic dunes. Why, for example, are some U-shaped and some V-shaped? What factors control the differences between erosional and constructional activity (McKee 1979c)? What controls the size of parabolic dunes? What are the thresholds of vegetation cover, sand wetness, sand supply or sand size that trigger parabolic dune development? Most large vegetated parabolic dunes now seem to be inactive; if so, why?

25.3 Dunes stabilized by cementation

25.3.1 Introduction

Stabilized dunes fall into two distinct categories: those stabilized penecontemporaneously with formation (with no marked change in climate); and dunes stabilized following a change in climate. This section describes only the first of these categories. These are mostly dunes stabilized by cementation of carbonate- and clay-rich sediments, in which stabilization occurs soon after emplacement, as water penetrates and dis-aggregates or dissolves material and redeposits cements. Although many of these stabilized dune are palaeo-features, they are capable of formation in arid or semi-arid climates. This distinguishes them from dunes stabilized by vegetational colonization, which in most cases has followed a change in climate. These vegetationally stabilized dunes are described in Chapter 29.2.

25.3.2 Lunettes and related dunes

(a) Introduction
When the clay content in dune sediments rises above

15–20%, dunes can be stabilized by disaggregation and cementation, and therefore assume forms somewhat different from those of incohesive sand (Bowler 1973). Clay dunes can take several forms: long ridges parallel to the coast, similar to vegetated beach ridges (Coffey 1909); linear dunes; and "lunettes" (and related dunes), described in this section. Clay dune formation depends on the prior formation of sand-size pellets of clay, viz. pelletization.

(b) Pelletization

Primary particles of silt and clay travelling in wind are not deposited as dunes, except in a broad sense. Knowing this, early workers on clay dunes speculated that the clay in clay dunes had been carried by spray (Hills 1939), but this hypothesis was soon rejected in favour of "pelletization". This is the aggregation of silt and clay into sand-size particles, in which form they can be moved by the wind in reptation, saltation and creep (Chepil 1957a, Teller 1972). They then travel with companion sand in variable mixes, but the pellets seldom travel far before they are disaggregated by mechanical bombardment. Although the pellets have a density of about 1.7 (cf. 2.37 for quartz), threshold velocities are not very different from those of quartz sand (Skidmore 1986b).

Clay pellets can come from three sources: alluvial deposits, especially shortly after flooding; bare soil surfaces; and saline depressions. Pellets may even travel as sedimentary particles in low-energy arid-zone streams, from the alluvium of which they may then be blown onto dunes (Nanson et al. 1986), but most pellets come from sabkhas and playas.

Pelletization can be achieved in several ways:
- by chemical bonding in smectite clays;
- by disaggregation and then bonding with salts;
- by the blistering and break-up of a salt crust;
- by the cracking and curling of non-saline crusts (Price 1963);
- by ice;
- by algae or bacteria (Bowler 1983, Wasson 1986).

Some of these process may be sequential, as when the break-up of a crust releases pellets aggregated with salts. Salts play the dominant rôle in Australian pelletization, the most common salt being halite, although gypsum or thenardite (a sodium sulphate) also play a part (Bowler 1980, 1983). Carbonates dominate the process of pelletization in inland semi-arid areas; and sulphates and chlorides become more important in dryer areas, being replaced by magnesium and potassium salts in the driest areas (Bowler 1986). Because of the importance of salts, pelletization is active in many coastal environments, as in Texas (Huffman & Price 1949, Coffey 1909, Rhodes 1982). The pelletization process usually has a strong seasonal rhythm: it is most pronounced in dry months with strong winds.

(c) Clay-dune sediments

Once on the surface of a dune, pellets are soon disaggregated in dew or rain, and the clays and silts bind together as a surface crust. Sediments within the dune disaggregate more slowly, the rate depending partly on the permeability of the sediment, and partly on the calcite content, for calcite recrystallization can disaggregate pellets (Dare-Edwards 1983). The sediment is then slowly bound together *en masse* by salts (Andrews 1981) or chemical bonds between clay particles. Mobile salts and ions, particularly chloride, are rapidly leached out of the dune, leaving less mobile salts such as the carbonates (Boulaine 1954, Macumber 1969).

The grain size and mineral composition of lakeshore dunes depend on local parent materials, be they clay, silt or sand, halite or gypsum. In the Kalahari and in Queensland, for example, most lakeshore dunes of lunette form are sandy (Lancaster 1978b, Pye 1982b). In the Murray–Darling lowlands of southeastern Australia, where parent materials are finer, so are most of the lakeshore dunes (although portions of many of even these are sandy). Sandy lakeshore dunes behave much as coastal dunes anywhere. They may be broken up by blowouts, and on the leeward side by a series of linear dunes trailing down wind.

(d) Lunette morphology

As clay content increases, mobility and dune- and bedding-slope angles decline. Large transverse dunes down wind of saline alluvial washes in the Strzelecki Desert in Australia with sediments composed of a mixture of quartz sand and up to 60% clay pellets have indistinct plan shapes and subdued slopes. When the clay content rises even further, adhesion of the clay does not allow high-angle bedding, except in high winds and seasonally dry conditions (Wasson 1983a). Dune bedding is destroyed by soil formation (Jauzein 1958, Trichet 1963, 1968, Benazzouz 1987).

Clay contents in clay dunes vary from 23% to 77% (Bowler 1973). The clay is commonly mixed with salts, particularly gypsum in the North African lakeshore dunes. The gypsum may be translocated during the consolidation of the lakeshore dune and form a dense crust or gypcrete (Benazzouz 1987).

In the cross section of clayey lakeshore dunes, the upwind slope is commonly steeper (8°–9°) than the downwind one (2°) (Bowler 1973), although symmetry is also common (Benazzouz 1987). Steepening of the upwind slope is commonly achieved by wave action. Bowler described many lakeshore dunes that had composite facies: there were phases of sandy bedding and others of clay, and symmetry altered accordingly. Many clay- and salt-cemented dunes, such as those in Colorado and in the Mojave, are eroded into yardangs after cementation (Blackwelder 1934).

Lakeshore dunes vary vastly in size. Some reach 50m high, and 20km long in parts of Tunisia (Coque 1979b). In plan, the most distinctive lakeshore dune shape is the **lunette**, down wind of an eroding pan (Fig. 21.11; Hills 1940, Bowler 1983, Goudie & Thomas 1986). The classic half-moon shape of the lunette, seen well in southeastern Australia, is more to do with the smooth log-spiral form of the attendant pan's wave-formed shore than with any purely aeolian process (Ch. 21.3.3).

Not all lakeshore dunes are smoothly crescentic, for the lakes from which their sediment originates are not all pans, in the narrow sense of the word. Some southern and northern African lakeshore dunes, loosely called lunettes, are V-shaped or irregular (Fig. 21.12; Coque & Jauzein 1967, Goudie & Thomas 1986, Benazzouz 1987), although others form smooth curves, as in the classic Australian case (Jauzein 1958). The difference may be due to the origin of the lakes, some originating as wave-formed pans, others from tectonic depressions (Perthouisot & Jauzein 1975, Coque 1979b). In coastal areas, as on the Gulf coast of Texas, most "lakeshore" dunes hug the edges of irregularly shaped tidal lagoons (Coffey 1909, Price 1963). Smoothness of form is linked to more frequent filling of the lakes, giving longer periods of wave activity (Bowler 1986).

Clay and silt sheets are described in Chapter 22.1.5d. The reasons why some clay is deposited in lunettes and some in sheets is obscure. It may be that, where salinity is low, pelletization is incomplete and the clay travels further because more of it is carried as dust than as sand-size pellets. It could also be that aeolian clay, like silty loess, requires a vegetated surface to allow it to accumulate, so that it characterizes the dispersed vegetation pattern of semi-arid areas rather than concentrated vegetation patterns of more arid environments.

(e) Clay-dune environments

Clay dunes therefore have very precisely defined environmental requirements. There must be a lake, either on the sea coast or inland. The very form of many of the upwind sides of lunettes indicates wave action. Strong seasonality is helpful, particularly where there is a marked dry season in which there is a fairly constant wind, strong enough to entrain the pellets, followed by a wet season in which the lake refills, and the dune is bodily cemented together (Bowler 1968). If there are to be enough clay pellets to build a dune, lake drying can be no more than seasonal since, if it were more permanent, the lake floor would be colonized and stabilized by plants. On the other hand, the water table has to retreat far enough below the capillary fringe to allow the surface to dry out thoroughly. If the climate is too dry, some of the pellets may be rapidly broken down by abrasion. Even in a wet environment such as the Gulf Coast of Texas, Price (1963) estimated that three-quarters of material eroded from saline depressions travelled beyond the neighbouring clay dune as dust. Vegetation is needed to trap the saltating pellets. Only if the wind is fairly constant, and if the pellets are in sufficient quantity, can they then be blown onto a dune. Salts are another requirement, although their absence is not absolutely limiting.

Many features of clay dunes are clearly the result of multiple phases of formation, themselves commonly the result of differing environmental conditions. Many lunettes and other shore dunes are arranged in multiple ridges, with small lunettes nested within larger ones (Boulaine 1954, Bettenay 1962, Lancaster 1978b, Bowler 1983, Goudie & Thomas 1986). The different ridges often have different characteristics, some clayey, some gypseous, some sandy. Even sections through individual ridges show variations in character. Palaeosols, in many cases calcretes, may be found within the lunette. The alternation of environments may, in places, be no more than seasonal: in the wet season, when the lake is full, a sandy beach is formed and is deflated to supply a sandy shore dune, in many cases with a slightly different orientation to those formed in the dry season. In the dry season, pelletization proceeds and clay dunes are formed. But the alternation can also be on a longer rhythm, in which clay dunes, salts, and saline mollusca denote arid phases; sandy shore dunes and freshwater mollusca denote wet ones (Bowler 1983, Goudie & Thomas 1986).

These and other features of lunettes show that most

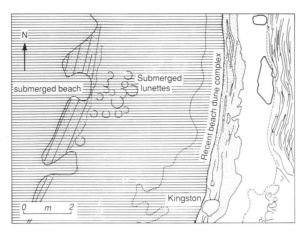

Figure 25.11 Lunette dunes submerged by rising post-glacial sea levels in southern Australia (after Short 1988).

are palaeo-features. In Australia, some have been buried by rising sea levels (Fig. 25.11; Short 1988), and many now have well developed soils (Stephens & Crocker 1946). The Australian lunettes began to form about 36,000yr ago, as a long lacustrine phase declined, and ceased to form about 17,000yr ago, as the climate became slightly wetter again (Bowler 1983, Wasson 1983a). Although some North African lunettes are active, Coque (1979b) believed that most of them too were inherited, but from an environment not unlike today's. In this region, clay dunes developed as drainage systems were disrupted when the climate became drier.

25.3.3 Aeolianite and cemented gypsum dunes

(a) Definition and distribution
Aeolianite (eolianite) is now the accepted term for aeolian sands that have been lithified (Fig. 25.12). Strictly speaking, aeolianites can be lithified with various cements, but in practice most aeolianites have been cemented by carbonates (calcium and to a lesser extent magnesium), and the term is often confined to these rocks, although "calcarenite" is more precise (Gardner 1983). Carbonate aeolianite is a formation almost, but not entirely, confined to high-energy tropical coasts. Tropical seas, especially those fed by upwelling, are necessary for the production of sufficient carbonate material, and high energy is required to move the carbonate sand shorewards and along shore, whence it can be blown onshore (Gardner 1983, McKee & Ward 1983). Carbonate sands survive

for only a few years in very high-rainfall, well vegetated areas.

The carbonate content of aeolianites has wide variations. In Israel aeolianites have a high quartz content (Yaalon 1967), but in Bermuda they have very little quartz (McKee & Ward 1983). The constituents of the carbonate sand itself are usually marine: mostly shell fragments, foraminifers and oöids. The Indian aeolianites, which consist largely of miliolid foraminifers, are beautifully evenly grained sandstones, much prized as a building material (Goudie & Sperling 1977). Sands may be mixed in varying quantities with dust, itself often carbonate-rich, added since deposition (McKee & Ward 1983).

Most aeolianites are further confined to a narrow coastal zone, because of either rapid lithification or the limited effectiveness of sea breezes. In the few places where onshore winds are strong, where the sea is productive and where aridity prevails (or prevailed), carbonate sands reach hundreds of kilometres inland, as in the Thar Desert in India (Goudie & Sperling 1977), and the Wahiba Sands of Oman (Glennie 1970, Gardner 1988, Warren 1988a). In the Wahiba Sands, sands with nearly 50% calcium carbonate are found 150km from the coast and form most of a sand body that is some 12,000km^2 in area (Goudie et al. 1987; Fig. 25.13).

When they move far inland, carbonate dunes adopt the full range of dune shapes. On vegetated coasts, however, they are either beach ridges, or, further inland, parabolic dunes (McKee & Ward 1983). Carbonate coastal dunes may interfinger with quartz sand dunes inland (Semeniuk & Glassford 1988).

Figure 25.12 Aeolianite in the Wahiba Sands of the Sultanate of Oman (photograph by R. Turpin).

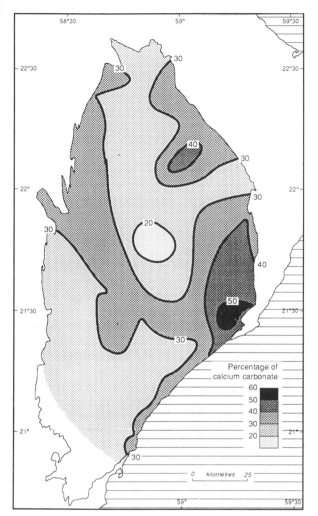

Figure 25.13 The distribution of carbonate in dune sands in the Wahiba Sands of the Sultanate of Oman (courtesy of D. K. C. Jones).

(b) Lithification

Understanding the lithification of aeolianite presents some problems. Although most of the cement must come from dissolved aragonite, the amount of cement and the density of aeolianite are not always consistent with this process. Nor do inputs of sea spray or carbonate dust entirely account for all the cement that is needed (Gardner 1983). There are also questions about the location of cementation: was it above, below or at the water table? If it was in the vadose zone, which seems most likely, then percolating rainwater must have dissolved the carbonate sand, and cemen-

tation must have occurred when the solution became supersaturated for short periods. Plant roots could have played an important rôle in this process (Ameil 1975). Cementation might have inhibited leaching, and this would have slowed down further weathering (Gardner 1983). Calcium carbonate may replace quartz in some high-pH environments, and total carbonate content can therefore increase in the early stages of weathering (Amiel 1975).

However, vadose cementation was cannot have been the only process, for lithification appears to have occurred at the water table itself, and complex layered zones of lithification (in effect calcretes) have formed in some profiles by repeated rises and falls of the water table (Schenk & Fryberger 1988, Semeniuk & Glassford 1988). In wet environments, there may be a later stage in which carbonate is slowly leached from the aeolianite. The rate of lithification is variable, depending on rainfall, carbonate content, water-table position and fluctuation, among other things (McKee & Ward 1983).

Many aeolianites were rapidly lithified after deposition of dunes in climatic phases with stronger winds or greater production of sand. The association of many deposits with cooler, drier climates in the Pleistocene is indicated by their presence below the present sea level, and their planation in places by marine terraces (Gardner 1983, 1988). The attribution of coastal aeolian phases to low or rising sea-level phases has been discussed in Chapter 22.2.5.

Because of rapid lithification, aeolianites are good sources of palaeo-environmental information, especially in wet tropical environments such as Bermuda, where aeolianite phases are separated by well developed ancient soils (McKee & Ward 1983). In much more arid Oman, there are fewer marker horizons, but it is nevertheless thought, on the evidence of degrees of lithification, that the aeolianites preserve evidence of at least three former major phases of sand movement before the Holocene (Gardner 1988). Some aeolianites, as in South Australia, may even be pre-Pleistocene (Milnes & Ludbrook 1986). Aeolianites generally preserve good evidence of palaeowind directions (Fig. 25.12; Glennie 1970, Yaalon & Laronne 1971, Gardner 1988).

Gypsum dune sands also fuse to form cemented deposits, as in southern Tunisia (Coque 1962, Watson 1983).

CHAPTER 26

Mobile dunes

This chapter examines the three-dimensional shapes of mobile dunes. While the explanations rely on an understanding of processes on individual dune slopes, as examined in Chapter 23, many new factors have to be considered as the scale of enquiry is enlarged.

Three primary groups of dune are described: **transverse dunes** (including barchans, dome dunes and continuous transverse dunes), **linear dunes** (including sayf dunes and sand ridges) and **star dunes** (including dune networks). There are also descriptions of three subsidiary dune types: **oblique dunes**, **sandsheets** and **zibars**. Distinctions are made according to form, growth and migration. Identification of the elemental types is uncomplicated in most cases, at least as regards form. Growth and migration data are harder to obtain. The great range in character of global wind climates, nevertheless, produces many intergrades.

26.1 Transverse dunes

26.1.1 Definition and distribution
Transverse dunes (Figs 23.5, Box Fig. 1, 24.1) have low length:width ratios and are conspicuously asymmetrical across their shorter axis, with gentler slopes on the windward than on the lee side. They are usually more densely spaced than linear dunes (smaller corridor-width:dune-width ratios).

A promising criterion for discriminating transverse and linear dunes is the "sinuosity ratio" of Hunter et al. (1983): the ratio of the distance along the crest to the straight-line distance between the ends of the dune. Hunter's Oregon dunes, which would here be termed "linear", had sinuosity values of 1.11. A transitional value between transverse and linear dunes might be about 1.3. Only Hunter et al. (1983) have measured this ratio, yet some form of quantified criterion is crucial to arguments about the behaviour

of linear and transverse dunes, for example to the claim that linear dunes migrate transversely (Hesp et al. 1989). The distinction needs to be more carefully drawn in future studies.

In a transverse dune, the mean direction of sand flow is at right angles to the crest, when averaged over the annual wind cycle. These dunes grow to and maintain equilibrium shapes and sizes; most migrate in a direction at right angles to their longer axis.

Transverse dunes are the fundamental dune form, and all unanchored, ephemeral dunes are transverse. Transverse meso-dunes survive only in relatively constant winds, although directional variability can be quite wide before a recognizably transverse form is destroyed (Fryberger & Dean 1979, Wasson & Hyde 1983b). The basic transverse dune becomes more and more overlain with small networks as the wind system becomes more complex. In moderately complex wind regimes, side winds create ephemeral dunes on top of the meso-dunes, and "aprons" at the base of the slip face (Fig. 23.1).

Transverse meso- and mega-dunes cover roughly 40% of all terrestrial sand seas (Breed et al. 1979b). They are common in the Northern Hemisphere, particularly in China (the Taklamakan and Ala Shan; Mainguet & Chemin 1988), stabilized dune fields in North America such as the Nebraska Sand Hills (Warren 1976a), and the Sahel, particularly northeast of Lake Chad (Peignot 1913). Transverse meso-dunes are also common on coasts, where the strongest winds come off the sea, and where there are reliable seabreezes (Lancaster 1982b, 1983a, 1985b). They dominate the sand seas of Mars (Lancaster & Greeley 1990). In large annual drift potentials, Fryberger & Dean (1979) found transverse dunes to be associated with more directionally concentrated winds than where drift potentials were lower, presumably because in highly mobile environments even cross winds are strong enough to distort the dune shape.

26.1.2 General three-dimensional characteristics

(a) Linguoid/barchanoid morphology

All transverse dunes adopt a linguoid/barchanoid three-dimensional shape (Fig. 23.1), a shape they share with other transverse bedforms, particularly subaqueous ripples (Allen 1968b). As with ripples, curvature is variable (Breed et al. 1979). The curvature of the lee slope of transverse dunes is undoubtedly related to dune height, being of larger radius in higher dunes, although there are no measurements of this relationship (Rubin & Hunter 1982). In some fields, as in the Jafurah Sand Sea in eastern Saudi Arabia, the linguoid segments are composed of coarse sand (they are essentially zibar parabolic dunes), and the barchanoid segments of fine sand.

No model of transverse dune behaviour has been tested in the field, although there have been tests in flumes of comparable subaqueous bedforms. Wilson's (1972b; Fig. 26.1) model of the three-dimensional behaviour of transverse dune pattern was based on these better tested models (Allen 1968b, Raudkivi 1976). Wilson's model produces three-dimensional shape with pairs of longitudinal (flow-parallel) vortices. These move some portions of the transverse ridges forwards to become linguoidal, and leave others as barchanoid sections. Wilson believed that the wavelengths of the transverse and longitudinal forms were about equal, creating equidimensional reticulation of dune landscapes, although this is not a necessary part of the model. In Wilson's model, the oblique steps created in the reticules induce strong oblique vortex flow in their lee, and these override the original vortex pattern, and create an *en echelon* arrangement of barchanoid and linguoid elements in subsequent transverse dunes (Fig. 26.1). Flow is convergent at the barchanoid sections and divergent at the linguoids.

In an alternative model, proposed by Yalin (1977), the barchanoid sections of transverse dunes are created by spheroidal eddies formed by the breakdown of cylindrical transverse eddies. If the analogy with subaqueous dunes is accepted, greater sinuosity should be associated with stronger winds (Allen 1968b, Rubin & McCulloch 1980).

(b) Equilibriation

The three-dimensional transverse pattern described above contains feedback mechanisms.

(i) All parts of the dune must migrate at the same rate, implying constant rates of erosion on slopes at different angles to the horizontal and to the oncoming wind (Howard et al. 1977). This is apparently achieved by variations in shear and in flow divergence or convergence.

(ii) If the height of a barchanoid element were to increase, there would be more flow-divergence, and less sand would be delivered to the summit, slowing its growth (Howard et al. 1977, 1978; Fig. 26.2).

(iii) If the barchanoid element were to become wider, and the plan-shape curve of the windward slope were to open up, there would be less divergence, and more sand would be deposited on the apex area, increasing the wind-wise length, decreasing the width:length ratio, and increasing divergence again.

(iv) A slipface on a dune of constant height that is straight across the wind, and which is being fed with sand from up wind, must always advance more quickly than the dune as a whole, because it must trap all the oncoming sand (Howard et al. 1977, 1978). However, curvature of the windward slope transmits some of the sand to the wings (or linguoid sections), and thus permits the dune to maintain an equilibrium shape. This implies feedback between the length, curvature and slope-angle of the barchanoid windward slope and the incoming rate of sand transport and the height of the slipface. Where there is little incoming sand, the windward slope should be steeper, with more divergence around the dune, and a more open curve (Howard et al. 1977). Where there is a strong supply of incoming sand, dunes should be elongated in the wind direction.

(v) The three-dimensional shape of the windward

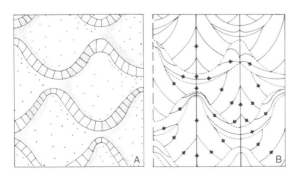

Figure 26.1 A model of transverse dune development (after Wilson 1970). A shows an idealized arrangement of barchanoid and linguoid elements. B shows hypothetical flow patterns over these ridges.

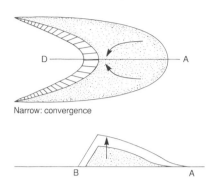

Narrow: convergence

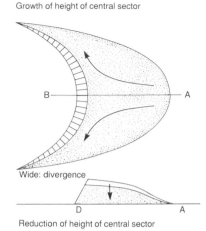

Growth of height of central sector

Wide: divergence

Reduction of height of central sector

Figure 26.2 Adjustment of the curvature and height of the windward slopes of transverse dunes.

slope of barchanoid elements is apparently very sensitive to flow direction. Howard et al. (1977, 1978) maintained that a barchan in a really unidirectional wind would be more elongated on the upwind side (more canoe-shaped) to give more flow-diversion to allow for equilibrium. They observed rapid adjustments of shape as wind directions changed.

(vi) The position of the brink is also very sensitive to changes in windspeed and sand supply (Howard et al. 1977). It reacts to variations in sand supply by adjusting its central curvature. Shear stress along the brinkline varies little, despite the variation in elevation, so that the lower portions on either side of the summit migrate faster than the summit. The resulting curvature of the central portion of the brink reduces the area available for deposition on the slipface, and this accelerates its movement (Fig. 26.3). The curve is also a delicate balance between the need to transmit oncoming sand around the dune and the slight inward convergence of flow that is consequent on curvature (Howard et al. 1978).

(vii) Apart from the adjustment of the central curve (vii, above), there appear also to be feedback adjustments to the angle of the slipfaces on the wings of barchanoid elements. These slipfaces and brinks are angled to the oncoming flow of sand (Fig. 26.4; Howard et al. 1977). Increasing obliquity reduces the amount of sand passing each unit length of brink, so that obliquity must decelerate slipface movement. There are also adjustments of flow velocity on brinks on the wings, for there is more marked crest/brink

separation on the wings than in the summit zone (Howard et al. 1978).

Negative feedback creates curvatures and angling of the brinks on the wings that are adjusted to ambient conditions. Thus, a tall barchanoid element, on which there is a strong contrast between the rate of migration of the wings and the crest must have stronger curvature (and brinks on the wings more angled to the oncoming wind) than a small one. This contrast in

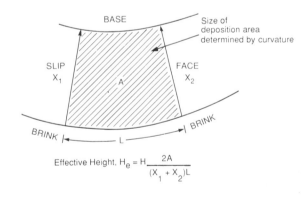

Effective Height, $H_e = H \dfrac{2A}{(X_1 + X_2)L}$

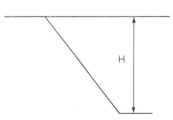

Figure 26.3 Equilibration of the curvature of the windward slopes of transverse dunes (after Howard et al. 1978).

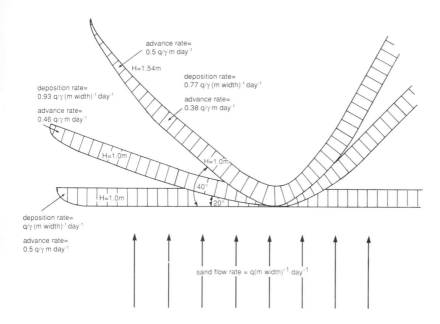

advance rate=
0.5 q/γ m day⁻¹

H=1.54m

deposition rate=
0.77 q/γ (m width)⁻¹ day⁻¹

deposition rate=
0.93 q/γ (m width)⁻¹ day⁻¹

advance rate=
0.38 q/γ m day⁻¹

advance rate=
0.46 q/γ m day⁻¹

H=1.0m

H=1.0m

40°

H=1.0m

20°

deposition rate=
q/γ (m width)⁻¹ day⁻¹

advance rate=
0.5 q/γ m day⁻¹

sand flow rate = q(m width)⁻¹ day⁻¹

Figure 26.4 Adjustment of the angles and heights of brinks on dune wings to oncoming sand flow.

curvature between large and small barchanoid elements is widely observed. These observations agree with those of early authorities reviewed by Bourcart (1928), who believed that curvature increased as windspeeds increased.

These remarks refer almost exclusively to adjustments of the barchanoid shape. Much less is known about the behaviour of linguoid elements.

26.1.3 Dome dunes

Dome dunes, as the term is used here (and by most authorities are dunes with no slip face, which are yet built of fairly fine-grained unimodal sand (McKee 1966, Mainguet & Callot 1978, Breed & Grow 1979). Most dome dunes are low, being only one or two metres high, although some are as high as nearby barchan and transversal dunes. They have also been called "ovoid", "whalebacks", "cake-like", "*galettes*" (griddle cakes), "*dunes en tas*", and "*boucliers*" (shields).

Defined in this way, dome dunes may not be completely distinct from zibars. The term "dome dune" has been used in Iraq for what some authorities would term zibar (Jawad Ali & Al-Ani 1983) and in Colorado (Fryberger et al. 1979), where dunes without slipfaces contain coarse sand. Tsoar (1986) found that low, "flat" dunes in Arizona and California could have a wide range of grain sizes; in other words, that they could be dome dunes or zibars. This is further

discussed in Sec.26.5.2. Although commonly used for meso-dunes only, the term "dome" has also been used for mega-dunes ("dome-shaped" dunes) in Saudi Arabia (Holm 1953, Breed & Grow 1979), and China (Zhu Zhenda 1984). These mega-dunes are assumed here to be a form of star dune.

The two-dimensional shape of dunes without slipfaces have been discussed in Chapter 23.3.5.

26.1.4 Barchans

A **barchan** is a crescentic dune isolated on a firm, coherent basement, such as a sabkha, pediment or desert pavement (Fig. 26.5). The word "*barchan*" is Turkic for "active dune", and is still used in this sense in the (former) Soviet Union, but it has been used in the present sense in western European languages for over 80 years. A barchan is called a "*khet*" in Qatarri Arabic (Johnstone & Wilkinson 1960), a "*rabadh*" in Saudi Arabia (Bagnold 1951) and a "*khord*" or "*zire*" in Mauritania (Dufour 1935).

Even isolated ripples and patches of sand only a few centimetres high, both in air and under water, adopt the crescentic shape (Fig. 23.7), suggesting that the barchan form is some universal expression of the dynamics of sand accumulation. Very small barchans have memories of the order of no more than hours. Not as common, but still widespread, are barchans about 1m high, with memories of a few months (before a seasonal change of winds). Meso-barchans,

Figure 26.5 Barchans in northern Chad.

3–10m high, with memories of 1–30yr, are far less common. I. G. Wilson (1973a) estimated that the sand contained in these barchans was less than 1% of all the dune sand on Earth. Mega-barchans, with memories of thousands of years, are even rarer. Meso-and mega-barchans are confined to directionally constant annual wind regimes (Bagnold 1941, Dubief 1952, Finkel 1959, Sharp 1964, Clos-Arceduc 1966a). Only these regimes allow the crescentic form to grow and to survive for more than a few months.

Isolation on a coherent base accentuates the fundamental barchanoid element of the transverse dune form (Bagnold 1941). In the saltation curtain, close to the surface, windspeeds are higher over the pebbly desert surface than over the dune itself, where they are retarded by saltation. This sweeps the wings forwards. Some of the sand arriving from up wind is channelled around the flanks, and leaves from the wings, trailing down wind, where it is shepherded into well defined streamers by pairs of vortices (Fig. 20.20). One of this pair of vortices is formed by the twisting around of the vortex in the lee of the slipface; the other is formed by a secondary component of flow towards the barchan flank from the faster-moving wind over the nearby desert floor. Small, short-memory barchans may be "calved" in these streamers.

The barchan "court" (Fig. 23.1) is kept partly clear of sand by the intensely turbulent wake in the lee of the slip face (Knott 1979). The turbulence is generated both by horizontal-axis roll-vortices shed at the brink, and by the vertical vortices shed from either flank

(Coursin 1964). Courts may be up to 5km long on barchans no more than 10m high.

In a barchan, the unavoidable loss of sand from the wings must be replaced by sand from up wind if the barchan is to maintain its size (Hastenrath 1987). The experiment reported in Chapter 23.2.1 shows that without sand a barchan slowly diminishes and then vanishes. Thus barchans are continuously renewing stores of sand. In a Peruvian barchan field, it has been estimated that an average 3m-high barchan gains and loses some $18m^3$ yr^{-1}, which means that it will have totally renewed its sand in 64 years, while travelling 1.7km (Lettau & Lettau 1969).

As in all dune fields, barchan fields contain dunes with different memories: ephemeral dunes created in a season, meso-memory dunes and, occasionally, mega-memory dunes inherited from former wind regimes. Unlike any other type of dune, however, some faster-moving ephemeral barchans travel for distances greater than the prevailing dune wavelength. Most of these small dunes end their lives by colliding with and being absorbed by larger neighbours, but while moving they are free to develop their forms slowly, unlike other short-memory dunes. Capot-Rey (1957b) and Verlaque (1958) observed the wings lengthened relative to the body as a barchan matured. In Peru, three-dimensional barchan characteristics show strong linear allometric relationships, for example between horn width and crest height, and windward slope angle and height (Finkel 1959, Hastenrath 1967).

There are still many unexplained features of barchans:

(i) The existence of mega-barchans in areas such as the Pleistocene Nebraska Sand Hills (Warren 1976a), parts of Peru (Simons 1956), the northern Rub' al Khali in Arabia, the Cherchen and Taklamakan Deserts in China (Breed & Grow 1979, Fryberger & Dean 1979) and the north-polar deserts of Mars (Tsoar et al. 1979). Most of these are over 100m high and 500m wide (Fig. 26.6). Moreover, in some of these areas mega-barchans occur in almost continuous sand cover, in which the accompanying linguoid elements are subdued. These features call into question any simple explanation of barchans as dunes formed where there are meagre supplies of sand.

(ii) Even more puzzling are "proto-mega-barchans" (Fig. 23.8). These features beg questions about secondary flow patterns.

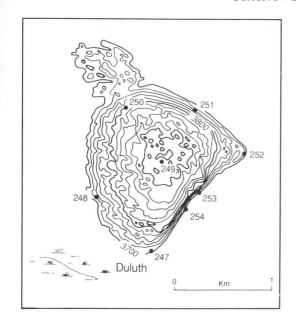

Figure 26.6 Mega-barchans in the Nebraska Sand Hills (sample sites referred to in Warren 1976b).

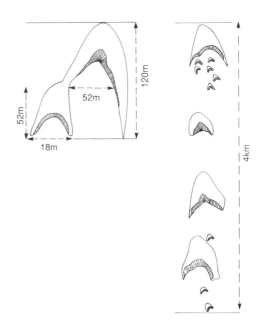

Figure 26.7 Barchan arrangements: *en echelon* and in convoy.

(iii) The regular size distribution of barchans in any one field (Figs 24.3, 26.5; Finkel 1959, Sarnthein & Walger 1974), and the regular distribution of sand within a field (Lettau & Lettau 1969). How are barchans first created and then marshalled to travel in such a regular pattern? The only available hypothesis for these processes is Wilson's (Sec. 26.1.2), although it has never been tested.

(iv) Variations in arrangement. Most barchans are arranged more or less *en echelon* (Figs 23.22, 26.5, 26.7), barchans down wind being fed by the sand streaming off the wings of ones up wind. But some barchans travel in "convoy", with barchans following each other in the lee zone (Fig. 26.6; Mainguet & Callot 1978).

(v) The asymmetry of many barchans (Fig. 23.22). Many cases of asymmetry can be satisfactorily explained by asymmetry in wind regime, in the sand supply or in the topography of the desert floor (Long & Sharp 1964, Lettau & Lettau 1969). But Clos-Arceduc (1969, 1971) pointed out that asymmetry cannot always be explained so easily (Fig. 26.8). He found groups of dunes in which closely neighbouring barchans were asymmetric in different senses. He and I. G. Wilson (1972b) believed that there was more to

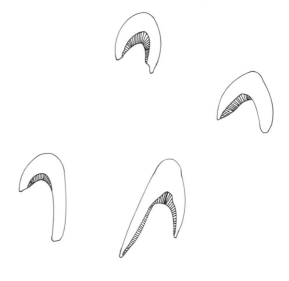

Figure 26.8 Asymmetrical barchans. Asymmetry is usually in the same sense in a group of barchans, but in this case it is in different senses in neighbouring dunes (after Clos-Arceduc 1967).

373

the asymmetry of barchans than wind regime, sand supply or topography.

26.1.5 Continuous transverse dunes

Perhaps just because they are so common, there is a dearth of vernacular names for transverse dunes: they are simply the "normal" dune condition. There is a richer terminology for the hollows in the lee of barchanoid sections: *"fuljes"* (*"faljes"*), *"bajirs"* and "cauldrons". Transverse dunes vary widely in size. Wing-to-wing widths of crescents range from a few metres to over 3km, and wind-wise lengths from a few metres to over 2km (Breed & Grow 1979).

Transverse dunes are formed in broadly the same kinds of wind regime as barchans, but where there is more sand supply. The difference between the two dune types is that in transverse dunes the barchanoid sections are joined by linguoid links that are usually lower and sometimes of coarser grain size than the barchans. Where there are variable wind velocities and sand-supply rates, as on coastal fields, barchans can be observed transmuting into transverse dunes and back again (Araya-Vergara 1987). As windspeed declines, for example on moving inland away from a coast, dunes grow, slow down and coalesce. Barchans reappear, from a transverse pattern, if sand supply declines or windspeed increases.

26.1.6 Lee projections

Low transverse dunes may, from time to time, develop "lee-projections" from the middle of the barchanoid slipface. These have been reported by Cholnoky (1902), Olson-Seffer (1908), Beadnell (1910), Cornish (1914), Bourcart (1928, after Harlé & Harlé 1919), Cooper (1958), Norris & Norris (1961) and Reid (1985). Most are ephemeral, and they develop only in certain wind conditions. Cooper believed that lee projections were formed by vertical eddies shed by either flank of barchanoid dunes, as described in Chapter 20.4.7, but the aerodynamics involved and the necessary wind or other meteorological conditions have not been discovered. Many lee-projections may be no more than the "T2" elements in a dune network (Sec. 26.4.2).

26.2 Linear dunes

26.2.1 Definition, description and classification

Earth has no landform more regular and extensive than linear dunes (Fig. 26.9). Estimates of the area of

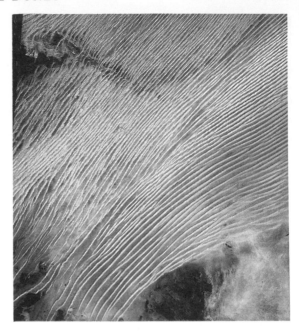

Figure 26.9 Linear dunes in the southwestern Rub' al Khali of Saudi Arabia (after Al-Hinai 1988). The dunes are spaced 1–2km apart.

sand seas in linear dunes vary from 30% to 60% (Fryberger & Goudie 1981, Lancaster 1982b). They dominate the Southern Hemisphere deserts, about half of the Rub' al Khali and many of the southern sand seas in the Sahara.

(a) Definition

Linear dunes are less curved (sinuosity ratio perhaps <1:3), have much greater length:width ratios (single dunes may reach 190km, although most are much shorter), and are more symmetrical across the shorter cross section than are transverse dunes. Slipfaces may take shape on either flank (but not simultaneously). Many, perhaps most, linear dunes grow by extension down wind (Tsoar 1978, Livingstone 1988), but some migrate obliquely. The two modes of migration are not mutually exclusive; they are likely to combine in different proportions according to wind regime.

(b) Terminology

"Linear" refers merely to the form of the dune, whereas "longitudinal" means near-parallelism with either a dominant wind or with the resultant of sand movement. The use of "linear" allows for explanations that involve complex wind regimes and avoids

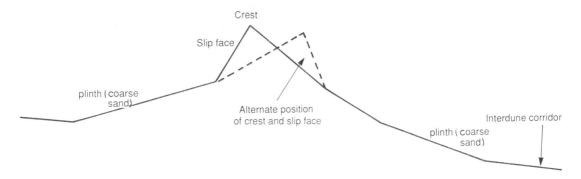

Figure 26.10 A typical cross section of a linear dune (after Lancaster 1982b).

the need to demonstrate longitudinality where it is difficulty to obtain wind data. These matters are discussed more fully below. Ignoring plurals, linear dunes have been called *sayf* (also spelled *seif, sief, saif,* or *sif*), *silk, fluq, irq, elb, draa, zemoul, slassel, ghurud, 'uruq, habl, cordon, bra, bhit, tibba, dune en vague,* longitudinal dune and sand ridge. The troughs between the ridges, which are usually more important to the local inhabitants than the ridges themselves, are called variously *goud, aftout, taieurt, feidj (fejj), sahane, shiqqa, gassi, straat, omouramba, bajir, ka'ar, couloire* or corridor.

(c) Description

Most linear dunes are asymmetric, sometimes in the same sense through the annual wind cycle, sometimes alternating with the seasons. Crests are steeper and more mobile than plinths (Fig. 26.10). Active linear meso-dunes have a succession of summits and cols along their length, and a meandering, sharp crest. Linear mega-dunes are less sinuous than linear meso-dunes, and may have multiple crests.

The most striking characteristic of many linear dunes (some would say only of sand ridges, defined below) is the regularity of their cross-sectional forms, orientations, dimensional ratios and spacing (Fig. 26.11). Orientation can be strikingly consistent. Within blocks of dunes of the order of 10km², there is generally strong correlation between corridor width and dune width, and between dune height and spacing (Aufrère 1935, Grove 1969, Breed & Grow 1979, Lancaster 1981b, 1983a, Mabbutt & Wooding 1983,

Wasson & Hyde 1983a, Thomas 1986a, 1988b). There is also an allometric increase of spacing with height and width (Aufrère 1935). Most striking, in the Australian and Kalahari sand ridges, is that when two linear dunes merge at a Y-junction, spacing readjusts down dune to the mean value (Fig. 26.11; Mabbutt & Wooding 1983, Thomas 1986a).

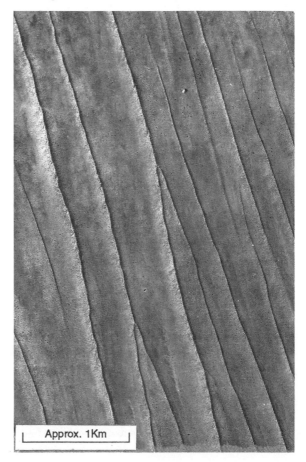

Figure 26.11 Regularly spaced linear dunes in the Simpson Desert of Australia.

Figures for variance of spacing between linear dunes in different parts of a sand sea can be quite large; but because they are a strong function of the size of the sample quadrat and the size and character of the area over which the sample data are aggregated, and the way in which spacing is defined, they do not always convey the regularity evident within small areas. Using $22km^2$ blocks and aggregating over whole sand seas, Breed & Grow's (1979) coefficients of variation ranged from 1.12 (very regularly spaced) in the Simpson Desert of Australia to 0.16 (very irregular) in the western Rub' al Khali. The regularity of linear dunes exhibited on the LANDSAT image of the southwestern Rub'al Khali in Figure 26.9 demonstrates just how misleading these statistics can be.

Although regular within small areas, linear dunes have a great range of size and form, both within and between sand seas. In 67 measured sample areas, Breed & Grow (1979) found that mean spacings ranged from 0.15km (in the Navajo Indian Reservation in Arizona) to 3.28km (in the southern Sahara). Height varied from 2m to 300m. Some linear dunes are tens of kilometres long, some a few hundred metres. Within individual sand seas, changes from one regular form to another can happen within a few kilometres (Mabbutt 1968, I. G. Wilson 1973a, McKee 1983, Mabbutt & Wooding 1983).

In Australia, spacing varies according to the character of the interdune surface. On clayey-sand substrates, linear dunes are less than 400m apart, on firm, silty surfaces they are 400–600m apart, and on stony surfaces they are more than 600m apart (Mabbutt 1968, 1980, Mabbutt & Wooding 1983). Thomas (1988b) compared height:spacing ratios from the Kalahari linear dunes with similar data for the Simpson and Great Sandy Deserts in Australia. There was regularity in each case, but of a different nature (Fig. 26.12). There is a tendency for dune spacing to increase down dune within any one sand sea (Mainguet 1983b, Mabbutt & Wooding 1983).

There is also great variation in other aspects of form. Linear mega-dunes are usually overlain by meso-dunes (linear or transverse), and these in turn are invariably overlaid by low-memory ephemeral dunes. Transverse or star mega-dunes are superimposed on some linear mega-dunes. Many linear dunes, particularly those in low-energy wind environments, have frequent Y-junctions, most of which open into the resultant wind direction (Fryberger & Dean 1979). Some dunes have extensive and complicated patterns of Y-junctions (Fig. 26.13; Goudie 1969, Mabbutt & Wooding 1983, Thomas 1986a). Some have none. Linear dunes in the Kalahari, for example, have many more Y-junctions than those in the Simpson Desert of Australia, and many more opening into the supposed dominant wind direction (Thomas 1986a).

(d) Classification

There have been several attempts to categorize the

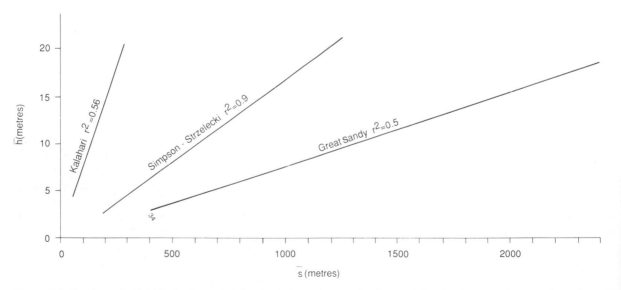

Figure 26.12 The relationship between height (h) and spacing (s) for linear dunes in three sand seas (after Thomas 1988b).

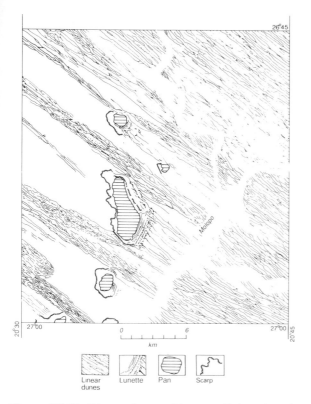

Figure 26.13 Linear dunes with many Y-junctions in the southwestern Kalahari (after Heine 1982).

variety of linear dune form. Price (1950), Breed & Grow (1979), Mainguet (1984a) and Tsoar (1989) distinguished:
– "sayfs" (short, sinuous, active over most of the dune slope and irregularly spaced)
from
– "sand-ridges" (long, straight, regularly spaced, and active only at the summit and the downwind tip, and in many cases vegetated).
McKee (1979b) saw three groups: "simple", with a single linear crest; "compound", with multiple linear crests; and "complex", with superimposed transverse or star dunes. Other authorities isolate oblique dunes (Hunter et al. 1983). It is difficult to define any of these categories with more precision, because there is a complete spectrum of types. They are discussed together in what follows.

(e) Features requiring explanation
The curious properties of linear dunes have attracted more attention than any other topic in dune geomorphology. Early conjecture about marine or tectonic genesis in North Africa, the Thar, the Namib and Australia can be dismissed, but a profusion of options remains. Recent reviews of the linear dunes include those of Twidale (1981a), Lancaster (1982b), McKee (1982), Glennie (1985), Livingstone (1988) and Tsoar (1989).

Five processes or characteristics of linear dunes need explanation:
   (i)   initiation and replication;
  (ii)   growth, maintenance and parallelism;
 (iii)   differences in size;
 (iv)   coexistence of linear dunes with other types of dune, and of linear dunes with different trend;
  (v)   regularity of form (spacing, height, width, and sand thickness).
Hypotheses to explain these features fall into five broad groups, which see linear dunes, variously: as obstacle dunes, as products of complex wind regimes, as products of secondary flow patterns, as erosional phenomena, or as products of vegetated landscapes. Some of these groups of hypotheses, particularly those concerning linear dunes as products of complex wind regimes, have themselves many forms.

26.2.2 Linear dunes as obstacle dunes
This group of hypotheses has answers to questions about initiation, coexistence with other dune forms, and perhaps to size and regularity. It has less to say about growth, maintenance or parallelism.

It is notable that anchored dunes are almost all linear, even in constant wind environments where neighbouring mobile dunes are transverse. This is true for ephemeral, meso-, mega- and macro-dunes, and even for some sand seas. Many linear dunes are indisputably obstacle dunes, and many authorities believe that obstacles are a general explanation for their initiation. The obstacles that have been invoked include hills and spurs (Melton 1940, Russell 1927), zibar dunes (Monod 1958, Tsoar 1978); gaps, endings or bends in transverse clay dunes (Wopfner & Twidale 1967, Twidale 1972 1981a), and trees or bushes (Cooper 1958, Tsoar & Møller 1986). Twidale's evidence for the formation of short linear dunes in the lee of large clay dunes in Australia is convincing, although the extension of the explanation to larger, more regularly spaced, linear dunes a few kilometres down wind is less so, for the two types of dune are not continuous and are at different scales (Mabbutt & Wooding 1983).

The obstacle hypothesis of linear dune formation can even explain something of size and regularity.

Melton's (1940) linear dunes, trailing off the Moen-kopi escarpment in Arizona, were spaced according to regular drainage-basin controlled fluvial dissection of the escarpment; the size of the linear dunes was related to the size of the spurs. Many desert plants occur at regular spacing, although the reasons are obscure (King & Woodell 1973, Noy-Meir 1973). If the plants are regularly spaced, so would be attached linear dunes. Connections between vegetation and linear dunes are further discussed in Sec. 26.2.6c.

Twidale's (1972, 1981a) hypothesis of linear dune initiation down wind of clayey transverse dunes also explains the coexistence of different dune types: the form of the transverse dunes depended on their clayey sediments, and that of the linear dunes on sands winnowed from among the clay pellets.

However, despite these undoubted connections, obstacles are not involved in the formation of most well developed linear dunes, for these appear to have developed on level, plant-free plains.

26.2.3 Linear dunes as products of complex wind regimes

This large group of hypotheses has little to say about initiation or spacing, but makes important contributions to the explanation of size, and the coexistence of different dune types. The major contribution is to the understanding of growth and maintenance. In these hypotheses, all linear dunes are seen as meso-memory features, retaining their shape through a complex annual regime of winds.

These hypotheses have the best empirical support: the global surveys of Fryberger & Dean (1979) and Wasson & Hyde (1983b) showed that most linear dunes, of whatever type, existed in more directionally variable wind regimes than did transverse dunes, and less directionally variable regimes than star dunes. Although strongly indicative, these findings need to be taken with caution. The exceptions noted by Fryberger & Dean are discussed below. Areas of major deviation between dune trend and resultant drift potential were rejected in Wasson & Hyde's analysis, and their assumptions, although explicit, can be called into question. Rubin (1984), for example, doubted their assumption that sand thickness was equated with sand supply; and the classification of parabolic dunes as transverse can also be challenged.

There are six, not mutually exclusive, although sometimes conflicting, variants of the multimodal regime model. The models are: barchan-to-linear, resultant, flow-diversion, dominant-skewed resultant, mobile crest, and gross bedform-normal.

(a) The barchan-to-linear model

This was one of Bagnold's (1941) hypotheses of linear dune formation. In the model, strong winds, carry sand over a pebble pavement and build a patch of sand into a dune, while gentle ones, unable to carry sand over the pavement to the dune, merely erode and elongate it. Bagnold elaborated the model so that the first dune was a barchan, one of whose wings eventually was elongated to become the linear dune (Fig. 26.14).

There is sound, although limited, field evidence for this model. Lancaster (1980b) reported active barchans in Namibia, with one wing elongated up to 1.5km, becoming, in effect, a linear dune. These linear dunes intercepted sand brought by a strong wind and were elongated by gentle persistent winds, as Bagnold predicted. A similar pattern was described in Peru by Hastenrath (1987). Warren (1976b) in stabilized mega-dunes in Nebraska, and Tsoar (1984) in active dunes in northern Sinai, also found elongated barchan wings, but maintained that they were elongated in the resultant rather than gentle-wind direction. If the regularity of transverse or barchan dunes is accepted, the barchan-to-linear dune model explains the regular spacing of the attached linear dunes.

There are two problems with this model. First, barchans and linear dunes of the same scale coexist in few areas. The second problem is the interpretation of coexistence. Clos-Arceduc (1969) found elongation of different wings in neighbouring barchans (Fig. 26.8), a pattern that cannot be explained by the simple barchan-to-linear model developed above.

(b) The resultant model

This was another of Bagnold's models (1951, 1953a), presaged by Aufrère's (1934) idea of a "parallelogram of forces". In this model, the linear dune extends in the direction of the resultant of sand movement: sand is blown back and forth by winds from different directions, travelling ultimately in the resultant direction. The model was apparently confirmed by Bagnold's own study of dune alignment in Arabia; by Cooper's (1958) work in Oregon; by Tsoar's and Warren's examples of barchans and linear dunes discussed above; by McKee & Tibbitts (1964) in Libya; by Warren (1972) for similar features in Niger; by Twidale's (1972, 1981a) observations in the Simpson Desert in Australia; and by the broad conclusions of Fryberger & Dean's (1979) global survey.

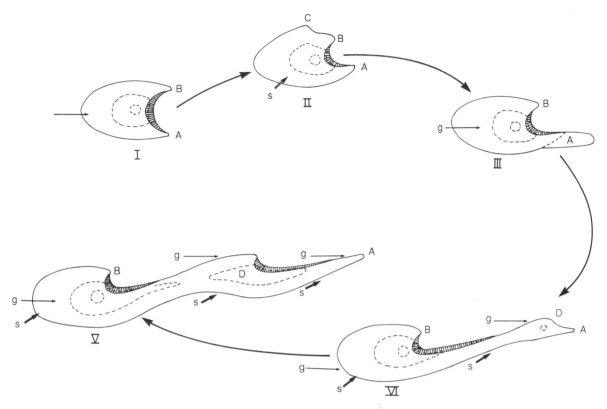

Figure 26.14 A model of the development of linear dunes (sayfs) from the wings of barchans in an asymmetrical wind regime. Arrows labelled "s" refer to strong winds; arrows labelled "g" to gentle winds (after Bagnold 1941).

The greatest problem with the "resultant" hypothesis is that, despite all the confirmations, there are many linear dunes that are not parallel to the apparent resultant. Bagnold (1953b) himself noted a discordance in the Western Desert of Egypt; Brookfield (1970) found consistent discrepancies of up to 20° in central Australia; and Besler (1980) and Lancaster (1983a) noted deviations of 20°–30° in the Namib. Fryberger & Dean's (1979) global survey found quite wide deviations in some places, such as the western Kalahari. Discrepancies may be no more than a problem with wind data, particularly if they are from distant weather stations (Fryberger & Dean 1979, Livingstone 1989a), although their persistence points to a more systematic problem.

Discrepancy could arise if the dunes had been aligned to ancient rather than modern winds, a notion discussed below. More problematic is the response of linear dunes, particularly sand ridges, to obstacles such as rock ridges, for one would expect a more complex response of the resultant in these cases (Fig.

26.15, Mainguet 1984c, Tsoar 1989). Finally, the resultant model has nothing to contribute to the discussion of spacing.

(c) The flow-diversion model

This was Tsoar's model (1978, 1983a, Tsoar & Yaalon 1983), and was the first to be based on detailed field monitoring. Tsoar accepted the resultant-drift model of linear dune alignment; his contribution was to an understanding of maintenance. His most important finding was that, when winds approached at certain angles to the dune crest, they created strong vortex-like flow in the lee, which ran parallel to the dune. Where the flow re-attached to the bed on the downwind side of the dune, velocities were higher than those on the crest (Fig. 26.16). These strong winds carried sand parallel to the dune. The existence of deflation in this zone was confirmed by grain-size distributions on linear dunes found by Hartmann & Christiansen (1988).

Tsoar (1978) found that along-dune sand movement

Figure 26.15 The response of sand ridges (linear dunes) in the vicinity of rock ridges in central Australia.

was at a maximum when wind approached the crest between 35° and 50°. This accords roughly with Seginer's (1975) and Mulhearn & Bradley's (1977) experimental findings with sand fences. The critical angle is likely, however, to depend on u_*.

Tsoar's model is shown in Figure 26.17. Sand is carried along those portions of the sinuous dune which are oblique to a particular wind, and deposited where the dune crest swings more nearly at right angles to it, building up these areas and giving a col/hill longitudinal profile. Since the locations of maximum erosion and deposition are different for different winds, depending on their angle of approach, winds

from one direction erode deposits laid down by those from an earlier one, always moving hills down the dune. The pattern of movement on any particular dune and the degree of contrast between hills and cols depends on the relative strengths of winds approaching from different directions. Along-dune sand transport also extends the dune.

Tsoar confirmed his model of flow around linear dunes in a wind tunnel (1985, Tsoar et al. 1985). His observations on oblique lee vortices broadly confirm experimental work on water flow in the lee of oblique steps by Allen (1968b), and Wilson's (1972b) ideas about the rôle of oblique vortices in desert dunes.

Although Twidale (1972) did not measure vortices over linear dunes, he speculated that they would be of the size of the interdune corridors, and therefore explain dune spacing. Tsoar's (1978) vortices, on the other hand, and even more so Livingstone's (1986), were much smaller than the width of the corridor. Tsoar (1978) suggested instead that spacing might be related to a wind-shadow effect, believing that windspeeds would need to pick up again before a new linear dune could form. He used the well known relationship between the height of an obstacle and its wind-shadow down wind to calculate that, if the formative winds blew at angles of 30°–45° to the linear dune crests, they should have height:width ratios of 1:6 to 1:8. Not all the evidence, however, supports this suggestion: Lancaster (1982c) found spacings at about 1:20 over a wide range of dune sizes.

(d) The dominant-skewed resultant model
This was Lancaster's (1982b) attempt to combine the resultant model and Tsoar's model, and to account for

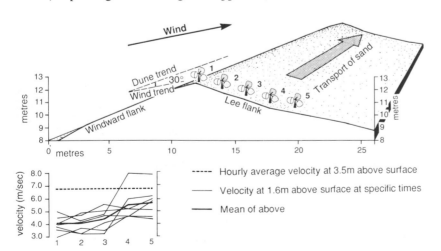

Figure 26.16 The generation of intense flow in the lee of a linear dune when the wind blows obliquely to the crest (data from Tsoar 1978).

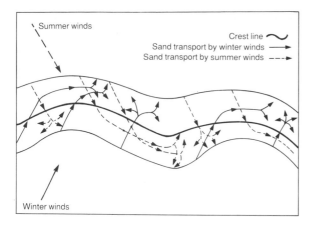

Figure 26.17 A model of linear (sayf) dune development in an acute bimodal wind regime (after Tsoar & Yaalon 1983).

the discrepancy between the trends of resultants of sand movement and linear dunes. Lancaster (1985a) noted that many linear dunes existed in wind regimes with a dominant wind from one direction and a number of weaker side winds. He therefore proposed that strong winds, blowing at angles of about 30° to a linear dune crest (Tsoar's optimal angle) would have greater influence on dune trend than their sand-moving power suggested. The model accounts for linear dune alignments in the Namib Sand Sea. A similar argument has been used for the Oregon coastal linear dunes by Carson & McLean (1986).

Lancaster's model is apparently supported by Brookfield's (1970) observations on the effects of strong winds, and is consistent with Fryberger & Dean's (1979) suggestion that, if strong winds at one season established a linear dune, more variable winds at other seasons would be forced to carry sand in this established direction. Despite some circumstantial evidence, Lancaster's model is not based on process studies, unlike Tsoar's or Livingstone's (below), and it needs to be better formulated before it can be tested.

(e) The mobile crest model

Like Tsoar, Livingstone (1986, 1989a) based his model on detailed field monitoring. It was developed on Namibian dunes that were much larger than Tsoar's and it existed in a wider multimodal wind regime. Although Livingstone confirmed that there was some along-dune transport of sand when winds blew obliquely to the dune crest, he found no lee-side

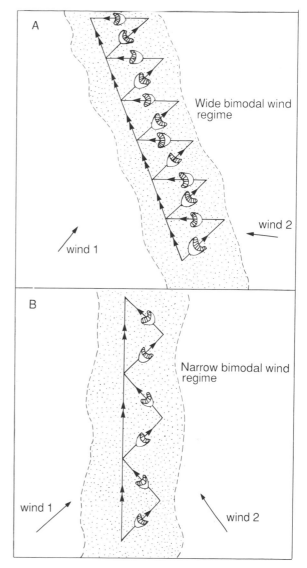

Figure 26.18 A model of linear dune development in a wide bimodal wind regime (after Livingstone 1986).

acceleration and a much smaller lee vortex, relative to the dune, than did Tsoar. Livingstone saw sand transport along linear dunes as the result of the back-and-forth, diagonal movement of ephemeral dunes on the crest of the mega-feature in response to different oblique winds in the annual cycle (Fig. 26.18). The model is similar to one developed by Streim (1954).

(f) Gross bedform-normal transport

It is a common observation that linear dunes are little more than frustrated transverse dunes. If the wind

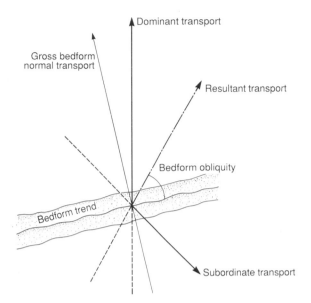

Labels on figure:
Dominant transport
Gross bedform normal transport
Resultant transport
Bedform obliquity
Bedform trend
Subordinate transport

Figure 26.19 A model of linear dunes adjusted to the "gross bedform-normal" transport of sand (after Rubin & Ikeda 1990).

explain initiation, growth, maintenance, parallelism and even obliquity. The starting point of their arguments is the belief that some process, more regular than fluvial or marine sediment supply, is building sand into regularly sized and regularly spaced ridges.

Like so many of the others, the first expression of this set of hypotheses was Bagnold's (1941). His model had two components:

(i) *Initiation* In 1941, Bagnold speculated that some "large-scale rotary movement in the air stream" produced the regularly spaced thin sand strips that he had observed in the Western Desert of Egypt. He later (1953b) linked linear dunes to the sorts of vortex motion then being discovered by meteorologists. Bagnold's strips were similar to the strips reported in Alaska by Swift et al. and in Libya by Hastings (Ch. 20.45b). This hypothesis is now usually termed the "roll-vortex" model.

(ii) *Maintenance* Bagnold (1941) observed in a wind tunnel that "transverse instability" developed where pre-existing longitudinal sand strips occurred on a pebbly surface. Strong winds were decelerated on the sand relative to the pebbly surface, resulting in transfer of sand to the strips. Bagnold suggested that this would enable the strips to grow into dunes.

were to blow from one direction throughout the year, they would develop into transverse dunes. This argument was taken further and tested on aeolian ripples formed on a rotating board in the open by Rubin & Hunter (1987), and later with a rotating board in a water flume by Rubin & Ikeda (1990). The hypothesis in these experiments was that a ripple would orientate itself to maximize bedform-normal (transverse) sand transport. Rubin & Hunter (1987) found that there was a sharp transition to linear-type features when two flows diverged by about 90° (Fig. 26.19). The gross bedform-normal trend was a better predictor of dune trend than the resultant of sediment transport.

Lancaster (1990) tested the hypothesis against dunes for which he had good wind data. The system worked well for transverse and star dunes, and for what appeared to be recent linear dunes, such as Tsoar's sayf dune in Sinai. But neither linear mega-dunes in the Namib Sand Sea, nor linear meso-dunes in the Strzelecki Desert in Australia, conformed to the bedform-normal rule (Lancaster 1990, Rubin 1990). Lancaster had to resort to the hypothesis that these were palaeofeatures, an hypothesis discussed in Chapter 29.1.1.

26.2.4 Linear dunes as products of secondary flow patterns

(a) Introduction
The champions of secondary-flow hypotheses are comprehensive in their ambitions. They use them to

(b) Initiation
The existence of roll-vortex flow in the planetary boundary layer is not in dispute (Ch. 20.4.5) and Bagnold's lead in associating it with linear dunes has been widely followed (Hanna 1969, Folk 1971a, b, Wilson 1972b, Gorycki 1973, Besler 1980), although Mainguet (1984b) and Tsoar & Møller (1986) believed that the secondary flow hypothesis applied only to "sand ridges", not to "sayfs". The roll-vortex mechanism has also been invoked on beaches (Tanner 1963, Norris 1956, Cloud 1966), in tidal seas (Houbolt 1968, Kenyon 1970, Stride 1982), and in the abyssal regions of the ocean (Lonsdale & Spiess 1977). The most highly developed meteorological argument in relation to linear dunes has come from Wippermann (1969) in an explanation of roll vortices and subsidiary transverse waves as a response to rotation in the planetary boundary layer.

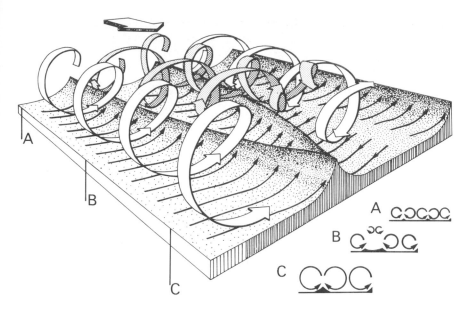

A

B

C

Figure 26.20 Hypothesis of linear dune formation by roll-vortex flow.

In brief, the roll-vortex hypothesis can be stated as follows (Figs 20.11 & 26.20). Roll vortices develop in the various ways outlined in Chapter 20.4.5. The pairs of counter-rotating vortices have wavelengths (for two counter-rotating rolls), two to four times the depth of the boundary layer. The secondary component in these vortices, being about 10% of the forward motion, encourages sand to move away from zones of downward divergent flow (where u_* values are high) towards those of upward convergent flow (where u_* is low). Saltating sand delivered obliquely to the new dune is diverted to flow along the dune (Howard 1977, 1985). The dunes grow in the zones of convergence and reach an elevation at which wind velocity is sufficient to move sand along the dune, maintaining uniformity of dune height (Wilson 1972b).

Apart from regularity, two features of linear dune morphology support the roll-vortex model:

 (i) Comparisons of the scale of vortices and dunes lend some support. For example, where the monsoon layer is only 1km thick, as in the Wahiba Sands in Oman (Pedgeley 1970), one might expect linear dunes to be spaced at about 2km, and this is indeed their spacing. More thorough research would have to investigate roll-vortex formation under "normal" as well as rare strong events.

 (ii) Y-junctions would be sites where pairs of roll vortices rise, and are replaced by vortex pairs from either side (Fig. 26.20; Folk 1971a). An

alternative explanation is that Y-junctions are forced by sidewinds (Thomas 1986b).

(c) Maintenance

Secondary-flow hypotheses are on stronger ground when they attempt to explain the maintenance and extension of linear dunes. Bagnold's transverse instability model was nicely corroborated by the discovery of linear meso-dunes which converged towards linear mega-dunes in the Wahiba Sands of Oman (Glennie 1970, 1985; Fig. 26.21). This kind of flow would be expected from meteorological theory and observations of flow over hills of the same scale as mega-dunes (Carruthers & Hunt 1989), even without the added influence of flow retardation by saltation on the dune.

Bagnold's (1953b) additional suggestion that unequal heating on dune and interdune surfaces, or on different sides of a dune, might encourage vortex formation was accepted by Tricart & Cailleux (1964) and Mabbutt et al. (1969). The suggestion was that upward convection would be encouraged over the dune, and might strengthen vortex flow. It is true that sand surfaces in tropical deserts can reach 69°C, enough to initiate quite strong upward convection (Howard 1985). A related suggestion is that unequal heating on the north and south sides of linear dunes causes cross-dune transport of sand, and so the observed cross-sectional asymmetry (Mabbutt et al. 1969).

Until recently there has been little confirmation of the dynamics of the roll-vortex model of linear dune

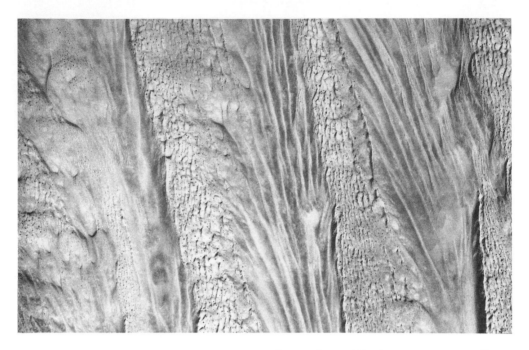

Figure 26.21 Meso-linear dunes rising up both flanks of a corridor between two mega-linear dunes in the Wahiba Sands of the Sultanate of Oman (after Warren 1988a).

formation, but the use of kites has shown that flow directions are towards a linear dune from both sides, when wind blows parallel to the dune (Tseo 1990). Tseo's measurements, although an important contribution, may indicate little more than the generation of vortex flow by an existing dune, and thus dune maintenance; they do not confirm linear-dune initiation by roll vortices. They also illustrate the difficulties of measuring flow on real linear dunes, where secondary flows, if they exist, are likely to be intermittent and short-lived.

(d) Objections to the roll-vortex model
There have been four major objections to the roll-vortex model (Lancaster 1982b, Livingstone 1988), each of which has a counter-argument. The first two (that the sizes of roll vortices and the spaces between linear dunes are not of the same order, and that there were no observations of roll vortices in operation) have been countered above, although these cases are not strong enough to be taken as proof of the vortex model.

The third objection is the association of linear dunes with multi-directional wind regimes. This argument can be countered in three ways. First, linear dunes may be formed only by occasional winds, perhaps only when strong winds and heating coincide (Steffan 1983). The dunes would then persist by channelling variable winds along their flanks. Secondly, linear dunes could have been formed by roll vortices in stronger more unidirectional winds of the past (Besler 1980, Glennie 1970, 1985). Here too, the argument is that, once formed, the dunes would become stable sand-transmitting forms, even when wind patterns changed. Palaeoclimatic models do show stronger and perhaps more unidirectional winds in glacial periods, and many desert dunes are undoubtedly paleoforms. Thirdly, there is the evidence of yardangs, whose geometry is not unlike that of linear dunes, and which could not have formed under any of the bi-directional models of linear dune formation. None of these three counterclaims has much supporting evidence. Livingstone's (1986, 1989a) demonstration that the Namib dunes are still very active points to the danger of tenuousness and circularity in the relic-dune argument. Steffan had little data to back his case, and the depth of ignorance about yardang spacing has been demonstrated in Chapter 21.3.2.

The fourth and last problem with the roll vortex model is the explanation of how mobile roll vortices arc fixed to the surface. Fixation might be accomplished by roughness or heating effects, but here again there are major problems of proof.

The roll vortex model, for all its attractive simplicity, still has a long way to go to achieve full credibility.

26.2.5 Linear dunes as erosional phenomena: the wind-rift model

(a) The hypothesis
The "wind-rift" (sometimes "wind-drift") proposal is that some linear dune fields are essentially collections of yardangs eroded from a once continuous cover of alluvial sand. In English, the idea can be traced to Meddlicott & Blanford's (1879) discussion of the Thar Desert, although the term itself was coined by Belknap (1928). In French, the model was proposed for Saharan dunes by Gautier & Chudeau (1908) and Aufrère (1928, 1929, 1930, 1932), and endorsed by Enquist & Frederick (1932), Brosset (1939), Monod (1958, for his "'alab") and Tricart & Cailleux (1964). Melton (1940) adopted it for some American linear dunes, and it was also used for the linear dune fields of central Australia (Madigan 1946, King 1956, 1960, Folk 1968, 1971). Barnard (1973) applied it to the Namib. It has been resurrected by Mainguet (1983b, 1984a), Mainguet & Chemin (1983) and El-Baz (1984a). The idea was explicitly linked to the secondary-flow hypothesis by Folk (1971b, 1976a), who suggested that the interdune erosion had been achieved by roll vortices.

(b) Wind rifting in semi-arid, clay-rich environments
The only substantial evidence for wind rifting comes from semi-arid environments. In the southeastern Australian Mallee, linear dunes are built of sediments with, on average, 7–10% clay content (Bowler & McGee 1978). The dunes reach 8m high and are about 600m apart. Up to five buried soils indicate dune accumulation in dry periods interspersed with wetter soil-forming periods (Hills 1939, Churchward 1961). It is claimed that the material for the dunes was excavated from saline interdunes. There may have been positive feedback: as deepening interdunes attracted more drainage and more salt, their sediments were more readily pelletized and eroded by wind (Bowler & McGee 1978).

Comparable linear features are found in an area of over 5,000km² in Oregon and Washington States, where aggrading soil horizons are also believed to

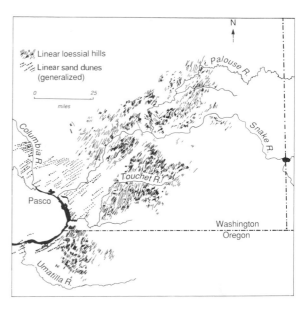

Figure 26.22 Linear depositional patterns in loess terrain in the Palouse area of Oregon and Washington states, USA (after Lewis 1960).

indicate a depositional–erosional origin (Fig. 26.22; Lewis 1960). Others occur in northwestern Senegal, where sediments with up to 80% of fine material occur in a regular linear dune topography, and where wind rifting has also been proposed (Tricart & Brochu 1955). These dunes do indeed seem to have been built from locally derived sediment (Barbey et al. 1975). Yet others are the linear "paha" ridges composed largely of fine sediment in the Midwest of North America (e.g. Flemal et al. 1972). Finally, well developed linear features near the Volga Delta in southern Russia, and composed of sediments rich in fine material, have also been seen as depositional–erosional dunes (Samojlov 1956, quoting Fedorovich 1946). Signs of flooding between these dunes, by higher stands of the Caspian Sea, had convinced Fedorovich that they were relics of stronger Pleistocene winds. All these linear features are comparable in size, spacing and extent to linear topography thought to be largely erosional and discussed in Chapter 21.3.2.

(c) Wind rifting in arid environments
Evidence of wind rifting in true sand deserts is more elusive. King (1960) discovered an alluvial core in ridges on the north side of Lake Eyre in Australia, proving yardang-like wind erosion at that point. He

extended the wind-rift explanation to all Australian sand ridges, but where clayey cores have been found in these, they have deemed by other authorities to be illuvial (Wopfner & Twidale 1967, Mabbutt 1980). There are multiple buried soil horizons in some sand ridges in the Strzelecki Desert (Wasson 1976, 1983b), but all the other evidence there points to dune construction on a level gravelly surface, rather than to the excavation of the interdunes. For example, the courses of ancient river channels extend beneath the Strzelecki sand ridges.

One of King's hypotheses was that the Y-junctions among Australian sand ridges were the ends of wind-eroded hollows (or parabolic dunes), but Mabbutt & Sullivan (1968) found that the level desert floor extended beneath Y-junctions, and Mabbutt & Wooding (1983) found that Y-junctions occurred more frequently where there were changes in the slope or character of the desert floor, both of which findings suggest that Y-junctions form by the convergence of a growing dunes rather than by excavation of the hollow. This appears to be confirmed by Thomas's (1986b) findings about Kalahari Y-junctions.

Until more conclusive proof is found in excavations across dunes, the wind-rift model must be held in doubt as a general explanation of linear dunes in deserts. It may, however, hold for areas like the one found by King near Lake Eyre, and for sediments with high clay contents in semi-arid areas, let alone for hard-rock mega-yardangs themselves.

26.2.6 Linear dunes and vegetation

Many, if not most, linear dunes are vegetated to a point where sand movement is severely restricted. This could have four explanations.

(a) Climatic change
Vegetated linear dunes could have experienced a secular change in rainfall or wind velocity. Many of the linear dunes in the West African Sahel have clearly been stabilized following climatic change, for they are now covered by dense vegetation and well developed soils. In Australia and the Kalahari, where soils and vegetation are not everywhere as well developed as in the Sahel, it has also been suggested that most linear dunes formed in a drier and windier environment than the present (Wasson 1984, Thomas & Goudie 1984). The association of linear dunes with climatic change may be fortuitous: the bimodal wind regimes that favour linear dune formation may be more characteristic of desert margins, and these are more likely to experience the critical climatic change between dune activity and inactivity, than are the desert cores.

(b) Stable surfaces for colonization
A change in climate cannot explain the coexistence of bare transverse dunes and vegetated linear dunes within the same rainfall regime (for example, in the Wahiba Sands in Oman; Warren 1988a). This prompts a second explanation: linear dunes, particularly on their lower slopes, have much more stable surfaces than transverse dunes and are therefore more readily colonized by plants (Bourcart 1928, Ash & Wasson 1983).

(c) Linear dunes as products of vegetated landscapes
Another suggestion is that linear dunes are actually encouraged by the presence of vegetation (Tricart & Cailleux 1964, Tsoar & Møller 1986, Thomas 1988d, 1989a, Tsoar 1989). In this model it is not only the genesis and spacing, but also the growth of linear dunes that is affected by plant cover. Plants protect the surface from all but the strongest winds, so that the linear dunes are orientated parallel only to strong winds. Greater protection of the flanks rather than crests, both with vegetation and soil crusts (Yair 1990) may restrict dune development only to the narrow crestal zone. Where, as in the southwestern Rub' al Khali, linear dunes are now found in vegetation-free environments, this model would have to maintain that there had been desiccation.

(d) Linear dunes and vegetated parabolic dunes
This proposal is that linear dunes are remnant wings of vegetated parabolic dunes, the transverse section of the parabolic curve having been blown away. There is good morphological evidence for this model in fields of parabolic dunes, where all stages between blowouts and linear dunes with no transverse section are found (Fig. 25.9). The model originated to explain some dune forms in Gascony in southwestern France (Capot-Rey & Capot-Rey 1948), and has acquired many supporters, most with experience of vegetated rather than truly desert conditions (Cornish 1908, Bourcart 1928, Hack 1941, Sprigg 1959, Nevyazhsky & Bidzhiev 1960, Verstappen 1968, 1972, Folk 1971, Twidale 1972, David 1981, Pye 1982a, Filion & Morisset 1983, Araya-Vergara 1987). The model has good support in semi-arid areas, such as the Thar (Fig. 26.23), but there is little evidence that linear dunes in deserts have formed in this way. The geo-

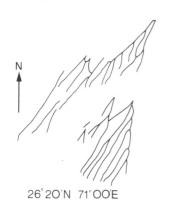

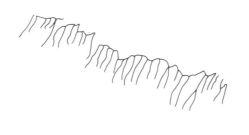

26°20'N 71°00'E   28°00'N 71°00'E

Figure 26.23 Linear dunes among vegetated parabolic dunes in the Thar Desert of India (after Mainguet 1984).

metry of most desert linear meso-dunes is much more regular than that of the linear dunes that have been explained with the parabolic model, and the size of linear mega-dunes removes them completely from the possibility of such an explanation.

26.2.7 Composite explanations

Few of the models outlined above have been developed for the same environment. Many may be valid only for their own special circumstances. Tsoar's flow-diversion model may apply to sayf-like dunes in a narrow bimodal wind regime; Livingstone's to large linear dunes in a wide bimodal wind regime; the "parabolic hypothesis" to areas with a sparse cover of vegetation; the wind-rift hypothesis to situations in which there is pelletization; and the secondary flow model to wide, hot plains.

Moreover, as explanations of linear dunes, some of the models may be more necessary than sufficient. There may have to be more than one necessary condition. A subsidiary wind, oblique to the dominant wind, may be necessary to create regularly spaced transverse dunes, which then determine the spacing of oblique linear dunes. Regularity could be supplied by dunes in the lee of regularly spaced large plants. These dunes might then be elongated depending on local meteorological conditions:

- in the resultant direction, as in Tsoar's or Livingstone's models, depending on the width of the bimodal wind regime;
- in the optimal direction for oblique vortices in unidirectional winds, after Wilson's or Lancaster's models; or
- parallel to the dominant wind on heated surfaces, after the roll vortex model.

Depending on sand-supply conditions and wind regimes, these processes would act individually or in combination, to give the variety of linear forms that exists.

26.3 Oblique dunes

26.3.1 Types of obliquity

Obliquity of morphologically linear and transverse dunes to dominant winds or to the resultant of sand movement has been widely reported (above and below, *passim*), yet there is confusion about the definition and meaning of obliquity. Obliquity exists in two dynamic and three morphological senses:

(i) Dynamic obliquity has two forms:
  – Obliquity of sand transport is the transport of sand obliquely across a transverse dune. It is, of necessity, a temporary phenomenon, and one that has little permanent effect on dune form.
  – Obliquity of dune movement was observed in apparently linear dunes on the Oregon coast by Hunter et al. (1983; Fig. 26.24),

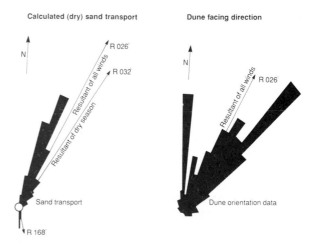

Figure 26.24 The relationship between sand movement and dune orientation in mobile coastal dunes on the Oregon coast, showing that the dunes are oblique to the direction of sand transport (after Hunter et al. 1983).

and inferred in apparently transverse mega-dunes in the Algodones dunefield by Havholm & Kocurek (1988).

Rubin (1990) noted that the classic linear dunes of the Strzelecki Desert in Australia were being eroded on their west flanks and were accreting on the east flanks.

(ii) Morphological obliquity has three forms:
- Oblique alignment on the Oregon coast has been described above (Hunter et al. 1983); these dunes were experiencing not only oblique dune movement, they were also aligned obliquely to the sand-movement resultant (Fig. 26.24). This has been termed "morphodynamic obliquity" by Rubin (1990).
- Obliquity of surface form: the most obvious sign of obliquity would be persistent cross-sectional asymmetry in dunes that might otherwise be seen as linear (Rubin 1990).
- Obliquity of internal structure: although obliquely moving linear dunes have structures like transverse dunes, there appear to be subtle differences (Rubin 1990). Obliquely moving transverse dunes, for example, are said to have distinctive internal structures (Havholm & Kocurek 1988).

Morphological obliquity must be a consequence of one form of dynamic obliquity, but the connections are obscure.

26.3.2 The occurrence of obliquity
There are four situations in which obliquity occurs: following climatic change, by diversion over oblique slopes in unidirectional winds, at roughness contrasts, or in multimodal wind regimes.

(a) Obliquity in changing climatic regimes
Obliquity could be the result of a secular change in wind regime since the formation of a dune. Obliquity of this sort is most likely to be maintained in mega-dunes. This was Hedin's (1903) explanation of obliquity in the Taklamakan, and comparable explanations have been proposed by Monod (1958), Warren (1970, 1988a), Glennie (1970), Besler (1982a) and others. In most of these cases obliquity was inferred from a discordance between the orientation of meso- and mega-dunes. Because so little is known of the relationships between dune form and wind regime, an explanation that invoked changing winds would need much more evidence than from form and alignment alone,

before it could be accepted.

(b) Obliquity in near-unidirectional wind regimes
Apparent obliquity of alignment in near-unidirectional wind regimes is quite common (Fig. 26.8). It should not be unexpected, because oblique forms are known to be quite stable in unidirectional subaqueous flow (Engelund & Fredsoe 1982). In aeolian dunes, obliquity is said to occur in the Fachi–Bilma Sand Sea in Niger, where there is a consistent angle of 30° between the axes of barchans and accompanying linear dunes and alignments of accompanying shadow dunes, barchans and yardangs (Durand de Corbiac 1958). The double obliquity here was compared by Mainguet (1984a, Mainguet & Callot 1978) to the wake of a boat.

In Figure 26.25 the trend of closely juxtaposed linear dunes is in some places to the left and in some places to the right of the axis of symmetry of the barchan dunes. I. G. Wilson's (1972b, 1973a) explanation of this kind of pattern was developed from his

Approx. 1Km

Figure 26.25 Linear dunes in two oblique directions in the Great Eastern Sand Sea of Algeria.

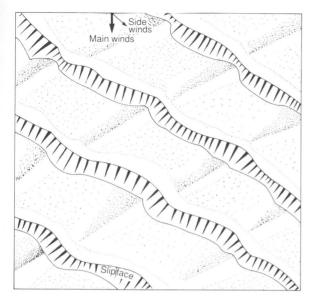

Side winds

Main winds

Slipface

Figure 26.26 A model for the development of linear dune obliquity in a near-unidirectional wind regime. The "main wind" creates a sinuous set of transverse dunes. The "side winds" elongate one "arm" of the pattern. The other arm remains as a subdued oblique element (after Wilson 1970).

models of transverse dune formation (Sec. 26.1.2). Overlapping flow-parallel vortices and flow-transverse secondary flows created dunes whose slipfaces presented oblique steps that were near to the optimal angle for the development of oblique flow (Fig. 26.26). At this angle, strong oblique vortices developed and created oblique dunes, and these could be either to the left or right of the prevailing wind direction, depending on chance factors. Where the obliquity at any one site is preferentially to the left or to the right of the evidently unidirectional flow, Wilson invoked a slight crosswind, tipping the scales in favour of the single oblique element. Observations on linear dunes in the lee of hills lend further support to Wilson's model of obliquity. Warren & Knott (1983) and Tsoar (1989) published the same air photograph of central Sinai in which linear dunes down wind of an escarpment have a consistent trend of about 30° to the trend of the unidirectional wind regime as indicated by accompanying barchans (Fig. 23.3). A similar phenomenon was noted on the Bilma scarp in the Ténéré Desert of Niger by Mainguet (1984a).

In another explanation, Wippermann (1969) suggested that morphological and dynamic obliquity arose

because roll vortices would be aligned at an angle to the flow. It must be said there is no field observation of the dynamics of any of these oblique dune forms or processes.

(c) Roughness-contrast obliquity
Roughness-contrast obliquity has been explained in Chapter 20.4.6. When winds flow over a positive roughness contrast (smooth to rough), they are deflected to the left in the Northern Hemisphere and to the right in the Southern Hemisphere (Fig. 20.13; Warren 1976b). A negative roughness contrast has the opposite effect. Such contrasts can occur when winds flow from transverse to linear dunes, or vice versa. This is well illustrated by Twidale's (1972) and Wasson's (1976) examples of transverse and linear dunes in Australia, and Lancaster's (1983b) examples from the Namib Sand Sea.

(d) Obliquity in multidirectional wind regimes
The discussion in Sec. 26.2.3b shows that obliquity of linear dunes to the resultant sand movement is not uncommon. Hunter et al. (1983) found that morphologically linear dunes in a multimodal wind regime were not only oblique to the resultant of sand transport (allowing for wet sand), but that they also migrated transversely. These discoveries prompted Rubin & Hunter (1985) to examine the sedimentary record, and to discover that there were far fewer records of linear dunes than their present-day distribution would suggest. One explanation would be that most linear dunes move obliquely to the resultant, for then their internal structures would be difficult to distinguish from those of transverse dunes.

Considering the great variety of wind environments, obliquity by movement and internal structure might be expected to be quite common, and a growing number of examples are being described. Carson & McLean's (1985) dunes in northern Canada are moving transversely to the resultant of sand movement, but also "bulking" (becoming larger) in the resultant direction, showing that significant amounts of sand are being taken in this direction. Hesp et al.'s (1989) dunes in northeastern China, while linear in form, were migrating transversely (linear by form, oblique by movement). Nevertheless, Rubin & Hunter's claim that most linear dunes were oblique by movement and internal structure may be rash, for records of the internal structure and movement of linear dunes are very few, and some have been found on good evidence to grow in the linear trend (Tsoar 1978).

It is not only linear dunes that have been found to be oblique in multidirectional wind regimes. Havholm & Kocurek (1988) found an obliquely moving mega-dune in southern California that had a conventional transverse form, at right angles to the resultant of sand movement. The smaller, lower-memory dunes on the flanks of the mega-dunes reflected the secondary flow patterns induced by the larger underlying feature, and many were moving obliquely to the resultant. This oblique movement of the meso-dunes induced oblique movement of the whole mega-dune, and indeed of the whole dunefield. Hesp et al. (1989) described other "oblique" transverse dunes on the South African coast.

26.3.3 Nomenclature for oblique dunes

There are three problems with oblique dunes as a category:

(i) Linear dunes that are oblique to the resultant are not unusual, so that if a classification depended only the basis of alignment with the resultant, many, if not most, dunes commonly classed as linear would be reclassified as oblique; but to do this one would have to establish oblique movement, and as Rubin (1990) found, this is not always easy; for

(ii) there are always problems in calculating the resultant, and in finding relevant wind data (Carson & McLean 1985); and

(iii) it is not easy, when mega-dunes are involved, to distinguish palaeo-dune trends from modern ones.

For obliquity in contemporary winds (where it can be established), the first step towards greater understanding is more precise nomenclature. It is suggested that, for example, the Oregon coastal dunes of Hunter et al. (1983) should be termed "resultant-oblique linear dunes by trend and movement". The dunes of Hesp et al. (1989), and those of Hedin would be "unidirectional-oblique transverse dunes by form". The dunes described by Havholm & Kocurek (1988) and Carson & McLean (1985) would be "resultant-oblique transverse mega-dunes by movement"; and so on.

26.4 Star dunes, dune networks and reversing dunes

Star dunes (Fig. 26.27) are mega-dunes whose cross sections are more symmetrical (along any axis) than other types of dune. Characteristic length:width ratios are near unity (although many star-like dunes are elongated). They may grow slowly to great heights; most move very slowly indeed.

Dune networks and *reversing dunes* are meso- or ephemeral-dune equivalents of star dunes.

26.4.1 Star dunes

(a) Description and distribution

Star dunes have been called sand mountains, stellate dunes, pyramidal dunes, sand massifs, "*oghrouds*", "*ghords*", "*tu'us*", "*barahis*", "*qa'ids*", "*medaños*", "*demkhas*" and probably by many other vernacular names, for they are striking features. They are the highest of mobile dunes, reaching to 400m in the Chinese deserts and in the Lut in Iran, 350m in the

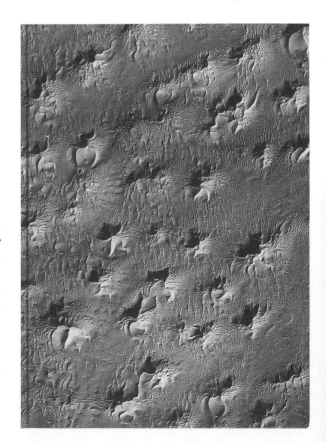

Figure 26.27 Star dune patterns in the Great Eastern Sand Sea of Algeria.

Issouane Sand Sea, and 320m in the Great Eastern Sand Sea in Algeria (Mainguet & Chemin 1984). Their formation has been reviewed by Lancaster (1989c). Star dunes, as defined here, are mobile dunes on open plains, as distinct from anchored echo dunes near major escarpments. This distinction may not always be easy to make, for disturbed wind regimes occur in a wide area around mountain massifs, and many star dunes are found in these zones. Moreover, even star dunes on open plains may be virtually immobile.

Star dunes, defined in this way, are much less common than either linear or transverse dunes (Breed & Grow 1979). They are found in the northwestern Sahara, the Fachi–Bilma Sand Sea in Chad/Niger, the Gran Desierto of Mexico, the margins of the northern Chinese sand seas, the northeastern Rub' al Khali, many small sand seas in the western USA, and the northeastern Namib Sand Sea. The greatest continuous area of star dunes is in the Great Eastern Sand Sea in Algeria, where they cover about 12,000km^2 (Mainguet & Chemin 1984).

Star dunes have several variants (Breed & Grow 1979): elongated but still peaked dunes, as in many small dune fields in California (Smith 1982), in the Fachi–Bilma Sand Sea (Mainguet 1984a) and in the Namib (Lancaster et al. 1987); dunes with three or more sharp-pointed, radiating arms as in the Great Eastern Sand Sea in Algeria; pyramidal dunes with four radiating arms as in the Gran Desierto (Lancaster 1989b); and rounder compact dunes with short arms as in Saudi Arabia (Holm's 1953 "dome dunes"). Many zones of star-dunes merge imperceptibly into zones with more clearly recognizable transverse or linear mega-dunes (I. G. Wilson 1973a, Breed & Grow 1979). Star dunes are found on the crests of many linear mega-dunes and in many sand seas star dunes are arranged in linear or quasi-regular patterns (Fig. 26.27). Star-like dunes, up to a kilometre in diameter, occur in central Australia, but do not reach the heights of star dunes in most other sand seas (Fig. 26.28). These may be "proto-star dunes" (cf. "proto-mega-barchans" and "proto-linear mega dunes" in Sec. 26.1.4 and Sec. 26.2.3).

The two-dimensional cross-sectional form of star dunes is not unlike that of linear mega-dunes: a stable, low-angle plinth usually of coarse sand, sometimes lightly vegetated; and steepening slopes capped by a series of slipfaces. Star-dune summits contain some of the finest and best sorted of aeolian sands (Folk 1971, Lancaster 1983b, 1989c). Star dunes are usually

Figure 26.28 A "proto-star dune" in central Australia (now stabilized).

covered by lower-memory dune networks. When the star dune is low, the meso-memory dunes in the network integrate the seasonal pattern of winds in the usual way (Sec. 26.4.2), but as the star dune grows, the network pattern begins to reflect flow around the mega-dune itself (Nielson & Kocurek 1987, Lancaster 1989c; Fig. 26.29). When the mega-dune reaches a

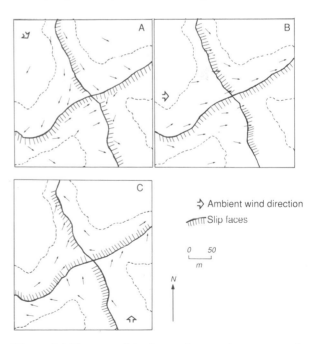

Figure 26.29 A model of star dune maintenance and growth. Once a star dune is in place, winds from different directions (open arrows) are channelled around the dune in such a way as to encourage growth (after Lancaster 1989c).

critical size, secondary, inward flows are generated in the lee of any wind, and these retain sand on the dune. This positive feedback should accelerate the growth of the star dune (Lancaster 1989c). Vertical growth, at the rate of 30cm yr^{-1}, has been observed on star dunes in the Great Sand Dunes in Colorado by Andrews (1981).

Although the main process on star dunes may be accretion, they are not necessarily stationary. Wilson (1972b) and Nielson & Kocurek (1987) argued that a completely balanced pattern of sand movement was unlikely, and therefore that most star dunes must be migrating, however slowly. In the Namib, star dunes have preferential alignments, and, therefore Lancaster (1983b) inferred, preferential directions of movement.

(b) Models of star-dune formation

Two hypotheses of star-dune formation have been proposed. The first can be termed the **standing-wave hypothesis**, a model which has been periodically resurrected since its first proposal by Hedin (1903), for example by Tricart & Cailleux (1964), Clos-Arceduc (1966b) and Folk (1971). In this model sand is swept onto the dunes by the standing atmospheric waves. The apparent regular spacing of some star dunes (Wilson 1972b, Mainguet 1984a) supports this model, but Breed & Grow's (1979) global survey found much less correlation between dune size and spacing in star dunes than in other forms of dune. The standing-wave hypothesis has been linked to the notion that many star dunes are erosional (as in the wind-rift hypothesis of linear dune formation).

The best authenticated hypothesis relates star dunes to regional wind regimes: the "multimodal wind regime hypothesis" (Mabbutt 1968, Fryberger & Dean 1979, Wasson & Hyde 1983b). Most star dunes exist in complex wind regimes (Fig. 26.30), and some star dunes appear to grow vertically as they accumulate sand brought in from a number of directions (Lancaster 1988c). Spacing in such a regime may be a matter of the average distance travelled by sand grains in a windy season (Allen 1982b).

Wilson's (1972b) hypothesis is a variant of the multimodal regime model, developed for the Great Sand Seas of Algeria (Fig. 26.31). He believed there to be major dune trends associated with winds from

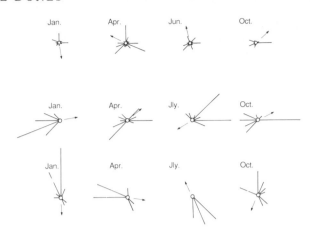

Figure 26.30 Wind regimes associated with star dunes in the eastern Great Eastern Sand Sea of Algeria (top), the Eastern Namib Sand Sea (centre) and the Gran Desierta in Mexico (foot) (after Lancaster 1989).

different directions. Where wind systems overlapped, so did the dune trends, and a star-dune pattern emerged.

(c) Maintenance

Whatever process initiates them, star dunes may be maintained, in part, by a further process. Masses of sand over 100m in height (as most star dunes are) are likely to develop their own thermally driven wind

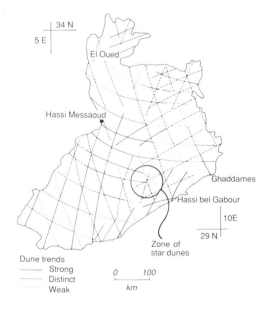

Figure 26.31 The relationship of star dunes to the intersection of major dune trends in the Great Eastern Sand Sea of Algeria (after Wilson 1970).

systems, which might encourage their growth, and control details of their morphology (Cornish 1897, Mainguet 1984a, Howard 1985).

26.4.2 Dune networks

Dune networks (Fig. 26.32) are the meso-dune, lower-memory equivalents of star dunes. The term "network" is used here more generally than it was by Mabbutt (1968) and Mabbutt & Wooding (1983): for them, networks were complex ring-like structures whose individual cells were hundreds of metres across. The cells in a dune network, as the term is used here, are a few tens of metres across. As such dune networks are what Melton (1940) in desperation called "complex", and what have by others been called "dune reticules", "*aklés*", "rhombic waffle patterns", "alveolar dunes" or "*dune quadrillage*".

As Aufrère noted (1935), dune networks (his "reticules") are formed by overlapping systems, usually of transverse dunes, each adjusted to a different seasonal wind. They occur only in continuous sandy terrain, for, where there is less sand, dunes are separated by bare desert floor, and the form they adopt is usually that of barchans constantly readjusting to seasonal changes in wind. Dune networks occur at several levels of complexity. At their simplest, they are no more than an overlap of a system ephemeral dunes, adjusted to a weak or short-period seasonal wind, on top of meso-dunes adjusted to the strong or persistent wind of another season. At their most complex they consist of the superimposition of two or more almost equally developed meso-dune systems.

Networks are the commonest form of mobile coastal and desert dune. Details of dune network behaviour have been described in the seasonally variable wind regimes of coastal Poland (Borówka 1980), and Oregon (Hunter et al. 1983), and in desert dunes at Kelso and the Algodones dunes in California (Sharp 1966, Havholm & Kocurek 1988). Sharp remarked on the vast quantities of sand that were moved back and forth on the summits of the large dunes in the Kelso field, despite the return of the pattern to more or less to the same place at the end of each annual cycle. These characteristics are common to most dune networks.

A system of nomenclature was developed for dune networks by Warren & Kay (1987; Fig. 26.33). In this example, a dominant transverse (meso-) dune system (T1) created by strong summer SSW winds is overlain by a smaller, secondary (ephemeral) transverse dune system (T2) adjusted at right angles to northeasterly winds. This in turn is overlain by tertiary and quaternary ephemeral dune systems (T3 and T4) adjusted to occasional winds from other directions. These lighter, short-duration winds activate only the upper slopes of the T1 system, so that the T3 and T4 systems are found only there. When light winds blow from the T1 direction, they create dunes with a shorter wavelength than the T1 ridges, and these are termed "T1a".

Each subtly different local regional wind regime

Figure 26.32 A dune network in the Wahiba Sands of the Sultanate of Oman.

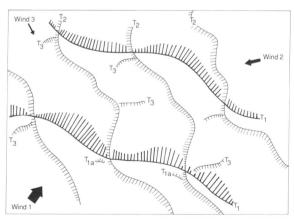

Figure 26.33 Terminology for dune networks. The strongest (wind 1) creates the most prominent set of transverse dunes ("T1"). Other winds produce less well developed systems of transverse dunes overlying these ("T2"; "T3" . . . etc.) (after Warren & Kay 1987). The evolution of this pattern is shown in Figure 23.9.

generates its own unique network pattern. Over time, only the longest-memory dunes (the "T1" dunes) survive from year to year; intermediate-memory dunes (T2 and perhaps T3) change through a yearly cycle; and minor details (short-memory dunes, T4 and so on) may change from day to day. In Warren & Kay's network, little more than the T1 system existed at the end of the summer season of constant strong winds, while some T4 dunes of the variable winter winds lasted barely a few hours. In other wind regimes there are different patterns of relative destruction and survival.

26.4.3 Reversing dunes

Reversing dunes are a special case of dune networks, in which the winds of two seasons are diametrically opposed. They were described by Hedin (1903) and Cornish (1897), and since then in many parts of the world, for example in the Great Sand Dunes of Colorado, where Andrews (1981) described how reversing dunes responded partly to anabatic and katabatic winds generated by the nearby Sangre de Cristo Mountains, and partly to regional westerlies (Fig. 26.34).

An element of dune reversal is common. It usually takes the form of the development of dunes with 1–2m-high slipfaces pushed by the reversed wind over the brinks of the pre-existing slipfaces. The topographic details of the process were followed in detail by Verlaque (1958), and the sedimentary ones by Reid (1985). Reid found that one wind winnowed fine sand from a windward face and, because this altered the roughness and threshold velocities of movement, the slope angle steepened. The same wind

deposited the finer sand in a lee position. The reverse wind quickly eroded this now inactive lee face, creating a new set of equilibrium slope angles on the new windward slope. The reverse wind piled fine sand onto the earlier-exposed coarse sand of the new lee face.

When reversal in a dune network in an anabatic/katabatic wind system is analyzed, it is seen that the system of wind pathways in the dominant direction through the network is much simpler than the system of pathways in the reverse (minor wind) direction (Burrell 1971). This suggests greater energy loss in the minor than in the major direction (to which the dune pattern is mainly adjusted), an observation which conforms to Yalin's ideas about dune formation in general, reported in Chapter 23.2.3c.

26.5 Sandsheets and zibars

26.5.1 Sandsheets

Sandsheets are areas of aeolian sand with very subdued dune forms. As used here, the term includes sand streaks or stringers (sharply defined, narrow sandsheets) and other features with low relief which were differentiated by Breed & Grow (1979). The largest known sandsheet, over 100,000km^2, is the Selima Sandsheet on the borders of Egypt, the Sudan and Libya. Other large sandsheets occur in the Ténéré desert in Niger (Warren 1971), in parts of Saudi Arabia (Holm 1960), in the southern Namib Sand Sea (Lancaster 1985b), and on the northeastern edges of the Australian sand deserts (Brookfield 1970). Most are much smaller than the Selima Sheet, many bordering sand seas. Many sandsheets act as stable bases over which other dunes migrate (Khalaf et al. 1984, Breed et al. 1987).

On the ground, most sandsheets appear to be virtually featureless, level plains underlain by near-horizontally bedded sands, only a few metres thick, commonly with a lag of coarser grains on the surface, and a significant clay fraction (Mabbutt 1968, Fryberger et al. 1979, Khalaf & Al Hashash 1983,

Figure 26.34 Reversing dunes in the Wahiba Sands of the Sultanate of Oman. The larger dunes are formed by the strong, summer southwest monsoon. The smaller dunes with slipfaces facing the camera are formed by reversed winds in the winter. A plot of changes in this dune field is shown in Figure 23.9.

Breed et al. 1987). However, in many of the Saharan and Arabian sandsheets, air photographs reveal zibar patterns on apparent sandsheets (below). In Australia, the sandsheets have subdued linear patterns (Brookfield 1970). Although the northern Australian and Sahelian sandsheets are evidently inactive, those in southern Egypt and northern Sudan seem to be active, even exhibiting change between LANDSAT scenes (Maxwell & Haynes 1989). Baked-bean cans, only about 40 years old, have been buried by a few centimetres of sand beneath the Selima Sandsheet (Breed et al. 1987). The great Selima Sandsheet has been found to be capped by a Neolithic soil, the production of which introduced some of the fine material (Haynes 1989).

Sandsheets can form in at least four situations:

(i) *In coarse sand* The most important control on the formation of the large desert sandsheets of the Sahara and Arabia is the coarseness of sediments (Khalaf et al. 1984, Breed et al. 1987, Khalaf 1989). This has also been cited as a control in Arctic Canada (McKenna-Neuman & Gilbert 1986), parts of the Mojave (Kocurek & Nielson 1986), and Australia (Mabbutt 1980; but see also Crocker 1946 and Mabbutt 1967). The clearly mobile sediment on the Selima Sandsheet has a mean size of 1.5mm (Breed et al. 1987).

(ii) *In the presence of vegetation* Some sandsheets are said to be kept fairly level because vegetation curtails sand movement and dune growth (Kocurek & Nielson 1986, Khalaf et al. 1984, Khalaf 1989). Many desert-margin sandsheets owe their origin to this kind of control. This may be part of the explanation of sandsheets in northern Australia, and on the north European Plain in the late Pleistocene (Schwan 1988).

(iii) *Nearness to the water table* In restricted zones, a water table near the surface (Stokes 1968) or periodic flooding (Fryberger et al. 1988) prevents the growth of dunes.

(iv) *Wind erosion* Some sandsheets have been shown to be erosional remnants of higher dunes (Tricart & Brochu 1955, Ahlbrandt & Fryberger 1981, Fryberger et al. 1984).

26.5.2 Zibars

(a) Description and distribution
Zibars (Figs 23.25, 26.35) are an extensive dune type, certainly more extensive than barchans or star dunes

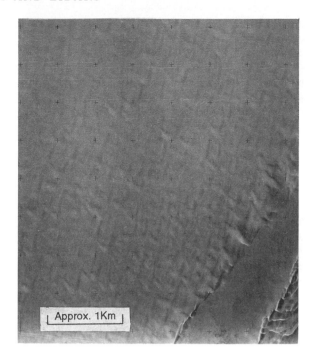

Approx. 1Km

Figure 26.35 A parabolic zibar from extremely arid southeastern Libya.

(even together), yet they have received little attention. The term was borrowed from the Arabic by Holm (1953) to describe "rolling transverse ridges . . . with a hard surface". Holm's zibars, and all others in the literature, conform to the more precise definition of Nielson & Kocurek (1986): dunes without slipfaces, built of coarse-grained sand. Most are of low relief, although some of the zibars in the Ténéré Desert are higher than nearby "sayf" dunes, reaching over 5m (Warren 1972).

Thus defined, zibars are synonymous with Bagnold's (1931) "whalebacks" in the Western Desert of Egypt (although Bagnold was not specific about the character of these), Greeley & Iversen's (1985: 171) "granule-armoured dunes", Monod's (1958) "mréyé" (or "giant ripples") in Mauritania, Mainguet & Callot's (1978) "giant undulations" in the Fachi–Bilma Sand Sea, and the "low rolling dunes without slip faces" of Lancaster (1983b), even though Lancaster described these as being themselves covered with zibars (this may be a zibar hierarchy). By this definition, zibars may also include "dome dunes", as in Iraq (Jawad Ali & Al-Ani 1983). The term may also extend to the "wanderrie banks" of central Australia, which have roughly the same dimensions and grain size as other zibars (Mabbutt 1963). The definition thus includes features with a wide range of heights, wavelengths and plan forms.

Zibars are common up wind of sand seas, in zones from which finer material has been winnowed (Capot-Rey 1947, Warren 1972, 1988a, Lancaster 1983b, Breed et al. 1987). They occur both in extreme deserts such as the central Sahara (Fig. 26.35) and in lightly vegetated areas such as southern California. Individual sheets of zibars are extensive: Monod's *mréyés* in Mauritania cover 10,000km².

Most zibars are straight in plan form, with little curvature, and are apparently aligned at right angles to the strongest winds, presumably the only ones able to move the coarse sand (Warren 1972, Lancaster 1983b). Nonetheless, many zibars have a "parabolic" form, called here **zibar parabolic dunes**. These have been described in the Jafurah sand sea of eastern Saudi Arabia (Anton & Vincent 1986), in Wyoming (Gaylord & Dawson 1987), in the Mallee region of New South Wales (Bowler & McGee 1978), in Western Australia (if "wanderrie banks" are zibars; Mabbutt 1963), in the southwestern Wahiba Sands in Oman (Goudie et al. 1987, Warren 1988a), in southwestern Egypt (Haynes 1989), and in the extremely arid Faya–Largeau area (Capot-Rey 1957b). Parabolic zibars in the extremely arid central Sahara are illustrated in Figure 26.35. The parabolic shape in zibars is less elongated and more symmetrical than in most vegetated parabolic dunes. Linear zibars have also been described (Warren 1972).

Three explanations have been proposed for zibar parabolic dunes. First, they are seen as no more than remnants of vegetated parabolic dunes. In south-western Egypt, Haynes (1989) found grass stems in their strata, and signs of a Neolithic soil on their surfaces, both indicating wetter conditions in the past; and the small amounts of vegetation in the Jafurah Sand Sea and Iraq convinced Anton & Vincent (1986) that these dunes were vegetated parabolic dunes. However, this proposal does not account for the very coarse sand in zibar parabolic dunes when compared to vegetated parabolic dunes.

A second and more complex mechanism, involving stream and wind action, was proposed by Mabbutt (1963) for his wanderrie banks. Finally, it is proposed here that zibar parabolic dunes are the linguoid sections of continuous transverse dunes, which have been exaggerated in areas with high concentrations of coarse sand. The coarse sand is channelled around the barchanoid sections and it collects in the linguoids, which are then enlarged. The fine sand moves quickly down wind (Warren 1972).

(b) Zibar dynamics

Many zibars are evidently active, even when covered by sparse vegetation (Nielson & Kocurek 1986, Breed et al. 1987). Finer material is deposited in the lee, and coarser material is left behind as the zibar moves forwards. Unlike conventional dunes there is thought to be no flow separation in the lee of zibars, only flow expansion. Theory suggests that coarse sand produces more gently sloping dunes (Ch. 23.5; Howard et al. 1977, Tsoar 1986). Simulation models of dune dynamics show that an increase in grain size or of windspeed creates dunes with gentler windward slopes (Howard et al. 1977).

CHAPTER 27

Aeolian sedimentary structures

27.1 Introduction

The internal structure of contemporary sand dunes reveals useful evidence about their mode of formation. The chief interest in aeolian sedimentary structures has come, however, from sedimentary geologists (McKee 1979d), whose research has made a major contribution to the understanding of desert dunes, from a very early date (Wade 1911). They have had two concerns: to find oil and water, for aeolian foreset beds are among the most porous of deposits, and therefore excellent underground reservoirs (Fryberger 1979); and to reconstruct past environments.

Ancient dune sandstones date from as far back as the Proterozoic (Ross 1983). They are particularly well represented in the Mesozoic in the western USA (Kocurek 1981a, Clemmensen & Blakey 1989), and in northwest Europe, particularly beneath the North Sea, where aeolian sandstones contain much of the oil and gas (Glennie 1983). In onshore Britain alone, there are extensive ancient aeolian sandstones in the Midlands (Shotton 1956), northeast England (Steele 1983), northwest England and southwest Scotland (Brookfield 1979) and northeast Scotland (Clemmensen 1987).

Sandstones are not always unmistakably aeolian, and controversy surrounds the environmental interpretation of some of them, notably those in the southwestern USA. This has motivated sedimentologists to enquire very carefully into aeolian sedimentary structures, and this has generated a large literature, which is discussed only briefly in this chapter.

27.2 Fundamental structures

Windblown sand accumulates in three ways: by grainfall, in climbing ripples, and on slipfaces (Fig. 27.1; Bagnold 1941, Hunter 1977a, b, McKee 1979c). Individual "strata" (deposits of short, internally uniform episodes, of the order of hours long) are arranged in "sets" (deposits of longer episodes within which the sets represent cyclic variation about a mean condition), themselves separated by "truncation surfaces" (breaks between episodes at various scales) (Hunter 1977b).

27.2.1 Grainfall strata

The processes of grainfall have been described in Chapter 23.8. Grainfall strata are preserved first at the top of slip faces, where they interdigitate with sandflow deposits, and dominate the deposit. The dominance of grainfall strata in these positions distinguishes the deposits of aeolian dunes from those of subaqueous dunes. Grainfall strata pinch out below about a metre from the brink, giving way to sandflow or

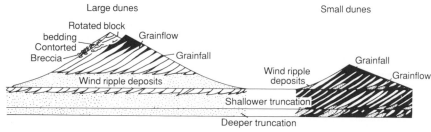

Figure 27.1 Grainfall, grainflow and ripple deposits related to position on transverse dunes (after Kocurek & Dott 1981).

Large dunes

Rotated block
bedding
Contorted
Breccia
Grainflow
Grainfall
Wind ripple deposits

Small dunes

Grainfall
Wind ripple deposits
Grainflow
Shallower truncation
Deeper truncation

397

slump strata. They are better preserved on lee slopes with lower angles than those of the slip face, which experience grainfall in certain wind conditions, and these may make up quite a high proportion of a total aeolian sedimentary deposit. Another site where grainfall strata accumulate is the narrow apron at the base of the slipface, which is created partly from material projected beyond the slipface in high winds. These "bottom-set" deposits in aprons may make up about 20% of a total aeolian deposit (Hunter 1985).

Although the angles of grainflow strata are gentler than those of sandflow and slump strata, they are generally steeper than those on which most ripple strata accumulate and they can reach 28°. Individual grainfall strata are on average thin (seldom 0.5cm), although very variable in thickness, parallel, and with indistinct boundaries. They become finer-grained down slope. Texture depends on wind velocity. Porosity ranges from 30% to 42% (Hunter 1977b, Schenk 1983).

27.2.2 Ripple strata

Ripple, "tractional" or "accretion" strata (Fig. 19.6) are the most diagnostically aeolian of sedimentary structures. Unlike most subaqueous ripple strata, most aeolian ripple strata do not preserve either the foreset strata deposited on the lee slopes of the ripples themselves, nor wavy rippled surfaces. They thus appear as parallel laminations, which are more finely laminated than most subaqueous ripple strata, each stratum usually being only a few millimetres thick (Fig. 19.6; Hunter 1977b, Kocurek & Dott 1981). Field experiments show that each stratum is deposited by a migrating ripple whose upper part has been stripped off by bombardment, to leave only the lower part for preservation (Hunter 1977b). Coarse grains of the ripple crests overlie finer grains of troughs. Individual strata are separated by sharp, planar "ripple-truncation" surfaces, formed as successive ripples climb over each other. Most structures are simple, but on rare steep slopes aeolian ripple structures can be complex (Hunter 1977a, b). Thin "pin-stripe" strata, formed by the accumulation of silt and clay in ripple troughs are sometimes present (Fryberger & Schenk 1988).

Climbing is in the wind direction, and either upwards or downwards depending on the local slope of the underlying dune, so that the direction of climb may diverge quite considerably from the ambient wind direction. In general, steeper down-current angles of climb occur when there is a higher deposition rate.

Most ripple strata are extensive laterally and have low angles (<6°), but in rare, steeply climbing situations may reach high angles (as when the wind reverses). Strata are thicker when the rate of sedimentation is higher. Ripple strata have porosities of about 40% (Hunter 1977b, Fryberger & Schenk 1981, Schenk 1983). They have coarser, more poorly sorted sand, with more marked fine-skewness than other kinds of aeolian strata; and many of the bulk sands are bimodal (Folk 1968; Chs 18.8.1c, 22.2.1c). Ripple strata are found in several different positions in a dune field:

(i) *Interdune corridors and the margins of sand seas and dune fields, sandsheets and zibar fields* provide the thickest sets of ripple strata, reaching several metres, especially between stable dunes such as star dunes or on sandsheets; these ripple strata are commonly disturbed by biogenic activity, and may intercalate with silt and clay lenses or evaporites from permanent or ephemeral lakes or puddles (Ahlbrandt & Fryberger 1981); if they are continually overridden, as by advancing transverse dunes, they may be diachronous and laterally fragmented. Ripple strata from sandsheets have been described by Nielson & Kocurek (1987) and Breed et al. (1987).

(ii) *Plinths* (defined in Fig. 23.1) provide sets that are generally thin and merge or interdigitate up slope with slipface strata, whose angles they approach (McKee & Douglass 1971); these sets are separated by many minor truncation surfaces (below).

(iii) *Crests of transverse dunes* (or "topset strata"; Ch. 23.3.5). Because of dune migration, these strata are seldom preserved.

27.2.3 Slipface strata

Slipfaces have been discussed in Chapter 23.3.7. Massive sets of slipface strata, also called "cross stratification", "encroachment strata" or "foreset strata", make up the greater part of most aeolian sandstones (Fig. 27.2). Individual strata are commonly a few centimetres thick and they lie at 31°–34°; sets may reach 35m thick. Slipface strata are very loose, with porosities of about 45% (Hunter 1977b). In horizontal cross section, sets of slipface strata are generally curved, reflecting the curves of the lee slopes of their parent dunes (Fig. 27.3). Thickness, curvature of sets and perhaps the thickness of sets, between second-order truncation surfaces, are positively correlated with the height of the parent slipface (Kocurek & Dott

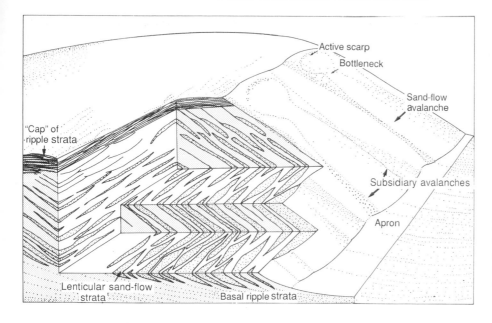

Figure 27.2 Structure of grainflow slipface deposits, also showing structures of avalanches (redrawn after Hunter 1977b).

1981, Rubin & Hunter 1982).

There are two subtypes of slipface strata:

(i) *Sandflow strata or "avalanche" strata* are formed by sand flows. On small slipfaces, they are lenticular in horizontal and vertical cross section (Fig. 27.2). On large slipfaces, sand-flow strata are almost tabular–planar except near the brink. Contacts between strata are generally sharp except near the brink (Hunter 1977b). Sands generally coarsen down slope. Fluctuations in wind velocity produce strata with different textures or mineral composition (Hunter & Richmond 1988). Sandflow strata interdigitate with grainfall deposits near the brink and ripple strata at the base of the slope.

(ii) *Slump strata* ("block-slide" strata) are formed when coherent bodies of sand slide down a slipface. They are tabular–planar or wedge–planar, pinching out at the top (Fryberger & Schenk 1981, Schenk 1983). Minor faults and folds may be formed in the block-sliding processes, and the slump may break into loose breccia at the base (McKee et al. 1971, Borówka 1979; Figs 23.20 & 21).

27.2.4 Truncation surfaces

Aeolian sand strata, being much more subject to changes in flow strength and direction than fluvial or marine deposits, are characteristically separated by many truncation (or bounding) surfaces (or planes). These are generally low-angle and they dip down wind (McKee 1979c). Truncation surfaces can be classified as follows (Brookfield 1977, 1979, Talbot 1985, Kocurek 1981b, 1988):

(i) *Super-truncation surfaces*, commonly characterized by palaeosols, formed by hiatuses in dune formation, as when there is a change in climate, or by deflation to a water table. They are commonly recognized by a surface lag of coarse sand.

(ii) *First-order truncation surfaces*, which are extensive and low angle, and formed by the passage of large (mega-) dunes.

Figure 27.3 Curving slipface deposits preserved in aeolianite in the southern Wahiba Sands of the Sultanate of Oman.

Figure 27.4 Second- and third-order truncation surfaces on a cliff exposure of aeolianite in the Wahiba Sands of the Sultanate of Oman.

(iii)   *Second-order truncation surfaces* (Fig. 27.4) which are gently to moderately dipping, and formed by the passage of moderately sized (meso-) dunes.

(iv)   *Third-order truncation surfaces* (Fig. 27.4) formed by local erosion and deposition in response to seasonal or daily winds (the formation and destruction of low-memory, ephemeral dunes). The way in which daily cycles of wind velocity and direction can create third-order truncation surfaces is beautifully illustrated by Hunter & Richmond (1988).

(v)   *Ripple truncation surfaces*, as described above. This series has parallels with the dune-memory hierarchy outlined in Chapter 24. The central three truncation-surface groupings correspond to the mega-, meso- and ephemeral dune categories of that hierarchy.

27.3 Sedimentary structures of common dune types

27.3.1 Transverse dunes

On transverse dunes, slipface strata have a sharply peaked orientation and are separated by second-order truncation surfaces (Bagnold 1941, McKee 1979c). The sets are more curved in barchans than in continuous transverse dunes. On active dunes, the curves are revealed when rain moistens the sand, or when a sabkha water table seeps up into the lower layers of a dune that has migrated over its surface (Fig. 23.22). The same patterns appear in aeolianites (Fig. 27.3).

27.3.2 Linear dunes

The most comprehensive analysis of linear dune stratification was undertaken by Tsoar (1982) on a "sayf" dune in Sinai. A rare heavy rainstorm had saturated the whole dune, and succeeding winds had stripped off layers of sand as it dried to reveal the dune structure (Fig. 27.5). Most of the deposition was in low-angle strata, up to about 25°, which occurred in crescentic units opening down wind. In the annually reversing wind regime of northern Sinai, these strata are truncated by second-order truncation surfaces created when the wind changes and erodes part of the previous season's strata. The ripple strata become increasingly steep towards the middle, or highest part of the dune and contain finer sand up slope (Sneh & Weissbrod 1983). Slipface strata occur only near the crest of linear dunes, dipping alternately in different directions. This pattern conforms more or less to the hypothesis of Bagnold (1941) and the findings of McKee & Tibbitts (1964) and McKee (1982).

Rubin (1990) believed that stratification in linear dunes would vary according to whether they were: (a)

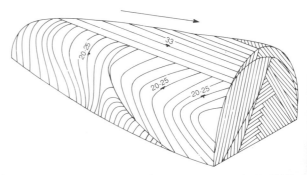

Figure 27.5 Bedding in a linear dune in northern Sinai, revealed by the stripping of dry sand from a deeply wetted sayf dune (after Tsoar 1982).

erosional (the wind-drift hypothesis); (b) accreting, but extending down dune (the dunes staying in the same place, separated by interdune deposits); or (c) oblique in the sense of Hunter et al. (1983), in which case they would be composed mostly of slipface strata, with some interdune strata near the base, and perhaps some superimposed along-dune structures as predicted by Glennie (1970).

27.3.3 Star dunes

Star dunes are composed mostly of low- to moderate-angle ripple strata, with many gently sloping truncation surfaces (Nielson & Kocurek 1987). Like linear dunes, slipface strata are preserved only near the summit. In star dunes, the dip directions are more variable.

27.3.4 Vegetated dunes

The characteristic stratification in vegetated dunes is a bimodal distribution of inclinations, specially of steeper-angle strata, and minor truncation surfaces (McBride & Hayes 1962, Embabi 1970/71, Goldsmith 1973, 1985, Yaalon 1975, Clemmensen 1986, Gunatilaka & Mwango 1987, 1989). Most dips are low, at about 12° (Goldsmith 1989). The pattern is derived from shadow dunes.

Not all coastal or streamside dunes, however, show this bimodal distribution, for when there is a large unvegetated transverse dune parallel to a beach or a stream bank (in areas of strong sand supply), dips are more unimodally distributed (Goldsmith 1973, McKee 1982). Many sets of aeolianite strata are extensive tabular–planar and wedge–planar sets derived from coastal dunes of this type; gently dipping truncation surfaces separate sets groups of strata (McKee & Ward 1983).

McKee (1966) found vegetated parabolic dunes to be predominantly composed of slipface strata, much like any other transverse dune, but with more lower-angle sets in the upper part. Many slipface strata were concave down wind and they showed a wide directional spread. There were, unsurprisingly, many plant remains. Some parabolic strata form characteristic concave-upward sets deposited in hollows (McKee 1979c, Bigarella 1979). Parabolic dune strata comprise quite a high proportion of some aeolianite deposits (Gardner 1983, McKee & Ward 1983, Semeniuk & Glassford 1988).

Since most vegetated dunes have formed in at least semi-arid climates, structures associated with wet sand are common in their strata. These include slump structures and brecciated material (McKee & Ward 1983)

27.3.5 Sandsheets and extensive interdune areas

In sandsheets and interdunes, ripple strata are at their thickest and most dominant. Most are very low-angle, and may be of very coarse sand. In the Selima Sandsheet in Sudan, these have evidently accumulated over more than 250,000yr (Breed et al. 1987). However, some sandsheets have been eroded across dune sediments, and are underlain by cross-bedded dune strata (McKee 1982, Fryberger et al. 1984). Many sandsheets contain zibar deposits which are climbing sets of low-angle lee-slope strata in coarse sand, interdigitated with even coarser inter-zibar deposits (Nielson & Kocurek (1986). The sands of sandsheets and zibar are characteristically coarser and more markedly bimodal or multimodal than those of other dune strata.

Low-angle strata are formed when sand blows over a sabkha or playa and is incorporated into the surface, mainly in the form of adhesion ripples (Talbot 1980).

27.4 Preservation

If the sediment transport rate decreases (over space or over time), there is deposition. Dunes then climb over preceding ones (Fig. 27.6), preserving parts of the earlier ones. The parts preserved, which are an order of magnitude smaller than the parent dune, are known as a "climbing translated strata" (Hunter 1977a).

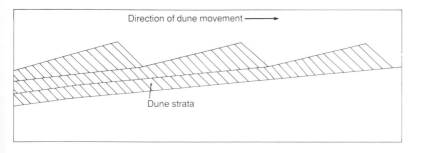

Direction of dune movement ⟶

Dune strata

Figure 27.6 Dune climbing, and the preservation of dune deposits (after Rubin & Hunter 1982).

Except in some aeolian ripples, the angle of climb is usually very low. Because the thickness of these strata is related to the ambient sand transport rate, the dune migration rate and the wavelength and height of the parent dunes, successive sets of strata between truncation surfaces are usually of roughly the same thickness (Rubin & Hunter 1982).

If linear dunes are features built only by linear extension (Tsoar's model), then their preservation in the geological record would be unlikely, because (a) they are formed where sand is sparse, and (b) they do not migrate laterally, and therefore cannot climb up on each other (Steele 1985). If, however, they do migrate as transverse dunes (or oblique dunes), their strata would be preserved, but would be difficult to distinguish from those of transverse dunes (Rubin & Hunter 1982, Rubin 1990).

CHAPTER 28

Sand seas and dunefields

85% of all unconsolidated windblown sand is said to be held in about 58 bodies over 32,000km² in area (I. G. Wilson 1973a). Even if, as is likely, there were errors in Wilson's cartographic data, there is little doubt that most aeolian sand is held in large accumulations. About 45% of the total area of these bodies is in Asia, 34% in Africa and 20% in Australia (Mainguet & Callot 1978). Wilson's data show a modal size of sand body at 188,000 km² and a skew towards smaller bodies (Fig. 28.1).

28.1 Definitions

"Sand seas" is a term used loosely by many author-ities to describe large bodies of sand, also often called **ergs** (or **irqs**). Other terms are: sand deserts, sands, sand hills, "*idihan*", "*qoz*", "*ramlat*", "*ghard*", "*nafud*", "*peski*", "*aklé*", "*kum*" and "*sha-mo*". To retain the size connotation of a sea and to add some precision to the subsequent discussion, the definition used here is as follows:

- a **sand sea** is an area of windblown sand of over 30,000km², where breaks in the cover are no wider than the natural wavelength of the local dunes; as a lower limit, 30,000km² comes near an inflection in the curve on Figure 28.1;
- a **dunefield** is a collection of more than 10 dunes in an area less than 30,000km², between which the breaks are no greater than the local dune wavelength.

The lower limit is arbitrary but convenient.

28.2 Location and extent

The areas of major active sand seas are given in Table 28.1. The largest are shown on Figure 16.1. Stabilized sand seas are listed and discussed in Chapter 29. The distinction between activity and stabilization is somewhat arbitrary, given the gradational nature of the boundary and the poor data in most deserts.

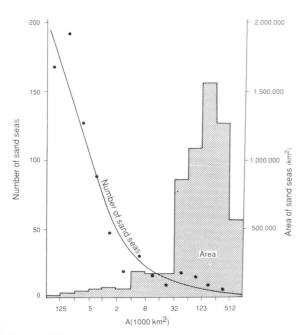

Figure 28.1 The size distribution of some major sand seas (after I. G. Wilson 1973a).

Table 28.1 Sizes of major active sand seas.

Name	Area ('000km²)	Source
Sahara		
Great Sand Sea, Egypt	72	El Baz (1984a)
Uwinat	31.5	El Baz (1984a)
Calanscio	62	Wilson (1973a)
Ribiana Sand Sea	65	Wilson (1973a)
Idihan Marzuq	58	Capot-Rey (1947)
Idihan Awbari	62	Wilson (1937a)

Table 28.1 Sizes of major active sand seas.

Name	Area ('000km²)	Source
Issaouane n'Irararen/ Tifernine	38	Wilson (1973a)
Great Eastern Sand Sea	192	Wilson (1973a)
Great Western Sand Sea	103	Wilson (1973a)
Irq Iguidi	68	Wilson (1973a)
Irq Chech-Adrar	319	Wilson (1973a)
Azefal–Tijirit–Makteir– Akchar	85	Wilson (1973a)
Amoukrouz–Aoukar– Ouarane	65	Wilson (1973a)
Majabât al Koubrâ	250	Monod (1958)
Fachi–Bilma	150	Mainguet & Callot (1978)
Selima Sandsheet	100	Breed et al. (1987)
Southern Africa		
Namib	34	Lancaster (1983b)
Kalahari (active zone)	100	Breed et al. (1979b)
Arabia		
Rub' al Khali	550	Anton (1984)
Al Dahna	51	Wilson (1973a)
Al Jafurah	57	Wilson (1973a)
Al Nafud	72	Wilson (1973a)
(former) Soviet Union		
Peski Karakum	380	Wilson (1973a)
Peski Kazylykum	276	Wilson (1973a)
Peski Priaralskye	56	Wilson (1973a)
Peski Mujunkum	38	Wilson (1973a)
Peski Sary Isnikotrav	65	Wilson (1973a)
Peski Dzosotin	47	Wilson (1973a)
China		
Taklamakan	337	Walker (1982)
Gurbantunggut Sha-mo	48.8	Walker (1982)
Qaidam Sha-mo	34.9	Walker (1982)
Badain Jaran Sha-mo	44	Walker (1982)
Tengger Sha-mo	42.7	Walker (1982)
Australia		
Simpson	80	Sprigg (1963)
Mars		
Four north-polar sand seas	110	Lancaster & Greeley (1990)
	500	
	470	
	330	

28.3 Characteristics

28.3.1 Borders

Most sand seas and dunefields are sharply defined. This comes about in several ways:

(i) *Advancing edges* The sharpness of downwind margins is the most easily explained, for here the sand sea is advancing over the desert surface. If there is a scarp next to the downwind edge, sharpness is maintained at an echo dune.

(ii) *Highland/lowland boundaries* Upwind borders abutting mountain massifs are sharpened by the sudden change from an undersaturated sand flow over the highlands to an oversaturated flow in the lowlands (Fig. 28.2; Wilson 1973a). Undersaturatedness over highlands results from acceleration and deflection; over-saturation on return to the lowlands from deceleration and convergence.

(iii) *Termination in rivers or the sea* Some sand seas and dunefields terminate down wind abruptly in seas or rivers. The most impressive coastal terminus is where the western Saharan sand seas debouch into the Atlantic on the Mauritanian coast, sustaining massive sub-marine sand turbidites offshore (Sarnthein & Walger 1974, Sarnthein & Diester-Haass 1977). There are also sharp termini in major rivers, as at the Nile in northern Sudan (Wilson 1970), at the Wadi al Batha on the northern side of the Wahiba Sands in Oman (Warren 1988a), at the Kuiseb river on the north of the Namib Sand Sea (Goudie 1972), and where the Algodones dunes in southern California meet the Colorado River (Sweet et al. 1988). As I. G. Wilson noted, such termini indicate that the river is carrying more sand than the wind. The implications of this in terms of magnitude–frequency relations is discussed in Chapter 18.7.

(iv) *Sand shepherding* Even without termini in scarps, seas or rivers, the borders of most sand seas and dunefields are sharp. The Abu Moharik dunefield in Egypt retains sharp boundaries for over 650km, although never more than 7km wide (Beadnell 1910, King 1918). Some of the borders of the Marzuq Sand Sea in Libya and of the Great Eastern Sand Sea in Tunisia are also very well defined (Fig. 28.3; Hagedorn 1979). This sharpness can be explained by the "sand-shepherding" process;

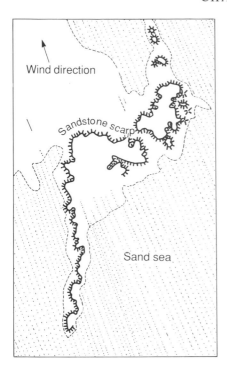

Figure 28.2 A sand sea is broken up where wind is accelerated over a scarp, and reconstituted in some places down wind where the wind decelerates (after Wilson 1973a).

sand moves quickly over stony desert surfaces, but slowly over sandy dune surfaces (Wilson 1971). Sand moving onto the sand sea or dunefield from the open desert is suddenly decelerated, and sand leaving the dune mass is suddenly accelerated.

(v) *Framing ridges* In some cases "framing ridges" develop at the edges of sand seas and dunefields (Fig. 28.3). Breed (1977) noted examples on the inland edge of the Namib, in Tunisia and Libya. Monod's (1958) *mréyé* in Mauritania is bounded by a conspicuous, named dune on the northwest side. Prominent framing ridges also occur on Martian sand seas (Tsoar et al. 1979). Three mechanisms might explain framing ridges. First, they could develop on the upwind margins of sand seas, where sand taken to the margin over a desert pavement, cannot be moved away as quickly over the sandy surface down wind. Secondly, they could form on wind-parallel margins where Bagnold's transverse instability process might be in operation. Thirdly, if there is vegetation,

they could be *randwallen* (Ch. 25.2.3a).

(vi) *Active/stabilized edges* Where active dunefields exist among stabilized dunes, a further mechanism creates sharp boundaries. This is a result of the delicate ecological balance experienced by plants on sandy soils: on the one hand there are the attractive moisture-holding capacities of sands, and on the other their repellant instability as a rooting medium. Once sand movement begins, moreover, any fine material held in the soil is winnowed away, and this alters soil moisture-holding capacity, perhaps even increasing it, possibly creating another sharp threshold.

28.3.2 Volume

If there are contoured maps of a sand sea or a dunefield it is a simple matter to produce a map of the

Figure 28.3 A sharply defined edge of a sand sea and a system of "framing ridges" around the northeastern end of the Great Eastern Sand Sea in southern Tunisia (an edge-enhanced LANDSAT MSS image).

spread-out sand volume. In the Great Algerian Sand Seas, I. G. Wilson (1973a) found that mean spread-out sand thickness was 14.5m, but that the sand depth was very uneven, with strong concentrations of sand in some areas. These concentrations are probably related to zones of intense sand supply, or accumulation, although it is difficult to interpret them. The pattern is clearer in the Nebraska Sand Hills, where there was a more unimodal wind regime at the time of formation. In Nebraska there are narrow zones of high volume, elongated in the dominant wind direction, presumably trailing from areas of greater sand supply (Fig. 28.4; Warren 1976a). By extrapolating dune-spacing and sand-volume patterns derived for terrestrial sand seas, Lancaster & Greeley (1990) estimated that the volume of the north-polar sand seas on Mars was 1158km^3, 78% of it held in one sand sea

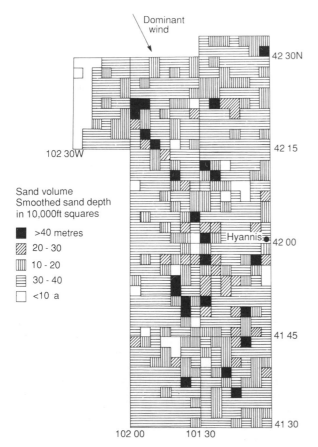

Figure 28.4 Patterns of sand volume in the Nebraska Sand Hills. Zones of thick sand have been elongated in the direction of the dominant wind. Prepared from topographic maps (after Warren 1976b).

(Fig. 23.26).

28.3.3 Dune assemblages
Only the smallest sand seas contain dunes of only one type. Even the Australian and Kalahari sand seas, which are dominated by linear dunes, do not retain the same pattern throughout. In the major sand seas of the Sahara, Arabia and China there are many dune types, some separated by abrupt and some by gradual boundaries (Fig. 28.5). It is difficult to explain these variations in dune pattern in more than general terms. In most of the cases the principal determinant must be spatial changes in wind pattern, but these are poorly documented. For example, the change from complex patterns in the northern parts of the Saharan and main Arabian sand seas, towards simpler linear patterns in the south, must be associated with the change from complex wind regimes dominated by the Westerlies to simpler ones of the Trades.

28.4 Growth and migration

The controls on the thickness, areal extent and movement of sand seas are somewhat different from those that control the size and movement of dunes. For large sand bodies, the controls operate at large spatial scales and over longer time periods. Furthermore, the concept of equilibrium size in dunes has no strict parallel for sand seas, for when dunes in a sand sea or dunefield have reached their equilibrium size, the sand sea can continue to extend in area or volume, given the necessary topography, wind systems and sand supply.

There are five main controls on the growth and movement of sand seas: sand supply, wind power, wind directionality, trap efficiency and time.

28.4.1 Sedimentary supply
By far the most important control on the size of a sand sea or dunefield is the supply of sand. This is because, at this scale, supply systems such as stream channels, beaches or upwind sources of aeolian sand are spatially more variable than any other factor.

(a) Non-aeolian supply
Many sand seas occur close to their fluvial and marine sources. The large sand seas of the northwestern Sahara are close to alluvium derived from the Atlas or the central Saharan mountains (I. G. Wilson 1971, Mainguet & Chemin 1983). The largely carbonate

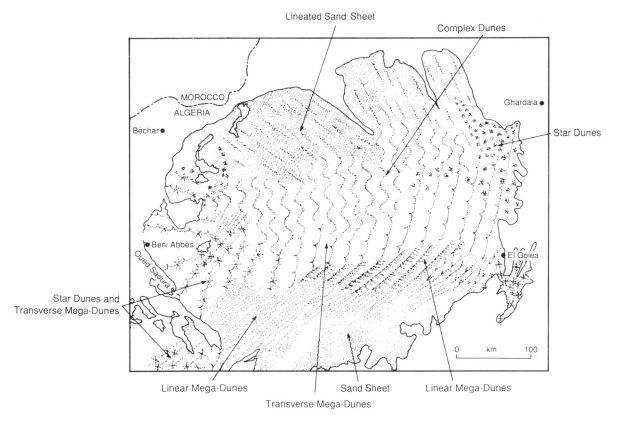

Lineated Sand Sheet

Complex Dunes

Star Dunes

Star Dunes and Transverse Mega-Dunes

Linear Mega-Dunes

Transverse Mega-Dunes

Sand Sheet

Linear Mega-Dunes

MOROCCO
ALGERIA

Bèchar

Ghardaia

Beni Abbès

Oued Saoura

El Golea

0   km   100

Figure 28.5 The dune assemblage in the Great Western Sand Sea in Algeria, showing great variation of form (redrawn after Breed et al. 1979).

Wahiba Sands of Oman are close to a biologically productive marine shelf and high wave-energy beaches (Warren 1988a). It has been maintained, on the evidence of mineralogy, that many sand seas are still very close to their non-aeolian sources. In the southern Sahara, Mainguet et al. (1983) found that, between Termit and Ader Doutchi, sand seas were autochthonous, and it is claimed that some Chinese sand seas have not travelled far from parent alluvial fans (Ma Chengyuan 1982, Chao Sung-Chiao 1984, Zhu Zhenda 1984, Zhu Zhenda et al. 1987). The confinement of many sand seas to large alluvial basins is a further indication of proximity to alluvial sources.

(b) Aeolian supply systems
In open deserts, however, sand seas and dunefields can break away from their non-aeolian sources and participate in input–output systems of their own. In the Sahara, aeolian sand streams feed sediment from one sand sea or dunefield to the next (Wilson 1973a,

Guy & Mainguet 1975; Fig. 28.6). The connectedness of sand seas along these sand streams means that upwind changes in the sand supply are ultimately transmitted to sand seas down wind (I. G. Wilson 1971). Wilson calculated that the minimum potential time for a disturbance to be transmitted through the Sahara was 200yr, but the actual residence time is probably much greater than the potential, because of sand trapping in mega-dunes, and disturbance probably takes much longer to transmit. The sand seas of the Sahel and southern Sahara are the termini of most of these sand streams and, over many thousands of years, have accumulated to great depths (Mainguet & Chemin 1983).

(c) Migration
The evidence that sand seas and dunefields have migrated far from their sources is contained mainly in their shape. Many sand seas extend over 100km parallel to the main drift direction, and many cross

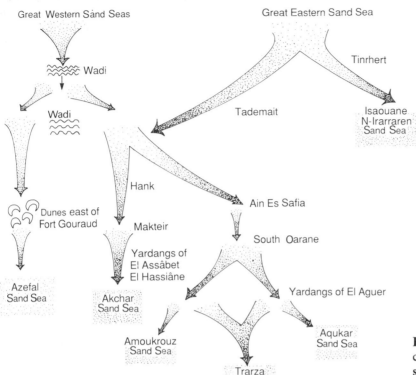

Figure 28.6 Sandflow patterns connecting some western Saharan sand seas (data from Mainguet et al. 1980).

fluvial divides and drainage lines. Elongation is most marked in strongly directionally concentrated wind regimes (Fryberger & Ahlbrandt 1979). Migration is further indicated by the long sand-free zones down wind of hill massifs that intrude into sand seas. A well known example is filled with Lake Faguebine in Mali. Others were illustrated by Urvoy (1933b). Yet another area is shown in Figure 25.3. A further indication of migration is the elongation of zones of deep sand within sand seas (Fig. 28.4).

It is usually assumed that sand seas and dunefields migrate down-resultant, but Sweet et al. (1988) believe that the Algodones dunefield in California was migrating obliquely as a consequence of the modification of the local wind regime by mega-dunes.

28.4.2 Wind power
Spatially extensive sand seas occur where there are strong winds, but high volumes of sand are found down wind of zones of high wind energy (Fryberger & Ahlbrandt 1979, Lancaster 1983b, 1984c, 1985b). Along one drift line $1.75 \times 10 \text{km}^2$ long in Mauritania, the drift potential falls to 3.6 from $4.9 \text{m}^3 \text{ m}^{-1} \text{ width}^{-1}$

yr^{-1}. If the wind had been saturated as it entered this stretch, then 0.2m of sand would have accumulated every 10^4yr (Breed et al. 1979b, Fryberger 1980). Sand seas in strong winds presumably migrate faster than those in gentle winds, but there are no comparative data in this area.

28.4.3 Wind directionality
In general, sand seas develop where saturated sand flows converge, or where wind energy declines (as where divergence of windflow reduces the sand-carrying capacity of the wind) (Fig. 28.7; Wilson 1971). Wind directionality plays much the same rôle for sand seas and dunefields as it plays for dunes: sand seas grow upwards more quickly in places where winds bring in sand from different directions than in zones where there is well directed throughput. This is the case in the northern Namib Sand Sea, which lies down wind of a zone of directionally concentrated winds, but which has itself accumulated in a zone of complex wind regimes (Lancaster 1985b). A similar process has happened in the Gran Desierto of north-western Mexico (Lancaster et al. 1987). Sand seas

presumably migrate more quickly in unidirectional wind regimes.

28.4.4 Trap efficiency

Trap efficiency also plays much the same rôle in sand seas as in dunes: a sand sea composed of transverse dunes is a better trap than one composed of linear dunes (Wilson 1971, Mainguet & Chemin 1983).

Traps around topographic obstacles occur at all scales from pebbles to mountain massifs. Just as there are anchored dunes, so there are **anchored sand seas or dunefields** (Wilson 1971, Fryberger & Ahlbrandt 1979). An example of anchoring is the Fachi–Bilma Sand Sea in the central Sahara, which has gathered sand swept around and over the Tibesti Mountains (Mainguet & Callot 1978, Mainguet 1985).

28.4.5 Rate of growth and age

Older sand seas, such as the Namib, contain more sand than young ones such as those in Australia. It is a simple matter to use measurements of sand or dune movement and of the volumes of sand in a dunefield or sand sea to arrive at estimates of age. Ball (1927) calculated that if dunes in the Abu Moharik dunefield in Egypt moved at 15m yr^{-1}, the dune belt itself would have been 35,000yr old. Using the same estimate for dune movement (15m yr^{-1}), and the known distance of advance from the coastal sand source, Warren (1988b) calculated that the Holocene dunefield

in the Wahiba Sands was about 8,000yr old. In coastal Baja California, Inman et al. (1966) estimated the advance of dunes to be 18m yr^{-1}, which gave an age for the 12km-wide dunefield of about 700yr. Sweet et al. (1988) calculated the migration rate for the Algodones dunefield at 0.0135m yr^{-1}. They maintained that the dunefield migrated more slowly than its constituent dunes. Other calculations of dunefield ages have been made by Shinn (1973) and Sarnthein & Walger (1974).

Using estimates of dune bulk transport, Fryberger & Ahlbrandt (1979) calculated that a sand sea of 10^4km^2 area and 100m depth would need between 13.5 million and 4.17 million yr to grow. More specifically, Ahlbrandt et al. (1983) suggested that the Nebraska Sand Hills had taken 6,000–8,000yr to grow. Rates of bulk transport may not be the most appropriate for this purpose, because sand seas grow as much by the accretion of sand to existing bedforms as by the inward migration of new ones. Using estimates of sand, rather than dune movement, I. G. Wilson (1971) estimated the age of the Great Eastern Sand Sea in Algeria to be 1,350,000yr.

An additional temporal requirement for the growth of large bodies of aeolian sand is continued or repeated aridity. Continued aridity appears to be one of the reasons for the growth of the large Namib Sand Sea (Lancaster 1989a). In the northern Sahara, on the other hand, the great volume of sand in the great sand

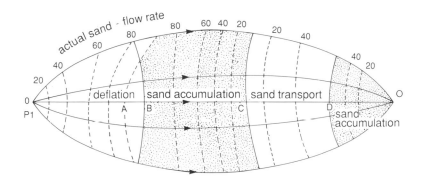

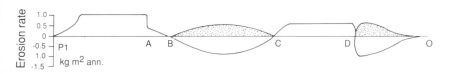

Figure 28.7 A model of the position and growth of sand seas (after Wilson 1971).

seas is more a function of the repeated alternation of phases of humidity and water erosion (to supply sand), and aridity and aeolian activity (for aeolian winnowing and spatial concentration).

When the bedforms reach an equilibrium size within a growing sand sea or dunefield, further growth is by bedform climbing (Ch. 27.4; Wilson 1971). Fryberger et al. (1984) estimated that accumulation in the Jafurah Sand Sea was of the order of 0.18mm yr^{-1}. A much higher rate of accumulation, 1.5mm yr^{-1}, was estimated for a coastal dunefield in South Africa (Illenberger & Rust 1988; Fig. 28.8).

28.5 Sedimentary patterns

Given some of the processes of development of sand seas and dunefields described above, one would expect a sequence of dune types and sediments to follow one another as a sand sea accreted and migrated (Kocurek 1981a, Porter 1986; Fig. 24.2). The pattern would only apply where there was a clear migration path. Ahead of a sand sea, as it migrated forwards, there would be a "fore sand sea" of low dunes, perhaps intercalated with fluvial deposits. This would be followed by the "central sand sea" of large dunes, leaving thick cross strata in the sedimentary record, and finally by a "back sand sea" dominated by coarse-grained sands in zibars. This sequence is more or less confirmed in the Algodones dunes of southern California (Fig. 28.9; Sweet et al. 1988). The upwind, trailing edges of the Idihan Marzuq in Libya, the Jafurah Sand Sea in eastern Saudi Arabia, the Wahiba Dunefield in Oman and the Namib Sand Sea are also composed of zibar-like dunes in coarse sand

(Capot-Rey 1947, Lancaster 1982c, Fryberger et al. 1984, Goudie et al. 1987)

28.6 Associated landforms

When sand seas or dunefields envelope pre-existing fluvially dissected topography, they block low- and medium-order stream channels, creating swamps, lakes or playas (Fig. 28.10). There are many examples: where the Saoura and other rivers empty into the Great Western Sand Sea in Algeria (Capot-Rey 1945), in the Rub' al Khali (McClure 1976), the Namib (Selby et al. 1979, Lancaster & Teller 1988), and the Tihodaine Sand Sea in Libya (DeVilliers 1948). Groundwater seepages are another way in which lakes can form. This has happened in some of the Libyan sand seas (DeVilliers 1948) and in the Mujunkum in central Asia (Suslov 1961). It is characteristic of humid-zone coastal sand dunes (Brothers 1954, Jennings 1957). The interaction of fluvial, lacustrine and aeolian systems creates distinctive landforms and deposits (Langford 1989). Alluvium is deposited between and over the lower slopes of dunes, which may also be eroded by streams. Alluvium is later eroded by wind or covered by advancing dunes. Wave-cut terraces are rapidly cut into soft dune sands, but are as rapidly degraded by wind erosion when the lake retreats (Inman et al. 1966).

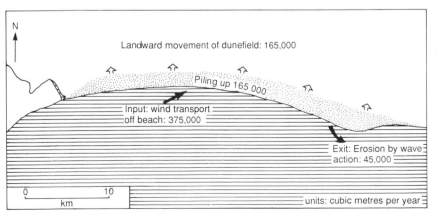

Figure 28.8 Sand accumulation in a coastal dune field on the South African coast, related to the wind direction and sand transport on the beach (after Illenbergen & Rust 1988).

410

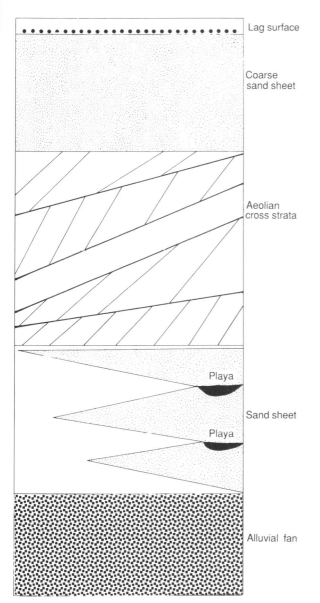

Lag surface

Coarse
sand sheet

Aeolian
cross strata

Playa

Sand sheet

Playa

Alluvial fan

Figure 28.9 A model of sedimentary accumulation following the passage of a sand sea (after Sweet et al. 1988).

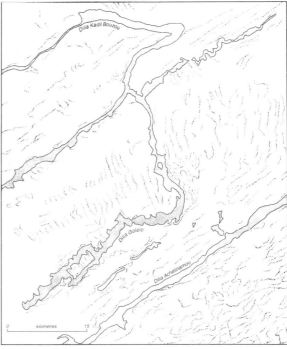

Figure 28.10 The diversion of major drainage lines by now-stabilized dunes in the Nigerian Sahel.

CHAPTER 29

Palaeo-aeolian deposits and landforms

29.1 Palaeo-aeolian geomorphology in deserts

29.1.1 Palaeo-dunes in deserts

In addition to the mere size of desert dunes, there are other indications that many are very ancient. First, and most useful for dating purposes, there are ancient lake deposits between some dunes, as in Arabia (McClure 1976, Gardner 1988). A second, unmistakable indicator of antiquity is the flooding of dunes by rising sea levels, as in the Eastern Province of Saudi Arabia, where large mega-barchans now lie beneath the waters of the Arabian Gulf (Al-Hinai et al. 1987), in Australia (Sprigg 1979) and in Oman (Gardner 1988). Thirdly, there are palaeosols and remains of palaeolithic man in many sand seas, particularly the great sand seas of the northwestern Sahara (Aufrère 1928, Gabriel 1979). The fourth indicator is more difficult to interpret; it is the apparent mismatch between the orientation of mega-dunes and meso-dunes (Hedin 1903, Aufrère 1932, Sevenet 1943, Monod 1958, Holm 1960, Reid 1985, Warren 1988a). The problem here is that, as has been shown above, little is known about the relations of many dune types to wind regimes (e.g. Brookfield 1970).

Three processes, usually acting together, may preserve palaeo-features in desert dunefields.

(i) *Secularly declining wind velocity* might induce a change from mega-dune to meso-dune formation, as suggested for the Wahiba Sands by Glennie (1970, 1985). More simply, it might mean a change from dune activity to inactivity, as proposed by Wasson (1986) for Australia, by Fryberger & Dean (1979) and Talbot (1984) for the south side of the Sahara, and by Thomas & Goudie (1984) for the Kalahari.

(ii) *Secular changes in wind direction* were invoked as explanations of complex dune patterns in hyper-arid deserts by Monod (1958), Besler (1982a) and Warren (1988a), although others (for example Sarnthein et al. 1981) believe that climatic changes during the last glaciation occurred without changes in wind direction.

(iii) *Changes in rainfall* would allow soil formation and surface stabilization, which would resist surface reactivation in a later return to drier or windier conditions. This was the sequence proposed to explain the survival of some of the features in the Majabât al Koubrâ by Monod (1958) and for those in the Nafud by Whitney et al.(1983).

29.1.2 Ancient wind-eroded features in deserts

The Lut mega-yardangs are thought by Krinsley (1970) to have formed in a drier period of the past, while the Borkou mega-yardangs around the Tibesti massif in the central Sahara, and some of those in Egypt may be as old as the Miocene. Southern Californian ventifacts were apparently formed by strong Pleistocene winds rather than modern ones (Sharp 1964, Dorn 1986, Laity 1987).

29.2 Palaeo-aeolian geomorphology on desert margins

29.2.1 The extent and location of stabilized dunes

Table 29.1 and Figure 29.1 show the extent and location of the major stabilized sand seas.

29.2.2 Ancient dunes in high latitudes

Active dunes covered large areas of the now-temperate lands during the last major glacial period.

Table 29.1 Major areas stabilized sands over 10^4km^2.

Sand sea	Area ('000km^2)	Source
Sahel		
Trarza/Cayor	57	Wilson (1972a)
Azaouad	69	" "
Azaouak	69	" "
Kano	294	" "
Sudanese Qoz	240	" "
Southern Africa		
Kalahari	1,000	Thomas (1987a)
Indian Subcontinent		
Thar	214	Wilson (1973a)
(former) USSR		
Pre-Kara-Kum (NW of the Aral Sea)	300	Suslov (1961)
Big and Little Baruski	100	" "
China		
Mu Us	630	Walker (1982)
Ortindag	21.4	" "
Horqin	42.3	" "
Australia		
Great Sandy	630	Wilson (1973a)
Great Victoria	300	" "
Strzelecki	140	present estimate
Europe		
Hungarian Plain	33	present estimate
North America		
Nebraska Sand Hills	34.4	Ahlbrandt & Fryberger (1980)
South America		
Pantanal	50	Klammer (1982)
Llanos, Venezuela	50	Tricart (1984)

Koster's (1988) list of stabilized dune areas in Europe and North America shows that they covered over 10^5km^2. On the northwestern European plain, discontinuous stabilized dunefields enveloped tens of thousands of square kilometres, between Belgium and Russia. There are outliers in eastern England, Sweden and Finland, and stabilized dunes cover 20% of Hungary (Seppälä 1971, 1972).

In North America, ancient dunes cover an even greater area. They occur in the western plains of Canada and the USA, in Alaska (where some are still active), and in parts of New York State (about 2,700km^2, Connally et al. 1972), Delaware (Denny & Owens 1979) and Quebec. The largest North American ancient sand sea, the Nebraska Sand Hills, covers 57,000km^2 (Fig. 29.2; Warren 1976a). Further large stabilized dunefields occur in the (former) USSR and China (Nevyazhskiy & Bidzhiev 1960, Kurbanmuradurov 1968). In the temperate Southern Hemisphere, there are stabilized dunefields in northern Tasmania (Bowden 1983), the Mallee country of Victoria, New South Wales and South Australia (Bowler & McGee 1978), and in parts of Argentina (Tricart 1969).

The formation of sand seas and dunefields in these now-temperate zones was stimulated by three features of the glacial period in high latitudes: aridity, plentiful supplies of sand from glacial outwash, and strong winds in the glacial margins, both katabatic winds off the glaciers themselves and intensified Westerlies in the zone towards the Equator (Wells 1983, Kutzbach & Wright 1985, Broccoli & Manabe 1987, Koster 1988). At the western end of the European ancient dunefields, relief is subdued (sand sheets with low dome dunes), perhaps because there was some vegetation cover. Farther east, perhaps where it was drier, dune forms were more accentuated, including large areas of vegetated parabolic dunes near the river valleys, and some fields of transverse dunes perhaps where sand supply was particularly copious (Schwan 1988). Longitudinal dunes seem mainly to be of "parabolic-arm" origin (Koster 1988). Some of the sedimentary structures in cold climate dunes are the product of synchronous deposition of sand and snow (review by Koster & Dijkmans 1988). The relief in most of the North American ancient dunefields and sand seas is also subdued, but there are mega-dunes in the Nebraska Sand Hills (Fig. 26.6), perhaps reflecting particularly strong winds, greater aridity and very generous supplies of sand (Warren 1976a).

29.2.3 Ancient dunes in the tropics and subtropics

In low latitudes, the expansion of the arid zone at the height of the glaciations brought dune-forming conditions into areas that were previously, and have since

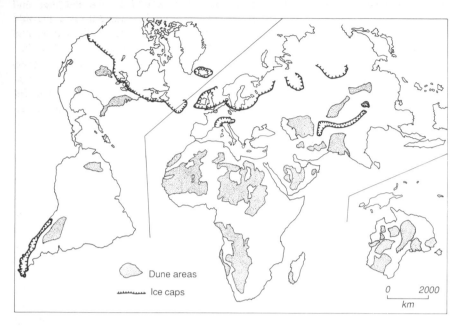

Figure 29.1 The extent of major sand seas at the height of the last glaciation. The map includes many sand seas now stabilized by vegetation (modified after Sarnthein 1978a).

become, less arid (Kolla et al. 1979, Goudie 1983b). Stabilized sand seas cover as much as 50% of the land surface between 30°N and 30°S, which is more than the surface they cover in modern deserts, for reasons discussed above (Fig. 29.1; Sarnthein 1978). Dunes were activated on both the equatorial and poleward margins of the great deserts in places such as northern Sinai, northern Saudi Arabia, the Thar Desert in India and Pakistan, the Sahel, northeastern Kenya and parts of Somalia, and parts of northern Australia. The Kala-

hari Sand, thought by many to be largely aeolian, covers 2,500,000km² of southern and central Africa, in Zambia, Zimbabwe, Namibia, Angola, Zaire and South Africa (Thomas 1987a). Further ancient sand seas are found in the Orinoco basin (Tricart 1984) and the Pantanal in Brazil (Klammer 1982).

There are also some large, now stabilized, dune-fields in semi-arid coastal areas. Large ones occur on the Queensland coast, where the Fraser Island field covers over 8×10^2km², and many others cover over

Figure 29.2 Stabilized dune sands in the northern High Plains of the USA (after Warren 1976a).

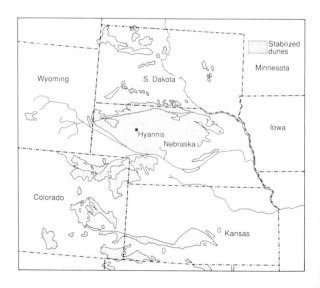

414

$4 \times 10^2 \text{km}^2$ (Lees 1989). Other stabilized coastal dune-fields are found on the California coast (Orme & Tchakerian 1986).

These now-stabilized dunefields in semi-arid and even humid tropical areas show that wind was much more active during the last glaciation, and probably at other times in the Pleistocene, than it is today. To what extent this was because velocities were higher, as they almost certainly were locally, or because rainfall was lower (as it too almost certainly was in some places) is uncertain. In Australia, Wasson (1986) speculated that wind velocities alone were sufficient to account for the greater dune activity, although else-where the received wisdom is that it was principally due to decreased rainfall. In most cases, dune activity may have followed changes of both rainfall and wind velocity. In the Sahel, Talbot's (1984) study of the limits of modern wind activity led him to conclude that the 50% increase in windspeeds and the 5°C drop in temperatures 19,000–17,000yr BP would not have increased C values (Fig. 16.8, Ch. 16.2a) to the point at which fully active dune formation could have occurred (which is known to have occurred at that time). He concluded that there must have been a reduction of rainfall of 20–50%, although he did not take into account increased evapotranspiration asso-ciated with the stronger winds.

29.2.4 Stabilization processes
This section is confined to stabilization consequent upon changes in rainfall and perhaps lowered wind velocities. Stabilization by cementation has been dis-cussed in Chapter 25.3.

(a) Ecological succession
Colonization by plants and animals occurs in small patches in the desert, as on an interdune that is seldom invaded by active sands, or where the sand supply has declined. It may also happen in response to some wider environmental change in rainfall or windspeeds. In either case there is a succession of plant species, perennials following annuals, giving increasing cover. This process has several stabilizing effects:

(i) Surface roughness increases, which in turn decreases sediment entrainment and transport. Large pieces of plant litter add to surface roughness (Otterman et al. 1975).

(ii) Silt and clay are trapped and, if occasionally wetted, help to bind the surface and form a crust.

(iii) Organic matter is added to the surface, and further helps to bind it; filamentous algae and fungi have much the same effect (Bond & Harris 1964, Daveau 1965, Barbey & Couté 1976); organic anti-wetting or water-repellent agents may be included (Bond 1968, Foggin & DeBano 1971, Bridge & Ross 1983, Jungerius & van der Muelen 1988, Agnew 1988). Anti-wetting agents are more characteristic of earlier than later successional stages, and affect sands more than finer soils (Foggin & DeBano 1971).

(iv) Roots penetrate the soil, and may build dense networks of root pipes (Barbey & Carbonel 1972). These pipes, rhizoconcretions, or "di-kaka" (Glennie & Evamy 1968), can become lithified when salts (particularly gypsum and calcium carbonate) are drawn towards the roots in the soil solution and concentrated in the rhizosphere (Gimmingham 1955). In high-rainfall areas, roots are replaced by silica, as well as by calcium carbonate, giving hard soil horizons (Hendry 1987). Root pipes may reach 40cm in diameter, if there is a good supply of solutes, as in calcareous dune sands (Amiel 1975). Net-works of root pipes may completely destroy the aeolian bedding pattern and penetrate up to 5m below the surface (Ahlbrandt et al. 1978).

There is also colonization by animals, particularly invertebrates (especially arthropods) and these also help to mix dust and organic matter through the profile, destroy bedding and bind the surface (Ahl-brandt et al. 1978).

(b) Additions
Fine sediments may be added from several sources, by far the most important being aeolian dust. In a sandy soil on ancient dunes in southern Israel the silt and clay content of the surface 5cm is distinctly high-er than the subsurface. This form of accumulation is common in semi-arid areas, where most stabilized dune sediments contain more silt and clay than active dunes. Silt and clay contents range between 15% and 30% (Huffington & Albritton 1941, Goudie et al. 1973, Goudie & Thomas 1986, Orme & Tchakarian 1986, Tsoar & Møller 1986). Much of the dust is composed of clay minerals and calcium carbonate. The calcium carbonate, when redissolved and redis-tributed, forms nodules or calcretes. In sufficient quantity it can cement the dunes into aeolianite.

(c) Weathering

Once dunes are no longer active, a number of processes may alter the physical properties of dunes. The weathering of clays from dune sediments (feldspars, micas and other unstable minerals), is faster in wetter climates, other things being equal (Gardner 1981a, Pye 1983c). Iron in rock minerals may be released and may stain the upper horizons of sandy soils red (Ch. 22.2.5a). If rainfall and temperature are high, even quartz grains may be weathered and split, as in humid tropical stabilized coastal dune sands (Pye 1983c). Weathering of the quartz grain surfaces produces microscopic pitting (Ch. 22.2.5). Compaction in shallow deposits of dune sand is not an important process, and although there may be some salt weathering, this cannot be active where rain has leached salts. Nevertheless, it may affect some coastal dune sands, where salt inputs are high. Frost weathering of quartz sands may happen in high latitudes (Pye 1983c).

(d) Soils

As dune soils develop, there is slow translocation of fine clay and silt down the profile to subsurface horizons. Depending on the ambient climate, carbonate, silica or iron dissolved farther up the soil profile are also redeposited lower down in the soil, or removed in the drainage water. Phosphorus first becomes more available and then progressively more fixed, so that soil fertility declines, in places to a point at which even vegetative cover may be reduced (Syers & Walker 1969)

In the early stages of soil formation, the translocation of carbonate and iron may form pisoliths or thin iron-cemented layers ("petroferric layers") along the bedding planes in the original sediment (Pye 1983c). In warm wet conditions silica removal can cause reddening or ferralitization of the soil profile (MacVicar et al. 1985). Given more time in these conditions, percolation can produce spectacular soil formation. Some podzolic, bleached A_2 horizons in eastern Australia are 18m thick (Thompson & Bowman 1984). Below the A_2, there may be a tough, humus-cemented B horizon or "alios" (Bourcart 1928, Simonett 1949, Bowden 1983, Pye 1983c). "Infiltration structures" may develop as flushes of water destroy the original bedding and create new contorted banding (Bigarella 1975, Ahlbrandt & Fryberger 1980).

In less humid areas, ferallitic soil profiles develop on ancient dune sands, reddened down to one or two metres, some with clay-rich B horizons (Daveau 1965, Warren 1968, Gardner 1981a, Felix-Henningson 1984). Horizons cemented by gypsum and halite can develop, derived from groundwaters (Glennie 1970). The most common horizon of this kind is cemented by calcium carbonate (calcretes).

(e) Slopes

In high-rainfall areas, silty or clayey surface soil horizons decrease permeability (Barbey & Couté 1976). Infiltration capacity is further decreased when rain-splash plugs pores between the sand grains with silt to create crusts and when anti-wetting, or water-repellent, properties are imparted to the soil surface by plants ((a), above). Decreased permeability allows runoff in storms, and this can be concentrated into rills and even gullies, whose sides, strengthened by the silt and clay, can maintain steep angles, at least in their early stages (Huffington & Albritton 1941, Daveau 1965, Bridge & Ross 1983, Barbey & Couté 1976, Yair 1990). Where the gullies empty into ancient interdune areas they may deposit sandy alluvial fans (Barth 1982, Thompson & Bowman 1984). Over many years these processes reduce slope angles from the 15°–30° range of the original dunes to a 3°–4° range (Talbot & Williams 1978, Pye 1983c). On Sahelian stabilized dunes the process appears to be cyclic: slopes are stabilized in wet years and gullied in droughts (Talbot & Williams 1979).

When aeolianite is exposed to subaerial weathering, solution of the carbonate produces karst topography, with pipes, karren and caves, notably on Bermuda (Gardner 1983). Slopes can be steep in this tough material, specially where they are undercut by waves. Exposed surfaces can develop a tough re-cemented patina or crust (Gardner 1983, McKee & Ward 1983, Pye 1983c).

(e) Stabilized wind-erosion features in semi-arid areas

The discussion above has shown that there are many signs of wind erosion preserved in semi-arid areas. Most occur in high latitudes where winds were particularly strong during glacial periods. They include ventifacts, lineated topography and pans.

PART 5

THE EVOLUTION OF DESERT LANDFORMS
A global perspective

CHAPTER 30

Aridity and its development

The purpose of this concluding part is to bring together the ever-increasing body of evidence that relates to the history and evolution of individual major desert regions, with particular reference to late Cenozoic events. It has become apparent that many desert areas are both old and characterized by an extremely varied environmental history. Space limitations preclude a fully comprehensive treatment, so that this account will concentrate on the pieces of evidence and landform effects that are especially significant in any particular area. The importance of a long-term perspective has become clear in recent decades because of the realization of the implications of plate tectonics and the increasing availability of evidence for the frequency and magnitude of Quaternary events. Of particular note has been the accumulation of large numbers of "absolute" dates derived from radiometric, palaeomagnetic and thermoluminescence techniques (Goudie 1983b, 1992).

As shown in Part 1, there are essentially two prime factors responsible for the development of climatic aridity: the relief and distribution of the land and sea, and the nature of the general atmospheric circulation and associated ocean currents. When such factors change, so will the nature and distribution of the world's arid lands.

High mountain chains running perpendicular to the prevailing winds can produce a "rain-shadow effect" on leeward slopes and in interior basins. Likewise, continental interiors, far removed from warm oceans and moist air masses, are generally dry. Thus, the nature of relief and the distribution of land and sea are responsible for the present existence of many of the arid lands, especially of middle and higher latitudes: the Great Basin of the USA and the High Plains, the Patagonian Desert, the great belt of steppe from southern Russia to Mongolia, and central Anatolia and interior Iran. Plainly, therefore, the growth and decay

of mountain ranges and continents under the influence of plate-tectonic induced movements is fundamental to an understanding of the history of desert terrains.

The second major influence, the nature of the general circulation of the atmosphere and associated ocean currents, relates to the positions of the major trade-wind belts, associated air subsidence and temperature inversions, together with the presence of cool offshore currents and areas of upwelling (Ch. 1.2). The great Saharan–Arabian zone, the Thar, the Namib, the Kalahari, the Atacama, the Californian deserts and the great arid expanses of Australia are all primarily "climatic" deserts situated in the heart of trade-wind belts. Thus, any change in the temperature regime of the world, which is responsible for driving both the atmospheric and oceanic circulation patterns at the macro-scale, may cause a change in the pattern of deserts, and latitudinal shifts of the continents may serve to change their positions with respect to the global pattern of trade-wind belts.

Given the recognition of the importance of plate movements in the Cenozoic and of global climatic changes of great magnitude and frequency in the Quaternary, it is unlikely that any desert area has developed its landforms purely under conditions of aridity. However, exceptions to this general rule are possible. For example, the thickness, extent and solubility of the nitrate crusts of the Atacama Desert suggest that, in that particular area, extreme aridity has been relatively long continued, while Meckelein (1959) has suggested that certain parts of the core of the Sahara may have had permanent aridity since the Pliocene. However, in some of the driest parts of the Sahara, with mean annual rainfall of no more than 25mm, other workers have found indisputable evidence for the importance of fluvial activity in pluvial times (e.g. Pachur & Kröpelin 1987).

Table 30.1 Evidence of climatic change in deserts.

Evidence	Inference	Example
Fossil dune systems	Past aridity	Thomas (1984a)
Breaching of dunes by river systems	Increased humidity	Daveau (1965)
Discordant dune trends	Changed wind direction	Warren (1970)
Lake shorelines	Balance of hydrological inputs and outputs	Street & Grove (1979)
Lake-floor sediments	Degree of water salinity, etc.	Gasse et al. (1980)
Lunette sediments	Hydrological status of lake basin	Bowler (1973)
Spring deposits and tufas	Groundwater activity	Butzer et al. (1978)
Duricrusts (lateritic) and related palaeosols	Intense chemical weathering under humid conditions	Goudie (1973)
Old drainage lines	Integrated hydrological network	Van der Graaf et al. (1977)
Fluvial sediments in ocean cores	Quantity of river flow	Sarnthein & Diester-Haass (1977)
Aeolian dust in ocean cores	Degree of aeolian deflation	Lever & McCave (1983)
Macro-plant remains (including charcoal), e.g. in pack rat middens	Nature of vegetation cover	Wells (1976)
Pollen analysis of terrestrial sediments	Nature of vegetation cover	Maley (1977)
Pollen analysis of marine sediments	Nature of vegetation cover	Hooghiemstra & Agwu (1988)
Fluvial aggradation and siltation	Desiccation	Mabbutt (1977)
Colluvial deposition	Reduced vegetation cover and stream flushing	Price-Williams et al. (1982)
Faunal remains	Biomes	Klein (1980)
Karstic phenomena	Increased hydrological activity	Cooke (1975)
Isotopic composition of groundwaters and speleothems	Palaeotemperatures and recharge rates	Sonntag et al. (1980)
Distribution of archaeological sites	Availability of water	Wendorf et al. (1976)
Drought and famine record	Aridity	Nicholson (1981)
"Frost" screes	Palaeotemperature	McBurney & Hey (1955)
Loess profiles and palaeosols	Aridity and stability	Pye (1984)

30.1 The evidence for climatic change

Palaeoforms abound in deserts, and so does the evidence for climatic change. Table 30.1 provides a summary of the types of evidence that have been used, and the inferences that can be gained from such evidence.

30.2 Aeolian sediments and landforms

The existence of large areas of dunes provides unequivocal evidence of aridity, so that the existence of areas of heavily vegetated, deeply weathered, gullied and degraded dunes indicates that there has been a shift towards more humid conditions. There is dispute about the precise precipitation thresholds that control major dune development. Undoubtedly wind strength and the nature of the vegetation cover have to be considered (Ash & Wasson 1983), but systems of palpably inactive dunes are today widespread in areas of relatively great humidity (Thomas 1984a). They indicate that in the Pleistocene there were times when sand deserts were much more extensive than today, as a result of lower precipitation conditions.

Dunes may also provide evidence for multiple

419

periods of sand movement, because of the presence of palaeosols in them, and comparison of modern winds and potential sandflows with those indicated by fixed dunes may suggest that there have been changes in wind regimes and atmospheric circulation patterns.

Dunes developed from lake basins by deposition on lee sides may indicate the nature of changing hydrological conditions within the basin, with dunes comprised of clay pellets tending to form from desiccated saline surfaces, and those with a predominance of quartzose sand-size material tending to form from lake beaches developed at times of higher lake levels (Bowler 1973).

The interrelationships of dunes and river systems may indicate the alternating significance of aeolian (dry) and fluvial (wet) dominance, as is made evident by a consideration of the courses of rivers such as the Niger and the Senegal, which have in their histories been ponded up or blocked by dunes.

Other aeolian features which have potential palaeoclimatic significance are loess sheets, and dust deposits in ocean-core sediments. The former are the product of periods of deflation from relatively unvegetated surfaces, and within loess profiles there may be palaeosols which resulted from stabilization of land surfaces. Dust deposits in ocean cores may also vary in their concentrations through a section, and thereby indicate the prevalence of deflational activity from adjacent land surfaces. A detailed treatment of the significance of such fine-grained aeolian materials is provided by Pye (1987).

30.3 Palaeolakes

The identification of former high shorelines around closed lake basins, and the study of lacustrine deposits from cores put down through lake floors, provide excellent evidence for hydrological changes in lake basins. These changes may in turn be related to estimates of former higher precipitation and evaporation. There are problems in assessing the relative importance of temperature and precipitation in affecting the water balance of a lake basin, and some changes can be caused by non-climatic factors (e.g. anthropogenic activity, tectonic disturbance, etc.). Nonetheless, correlations of absolutely dated shorelines and deposits have done much in recent decades to increase and improve knowledge of the extent and duration of humid phases in the Quaternary (Street & Grove 1979).

30.4 Fluvial landforms

Changes in effective precipitation may be reflected in the frequency and range of floods in desert rivers and in their load:discharge ratios. In particular, in sensitive piedmont zones, marked alternations may occur between planation and incision. As Mabbutt (1977) has remarked:

> There is no general agreement on the climatic regimes involved, however, still less on the climatic controls; nor need these be similar under all desert conditions. In extreme deserts an increase in the sediment yield of an upland catchment might indicate increased rainfall, whereas in a marginal desert with significant surface control by vegetation a similar result might follow from diminution of rainfall . . .

Moreover, although phases of fan aggradation and entrenchment may be a response to changes in precipitation, they can also result from tectonic changes, from changes in sediment supply resulting from changes in the operation of such temperature-controlled processes as frost-shattering, or from the inherent instability of fan surfaces. This is discussed at greater length in Chapter 12.4. Similar uncertainties surround the reasons for periods of arroyo cut and fill (Cooke & Reeves 1976).

Rather less contentious are the remains of major fluvial systems in areas where under present conditions flow is either very localized or non-existent. For example, large wadi systems, traced by the presence of both palaeochannels and associated coarse gravel deposits, occur in hyper-arid areas such as the southeastern Sahara (Pachur & Kröpelin 1987).

30.5 Caves, karst, tufas and groundwaters

Severe solutional attack on limestones to produce karstic phenomena (e.g. major cave systems) requires the presence of water, and may thus give some general indication of past humidity conditions. However, of rather more significance than this is the fact that caves are major sites for deposition, and their speleothems often provide a record of environmental change that can be studied by sedimentological and isotopic means. Other sediments that may be preserved in cave systems are frost spall deposits (the identification of which can be problematical, and confused with breccias caused by processes such as salt weathering) and aeolian materials.

Also of considerable significance for environmental reconstruction are the products of limestone precipitation that occur on the margins of limestone areas in the form of tufas and travertines. The tufas may contain plant and animal remains, palaeosols, etc., which may yield environmental information, and the tufas themselves may be indicative of formerly more active hydrological systems (Butzer et al. 1978).

In recent decades the isotopic dating of groundwater reserves has provided new information on those periods when groundwater was being recharged and those when it was not. For example, in the Sahara the studies of Sonntag and collaborators (1980) have indicated that very limited groundwater recharge occurred during the period of the last glacial maximum (20,000–14,000 BP), confirming that this was a period of relative aridity.

30.6 Miscellaneous geomorphological indicators

Much geomorphological evidence is often suggestive of significant change in environmental conditions, but it is often difficult to be precise about the degree of change involved, and to separate the rôle of climate from other influences.

In some arid areas there are weathering crusts and palaeosols that exist outside what are thought to be their normal formative climatic ranges, and their current state of breakdown may indicate that they are out of equilibrium with the present climate. The iron-rich duricrusts of the southern margins of the Sahara and of large areas of arid Australia may come into this category.

Elsewhere, there is evidence in slope deposits of changes in precipitation. For example, massive slumping along the escarpments of the plateaux of Colorado has been attributed to cooler and wetter conditions which caused shales to become saturated and unstable. Likewise, in southern Africa, there are extensive sheets of colluvium, which have caused the infilling of old drainage lines, and which may have formed when steep slopes behind the pediments on which the colluvium was deposited were destabilized by vegetation sparsity under arid conditions, providing more sediment supply than could be removed by pediment and throughflowing stream systems (Price-Williams et al. 1982).

The presence of extensive relict frost screes has also been interpreted as having climatic significance, having been produced under colder, but probably moist, conditions (McBurney & Hey 1955).

30.7 Faunal and floral remains

Pollen analysis may provide evidence of past vegetational conditions, from which past climatic circumstances may, with care, be unravelled. However, biases in pollen occurrence and preservation may occur as a result of differences in depositional history, soil matrix and position of site, and because of differences in the type of pollen preserved. In particular, care has to be taken in assessing the significance of any pollen type which is known to be produced in large quantities and which can also be preferentially transported and deposited over large distances.

Other evidence of past vegetational conditions is provided by charcoal deposits which, with the help of scanning electron microscopy, can enable the identification of woody plants, sometimes to the species level. However, it needs to be remembered that charcoal from archaeological sites may reflect human selection of firewood as much as it reflects vegetational conditions.

In the southwestern USA there are some remarkable floral deposits that were accumulated by pack rats – rodents of the genus *Neotama*. The middens of these animals contain extensive plant debris, and ancient middens sheltered in caves or overhangs may remain intact for tens of thousands of years (Wells 1976).

Inferences of palaeoclimates may also be drawn from the relative proportions of different faunal remains and knowledge about their modern habitat preferences. For example, the relative proportions of grazers and browsers may indicate the relative importance of grassland or predominantly bushy vegetation respectively (Klein 1980). In addition to the remains of large mammals, cave sites may preserve the remains of small rodents and insectivores, many of which are the result of deposition of regurgitated material by birds such as owls. These can give an indication of environment both through their species composition and through the size of bones, which may be related to palaeotemperature (Avery 1982).

30.8 Historical and archaeological evidence

The changing distributions and fortunes of different human groups has for long been used to infer changes in the suitability of the desert environment for human habitation, persuading such workers as Huntington to postulate pulses of climatic change in areas such as central Asia (Huntington 1907). In general, the absence of human habitation has been used to infer increased aridity, while the existence of human habitations in hyper-arid areas has been used to infer the existence of more clement conditions. While considerable care needs to be exercised in attributing the rise and fall of particular human cultures to climatic stimuli of this type, given the range of other factors that could be important (e.g. political, military, etc.), there are some examples where the evidence is relatively unambiguous. For example, in the hyper-arid eastern Sahara, artefactual materials occur in areas which are far too dry for human habitation today (Wendorf et al. 1976).

In historical times other types of evidence become available. Nicholson (1981), for instance, has attempted to reconstruct Saharan climates on the basis of famine and drought chronologies, and geographical and climatic descriptions in travellers' reports and diaries.

30.9 The evidence from the oceans

One of the most important developments in palaeo-environmental reconstruction over the past two decades has been the use of information derived from deep-sea cores. Such information has the advantage of covering long time spans, and has been less fragmented by past depositional erosion and diagenesis than most terrestrial evidence.

Ocean floors off the world's major deserts have acted as depositories for sediments derived from neighbouring landmasses by aeolian and fluvial inputs (Sirocko & Lange 1991). These sediments contain a whole range of information for environmental reconstruction, and have the added advantage that they may be susceptible to relatively accurate dating by radiocarbon, palaeomagnetism, uranium series and the like. They record the relative importance of fluvial and aeolian inputs (Sarnthein & Diester-Haass 1977), the degree of weathering of river-borne feldspars (Bonatti & Gartner 1973), the pollen and phytolith rain (Rossignol-Strick & Duzer 1980), the salinity of the ocean or sea (Deuser et al. 1976), the intensity of monsoonal winds and of upwelling activity (Prell et al. 1980), and the temperature of offshore waters recorded by oxygen isotope ratios in foraminiferal tests.

CHAPTER 31

Deserts of the world

31.1 The Sahara and its margins

The Sahara is the world's largest desert (covering *c.* 7 million km²), and the region comprising the Sahara and the Nile occupies about half of the entire African continent (Williams & Faure 1980). The greater part of the region is free of surface water and is sparsely vegetated; and, being exposed to dry, descending, northeasterly airstreams its mean annual rainfall is less than 400mm, and over vast areas less than 100mm (Grove 1980). Temperatures are also high, and evaporation losses from free water surfaces and transpiration losses from vegetated areas are greater than anywhere else on the globe. General discussions of the Saharan environments are provided by Gautier (1935), Cloudsley-Thompson (1984) and Capot-Rey (1953).

The general morphology of the Sahara has been discussed by Mainguet (1983a), who suggests that its most distinctive characteristic, save only the relief provided by the Hoggar and Tibesti massifs, is its flatness. This flatness is associated with great sandstone plateaux, a series of broad, closed basins (of which Chad is the most notable), and a series of wind-moulded landscapes which include deflational regs, corrasional fields of yardangs and areas of sand deposition (ergs). Some of the major geological, geomorphological and climate features of the area are shown in Figure 31.1.

Like most other deserts, the Sahara shows the imprint of a long evolution. For example, the Sahara was glaciated in Ordovician times, when from palaeomagnetic evidence it is apparent that the South Pole was located about the centre of this region. Well preserved striations, crescentic gouges, erratic boulders and glacial lineations are still evident in the present landscape (Beruf et al. 1971, Fairbridge 1970). In the early Tertiary there was a long interval of intense weathering, producing duricrusts on the

southern side of the Sahara in tectonically stable lowlands (Faure 1962), while Tertiary and Quaternary uplift, with associated volcanism, produced major Saharan massifs such as the Hoggar and Tibesti. In the late Tertiary, climatic deterioration and tectonic movements between Africa and Europe led to the gradual diminution of the Tethys Sea, and the formation in the so-called Messinian salinity crisis (*c.* 6 million yr BP) of a large closed depression in the vicinity of the present Mediterranean basin. The northern Sahara must have been arid at that time, and large spreads of evaporites formed before a marine transgression through the Straits of Gibraltar caused a re-establishment of marine conditions (Hsu et al. 1977). The Nile cut down deeply to the low base level, forming a canyon 2500m deep, 1300km long and 10–20km wide, dimensions which comfortably exceed those of the present-day Grand Canyon in Colorado (Said 1982).

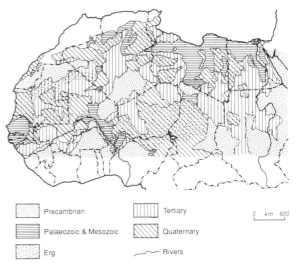

Precambrian	Tertiary	0 km 600
Palaeozoic & Mesozoic	Quaternary	
Erg	Rivers	

Figure 31.1 Geological and morphological features of the Sahara (after Mainguet 1983a).

423

In late Cenozoic times (van Zinderen Bakker 1984) aridity became a prominent feature of the Saharan environment, probably because of the occurrence of several independent but roughly synchronous geological events (Williams 1984):

- As the African plate moved northwards, there was a migration of northern Africa from wet equatorial latitudes (where the Sahara had been at the end of the Jurassic) into drier subtropical latitudes.
- During the late Tertiary and Quaternary, uplift of the Tibetan plateau had a dramatic effect on world climates, helping to create the easterly jet stream which now brings dry subsiding air to the Ethiopian and Somali deserts.
- The progressive build-up of polar ice caps during the Cenozoic climatic decline created a steeper temperature gradient between the Equator and the Poles, and this in turn led to an increase in Trade Wind velocities and their ability to mobilize sand into dunes.
- Cooling of the ocean surface may have reduced the amount of evaporation and convection in low latitudes, thus reducing the amount of tropical and subtropical precipitation.

Thus, although the analysis of deep-sea cores in the Atlantic offshore from the Sahara indicates that some aeolian activity dates back to the early Cretaceous (Lever & McCave 1983), it was probably around 2–3 million yr BP that a high level of aridity became established. From about 2.5 million yr BP the great tropical inland lakes of the Sahara began to dry out, and this is more or less contemporaneous with the onset of mid-latitude glaciation. Aeolian sands become evident in the Chad basin at this time, and such palynological work as there is indicates substantial changes in vegetation characteristics (Servant-Vildary 1973, Street & Gasse 1981).

In the Pleistocene a clearer pattern of climatic oscillations became apparent, with alternations of aridity and greater humidity, although dates are a matter of controversy, especially in North Africa, (Fontes & Gasse 1989). Each cycle may have been of the order of 100,000yr in duration, with nine-tenths of the cycle involving a gradual build-up towards peak aridity, coldness and aeolian activity, followed by a rapid but short-lived return to milder and wetter conditions (Williams 1984). During dry phases, ocean cores demonstrate that large quantities of dust were generated by the Sahara (Parmenter & Folger 1974, Parkin & Shackleton 1973), and there was an equatorward spread of dune fields in the Sahelian–Soudano

zone (Grove & Warren 1968, Michel 1973, Sarnthein 1978, Nichol 1991). The situation becomes especially clear at the time of the maximum of the last glaciation (the "Ogolian" of French workers), when aeolian turbidites were deposited on the Atlantic continental shelf, there was a substantial increase in dust output, and fluvial inputs from rivers such as the Senegal were minimal (Sarnthein & Diester-Haass 1977). Desiccation of lakes took place on the southern side of the Sahara between 23,000 and 16,000yr BP (Fig. 31.2A, B), and active dunes extended up to 450km southwards into the Sahel, blocking the courses of the Senegal and Niger rivers and probably also crossing the Nile (Rognon 1976, Williams & Adamson 1974; Fig. 31.1B).

Towards the end of the last glacial, at around 12,500yr BP, there was a major change in environmental conditions, characterized by a redevelopment of lakes in the Chad basin and eastern Africa. Extensive lakes and swamps also formed along the Blue and White Niles in the Gezira area. A peak of humidity may have occurred around 9,000–8,000yr BP (Pachur & Hoelzmann 1991) and, if one assumes that temperatures were broadly the same as today, the rainfall increase may have been as much as 65%. There were other lake oscillations during the course of the Holocene, but from about 5,000yr BP conditions began to deteriorate irregularly towards the present situation. The general trend towards aridity is reflected in a sharp drop in the influx of montane and Sudano–Guinean pollen into the Chad area between 5,000 and 4,000yr BP (Maley 1977). Saharan climates have also undergone significant changes in recent centuries, as Nicholson (1981) has shown from an analysis of historical sources, meteorological records, and studies of lake and river fluctuations (Fig. 31.2C). For example, from the early sixteenth century until the eighteenth century the Sahel experienced increased rainfall in comparison with the present century; in the eighteenth century there were frequent famines and droughts; there was increased precipitation in the late nineteenth century; and in the early twentieth century there was marked aridity along the Saharan margins. The rivers and lakes showed low positions around 1912, again in the early 1940s, and in the period since the late 1960s. This last fluctuation caused a marked diminution in the area and volume of Lake Chad, and it led to a marked increase in the level of dust-storm activity in a great belt from Mauritania to Sudan (Middleton 1985).

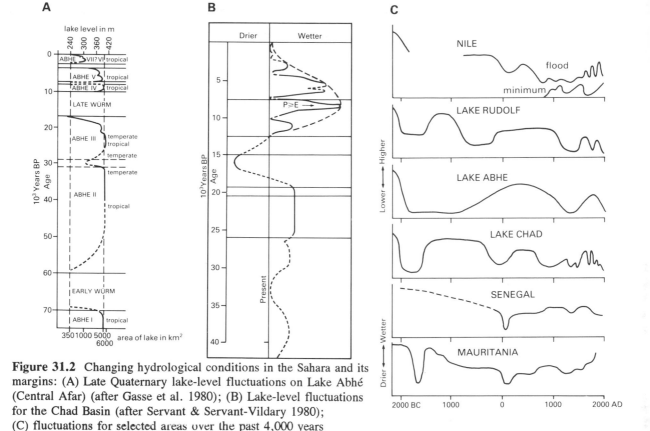

Figure 31.2 Changing hydrological conditions in the Sahara and its margins: (A) Late Quaternary lake-level fluctuations on Lake Abhé (Central Afar) (after Gasse et al. 1980); (B) Lake-level fluctuations for the Chad Basin (after Servant & Servant-Vildary 1980); (C) fluctuations for selected areas over the past 4,000 years (after Nicholson 1981).

31.2 Southern Africa

There are three main desert areas in southern Africa. In the west there is the coastal desert of the Namib (Lancaster 1989) which extends some 2,800km from south of Luanda in Angola to St Helena bay in the Republic of South Africa. It is bounded on the east by the Great Escarpment, and forms a narrow strip, generally less than about 150km wide. In the central and coastal parts of the Namib, hyper-aridity prevails (the mean annual rainfall at Walvis Bay is around 23mm), although fog and dew are of frequent occurrence. The desert contains two major dune fields separated by inselberg-studded gravel plains – the Skeleton Coast erg (Lancaster 1982a), and the great Namib erg to the south of the Kuiseb Valley (Lancaster 1981b). Good general reviews of the desert are provided by Logan (1960) and Beaudet & Michel (1978).

The most striking feature of the Namib is the 34,000km^2 sand sea that stretches for over 300km

between Luderitz and the Kuiseb River (Figs 31.3, 31.4). Three major dune types occur (Fig. 31.3B): transverse and barchanoid dunes which occur aligned normal to SSW to SW winds (Fig. 31.3C) in a 20km wide strip along the coast; linear dunes on N–S to NW–SE alignments, reaching heights of 150–170m in the centre of the erg (Fig. 31.3D); and dunes of star form which occur along the eastern margins of the erg. Detailed morphometric and sedimentological data are provided by Lancaster (1983a).

The Namib seems to be a desert of considerable antiquity, for the character of most Tertiary sediments in the Namib is suggestive of arid or semi-arid conditions. The cross-bedded Tsondab Sandstone Formation represents the accumulation of a major sand sea in the central and southern Namib over a period of 20–30 million yr prior to the mid- to late Miocene (Lancaster 1984a). In addition, extensive calcrete formation seems to have occurred at the end of the Miocene, while in the Pliocene a climate of modern affinities was developing in the region.

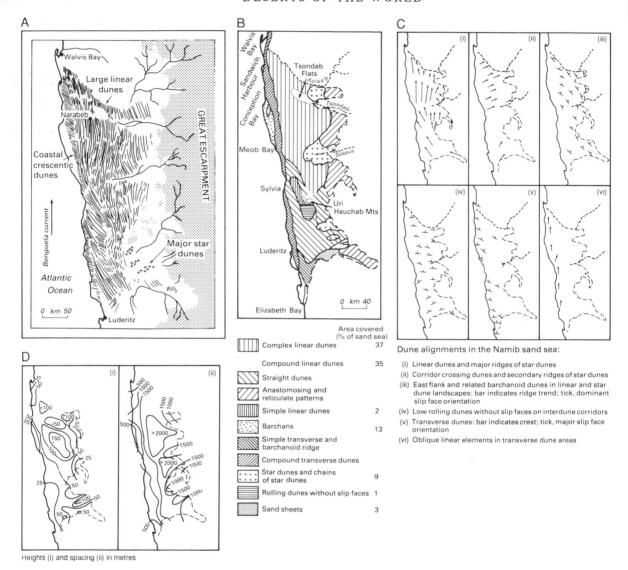

Figure 31.3 Dunes of the Central Namib Desert (after Lancaster 1980a, 1983a). (A, B) Distribution of dune types from LANDSAT imagery and air-photo interpretation; (C) dune alignments; (D) spatial variation in dune height and spacing.

Seisser's investigation of offshore sediments (Seisser 1978, 1980) has indicated that upwelling of cold waters intensified significantly from the late Miocene (7–10 million yr BP) and that the Benguela Current developed progressively thereafter. Pollen analysis of such sediments indicates that hyper-aridity occurred throughout the Pliocene, and that the accumulation of the main Namib erg started at that time. For much of the Pleistocene, aridity was also the norm, and although there have been periods of increased fluvial and lacustrine activity, most of the

sedimentological and faunal record suggests that moist phases were relatively short-lived and of limited intensity. However, fossil silts caused by ponding of river waters in the dune field occur in the Kuiseb Gorge, the Sossus Vlei and the Tsondab Valley (Butzer 1984), but no coherent picture emerges as yet from the few dates which are available. There are also speleothems in the Rössing cave of the central Namib (Heine & Geyh 1984) which indicate phases of greater hydrological activity in the late Pleistocene before 25,000yr BP. The last glacial maximum (c. 18,000yr

Figure 31.4 Sand encroachment on the abandoned mining settlement of Kolmansdorp, near Luderitz in Namibia.

BP) may have been dry (Vogel 1989).

To the east of the Great Western Escarpment is a second major desert: the interior desert of the Kalahari (Passarge 1904, Thomas & Shaw 1991), a word derived from the Setswana word "Kgalagale" which means "always dry". However, the area covered by the term is far from clear. Thomas (1984c) has identified three main regions (Fig. 31.5C):

(i) The Kalahari dune desert in the arid southwest interior of Botswana and adjoining parts of Namibia and South Africa. The primarily summer rainfall is less than 200mm per annum (Fig. 31.5B) and is just sufficient to stabilize the plinths of a major field of dominantly linear dunes.

(ii) The Kalahari region (or thirstland) approximately delineated in the north by the Okavango Swamps and in the south by the Orange and Limpopo Rivers. This is an area of little or no surface drainage despite a relatively higher rainfall (*c.* 600mm per annum). It is almost entirely covered with grass and woodland, and has extraordinarily low relief.

(iii) The Mega-Kalahari, which is an extensive area consisting of a basin infilled by continental sediments of the Kalahari Beds. This extends from beyond the Congo River in Zaire to the Orange River in South Africa. Precipitation may be as high as 1500mm and vegetation may range from savanna to tropical moist forest. Nonetheless, it displays evidence for formerly more extensive aridity, in terms of both the development of ancient dune systems and of the widespread distribution of closed depressions, called **pans** (Goudie & Thomas 1985).

The Kalahari is another long-continued area of terrestrial sedimentation, with sequences of marls, sands, lake deposits, calcretes, silcretes, etc., dating back to the Cretaceous. These beds, which are called the Kalahari Group, are ill exposed, and much of the information comes from borehole records. Dating evidence is still slender, but there is general agreement that the sediments accumulated under arid to semi-arid environments. The Kalahari Group is especially noted for the extensive development of calcretes (Netterberg 1978) and of silcretes (Summerfield 1983b), while the Kalahari Sands (which are either *in situ* or reworked aeolian materials for the most part) stretch over large tracts of country (*c.* 2.5 million km^2) between the Orange River in the south and the Congo River in the north. The river systems of parts of Zambia and Zimbabwe inherit their alignments from a previously more extensive aeolian cover (Thomas 1984b), and throughout the Kalahari there are extensive fossil drainage networks which are now either ephemeral or dry (Boocock & Van Straten 1962).

The internal basin of the Kalahari, in which these sediments accumulated, was created by tectonic processes in mid-Jurassic times at the final division of Gondwanaland. There are various sub-basins and

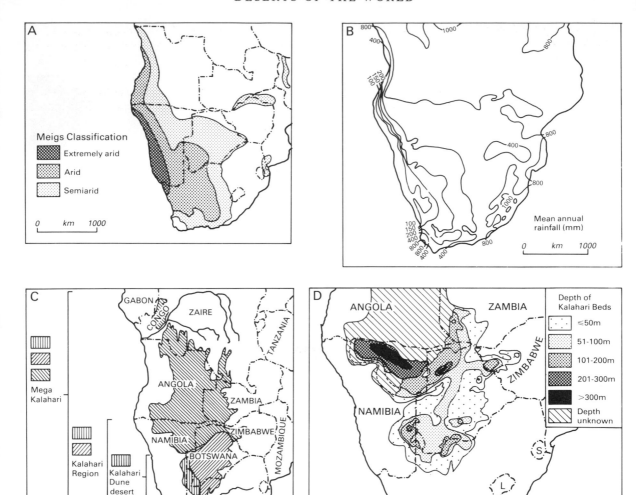

Figure 31.5 (A) Aridity in southwest Africa (after Meigs 1953); (B) precipitation in southern Africa; (C) the location of the three Kalahari areas (after Thomas 1984c); (D) the depths of the Kalahari beds in southern Africa (after Thomas 1984c).

graben structures within the area, and the greatest depths of Kalahari sediments occur in the Etosha Pan area of northern Namibia, locally exceeding 370m in thickness (Fig. 31.5D).

There is abundant evidence for environmental changes in the Pleistocene within the Kalahari (Fig. 31.5) with ancient dunes and palaeolakes being the most striking manifestations (Grove 1969). The dune systems of the Mega-Kalahari occur in areas where mean annual rainfall currently exceeds 800mm and they have been mapped in the Hwange and Victoria Falls areas of Zimbabwe (Fig. 31.6B), Zambia and Angola by Thomas (1984a), and in Botswana by Lancaster (1981a). They indicate formerly very extensive late Pleistocene aridity (with rainfalls less than 150mm per annum), and there is also some suggestion that wind directions may have been different from those pertaining in the region today (Heine 1982). The presence of sandy and relatively clayey lunettes (Fig. 31.6D) in association with small closed depressions (pans) has also been used to indicate the hydrological fluctuations of the late Pleistocene (Lancaster 1978a).

The greatest palaeolake in the area, however, is that of the Makgadikgadi Depression in northern Botswana (Fig. 31.6A). Strandlines extend to an altitude of 940–50m (*c.* 50m above current pan floor level), and the maximum extent of the lake was probably around 60,000km², about the size of today's Lake Victoria and larger than palaeolake Bonneville in the USA (Cooke 1980).

The palaeoclimatic significance of the Makgadikgadi lakes is, however, uncertain given the increasing body of evidence for tectonic instability in the Okavango delta region.

The third desert of southern Africa is sometimes called the Great Karoo semi-desert. It occurs as a plateau at an altitude of 600–1000m, tends to be underlain by horizontally bedded Palaeozoic sediments of the Beaufort Series, is bounded on the north and south by mountain ranges, and has a primarily winter rainfall regime that produces 130–400mm per annum. Information on the evolution of this arid region is sparse.

The dating of the various fluctuations of climate in the southern African region is still highly uncertain. Lancaster (1984a) remarks that "there is no reliable chronology of events, nor any agreement upon the nature of the changes in regional climatic patterns", while Butzer (1984: 51) finds "the Kalahari–Namib evidence is both patterned and ichoate. No distinctive

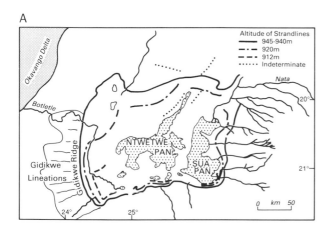

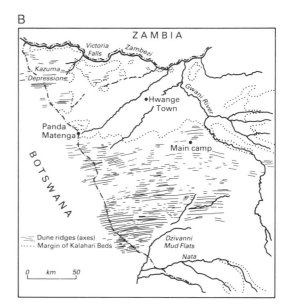

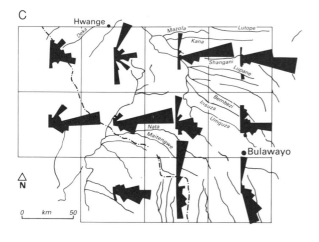

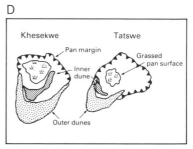

Figure 31.6 Examples of landforms resulting from past climatic events in the Kalahari (after Thomas 1984a, Lancaster 1978a). (A) The palaeostrandlines of the Makgadikgadi Depression, Botswana; (B) the palaeodunes of southwest Zimbabwe; (C) a drainage pattern derived from development along palaeodune swales in Zimbabwe; (D) pans and two different ages of lunette in Botswana.

A

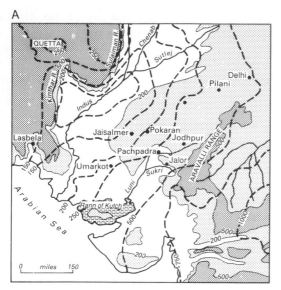

B

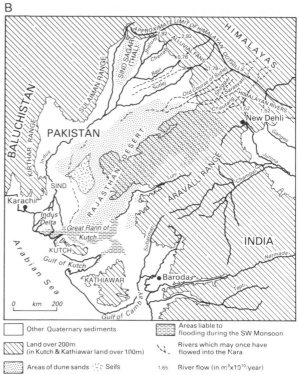

Figure 31.7 (A) Generalized map of the arid and semi-arid parts of India and Pakistan, with isohyets in millimetres and contour lines at 200m and 500m (after Verstappen 1970). (B) The Ranns of Kutch, the Rajasthan Desert and the Indus River system (after Glennie & Evans 1976).

interregional contrasts emerge, and different categories of data are often difficult to reconcile within one area".

31.3 The Great Indian Desert or Thar

The arid zone of the northwestern part of the Indian subcontinent extends from the Aravalli Range in the east to the Indus Plain and the mountains of Baluchistan in the west (Fig. 31.7). In Rajasthan it is traditionally called *Marwar* or "place of death", but desert conditions are not especially extreme, for only locally does mean annual precipitation fall below 100mm. The southwest monsoon manages to produce summer rainfall in the area, and rural population densities are quite high. In the driest parts of the area, in the west, the Indus and its tributaries are through-flowing rivers that are now much used for irrigation. General background information on the area is provided by Allchin et al. (1978).

An important feature of the Great Indian Desert is the presence of the snow-fed Indus and its tributaries, both past and present (Fig. 31.7). The discharge

pattern is highly seasonal and floods can be very destructive. Its average annual flow is about twice that of the Nile, and during floods the river in the plains of Sind can be over 16km wide. The river carries a huge sediment load, so that the Indus Delta is thought to have extended 80km in the past 2,000yr, and some 10m of aggradation has taken place in the past 5,000yr (Lambrick 1964). However, perhaps the most interesting hydrological features of the Thar rivers are their propensity to change course or to disappear, and the history of competition between the Ganga and Indus systems in the northern Punjab. Ancient river courses of different ages have been located (Wilhelmy 1969) and the geomorphology of ancient Indus courses in the Thal doab (between Chenab and Indus) has been described by Higgins et al. (1973).

Another distinctive feature of the Thar is the nature of its dunes. The coastline of the Arabian Sea, the alluvial plain of the Indus and the weathering of widespread outcrops of sandstones and granites provide plentiful sand for aeolian reworking. The dominant sand-moving winds come from the southwest, and this accounts for both the long-distance transport of foraminiferal tests from the coast and for the overall alignments of the dunes (Goudie & Sperling 1977). In the coastal regions of Saurashtra and Kutch the dunes are composed of calcareous aeolianites, which are locally called **miliolites** (Biswas 1971, Sperling & Goudie 1975, Marathe et al. 1977), but as one moves inland they become generally quartzose (Fig. 31.8C). The general pattern of the dunes has been mapped by Breed et al. (1979b), who pointed out the particular importance of clustered parabolics of rake-like form in much of the desert (Fig. 31.8B). This may reflect

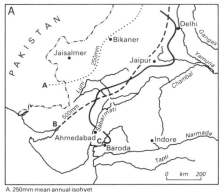

A. 250mm mean annual isohyet
B. 500mm mean annual isohyet
C. Former extension of sand desert

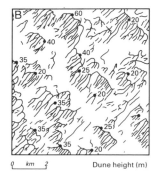

Dune height (m)

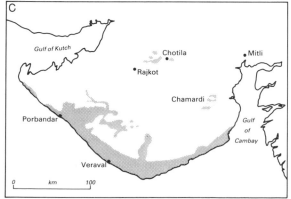

the relatively high rainfall levels with the concomitantly relatively dense vegetation cover, or it may be one response to the fact that in terms of wind-energy levels the Thar is one of the least energetic of the world's deserts. Large sayf dunes which occur in the west of the desert, especially near Umarkot in Pakistan, have some seasonally inundated lakes in their swales (*dhands*) and may be derived as blown-out parabolics (Verstappen 1968).

Locally within the Thar there are closed basins and salt deposits. Some may be the result of the blocking of drainage systems by dunes as at Sambhar or on some of the tributaries of the Luni River (Ghose 1964), while others, such as the Jaisalmer and Pokaran Ranns, may be related to faulting (Pandey & Chatterji 1970). Coastal deposits are also important areas of salt, and the Rann of Kutch is a major sabkha area (Glennie & Evans 1976).

Possibly the most contentious issue surrounding the Thar Desert is its age and origin. Fossil evidence for pre-Pleistocene climates is scanty in Gujarat, Sind and Rajasthan, but the records of *Dipterocarpoxylon malavii*, *Cocos*, *Mesua tertiara* and *Garcinia borocahii* from the Tertiary beds of Kutch and Barmer, and the Eocene lignite at Palna near Bikaner may be suggestive of conditions rather similar to those currently pertaining in eastern Bengal, upper Burma and Assam (Singh 1969). However, it is far from clear when the desert became established. Many authors have maintained that the desert is only of Holocene age and is the result of postglacial progressive desiccation (see, for example, Krishnan 1952). The arguments have been reviewed by Meher-Homji (1973) and dismissed by Allchin et al. (1978).

The stratigraphy of the Rajasthan lakes at Sambhar, Didwana and Lunkaransar has shown conclusively that a major aeolian layer pre-dates Holocene freshwater deposits (Singh et al. 1972), and there are also hypersaline evaporite layers that date back to the last glacial maximum (Wasson et al. 1984). The researches of Goudie et al. (1973) have shown that there are several phases of dune formation in the late Pleistocene, and that the dune fields were formerly more extensive and

Figure 31.8 (A) The former extension of the Great Indian Sand Desert in the late Pleistocene (after Goudie et al. 1973); (B) clustered parabolic dunes in the Barmer District (26°N 71°41′E), from the 1:50,000 Survey of India sheets; (C) aeolianite (miliolite) distribution in Saurashtra (Kathiawar) (after Allchin et al. 1978).

Figure 31.9 Gullied and vegetated sand dunes in Gujarat, northwest India, formed under more arid conditions in the late Pleistocene. In the background is the Great Basalt Escarpment of peninsular India.

active than they are today (Figs 31.8A, 31.9). Many of the dunes are now stable, vegetated, gullied by fluvial action, and overlain by slopewash deposits and, in the case of the coastal and near-coastal miliolites, they have been strongly cemented into material used for building. The lake stratigraphy also shows evidence for fluctuations of humidity in the Holocene, and this may have influenced the fortunes of the Harrappan civilization in the Indus Valley and its margins. The late Pleistocene aridity of the area may be confirmed by the presence of loess layers in river terrace deposits dated to *c.* 20,000–10,000yr BP in the Allahabad region (Williams & Clarke 1984) and by high dust loadings in Indian Ocean cores (Kolla & Biscaye 1977).

Furthermore, detailed investigations of planktonic foraminifera in the Bay of Bengal (Cullen 1981) indicate high levels of salinity at the time of the Last Glacial Maximum, suggesting that there was reduced runoff in the Ganga–Brahmaputra at that time, probably as a result of a less vigorous monsoonal circulation. High global radiation receipts at around 9,000yr BP caused aridity to be reduced as the vigour of the monsoonal circulation returned (Kutzbach 1981).

31.4 Arabia and the Middle East

The Middle East is an area of sometimes great aridity, and also of great topographic diversity. On the one

hand there are major mountain ranges: the Zagros and Elburz mountains of Iran, the Taurus mountains of Turkey, the Asir Mountains of Arabia, and the Jebel al Akhdar of Oman. On the other there are the extensive inland plains and plateaux of Arabia, with their two great sand seas, An Nafud and Rub' al Khali, and the large intermontane basins in which lie the kavirs (salt plains) of Iran.

This topographic diversity owes much to the tectonic history and plate-tectonic setting of the region. Much of Arabia represents the remnant of part of the ancient landmass of Gondwanaland, while the mountain ranges are associated with the interaction of three great plates: the African, the Eurasian and the Arabian. The Red Sea and the Gulf of Aden have been formed as the result of seafloor spreading as Arabia moved away from Africa, and there has also been about 100km of left-lateral movement along the Dead Sea Fault system since Miocene times, as the Arabian plate has moved northwards relative to the micro-plate of Sinai. In Iran the same northward movement of Arabia towards Eurasia has caused widespread overthrusting, and sediments have been folded into a series of major synclines and anticlines. The underthrusting of Iran by the Arabian plate resulted in the complex folding of the Zagros mountains. Eruptive rocks occur in zones of structural weakness, in the highland zones of Turkey and Iran, and also adjacent to the major faulting zones of the Dead Sea Lowlands and the Red Sea.

Such structural considerations thus underlie the gross morphology of the region. The detailed morphology owes much to environmental changes of the Tertiary and Pleistocene. Widespread humid conditions in the late Tertiary may have been of particular importance (Hötzl et al. 1978). The evidence for this is provided by the development of erosional and weathering features on basalt lava flows of known age. Intense lateritic weathering is evident on a basalt flow 3.5 million yr old, as is extensive fluvial dissection. By contrast, a younger flow, dated to the early Pleistocene (*c.* 1.2 million yr old), shows no such features. Blocked-up linear drainage systems may date back to this stage, and associated gravel fans interfinger with the deposits of the regressing late

Pliocene/early Pleistocene sea.

During the Quaternary, long-continued humid periods appear to have been much less significant, although the evidence for pluvials and interpluvials is well known. Indeed, it was in the Dead Sea trough that some of the first evidence for pluvials was identified (Lartet 1865). Nonetheless, aridity has probably been a dominant feature for much of the Pleistocene (Anton 1984). Unfortunately, precisely dated, reliable palaeo-environmental information is not readily available, especially before about 40,000yr BP, and attempts at correlation across the very varied climatic environments of the Near East have so far produced results which are confused both temporally and spatially. So, for example, in their analysis of palynological data for a range of sites, Bottema & van Zeist (1981) have found for the Pleniglacial (*c.* 50,000–14,000yr BP) that not only are there striking differences in vegetational and climatic history between the Levant and western Iran, but that even within the Levant the climatic history deduced for northern Israel cannot be brought into line with that of northwestern Syria. For that reason, at present it is probably prudent to provide some results from relatively well dated situations from selected sites across the area, rather than to try to produce premature correlations.

Whitney (1983) has undertaken an analysis of information for Saudi Arabia and, on the basis of a large number of radiocarbon dates, finds the pattern shown in Figure 31.10A. He believed that the late Pleistocene alluvium, calcrete and lake deposits define a clear pluvial episode between about 33,000yr and 24,000yr BP, with the most intense phase of pluvial conditions probably being between 28,000yr and 26,000yr BP. Quite large lakes occurred in the Rub' al Khali at this time (McClure 1976; see also Fig. 12.15), and there seems to be a broad correspondence in climatic history with that encountered in Africa (Street-Perrott et al. 1985). From about 19,000yr to 10,000yr BP aeolian activity prevailed through Arabia, and dune development occurred at times of low sea level on the floor of the Arabian Gulf (Sarnthein 1972), permitting Saudi sand to enter Bahrain Island (Doornkamp et al. 1980). There are also extensive spreads of cemented aeolianites which extend below the present sea level in the Gulf States and Oman, and which formed at the time of low glacial sea levels.

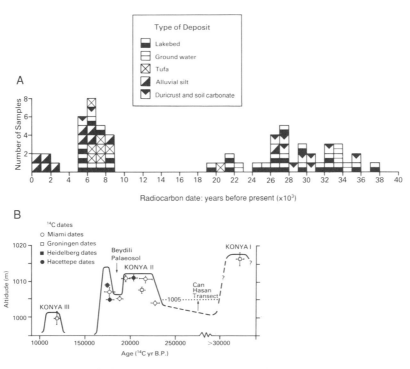

Figure 31.10 Environmental changes in the Middle East since *c.* 40,000yr ago. (A) The distribution of ages of isotopically dated superficial deposits in Saudi Arabia (after Whitney 1983); (B) a generalized lake level curve for the Konya Basin, Anatolia, Turkey, showing ^{14}C dates from shoreline deposits (after Roberts 1983).

Ocean core deposits in the Arabian Sea also indicate that large quantitites of silt were being exported around 18,000yr BP (Kollya & Biscaye 1977). A second major cluster of dates suggests that pluvial conditions began again in Arabia about 9,000yr BP or slightly later, causing the deposition of tufa, lake deposits in the Rub' al Khali, soil carbonates and spreads of fine alluvial silts along major wadis. Pluvial conditions lasted until c. 5,000yr BP.

The Konya Basin, in one of the more arid parts of the Anatolian Plateau, Turkey, has a series of dated shorelines which permit the reconstruction of hydrological changes in the northwestern part of the Near East. Figure 31.10B shows the lake-level curve that has been established by Roberts (1983), with three phases of high level: Konya I prior to 30,000yr BP, Konya II between 23,000yr and 17,000yr BP, and Konya III around 12,000–11,000yr BP. The important Konya II event appears to have occurred at the same time as the build-up of the last major Northern Hemisphere ice sheets, while the dramatic fall of palaeolake Konya around 17,000yr BP occurred well before the northward retreat of the Laurentide and Fennoscandian ice sheets. The early Holocene lacustral phase, noted in Saudi Arabia, is not represented in the Konya sequence. Roberts believed that the palaeolakes of Iran and Anatolia were more a product of reduced evaporation brought about by temperature depression than of changes in precipitation.

Nonetheless, when one considers another area, comprising the high plateaux of western Iran, detailed pollen analytical work from Lake Zeribar (Bottema & van Zeist 1981) indicates that vegetation conditions have changed very markedly in response to changes in precipitation levels over the past 40,000 years. Until c. 14,000yr BP, during a period broadly coincidental with the Pleniglacial of the European Würm glacial chronology, the vegetation was predominantly open, with steppe or desert-steppe in which *Artemisia*, *Chenopoiaceae* and *Umbelliferae* were important. This is seen to be the result not so much of coldness but of accentuated aridity. Conditions for tree growth improved after 14,000yr BP, but it was not until well into the Holocene that *Quercus* and *Pistacia* forests became established over wide areas.

The differences that have been observed in the pollen and lacustrine records for the late Pleistocene in the Middle East at different sites may be susceptible to a climatological explanation based on the varying importance of major systems (e.g. the summer monsoon, cyclonic westerlies, and cold Eurasian air

masses) at different times and in different places. For example, Farrand (1979) has argued that, during the late Pleistocene glacial, the northern shores of the Mediterranean Basin may have been under the influence of the cold, dry air masses generated by the giant ice sheets of Eurasia, while the southern shores (including the southern Levant) would have received precipitation from westerly cyclonic systems that were compressed between the northern cold air and the more or less fixed subtropical high-pressure belt. Southern Arabia would have been influenced by the varying strength of the monsoons, and thus it shows a certain similarity of climatic trends to those observed in East Africa and the Thar.

31.5 China and the (former) USSR

The deserts of China cover an area of around 1.1 million km^2 and occupy about 11.5% of the total land area of the country. They are located in the temperate zone, stretching from 75°E to 125°E, and from 35°N to 50°N. Within this huge area, extreme aridity characterizes the Taklamakan of the Tarim basin. The locations of the main desert zones are shown in Figure 31.11. They are positioned in the great inland basins and high plateaux, with elevations generally lying between 500m and 1500m, although there are some areas, such as the Turfan Depression, which lie below sea level. A distinction is often drawn between rocky and gravel deserts, termed "gobi", and the sandy deserts, termed "shamo".

The Chinese deserts appear to be very old (Chao Sung-chiao 1984), being formed as early as the late Cretaceous and early Tertiary. At this time the area was mostly under a subtropical high-pressure belt. Late in the Tertiary the Tibetan Plateau was uplifted and the great Himalayan orogeny occurred. The continentality of the climate was greatly strengthened, the monsoon system became well established, and northwestern China became even more arid. Ancient lakes in the Tarim and other inland basins diminished or dried out gradually, and the Taklamakan and other sandy deserts probably enlarged considerably at that time. Continued uplift of the mountains during the Pleistocene and Holocene have further accentuated the aridity.

There is abundant evidence of climatic fluctuations during the Quaternary in the form of very complex loess profiles, suites of old lake shorelines (most notably around Lop Nor), and miscellaneous historical

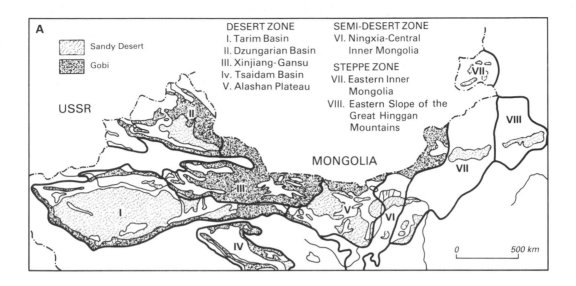

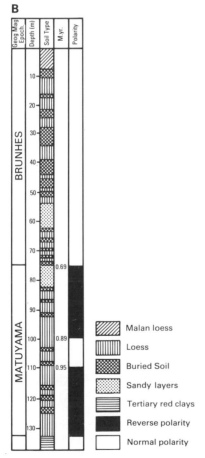

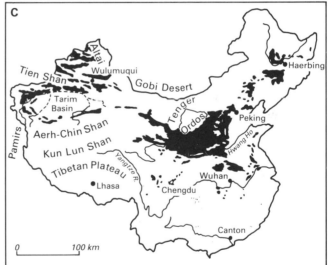

Figure 31.11 (A) The main desert areas of China (after Chao Sung-chiao 1984); (B) geomagnetic stratigraphy of the loess section in Luochuan in Shaanxi Province (after An et al., in Derbyshire 1983); (C) the distribution of loess in China (after Pye 1984a).

435

and archaeological evidence of the type employed by Huntington (1907) in his *Pulse of Asia*. Of these the loess profiles give the longest record of environmental change, for maximum thicknesses of 335m have been observed (Derbyshire 1983). The materials have also proved susceptible to dating by palaeomagnetic and thermoluminescence methods. The oldest loess in the Central Loess Plateau has been dated palaeomagnetically at about 2.4 million yr (Heller & Liu Tong Sheng 1982), and this confirms that aridity has a lengthy history, for much of the silt is derived from the Gobi and Ordos deserts (Pye 1984a). Seventeen periods of loess sedimentation, separated by periods of non-deposition and soil development, have been recognized in the classic section of Luochuan for the period since 1.67 million yr BP, and the depositional episodes have been correlated with glacial periods (Fig. 31.11B). It has been suggested that glacial maxima in Tibet, Tien Shan and the Kun Lun mountains were accompanied by a greater frequency of

cyclonic depressions and sandstorms in the Gobi Desert and by more effective easterly transport of dust by a westerly jet stream centred north of the Tibetan anticyclone (Liu Tung Sheng et al. 1982; Fig 31.11C).

The Turkestan desert of the (former) USSR lies between 36°N and 48°N and between 50°E and 83°E. It is bounded on the west by the Caspian Sea, on the south by the mountains bordering Iran and Afghanistan, on the east by the mountains bordering Sinkiang, and on the north by the Kirghiz Steppe. Two great ergs are included: the Kara-Kum ("black sands") and the Kyzyl-Kum ("red sands").

As in China (Goudie et al. 1984), loess deposits are both extensive and thick (up to 200m), and they have been dated in a similar manner. The loess record in Uzbekistan extends as far back as 2 million yr, there are at least nine major soils in loess above the Brunhes/Matuyama boundary (*c.* 690,000yr BP), and loess deposition appears to have been relatively slight

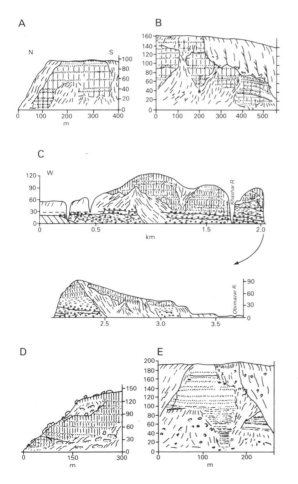

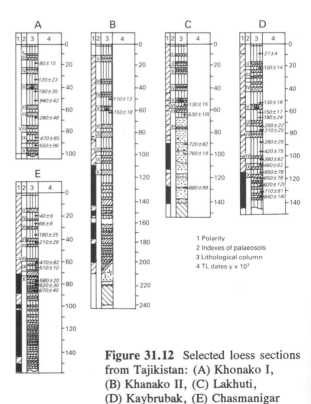

Figure 31.12 Selected loess sections from Tajikistan: (A) Khonako I, (B) Khanako II, (C) Lakhuti, (D) Kaybrubak, (E) Chasmanigar (after Dodonov and others, in Goudie et al. 1984).

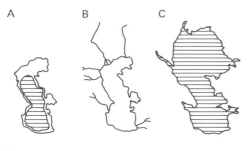

Figure 31.13 The changing nature of the Caspian Sea. (A) The extent of the sea during the Mikulino interglacial and (B) at the present day; (C) the greatly expanded sea during the Early Valdai glaciation; (D) the transgressions and regression of the Caspian since the last interglacial. (A–C from Goudie 1983b, D modified from Chepalyga 1984).

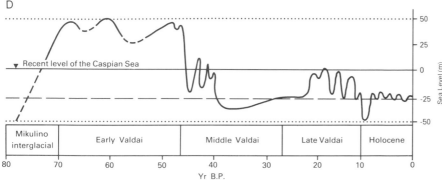

during the Holocene. Likewise, in Tajikistan the loess record goes back to the Pliocene, and impressive sections contain more than 20 loess units with intervening palaeosols, many of which show heavy calcification (Fig. 31.12).

The loess horizons themselves appear to have formed under more arid conditions than today (Lazarenko 1984), for they contain large amounts of carbonates and soluble salts, have a xerophilous mollusc fauna, and show few indications of waterlogging. Rates of accumulation in the late Pleistocene appear to have been about four times faster than in the early Pleistocene, and may have reached a rate of 1.5m 10^{-3} yr^{-1}. Shortly before the Holocene, loess accumulation ceased almost everywhere in Central Asia.

A progressive trend towards greater aridity through the Quaternary is evident from soil and pollen evidence within the profiles, and may be related to progressive uplift of the Ghissar and Tien Shan Mountains. Detailed analyses are provided by Davis et al. (1980) and Lazarenko et al. (1981).

The Aral-Caspian basin in the western part of the Turkestan Desert shows dramatic evidence of marked hydrological fluctuations, and during glacial times may have been occupied by the greatest known pluvial lake, covering an area of around 1.1 million km^2 (Fig. 31.13A–C). The highest shoreline was 76m above present Caspian level, and the Aral and the Caspian were united, and extended some 1,300km up the

Volga River from its present mouth. The largest transgressions occurred during early glacial phases, partly because reduced temperatures caused less evaporative loss, partly because of inputs of glacial meltwater, and partly because the surface over a large part of the catchment was sealed by permafrost (Chepalyga 1984). Interglacials were times of regression (Fig. 31.13B).

31.6 North America

The deserts of North America occupy much of western USA and northern Mexico between 44°N and 22°N (Fig. 31.14A). They extend southwards from central and eastern Oregon, embracing nearly all of Nevada and Utah, into southwestern Wyoming and western Colorado, reaching westwards in southern California to the eastern base of the Sierra Nevada, the San Bernadino mountains and the Cuyamaca mountains. From southern Utah the desert extends into Arizona, and on into the Chihuahua Desert of Mexico. The Sonoran desert of California extends into Baja California and to the eastern side of the Gulf of California. These deserts owe their aridity to a variety of conditions. Orographic barriers are especially important in the north and in parts of California, while the southern portion comes under the influence of a subtropical high-pressure cell, and has a summer

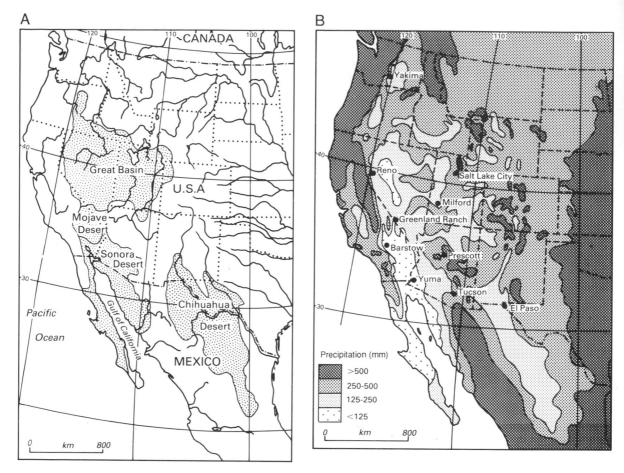

Figure 31.14 (A) The deserts of North America (after Petrov 1976); (B) precipitation in the deserts of North America (from Petrov 1976).

maximum of precipitation. Extreme aridity is relatively limited in extent, with the most arid regions lying along the Gulf of California and in the Mojave. Even in such hyper-arid areas, fluvial activity is probably significant because of the close proximity of high mountain ranges.

A major control of the development and climatic evolution of the deserts of North America was the Cordilleran orogeny, which began at the close of the Cretaceous period and continued into the Cenozoic. This involved the intrusion of massive granite batholiths in the Sierra Nevada and Idaho, and low-angle thrusting of immense slabs of rock eastwards over one another along a line extending from Mexico to north-western Canada. This thrusting ceased in Miocene times (c. 12 million yr BP), but uplift and tectonic processes continued thereafter. Structural domes were

uplifted, volcanism occurred, and rivers cut spectacular canyons into the uplifted masses. The Basin–Range province started to develop in the late Oligocene, creating a landscape of fault-bounded blocks and troughs. This was caused by crustal extension (Fig. 31.15), either because high heat flow above the subducted plate may have caused doming and extension of the crust, or because the subducting plate may have broken, causing very high heatflow and extension of the crust for a limited period.

These tectonic events provided the setting for some of the most important geomorphological features of the arid west – the playas – for, although deflation and other processes have contributed to their development and form, they are essentially products of a particular tectonic setting. That these mainly dry playa basins formerly contained large lakes was documented

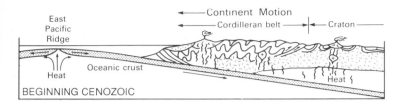

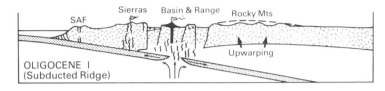

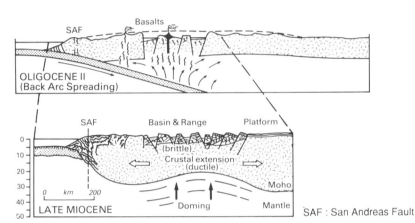

Figure 31.15 Alternative hypotheses for the Cenozoic stretching of the western USA to form the Basin–Range Province. It is assumed that a low-angle subduction caused widespread magmatism with the formation of granite batholiths and andesite volcanoes. Rifting and basalt volcanism may have resulted from either (1) subduction of the East Pacific Ridge with great heat flow above its central rift or (2) high heat flow followed by mantle doming and crustal extension after subduction had ceased.
SAF = San Andreas Fault.

around a century ago by such workers as Russell (1883) and Gilbert (1890). It is now recognized that more than a hundred closed basins in the western USA contained lakes during the late Wisconsin, but only about 10% of them are perennial and of substantial size today. The largest of the basins was Bonneville, which had a length of around 500km, a maximum area of 51,640km^2 and a depth of about 335m. Lahontan, the second largest lake, had an area of 22,900km^2 and a 280m maximum depth.

The dating of hydrological changes in these basins over the past 40,000yr is relatively precise (Smith & Street-Perrott 1983), and the lake-level curves from a number of basins show a considerable degree of similarity (Figs 31.16, 31.17). It is strikingly evident that from c. 24,000–14,000yr BP (i.e. more or less contemporaneous with the maximum of the last, Wisconsin, glaciation) that lake levels were high. There was then a phase when lake levels showed marked fluctuations, before a period of drought between 10,000yr and 5,000yr BP, which culminated between 6,000yr and 5,000yr BP. While the temporal patterns

of the fluctuations may be relatively well known, the explanation for the greatly expanded pluvial lakes is more controversial, and there has been a longstanding debate as to whether the crucial control was the diminished evapotranspirational losses brought about by late Pleistocene temperature reductions, or some increase in precipitation levels (see, for example, Brakenridge 1978).

Although the pluvial legacy is widely evident in the presence of old lake beds, high lake shorelines, and formerly integrated fluvial systems, elsewhere there is abundant evidence for the formerly greater power and distribution of aeolian processes, as revealed by the presence of aligned drainage systems, yardangs, shaped depressions and associated lunettes and, most importantly of all, of palaeodune systems. These were first recognized by Price (1944) and have recently been mapped in a great belt of country in the lee of the western cordillera between the Canadian Line and the Gulf of Mexico (Wells 1983). The dating of the largely relict dune fields, of which the Nebraskan Sandhills are the largest example (Warren 1976a), is

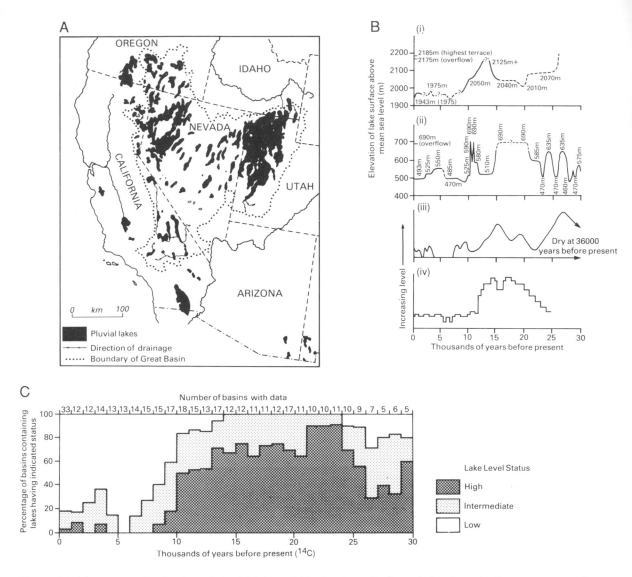

Figure 31.16 (A) The distribution of pluvial lakes within the Great Basin and elsewhere in the western USA that are known or inferred to have expanded during the period 10,000–25,000yr BP (after Smith & Street-Perrott 1983); (B) lake-level histories of selected basins since 30,000yr BP from various sources (after Smith & Street-Perrott 1983): (i) Lake Russell (Mono Basin); (ii) Searles Lake; (iii) Lake Lahontan; (iv) Lake Bonneville. (C) Relative percentages of lake basins containing dated materials at elevations indicative of low, intermediate or high stages versus their radiocarbon ages (after Smith & Street-Perrott 1983).

controversial. Debate surrounds the relative importance of late Pleistocene and mid-Holocene (altithermal) arid phases (Ahlbrandt et al. 1983). The stratigraphy, sedimentology and arrangement of lunettes in the lee of large deflation basins cut into palaeodrainage systems on the High Plains of Texas indicates that there were multiple phases of aeolian activity. Indeed, aeolian activity in the High Plains, as represented by the Blackwater Draw Formation, dates back to beyond 1.4 million yr BP (Holliday 1989).

Figure 31.17 Well developed shorelines on slopes in the Great Salt Lake Basin, Utah, formed by late Pleistocene Lake Bonneville.

31.7 South America

The main desert areas of South America are inextricably associated with the Andean cordillera. The most extensive zone of aridity includes the coastal Peruvian and Atacama deserts to the west of the mountains from *c.* 5°S to *c.* 30°S; to the east of the cordillera lie the Monte and Patagonian deserts of Argentina (Fig. 31.18).

The Monte and Patagonian deserts both lie essentially in the lee of the Andes. The Monte Desert, which is more or less continuous with the deserts to the west, is composed of basin–range topography, including mountain blocks, extensive piedmont surfaces, and largely internal drainage. Volcanic features are also to be found. The evolution of the region is not well understood, but Walter Penck was amongst those who have contributed towards the description of slope evolution and piedmont development. The Patagonian Desert stretches for over 500km between the Andes and the sea. It owes its aridity to the mountains, which block the rain-bearing winds from the west, and to the cold Falkland Current off the coast. The region is dominated by piedmont plains that slope eastwards towards the Atlantic, where they are terminated by marine surfaces, by ephemeral rivers that are entrenched into them, and by enclosed drainage basins. Volcanic, glacial and fluvial deposits occur extensively in the region. Several glacial episodes

Figure 31.18 Arid lands of South America (after Meigs 1953).

441

during the Quaternary in the Patagonian Andes certainly influenced the evolution of this arid area strongly, especially in feeding fluvioglacial gravels into the desert (Mercer 1976).

West of the Andes, the Peru–Chile desert has several distinctive features. Climatically, the aridity is created by subtropical atmospheric subsidence reinforced by the upwelling of cold coastal waters associated with the north-flowing Peru current. As a result, it is one of the world's driest areas, although precipitation does increase eastwards with elevation in the Andes. The coastal zone is characterized by fogs (*camanchaca*) that roll in from the Pacific on many days, providing unexpectedly high humidities. The region has many local and regional winds but there are very few areas of sand dunes. Nevertheless, mobile barchans and yardangs are well developed in southern Peru.

Geologically, the region is dominated by the Andean cordillera (up to 7,000m asl) and, offshore, the Peru–Chile trench (up to 7,600m bsl). Both features are associated with the westward migration of the South American plate over the eastward-moving Nazca plate. The region has abundant evidence of volcanic activity, including volcanoes, calderas, enormous pyroclastic flow deposits (ignimbrites), lava flows, geysers and solfatara (Fig. 31.19; Naranjo &

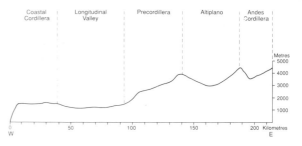

Figure 31.20 Topographic profile of the Atacama Desert (22°S) in Chile.

Cornejo 1989). Andesitic rocks are typical of the volcanic region, and contemporary activity is confined to a few moderately explosive volcanoes. Tectonic activity, in the form of extensive folding and faulting, is widespread and is responsible, *inter alia*, for the longitudinal differentiation of the topography.

The principal longitudinal features of the Atacama topography (Fig. 31.20) are, from west to east, the Coastal Cordillera (up to *c.* 2,000m asl); the Longitudinal Valley, the pre-cordillera (an area of basin–range topography, salt domes and slopes rising up to the Andes), and the Altiplano (an extensive, arid plateau at over 4,000m asl where volcanic activity is widespread). The Andean flanks, and the Longitudinal Valley are dominated by massive aggradation during the Oligocene and, possibly as a consequence of the onset of aridity, by subsequent extensive, bedrock-dominated pediment surfaces developed across a variety of rocks, including granite (e.g. Naranjo & Paskoff 1985). Segerstrom (1963) referred to the pediment topography as a "matureland", although in reality it probably consists of several uplifted and dissected erosional surfaces. There are extensive and probably complementary piedmont deposits within the Longitudinal Valley, the internal drainage basins and along the perennial rivers.

The drainage of the region, especially in Chile, is dominated by enclosed drainage basins that focus on salars. In addition, snowmelt in the Andes nourishes a few perennial streams that cross the desert to the Pacific. Only the Rio Loa crosses the driest zone of the Norte Grande, but to the south, as the precipitation rises and the snowline falls in the Norte Chico, the rios Copiapó, Huasco and Elqui all cut across the major zones of longitudinal relief. Salts, many of which may be of volcanic origin, are widespread. They are found in many salars; nitrates and iodates are distinctively concentrated, especially on the eastern side of the coastal cordillera (see Ch. 5.2);

Figure 31.19 Volcanic activity, including a small geyser, in a salt lake at El Tatio, in the high Andes of the Atacama Desert, Chile.

and the salt domes occur in the precordillera of the Norte Grande.

The evolution of the Peru–Chile deserts is still a matter of considerable speculation, but a few observations are in order. First, and contrary to common opinion, it seems likely that there has not been enormous Andean uplift during the Quaternary. One study, based on K-Ar dating of ignimbite flows, suggested that in the high Andes of the Norte Chico the pediplain topography has suffered remarkably little erosion since the Upper Miocene, and that over the past 9–12 million yr there has been entrenchment only in the canyons of *c.* 100–200m (Clark et al. 1967). This view is confirmed by Naranjo & Paskoff (1985).

Secondly, the Atacama Desert is probably very old. It is generally believed that it has been arid since at least the late Eocene, with hyper-aridity since the middle to late Miocene (Alpers & Brimhall 1988). The uplift of the Central Andes cordillera during the Oligocene and early Miocene was a critical palaeoclimatic factor, providing a rain-shadow effect and also stabilizing the southeastern Pacific anticyclone. However, also of great significance (and analogous to the situation in the Namib) was the development 15–13 million yr BP of cold Antarctic bottom waters and the cold Humboldt current as a result of the formation of the Antarctic ice sheet.

Thirdly, it is a widely held view that much of the region is a "core desert", where climatic change has been quite limited during the Quaternary. Certainly much of the Atacama gives this impression. Although there is some evidence of glaciation in the high Andes of the desert during the Quaternary, it was very local and only on the highest mountains. Morainic deposits showed that short glaciers extended to *c.* 4,200m. Morainic deposits and associated features have been identified in the northern Atacama, east of the Salar de Atacama (e.g. Hollingworth & Guest 1967). Farther south, in the Norte Chico, Caviedes & Paskoff (1975) and Weischet (1969) have described glacial features that suggest three or four glacial "periods", although their ages are uncertain.

Clapperton (1983), Mercer (1976) and others leave no doubt that there has been a series of roughly synchronous glacier fluctuations of similar magnitude in the Andes during the late Pleistocene, and that these events were preceded by glaciations of similar magnitude during the past 3.5 million yr BP. Clapperton recognized significant glacial episodes in the Holocene (*c.* 16,000–10,000yr BP), the last glaciation (*c.* 18,000–16,000yr BP), the penultimate glaciation

(*c.* 170,000–140,000yr BP), the pre-penultimate glaciations (< 1.8 million yr BP) and the pre-Pleistocene glaciations (1.8–5 million yr BP). The glacial episodes undoubtedly reflect climatic changes. In the Atacama region, the glacial features were minor, but the impact of the climatic changes on the evolution of landforms is not yet clear, except in so far as there is evidence of fluctuations in lake levels in the salars.

Evidence of precipitation changes is provided by lake basins in the Altiplano. Hastenrath and Kutzbach (1985), for example, have shown that in the late Pleistocene (before 28,000yr BP and from 12,500 to 11,000yr BP), lakes in the Peruvian–Bolivian altiplano were four to six times more extensive than today, and that this implies rainfall increases of around 50–75%. By contrast, in the mid-Holocene (*c.* 7,700–3,650yr BP) Lake Titicaca was at a very low level (Wirrman & Almeida 1987).

Along the perennial river valleys and the coast there is also evidence of Quaternary evolution in the form of marine and fluvial terraces and their associated deposits. Because the number of these surfaces and deposits varies from sector to sector along the coast, it seems probable that the sequences may reflect differential tectonic movements, as well as fluctuating sea levels. As a result, clear generalizations are difficult, but is seems likely that the highest major surface of erosion-aggradation is at least Pliocene in age, and possibly much older. Paskoff (1978–9) has suggested that the high cliff, which is up to 800m high and is so characteristic of the Chilean desert coast, probably originated as a Miocene fault scarp which has retreated and been embroidered by an oscillating sea level ever since.

31.8 Australia

With the exception of Antarctica, Australia is the driest of the continents, with a total of around 5.4 million km^2 experiencing appreciable aridity. Paradoxically, however, aridity is not especially extreme in its intensity, and mean annual precipitation levels do not fall below 100–125mm. Indeed, because of another major control on its landscape evolution, its relatively long history of tectonic stability over large areas, many of the present features of its geomorphology are inherited from a great variety of climates that may go back to the Jurassic or earlier. Dunes as old as the Eocene are still preserved (Benbow 1990). The importance of such relict features has been succinctly summarized thus by Langford-Smith (1982):

> The fundamental base for much of the flat or gently undulating desert landscape is Cretaceous; most of the macro-forms such as plateaux and mesas, and structural features such as the larger lake depressions are Tertiary; while meso-forms like sand dunes, prior stream formations, and the many small playas are Pleistocene. The Holocene has had little bearing on the deserts of today, apart from European man's contribution to the degeneration of ecosystems on some semi-arid/arid margins.

Useful general reviews of the distinctive landscapes of the Australian Deserts are provided by Mabbutt (1969, 1984).

The impact of post-Tertiary climates can be appreciated only within the context of plate tectonics and continental drift. Around 50–60 million yr BP, Australia began to drift apart from Antarctica, and it migrated northwards, drifting to within 4° of its present latitude by the early Miocene (c. 25 million yr BP). Thus, not only has Australia been subjected to climatic changes resulting from changes in latitude, but it has also been subjected to the climatic effects that such movement had on the nature of oceanic and climatic circulation systems in the Southern Hemisphere, and to the global climatic changes associated with the so-called Cenozoic climatic decline.

Among the important relict Tertiary features of the Australian desert are the widespread duricrusts (Woolnough 1927), which include silcretes (Langford-Smith 1978a, b) and laterites with associated deep weathering profiles. There are also widespread relict palaeodrainage systems on vast erosion surfaces (Van der Graaf et al. 1977), and there was a virtual inland sea in the Lake Eyre depression.

However, in the late Miocene–Pliocene, there was a gradual transition to a more arid climate. There is, however, scant evidence about the nature of climates for much of the Pleistocene, and it is only for the past 40,000–50,000yr that there is much information. The importance of major Pleistocene climatic and hydrological changes is clearly evident in the riverine plains of the east of the arid zone, which are part of the Murray system (e.g. Schumm 1968; Ch. 11.7). There are large spreads of alluvium deposited by ancient "prior streams", which were sinuous bedload channels indicative of coarse sediment transport in flash floods. These plains are mantled by aeolian clays, called parna, and the active floodplains are slightly entrenched beneath the prior stream deposits and are occupied by meandering, suspended-load channels. Cooper Creek, which flows into Lake Eyre, shows a comparably varied history in the late Pleistocene

Figure 31.21 (A) Physiographic desert types in Australia (after Mabbutt 1984). (B) Boundaries of dune fields, average directions of dune elongation judged from Y-junctions and crestal orientations, and dunefield names (after Wasson 1984). (C) Changes in precipitation at Lynch's Crater, near Atherton, Queensland, derived from pollen analysis (according to A. P. Kershaw). The dashed line indicates the direction of change only (after Bowler et al. 1976). (D, E) movements of lake levels in the Willandra Lakes of New South Wales and Lake Keilambete in western Victoria, derived from sedimentological analyses of lake deposits. Thickened lines indicate high lake levels of low salinity (after Bowler et al. 1976).

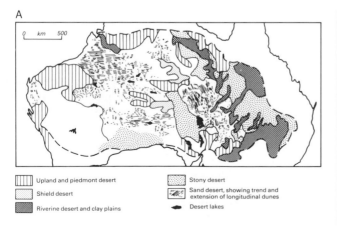

A

Upland and piedmont desert

Shield desert

Riverine desert and clay plains

Stony desert

Sand desert, showing trend and extension of longitudinal dunes

Desert lakes

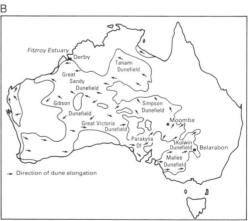

B

Fitzroy Estuary

Derby

Tanami Dunefield

Great Sandy Dunefield

Gibson Dunefield

Simpson Dunefield

Moomba

Great Victoria Dunefield

Parakylia Df

Kulwin Dunefield

Belarabon

Mallee Dunefield

→ Direction of dune elongation

(Bowler & Wasson 1984).

The widespread sand deserts of Australia also display the impact of Pleistocene changes (Wasson 1984), and a striking feature of the linear dune fields of the Australian deserts is that they extend well beyond the present desert areas (Fig. 31.21). They are relict features of Pleistocene aridity, and some may be of late Pleistocene age. This applies to the stable dunes of the northern part of the Great Sandy Desert which extends beneath the Holocene alluvium of the Fitzroy Estuary (Jennings 1975), and which thus formed at a time of low sea levels. Lacustrine and lunette sediments in the Willandra lakes area of New South Wales also afford evidence for a period of dune encroachment and lake desiccation in the late Pleistocene (Bowler et al. 1976; Fig. 31.21D). This glacial aridity may have become prevalent after 25,000yr BP, before which lake levels in the southern parts of Australia appear to have been relatively high. Northeastern Australia, however, was significantly

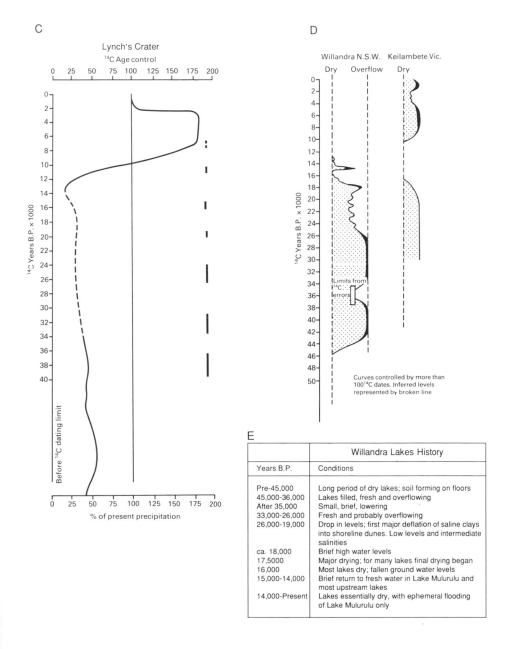

E	Willandra Lakes History	
Years B.P.	Conditions	
Pre-45,000	Long period of dry lakes; soil forming on floors	
45,000-36,000	Lakes filled, fresh and overflowing	
After 35,000	Small, brief, lowering	
33,000-26,000	Fresh and probably overflowing	
26,000-19,000	Drop in levels; first major deflation of saline clays into shoreline dunes. Low levels and intermediate salinities	
ca. 18,000	Brief high water levels	
17,5000	Major drying; for many lakes final drying began	
16,000	Most lakes dry; fallen ground water levels	
15,000-14,000	Brief return to fresh water in Lake Mulurulu and most upstream lakes	
14,000-Present	Lakes essentially dry, with ephemeral flooding of Lake Mulurulu only	

less humid than today well before 25,000yr BP, for the pollen spectra from Lynch's Crater in Queensland indicate a drastically reduced rainfall (2,500–500mm yr^{-1}) in the interval between about 80,000yr and 20,000yr BP (Fig. 31.21C). Overall, the combined evidence from marine cores on the Timor continental shelf, pollen analysis and the study of lake and dune deposits, indicates that at the time of the last glacial maximum (*c.* 25,000–18,000yr BP) much of Australia was drier and windier than today, and was surrounded by a much broader continental shelf (Williams 1985). Soon after 17,000yr BP, temperatures, precipitation and sea levels began to rise, and most of the desert dunes were probably becoming stabilized by *c.* 13,000yr BP. In the early Holocene, rainfall levels were higher and forests became more widespread. There is some evidence for renewed dune activity in western New South Wales beginning *c.* 2,500yr BP (Wasson 1976), and this may signify the passing of the mid-Holocene humid phase and a brief return to relative aridity.

31.9 Conclusion

The world's deserts show the imprint of climatic changes at many different time scales. These range from runs of a few dry or wet years, through to phases of the duration of a decade or decades (e.g. the Dust Bowl years of North America in the 1930s, or the persistent drought of the Sudano-Sahel belt since the late 1960s), through to extended periods of some centuries or millennia (e.g. the dry hypsithermal phase of the American High Plains in the mid-Holocene or the intense pluvial or lacustral phase of tropical Africa in the early Holocene), through to major Pleistocene events related to the glacial and interglacials of higher latitudes (which may have had a duration of the order of 100,000yr), and through to the long-term "geological changes" of the order of millions of years associated with major shifts in the positions of the continents, major tectonic and orogenic events, and the configurations of the ocean basins and their associated circulation systems.

The short-term fluctuations are related to changes in the general atmospheric circulation, with such changes as the water temperatures of the Pacific Ocean associated with the El Niño effect or the zonality or meridionality of the Rossby waves playing a major rôle (Winstanley 1973). Medium-term fluctuations, such as the early Holocene pluvial of the tropics, may be related to changes in Earth geometry, and Kutzbach (1981) has indicated that, at around 9,000yr BP, theoretical insolation receipt during the Northern Hemisphere summer may have been larger than now (by about 7%), thereby leading to an intensification of the monsoonal circulation and associated precipitation. At a longer timescale, related to the interglacial and glacial fluctuations of the Pleistocene, a variety of factors may have been involved. For example, during a major glacial phase, tropical aridity may have been heightened by an increased continentality of climate resulting from the withdrawal of the sea from the continental shelves as a consequence of glacial eustasy. Such a drop in sea level, together with an extension of sea ice, would result in less evaporation from the ocean surface, leading to less rain. Moreover, the worldwide cooling of the oceans would lead to less evaporation and convection, and the generation of fewer tropical cyclones.

Furthermore, thermal variations provoked by the growth and decay of the great ice sheets decisively influenced the patterns of the general atmospheric circulation. For example, in theory, an increased temperature gradient resulting from the presence of the great Scandinavian and Laurentide ice sheets would result in stronger westerlies, an equatorial displacement of major circulation features, and an intensification and shrinking of the Hadley cell zone and the zone of extratropical Rossby wave circulation. The result would, according to Nicholson & Flohn (1981), have been a greater degree of westerly flow and mid-latitude cyclogenesis over an area such as North Africa. A corresponding displacement of the subtropical high-pressure zone would have displaced the aridity maximum into West Africa. The decreased thermal contrast between the two hemispheres compared to the situation today would have had an impact. At present the Southern Hemisphere, in comparison with the Northern, is much cooler and its temperature gradient greater. This is because of the varying amounts and distribution of ocean and land in the two hemispheres. In the Southern Hemisphere the stronger temperature gradient produces a more intense atmospheric circulation, and is probably largely responsible for the asymmetry that exists, whereby the meteorological equator lies in the Northern Hemisphere. During a glacial phase intense continental glaciation should have led to a displacement of this meteorological equator to a position more coincident with the geographical Equator (i.e. southwards). This would disrupt the annual march of the monsoon, and thereby reduce precipitation amounts in such areas as the Thar and southeast Arabia.

The trigger for the glacial/interglacial fluctuations themselves is now thought to lie in the Milankovitch mechanism of Earth orbital fluctuations, with amplitudes of the order of about 100,000yr, 43,000yr and 19,000–24,000yr (Imbrie & Imbrie 1979). Spectral analysis has shown such wavelengths in ocean-floor sediments and in Chinese loess profiles (Lu 1981).

The longest scales of change, which include the establishment of the world's arid areas, are related to variations in the configurations of the continents and oceans, occasioned by global tectonics. This mechanism operated in a variety of ways. First, the drift of the continents changed their latitudinal positions and thus their position with respect to major climatic belts. Secondly, because of the rise of the great Cenozoic mountain systems caused by plate collision, rain-shadow deserts were created on the continents, especially in Asia, northern Argentina, the southwestern USA and the Atlas. Thirdly, changes in the configurations of the oceans and continents caused a general world cooling – the so-called Cenozoic climatic decline. Such cooling would have affected the nature and intensity of the general atmospheric circulation and of ocean currents. The significance of this last point is that in the Southern Hemisphere the continents all have dry zones along their western margins. Each of these deserts is to a large degree caused by an upwelling of cold water off the coast, and such cold water is mostly of Antarctic origin. Such cold currents would, therefore, not exist until Antarctica was producing large quantities of cold meltwater and sea ice. For this to happen, Antarctica had to have moved to a suitable latitudinal position following its separation from Australia (Selby 1985). These sorts of influences are seen in the history of the Atacama Desert, an area which has been arid since at least the late Eocene. The uplift of the Central Andes cordillera during the Oligocene and early Miocene was one critical palaeoclimatic factor, providing a rain shadow for precipitation from the Amazon basin, and stabilizing the southeastern Pacific anticyclone. However, it was the dramatic cooling of Antarctic bottom waters and of the Humboldt Current around 13–15 million yr BP, associated with the formation of the Antarctic ice sheet, that enabled hyper-aridity to be established for the first time in the area (Alpers & Brimhall 1988).

BIBLIOGRAPHY

Entries *entirely* in **boldface type** are edited volumes of multiple contributions referred to more than once.

Abou Al Heya, M. K. & W. M. Shehata 1986. Engineering properties of Al-lith Sabkha, Saudi Arabia, *5th Congress International Association Engineering Geologists, Proceedings* **3.5.1**, 935–41.

Abrahams, A. D. 1972. Drainage densities and sediment yields in eastern Australia. *Australian Geographical Studies* **10**, 19–44.

Abrahams, A. D. (ed.) 1986. *Hillslope processes*. Boston: Allen & Unwin.

Abrahams, A. D., S-H. Luk, A. J. Parsons 1988. Threshold relations for the transport of sediment by overland flow on desert hillslopes. *Earth Surface Processes and Landforms* **13**, 407–19.

Abrahams, A. D., A. J. Parsons, R. U. Cooke, R. W. Reeves 1984. Stone movement on hillslopes in the Mojave Desert, California: a 16-year record. *Earth Surface Processes and Landforms* **9**, 365–70.

Abrahams, A. D., A. J. Parsons, P. J. Hirsh 1985. Hillslope gradient – particle size relations: evidence for the formation of debris slopes by hydraulic processes in the Mojave Desert. *Journal of Geology* **93,** 347–57.

Abrahams, A. D., A. J. Parsons, S-H. Luk 1988. Hydrologic and sediment responses to simulated rainfall on desert hillslopes in southern Arizona. *Catena* **15**, 103–17.

Adams, R. McC. 1981. *Heartland of cities*. Chicago: University of Chicago Press.

Agnew, C. T. 1988. Soil hydrology in the Wahiba Sands. *Journal of Oman Studies*, Special Report 3, 191–200.

Ahlbrandt, T. S. 1974. The source of sand for the Killpecker sand dune field, southwestern Wyoming. *Sedimentary Geology* **11**, 39–57.

Ahlbrandt, T. S. 1979. Textural parameters of eolian deposits. See McKee (1979), 21–51.

Ahlbrandt, T. S. & S. Andrews 1978. Distinctive sedimentary features of cold-climate eolian deposits, North Park, Colorado. *Palaeogeography, Palaeoclimatology, Palaeoecology* **25**, 327–51.

Ahlbrandt, T. S., S. Andrews, D. T. Gwynne 1978. Bioturbation in eolian deposits. *Journal of Sedimentary Petrology* **48**, 839–48.

Ahlbrandt, T. S. & S. G. Fryberger 1980. Eolian deposits in the Nebraska Sand Hills. USGS Professional Paper 1120A–C.

Ahlbrandt, T. S. & S. G. Fryberger 1981. Sedimentary features and significance of interdune deposits. See *Recent and ancient nonmarine depositional environments: models for exploration*, F. G. Ethridge & R. M. Flores (eds), SEPM Special Publication 31, 293–314.

Ahlbrandt, T. S., J. B. Swinehart, D. G. Maroney 1983. The dynamic Holocene dune fields of the Great Plains and Rocky Mountain Basins. See Brookfield & Ahlbrandt (1983), 379–406.

Ahlcrona, E. 1986. *Monitoring the impact of climate and man on land transformation – a study in an arid and semi-arid environment in central Sudan*. Lunds Universitet Naturgeografiska Institution, Rapporter och Notiser 66.

Ahnert, F. 1960. The influence of Pleistocene climates on the morphology of cuesta scarps on the Colorado Plateau. *Annals of the AAG* **50**, 139–56.

Aimé, S. & M. J. Penven 1982. Le complexe dunaire du Cap Falcon (Oran): étude morphodynamique appliqué et perspectives d'aménagement. *Méditerranée* **45**, 3–16.

Aires-Barros, L., R. C. Graça, A. Veloz 1975. Dry and wet laboratory tests and thermal fatigue of rocks. *Engineering Geology* **9**, 249–65.

Akagi, Y. 1972. Pediment morphology in Japan. *Fukuoka University of Education, Bulletin* **21**, 52–76.

Akagi, Y. 1980. Relations between rock type and the slope form in the Sonora Desert, Arizona. *Zeitschrift für Geomorphologie* **24**, 129–40.

Akili, W. & J. K. Torrance 1981. The development and geotechnical problems of sabkha, with preliminary experiments on the static penetration resistance of cemented sands. *Quarterly Journal of Engineering Geology* **14**, 59–73.

Akpokodje. E. G. 1984. Influence of rock weathering on the genesis of gypsum and carbonate in some Australian arid zone soils. *Australian Journal of Soil Research* **22**, 243–51.

Al-Hinai, K. G. 1988. *Quaternary aeolian sand mapping in Saudi Arabia, using remotely-sensed imagery*. PhD thesis, Imperial College, University of London.

Al-Hinai, K. G. & J. M. Moore 1987. Monitoring of sand migration in eastern Arabia by remote sensing. *20th International Symposium on Remote Sensing of Environment, Proceedings* (Environment Research Institute of Michigan, Ann Arbor, Michigan).

Al-Hinai, K. G., J. M. Moore, P. R. Bush 1987. LANDSAT image enhancement study of possible submerged sand dunes in the Arabian Gulf. *International Journal of Remote Sensing* **2**, 251–8.

Al-Saleh, S. & F. I. Khalaf 1982. Surface textures of quartz grains from various recent sedimentary environments in Kuwait. *Journal of Sedimentary Petrology* **52**, 215–26.

Al-Sayari, S. & J. G. Zötl (eds) 1978. *Quaternary period* *in Saudi Arabia*, volume 1. Wien: Springer.

Alexander, E. B. & W. D. Nettleton 1977. Post-Mazama natrargids in Dixie Valley, Nevada. *Soil Science Society of America, Journal* **41**, 1210–12.

Ali, Y. A. & I. West 1983. Relationships of modern gypsum nodules in sabkhas of loess to compositions of brines and sediments in northern Egypt. *Journal of Sedimentary Petrology* **53**, 1151–68.

Alimen, M-H. 1953. Variations granulométriques et morphoscopiques du sable le long de profiles dunaires au Sahara occidental. In *Actions éoliennes*, CNRS Colloques Internationaux 35, 219–35.

Alimen, M-H., M. Buron, J. Chavaillon 1958. Caractères granulométriques de quelques dunes d'ergs du Sahara nord-occidental. *Academie des Sciences à Paris, Comptes Rendus* **247**, 1758–61.

Alizai, H. V. & L. C. Hurlbert 1970. Effects of soil texture on evaporative loss and available water in semi-arid climates. *Soil Science* **110**, 328–32.

Allan, J. A. (ed.) 1981. *The Sahara: ecological change and early economic history*. London: Menas Press.

Allchin, B., A. S. Goudie, K. Hegde 1978. *The prehistory and palaeogeography of the Great Indian Desert*. London: Academic Press.

Allen, J. R. L. 1965. Scour marks in snow. *Journal of Sedimentary Petrology* **35**, 331–8.

Allen, J. R. L. 1968a. The nature and origin of bedform hierarchies. *Sedimentology* **10**, 161–82.

Allen, J. R. L. 1968b. *Current ripples: their relation to patterns of water and sediment motion*. Amsterdam: Elsevier/North Holland.

Allen, J. R. L. 1968c. The diffusion of grains in the lee of ripples, dunes and sand deltas. *Journal of Sedimentary Petrology* **38**, 621–33.

Allen, J. R. L. 1969a. The maximum slope angle attainable by surfaces underlain by bulked equal spheroids with variable dimensional ordering. *Geological Society of America, Bulletin* **80**, 1923–30.

Allen, J. R. L. 1969b. On the geometry of current ripples in relation to the stability of flow. *Geografiska Annaler* **51**, 61–96.

Allen, J. R. L. 1970a. The avalanching of granular solids on dunes and similar slopes. *Journal of Geology* **78**, 326–51.

Allen, J. R. L. 1970b The angle of initial yield of haphazard assemblages of equal spheres, in bulk. *Geologie en Mijnbouw* **49**, 13–22.

Allen, J. R. L. 1971. Intensity of deposition from avalanches and loose packing of avalanche deposits. *Sedimentology* **18**, 105–11.

Allen, J. R. L. 1974. Reaction, relaxation and lag in natural sedimentary systems: general principles, examples and lessons. *Earth Science Reviews* **10**, 263–342.

Allen, J. R. L. 1976. Bedforms and unsteady processes: some concepts of classification and response illustrated by common one-way types. *Earth Surface Processes* **1**, 361–74.

Allen, J. R. L. 1980. Large transverse bedforms and the character of boundary layers in shallow-water environments. *Sedimentology* **27**, 317–24.

Allen, J. R. L. 1982a. *Sedimentary structures* [2 volumes]. Amsterdam: Elsevier.

Allen, J. R. L. 1982b. Simple models for the shape and symmetry of tidal sand waves: (1) statically stable equilibrium forms. *Marine Geology* **48**, 31–49.

Allen, J. R. L. 1985. *Principles of physical sedimentology*. London: Allen & Unwin.

Allen, J. R. L. & J. D. Collinson 1974. The superimposition and classification of dunes formed by unidirectional aqueous flows. *Sedimentary Geology* **12**, 169–78.

Allenby, R. J. 1988. Origin of rectangular and aligned lakes in the Beni basin of Bolivia. *Tectonophysics* **145**, 1–20.

Allison, R. J. 1988. Sediment types and sources in the Wahiba Sands, Oman. *Journal of Oman Studies*, Special Report **3**, 161–8.

Allison, R. J. 1990. Rock control and slope profiles in a tropical limestone environment: the Napier Range of Western Australia. *Geographical Journal* **156**, 200–11.

Alpers, C. N. & G. H. Brimhall 1988. Middle Miocene climatic change in the Atacama Desert, northern Chile: evidence from supergene mineralisation at La Escondida. *Geological Society of America, Bulletin* **100**, 1640–56.

Ambrose, G. J. & R. B. Flint 1981. A regressive Pliocene lake system and silicified strandlines in northern South Australia: implications for regional stratigraphy and silcrete genesis. *Journal of the Geological Society of South Australia* **28**, 81–94.

Amiel, A. J. 1975. Progressive pedogenesis of eolianite sandstone. *Journal of Sedimentary Petrology* **45**, 513–9.

Amit, R. & R. Gerson 1986. The evolution of Holocene reg (gravelly) soils in deserts – an example from the Dead Sea region. *Catena* **13**, 59–79.

Amstutz, G. C. & R. Chico 1958. Sand size fractions of south-Peruvian barchans and a brief review of the genetic grain shape function. *Vereinigung Schweizerischer Petroleum-Geologen-und-Ingeniuren, Bulletin* **24**, 47–52.

Amundson, R. G., O. A. Chadwick, J. M. Sowers, H. E. Dower 1988. Soil evolution along an altitudinal transect in the eastern Mojave Desert, California. *Geoderma* **43**, 349–72.

Amundson, R. G., O. A. Chadwick, J. M. Sowers, H. E. Dower 1990. The stable isotope chemistry of pedogenic carbonates at Kyle Canyon, Nevada. *Soil Science Society of America, Journal* **53**, 201–10.

Anderson, E. W. & W. Büttiker 1988. Water-extracting tenebrionid beetles from the Wahiba Sand Desert, Oman. *Journal of Oman Studies*, Special Report **3**, 321–4.

Anderson, R. S. 1986. Erosion profiles due to particles entrained by wind: application of an eolian sediment transport model. *Geological Society of America, Bulletin* **97**, 1270–8.

Anderson, R. S. 1987a. A theoretical model for aeolian impact ripples. *Sedimentology* **34**, 943–56.

Anderson, R. S. 1987b. Eolian sediment transport as a

stochastic process: the effects of a fluctuating wind on particle trajectories. *Journal of Geology* **95**, 497–512.

Anderson, R. S. 1988. The pattern of grainfall deposition in the lee of aeolian dunes. *Sedimentology* **34**, 175–88.

Anderson, R. S. 1989. Saltation of sand: a qualitative review with biological analogy. See Gimmingham et al. (1989), 149–65.

Anderson, R. S. & P. K. Haff 1988. Simulation of eolian saltation. *Science* **241**, 820–3.

Anderson, R. S. & P. K. Haff 1991. Wind modification and bed response during saltation of sand in air. *Acta Mechanica Supplementum* **2**, 21–52.

Anderson, R. S. & B. Hallet 1986. Sediment transport by wind: toward a general model. *Geological Society of America, Bulletin* **97**, 523–35.

Anderson, R. S., M. Sorensen, B. B. Willetts 1990. A review of recent progress in our understanding of aeolian sediment transport. *Acta Mechanica Supplementum* **2**, 1–20.

Anderson, R. V. 1947. Origin of the Libyan oasis basins [abstract]. *Geological Society of America, Bulletin* **58**, 1163.

Andres, W. 1980. On the paleoclimatic significance of erosion and deposition in arid regions. *Zeitschrift für Geomorphologie*, Supplementband 36, 113–22.

Andrews, S. 1981. Sedimentology of Great Sand Dunes, Colorado. SEPM Special Publication 31, 279–91.

Anstey, R. L. 1965. *Physical characteristics of alluvial fans*. US Army Natick Laboratory, Technical Report ES-20.

Antevs, E. 1952. Arroyo cutting and filling. *Journal of Geology* **60**, 375–85.

Anton, D. 1984. Aspects of geomorphological evolution: palaeosols and dunes in Saudi Arabia. See *The Quaternary of Saudi Arabia, 2: sedimentological, hydrogeological, hydrochemical, geomorphological geochronological and climatological investigations in western Saudi Arabia*, A. R. Jado & J. G. Zotl (eds), 275–96. Berlin: Springer.

Anton, D. & F. Ince 1986. A study of sand color and maturity in Saudi Arabia. *Zeitschrift für Geomorphologie* **30**, 339–56.

Anton, D. & P. Vincent 1986. Parabolic dunes of the Jafurah Desert, Eastern province, Saudi Arabia. *Journal of Arid Environments* **11**, 187–98.

Arakel, A. V. 1982. Genesis of calcite in Quaternary soil profiles, Hutt and Leeman Lagoons, Western Australia. *Journal of Sedimentary Petrology* **52**, 109–25.

Arakel, A. V. 1985. Vadose diagenesis and multiple calcrete soil profile development in Hutt Lagoon area, Western Australia. *Revue Géographie Physique et de Géologie Dynamique* **26**, 243–54.

Arakel, A.V. & D. McConchie 1982. Classification and genesis of calcrete and gypsite lithofacies in palaeodrainage systems of inland Australia and their relationship to carnotite mineralization. *Journal of Sedimentary Petrology* **52**, 1149–70.

Araya-Vergara, J. F. 1987. The evolution of modern coastal dune systems in central Chile. See Gardiner (1987), 1231–9.

Arkley, R. J. 1963. Calculation of carbonate and water movement in soil from climatic data. *Soil Science* **96**, 239–48.

Ascher R. & M. Ascher 1965. Recognizing the emergence of Man. *Science* **147**, 243–50.

Ash, J. E. & R. H. Wasson 1983. Vegetation and sand mobility in the Australian desert dunefield. *Zeitschrift für Geomorphologie* **45**, 7–25.

Ashour, M. M. 1985. Textural properties of Qatar dune sands. *Journal of Arid Environments* **8**, 1–14.

Ashour, M. M. & S. M. Abd-el-Mogheith 1983. Calcareous duricrusts in north-western Libya. *Journal of Arid Environments* **6**, 203–14.

Atkinson, K. & B. Waugh 1979. Morphology and mineralogy of red desert soils in the Libyan Sahara. *Earth Surface Processes* **4**, 103–15.

Aufrère, L. 1928. L'orientation des dunes et la direction des vents. *Academie des Sciences à Paris, Comptes Rendus* **187**, 833–5.

Aufrère, L. 1929. Le problème géologique des dunes dans les déserts chauds du nord de l'ancien monde (Sahara, Arabie, Inde). *Association Français pour l'Avancement de Science*, 53ᵉ session, 393–7.

Aufrère, L. 1930. L'orientation des dunes continentales. *12th International Geographical Congress, Proceedings*, 220–31.

Aufrère, L. 1931. Le cycle morphologique des dunes. *Annales de Géographie* **40**, 362–85.

Aufrère, L. 1932. Morphologie dunaire et météorologie saharienne. *Association de Géographes Français, Bulletin* **56**, 34–47.

Aufrère, L. 1934. Les dunes du Sahara algérien. *Association des Géographes Français, Bulletin* **183**, 130–42.

Aufrère, L. 1935. Essai sur les dunes du Sahara algérien. *Geografiska Annaler* **18**, 481–550.

Avery, D. M. 1982. Micromammals as palaeoenvironmental indicators and an interpretation of the late Quaternary in the Southern Cape Province, South Africa. *Annals of the South African Museum* **85**, 183–374.

Ayyad, M. A. 1973. Vegetation and environment of the western Mediterranean coastal land of Egypt, I: the habitat of sand dunes. *Journal of Ecology* **61**, 509–23.

Azizov, A., I. M. Ismailov, K. G. Kadyrov, K. M. Mirzazhanov, B. R. Toshov 1979. Relation between the amount of loamy sand soil removed by wind and soil moisture [in Russian]. *Pochvovedeniye* **4**, 105–7.

Babikir, A. A. A. 1986. The vegetation of natural depressions in Qatar in relation to climate and soil. *Journal of Arid Environments* **10**, 165–73.

Babikir, A. A. A. & C. C. E. Jackson 1985. Ventifacts distribution in Qatar. *Earth Surface Processes and Landforms* **10**, 3–16.

Bagnold, R. A. 1931. Journeys in the Libyan Desert. *Geographical Journal* **78**, 13–39, 524–33.

Bagnold, R. A. 1936. The movement of desert sand. *Royal

Society of London, Proceedings **A157**, 594–620.

Bagnold, R. A. 1937. The size grading of sand by wind. *Royal Society of London, Proceedings* **A163**, 250–64.

Bagnold, R. A. 1938. Grain structure of sand dunes in relation to water content. *Nature* **142**, 403.

Bagnold, R. A. 1941. *The physics of blown sand and desert dunes.* London: Methuen.

Bagnold, R. A. 1951. The sand formations in southern Arabia. *Geographical Journal* **127**, 78–86.

Bagnold, R. A. 1953a. Forme des dunes de sables et régime des vents. *Colloque International du Centre Nationale de la Recherche Scientifique* **35**, 23–32.

Bagnold, R. A. 1953b. The surface movement of blown sand in relation to meteorology. Research Council of Israel, Special Publication 2, 89–93.

Bagnold, R. A. 1954a. The physical aspects of dry deserts. See *Biology of deserts*, J. L. Cloudsley-Thompson (ed.), 7–12. London: Edward Arnold.

Bagnold, R. A. 1954b. Experiments on a gravity-free dispersion of large solid spheres in a Newtonian fluid under shear. *Royal Society of London, Proceedings* **A225**, 49–63.

Bagnold, R. A. 1956. The flow of cohesionless grains in fluids. *Royal Society of London, Philosophical Transactions* **A249**, 235–97.

Bagnold, R. A. 1960. The re-entrainment of settled dust. *International Journal of Air Pollution* **2**, 357–63.

Bagnold, R. A. 1966. The shearing and dilation of dry sand and the "singing mechanism". *Royal Society of London, Proceedings* **A295**, 219–32.

Bagnold, R. A. 1985. Transport of granular solids by wind and water compared. See Barndorff-Nielsen et al. (1985), 1–8.

Bagnold, R. A. & O. E. Barndorff-Nielsen 1980. The pattern of natural size distributions. *Sedimentology* **27**, 199–207.

Baillie, I. C., P. H. Faulkner, G. P. Espin, M. J. Levett, B. Nicholson 1986. Problems of protection against piping and surface erosion in central Tunisia. *Environmental Conservation* **13**, 27–32.

Baitulin, I. O. 1972. Depth of penetration of the roots of some edifactors in natural phytocenoses of the desert in the Alma–Ata oblast [in Russian]. *Problems of Desert Development* **5**, 54–5.

Baker Jr, H. W. 1976. Environmental sensitivity of sub-microscopic surface textures on quartz sand grains – a statistical evaluation. *Journal of Sedimentary Petrology* **46**, 871–80.

Baker, V. R. 1977. Stream-channel response to floods, with examples from central Texas. *Geological Society of America, Bulletin* **88**, 1057–71.

Baker, V. R. 1978. The Spokane flood controversy and the Martian outflow channels. *Science* **202**, 1249–59.

Baker, V. R., R. C. Kochel, P. C. Patton, G. Pickup 1983a. Palaeohydrologic analysis of Holocene flood slack-water sediments. *Special Publications of the International Association of Sedimentologists* 6, 229–39.

Baker, V. R., G. Pickup, H. A. Paloch 1983b. Desert

palaeofloods in central Australia. *Nature* **301**, 502–4.

Bakhit, A. & F. N. Ibrahim 1982. Geomorphological aspects of the process of desertification in western Sudan. *Geojournal* **6**, 19–24.

Balchin, W. G. V. & N. Pye 1955. Piedmont profiles in the arid cycle. *Geologists' Association, Proceedings* **66**, 167–81.

Ball, J. 1899/1900. *Kharga Oasis: its topography and geology.* Egyptian Geological Survey Report. Cairo: Government Printing Department.

Ball, J. 1927. Problems of the Libyan Desert. *Geographical Journal* **70**, 21–38, 105–28, 209–24.

Barbey, C. & J-P. Carbonel 1972. Le ravinement dunaire en milieu sahélien. Observations lors de pluies récents dans la région de Naoukchott. *Academie des Sciences à Paris, Comptes Rendus* **D274**, 2933–5.

Barbey, C., J. P. Carbonel, S. Duplaix, L. Le Ribault, J. Tourenq 1975. Étude sedimentologique de formations dunaires en Mauritanie occidentale. *Institut Fondemental d'Afrique Noire, Bulletin* Série **A37**, 255–81.

Barbey, C. & A. Couté 1976. Croûtes à cyanophycères sur les dunes du Sahel mauritanéen. *Institut Fondemental d'Afrique Noire* **A38**, 732–6.

Bariss, N. 1967. Effects of source relations upon loess topography [abstract]. *Nebraska Academy of Sciences, Proceedings* **78**, 35.

Barnard, W. S. 1973. Duinformasies in die Sentrale Namib. *Tegnikon*, 2–13.

Barndorff-Nielsen, O. E. 1989. Sorting, texture and structure. See Gimmingham et al. (1989), 167–79.

Barndorff-Nielsen, O.E. & C. Christiansen 1985. The hyperbolic shape triangle and classification of sand sediments. See Barndorff-Nielsen et al. (1985), 649–76.

Barndorff-Nielsen, O. E., J-T. Moller, K. Rasmussen, B. B. Willetts (eds) 1985. *International workshop on the physics of blown sand*, **Proceedings, Aarhus, May 28–31, 1985. Department of Theoretical Statistics, Institute of Mathematics, University of Aarhus; Memoir 8.**

Barndorff-Nielsen, O. E., K. Dalsgaard, C. Halgren, G. Kuhlman, J-T. Moller, G. Schon 1982. Variations in particle size over a small dune. *Sedimentology* **29**, 53–65.

Barrett, P. J. 1980. The shape of rock particles, a critical review. *Sedimentology* **27**, 291–303.

Barry, R. G. 1978. Dry climates – past and present. *Progress in Physical Geography* **2**, 116–26.

Barth, H. K. 1976. Pediment generationen und Relief-entwicklung im Schichtstufenland Saudi-Arabiens. *Zeitschrift für Geomorphologie*, Supplementband 24, 111–19.

Barth, H. K. 1982. Accelerated erosion of fossil dunes in the Gourma region (Mali) as a manifestation of desertification. See Yaalon (1982), 211–9.

Bartowski, T. 1969. Relief linéaire – relief typique des régions de loess actuelles. *Biuletyn Periglacjalny* **20**, 213–6.

Baschin, O. J. 1899. Die enstehung wellenahnlicher Oberflochenformen. *Zeitschrift für Gessellschaft für Erdkunde zu Berlin*, III **34**, 408–24.

Baschin, O. J. 1903. Dünenstudien. *Zeitschrift für Gesellschaft für Erdkunde zu Berlin* **36**, 422–430 [abstracted in *Geographical Journal* **23**, 588].

Bauer, S. J. & B. Johnson, 1979. Effects of slow uniform heating on physical properties of the Westerly and Charcoal granites. *20th US Symposium on Rock Mechanics, Proceedings*, 7–18.

Bauman, A. J. 1976. Desert varnish and marine ferromanganese oxide nodules; congeneric phenomena. *Nature* **259**, 387–8.

Beadnell, H. J. L. 1909a. *An Egyptian oasis: an account of the oasis of Kharga in the Libyan desert*. London: John Murray.

Beadnell, H. J. L. 1909b. Lake basins created by wind erosion. *Journal of Geology* **3**, 47–9.

Beadnell, H. J. L. 1910. The sand dunes of the Libyan Desert. *Geographical Journal* **35**, 379–95.

Beaty, C. B. 1963. Origin of alluvial fans, White Mountains, California and Nevada. *Annals of the AAG* **53**, 516–35.

Beaty, C. B. 1974a. Debris flows, alluvial fans and a revitalized catastrophism. *Zeitschrift für Geomorphologie* **21**, 39–51.

Beaty, C. B. 1974b. Needle ice and wind in the White Mountains of California. *Geology* **2**, 565–7.

Beaudet, G., R. Coque, P. Michel, P. Rognon 1977a. Alterations tropicales et accumulations ferrugineuses entre la vallée du Niger et les massifs centraux Sahariens (Aïr et Hoggar). *Zeitschrift für Geomorphologie* **21**, 297–322.

Beaudet, G., R. Coque, P. Michel, P. Rognon 1977b. Y-a-t'il eu capture du Niger? *Association Géographique Français, Bulletin* **445–6**, 215–22.

Beaudet, G. & P. Michel 1978. *Recherches géomorphologiques en Namibie Centrale*. Strasbourg: Association Géographique d'Alsace.

Beaumont, P. 1972. Alluvial fans along foothills of the Elburz Mountains, Iran. *Palaeogeography, Palaeoclimatology, Palaeoecology* **12**, 251–73.

Beeson, R. 1984. The use of fine fractions of stream sediments in geochemical exploration in arid and semi-arid terrains. *Journal of Geochemical Exploration* **22**, 119–32.

Begin, Z. B. & M. Inbar 1984. A method for the determination of the discharge–frequency relationship in ungauged arid gravel streams. *Journal of Hydrology* **69**, 163–72.

Behiery, S. A. 1967. Sand forms in the Coachella valley, southern California. *Annals of the AAG* **57**, 25–48.

Belknap, R. L. 1928. *Some Greenland sand dunes*. Papers, Michigan Academy of Science, Arts and Letters 10.

Bell, J. W. 1981. Subsidence in Las Vegas Valley. *Nevada Bureau of Mines and Geology, Bulletin* **95**.

Bellair, P. 1952. Sables désertiques et morphologie éolienne. *19th International Geological Congress, Proceedings*. Algiers, 1952, Section 8, Fasc. 7, 113–8.

Bellair, P. 1958. Le sédimentation et la sélection différentielle des évaporites. *Eclogae Geologicae Helveticae* **51**,

495–500.

Belly, P-Y. 1964. *Sand movement by wind*. US Army, Coastal Engineering Research Center, Technical Memorandum 1.

Ben-Zvi, A. & D. Cohen 1975. Frequency and magnitude of flows in the Negev. *Catena* **2** 193–200.

Benazzouz, M. T. 1987. A comparative study of the lunettes in the Tarf and Hodna basins (Algeria). See Gardiner (1987), 1217–29.

Benbow, M. C. 1990. Tertiary coastal dunes of the Eucla Basin, Australia. *Geomorphology* **3**, 9–29.

Bendali, F., C. Floret, E. Le Floc'h, R. Pontanier 1990. The dynamics of vegetation and sand mobility in arid regions of Tunisia. *Journal of Arid Environments* **18**, 21–32.

Berg, A. W. 1990. Formation of mima mounds: a seismic hypothesis. *Geology* **18**, 281–4.

Berkey, C. P. & F. K. Morris 1927. *Geology of Mongolia* [2 volumes]. New York: American Museum of Natural History.

Beruf, S., B. Biju-Duval, O. de Charpal, P. Rognon, P. Gariel, O. A. Bennacef 1971. *Les grès du Paléozoique inférieur au Sahara*. Paris: Editions Technip.

Besler, H. 1977. Fluviale und äolische Formen zwischen Schott und Erg. See *Geographische Untersuchungen am Nordrand der tunisischen Sahara*, W. Meckelein (ed.), Stuttgarter Geographische Studien 91, 19–81.

Besler, H. 1980. *Die Dünen – Namib: Entstehung und Dynamik eines Ergs*. Stuttgarter Geographische Studien 96.

Besler, H. 1981. Surface structures on Namib dunes caused by moisture. *Namib und Meer* **9**, 11–7.

Besler, H. 1982a. The north-eastern Rub' al Khali within the borders of the United Arab Emirates. *Zeitschrift für Geomorphologie* **26**, 495–505.

Belser, H. 1982b. A contribution to the aeolian history of the Tanezrouft. *l'Association Géographes Français, Bulletin* **484**, 55–9.

Besler, H. 1983. The response diagram: distinction between aeolian mobility and stability of sands and aeolian residuals by grain size parameters. *Zeitschrift für Geomorphologie*, Supplementband 45, 287–301.

Besler, H. 1984. The development of the Namib dune field according to sedimentological and geomorphological evidence. See Vogel (1984), 445–53.

Bettenay, E. 1962. The salt lake systems and their associated aeolian features in the semi-arid regions of western Australia. *Journal of Soil Science* **13**, 10–18.

Bettenay, E. & F. J. Hingston 1964. Development and distribution of soils in the Merredin area, Western Australia. *Australian Journal of Soil Research* **2**, 173–86.

Beydoun, Z. R. 1980. Some Holocene geomorphological and sedimentological observations from Oman and their palaeogeographical implications. *Journal of Petroleum Geology* **2**, 427–37.

Bigarella, J. J. 1975. Structures developed by dissipation of dune and beach ridge deposits. *Catena* **2**, 107–52.

Bigarella, J. J. 1979. The Lagoa dune field, Brazil. See

McKee (1979a).

Binda, P. L. 1983. On the skewness of some eolian sands from Saudi Arabia. See Brookfield & Ahlbrandt (1983), 27–39.

Bird, E. C. F. 1968. *Coasts*. Canberra: Australian National University Press.

Bird, E. C. F. 1974. Dune stability on Fraser Island. *Queensland Naturalist* 21, 15–21.

Bird, E. C. F. & D. J. B. Jones 1988. The origin of foredunes on the coast of Victoria, Australia. *Journal of Coastal Research* 4, 181–92.

Birkeland, P. W. 1974. *Pedology, weathering and geomorphological research*. New York: Oxford University Press.

Birot, P. 1962. *Contributions à l'étude de la désegrégation des roches cristallines*. Paris: Centre Documentation Universitaire.

Birot, P. 1968. *The cycle of erosion in different climates*. London: Batsford.

Birot, P. & J. Dresch 1966. Pédiments et glacis dans l'Ouest des États-Unis. *Annales de Géographie* 411, 513–52.

Bisal, F. & W. Ferguson 1970. Effect of non-erodible aggregates and wheat stubble on initiation of soil drifting. *Canadian Journal of Soil Science* 50, 31–4.

Bisal, F. & J. Hsieh 1966. Influence of moisture on erodibility of sand by wind. *Soil Science* 102, 143–6.

Bisal, F. & K. F. Nielsen 1962. Movement of soil particles in saltation. *Canadian Journal of Soil Science* 42, 81–6.

Biswas, S. K. 1971. The miliolite rocks of Kutch and Kathiawar (western India). *Sedimentary Geology* 5, 147–64.

Black, R. F. & W. L. Barksdale 1949. Oriented lakes of northern Alaska. *Journal of Geology* 57, 105–18.

Blackwelder, E. 1925. Exfoliation as a phase of rock weathering. *Journal of Geology* 33, 793–806.

Blackwelder, E. 1926. Fire as an agent in rock weathering. *Journal of Geology* 35, 134–40.

Blackwelder, E. 1928. The origin of desert basins of southwestern USA [abstract]. *Geological Society of America, Bulletin* 39, 262–3.

Blackwelder, E. 1929. Cavernous rock surfaces of the desert. *American Journal of Science* 17, 393–9

Blackwelder, E. 1931a. The lowering of playas by deflation. *American Journal of Science* 21, 140–4.

Blackwelder, E. 1931b. Desert plains. *Journal of Geology* 39, 133–40.

Blackwelder, E. 1933. The insolation hypothesis of rock weathering. *American Journal of Science* 26, 97–113.

Blackwelder, E. 1934. Yardangs. *Geological Society of America, Bulletin* 24, 159–66.

Blackwelder, E. 1948. Historical significance of desert lacquer [abstract]. *Geological Society of America, Bulletin* 59, 1367.

Blackwelder, E. 1954. Pleistocene lakes and drainage in the Mojave Region, southern California. *California Division of Mines Bulletin* 170V, 35–40.

Blair, R. W. 1987. Development of natural sandstone

arches in south-eastern Utah. See Gardiner (1987), 597–604.

Blake, W. P. 1858. *Report of a geological reconnaissance in California*. New York: Balière.

Blanford, W. T. 1877. Geological notes on the great desert between Sind and Rajputana. *Geological Survey of India, Records* 10, 10–21.

Blank, H. R. & E. W. Tynes 1965. Formation of caliche in situ. *Geological Society of America, Bulletin* 76, 1387–92.

Blockhuis, W. A., H. Pape, S. Slager 1968. Morphology and distribution of pedogenic carbonate in some vertisols of the Sudan. *Geoderma* 2, 173–200.

Bluck, B. J. 1964. Sedimentation on an alluvial fan in southern Nevada. *Journal of Sedimentary Petrology* 34, 395–400.

Blume, H-P. 1987. Bildung sandgefülter Spalten unter periglaziären und warmariden Bedingungen. *Zeitschrift für Geomorphologie* 31, 443–8.

Blume, H-P. & H. K. Barth 1979. Laterische Krustenstufen in Australien. *Zeitschrift für Geomorphologie*, Supplementband 33, 46–56.

Blume, H-P., W-G. Vahrson, H. Meshref 1985. Dynamics of water, temperature, and salts in typical aridic soils after artificial rainstorms. *Catena* 12, 343–62.

Blümel, W. D. 1982. Calcretes in Namibia and SE Spain – relations to substratum, soil formation and geomorphic actors. *Catena Supplement* 1, 67–95.

Bobek, H. 1969. Zür Kenntnis der Südlichen Lut. *Mitteilungen der Österreicher geographischen Gesellschaft, Wien* 3, 155–92.

Bocquier, G. 1968. Biogéocénoses et morphogenèse actuelle de certaines pédiments du Bassin Tchadien. *Transactions 19th International Congress of Soil Science* 1964, 630–5.

Bogoch, R. & P. Cook 1974. Calcite cementation of a Quaternary conglomerate in southern Sinai. *Journal of Sedimentary Petrology* 44, 917–20.

Bolton, H. C. 1889. A new Mountain of the Bell. *Nature* 39, 607–8.

Bonatti, E. & S. Gartner 1973. Caribbean climate during Pleistocene ice ages. *Nature* 244, 565.

Bond, R. D. 1968. Water-repellent sands. *19th International Congress of Soil Science, Transactions* 1, 339–47.

Bond, R. D. & J. R. Harris 1964. The influence of microflora on the physical properties of soils, 1: effects associated with filamentous algae and fungi. *Australian Journal of Soil Research* 2, 111–22.

Boocock, E. & O. J. Van Straten 1962. Notes on the geology and hydrogeology of the central Kalahari, Bechuanaland Protectorate. *Geological Society of South Africa, Transactions* 65, 125–71.

Boorman, L. A. 1977. Sand dunes. See *The coastline*, R. S. K. Barnes (ed.), 161–97. New York: John Wiley.

Boorman, L. A. 1989. The grazing of British sand dune vegetation. See Gimmingham et al. (1989), 75–88.

Boothroyd, J. C. & D. K. Hubbard 1975. Genesis of

bcdforms in mesotidal estuaries. See *Estuarine research*, L. E. Cronin (ed.), 217–34. London: Academic Press.

Booysen, J. J. 1974. Alluvial fans in the middle Bree River basin: a morphometric comparison. *South African Geographer* **4**, 388–404.

Borówka, R. K. 1979. Accumulation and deposition of eolian sands on the lee slope of dunes and their influence on the formation of sedimentary structures. *Quaestiones Geographicae* **5**, 5–22.

Borówka, R. K. 1980. Present-day transport and sedimentation processes of eolian sands – controlling the factors and resulting phenomena on a coastal dune area. *Poznanskie Towarzystwo Przyjaciol Nauk* **20**, 123–6.

Borsy, Z., E. Csongor, I. Szabol 1982. Mobile sand phases in the north-east part of the Great Hungarian Plain. See *Quaternary studies in Hungary*, M. Pesci (ed.), 193–208. Budapest: INQUA Hungarian National Committee.

Bosworth, T. 1922. *Geology of the Tertiary and Quaternary periods in the northwest parts of Peru*. London: Macmillan.

Bottema, S. & W. van Zeist 1981. Palynological evidence for the climatic history of the Near East, 50,000–6,000 BP. *Colloques Internationaux du CNRS* **598**, 111–32.

Boulaine, J. 1954. La sebkha Ben Ziane et sa "lunette" ou bourrelet; éxample de complexe morphologique formé par la dégradation éolienne des sols salés. *Revue de Géomorphologie Dynamique* **5**, 102–23.

Boulaine, J. 1958a. Sur la présence de takyr au Sahara français. *Institut Recherches Sahariennes, Traveaux* **17**, 193–4.

Boulaine, J. 1958b. Sur la formation des carapaces calcailes. *Bulletin de la Service Géologiques de l'Algerie, Traveaux Collaboratifs*, **20**, 7–19.

Bourcart, J. 1928. L'action du vent à la surface de la terre. *Revue de Géographie Physique et Géologie Dynamique* **1**, 26–54.

Bourman, R. P. 1986. Aeolian sand transport along beaches. *Australian Geographer* **17**, 30–5.

Bourman, R. P., A. R. Milnes, J. M. Oades 1987. Investigations of ferricretes and related surficial ferruginous materials in parts of southern and eastern Australia. *Zeitschrift für Geomorphologie*, Supplementband 64, 1–24.

Bourne, J. A., C. R. Twidale, D. M. Smith 1974. The Corrobinnie Depression, Eyre Peninsula, South Australia. *Royal Society of South Australia, Transactions* **98**, 139–52.

Boussingault, M. 1882. Sur l'apparition du manganese à la surface des roches. *Annales de Chemie et de Physique* **27**, Série 5, 289–311.

Bowden, A. R. 1983. Relict terrestrial dunes: legacies of a forer climate in coastal northeastern Tasmania. *Zeitschrift für Geomorphologie*, Supplementband 45, 153–74.

Bowden, L. W., J. R. Huning, C. F. Hutchinson, C. W. Johnson 1974. Satellite photograph presents first comprehensive view of local wind: the Santa Ana. *Science* **184**, 1077–8.

Bowen, A. S. & D. Lindley 1977. A wind-tunnel investigation of the wind speed and turbulence characteristics close to the ground over various escarpment shapes. *Boundary-layer Meteorology* **12**, 259–72.

Bowler, J. M. 1968. Australian landform example: lunette. *Australian Geographer* **10**, 402–4.

Bowler, J. M. 1973. Clay dunes: their occurrence, formation and environmental significance. *Earth Science Reviews* **9**, 315–38.

Bowler, J. M. 1975. Deglacial events in southern Australia: their age, nature and palaeoclimatic significance. See *Quaternary studies*, R. P. Suggate & M. M. Creswell (eds), 75–82. Wellington: Royal Society of New Zealand.

Bowler, J. M. 1980. Quaternary chronology and palaeohydrology in the evolution of the mallee landscapes. See Storrier & Stannard (1980), 17–36.

Bowler, J. M. 1983. Lunettes as indices of hydrologic change: a review of Australian evidence. *Royal Society of Victoria, Proceedings* **95**, 147–68.

Bowler, J. M. 1986. Spatial variability and hydrologic evolution of Australian lake basins: analogue for Pleistocene hydrologic change and evaporite formation. *Palaeogeography, Palaeoclimatology, Palaeoecology* **54**, 21–41.

Bowler, J. M. , G. S. Hope, J. N. Jennings, G. Singh, D. Walker 1976. Late Quaternary climates of Australia and New Guinea. *Quaternary Research* **6**, 359–94.

Bowler, J. M. & J. W. McGee 1978. Geomorphology of the Mallee region in semi-arid northern Victoria and western New South Wales. *Royal Society of Victoria, Proceedings* **90**, 5–20.

Bowler, J. M. & H. A. Polach 1971. Radiocarbon analyses of soil carbonates from palaeosols in southeastern Australia. See Yaalon (1971), 97–108.

Bowler, J. M. & R. J. Wasson 1984. Glacial age environments of inland Australia. See Vogel (1984), 183–208.

Bowman, D. 1978. Determination of intersection points within a telescopic alluvial fan complex. *Earth Surface Processes* **3**, 265–76.

Bowman, D. 1982. Iron coating in a recent terrace sequence under extremely arid conditions. *Catena* **9**, 353–60.

Bowman, D., A. Karnieli, A. Issar, H. J. Bruins 1986. Residual colluvio-aeolian aprons in the Negev Highlands (Israel) as a palaeoclimatic indicator. *Palaeogeography, Palaeoclimatology, Palaeoecology* **56**, 89–101.

Bradley, E. F. 1980. An experimental study of the profiles of wind speed, shearing stress and turbulence at the crest of a large hill. *Quarterly Journal of the Royal Meteorological Society* **106**, 101–23.

Bradley, W. C. 1963. Large-scale exfoliation in massive sandstones of the Colorado Plateau. *Geological Society of America, Bulletin* **74**, 519–27.

Bradley, W. C., J. T. Hutton, C. R. Twidale 1978. Rôle of Knihovna salts in development of granitic tafoni, South Australia. *Journal of Geology* **86**, 647–54.

Braithwaite, C. J. R. 1983. Calcrete and other soils in Quaternary limestones: structures, processes and applications. *Journal of the Geological Society of London* **140**, 351–63.

Brakenridge, G. R. 1978. Evidence for a cold, dry full-glacial climate in the American southwest. *Quaternary Research* **9**, 22–40.

Brakenridge, G. R. & J. Shuster 1986. Late Quaternary geology and geomorphology in relation to archaeological site locations, southern Arizona. *Journal of Arid Environments* **10**, 225–39.

Brazel, A. J. & W. G. Nickling 1986. The relationship of weather types to dust storm generation in Arizona (1955–1980). *Journal of Climatology* **6**, 255–75.

Breed, C. S. 1977. Terrestrial analogs of the Hellespontus dunes on Mars. *Icarus* **30**, 326–40.

Breed, C. S., N. S. Embabi, H. El-Etr, M. J. Grolier 1980. Wind deposits in the Western Desert. *Geographical Journal* **146**, 88–90.

Breed, C. S., M. J. Grolier, J. F. McCauley 1979a. Eolian features in the western desert of Egypt and some applications to Mars. *Journal of Geophysical Research* **84**, 8205–21.

Breed, C. S., S. C. Fryberger, S. Andrews, C. McCauley, F. Lennartz, D. Gebel, K. Horstman 1979b. Regional studies of sand seas using LANDSAT (ERTS) imagery. See McKee (1979a), 305–97.

Breed, C. S. & T. Grow 1979. Morphology and distribution of dunes in sand seas observed by remote sensing. See McKee (1979a), 253–303.

Breed, C. S., J. F. McCauley, P. A. Davis 1987. Sand sheets of the eastern Sahara and ripple blankets on Mars. See Frostick & Reid (1987), 337–60.

Breed, C. S., J. F. McCauley, M. J. Grolier 1982. Relic drainages, conical hills, and the eolian veneer in southwest Egypt – applications to Mars. *Journal of Geophysical Research* **87**, 9929–50.

Breed, C. S., J. F. McCauley, M. I. Whitney 1989. Wind erosion forms. See Thomas (1989c), 284–307.

Bresler, E., B. L. McNeal, D. L. Carter 1982. *Saline and sodic soils*. Berlin: Springer.

Bréssolier, C. & Y-F. Thomas 1977. Studies on wind and plant interactions on French Atlantic coastal dunes. *Journal of Sedimentary Petrology* **47**, 331–8.

Brewer, R. 1956. A petrographic study of two soils in relation to their origin and classification. *Journal of Soil Science* **7**, 268–79.

Brewer, R., E. Bettenay, H. M. Churchward 1972. *Some aspects of the origin and development of the red and brown hardpan soils of Bulloo Downs, Western Australia*. CSIRO Division of Soils, Technical Paper 13.

Brice, J. C. 1966. Erosion and deposition in the loess-mantled Great Plains, Medicine Creek drainage basin, Nebraska. USGS Professional Paper 352H, 255–339.

Bridge, B. J. & P. J. Ross 1983. Water erosion in vegetated sand dune at Cooloola, southeast Queensland. *Zeitschrift für Geomorphologie*, Supplementband 45, 227–44.

Bridges, E. M. & P. Bull 1981. The role of silica in the formation of compact indurated horizons in the soils of

South Wales. See Bullock & Murphy (1981), 605–13.

Briggs, D. J. & J. France 1982. Mapping soil erosion by wind for regional environmental planning. *Journal of Environmental Management* **15**, 158–68.

Briggs, L. I., D. S. McCulloch, F. Moser 1962. The hydraulic shape of sand particles. *Journal of Sedimentary Petrology* **32**, 645–56.

Broccoli, N. J. & S. Manabe 1987. The effects of the Laurentide ice-sheet on North American climate during the last glacial maximum. *Géographie Physique et Quaternaire* **41**, 291–9.

Brookfield, M. E. 1970. Dune trends and wind regime in central Australia. *Zeitschrift für Geomorphologie*, Supplementband 10, 121–53.

Brookfield, M. E. 1977. The origin of bounding surfaces in ancient eolian sandstones. *Sedimentology* **24**, 303–32.

Brookfield, M. E. 1979. Anatomy of a Lower Permian aeolian sandstone complex, southwestern Scotland. *Scottish Journal of Geology* **15**, 81–96.

Brookfield, M. E. & T. S. Ahlbrandt (eds) 1983. *Eolian sediments and processes***. Amsterdam: Elsevier.**

Brosset, D. 1939. Essai sur les ergs du Sahara occidental. *Institut Français de l'Afrique Noire, Traveaux* (Dakar) **1**, 657–90.

Brothers, R. N. 1954. A physiographic study of Recent sand dunes on the Auckland west coast. *New Zealand Geographer* **10**, 47–59.

Brown, A. J. 1983. Channel changes in arid badlands, Borrego Springs, California. *Physical Geography* **4**, 82–102.

Brown, C. N. 1956. The origin of caliche in the north-eastern Llano Estacado. *Journal of Geology* **64**, 1–15.

Brown, G. F. 1960. Geomorphology of western and central Saudi Arabia. *Report, 21st International Geological Congress* **21**, 150–9.

Brown, R. A. 1983. The flow in the planetary boundary layer. See Brookfield & Ahlbrandt (1983), 291–310.

Brüggen, J. 1950. *Fundamentos de la geologia de Chile*. Santiago: Instituto Géografico Militar.

Brugmans, F. 1983. Wind ripples in an active drift sand area in the Netherlands: a preliminary report. *Earth Surface Processes and Landforms* **8**, 527–34.

Bruins, H. J. 1976. *The origin, nature and stratigraphy of palaeosols in the loessal deposits of the north-west Negev (Netivot, Israel)*. MSc thesis, Hebrew University of Jerusalem.

Bryan, K. 1923a. Wind erosion near Lees Ferry, Arizona. *American Journal of Science* **6**, 291–307.

Bryan, K. 1923b. Pedestal rocks in the arid South-West. USGS Bulletin 760-A, 1–11.

Bryan, K. 1925. The Papago country. USGS Water-Supply Paper 499.

Bryan, K. 1927. Pedestal rocks formed by differential erosion. USGS Bulletin 790, 1–15.

Bryan, K. 1936. The formation of pediments, *International Geological Congress Report, 16th Session* **2**, 765–75.

Bryan, R. & A. Yair 1982. *Badlands geomorphology and piping*. Norwich: Geobooks.

Buckley, R. C. 1982. Soil and vegetation of the central Australian sand ridges. *Australian Journal of Ecology* **7**, 187–200.

Buckley, R. C. 1987. The effect of sparse vegetation on the transport of dune sand by wind. *Nature* **325**, 426–8.

Buckley, R. C., W. Chen, Y. Liu, Z. Zhu 1986. Characteristics of the Tengger dunefield, north-central China, and comparison with the central Australian dunefields. *Journal of Arid Environments* **10**, 97–101.

Büdel, J. 1957. Die "Doppelten Einebnungsflächen" in den feuchten Tropen. *Zeitschrift für Geomorphologie* **1**, 201–28.

Büdel, J. 1970. Pedimente, Rumpfflächen und Rückland-Steilhänge. *Zeitschrift für Geomorphologie* **14**, 1–57.

Budyko, M. I. 1974. *Climate and life*. New York: Academic Press.

Bui, E. N., J. B. Dixon, H. Shadfan, L. P. Wilding 1990. Geomorphic features and associated iron oxides of the Dallol Bosso of Niger (West Africa). *Catena* **17**, 41–54.

Bui, E. N., J. M. Mazullo, L. P. Wilding 1989. Using quartz grain size and shape analysis to distinguish between aeolian and fluvial deposits in Dallol Bosso of Niger (West Africa). *Earth Surface Processes and Landforms* **14**, 157–66.

Bui, E. N. & L. P. Wilding 1988. Pedogenesis and mineralogy of a Halaquept soil of Niger (West Africa). *Geoderma* **43**, 49–64.

Bull, W. B. 1962. Relation of textural (CM) patterns to depositional environment of alluvial-fan deposits. *Journal of Sedimentary Petrology* **32**, 211–16.

Bull, W. B. 1963. Alluvial fan deposits in western Fresno County, California. *Journal of Geology* **71**, 243–51.

Bull, W. B. 1964a. *Alluvial fans and near-surface subsidence in western Fresno County, California*. USGS Professional Paper 437A.

Bull, W. B. 1964b. Geomorphology of segmented alluvial fans in western Fresno County, California. USGS Professional Paper 352E, 85–129.

Bull, W. B. 1968. Alluvial fans. *Journal of Geological Education* **16**, 101–6.

Bull, W. B. 1972. Recognition of alluvial fan deposits in the stratigraphic record. See *Recognition of ancient sedimentary environments*, W. K. Hamblin & J. K. Rigby (eds), 63–83. SEPM Special Publication 16.

Bull, W. B. 1977. The alluvial-fan environment. *Progress in Physical Geography* **1**, 222–70.

Bull, W. B. 1979. Threshold of critical power in streams. *Geological Society of America, Bulletin* **90**, 453–64.

Bull, W. B. 1984. Tectonic geomorphology. *Journal of Geological Education* **32**, 310–24.

Bullock, P. & C. P. Murphy (eds) 1983. *Soil micromorphology***. Berkhamsted, England: A. B. Academic Press.**

Burkham, D. E. 1970a. *Precipitation, streamflow and major floods at selected sites in the Gila River drainage basin above Coolidge Dam, Arizona*. USGS Professional Paper 655B.

Burkham, D. E. 1970b. *A method for relating infiltration rates to streamflow rates in perched streams*. USGS Professional Paper 700D.

Burrell, G. J. 1971. The locational and spatial analysis of a dune system in the Ziz valley, southern Morocco. *Reading Geographer* **2**, 24–41.

Busche, D. 1972. Alluvial fan studies on the northern slope of the Tibesti Massif, Chad. See *Arbeitsberichte aus der Forschungsstation Bardai/Tibesti*; III *Feldarbeiten, 1966/ 67*, 95–104. Berlin: Berliner Geographische Abhandlungen.

Busche, D. 1979. Planation surface and the problem of scarp retreat on the western margin of the Murzut Basin, central Sahara. See van Zinderen Bakker & Coetzee (1979), 50–5.

Busche, D. 1980. On the origin of the Msak Mallat and the Hamadat Manghini escarpment. See Salem & Busrewil (1980), 837–48.

Busche, D. 1983. Silcrete in der zentralen Sahara (Murzuk–Bechen, Djado-Plateau und Kaouar; Süd-Libyen und Nord-Niger). *Zeitschrift für Geomorphologie*, Supplementband 48, 35–49.

Busche, D., M. Draga, H. Hagedorn 1984. *Les sables éoliens – modelés et dynamique – la menace éolienne et son contrôle, bibliographie annotée*. Rossdorf, Germany: Deutches Gesellschaft für technische Zusammenarbeit.

Busche, D. & W. Erbe 1987. Silicate karst landforms of the southern Sahara (northeastern Niger and southern Libya). *Zeitschrift für Geomorphologie*, Supplementband 64, 57–72.

Busche, D. & H. Hagedorn 1980. Landform development in warm deserts – the central Saharan example. *Zeitschrift für Geomorphologie*, Supplementband 36, 123–39.

Butler, B. E. 1956. Parna – an aeolian clay. *Australian Journal of Science* **18**, 145–51.

Butler, B. E. 1960. Riverine deposition during arid phases. *Australian Journal of Science* **22**, 451–2.

Butler, B. E. 1974. A contribution towards the better specification of parna and some other aeolian clays in Australia. *Zeitschrift für Geomorphologie*, Supplementband 20, 106–16.

Butler, B. E. & J. T. Hutton 1956. Parna in the Riverine Plain of southeastern Australia and the soils thereon. *Australian Journal of Agricultural Research* **7**, 536–53.

Butler, P. R. & J. F. Mount 1986. Corroded cobbles in southern Death Valley: their relationship to honeycomb weathering and lake shorelines. *Earth Surface Processes and Landforms* **11**, 377–87.

Butt, C. R. M. 1985. Granite weathering and silcrete formation on the Yilgarn block, Western Australia. *Australian Journal of Earth Sciences* **32**, 415–32.

Butterfield, G. R. 1991. Grain transport in steady and unsteady turbulent airflows. *Acta Mechanica* **1**, 97–122.

Butzer, K. W. 1965. Desert landforms at the Kurkur oasis, Egypt. *Annals of the AAG* **55**, 578–91.

Butzer, K. W. 1984. Archaeology and Quaternary environments in the interior of southern Africa. See *Southern African prehistory and paleoenvironments*, R. G. Klein (ed.), 1–64. Rotterdam: Balkema.

Butzer, K. W. & C. L. Hansen 1968. *Desert and river in Nubia*. Madison: University of Wisconsin Press.

Butzer, K. W., R. Stuckenrath, A. J. Bruzewick, D. M. Helgren 1978. Late Cenozoic palaeoclimates of the Gaop Escarpment, Kalahari margin, South Africa. *Quaternary Research* 10, 310–39.

Cailleux, A. 1960. Sur les marnes et lacs ronds des plaines aujourd'hui temperées. *Revue de Géomorphologie Dynamique* 11, 28–9.

Cailleux, A. & K. Wuttke 1964. Morphoscopie des sables quartzeux dans l'ouest des États Unis d'Amerique du Nord. *Boletim Paranaense de Geografia* 10–15, 79–87.

Calkin, P. E. & A. Cailleux 1962. A quantitative study of cavernous weathering (taffonis) and its implications to glacial chronology in the Victoria Valley, Antarctica. *Zeitschrift für Geomorphologie* 6, 317–24.

Calkin, P. E. & R. H. Rutford 1974. The sand dunes of Victoria Valley, Antarctica. *Geographical Review* 64, 189–216.

Callen, R. A. 1983. Late Tertiary "grey billy" and the age and origin of surficial silicifications (silcrete) in South Australia. *Journal of the Geological Society of Australia* 30, 393–410.

Cameron, R. E. 1969. Cold desert characteristics and problems relevant to other arid lands. See McGinnies & Goldman (1969), 169–205.

Campbell, E. M. 1968. Lunettes in southern South Australia. *Royal Society of South Australia, Transactions* 92, 85–109.

Campbell, S. E., J–S. Seeler, S. Golubic 1989. Desert crust formation and soil stabilization. *Arid Soil Research and Rehabilitation* 3, 217–28.

Camuti, P., G. Rodolfi, M. Schieppati 1986. Swelling soils in the Addis Ababa (Ethiopia) metropolitan area and their pedological and engineering properties. *5th International Association of Engineering Geology Congress* 3, 667–71.

Cantwell, B. J. 1981. Organized motion in turbulent flow. *Annual Review of Fluid Mechanics* 13, 457–515.

Capot-Rey, R. 1943. La morphologie de l'Erg occidental. *Institut de Recherches Sahariennes, Traveaux* 2, 69–103.

Capot-Rey, R. 1945. The dry and humid morphology of the western Erg. *Geographical Review* 35, 391–407.

Capot-Rey, R. 1947. L'Edeyen de Mourzouk. *Institut de Recherches Sahariennes, Traveaux* 4, 67–109.

Capot-Rey, R. 1948. Le déplacement des sables éoliens et la formation des dunes désertiques, d'après Bagnold. *Institut Recherches Sahariennes, Traveaux* 5, 47–80.

Capot-Rey, R. 1953. *Le Sahara français*. Paris: Presses Universitaires de France.

Capot-Rey, R. 1957a. Le vent et le modèle éolien au Borku. *Institut Recherches Sahariennes, Traveaux* 15, 155–7.

Capot-Rey, R. 1957b. Sur une forme d'érosion éolienne dans le Sahara français. *Tidjschrift Konligke Nederland Aardnjsk Genootschap* 74, 242–7.

Capot-Rey, R. 1963. Contribution à l'étude et la représentation des barkanes. *Institut Recherches Saharienne, Traveaux* 22, 37–60.

Capot-Rey, R. 1970. Remarques sur les ergs du Sahara. *Annales de Géographie* 79, 2–19.

Capot-Rey, R. & F. Capot-Rey 1948. Le déplacement des sables éoliens et la formation des dunes désertiques d'après R. A. Bagnold. *Institut Recherches Sahariennes, Traveaux* 5, 47–80.

Carlisle, W. J. & R. W. Mars 1982. Eolian features of the southern High Plains and their relationship to windflow patterns. Geological Society of America, Special Paper 192, 89–105.

Carlston, T. N. & J. M. Prospero 1972. Large-scale movement of Saharan air outbreaks over the northern equatorial Atlantic. *Journal of Applied Meteorology* 11, 283–97.

Carrigy, M. A. 1970. Experiments on the angles of repose of granular materials. *Sedimentology* 14, 147–58.

Carrol, D. 1944. The Simpson Desert Expedition, 1939; scientific reports 2: geology – desert sands. *Royal Society of South Australia, Transactions* 68, 49–59.

Carroll, J. J. & J. A. Ryan 1970. Atmospheric vorticity and dust devil rotation. *Journal of Geophysical Research* 75, 5179–84.

Carruthers, D. J. & T. W. Choularton 1982. Air flow over hills of moderate slope. *Quarterly Journal of the Royal Meteorological Society* 108, 603–24.

Carruthers, D. J. & J. C. R. Hunt 1989. Fluid mechanics of airflow over hills: turbulence, fluxes and waves in the boundary layer. See *Workshop*, American Meteorological Society, Boston.

Carson, C. E. & K. M. Hussey 1960. Hydrodynamics of three Arctic lakes. *Journal of Geology* 68, 585–600.

Carson, C. E. & K. M. Hussey 1962. The oriented lakes of Alaska. *Journal of Geology* 70, 417–39.

Carson, M. A. 1971. An application of the concept of threshold slopes to the Laramie Mountains, Wyoming. *Institute of British Geographers, Transactions* 3, 31–48.

Carson, M. A. & M. J. Kirkby 1972. *Hillslope form and process*. Cambridge: Cambridge University Press.

Carson, M. A. & P. A. McLean 1986. Development of hybrid aeolian dunes: the William River dune field, northwest Saskatchewan, Canada. *Canadian Journal of Earth Sciences* 23, 1974–90.

Carter, R. W. G. 1976. Formation, maintenance and geomorphological significance of an aeolian shell pavement. *Journal of Sedimentary Petrology* 46, 418–29.

Carter, R. W. G. 1977. The rate and pattern of sediment interchange between beach and dune. *Symposium on coastal sedimentology, proceedings*, 3–34. Talahassee: Florida State University.

Carter, R. W. G. & G. W. Stone 1989. Mechanisms associated with the erosion of sand dune cliffs, Magilligan, Northern Ireland. *Earth Surface Processes and Landforms* 14, 1–10.

Cas, R. A. F. & J. V. Wright 1987. *Volcanic successions: modern and ancient*. London: Unwin Hyman.

Castel, I. I. Y., E. Koster, R. Slotboom 1989. Morphogenetic aspects and age of Late Holocene eolian drift sands in Northwest Europe. *Zeitschrift für Geomorphologie* **33**, 1–26.

Castellani, V. & W. Dragoni 1987. Some considerations regarding karstic evolution of desert limestone plateaus. See Gardiner (1987), 1199–206.

Caviedies, C. N. & R. Paskoff 1975. Quaternary glaciations in the Andes of north-central Chile. *Journal of Glaciology* **14**, 155–70

Cayeux, L. 1928. Origines de sables des dunes sahariennes. *14th International Geological Congress, Comptes Rendus*, 783–8.

Chadwick, O. A. & J. O. Davis 1990. Soil-forming intervals caused by eolian sediment pulses in the Lohontan Basin, northwestern Nevada. *Geology* **18**, 243–6.

Chadwick, O. A., D. M. Hendricks, W. D. Nettleton 1989. Silicification of the Holocene soils in the Moniter Valley, Nevada, *Soil Science Society of America, Journal* **53**, 158–64.

Chakrabarti, C. & D. R. Lowe 1981. Diffusion of sediment on the lee of dune-like bedforms: theoretical and numerical analysis. *Sedimentology* **28**, 531–45.

Chamberlain, A. C. 1983. Roughness length of sand sea and snow. *Boundary-layer Meteorology* **25**, 405–9.

Chamley, H., G. Coudé-Gaussen, P. Debrabant, P. Rognon 1987. Contribution autochtone et allochtone à la sédimentation Quaternaire de l'île de Fuerteventura (Canaries): altération ou apport éoliens. *Société Géologique de France, Bulletin* **8**, III, 939–52.

Chao Sung-Chiao 1984. The sandy deserts and the Gobi of China. See El-Baz (1984b), 95–113.

Chao Sung-Chiao & Xing Jiaming 1982. Origin and development of the Shamo (sandy deserts) and the Gobi (stony deserts) of China. *Striae* **17**, 79–91.

Chapman, R. W. 1974. Calcareous duricrusts in Al-Hasa, Saudi Arabia. *Geological Society of America, Bulletin* **85**, 119–30

Chapman, R. W. 1980. Salt weathering by sodium chloride in the Saudi Arabian desert. *American Journal of Science* **280**, 116–29

Charley, J. L. & S. L. Cowling 1968. Changes in soil nutrient status resulting from overgrazing and their consequences in plant communities of semi-arid areas. *Ecological Society of Australia, Proceedings* **3**, 28–38.

Charney, J. G. 1975. Dynamics of deserts and drought in the Sahel. *Quarterly Journal of the Royal Meteorological Society* **101**, 193–202.

Chartres, C. J. 1982. The pedogenesis of desert loam soils in the Barrier Range, western New South Wales, I: soil parent materials. *Australian Journal of Soil Research* **20**, 269–81.

Chartres, C. J. 1983a. The pedogenesis of desert loam soils in the Barrier Range, western New South Wales, II: weathering and soil formation. *Australian Journal of Soil Research* **21**, 1–13.

Chartres, C. J. 1983b. The micromorphology of desert loam soils and implications for Quaternary studies in western New South Wales. See Bullock & Murphy (1983), 273–9.

Chatterji, S. & J. W. Jeffery 1963. Crystal growth during the hydration of $CaSO_4.H_2O$. *Nature* **200**, 463–4.

Chaudhri, R. S. & H. M. M. Khan 1981. Textural parameters of desert sediments – Thar Desert (India). *Sedimentary Geology* **28**, 43–62.

Chepalyga, A. L. 1984. Inland sea basins. See Velichko (1984), 229–47.

Chepil, W. S. 1945a. Dynamics of wind erosion, I: nature of movement of soil by wind. *Soil Science* **60**, 305–20.

Chepil, W. S. 1945b. Dynamics of wind erosion, III: the transport capacity of the wind. *Soil Science* **60**, 475–80.

Chepil, W. S. 1950. Properties of soil which influence wind erosion, I: the governing principle of surface roughness. *Soil Science* **69**, 149–62.

Chepil, W. S. 1956. Influence of moisture on erodibility of soil by wind. *Soil Science Society of America, Proceedings* **20**, 288–92.

Chepil, W. S. 1957a. Sedimentary characteristics of dust storms, I: sorting of wind-eroded soil material. *American Journal of Science* **255**, 12–22.

Chepil, W. S. 1957b. *Width of field strips to control wind erosion*. Technical Bulletin 92, Kansas State College of Agriculture and Applied Science.

Chepil, W. S. 1959. Equilibrium of soil grains at the threshold of movement by wind. *Soil Science Society of America, Proceedings* **23**, 422–8.

Chepil, W. S. & R. A. Milne 1939. Comparative study of soil drifting in the field and in a wind tunnel. *Scientific Agriculture* **19**, 249–57.

Chepil, W. S. & F. H. Siddoway 1959. Strain-gauge anemometer for analyzing various characteristics of wind turbulence. *Journal of Meteorology* **16**, 411–8.

Chepil, W. S., F. H. Siddoway, D. V. Armbrust 1962. Climatic factor for estimating wind erodibility of farm fields. *Journal of Soil and Water Conservation* **17**, 162–5.

Chepil, W. S. & N. P. Woodruff 1954. Estimations of wind erodibility of field surfaces. *Journal of Soil and Water Conservation* **9**, 257–65.

Chepil, W. S. & N. P. Woodruff 1963. The physics of wind erosion and its control. *Advances in Agronomy* **15**, 211–302.

Chhotani, O. B. 1988. The termites of Oman. *Journal of Oman Studies*, Special Report 3, 363–71.

Chico, R. J. 1963. *Playa mud cracks: regular and king size*. Geological Society of America, Special Paper 76.

Chivas, A. R., T. Torgersen, J. M. Bowler (eds) 1986. Palaeoenvironments of salt lakes. *Palaeogeography, Palaeoclimatology, Palaeoecology* **54**.

Cholnoky, E. von 1902. Die Bewegungsgesetze des Flugsandes. *Foldtani Kozlony* **32**, 106–43.

Chong, G. 1984. Die Salare in Nordchile – Geologie, Struktur und Geochemie. *Geotektonische Forschungen* **67**, 1–146.

Chopra, K. & L. F. Hubert 1964. Karman vortices in the Earth's atmosphere. *Nature* **203**, 1341–3.

Chorley, R. J., S. A. Schumm, D. E. Sugden 1984. *Geomorphology*. London: Methuen.

Chow, V. T. (ed.) 1964. *Handbook of applied hydrology.*

New York: McGraw-Hill.

Christenson, G. E. & C. Purcell 1985. Correlation and age of Quaternary alluvial-fan sequences, Basin and Range Province, southwestern United States. Geological Society of America Special Paper 203, 115–22.

Christian, L. B. 1970. Ancient windblown terrain of central California. *Mineral Information Bulletin* (California Division of Mines and Geology) 23, 175–9.

Christiansen, C., P. Blaesild, K. Dalsgaard 1984. Reinterpreting "segmented" grain-size curves. *Geological Magazine* 121, 47–51.

Christiansen, C. & D. Hartmann 1988. Sahara: a package of PC computer programs for estimating both log-hyperbolic grain-size parameters and standard moments. *Computers and Geosciences* 14, 557–625.

Chu Chen-Ta (Zhu Zhenda) 1963. A preliminary study of the problems of the dynamic changing process of sand dunes under the pressure of the wind. K'au Geographical Monographs 58–78. Beijing: Scientific Press, Beijing. (Translated by Department of Commerce, Clearing House for Federal Scientific and Technical Information, US Joint Publications Research Service, National Technical Information Service, AD 299–315).

Chudeau, R. 1911. Remarques sur les dunes, à propos d'une étude de M. H. J. Llewellyn-Beadnell. *La Géographie* 24, 153–60.

Church, M. & D. M. Mark 1980. On size and scale in geomorphology. *Progress in Physical Geography* 4, 342–90.

Churchward, H. M. 1961. Soil studies at Swan Hill, Victoria, II: dune moulding and parna formation. *Australian Journal of Soil Research* 1, 103–16.

Clapperton, C. M. 1983. The glaciation of the Andes. *Quaternary Science Reviews* 2, 83–155.

Clark, A. H., R. U. Cooke, C. Mortimer, R. H. Sillitoe 1967. Relationships between supergene mineral alteration and geomorphology, southern Atacama Desert, Chile – an interim report. *Institute of Mining and Metallurgy, Transactions* B76, 89–96.

Clark, A. H., A. E. S. Maya, C. Mortimer, R. H. Sillitoe, R. U. Cooke, N. J. Snelling 1967. Implications of the isotopic ages of ignimbrite flows, southern Atacama Desert, Chile. *Nature* 215, 723–4.

Clayton, P. A. & L. J. Spencer 1933. Silica glass in the Libyan Desert. *Mineralogical Magazine* 23, 503–8.

Clements, T., R. H. Merriam, R. O. Stone, J. F. Mann Jr, J. L. Eyman 1957. *A study of desert surface conditions*. Headquarters Quartermaster Research and Development Command, Environmental Protection Research Division Technical Report EP53.

Clements, T., J. F. Mann, Jr, R. O. Stone, J. L. Eymann 1963. *A study of windborne sand and dust in desert areas*, United States Army Natick Laboratories, Technical Report ES-8.

Clemmensen, L. B. 1986. Storm-generated eolian sand shadows and their sedimentary structures, Vejers Strand, Denmark. *Journal of Sedimentary Petrology* 56, 520–7.

Clemmensen, L. B. 1987. Complex star dunes and associated aeolian bedforms, Hopeman Sandstone (Permo-Triassic), Moray Firth Basin, Scotland. See Frostick & Reid (1987), 213–31.

Clemmensen, L. B. & R. C. Blakey 1989. Erg deposits in the lower Jurassic Wingate Sandstone, northwestern Arizona: oblique dune sedimentation. *Sedimentology* 36, 449–70.

Clifton, H. E. 1977. Rain impact ripples. *Journal of Sedimentary Petrology* 47, 678–9.

Clos-Arceduc, A. 1966a. L'application des methodes d'interprétation des images à des problèmes géographiques: examples et résultats; méthodologie. *Institut Français de Pétrole, Revue* 21, 1783–800.

Clos-Arceduc, A. 1966b. Le rôle déterminant des ondes aériennes stationaire dans la structure des ergs sahariens et les formes des érosion avoisinantes. *Academie des Sciences à Paris, Comptes Rendus* D262, 2673–6.

Clos-Arceduc, A. 1967. La direction des dunes et ses rapports avec celle du vent. *Academie des Sciences à Paris, Comptes Rendus* D264, 1393–6.

Clos-Arceduc, A. 1969. *Essai d'éxplication des formes dunaires sahariennes*. Études de Photo-Interprétation, Institut Géographique National.

Clos-Arceduc, A. 1971. Disposition des structures éoliennes au voisinage d'une groupe de barkhanes à parcours limité. Étude d'une groupe isolé de barkhanes au sud du Tibesti. Evolution des barkhanes sur un parcours limité par deux aires où la déflation éolienne interdit la presence des dunes. In *Photo-interprétation*, 71–2. Paris: Editions Technip.

Cloud Jr, P. E. 1966. Beach cusps: response to Plateau's rule? *Science* 154, 890–1.

Coaldrake, J. E. 1954. The sand dunes of the Ninety-Mile Plain, southeastern Australia. *Geographical Review* 44, 394–407.

Coffey, G. N. 1909. Clay dunes. *Journal of Geology* 17, 754–5.

Colclough, J. D. 1973. Soil erosion and soil erosion control in Tasmania: vegetation to control tunnel erosion. *Tasmanian Agricultural Journal* 44, 65–70.

Conca, J. L. & G. R. Rossman 1982. Case hardening of sandstone. *Geology* 10, 520–3.

Connacher, A. J. 1971. The significance of vegetation, fire and man in the stabilization of sand dunes near the Warburton Ranges, central Australia. *Earth Science Journal* 5, 92–4.

Connacher, A. J. 1975. Throughflow as a mechanism responsible for excessive soil salinisation in non-irrigated, previously arable lands in the Western Australian wheatbelt: a field study. *Catena* 2, 31–67.

Connally, G. G., D. H. Krinsley, L. A. Sirkin 1972. Late Pleistocene erg in the upper Hudson Valley, New York. *Geological Society of America, Bulletin* 83, 1537–42.

Cooke, C. W. 1943. Elliptical bays. *Journal of Geology* 51, 419–27.

Cooke, H. J. 1975. The palaeoclimatic significance of caves and adjacent landforms in Western Ngamiland, Botswana. *Geographical Journal* 141, 430–44.

Cooke, H. J. 1980. Landform evolution in the context of climatic change and neo-tectonism in the middle Kalahari of northern central Botswana. *Institute of British Geographers, Transactions* **5**, 80–99.

Cooke, R. U. 1970a. Morphometric analysis of pediments and associated landforms in the western Mojave Desert, California. *American Journal of Science* **269**, 26–38.

Cooke, R. U. 1970b, Stone pavements in deserts. *Annals of the AAG* **60**, 560–77.

Cooke, R. U. 1979. Laboratory simulation of salt weathering processes in arid environments. *Earth Surface Processes* **4**, 347–59.

Cooke, R. U. 1981. Salt weathering in deserts. *Geologists' Association, Proceedings* **92**, 1–16.

Cooke, R. U. 1984. *Geomorphological hazards in Los Angeles*. London: Allen & Unwin.

Cooke, R. U. 1986. Surface forms of north Panamint Valley. See *Environmental and paleoenvironmental studies in Panamint Valley*, E. L. Davis & C. Raven (eds), 8–27. San Diego: Great Basin Foundation.

Cooke, R. U., D. Brunsden, J. C. Doornkamp, D. K. C. Jones 1982. *Urban geomorphology in drylands*. Oxford: Oxford University Press.

Cooke, R. U. & J. C. Doornkamp 1990. *Geomorphology in environmental management*, 2nd edn. Oxford: Oxford University Press.

Cooke, R. U., A. S. Goudie, J. C. Doornkamp 1978. Middle East – review and bibliography of geomorphological contributions. *Quarterly Journal of Engineering Geology* **11**, 9–18.

Cooke, R. U. & P. F. Mason 1973. Desert knolls, pediment and associated landforms in the Mojave Desert, California. *Revue de Géomorphologie Dynamique* **22**, 49–60.

Cooke, R. U. & C. Mortimer 1971. Geomorphological evidence of faulting in the southern Atacama Desert. *Revue de Géomorphologie Dynamique* **20**, 71–8.

Cooke, R. U. & R. W. Reeves 1972. Relations between debris size and the slope of mountain fronts and pediments in the Mojave Desert, California. *Zeitschrift für Geomorphologie* **16**, 76–82.

Cooke, R. U. & R. W. Reeves 1976. *Arroyos and environmental change in the American Southwest*. Oxford: Oxford University Press.

Cooke, R. U. & I. J. Smalley 1968. Salt weathering in deserts. *Nature* **220**, 1226–7.

Cooke, R. U. & A. Warren 1973. *Geomorphology in deserts*. London: Batsford.

Cooper, W. S. 1958. *Coastal sand dunes of Oregon and Washington*. Geological Society of America, Memoir 72.

Cooper, W. S. 1967. *Coastal dunes of California*. Geological Society of America, Memoir 104.

Coque, R. 1960. L'évolution des versants en Tunisie présaharienne. *Zeitschrift für Geomorphologie*, Supplementband 1, 172–7.

Coque, R. 1962. *La Tunisie présaharienne: étude géomorphologique*. Paris: Colin.

Coque, R. 1969. Recherches sur la géomorphologie et la géologie du Quaternaire de l'Afrique aride. See van Zinderen Bakker (1969), 32–6.

Coque, R. 1972a. Premiers résultats de recherches géomorphologiques dans le piédmont de Thèbes (Haute-Egypte). *Association de Géographes Français, Bulletin* **399**, 227–34.

Coque, R. 1972b. Cartes géomorphologiques à petite échelle de la Tunisie présaharienne. *Mémoires et Documents de Service du Documents et de Cartographie Géographiques* **12**, 155–7.

Coque, R. 1976. Observations sur la limite séptentrionale des accumulations ferrugineuses de l'Afrique de l'Ouest. See *Géomorphologie de relief dans les pays tropicaux chauds et humides*, CEGET, Bordeaux, 69–80.

Coque, R. 1979a. Problèmes de corrélations des niveaux Quaternaires dans les piédmonts Nord-Sahariens. *Association de Géographes Français, Bulletin* **462**, 215–22.

Coque, R. 1979b. Sur la place du vent dans l'érosion en milieu aride: l'éxample des lunettes (bourrelets éoliens) de la Tunisie. *Méditerranée* **35**, 15-21.

Coque, R. & A. Jauzein 1967. Geomorphology and Quaternary Geology of Tunisia. *9th Annual Field Conference of the Petroleum Exploration Society of Libya*, 227–57.

Coque, R., M. Mainguet, P. Rognon 1980. Les grandes orientations de la recherche sur les régions arides. *Recherches Géographiques en France* **15**, 99–107.

Corbel, J. 1963. Pédiments d'Arizona. *Centre de Documentation Cartographique et Géographique, Mémoires et Documents* **9**, 33–95.

Corbel, J. 1964. L'érosion terrestre, étude quantitative (méthodes—techniques—résultats). *Annales Géographie* **73**, 385–412.

Cornish, V. 1897. On the formation of sand dunes, *Geographical Journal* **9**, 278–309.

Cornish, V. 1908. On the observation of desert sand dunes, *Geographical Journal* **31**, 400–2.

Cornish, V. 1914. *Waves of sand and snow*. London: T. Fisher-Unwin.

Cornish, V. 1927. Waves in granular material formed and propelled by winds and currents. *Royal Astronomical Society, Geophysics Supplement*, Monthly Notes 1, 447–67.

Correns, C. W. 1949. Growth and dissolution of crystals under linear pressure. *Discussions of the Faraday Society* **5**, 267–71.

Corte, A. E. 1963. Particle sorting by repeated freezing and thawing. *Science* **142**, 499–501.

Corte, A. E. & A. Higashi 1964. *Experimental research on desiccation cracks in soil*. US Army Material Command, Cold Regions Research and Engineering Laboratory Research Report 66.

Cote, H. 1957. Quelques aspects de la morphologie de l'Ahaggar., *Revue Géographie Lyon* **23**, 321–32.

Cotton, C. A. 1947. *Climatic accidents in landscape-making*. Christchurch, New Zealand: Whitcombe & Tombs.

Coudé-Gaussen, G. 1984a. Le cycle des poussières éoliennes désertiques actuelles et la sédimentation des loess péridésertiques Quaternaires. *ELF Aquitaine, Bulletin* **8**, 167–82.

Coudé-Gaussen, G. 1984b. Mise en place des basses terrasses Holocènes dans les Matmata et leurs bordures (sud-Tunisien). *Association Français pour l'Étude du Quaternaire*, Paris 1-2-3, 173–80.

Coudé-Gaussen, G. 1987. The pre-Saharan loess: sedimentological characterisation and palaeoclimatological significance. *Geojournal* **15**, 177–83.

Coudé-Gaussen, G. & S. Balescu 1987. Étude comparé de loess périglaciare et prédésertiques: premiers résultats d'un examen des grains de quartz au microscope électrique à balayage. *Catena*, Supplement 9, 129–44.

Coudé-Gaussen, G. & P. Blanc 1985. Présence des grains éolisés de palygorskite dans les poussières actuelles et des sédiments récents d'origine désertique. *Société Géologique de France, Bulletin* **1**, 571–9.

Coudé-Gaussen, G., M-N. Le Coustumer, P. Rognon 1984a. Paléosols d'âge Pléistocène superieur dans les loess des Matmata (sud-Tunisien). *Sciences Géologiques, Bulletin* **37**, 359–86.

Coudé-Gaussen, G. & P. Rognon 1983. Les poussières sahariennes. *La Recherche* **147**, 1050–61.

Coudé-Gaussen, G., P. Rognon, N. Federoff 1984b. Piegeage de poussières éoliennes dans des fissures de granitoides du Sinai oriental. *Compte Rendus de l'Academie des Sciences de Paris* II, **298**, 369–74.

Coudé-Gaussen, G. & P. Rognon 1988a. Charactérisation sédimentologique et conditions paléoclimatiques de la mis en place de loess au Nord du Sahara à partir de l'example du sud-Tunisien. *Société Géologique de France* 4, 1081–90.

Coudé-Gaussen, G. & P. Rognon 1988b. Origine éolienne de certains encroûtements calcaires sur l'île de Fuerteventura (Canaries orientales). *Geoderma* **42**, 271–93.

Coudé-Gaussen, G., P. Rognon, A. Weisrock 1982. Évolution du matériel sableux au cours de son déplacement dans un système dunaire; les barkhanes du Cap Sim au sud d'Essaouira, Maroc. *Academie des Sciences à Paris, Comptes Rendus* D295, 621–4.

Coulson, A. 1940. The sand dunes of the Portland District and their relation to post-Pliocene uplift. *Royal Society of Victoria, Proceedings*, **52**, 314–32.

Coursin, A. 1964. Observation et expériences faites en Avril et Mai 1956 sur les barkhanes du Souhel et Abiodh (région est de Port Étienne). *Institut Française de l'Afrique Noire, Bulletin*, Série A **26**, 989–1022.

Cox, G. W. 1984. The distribution and origin of mima mound grasslands in San Diego County, California. *Ecology* **65**, 1397–1405.

Cox, G. W. & C. G. Gakahu 1986. A latitudinal test of the fossorial rodent hypothesis of mima mound origin. *Zeitschrift für Geomorphologie* **30**, 485–501.

Cox, G. W., B. G. Lovegrove, W. R. Siegfried 1987. The small stone content of mima-like mounds in the South African Cape region: implications for mound origin. *Catena* **14**, 165–76.

Crandell, D. R. 1958. *Geology of the Pierre area, South Dakota*. USGS Professional Paper 307.

Crawford, C. S. & J. R. Gosz 1982. Desert ecosystems: their resources in space and time. *Environmental Conservation* **9**, 181–96.

Criswell, D. R., J. F. Lindsay, D. L. Reasoner 1975. Seismic and acoustic emissions of a booming dune. *Journal of Geophysical Research* **80**, 4963–74.

Crittenden, M. D. 1963. Effective viscosity of the earth derived from isostatic loading of Pleistocene Lake Bonneville. *Journal of Geophysical Research* **68**, 5517–30.

Crocker, R. L. 1946. The soil and vegetation of the Simpson Desert and its borders. *Royal Society of South Australia, Transactions* **70**, 235–58.

Crompton, E. 1960. The significance of the weathering: leaching ratio in the differentiation of major soil groups, with particular reference to some strongly leached brown earths in the hills of Butan. *7th International Congress of Soil Scientists, Transactions* **4**, 406–12.

Cui, B., P. Komar, J. Baba 1983. Settling velocities of natural sand grains in air. *Journal of Sedimentary Petrology* **53**, 1205–11.

Cullen, J. L. 1981. Microfossil evidence for changing salinity patterns in the Bay of Bengal over the last 20,000 years. *Palaeogeography, Palaeoclimatology, Palaeoecology* **35**, 315–56.

Curtiss, B., J. B. Adams, M. S. Ghiorso 1985. Origin, development and chemistry of silica–alumina rock coatings from the semi-arid regions of the island of Hawaii. *Geochimica et Cosmochimica Acta* **49**, 49–56.

Curzon, Lord 1923. *Tales of travel*. London: Hodder & Stoughton.

Dacey, M. F. & W. C. Krumbein 1979. Model of breakage and selection for particle size distributions. *Mathematical Geology* **11**, 193–222.

Dalquert, W. W. & V. B. Sheffer 1942. The origin of mima mounds of western Washington. *Journal of Geology* **50**, 68–84.

Dan, J., R. Moshe, N. Alperovitch 1973. The soils of Sede Zin. *Israel Journal of Earth Sciences* **22**, 211–27.

Dan, J. & D. H. Yaalon 1966. Trends in the development with time in the Mediterranean environments of Israel. *Conference on Mediterranean Soils, Transactions*, Madrid, 139–45.

Dan, J. & D. H. Yaalon 1971. On the origin and nature of paleo-pedological formations in coastal fringe areas of Israel. See Yaalon (1971), 245–60.

Dan, J. & D. H. Yaalon 1982. Automorphic saline soils in Israel. See Yaalon (1982), 103–15.

Dan, J., D. H. Yaalon, H. Koyumdjisky 1972. Catenary soil relationships in Israel, I: the Netanya catena on coastal dunes of the Sharon. *Geoderma* **2**, 95–120.

Dan, J., D. H. Yaalon, R. Moshe, S. Nissim 1982. Evolution of reg soils in southern Israel and Sinai. *Geoderma* **28**, 173–202.

Danin, A. 1983. Weathering of limestone in Jerusalem by

cyanobacteria. *Zeitschrift für Geomorphologie* **27**, 413–21.

Danin, A., R. Gerson, K. Marton, J. Garty 1982. Patterns of limestone and dolomite weathering by lichens and blue–green algae and their palaeoclimatic significance. *Palaeogeography, Palaeoclimatology, Palaeoecology* **37**, 221–33.

Danin, A. & D. H. Yaalon 1982. Silt and clay sedimentation and decalcification during plant succession in sand of the Mediterranean coastal plain of Israel. *Israel Journal of Earth Sciences* **31**, 101–9.

Dare-Edwards, A. J. 1983. *Loessic clays of south Eastern Australia*. Loess Letter Supplement 2.

Daveau, S. 1965. Dunes ravinées et dépôts du Quaternaire récent dans le Sahel mauretanien. *Revue de Géographie de l'Afrique occidental* **1–2**, 7–48.

Daveau, S. 1966. *Le relief du Baten d'Atar (Adrar mauritanien)*. Mémoires et Documents, Centre de Recherches et Documents Cartographique Géographique, 2.

Daveau, S., R. Musinho, C. Toupet 1967. Les grands dépressions fermées de l'Adrar mauritanien, Sabkha de Chemchane et Richat. *Institut Français de l'Afrique Noire, Bulletin* **A29**, 414–46.

David, P. P. 1981. Stabilized dune ridges in northern Saskatchewan. *Canadian Journal of Earth Sciences* **18**, 286–310.

Davidson, A. P. 1986. An investigation into the relationship between salt weathering debris production and temperature. *Earth Surface Processes and Landforms* **11**, 335–41.

Davis, E. L. & S. Winslow 1965. Giant ground figures of prehistoric deserts. *American Philosophical Society, Proceedings* **109**, 8–21.

Davis, R. S., V. A. Ranov, A. E. Dodonov 1980. Early man in Soviet Central Asia. *Scientific American* **243**, 92–102.

Davis, W. M. 1905. The geographical cycle in an arid climate. *Journal of Geology* **13**, 381–407.

Davis, W. M. 1930. Rock floors in arid and in humid climates. *Journal of Geology* **38**, 1–27, 136–8.

Davis, W. M. 1933. Granite domes of the Mojave Desert, California. *San Diego Society Natural History, Transactions* **7**, 211–58.

Davis, W. M. 1954. The geographical cycle in an arid climate. See *Geographical essays*, D. W. Johnson (ed.), 296–322. London: Constable.

Davis, W. M. 1980. *The physical geography (geomorphology) of William Morris Davis* [compiled, illustrated, edited and annotated by P. K. King & S. A. Schumm]. Norwich: Geobooks.

Davy, E. G., F. Mattei, S. Solomon 1976. *An evaluation of climate and water resources for development of agriculture in the Sudano-Sahelian zone of West Africa*. World Meteorological Organisation Publication 459.

Dawson, P. J. & J. D. Marwitz 1982. *Wave structures and turbulent features of the winter airflow in southern Wyoming*. Geological Society of America Special Paper

192, 55–63.

DeBano, L. F. 1981. *Water-repellent soils: a state of the art*. USDA Forest Service, Forest and Range Experiment Station General Technical Report, PSW–46.

Defant, F. 1951. Local winds. See *Compendium of meteorology*, T. F. Malone (ed.), 655–72. Boston: American Meteorological Society.

De Félice, P. 1956. Processus de soulèvement des grains de sables par le vent. *Academie des Sciences à Paris, Comptes Rendus* **242**, 920–3.

De Félice, P. 1968. Étude des échanges de chaleur entre l'air et le sol sur deaux sols de nature différentes. *Archive für Meteorologie, Geophysik und Bioklimatologie* **B16**, 70–80.

DeGraff, J. V. 1978. Regional landslide evaluation: two Utah examples. *Environmental Geology* **2**, 203–14.

Delo, D. M. 1930. Dreikanter in Wyoming and Montana. *Science* **72**, 604.

de Martonne, E. & L. Aufrère 1928. L'extension des régions privées d'écoulement vers l'océan. *Annales de Géographie* **38**, 1–24.

Dendy, F. E. & G. C. Boulton 1976. Sediment-yield-runoff–drainage-area relationships in the United States. *Journal of Soil and Water Conservation* **32**, 264–6.

Denny, C. S. 1965. *Alluvial fans in the Death Valley region, California and Nevada*. USGS Professional Paper 466.

Denny, C. S. 1967. Fans and pediments. *American Journal of Science* **265**, 81–105.

Denny, C. S. & J. P. Owens 1979. *Surface and shallow subsurface geologic studies of the emerged coastal plain of the middle Atlantic states*. USGS Professional Paper 1067C.

De Ploey, J. 1977. Some experimental data on slopewash and wind action with reference to Quaternary morphogenesis in Belgium. *Earth Surface Processes* **2**, 101–15.

De Ploey, J. 1980. Some field measurements and experimental data on wind-blown sands. See *Assessment of erosion*, M. de Boodt & D. Gabriels (eds), 541–52. New York: John Wiley.

De Ploey, J. & A. Yair 1985. Promoted erosion and controlled colluviation: a proposal concerning land management and landscape evolution. *Catena* **12**, 105–10.

Depuydt, F. 1972. *De Belgische strand – en duin formaties in het kader van de geomorphologie der zuidoostelijke Norrdseekust*. Verhandelingen van Koninklijke Academie voor Wettenschappen, Litteren en Schone Kunsten van Belgie, Klasse der Wettenschappen, 34.

Derbyshire, E. 1978. The middle Hwang He loess lands. *Geographical Journal* **144**, 191–4.

Derbyshire, E. 1983. On the morphology, sediments, and origin of the Loess Plateau of central China. See Gardner & Scoging (1983), 172–94.

Desio, A. 1968. *Short history of the geological, mining and oil exploration in Libya*. Istituto Geologia dell'Università degli Studi di Milano, Series G, 250.

Deuser, W. G., E. H. Ross, L. E. Waterman 1976. Glacial and pluvial periods; their relationship revealed by

Pleistocene sediments of the Red Sea and Gulf of Aden. *Science* **191**, 1168–70.

DeVilliers, C. 1948. Les dépôts Quaternaire de l'Erg Tihodaine (Sahara centrale). *Société Géologique de la France, Bulletin*, Série 5 **18**, 189–91.

Dewolf, Y. & M. Mainguet 1976. Une hypothèse éolienne et tectonique sur l'alignement et l'orientation des buttes Tertiares du Bassin de Paris. *Revue de Géographie Physique et de Géologie Dynamique* **18**, 415–26.

Dietrich, R. V. 1977. Wind erosion by snow. *Journal of Glaciology* **18**, 148–9.

Dijkmans, J. W., A. E. Koster, J. P. Galloway, G. Mook 1986. Characteristics and origin of calcretes in a subarctic environment, Great Kobuk sand dunes, northwestern Alaska. *Arctic and Alpine Research* **18**, 377–87.

Dincer, T., A. Al-Mugrin, U. Zimmerman 1974. Study of the infiltration and recharge through the sand dunes in arid zones with special reference to the stable isotopes and thermonuclear tritium. *Journal of Hydrology* **23**, 79–109.

Doehring, D. O. 1970. Discrimination of pediments and alluvial fans from topographic maps. *Geological Society of America, Bulletin* **81**, 3109–16.

Doehring, D. O. (ed.) 1977. *Geomorphology in arid regions.* **Boston: Allen & Unwin.**

Dohrenwend, J. C., S. G. Wells, L. D. McFadden, B. D. Turrin 1987. Pediment dome evolution in the eastern Mojave desert, California. See Gardiner (1987), 1047–62.

Dohrenwend, J. C., S. G. Wells, B. D. Turrin 1986. Degradation of Quaternary cinder cones in the Cima volcanic field, Mojave desert, California. *Geological Society of America, Bulletin* **97**, 421–7.

Dokka, R. K. 1978. A method of determining the direction of strong winds, northwestern Coachella Valley, California. *Association of Engineering Geologists, Bulletin* **15**, 375–81.

Dolan, R., P. J. Godfrey, W. E. Odum 1973. Man's impact on the barrier islands of North Carolina: a case study of the implications of large-scale manipulations of the natural environment. *American Scientist* **61**, 152–62.

Doornkamp, J. C., D. Brunsden, D. K. C. Jones (eds) 1980. *Geology, geomorphology and pedology of Bahrain.* Norwich: Geobooks.

Dorn, R. I. 1982. Enigma of the desert. *Environment Southwest* **497**, 3–5.

Dorn, R. I. 1983. Cation-ratio dating: a new rock-varnish age-determination technique. *Quaternary Research* **20**, 49–73.

Dorn, R. I. 1986. Rock varnish as an indicator of aeolian environmental change. See Nickling (1986), 291–307.

Dorn, R. I. 1988. A rock varnish interpretation of alluvial-fan development in Death Valley, California. *National Geographic Research* **4**, 56–73.

Dorn, R. I. 1989a. Cation-ratio dating of rock varnish: a geographic assessment. *Progress in Physical Geography* **13**, 559–96.

Dorn, R. I. 1989b. A comment on "A note on the charac-

teristics and possible origins of desert varnish from southeast Morocco" by Drs Smith & Whalley. *Earth Surface Processes and Landforms* **14**, 167–70.

Dorn, R. I., D. B. Bamforth, T. A. Cahill, J. C. Dohrenwend, B. D. Turrin, D. J. Donahue, A. J. Jull, A. Long, M. E. Macko, E. B. Weil, D. S. Whitty, T. H. Zabel 1986. Cation-ratio and accelerator radiocarbon dating of rock varnish on Mojave artifacts and landforms. *Science* **231**, 830–3.

Dorn, R. I. & M. J. DeNiro 1984. Stable carbon isotope ratios of rock varnish organic matter: a new palaeo-environmental indicator. *Science* **227**, 1472–4.

Dorn, R. I., M. J. DeNiro, H. O. Ajie 1987. Isotopic evidence for climatic influence on alluvial-fan development in Death Valley. *Geology* **15**, 108–10.

Dorn, R. I. & T. M. Oberlander 1981a. Microbial origin of desert varnish. *Science* **213**, 1245–7.

Dorn, R. I. & T. M. Oberlander 1981b. Rock varnish origin, characteristics and usage. *Zeitschrift für Geomorphologie* **25**, 240–436.

Dorn, R. I., D. L. Tanner, B. D. Turrin, J. C. Dohrenwend 1987. Cation-ratio dating of Quaternary materials in the east-central Mojave Desert, California. *Physical Geography* **8**, 72–81.

Douglas, G. R. 1987. Manganese rock coatings from Iceland. *Earth Surface Processes and Landforms* **12**, 301–10.

Douglas, I. 1967. Man, vegetation and sediment yield of rivers. *Nature* **215**, 925–8.

Douglas, I. & T. Spencer (eds) 1985. *Environmental change and tropical geomorphology.* **London: Allen & Unwin.**

Douglass, A. E. 1909. The crescentic dunes of Peru. *Appalachia* **12**, 34–45.

Downes, R. G. 1946. Tunnelling erosion in north-eastern Victoria. *Journal CSIRO* **19**, 283–91.

Draga, M. 1983. Eolian activity as a consequence of beach nourishment – observations at Westerland (Sylt), German North Sea Coast. *Zeitschrift für Geomorphologie*, Supplementband 45, 303–19.

Dragovich, D. 1969. The origin of cavernous surfaces (tafoni) in granitic rocks of southern South Australia. *Zeitschrift für Geomorphologie* **13**, 163–81.

Dragovich, D. 1984. The survival of desert varnish in subsurface positions, western New South Wales, Australia. *Earth Surface Processes and Landforms* **9**, 425–34.

Dragovich, D. 1986. Minimum age of some desert varnish near Broken Hill, New South Wales. *Search* **17**, 149–51.

Dragovich, D. 1988. A preliminary electron probe study of microchemical variations in desert varnish in western New South Wales, Australia. *Earth Surface Processes and Landforms* **13**, 259–70.

Dregne, H. E. 1976. *Soils of arid regions.* Amsterdam: Elsevier.

Dregne, H. E. 1977. *Map of status of desertification in the hot arid regions.* UN Conference on Desertification, Nairobi: A/CONF/74/31.

Dresch, J. 1953. Morphologie de la chaîne d'Ungarta.

Institut de Recherches Sahariennes, Traveaux **9**, 25–38.

Dresch, J. 1957. Pediments et glacis d'érosion, pédiplains et inselbergs. *Information Géographique* **22**, 183–96.

Dresch, J. 1982. *Géographie des régions arides*. Paris: Presses Universitaires de France.

Drever, J. I. & C. L. Smith 1978. Cyclic wetting and drying of the soil zone as an influence on the chemistry of groundwater in arid terrains. *American Journal of Science* **278**, 1448–52.

Drury, S. A. 1987. *Image interpretation in geology*. London: Allen & Unwin.

Dubief, J. 1952. Le vent et le déplacement du sable au Sahara. *Institut de Recherches Sahariennes, Traveaux* **8**, 123–64.

Dubief, J. 1953. *Essai sur l'hydrologie superficielle au Sahara*. Birmandreis: Service des Études Scientifiques, Gouvernement Général de l'Algérie.

Dubief, J. 1959. *Le climat du Sahara*. Mémoire de l'Institut de Recherches Sahariennes, Université d'Alger.

Dubief, J. 1979. Review of the North African climate with particular emphasis on the production of eolian dust in the Sahel Zone and in the Sahara. See Morales (1979), 27–48.

Dudal, R. & M. F. Purnell 1986. Land resources: salt-affected soils. *Reclamation and Revegetation Research* **5**, 1–9.

Dufour, A. 1935. Observations sur les dunes du Sahara méridional. *L'Association de Géographes Français, Bulletin* **88**, 84–8.

Dunkerley, D. L. 1978. Hillslope form and climate: discussion and reply. *Geological Society of America, Bulletin* **89**, 1111–14.

Dunkerley, D. L. 1979. The morphology and development of *Rillenkarren*. *Zeitschrift für Geomorphologie* **23**, 332–48.

Durand, A., M. Icole, S. Bieda 1982. Sédiments et climats Quaternaire du Sahel central: exemple de la vallée de Maradi (Niger meridional). *Cahiers d'ORSTOM, Série Géologie* **12**, 77–90.

Durand de Corbiac, H. D. 1958. "Autant en emporte le vent", ou l'érosion et l'accumulation autour du Tibesti. *Association Ing. Géog., Bulletin* **4**, 147–56.

Dury, G. H. 1966a. Duricrusted residuals on the Barrier and Cobar pediplains of New South Wales, *Journal of the Geological Society of Australia* **13**, 299–307.

Dury, G. H. 1966b. Pediment slope and particle size at Middle Pinnacle, near Broken Hill, New South Wales. *Australian Geographical Studies* **4**, 1–17.

Dury G. H. 1968. Gibber. See Fairbridge (1968), 424.

Dury, G. H. & G. M. Habermann 1978. Australian silcretes and Northern Hemisphere correlates. See Langford-Smith (1978), 223–60.

Dury, G. H. & T. Langford-Smith 1964. The use of the term peneplain in descriptions of Australian landscapes. *Australian Journal of Science* **27**, 171–5.

Dury, G. H., E. D. Ongley, V. A. Ongley 1967. Attributes of pediment form on the Barrier and Cobar pediplains of New South Wales. *Australian Journal of Science* **30**, 33.

Dutton, R. W. (ed.) 1988. Scientific results of the Royal Geographical Society's Oman Wahiba Sands project. *Journal of Oman Studies*, Special Report 3.

Dzulynski, St. & A. Kotarba 1979. Solution pans and their bearing on the development of pediments and tors in granite. *Zeitschrift für Geomorphologie* **23**, 172–91.

ERIM 1982. *Remote sensing of arid and semi-arid lands*. Ann Arbor, Michigan: Environmental Research Institute of Michigan.

Eckert, R. E., M. K. Wood, W. H. Blackburn, F. F. Peterson 1979. Impacts of off-road vehicles on infiltration and sediment production of two desert soils. *Journal of Range Management* **32**, 394–7.

Eckis, R. 1928. Alluvial fans of the Cucamonga District, southern California. *Journal of Geology* **36**, 111–41.

Eggler, D. H., E. E. Larson, W. C. Bradley 1969. Granites, gneisses and the Sherman erosion surface, southern Laramie Range, Colorado–Wyoming. *American Journal of Science* **267**, 510–22.

Ehlen, J. 1976. *Photo analysis of a desert area*. US Army Engineer, Topographic Laboratories, Fort Belvour, Virginia.

Ehlers, E. 1982. Man and environment – problems in rural Iran. *Applied Geography and Development* **19**, 108–25.

Eisenberg, J., J. Dan, H. Koyumdjisky 1982. Relationships between moisture penetration and salinity in soils of the northern Negev (Israel). *Geoderma* **28**, 313–43.

El Aref, M., S. Abdel Wahab, S. Ahmed 1985. Surficial calcareous crust of caliche type along the Red Sea coast, Egypt. *Geologische Rundschau* **74**, 155–63.

El-Baz, F. 1979. Color of desert surfaces in the Arabian Peninsula. See *Apollo–Soyuz test project*, vol. 2, F. El-Baz & D. M. Warner (eds), 285–99. Washington DC: NASA.

El-Baz, F. 1984a. The desert in the space age. See El-Baz (1984b), 1–29.

El-Baz, F. (ed.) 1984b. *Deserts and arid lands*. Den Haag: Martinus Nijhoff.

El-Baz, F. 1986, Monitoring the sand-moving desert winds. See El-Baz & Hassan (1986), 141–58.

El-Baz, F. & M. H. A. Hassan (1986). *Physics of desertification*. Den Haag: Martinus Nijhoff.

El-Baz, F. & T. A. Maxwell (eds) 1982. *Desert landforms of southwest Egypt: a basis for comparison with Mars*. (CR-3611) Washington DC: NASA.

El-Baz, F. & D. M. Warner (eds) 1979. *Apollo–Soyuz test project: summary science report*. Washington DC: NASA.

El-Baz, F., et al. 1980. Journey to the Gilf Kebir and Uweinat, southwest Egypt 1978. *Geographical Journal* **146**, 51–93.

El Hinawi, E., S. Loukina, M. Mandour 1990. Isotopic composition of calcrete from the Mediterranean coastal plain west of Alexandria, Egypt. *Journal of Arid Environments* **19**, 39–44.

Ellis, C. I. & R. B. C. Russell 1974. The use of salt-laden soils (sabkha) for low-cost roads. *Conference on low-cost roads, Kuwait, 25–28 November 1974*, 1–25.

Ellwood, B. B. & J. H. Howard 1981. Magnetic fabric development in an experimentally produced barchan dune. *Journal of Sedimentary Petrology* **51**, 97–100.

Ellwood, J. M., P. D. Evans, I. G. Wilson 1975. Small scale aeolian bedforms. *Journal of Sedimentary Petrology* **45**, 554–61.

Elvidge, C. D. & R. M. Iverson 1983. Regeneration of desert pavement and varnish. See Webb & Wilshire (1983), 225–43.

Ely, L. L. & V. R. Baker 1985. Reconstructing paleoflood hydrology with slackwater deposits: Verde River, Arizona. *Physical Geography* **5**, 103–26.

Embabi, N. S. 1970/71. Structures of barchan dunes at the Kharga Oasis Depression, the Western Desert, Egypt (and a comparison with structures of two aeolian microforms from Saudi Arabia). *Société de Géographie d'Egypte, Bulletin* **43/44**, 53–71.

Embabi, N. S. 1986/87. Dune movement in the Kharga and Dhakla Oases depressions, the Western Desert, Egypt. *Société de Géographie d'Egypte, Bulletin* **59/60**, 35–70.

Emmanuel, C. B. & B. R. Bean 1972. Observations of Helmholtz waves in the lower atmosphere with an acoustic sounder. *Journal of Atmospheric Sciences* **29**, 886–92.

Emmett, W. W. 1970. *The hydraulics of overland flows on hillslopes*. USGS Professional Paper 662A.

Engel, C. E. & R. P. Sharp 1958. Chemical data on desert varnish. *Geological Society of America, Bulletin* **69**, 487–518.

Engel, P. 1981. Length of flow separation over dunes. *Proceedings of the American Society of Civil Engineers, Journal of the Hydraulics Division* **107**, 1133–43.

Engelund, F. & J. Fredsoe 1982. Sediment ripples and dunes. *Annual Review of Fluid Mechanics* **14**, 13–37.

Englebert, V. 1970. The Danakil, nomads of Ethiopia's wasteland. *National Geographic Magazine* **137**, 186–211.

Enquist, F. & E. Frederick 1932. The relation between dune form and wind direction. *Geologiska Föreningens i Stockholm Föhandlingar* **54**, 19–59.

Ericksen, G. E. 1981. *Geology and origin of the Chilean nitrate deposits*. USGS Professional Paper 1188.

Eriksson, E. 1958. The chemical climate and saline soils in the arid zone. *Arid Zone Research* **10**, 147–88.

Evans, G., C. G. St. C. Kendall, P. Skipwith 1964. Origin of the coastal flats, the sabkha, of the Trucial Coast, Persian Gulf. *Nature* **202**, 759–61.

Evans, I. S. 1970. Salt crystallization and rock weathering, a review. *Revue de Géomorphologie Dynamique* **19**, 153–77.

Evans, I. S. 1971. Physical weathering in hot deserts: the processes of haloclasty and thermoclasty. *British Geomorphological Research Group conference proceedings (Durham, 25th September)*.

Evans, J. R. 1962. Falling and climbing sand dunes in Cronese ("Cat") Mountains, San Bernardino County, California. *Journal of Geology* **70**, 107–13.

Evenari, M. L. 1981. Ecology of the Negev Desert, a critical review of our knowledge. See *Developments in arid zone ecology and environmental quality*, H. Shuval (ed.), 1–33. Philadelphia: Balban.

Evenari, M. L. 1985. The desert environment. See Evenari et al. (1986), 1–22.

Evenari, M. L., I. Noy-Meir, D. W. Goodall (eds) 1986. *Hot deserts and arid shrublands, B*. Amsterdam: Elsevier.

Evenari, M. L., L. Shanan, N. Tadmor 1968. Runoff farming in the desert, 1: experimental layout. *Agronomy Journal* **60**, 29–32.

Evenari, M., L. Shanan, N. H. Tadmor 1971. *The Negev: the challenge of a desert*. Cambridge, Mass.: Harvard University Press.

Evenari, M., D. L. Yaalon, Y. Gutterman 1974. Note on soils with vesicular structure in deserts. *Zeitschrift für Geomorphologie* **18**, 162–72.

Everard, C. E. 1963. Contrasts in the form and evolution of hill-side slopes in central Cyprus. *Institute of British Geographers, Transactions* **32**, 31–47.

Exner, F. M. 1920. Zür Physik der Dünen. *Akademie der Wissenschaften, Wien, Mathematisch–Naturwissenschaftliche Klasse, Abteilung* **1**, 129, 929–52.

Exner, F. M. 1927. Über Dünen und Sandwellen. *Geografiska Annaler* **9**, 81–9.

Fahey, B. D. 1983. Frost action and hydration as rock weathering mechanisms on schist: a laboratory study. *Earth Surface Processes and Landforms* **8**, 535–45.

Fahey, B. D. 1986. A comparative study of salt crystallisation and salt hydration as potential weathering agents in deserts. *Geografiska Annaler* **68A**, 107–11.

Fairbridge, R. W. (ed.) 1968. *Encyclopedia of geomorphology*. (2nd edn revised with J. Bourgeois 1978). New York: Van Nostrand Reinhold.

Fairbridge, R. W. 1970. An ice age in the Sahara. *Geotimes* **5**, 18–20.

Fan, L. T. & L. A. Disrud 1977. Transient wind erosion: a study of the nonstationary effect on the rate of wind erosion. *Soil Science* **124**, 61–5.

Farr, T. G. & J. B. Adams 1984. Rock coatings in Hawaii. *Geological Society of America, Bulletin* **95**, 1077–83.

Farrand, W. R. 1979. Blank on the Pleistocene map. *Geographical Magazine* **51**, 548–54.

Farres, P. 1978. The role of time and aggregate size in the crusting process. *Earth Surface Processes* **3**, 243–54.

Faure, H. 1962. *Reconnaissance géologique des formations post-Paléozoiques du Niger oriental*. Mémoires BRGM 47.

Fehrenbacher, J. B., K. R. Olson, I. J. Jansen 1986. Loess thickness in Illinois. *Soil Science* **141**, 423–31.

Felix-Henningson, P. 1984. Zür Relief und Bödentwicklung der Goz-Zone Nordkordofans im Sudan. *Zeitschrift für Geomorphologie* **28**, 285–304.

Filion, L. 1984. A relationship between dunes, fire and climate recorded in the Holocene deposits of Quebec. *Nature* **309**, 543–6.

Filion, L. 1987. Holocene development of parabolic dunes in the central St Lawrence Lowland, Quebec. *Quaternary Research* **28**, 196–209.

Filion, L. & P. Morisset 1983. Eolian landscapes along the eastern coast of Hudson Bay, Northern Quebec. *Nordicana* **47**, 73–94.

Findlater, J. 1969. A major low-level air current near the Indian Ocean during the northern summer. *Quarterly Journal of the Royal Meteorological Society* **95**, 362–80.

Finkel, H. J. 1959. The barchans of southern Peru. *Journal of Geology* **67**, 614–47.

Finlayson, A. A. 1984. Land surface evaluation for engineering practice: applications of the Australian PUCE system for terrain analysis. *Quarterly Journal of Engineering Geology* **17**, 149–58.

Fisher, O. 1866. On the disintegration of a chalk cliff. *Geological Magazine* **3**, 354–6.

Fisher, P. F. & P. Galdies 1988. A computer model for barchan dune movement. *Computers and Geosciences* **14**, 229–53.

Fisk, E. P. 1971. Desert glaze. *Journal of Sedimentary Petrology* **41**, 1136–7,

Flach, K. W., W. D. Nettleton, L. H. Gile, J. G. Cady 1969. Pedocementation: induration by silica, carbonate and sesquioxides in the Quaternary. *Soil Science* **107**, 442–53.

Flammand, G. B. M. 1899. La traversée de l'erg occidental (grands dunes du Sahara oranais). *Annales de Géographie* **9**, 231–41.

Flemal, R. C., I. E. Odom, R. G. Vail 1972. Stratigraphy and origin of the paha topography of northwestern Illinois. *Quaternary Research* **2**, 232–43.

Flenley, E. C., N. R. J. Fieller, D. D. Gilbertson 1987. The statistical analysis of "mixed" grain size distributions from aeolian sands in the Libyan Pre-Desert using log skew Laplace models. See Frostick & Reid (1987), 271–80. Oxford: Blackwell Scientific.

Fletcher, B. 1976. The incipient motion of dust by an airflow. *Journal of Physics D: Applied Physics* **9**, 913–24.

Fletcher, S. E. & W. P. Martin 1948. Some effects of algae and moulds in the rain-crust of desert soil. *Ecology* **29**, 95–100.

Flint, R. F. 1955. *Pleistocene geology of eastern South Dakota*. USGS Professional Paper 262.

Flint, R. F. & G. Bond 1968. Pleistocene sand ridges and pans in western Rhodesia. *Geological Society of America, Bulletin* **79**, 299–314.

Flint, S. 1985. Alluvial fan and playa sedimentation in an Andean arid closed basin: the Paciencia Group, Antofagasta Province, Chile. *Journal of the Geological Society of London* **142**, 533–46.

Flohn, H. 1969. Local wind systems in general climatology. See *General climatology*, vol. 2, H. Flohn (ed.), 139–72. Amsterdam: Elsevier.

Flohn, H. 1984. Climatic evolution in the southern hemisphere and the equatorial region during the Late Cainozoic. See Vogel (1984), 5–20.

Flohn, H. & S. Nicholson 1980. Climatic fluctuations in the arid belt of the "Old World" since the last glacial maximum: possible causes and future implications. See Sarnthein et al. (1980), 3–21. Rotterdam: Balkema.

Foggin III, G. J. & L. F. DeBano 1971. Some geographic implications of water-repellent soils. *Professional Geographer* **23**, 347–51.

Folk, R. L. 1968. Bimodal supermature sandstone: product of the desert floor. *23rd International Geological Congress* **8**, 9–32.

Folk, R. L. 1969. Grain shape and diagenesis in the Simpson desert, Northern Territory, Australia. *Geological Society of America, Abstracts with Programs for 1969* **7**, 68–9.

Folk, R. L. 1971a. Longitudinal dunes of the northwestern edge of the Simpson Desert, Northern Territory, Australia, 1: geomorphology and grain size relationships. *Sedimentology* **16**, 5–54.

Folk, R. L. 1971b. Genesis of longitudinal and oghrourd dunes elucidated by rolling upon grease. *Geological Society of America, Bulletin* **82**, 3461–8.

Folk, R. L. 1976a. Rollers and ripples in sand, streams and sky; rhythmic alteration of transverse and longitudinal vortices in three orders. *Sedimentology* **23**, 649–69.

Folk, R. L. 1976b. Reddening of desert sands, Simpson Desert, Northern Territory, Australia. *Journal of Sedimentary Petrology* **46**, 604–15.

Folk, R. L. 1977a. Longitudinal ridges with tuning-fork junctions in the laminated interval of flysch beds: evidence of low-order helical flow in turbidites. *Sedimentary Geology* **19**, 1–6.

Folk, R. L. 1977b. Folk's bedform theory – reply. *Sedimentology* **24**, 864–74.

Folk, R. L. 1978. Angularity and silica coatings of Simpson Desert sand grains, Northern Territory, Australia. *Journal of Sedimentary Petrology* **48**, 611–24.

Folk, R. L. & W. C. Ward 1957. Brazos River bar: a study in the significance of grain-size parameters. *Journal of Sedimentary Petrology* **27**, 3–26.

Fontes, J-C. & F. Gasse 1989. On the ages of humid Holocene and late Pleistocene phases in North Africa – remarks on "Late Quaternary climatic reconstruction for the Maghreb (North Africa)" by P. Rognon. *Palaeogeography, Paleoclimatology, Palaeoecology* **70**, 393–8.

Fontes, J-C. & P. Pouchan 1975. Les cheminées du lac Abbé (TFAI): stations hydroclimatiques de l'Holocène. *Academie des Sciences à Paris, Comptes Rendus*, **E280**, Series D, 383–6.

Fookes, P. G. & L. Collis 1977. *Concrete in the Middle East*. London: Cement and Concrete Association.

Fookes, P. G., W. J. French, S. M. Rice 1985. The influence of ground and groundwater geochemistry on construction in the Middle East. *Quarterly Journal of Engineering Geology* **18**, 101–28.

Fookes, P. G. & J. L. Knill 1969. The application of engineering geology in the regional development of northern and central Iran. *Engineering Geology* **3**, 81–120.

Forster, S. M. & T. H. Nicolson 1981. Aggregation of sand from a maritime embryo sand dune. *Soil Biology and Biochemistry* **13**, 199–203.

Foureau, F. 1903. Quelques considérations sur les dunes et les phenomènes éoliens. *Mission Saharien, Documents Scientifiques* **2**, 213–37.

Fournier, F. 1960. *Climat et érosion: la relation entre l'érosion du sol par l'eau et les précipitations atmosphériques.* Paris: Presses Universitaires de France.

Frank, T. D. 1984. The effect of change in vegetation cover and erosion patterns on albedo and texture of LANDSAT on images in a semiarid environment. *Annals of the AAG* **74**, 393–407.

Fredsoe, J. 1982. Shape and dimensions of stationary dunes in rivers. *American Society of Civil Engineers, Proceedings; Journal of the Hydraulics Division* **108**, 932–47.

Fredsoe, J. 1986. Shape and dimensions of dunes in open channel flow. See El-Baz & Hassan (1986), 385–97.

Free, E. E. (ed.) 1911. *The movement of soil material by the wind.* USDA Bureau of Soils Bulletin 68.

French, R. H. 1987. *Hydraulic processes on alluvial fans.* Amsterdam: Elsevier.

French, W. J. & A. B. Poole 1976. Alkali–aggregate reactions and the Middle East. *Concrete* **10**, 18–20.

Friedman, G. M. 1961. Distinction between dune, beach and river sands from their textural characteristics. *Journal of Sedimentary Petrology* **31**, 514–29.

Friedman, G. M. 1979. Differences in size distribution of populations of particles among sands of various origins. *Sedimentology* **26**, 3–32.

Friedmann, E. J. 1980. Endolithic microbial life in hot and cold deserts. *Origins of Life* **10**, 223–35.

Friedmann, E. J. 1982. Endolithic micro-organisms in the Antarctic cold desert. *Science* **215**, 1045–53.

Friedmann, E. J. & M. Galun 1974. Desert algae, lichens and fungi. See *Desert biology*, vol. 2, G. W. Brown (ed.), 165–212. New York: Academic Press.

Froidefond, J. M. & R. Prud'homme 1991. Coastal erosion and aeolian sand transport on the Aquitaine coast, *Acta Mechanica, Supplementum* **2**, 147–60.

Frostick, L. E. & I. Reid 1979. Drainage-net control of sedimentary parameters in sand-bed ephemeral streams. See *Geographical approaches to fluvial processes*, A. Pitty (ed.), 173–201. Norwich: Geobooks.

Frostick, L. E. & I. Reid 1982. Talluvial processes, mass washing and slope evolution in arid environments. *Zeitschrift für Geomorphologie*, Supplementband 44, 53–67.

Frostick, L. E. & I. Reid 1983. Taphonomic significance of sub-aerial transport of vertebrate fossils on steep semi-arid slopes. *Lethaia* **16**, 157–64.

Frostick, L. E. & I. Reid (eds) 1987. *Desert sediments: ancient and modern.* Oxford: Blackwell Scientific.

Frostick, L. E., I. Reid, J. T. Layman 1983. Changing size distribution of suspended sediment in arid-zone flash floods. *Special Publications of the International Association of Sedimentologists* **6**, 97–106.

Fry, E. J. 1924. A suggested explanation of the mechanical action of lithophytic lichens on rocks (shale). *Annals of Botany* **38**, 175–96.

Fry, E. J. 1927. The mechanical action of crustaceous lichens on substrata of shale, schist, gneiss, limestone and obsidian. *Annals of Botany* **41**, 437–60.

Fryberger, S. G. 1979. Eolian–fluviatile (continental) origin of ancient stratigraphic trap for petroleum in Weber Sandstone, Rangeley Oil Field, Colorado. *Mountain Geologist* **16**, 1–36.

Fryberger, S. G. 1980. Dune forms and wind regime, Mauritania, West Africa: implications for past climate. See Sarnthein et al. (1980), 79–96.

Fryberger, S. G. & T. S. Ahlbrandt 1979. Mechanisms for the formation of eolian sand seas. *Zeitschrift für Geomorphologie* **23**, 440–60.

Fryberger, S. G., T. S. Ahlbrandt, S. Andrews 1979. Origin, sedimentary features and significance of low-angle eolian "sand sheet" deposits, Great Sand Dunes National Monument and vicinity, Colorado. *Journal of Sedimentary Petrology* **49**, 733–46.

Fryberger, S. G., A. M. Al-Sari, T. J. Clisham, S. A. R. Rizvi, K. G. Al-Hinai 1984. Wind sedimentation in the Jafurah sand sea, Saudi Arabia. *Sedimentology* **31**, 413–31.

Fryberger, S. G. & G. Dean 1979. Dune forms and wind regime. See McKee (1979a), 305–97.

Fryberger, S. G. & A. S. Goudie 1981. Arid geomorphology. *Progress in Physical Geography* **5**, 420–8.

Fryberger, S. G. & C. Schenk 1981. Wind sedimentation tunnel experiments on the origins of aeolian strata. *Sedimentology* **28**, 805–21.

Fryberger, S. G. & C. Schenk 1988. Pin stripe lamination: a distinctive feature of modern and ancient eolian sediments. *Sedimentary Geology* **55**, 1–16.

Fryberger, S. G., C. Schenk, L. F. Krystinik 1988. Stokes surfaces and the effects of near-surface groundwater table on aeolian deposition. *Sedimentology* **35**, 21–41.

Frye, J. C. 1950. Origin of the Kansas Great Plains depressions. *Kansas State Geological Survey, Bulletin* **86**, (1).

Fuchs, M. 1979. Atmospheric transport processes above arid land vegetation. See Goodall & Perry (1979), vol. 1, 393–408.

Fuller, W. H., R. E. Cameron, A. Raica Jr 1960. Fixation of nitrogen in desert soils by algae. *7th International Congress of Soil Science* **2**, 617–24.

Gabriel, A. 1938. The southern Lut and Iranian Baluchistan. *Geographical Journal* **92**, 193–210.

Gabriel, A. 1964. Die Auswirkung vertikaler Luftströmingen und elektrikischer Spannungsfelder in kahlen sanden; neue Gedanken zür Dünenmorphologie als Diskussionsarbeiträge. *Österreichische Geographische Gesellschaft, Mitteilungen* **107**, 125–37.

Gabriel, B. 1979. Ur und Gruehgeschichte als Hilfwissenschaft der Geomorphologie um ariden Nord-afrika. See *Festschrift für Wolfgang Meckelein*, C. Borcherdt et al. (eds). *Stuttgarter Geographische Studien* **93**, 135–48.

Gale, H. S. 1914. Salines in the Owens, Searles and Panamint basins, south-eastern California. *USGS Bulletin* **580-L**, 251–323.

Galloway, W. E. & D. K. Hobday 1983. *Terrigenous clastic depositional systems*. New York: Springer.

Galon, R. (ed.) 1969. *Dune processes and forms in Poland* [in Polish]. Polska Academia Nauk, Istytut Geografii, *Prace Geograficzne*, Warsaw **75**.

Ganor, E. & Y. Mamane 1982. Transport of Saharan dust across the eastern Mediterranean. *Atmospheric Environment* **16**, 581–7.

Gardiner, V. (ed.) 1987. *International geomorphology 1986*. Chichester, England: John Wiley.

Gardner, L. R. 1972. Origin of the Mormon Mesa caliche, Clark County, Nevada. *Geological Society of America Bulletin* **83**, 143–56.

Gardner, R. A. M. 1981a. Reddening of dune sands – evidence from southeast India. *Earth Surface Processes and Landforms* **6**, 459–68.

Gardner, R. A. M. 1981b. Reddening of tropical coastal dune sands. See R. C. L. Wilson (1983), 103–15.

Gardner, R. A. M. 1983. Aeolianite. See Goudie & Pye (eds), 265–300. London: Academic Press.

Gardner, R. A. M. 1988. Aeolianites and marine deposits of the Wahiba Sands: character and palaeoenvironments. *Journal of Oman Studies*, Special Report 3, 75–94.

Gardner, R. A. M. & K. Pye 1981. Nature, origin and palaeoenviromental significance of red coastal and desert sand dunes. *Progress in Physical Geography* **5**, 514–34.

Gardner, R. A. M. & H. Scoging (eds) 1983. *Megageomorphology*. Oxford: Oxford University Press

Garner, H. F. 1976. Insolation warmed over: comment. *Geology* **4**, 264.

Gasse, F., P. Rognon, F. A. Street 1980. Quaternary history of the Afar and Ethiopian Rift lakes. See Williams & Faure (1980), 361–400.

Gautier, E. F. 1935. *Sahara, the great desert*. New York: Columbia University Press. [Reprinted 1970, by Octagon Press, New York.]

Gautier, E. F. & R. Chudeau 1909. *Missions au Sahara, I: Sahara algérien*. Paris: Armand Colin.

Gautier, M. 1975. Actions naturelles et actions humaines dans l'évolution d'une flèche littorale: le cordon dunaire des Moutiers-en-Retz (LA). *Norois* **22**, 549–62.

Gaylord, D. R. 1982. Geologic history of the Ferris dune field, south-central Wyoming. *Geological Society of America Special Paper* **192**, 65–82.

Gaylord, D. R. & P. J. Dawson 1987. Airflow–terrain interactions through a mountain gap, with an example of eolian activity beneath an atmospheric hydraulic jump. *Geology* **15**, 789–92.

Gerety, K. M. & R. Slingerland 1983. Nature of the saltating population in wind tunnel experiments with heterogeneous size density sands. See Brookfield & Ahlbrandt (1983), 133–48.

Gerson, R. 1977. Sediment transport for desert watersheds in erodible materials. *Earth Surface Processes* **2**, 343–61.

Gerson, R. 1982. Talus relicts in deserts: a key to major climatic fluctuations. *Israel Journal of Earth Sciences* **31**, 123–32.

Gerson, R. & R. Amit 1987. Rates and modes of dust accretion and deposition in an arid region – the Negev, Israel. See Frostick & Reid (1987), 157–69.

Ghose, B. 1964. Geomorphological aspects of the formation of salt basins in the Lower Luni Basin. *Papers from Symposium on Problems of Indian Arid Zone, Jodhpur*, 169–78.

Gibbens, R. P., J. N. Tromble, J. T. Hennessey, M. Cardenas 1983. Soil movement in mesquite dunelands and former grasslands of southern New Mexico from 1933 to 1980. *Journal of Range Management* **36**, 145–8.

Gibson, E. S. H. 1946. Singing sands. *Royal Society of South Australia, Transactions* **70**, 35–44.

Gilbert, G. K. 1875. Report on the Geology of Portions of Nevada, Utah, California and Arizona, part 1: of geographical and geological explorations and surveys west of the 100th meridian. Engineers Department, US Army, 3, 21–187.

Gilbert, G. K. 1877. *Report on the geology of the Henry Mountains*. Washington DC: Government Printing Office.

Gilbert, G. K. 1890. *Lake Bonneville*. US Geographical Survey Monograph 1.

Gilbert, G. K. 1895. Lake basins created by wind erosion. *Journal of Geology* **3**, 47–9.

Gilbert, G. K. 1914. *Transportation of debris by running water*. USGS Professional Paper 86.

Gile, L. H. 1966. Coppice dunes and the Rotura soil. *Soil Science Society of America, Proceedings* **30**, 657–60.

Gile, L. H. 1967. Soils of an ancient basin floor near Las Cruces, New Mexico. *Soil Science* **103**, 265–76.

Gile, L. H. 1975. Holocene soils and soil–geomorphic relations in an arid region of southern New Mexico. *Quaternary Research* **5**, 321–60.

Gile, L. H. 1977. Holocene soils and soil–geomorphic relations in an arid region of southern New Mexico, *Quaternary Research* **7**, 112–32.

Gile, L. H. & R. B. Grossman 1968. Morphology of an argillic horizon in desert soils of southern New Mexico. *Soil Science* **106**, 6–15.

Gile, L. H. & J. W. Hawley 1966. Periodic sedimentation and soil formation in an alluvial fan–piedmont in southern New Mexico. *Soil Science Society of America, Proceedings* **30**, 261–8.

Gile, L. H. & J. W. Hawley 1968. Age and comparative development of desert soils at the Gardner Spring radiocarbon site, New Mexico. *Soil Science Society of America, Proceedings* **32**, 709–16.

Gile, L. H., F. F. Peterson, R. B. Grossman 1965. The K-horizon – a master horizon of $CaCO_3$ accumulation. *Soil Science* **99**, 74–82.

Gile, L. H., F. F. Peterson, R. B. Grossman 1966. Morphological and genetic sequences of carbonate accumulation in desert soils. *Soil Science* **101**, 347–60.

Gill, E. W. B. 1948. Frictional electrification of sand. *Nature* **162**, 568–9.

Gillette, D. A. 1977. Fine particle emissions due to wind erosion. *American Society of Agricultural Engineering, Transactions* **20**, 890–7.

Gillette, D. A. 1978. A wind-tunnel simulation of the

erosion of soil; effect of soil moisture, sandblasting, wind speed and soil consolidation on dust production. *Atmospheric Environment* **12**, 1735–43.

Gillette, D. A. 1979. Environmental factors affecting dust emission by wind erosion. See Morales (1979), 71–91.

Gillette, D. A. 1980. Major contributions of natural primary continental aerosols: source mechanisms. *Annals of the New York Academy of Sciences* **338**, 348–58.

Gillette, D. A. 1981. Production of dust that may be carried great distances. *Geological Society of America, Special Paper* **186**, 11–26.

Gillette, D. A. 1986a. Production of dust. See El-Baz & Hassan (1986), 251–60.

Gillette, D. A. 1986b. Threshold velocities for dust production. See El-Baz & Hassan (1986), 321–6.

Gillette, D. A., J. Adams, A. Endo, D. Smith, R. Kihl 1980. Threshold velocities for input of soil particles into the air by desert soils. *Journal of Geophysical Research* **85**, (C10), 5621–30.

Gillette, D. A., J. Adams, D. Muhs, R. Kihl 1982. Threshold friction velocities and rupture moduli for crusted desert soils for the input of soil particles to the air. *Journal of Geophysical Research* **87C**, 9003–15.

Gillette, D. A., I. H. Blifford, D. W. Fryear 1974. The influence of wind velocity on size distributions of soil wind aerosols. *Journal of Geophysical Research* **79**, 4068–75.

Gillette, D. A. & P. H. Stockton 1986. Mass momentum and kinetic energy fluxes of saltating particles. See Nickling (1986), 35–56.

Gillette, D. A. & P. H. Stockton 1990. Effect of non-erodible elements on wind movement of soil material. In *Extended abstracts, NATO Advanced Workshop on sand, dust and soil in their relation to aeolian and littoral processes.* 14–18 May, 1990, Sandbjerg, Denmark.

Gillette, D. A. & T. R. Walker 1977. Characteristics of air-borne particles produced by wind erosion of sandy soil, High Plains of West Texas. *Soil Science* **123**, 97–110.

Gilluly, J. 1937. Physiography of the Ajo Region, Arizona. *Geological Society of America, Bulletin* **48**, 323–48.

Gimmingham, C. H. 1955. A note on the water table, sand movement and plant distribution in a North African oasis. *Journal of Ecology* **43**, 22–5.

Gimmingham, C. H., W. Ritchie, B. B. Willetts, A. J. Willis (eds) 1989. *Symposium: coastal sand dunes. Royal Society of Edinburgh, Proceedings* B96.

Glasby, G. P., J. K. McPherson, B. P. John, J. H. Johnson, J. R. Keys, A. G. Freeman, M. J. Tricker 1981. Desert varnish in southern Victoria Land, Antarctica. *New Zealand Journal of Geology and Geophysics* **24**, 389–97.

Glasovskaya, M. A. 1968. Geochemical landscapes and types of geochemical soil sequences. *19th International Congress of Soil Science, Proceedings* **4**, 303–12.

Glennie, K. W. 1970. *Desert sedimentary environments.* Amsterdam: Elsevier.

Glennie, K. W. 1983. Lower Permian Rotliegende desert sedimentation in the North Sea area. See Brookfield & Ahlbrandt (1983), 521–41.

Glennie, K. W. 1985. Early Permian (Rotliegendes) palaeo-winds of the North Sea – reply. *Sedimentary Geology* **45**, 297–313.

Glennie, K. W. & B. D. Evamy 1968. Dikaka: plants and plant root structures associated with aeolian sand. *Palaeogeography, Palaeoclimatology, Palaeoecology* **4**, 77–87.

Glennie, K. W. & G. Evans 1976. A reconnaissance of the recent sediments of the Ranns of Kutch, India. *Sedimentology* **23**, 625–47.

Goddard, D. & Y. X. Picard 1973. *Los Medaños de Coro: composicion, granulometria y migracion de las arenas.* Caracas, Venezuela: Division de Geologia Marina, Ministerio de Energia y Minas.

Goldsmid, J. G. 1897. Singing and drifting sand. *Geographical Journal* **9**, 454–5.

Goldsmith, V. 1973. Internal geometry and origin of vegetated coastal sand dunes. *Journal of Sedimentary Petrology* **43**, 1128–42.

Goldsmith, V. 1985. Coastal dunes. See *Coastal sedimentary environments*, R. A. Davis (ed.), 171–236. New York: Springer.

Goldsmith, V. 1989. Coastal sand dunes as geomorphological systems. See Gimmingham et al. (1989), 3–15.

Goldsmith, V., P. Rosen, Y. Gertner 1988. *Eolian sediment transport on the Israeli coast.* Final report to US–Israel Binational Science Foundation, Jerusalem.

Goodall, D. W. & R. A. Perry 1979. *Arid-land ecosystems* [2 vols]. Cambridge: Cambridge University Press.

Goosens, D. 1985. The granulometric characteristics of a slowly moving dust cloud. *Earth Surface Processes and Landforms* **10**, 353–62.

Goosens, D. 1988a. Sedimentary characteristics of natural dust in the wake of symmetrical hills. *Zeitschrift für Geomorphologie* **32**, 499–502.

Goosens, D. 1988b. The effect of surface curvature on the deposition of loess: a physical model. *Catena* **15**, 179–94.

Goosens, D. 1988c. Scale model simulations of the deposition of loess in hilly terrain. *Earth Surface Processes and Landforms* **13**, 533–44.

Goosens, D. & Z. I. Offer 1990. A wind-tunnel simulation and field verification of desert dust deposition (Avdat Experimental Station, Negev Desert). *Sedimentology* **37**, 7–22.

Gorycki, M. A. 1973. Sheetflood structure: mechanism of beach cusp formation and related phenomena. *Journal of Geology* **81**, 109–17.

Goudharzi, G. H. 1970. *Nonmetallic mineral resources: saline deposits, silica sand, sulfur and Trona.* USGS Professional Paper 660.

Goudie, A. S. 1969. Statistical laws and dune ridges in southern Africa. *Geographical Journal* **135**, 404–6.

Goudie, A. S. 1970. Notes on some major dune types in southern Africa. *South African Geographical Journal* **52**, 93–101.

Goudie, A. S. 1972. Climate, weathering, crust formation,

dunes and fluvial features of the central Namib Desert near Gobabeb, South-West Africa. *Madoqua*, series II **1**, 15–31.

Goudie, A. S. 1973. *Duricrusts in tropical and sub-tropical landscapes*. Oxford: Oxford University Press.

Goudie, A. S. 1974. Further experimental investigation of rock weathering by salt and other mechanical processes. *Zeitschrift für Geomorphologie,* Supplementband 21, 1–12.

Goudie, A. S. 1978. Dust storms and their geomorphological implications. *Journal of Arid Environments* **1**, 291–310.

Goudie, A. S. 1983a. Dust storms in space and time. *Progress in Physical Geography* **7**, 502–30.

Goudie, A. S. 1983b. The arid earth. See Gardner & Scoging (1983), 152–71.

Goudie, A. S. 1983c. Calcrete. See Goudie & Pye (1983), 93–131.

Goudie, A. S. 1984. Salt efflorescences and salt weathering in the Jurza Valley, Karakoram Mountains, Pakistan. See Miller (1984), 607–15.

Goudie, A. S. 1985a. Duricrusts and landforms. See *Geomorphology and soils*, K. S. Richards, R. R. Arnett, S. Ellis (eds), 37–57. London: Allen & Unwin.

Goudie, A. S. 1985b. *Salt weathering*. School of Geography, University of Oxford, Research Paper 33.

Goudie, A. S. 1986. Laboratory simulation of the "wick effect" in salt weathering of rock. *Earth Surface Processes and Landforms* **11**, 275–85.

Goudie, A. S. 1989. Weathering processes. See Thomas (1989c), 11–24.

Goudie, A. S. 1992. Environmental change, 3rd edn. Oxford: Oxford University Press.

Goudie, A. S., B. Allchin, K. T. M. Hegde 1973. The former extensions of the Great Indian Sand Desert. *Geographical Journal* **139**, 243–57.

Goudie, A. S & K. Pye (eds) 1983. *Chemical sediments and geomorphology*. London: Academic Press.

Goudie, A. S. & P. A. Bull 1984. Slope process change and colluvium deposition in Swaziland: an SEM analysis. *Earth Surface Processes and Landforms* **9**, 289–99.

Goudie, A. S., R. U. Cooke, I. Evans 1970. Experimental investigation of rock weathering by salts. *Area 1970*, 42–8.

Goudie, A. S., R. U. Cooke, J. C. Doornkamp 1979. The formation of silt from quartz dune sand by salt-weathering processes in deserts. *Journal of Arid Environments* **22**, 102–12.

Goudie, A. S. & R. U. Cooke 1984. Salt efflorescences and saline lakes: a distributional analysis. *Geoforum* **15**, 563–82.

Goudie, A. S. & M. J. Day 1980. Disintegration of fan sediments in Death valley, California, by salt weathering. *Physical Geography* **1**, 126–37.

Goudie, A. S., H. M. Rendell, P. A. Bull 1984. The loess of Tajik SSR. See Miller (1984), 399–412.

Goudie, A. S. & M. J. S. Sands 1989. The Kimberley Research Project, Western Australia 1988: a report.

Geographical Journal **155**, 161–6.

Goudie, A. S. & C. H. B. Sperling 1977. Long distance transport of foraminiferal tests by wind in the Thar Desert, north-west India. *Journal of Sedimentary Petrology* **47**, 630–3.

Goudie, A. S. & D. S. G. Thomas 1985. Pans in southern Africa with particular reference to South Africa and Zimbabwe. *Zeitschrift für Geomorphologie* **29**, 1–19.

Goudie, A. S. & D. S. G. Thomas 1986. Lunette dunes in southern Africa. *Journal of Arid Environments* **10**, 1–12.

Goudie, A. S. & A. Watson 1981. The shape of desert sand dune grains. *Journal of Arid Environments* **4**, 185–90.

Goudie, A. S. & A. Watson 1984. Rock block monitoring of rapid salt-weathering in southern Tunisia. *Earth Surface Processes and Landforms* **9**, 95–8.

Goudie, A. S., A. Warren, D. K. C. Jones, R. U. Cooke 1987. The character and possible origins of the aeolian sediments of the Wahiba Sand Sea, Oman. *Geographical Journal* **153**, 231–56.

Goudie, A. S. & G. L. Wells 1990. *The nature, distribution and formation of pans in arid zones* [MS].

Goudie, A. S. & J. Wilkinson 1978. *The warm desert environment*. Cambridge: Cambridge University Press.

Government of Qatar 1977. *Technical report: sand dune movement study south of Umm Said, 1963–1976* (Huntings Surveys Laboratory, Herts, UK).

Gozdzik, J. S. 1981. Les changements de processus éoliens dans la Pologne centrale au cours du Vistulian (Wurm). *Recherches Géographiques à Strasbourg* **16–17**, 115–20.

Graf, W. H. 1971. *Hydraulics of sediment transport*. New York: McGraw–Hill.

Graf, W. L. 1977. The rate law in fluvial geomorphology. *American Journal of Science* **277**, 178–91.

Graf, W. L. 1978. Fluvial adjustments to the spread of tamarisk in the Colorado Plateau region. *Geological Society of America, Bulletin* **89**, 1491–1501.

Graf, W. L. 1979a. Mining and channel response. *Annals of the AAG* **69**, 262–75.

Graf, W. L. 1979b. The development of montane arroyos and gullies. *Earth Surface Processes* **4**, 1–14.

Graf, W. L. 1979c. Rapids in canyon rivers. *Journal of Geology* **87**, 533–51.

Graf, W. L. 1980. The effect of dam closure on downstream rapids. *Water Resources Research* **16**, 129–36.

Graf, W. L. 1981. Channel instability in a sand-river bed. *Water Resources Research* **17**, 1087–94.

Graf, W. L. 1983a. Downstream changes in stream power in the Henry Mountains, Utah. *Annals of the AAG* **73**, 373–87.

Graf, W. L. 1983b. Flood-related channel change in an arid-region river. *Earth Surface Processes and Landforms* **8**, 125–39.

Graf, W. L. 1983c. Variability of sediment removal in a semi-arid watershed. *Water Resources Research* **19**, 643–52.

Graf, W. L. 1988a. *Fluvial processes in dryland rivers*. New York: Springer.

Graf, W. L. 1988b. Definition of flood plains along arid-region rivers. See *Flood geomorphology*, V. R. Baker, R. C. Kochel, P. C. Patton (eds), 231–42. Chichester, England: John Wiley.

Graf, W. L. 1989. Holocene lacustrine deposits and sediment yield in Lake Canyon, southeastern Utah. *National Geographic Research* **5**, 146–60.

Grandet, C. 1957. Sur la morphologie dunaire de la rive sud du Lac Faguébine. *Institut des Recherches Sahariennes, Traveaux* **16**, 171–80.

Grass, A. J. 1970. Initial instability of fine bed sand. *American Society of Civil Engineers, Proceedings; Journal of the Hydraulics Division* **96**, 619–32.

Grass, A. J. 1971. Structural features of turbulent flow over smooth and rough boundaries. *Journal of Fluid Mechanics* **50**, 233–55.

Grass, A. J. 1982. The influence of boundary layer turbulence on the mechanics of sediment transport, *Euromech 156: mechanics of sediment transport*, Istanbul, 3–17.

Greeley, R. 1979. Silt–clay aggregates on Mars: a model for the formation of "sand". *Journal of Geophysical Research* **84**, 6248–54.

Greeley, R. 1986. Aeolian landforms: laboratory simulations and field studies. See Nickling (1986), 195–211.

Greeley, R. & J. D. Iversen 1985. *Wind as a geological process on Earth, Mars, Venus and Titan*. Cambridge: Cambridge University Press.

Greeley, R. & J. D. Iversen 1986. Aeolian processes and features at Amboy Lava Field, California. See El-Baz & Hassan (1986), 290–317.

Greeley, R., N. Lancaster, L. Gaddis, K. Rasmussen, B. R. White, R. S. Saunders, S. Wall, A. Dobrovolskis, J. D. Iversen 1990. Relationships between topographic roughness and aeolian processes. *Acta Mechanica* **2**, 77–88.

Greeley, R. & R. Leach 1978. A preliminary assessment of the effects of electrostatics on aeolian processes. *Reports, Planetary Geology Program, 1977–8*, NASA, TM 79729, 236–7.

Greeley, R., R. Leach, S. H. Williams, B. R. White, J. B. Pollack, D. H. Krinsley, J. R. Marshall 1982. Rate of wind abrasion on Mars. *Journal of Geophysical Research* **87**, 10,009–24.

Greeley, R., J. R. Marshall, J. B. Pollack 1987. Physical and chemical modification of the surface of Venus by windblown sand particles. *Nature* **327**, 313–31.

Greeley, R., S. H. Williams, J. R. Marshall 1983. Velocities of windblown particles in saltation: preliminary laboratory and field measurements. See Brookfield & Ahlbrandt (1983), 133–48.

Gregory, J. W. 1914. The lake system of Western Australia. *Geographical Journal* **43**, 656–64.

Gregory, K. J. (ed.) 1977. *River channel changes*. Chichester, England: John Wiley.

Gregory, K. J. & D. E. Walling 1973. *Drainage basin form and process: a geomorphological approach*. London: Edward Arnold.

Grenier, M. P. 1968. Observations sur les taffonis du désert Chilien. *Association Géographes Français, Bulletin* **364–5**, 193–211.

Griffiths, J. S. 1978. *Flood assessment in ungauged semi-arid catchments as a branch of applied geomorphology*. King's College, London Department of Geography Occasional Papers 8.

Griggs, D. T. 1936. The factor of fatigue in rock exfoliation. *Journal of Geology* **44**, 781–96.

Grolier, M. J., J. F. McCauley, C. S. Breed, N.S. Embabi 1980. Yardangs of the Western Desert. *Geographical Journal* **135**, 191–212.

Gross, G. 1987. A numerical study of the air flow within and around a single tree. *Boundary-layer Meteorology* **40**, 311–27.

Grove, A. T. 1958. The ancient erg of Hausaland and similar formations on the south side of the Sahara. *Geographical Journal* **124**, 526–33.

Grove, A. T. 1960. Geomorphology of the Tibesti region with special reference to the western Tibesti. *Geographical Journal* **126**, 18–31.

Grove, A. T. 1969. Landforms and climatic change in the Kalahari and Ngamiland. *Geographical Journal* **135**, 191–212.

Grove, A. T. 1972. The dissolved and solid load carried by some West Africa rivers: Senegal, Niger, Benue and Shari. *Journal of Hydrology* **16**, 277–300.

Grove, A. T. 1980. Geomorphic evolution of the Sahara and the Nile. See Williams & Faure (1980), 7–16.

Grove, A. T. & A. Warren 1968. Quaternary landforms and climate on the south side of the Sahara. *Geographical Journal* **134**, 194–208.

Grünert, J. 1987. Landslides in the central Sahara. See Gardiner (1987), 487–98.

Grünert, J. & D. Busche 1980. Large-scale fossil landslides at the Msak Mallat and Hamadat Manghini escarpment. See Salem & Busrewil (1980), 849–60.

Grünert, J. & H. Hagedorn 1976. Beobachtungen an Schichtstufen der Nubischen Serie (Zentralsahara). *Zeitschrift für Geomorphologie*, Supplementband 24, 99–110.

Gunatilaka, A. & S. Mwango 1987. Continental sabkha pans and associated nebkhas in southern Kuwait, Arabian Gulf. See Frostick & Reid (1987), 187–203.

Gunatilaka, A. & S. B. Mwango 1989. Flow separation and the internal structure of shadow dunes. *Sedimentology* **61**, 125–34.

Gunatilaka, A., A. Saleh, A. Al-Temeemi 1980. Plant-controlled supratidal anhydrite from Al-Khiran, Kuwait. *Nature* **288**, 257–60.

Gupta, J. P. 1979. Observations on the periodic variations of moisture in stabilised and unstabilised dunes of the Indian Desert. *Journal of Hydrology* **41**, 153–6.

Gupta, J. P., R. K. Agarwal, N. P. Raikhy 1981. Soil erosion by wind from bare sandy plains in western Rajasthan. *Journal of Arid Environments* **4**, 15–20.

Guy, H. P., D. B. Simons, E. V. Richardson 1966. *Summary of alluvial channel data from flume experiments from 1956 to 1961*. USGS Professional Paper 462I.

472

Guy, M. & M. Mainguet 1975. Les courants de transport éolien au Sahara et leurs manifestations au sol. *Academie des Sciences à Paris, Comptes Rendus* **281**, 103–6.

Haberland, W. & H-J. Pachur 1980. Über Deflationsformen in der zentralen Sahara. See *Beiträge zür Geomorphologie und Länderkunde, Prof. Hartmut Valentin zum Gedächtnis*, B. Hofmeister & A. Steinecke (eds), *Berliner Geographische Studien* **7**, 309–22.

Hack, J. T. 1941. The dunes of western Navajo county. *Geographical Review* **31**, 240–362.

Hadley, R. F. 1961. *Influence of riparian vegetation on channel shape, northeastern Arizona*. USGS Professional Paper 424C, 30–31.

Hadley, R. F. 1967. Pediments and pediment-forming processes. *Journal of Geological Education* **15**, 83–9.

Hadley, R. F. 1977. Some concepts of erosional processes and sediment yield in a semiarid environment. See *Erosion*, T. J. Toy (ed.), 78–82. Norwich: Geobooks.

Haff, P. K. 1986. Booming sands. *American Scientist* **74**, 376–81.

Haff, P. K. 1990. Computer simulation of saltation impact. *Acta Mechanica*.

Haff, P. K. & B. T. Werner 1986. Computer simulation of the mechanical sorting of grains. *Powder Technology* **48**, 239–45.

Hagedorn, H. 1968. Über äolische Abtragung und Formung in der Südest-Sahara. *Erdkunde* **22**, 257–69.

Hagedorn, H. 1979. Das Verbeitungsmuster der Dünen am Westrand Murzuk-Bekens, Zentrale Sahara. See *Natur und Wirtschaftsgeographische Forschungen in Afrika*, J. Budel (ed.), *Würtzburg geographische Abhandelungen* **49**, 103–23.

Hagedorn, H. & H. J. Pachur 1971. Observations on climatic geomorphology and Quaternary evolution of landforms in south-central Libya. See *Symposium on the geology of Libya*, C. Gray (ed.), 387–400. Faculty of Science, University of Libya.

Hagen, L. J. 1984. Soil aggregate abrasion by impacting sand and soil particles. *American Society of Agricultural Engineering, Transactions* **27**, 805–8.

Hagen, L. J. & D. V. Armbrust 1985. Effects of field ridges on soil transport by wind. See Barndorff-Nielsen et al. (1985), 563–86.

Hails, J. R. & J. Bennet 1980. Wind and sediment movement in coastal dune areas. See *17th coastal engineering conference*, 1565–75. New York: American Society for Civil Engineers.

Haines, D. A. & M. C. Smith 1983. Wind-tunnel generation of roll vortices over a differentially heated surface. *Nature* **306**, 351–3.

Halevy, G. & E. H. Steinberger 1974. Inland penetration of the summer inversion from the Mediterranean coast of Israel. *Israel Journal of Earth Science* **223**, 47–54.

Halitim, A., M. Robert, G. Pedro 1983. Étude experimentale de l'épigenie calcaire des silicates en milieu confine – caractérisation des conditions, de son developpement et des modalités de sa mis en jeu.

Sciences Géologiques Memoires **71**, Université Louis Pasteur de Strasbourg, Institut de Géologie, 63–73.

Hall, K. 1979. Sorted stripes orientated by wind action: some observations from sub-Arctic Marion Island. *Earth Surface Processes* **4**, 281–90.

Hallbauer, D. K. & H. M. Jahns 1977. Attack of lichen on quartzitic rock surfaces. *Lichenologist* **9**, 119–22.

Hallet, J. & T. Hoffer 1971. Dust devil systems. *Weather* **224**, 247–50.

Hallsworth, E. G. & G. G. Beckmann 1969. Gilgai in the Quaternary. *Soil Science* **107**, 409–20.

Hallsworth, E. G., G. K. Robertson, F. R. Gibbons 1955. Studies of pedogenesis in New South Wales, VII: the "gilgai" soils. *Journal of Soil Science* **6**, 1–31.

Hallsworth, E. G. & H. D. Waring 1964. Studies of pedogenesis in New South Wales, VIII: an alternative hypothesis for the formation of solodized solometz of the Pilliga district. *Journal of Soil Science* **15**, 158–77.

Hammond, E. 1954. A geomorphic study of the Cape Region of Baja California. *University of California Publications in Geography* **10**, 45–112.

Hammond, F. D. C. & A. D. Heathershaw 1981. A wave theory for sandwaves in shelf seas. *Nature* **293**, 208–10.

Han, Mukang 1985. Tectonic geomorphology and its application to earthquake prediction in China. See Morisawa & Hack (1985), 367–86.

Hanawalt, R. B. & R. H. Whittaker 1977. Altitudinal patterns of Na, K, Ca and Mg in soils and plants in the San Jacinto Mountains, California. *Soil Science* **123**, 25–36.

Hanna, S. R. 1969. The formation of longitudinal sand dunes by large helical eddies in the atmosphere. *Journal of Applied Meteorology* **8**, 874–83.

Hardan, A. 1971. Archaeological methods for dating of soil salinity in the Mesopotamian plain. See Yaalon (1971), 181–7.

Hardie, L. A., J. P. Smoot, H. P. Eugster 1978. Saline lakes and their deposits: a sedimentological approach. See Matter & Tucker (1978), 7–42.

Hardisty, J. & R. J. S. Whitehouse 1988. Evidence for a new sand transport process from experiments on Saharan dunes. *Nature* **332**, 532–4.

Harlé, E. & J. Harlé 1919. Mémoire sur les dunes de Gascogne, avec observations sur le formation des dunes. Extrait du *Sec. de Géographie du Comité des Traveaux Historiques et Scientifiques, Bulletin* **34**, 145.

Harmse, J. T. & C. J. Swanevelder 1988. Further evidence for the applicability of Besler's "response diagram". *Zeitschrift für Geomorphologie* **32**, 471–9.

Harrington, C. D. & J. W. Whitney 1987. Scanning electron microscope method for rock-varnish dating. *Geology* **15**, 967–70.

Harris, S. A. 1957. The mechanical constitution of certain present-day Egyptian sand dunes. *Journal of Sedimentary Petrology* **27**, 421–34.

Harris, S. A. 1959. The classification of gilgaied soils: some evidence from northern Iraq. *Journal of Soil Science* **10**, 27–33.

Harrison, J. V. & N. L. Falcon 1938. An ancient landslip at Saidmarreh in southwestern Iran. *Journal of Geology* **46**, 296–309.

Harrison, S. P., S. E. Metcalfe, F. A. Street-Perrot, A. B. Pittock, C. N. Roberts, M. J. Salinger 1984. A climatic model of the last glacial/interglacial transition based on palaeotemperature and palaeohydrological evidence. See Vogel (1984), 21–34.

Hartmann, D. & C. Christiansen 1988. Settling-velocity distribution and sorting processes on a longitudinal dune. *Earth Surface Processes and Landforms* **13**, 649–56.

Harvey, A. M. 1978. Dissected alluvial fans in southeast Spain. *Catena* 5, 177–211.

Harvey, A. M. 1982. The role of piping in the development of badlands and gulley systems in south-east Spain. See *Badland geomorphology and piping*, R. Bryan & A. Yair (eds), 317–35. Norwich: Geobooks.

Harvey, A. M. 1984a. Geomorphological response to an extreme flood: a case from southeast Spain. *Earth Surface Processes and Landforms* 9, 267–79.

Harvey, A. M. 1984b. Aggradation and dissection sequences on Spanish alluvial fans: influence on morphological development. *Catena* **11**, 289–304.

Harvey, A. M. 1987. Alluvial fan dissection: relationships between morphology and sedimentation. See Frostick & Reid (1987), 87–103.

Harvey, C. P. D. & A. T. Grove 1982. A prehistoric source of the Nile. *Geographical Journal* **148**, 327–36.

Hassanein Bey 1925. *The lost oasis. London:* Butterworth.

Hassouba, H. & H. F. Shaw 1980. The occurrence and origin of sedimentary palygorskite in Quaternary sediments of the coastal plain of north-west Egypt. *Clay Minerals* **15**, 77–83.

Hastenrath, S. L. 1967. The barchans of the Arequipa region, southern Peru. *Zeitschrift für Geomorphologie* **11**, 300–31.

Hastenrath, S. L. 1978. Mapping and surveying – dune shape and multiannual displacement. See Lettau & Lettau (1978b), 74–88.

Hastenrath, S. L. 1987. The barchan dunes of southern Peru revisited. *Zeitschrift für Geomorphologie* **31**, 167–78.

Hastenrath, S. L. & J. Kutzbach 1985. Late Pleistocene climate and water budget of the South American altiplano. *Quaternary Research* **24**, 249–56.

Hastings, J. D. 1971. Sand streets. *Meteorological Magazine* **100**, 155–9.

Hastings, J. R. & R. M. Turner 1965. *The changing mile.* Tucson: University of Arizona Press.

Havholm, K. G. & G. Kocurek 1988. A preliminary study of the dynamics of a modern draa, Algodones, southeastern California. *Sedimentology* **35**, 649–69.

Hay, R. L. & R. J. Reeder 1978. Calcretes of Olduvai Gorge and the Ndolanya Beds of northern Tanzania. *Sedimentology* **25**, 649–73.

Haynes Jr, C. V. 1968. Geochronology of late-Quaternary alluvium. See *Means of correlation of Quaternary successions*, R. B. Morrison & H. E. Wright Jr (eds),

591–615. Salt Lake City: University of Utah Press.

Haynes Jr, C. V. 1989. Bagnold's barchan: a 57-year record of dune movement in the eastern Sahara and implications for dune origin and palaeoclimate since Neolithic times. *Quaternary Research* **32**, 153–67.

Hedin, S. 1903. *Central Asia and Tibet* [2 volumes]. New York: Scribners.

Hedin, S. 1904–5. *Scientific results of a journey in Central Asia* (Lithographic Institute of the General Staff of the Swedish Army, Stockholm), 1, The Tarim Basin.

Hefley, H. M. & R. Sidwell 1945. Geological and ecological observations on some High Plains dunes. *American Journal of Science* **243**, 361–76.

Hegde, K. T. M. & S. P. Sychantharong 1982. The Great Indian Desert. *Striae* **17**, 92–100.

Heidinga, H. A. 1984. Indications of severe drought during the 10th century AD from an inland dune area in the central Netherlands. *Geologie en Mijnbouw* **63**, 241–8.

Heine, K. 1981. Aride und pluviale Bedingungen Während der letzten Altzeit in der Südwest-Kalahari, Südliches Africa: ein Beiträge zür klimagenetischen Geomorphologie der Dünen, Pfanen und Täler. *Zeitschrift für Geomorphologie*, Supplementband 38, 1–37.

Heine, K. 1982. The main stages of the Late Quaternary evolution of the Kalahari region, southern Africa. See Vogel et al. (1982).

Heine, K. & M. A. Geyh 1984. Radiocarbon dating of speleothems from the Rötting cave, Namib desert, and palaeoclimatic implications. See Vogel (1984), 465–70.

Helldén, U. 1984. *Drought impact monitoring – a remote sensing study of desertification in Kordofan, Sudan.* Lunds Universitets Naturgeografiska Institution, Rapporter och Notiser 61.

Heller, F. & Liu Tung Sheng 1982. Magneto-stratigraphical dating of loess deposits in China. *Nature* **300**, 431–3.

Hendry, D. A. 1987. Silica and calcium carbonate replacement in tropical dune sands, SE India. See Frostick & Reid (1987), 309–20.

Hennessey, J. T., B. Kies, R. P. Gibbens, J. M. Tromble 1986. Soil sorting by forty-five years of wind erosion on a southern New Mexico range. *Soil Science Society of America, Journal* **50**, 391–4.

Hereford, R. 1984. Climate and ephemeral-stream processes: twentieth-century geomorphology and alluvial stratigraphy of the Little Colorado River, Arizona. *Geological Society of America, Bulletin* **95**, 654–68.

Hesp, P. A. 1979. Sand trapping ability of culms of marram grass (*Ammophila arenaria*). *Journal of the Soil Conservation Service of New South Wales* **35**, 156–60.

Hesp, P. A. 1981. The formation of shadow dunes. *Journal of Sedimentary Petrology* **51**, 101–12.

Hesp, P. A. 1983. Morphodynamics of incipient foredunes in New South Wales, Australia. See Brookfield & Ahlbrandt (1983), 325–43.

Hesp, P. A. 1988. Morphology, dynamics and internal stratification of some established foredunes in southeast Australia. *Sedimentary Geology* **55**, 17–42.

Hesp, P. A. 1989. A review of biological and geomorphological processes involved in the initiation and development of incipient foredunes. See Gimmingham et al. (1989), 181–215.

Hesp, P. A., R. Hyde, V. Hesp, Q. Zhengyu 1989. Longitudinal dunes can move sideways. *Earth Surface Processes and Landforms* **14**, 447–52.

Hesp, P. A., W. Illenberger, I. Rust, A. McLaughlan, R. Hyde 1988. Some aspects of transgressive dunefields and transverse ridge geomorphology and dynamics, south coast, South Africa. *Zeitschrift für Geomorphologie*, Supplementband 73, 111–23.

Hesp, P. A. & A. D. Short 1980. Dune forms of the Younghusband Peninsula, SE Australia. See Storrier & Stannard (1980), 65–6.

Hewes, L. I. 1948. A theory of surface cracks in mud and lava and resulting geometric relations. *American Journal of Science* **246**, 138–49.

Hewett, D. G. 1970. The colonisation of sand dunes after stabilisation with marram grass (*Ammophila arenaria*). *Journal of Ecology* **58**, 653–68.

Hickox, C. F. 1959. Formation of ventifacts in a moist temperate climate. *Geological Society of America, Bulletin* **70**, 1489–90.

Higgins, C. G. 1956. Formation of small ventifacts. *Journal of Geology* **64**, 506–16.

Higgins, G. M., S. Baig, R. Brinkman 1974. The sands of the Thal: wind-regimes and sand ridge formations. *Zeitschrift für Geomorphologie* **18**, 272–90.

Higgins, G. M., A. Mushtuq, R. Brinkman 1973. The Thal interfluve, Pakistan: geomorphology and depositional history. *Geologie en Mijnbouw* **52**, 147–55.

Hill, D. R. & J. M. Tompkin 1953. *General engineering geology of the Wray area, Colorado and Nebraska*. USGS Bulletin 1001.

Hillel, D. & N. H. Tadmor 1962. Water regime and vegetation in the central Negev highlands of Israel. *Ecology* **43**, 33–41.

Hills, E. S. 1939. The physiography of western Victoria. *Royal Society of Victoria, Proceedings* **51**, 297–320.

Hills, E. S. 1940. The lunette, a new landform of aeolian origin. *Australian Geographer* **3**, 15–21.

Hinkle, M. E. 1988. *Geochemical sampling in arid environments by the USGS*. USGS Circular 997.

Hobbs, W. H. 1917. The erosional degradational processes of deserts, with special reference to the origin of desert depressions. *Annals of the AAG* **7**, 25–60.

Hobbs, W. H. 1942. Wind – the dominant transportation agent within extramarginal zone to continental glaciers. *Journal of Geology* **50**, 556–9.

Högbom, I. 1923. Ancient inland dunes of north and middle Europe. *Geografiska Annaler* **5**, 113–243.

Hogg, S. E. 1982. Sheetfloods, sheetwash, sheetflow, or . . .? *Earth Science Reviews* **18**, 59–76.

Holland, T. H. 1911. The origin of desert salt deposits. *Geological Society of Liverpool, Proceedings* **11**, 227–50.

Holliday, V. T. 1989. The Blackwater Draw Formation (Quaternary): a 1.4 ± Ma record of eolian sedimentation and soil formation on the southern High Plains.

Geological Society of America, Bulletin **101**, 1598–1607.

Hollingworth, S. E. & J. E. Guest 1967. Pleistocene glaciation in the Atacama Desert, northern Chile. *Journal of Glaciology* **6**, 749–51.

Holm, D. A. 1953. Dome-shaped dunes of the central Nejd, Saudi Arabia. *19th International Geological Congress, Comptes Rendus* **7**, 107–12.

Holm, D. A. 1960. Desert geomorphology of the Arabian Peninsula. *Science* **132**, 1369–79.

Holzer, T. L. 1981. Preconsolidated stress of aquifer systems in areas of induced land subsidence. *Water Resources Research* **17**, 693–704.

Honeycutt, C. W., R. P. Heil, C. V. Cole 1990. Climatic and topographic relations of three Great Plains soils, I: soil morphology. *Soil Science Society of America, Journal* **54**, 469–75.

Hooghiemstra, H. & C. O. C. Agwu 1988. Changes in the vegetation and trade winds in equatorial northwest Africa 140,000–70,000 yr BP as deduced from two marine pollen records. *Palaeogeography, Palaeoclimatology, Palaeo-ecology* **66**, 173–213.

Hooke, R. LeB. 1967. Processes on arid-region alluvial fans. *Journal of Geology* **75**, 438–60.

Hooke, R. LeB. 1968. Steady-state relationships on arid-region alluvial fans in closed basins. *American Journal of Science* **266**, 609–29.

Hooke, R. LeB. 1972. Geomorphic evidence for late-Wisconsin and Holocene tectonic deformation, Death Valley, California. *Geological Society of America, Bulletin* **85**, 2073–98.

Hooke, R. LeB. 1987. Mass movement in semi-arid environments and the morphology of alluvial fans. See *Slope stability*, M. G. Anderson & K. S. Richards (eds), 505–29. Chichester, England: John Wiley.

Hooke, R. LeB. & W. L. Rohrer 1977. Relative erodibility of source-area rock types, as determined from second-order variations in alluvial-fan size. *Geological Society of America, Bulletin* **88**, 1177–82.

Hooke, R. LeB. & W. L. Rohrer 1979. Geometry of alluvial fans: effect of discharge and sediment size. *Earth Surface Processes* **4**, 147–66.

Hooke, R. LeB., H. Yang, P. W. Weiblen 1969. Desert varnish an electron microprobe study. *Journal of Geology* **77**, 275–88.

Horikawa, K. & H. W. Shen 1960. *Sand movement by wind action (on the characteristics of sand traps)*. US Army Corps of Engineers Technical Memoir 119.

Hörner, N. G. 1932. Lop Nor: topographical and geological summary. *Geografiska Annaler* **14**, 297–321.

Hörner, N. G. 1933. Geomorphic processes in continental basins of central Asia. *International Geological Congress Report, 16th Session* **2**, 721–35.

Horton, R. E. 1945. Erosional development of streams and their drainage basins: hydrophysical approach to quantitative morphology. *Geological Society of America, Bulletin* **56**, 275–370.

Hossain, D. & K. M. Ali 1988. Shear strength and consolidation characteristics of Obhor sabkha, Saudi

Arabia. *Quarterly Journal of Engineering Geology* **21**, 347–59.

Hotta, S., S. Kubota, S. Katori, K. Horikawa 1984. Blown sand on a wet sand surface. *19th Coastal Engineering Conference, proceedings*, 1265–81. New York: American Society of Civil Engineers.

Hötzl, J., F. Kramer, V. Maurin 1978. Quaternary sediments. See *Quaternary period in Saudi Arabia*, volume 1, S. Al-Sayari & J. G. Zötl (eds), 264–311. Wien: Springer.

Houbolt, J. J. H. C. 1968. Recent sediments in the southern bight of the North Sea. *Geologie en Mijnbouw* **47**, 245–73.

Howard, A. D. 1977. The effect of slope on the threshold of motion and its application to the orientation of wind ripples. *Geological Society of America, Bulletin* **88**, 853–6.

Howard, A. D. 1985. Interaction of sand transport with topography and local winds in the northern Peruvian coastal desert. See Barndorff-Nielsen et al. (1985), 511–44.

Howard, A. & R. Dolan 1981. Geomorphology of the Colorado River in the Grand Canyon. *Journal of Geology* **89**, 269–98.

Howard, A. D., R. C. Kochel, H. E. Holt 1988. *Sapping features of the Colorado Plateau*. NASA Publication SP-491.

Howard, A. D., J. B. Morton, M. Gad-el-Hak, D. B. Pierce 1977. *Simulation model of erosion and deposition on a barchan dune*. Contractor Report, NASA CR-2838, contract NGR-47-005-172, Washington DC.

Howard, A. D., J. B. Morton, M. Gad-el-Hak, D. B. Pierce 1978. Sand transport model of barchan dune equilibrium. *Sedimentology* **25**, 307–38.

Howard, A. D. & J. L. Walmsley 1985. Simulation model of isolated dune sculpture by wind. See Barndorff-Nielsen et al. (1985), 377–92.

Howe, G. M. *et al.* 1968. *Classification of world desert areas*. US Army Natick Laboratories Technical Report 69-38-ES.

Hoyt, J. H. 1966. Air and sand movement to the lee of dunes. *Sedimentology* **7**, 137–43.

Hsü, K. J., L. Montadert, D. Bernoulli, M. B. Cita, A. Erickson, R. E. Garrison, R. B, Kidd, F. Mélières, C. Muller, R. Wright 1977. History of the Mediterranean salinity crisis. *Nature* **267**, 399–403.

Hsu, Shih-Ang 1970. Coastal air-circulation system: observation and empirical model. *Monthly Weather Review* **98**, 487–509.

Hsu, Shih-Ang 1971. Measurement of shear stress and roughness length on a beach. *Journal of Geophysical Research* **76**, 2880–5.

Huffington, R. M. & C. C. Albritton 1941. Quaternary sands of the High Plains. *American Journal of Science* **239**, 325–38.

Huffman, G. G. & W. A. Price 1949. Clay dune formation near Corpus Christi, Texas. *Journal of Sedimentary Petrology* **19**, 118–27.

Hume, W. F. 1925. *Geology of Egypt, 1*. Cairo: Government Press.

Hunt, C. B. 1975. *Death Valley: geology, ecology, archaeology*. Berkeley: University of California.

Hunt, C. B. 1983. Physiographic overview of our arid lands in the western US. See *Origin and evolution of deserts*, S. G. Wells & D. R. Haragan (eds), 7–63. Albuquerque: University of New Mexico Press.

Hunt, C. B. & D. R. Mabey 1966. *Stratigraphy and structure, Death Valley, California*. USGS Professional Paper 496A.

Hunt, C. B. & A. L. Washburn 1960. *Salt features that simulate ground patterns found in cold climates*. USGS Professional Paper 400B.

Hunt, C. B. & A. L. Washburn 1966. Patterned ground, in USGS Professional Paper 494B, 104-33.

Hunt, J. C. R. & C. F. Barrett 1989. *Wind loss from stockpiles and transport of dust*. Warren Spring Laboratory Report, LR 688 (PA). London: HMSO.

Hunt, J. C. R. & P. Nalpanis 1985. Saltating and suspended particles over flat and sloping surfaces, I: modelling concepts. See Barndorff-Nielsen et al. (1985), 9–36.

Hunt, J. C. R. & W. H. Snyder 1980. Effects of stratification on the flow structure and turbulent diffusion around three-dimensional hills, part I: flow structure. *Journal of Fluid Mechanics* **96**, 671–704.

Hunter, R. B., E. M. Romney, A. Wallace 1982. Nitrate distribution in Mojave Desert soils. *Soil Science* **134**, 22–30.

Hunter, R. E. 1973. Pseudo-cross lamination formed by climbing adhesion structures. *Journal of Sedimentary Petrology* **43**, 1125–7.

Hunter, R. E. 1977a. Terminology of cross-stratified sedimentary layers and climbing-ripple structures. *Journal of Sedimentary Petrology* **47**, 697–706.

Hunter, R. E. 1977b. Basic types of stratification in small eolian dunes. *Sedimentology* **24**, 361–87.

Hunter, R. E. 1980. Quasi-planar adhesion stratification – an eolian structure formed in wet sand. *Journal of Sedimentary Petrology* **50**, 263–6.

Hunter, R. E. 1985. A kinematic model for the structure of lee-side deposits. *Sedimentology* **32**, 409–422.

Hunter, R. E. & B. M. Richmond 1988. Daily cycles in coastal dunes. *Sedimentary Geology* **55**, 43–67.

Hunter, R. E., B. M. Richmond, T. R. Alpha 1983. Storm-controlled oblique dunes of the Oregon coast. *Geological Society of America, Bulletin* **94**, 1450–65.

Huntington, E. 1907. *The pulse of Asia*. Boston: Houghton Mifflin.

Huntington, E. 1914. *The climatic factor as illustrated in arid America*. Carnegie Institution of Washington, Publication 192.

Hurault, J. 1966. Étude photo-aérienne de la tendance à la remobilisation des sables éoliens sur la rive nord du Lac Tchad. *Institut Français du Petrole, Revue* **21**, 1837–46.

Hutton, J. T. 1968. The distribution of the more soluble chemical elements associated with soils indicated by analysis of rainwater, soils and plants. *9th International Congress of Soil Science, Transactions* **4**, 312–3.

Hutton, J. T. & J. C. Dixon 1981. The chemistry and mineralogy of some South Australian calcretes and associated soft carbonates and their dolomitization. *Geological Society of Australia, Journal* **28**, 71–9.

Hutton, J. T., C. R. Twidale, A. R. Milnes 1978. Characteristics and origin of some Australian silcretes. See Langford-Smith (1978b), 19–39.

Hyde, R. & R. Wasson 1983. Radiative and meteorological control on the movement of sand at Lake Mungo, New South Wales, Australia. See Brookfield & Ahlbrandt (1983), 311–24.

Ibrahim, F. N. 1982. The Aswan High Dam – serious human interference with the ecosystem. *Development and Cooperation* **6**, 12–15.

Idso, S. B. 1974. Tornado or dust devil: enigma of desert whirlwinds. *American Scientist* **62**, 530–41.

Idso, S. B. 1977. A note on some recently proposed mechanisms of genesis of deserts. *Royal Meteorological Society, Quarterly Journal* **103**, 369–70.

Illenberger, W. K. & I. C. Rust 1988. A sand budget for the Alexandria coastal dunefield, South Africa. *Sedimentology* **35**, 513–21.

Imbrie, J. & K. P. Imbrie 1979. *Ice ages: solving the mystery*. London: Methuen.

Inbar, M. 1982. Spatial and temporal aspects of man-induced changes in the hydrological and sedimentological regime of the Upper Jordan River. *Israel Journal of Earth Sciences* **31**, 53–66.

Inglis, D. R. 1965. Particle sorting and stone migration by freezing and thawing. *Science* **148**, 1616–17.

Inman, D. L., G. C. Ewing, J. B. Corless 1966. Coastal sand dunes of Guerrero Negro, Baja California, Mexico. *Geological Society of America, Bulletin* **77**, 787–802.

Iversen, J. D. 1983. Saltation threshold and deposition rate modelling. See Brookfield & Ahlbrandt (1983), 103–14.

Iversen, J. D. 1986a. Small-scale wind-tunnel modelling of particle transport – Froude number effect. See Nickling (1986), 19–33.

Iversen, J. D. 1986b. Saltation threshold mechanics. See El-Baz & Hassan (1986), 344–60.

Iversen, J. D., R. Greeley, J. R. Marshall, J. Pollack 1987. Aeolian saltation threshold: effect of density ratio. *Sedimentology* **34**, 699–706.

Iversen, J. D., R. Greeley, J. B. Pollack 1976a. Wind-blown dust on Earth, Mars and Venus. *Journal of Atmospheric Sciences* **33**, 2425–9.

Iversen, J. D., R. Greeley, B. R. White, J. B. Pollack 1976b. The effect of vertical distortion in the modelling of sediment phenomena: Martian crater wake streaks. *Journal of Geophysical Research* **81**, 4846–56.

Iversen, J. D., J. B. Pollack, R. Greeley, B. R. White 1976c. Saltation threshold on Mars: the effect of inter-particle force, surface roughness, and low atmospheric density. *Icarus* **29**, 381–93.

Iversen, J. D., W. P. Wang, K. R. Rasmussen, H. E. Mikkelsen, R. N. Leach 1991. Roughness element effect on local and universal saltation transport. *Acta Mechanica* **2**, 65–76.

Iversen, J. D. & B. R. White 1982. Saltation threshold on Earth, Mars and Venus. *Sedimentology* **29**, 111–19.

Iverson, R. M. 1980. Processes of accelerated pluvial erosion on desert hillslopes modified by vehicular activity. *Earth Surface Processes* **5**, 369–88.

Ives, R. L. 1946. Desert ripples. *American Journal of Science* **244**, 492–501.

Jachens, R. C. & T. L. Hölzer 1982. Differential compaction mechanism for earth-fissures near Casa Grande, Arizona. *Geological Society of America, Bulletin* **93**, 998–1012.

Jackson, P. S. & J. C. R. Hunt 1975. Turbulent wind flowover a low hill. *Royal Meteorological Society, Quarterly Journal* **101**, 929–55.

Jackson, R. G. 1976. Sedimentological and fluid dynamic implications of turbulent bursting phenomena in geophysical flows. *Journal of Fluid Mechanics* **77**, 531–60.

Jaeger, F. 1921. *Deutsch Sudwestafrika*. Breslau: Wissenschaftliche Gesellschaft.

Jäkel, D. 1974. Forschungsstation, Bardai, FU-Geologen in der Zentralsahara. *Pressedienst Wissenschaft FU Berlin* **5**.

Jäkel, D. 1980. Die Bildung von Barchanen in Faya-Largeau, Republique du Tchad. *Zeitschrift für Geomorphologie* **24**, 141–59.

Jänicke, R. 1979. Monitoring and critical review of the estimated source strength of mineral dust from the Sahara. See Morales (1979), 233–42.

Jarrel, W. M. & R. A. Virginia 1990. Soil cation accumulation in a mesquite woodland: sustained production and long-term estimates of water use and nitrogen fixation. *Journal of Arid Environments* **18**, 51–8.

Jauzein, A. 1958. Les formes d'accumulation éolienne liées à des zones inondables en Tunisie séptentrionale. *Academie des Sciences à Paris, Comptes Rendus* **247**, 2396–9.

Jawad, A. A. & R. A. Al-Ani 1983. Sedimentological and geomorphological study of sand dunes in the western desert of Iraq. *Journal of Arid Environments* **6**, 13–32.

Jennings, J. N. 1955. Le complexe des sebkhas: un commentaire provenant des antipodes. *Revue de Géomorphologie Dynamique* **6**, 69–72.

Jennings, J. N. 1957. Coastal dune lakes as exemplified from King Island, Tasmania. *Geographical Journal* **123**, 59–70.

Jennings, J. N. 1975. Desert dunes and estuarine fill in the Fitzroy Estuary (NW Australia). *Catena* **2**, 215–62.

Jennings, J. N. 1983. The disregarded karst of the arid and semiarid domain. *Karstologia* **1**, 61–73.

Jennings, J. N. 1985. *Karst geomorphology*, revised edition. Oxford: Basil Blackwell.

Jenny, H. 1941. *Factors of soil formation*. New York: McGraw-Hill.

Jensen, J. L., K. Rasmussen, M. Sorensen, B. B. Willetts 1984. *The Hanstholm Experiment, 1982: sand grain saltation on a beach*. Department of Theoretical

Statistics, Institute of Mathematics, University of Aarhus, Research Reports 125.

Jensen, J. L. & M. Sorensen 1986. Estimation of some aeolian saltation transport parameters: a re-analysis of Williams's data. *Sedimentology* **33**, 547–58.

Jensen, N-O. & O. Zeman 1985. Perturbations to mean wind and turbulence in flow over topographic forms. See Barndorff-Nielsen et al. (1985), 351–68.

Jensen, N-O. & O. Zeman 1990. Numerical modelling of transverse dune forms. In *Extended abstracts, NATO Advanced Workshop on sand, dust and soil in their relation to aeolian and littoral processes, 14–18 May, 1990, Sandbjerg, Denmark.*

Jerwood, L. C., D. A. Robinson, R. B. G. Williams 1987. Frost and salt weathering as periglacial processes: the results and implications of some laboratory experiments. See *Periglacial processes and landforms in Britain and Ireland*, J. Boardman (ed.), 135–43. Cambridge: Cambridge University Press.

Jessup, R. W. 1960. The Stony Tableland soils of the southeastern portion of the Australian arid zone and their evolutionary history. *Journal of Soil Science* **11**, 188–96.

Jessup, R. W. 1961. A Tertiary–Quaternary pedological chronology for the southeastern portion of the Australian arid zone. *Journal of Soil Science* **12**, 199–213.

Johannessen, C. L., J. J. Feiereisen, A. N. Wells 1982. Weathering of ocean cliffs by salt expansion in a mid-latitude coastal environment. *Shore and Beach*, 26–34.

Johnson, A. J. & D. M. O'Brien 1973. A study of an Oregon sea breeze event. *Journal of Applied Meteorology* **12**, 1267–83.

Johnson, D. L. 1977. Late Quaternary climate of coastal California: evidence for an ice age refugium. *Quaternary Research* **8**, 154–79.

Johnson, D. W. 1932. Rock planes in arid regions. *Geographical Review* **22**, 656–65.

Johnson, D. W. 1942. *The origin of the Carolina Bays.* New York: Columbia University Press.

Johnstone, W. M. & J. C. Wilkinson 1960. Some geographical aspects of Qatar. *Geographical Journal* **126**, 442–50.

Jones, A. R. 1987. An evaluation of satellite thematic mapper imagery for geomorphological mapping in arid and semi-arid environments. See Gardiner (1987), 343–58.

Jones, C. R. 1982. The Kalahari of southern Africa. *Striae* **17**, 20–34.

Jones, D., M. J. Wilson, J. M. Tait 1980. Weathering of a basalt by *Pertusaria corallina*. *Lichenologist* **12**, 277–89.

Jones, D. K. C., R. U. Cooke, A. Warren 1986. Geomorphological investigation, for engineering purposes, of blowing sand and dust hazard. *Quarterly Journal of Engineering Geology* **19**, 251–70.

Jones, J. R., H. Weier, P. R. Considine 1988. Geology and hydrogeology of the pre-dune sand deposits of the Wahiba Sands, Sultanate of Oman. *Journal of Oman Studies*, Special Report 3, 61–73.

Judson, S. 1950. Depressions of the northern portion of the southern High Plains of eastern New Mexico. *Geological Society of America, Bulletin* **61**, 253–74.

Junge, C. 1979. The importance of mineral dust as an atmospheric constituent. See Morales (1979), 49–60.

Jungerius, P. D. 1984. A simulation model of blowout development. *Earth Surface Processes and Landforms* **9**, 509–12.

Jungerius, P. D. & F. van der Meulen 1988. Erosion processes in a dune landscape along the Dutch coast. *Catena* **15**, 217–28.

Jungerius, P. D. & F. van der Meulen 1989. The development of dune blowouts, as measured with erosion pins and sequential air photos. *Catena* **16**, 369–76.

Jungerius, P. D., A. J. T. Verheggen, A. J. Wiggers 1981. The development of blowouts in "De Blink", a coastal dune area near Noordwijkerhout, The Netherlands. *Earth Surface Processes and Landforms* **6**, 375–96.

Jutson, J. T. 1918. The influence of salts in rock weathering in sub-arid Western Australia. *Royal Society of Victoria, Proceedings* **30**, 165–72.

Kahle, C. F. 1977. Origin of subaerial Holocene calcareous crust: role of algae, fungi and sparmicritisation. *Sedimentology* **24**, 413–35.

Kaimal, J. C. & J. A. Businger 1979. Case studies of a convective plume and a dust devil. *Journal of Applied Meteorology* **9**, 612–20.

Kaiser, K. 1970. Beobachtunger über Fließmarken an Leeseitigen Barchan-Hänger. *Kurt Kaiser Festschrift, Sonderband Kölner Geographischen Arbeiten*, 65–71.

Kaldi, J., D. H. Krinsley, D. Lawson 1978. Experimentally produced aeolian surface textures on quartz sand grains from various environments. See Whalley (1978), 261–74.

Kalu, A. E. 1979. The African dust plume: its characteristics and propagation across West Africa in winter. See Morales (1979), 95–118.

Kamra, A. K. 1972. Measurements of electrical properties of dust storms. *Journal of Geophysical Research* **77**, 5856–69.

Kamra, A. K. 1977. Effect of dust-raising winds on the atmospheric electric field. *Electrical processes in the atmosphere* (5th International Conference on Atmospheric Electricity, Proceedings 1974), 168–74. Darmstadt: Steinkopf.

Kappus, U., J. M. Bleck, S. H. Blair 1978. Rainfall frequencies for the Persian Gulf Coast of Iran. *Hydrological Sciences Bulletin* **23**, 119–29.

Kar, A. 1984. Morphology and evolution of some sandstone pediments in Rajasthan Desert. *Geographical Review of India* **46**, 67–74.

Kar, A. 1987. Origin and transformation of longitudinal sand dunes in the Indian Desert. *Zeitschrift für Geomorphologie* **31**, 311–37.

Kar, A. 1990. The megabarchanoids of the Thar: their environment, morphology and relationship with longitudinal dunes. *Geographical Journal* **156**, 51–61.

Kästner, M. & W. Kästner 1978. Saharasaubwolke über dem Nordatlantik beobachtet mit METEOSAT Bildern. *Promet* **8**, 59–61.

Kedar, Y. 1967. *The ancient agriculture in the Negev* [in Hebrew]. Jerusalem: Bialik Foundation.

Keller, C. 1946. *El Departamento de Arica*. Min. Econ. y Com., Santiago de Chile.

Kelletat, D. 1980. Studies in the age of honeycombs and tafoni features. *Catena* **7**, 317–25.

Kelly, R. D. 1984. Horizontal roll and boundary-layer inter-relationships observed over Lake Michigan. *Journal of Atmospheric Sciencess* **41**, 1816–26.

Kelts, K. & M. Shahrabi 1986. Holocene sedimentology of hypersaline Lake Urmia, northwestern Iran. *Palaeogeography, Palaeoclimatology, Palaeoecology* **54**, 105–30.

Kennedy, J. F. 1969. The formation of sediment ripples, dunes and antidunes. *Annual Review of Fluid Mechanics* **1**, 147–69.

Kenyon, N. H. 1970. Sand ribbons of European tidal seas. *Marine Geology* **9**, 25–35.

Kerr, A., B. J. Smith, B. Whalley, J. P. McGreevy 1984. Rock temperatures from southeast Morocco and their significance for experimental rock-weathering studies. *Geology* **12**, 306–9.

Kerr, R. C. & J. O. Nigra 1952. Eolian sand control. *American Association of Petroleum Geologists, Bulletin* **36**, 1541–73.

Kesel, R. H. 1973. Problèmes de recherches: les éléments du modèle des inselbergs: definition et sythèse. *Revue de Geomorphologie Dynamique* **22**, 97–108.

Kesel, R. H. 1977. Some aspects of the geomorphology of inselbergs in central Arizona, USA. *Zeitschrift für Geomorphologie* **21**, 119–46.

Kesseli, J. E. & C. B. Beaty 1959. *Desert flood conditions in the White Mountains of California and Nevada*. Headquarters of Quartermaster Research Engineering Command, US Army, Environmental Protection Research Division, Technical Report, EP-108.

Keyes, C. R. 1909. Base level of eolian erosion. *Journal of Geology* **17**, 659–63.

Keyes, C. R. 1912. Deflative scheme of the geographic cycle in an arid climate. *Geological Society of America, Bulletin* **23**, 537–62.

Kezao, C. & J. M. Bowler 1986. Late Pleistocene evolution of salt lakes in Qaidam basin, Qinghai Province, China. *Palaeogeography, Palaeoclimatology, Palaeoecology* **54**, 87–104.

Khalaf, F. I. 1989. Textural characteristics and genesis of aeolian sediments in the Kuwait desert. *Sedimentology* **36**, 253–71.

Khalaf, F. I., D. Al-Ajmi, N. Saleh, M. Al-Hashash 1990. Aeolian processes and sand encroachment problems in Kuwait. In *Extended abstracts, NATO Advanced Workshop on sand, dust and soil in their relation to aeolian and littoral processes*. 14–18 May, 1990, Sandbjerg, Denmark.

Khalaf, F. I. & M. Z. Al-Hashash 1983. Aeolian sedimentation in the north-western part of the Arabian Gulf. *Journal of Arid Environments* **6**, 319–32.

Khalaf, F. I., I. M. Gharib, M. Z. Al-Hashash 1984. Types and characteristics of the recent surface deposits of Kuwait, Arabian Gulf. *Journal of Arid Environments* **7**, 9–33.

Khan, I. H. & S. I. Hasnain 1981. Engineering properties of sabkha soils in the Benghazi plain and construction problems. *Engineering Geology* **17**, 175–83.

Khawlie, M. R. & H. I. Hassanain 1984. Engineering geology of the Hammana landslides, Lebanon. *Quarterly Journal of Engineering Geology* **17**, 137–48.

Kidson, C., R. L. Collin, N. W. T. Chisholm 1989. Surveying a major dune system – Braunton Burrows, northwest Devon. *Geographical Journal* **155**, 94–105.

Killian, Ch. 1945. Un cas très particulier d'humification au désert due à l'activité de micro-organismes dans le sol des "nebkas". *Revue Canadienne de Biologie* **4**, 3–36.

Killigrew, L. P. & R. J. Gilkes 1974. Development of playa lakes in southwestern Australia. *Nature* **247**, 454–5.

Kindle, E. M. 1917. Some factors affecting the development of mud-cracks. *Journal of Geology* **25**, 135–44.

King, C. A. M. 1972. *Beaches and coasts,* 2nd edition. London: Edward Arnold.

King, D. 1956. The Quaternary stratigraphic record at Lake Eyre North and the evolution of existing topographic forms. *Royal Society of South Australia, Transactions* **79**, 97–103.

King, D. 1960. The sand-ridge deserts of Australia and related aeolian landforms of the Quaternary arid cycles. *Royal Society of South Australia, Transactions* **83**, 99–108.

King, L. C. 1936. Wind-faceted stones from Marlborough, New Zealand. *Journal of Geology* **44**, 201–13.

King, L. C. 1939. *South African scenery*. Edinburgh: Oliver & Boyd.

King, L. C. 1951. The geomorphology of the eastern and southern districts of Southern Rhodesia. *Geological Society of South Africa, Transactions* **54**, 33–64.

King, L. C. 1953. Canons of landscape evolution. *Geological Society of America, Bulletin* **64**, 721–52.

King, L. C. 1962. *The morphology of the Earth.* Edinburgh: Oliver & Boyd.

King, T. J. & S. R. J. Woodell 1973. The causes of regular pattern in a desert perennial. *Journal of Ecology* **61**, 761–5.

King, W. H. J. 1916. The nature and formation of sand ripples and dunes. *Geographical Journal* **47**, 189–209.

King, W. H. J. 1918. Study of a dunebelt. *Geographical Journal* **51**, 16–33.

Kirk, L. G. 1952. Trails and rocks observed on a playa in Death Valley National Monument, California. *Journal of Sedimentary Petrology* **22**, 173–81.

Kirkby, A. & M. J. Kirkby 1974. Surface wash at the semi-arid break of slope. *Zeitschrift für Geomorphologie* **21**, 151–76.

Kirkby, A. & M. J. Kirkby 1976. Geomorphic processes and the surface survey of archaeological sites in semi-arid areas. See *Geoarchaeology: Earth science and the past,*

D. Davidson & M. Shackley (eds), 229–53. London: Duckworth.

Kirkby, M. J. & R. J. Chorley 1967. Throughflow, overland flow and erosion. *Bulletin of the International Association of Scientific Hydrology* 12, 5–21.

Kiya, M. & K. Sasaki 1983. Structure of turbulent separation bubble. *Journal of Fluid Mechanics* 137, 83–113.

Klammer, C. 1982. Die Palaeowüste des Pantannal von Mato Grosso und die Pleistozane Klimageschichte der Brazilianischen Randtropen. *Zeitschrift für Geomorphologie* 26, 393–416.

Klappa, C. F. 1978. Biolithogenesis of *Microcodium*. *Sedimentology* 25, 489–522.

Klappa, C. F. 1979. Lichen stromatolites: criterion for subaerial exposure and a mechanism for the formation of laminar calcretes (caliche). *Journal of Sedimentary Petrology* 49, 387–400.

Klappa, C. F. 1980. Brecciation textures and tepee structures in Quaternary calcrete (caliche) profiles from eastern Spain: the plant factor in their formation. *Geological Journal* 15, 81–9.

Klein, M. 1984. Weathering rates of limestone tombstones measured in Haifa, Israel. *Zeitschrift für Geomorphologie* 28, 105–11.

Klein, R. G. 1980. Environmental and ecological implications of large mammals from Upper Pleistocene and Holocene sites in southern Africa. *Annals of the South African Museum* 81, 223–83.

Kline, S. J., W. C. Reynolds, F. A. Schraub, P. W. Runstadler 1967. The structure of turbulent boundary layers. *Journal of Fluid Mechanics* 30, 741–73.

Knauss, K. G. & T. L. Ku 1980. Desert varnish: potential for age-dating via uranium-series isotopes. *Journal of Geology* 88, 95–100.

Knetsch, G. & M. Yallouze 1955. Remarks on the origin of the Egyptian oasis depressions. *Société Géographique d'Egypte, Bulletin* 28, 21–33.

Knighton, D. 1984. *Fluvial forms and processes*. London: Edward Arnold.

Knott, P. 1979. *The structure and pattern of dune-forming winds*. PhD dissertation, University of London [2 volumes].

Knottnerus, D. J. C. 1980. Relative humidity of the air and critical wind velocity in relation to erosion. See *Assessment of erosion*, M. De Boodt & M. D. Gabriels (eds), 531–9. Chichester, England: John Wiley.

Kobayshi, S. & T. Ishihara 1979. Interaction between wind and snow surface. *Boundary-layer Meteorology* 16, 35–47.

Kochel, R. C. & R. A. Johnson 1984. Geomorphology, sedimentology and depositional processes in humid-temperate alluvial fans in central Virginia, USA. See *Sedimentology of gravels and conglomerates*, Canadian Society Petrology Geology Memoir 10, E. H. Koster & R. J. Steel (eds), 109–22.

Kocurek, G. 1981a. Erg reconstruction: Entrada Sandstone (Jurassic) of northern Utah and Colorado. *Palaeogeography, Palaeoclimatology, Palaeoecology* 36, 125–53.

Kocurek, G. 1981b, Significance of interdune deposits and bounding surfaces in eolian dune sands. *Sedimentology* 28, 754–80.

Kocurek, G. 1986. Origins of low-angle stratification in eolian deposits. See Nickling (1986), 176–93.

Kocurek, G. 1988. First-order and super bounding surfaces in eolian sequences – bounding surfaces revisited. *Sedimentary Geology* 56, 193–206.

Kocurek, G. & R. H. Dott 1981. Distinctions and uses of stratification types in the interpretation of eolian sand. *Journal of Sedimentary Petrology* 51, 579–95.

Kocurek, G. & G. Fielder 1982. Adhesion structures. *Journal of Sedimentary Petrology* 52, 1229–42.

Kocurek, G. & J. Nielson 1986. Conditions favourable for the formation of warm-climate eolian sand sheets. *Sedimentology* 33, 795–816.

Kocurek, G., M. Townsley, E. Yeh, M. Sweet, K. Havholm 1990. Dune and dune-field development stages on Padre Island, Texas: effects of lee airflow and sand saturation levels and implications for interdune deposition. *Acta Mechanica. In Extended abstracts, NATO Advanced Workshop on sand, dust and soil in their relation to aeolian and littoral processes.* 14–18 May, 1990, Sandbjerg, Denmark.

Kolla, V. & P. C. Biscaye 1977. Distribution and origin of quartz in the sediments of the Indian Ocean. *Journal of Sedimentary Petrology* 47, 642–9.

Kolla, V., P. E. Biscaye, A. F. Hanley 1979. Distribution of quartz in Lower Quaternary Atlantic sediments in relation to climate. *Quaternary Research* 11, 261–77.

Kolm, K. E. 1982. *Predicting surface wind characteristics of southern Wyoming from remote sensing and eolian geomorphology.* Geological Society of America, Special Paper 192, 25–53.

Kolm, K. E. 1985. Predicting surface wind characteristics of southern Wyoming from remote sensing and eolian geomorphology. See Barndorff-Nielsen et al. (1985), 421–81.

Kosmowska-Suffczynska, D. 1980. Dune forms in Sebkhet El-Muh in Palmyra Region [in Polish, English summary]. *Prace i Studia Geograficzne* 2, 177–88.

Kostaschuk, R. A., G. M. MacDonald, P. E. Putnam 1986. Depositional processes and alluvial-fan–drainage-basin morphometric relationships near Banff, Alberta, Canada. *Earth Surface Processes and Landforms* 11, 471–84.

Koster, E. A. 1968. De invloed van markebossen op de vorming van zeer hoge stuifzanderuggen ("Randwallen") op de veluwe. *Boor en Spade* 16, 66–73.

Koster, E. A. 1978. *De stuifzanden van Veluwe; een fysisch–feografische studie*. Publicaties van het Fysisch–Geografish en Bodemkundig Laboratorium van de Universiteit van Amsterdam, 27.

Koster, E. A. 1988. Ancient and modern cold-climate aeolian sand deposition: a review. *Journal of Quaternary Science* 3, 69–83.

Koster, E. A. & J. W. A. Dijkmans 1988. Niveo-aeolian deposits and denivation forms with special reference to the Great Kobuk sand dunes, northwestern Alaska. *Earth Surface Processes and Landforms* **13**, 153–70.

Kotwicki, V. 1986. *Floods of Lake Eyre.* Adelaide: South Australia Government Printer.

Krinitzky, E. L. 1949. Origin of pimple mounds. *American Journal of Science* **247**, 706–14.

Krinsley, D. H. 1970. *A geomorphological and palaeo-climatological study of the playas of Iran.* USGS Final Scientific Report, Contract, PRO CP70-800, US Air Force Cambridge Research Laboratories, Hanscom Field, Bedford Massachusetts [2 volumes].

Krinsley, D. H. 1978. The present state and future prospects of environmental discrimination by scanning electron microscopy. See Whalley (1978), 169–79.

Krinsley, D. H., P. F. Friend, R. Klimentidis 1976. Eolian transport textures on the surfaces of sand grains of early Triassic age. *Geological Society of America, Bulletin* **87**, 130–2.

Krinsley, D. H. & I. J. Smalley 1972. Sand. *American Scientist* **60**, 286–91.

Krinsley, D. H. & I. J. Smalley 1973. Shape and nature of small sedimentary quartz particles. *Science* **180**, 1277–9.

Krishnan, M. S. 1952. Geological history of Rajasthan and its relation to present-day conditions. *Proceedings of symposium on the Rajputana Desert*, 19–31.

Krumbein, W. E. 1969. Über den Einfluss der Mikroflora auf die exogene Dynamik (Verwitterung und Krusten-bildung). *Geologische Rundschau* **58**, 333–65.

Krumbein, W. E. & K. Jens 1981. Biogenic rock varnishes of the Negev Desert (Israel): an ecological study of iron and manganese transformation by cyano-bacteria and fungi. *Oecologia* **50**, 25–28.

Ku, T-L., W. B. Bull, S. T. Freeman, K. G. Knauss 1979. Th^{230}–U^{234} dating of pedogenic carbonates in gravelly desert soils of Vidal Valley, southeastern California. *Geological Society of America, Bulletin* **90**, 1063–73.

Kuenen, Ph. 1960. Experimental abrasion, 4: eolian action. *Journal of Geology* **68**, 427–49.

Kuenen, Ph. 1969. The origin of quartz silt. *Journal of Sedimentary Petrology* **39**, 1631–3.

Kuenen, Ph. & W. G. Perdok 1962. Experimental abrasion, 5: frosting and defrosting of quartz grains. *Journal of Geology* **70**, 648–58.

Kuettner, J. P. 1971. Cloud bands in the Earth's atmo-sphere. *Tellus* **23**, 404–26.

Kuhlman, H. 1958. Quantitative measurements of aeolian sand transport. *Geografisk Tidsskrift* **57**, 51–74.

Kurbanmuradov, K. 1968. Types of old overgrown aeolian sands of the western part of central Karakum [in Russian]. *Problems of Desert Development* **6**, 57–62.

Kutzbach, J. E. 1981. Monsoon climate of the early Holocene: climatic experiment with the Earth's orbital parameters for 9,000 years ago. *Science* **214**, 59–61.

Kutzbach, J. E. & H. E. Wright 1985. Simulation of the climate of 18,000 years BP: results for the North Amer-ican / North Atlantic / European sector and comparison with the geologic record of North America. *Quaternary Science Reviews* **4**, 147–87.

Kwaad, F. J. P. M. 1970. Experiments on the granular disintegration of granite by salt action. In *From field to laboratory*, 67–80 (*Fys. Geogr. en Bodemkundig Lab.*, **16**).

Lachenbruch, A. H. 1962. *Mechanisms of thermal contrac-tion cracks and ice-wedge polygons in permafrost.* Geological Society of America, Special Paper 70.

Lai, R. J. & J. Wu 1978. *Wind erosion and deposition along a coastal sand dune.* Sea Grant Program, Univer-sity of Delaware, Report DEL-SG-10-78.

Laity, J. E. 1987. Topographic effects on ventifact develop-ment, Mojave Desert, California. *Physical Geography* **8**, 113–32.

Laity, J. E. & M. C. Malin 1985. Sapping processes and the development of theatre-headed valley networks on the Colorado Plateau. *Geological Society of America, Bulletin* **96**, 203–17.

LaMarche, V. C. 1967. *Speroidal weathering of thermally metamorphosed limestone and dolomite, White Mountains, California.* USGS Professional Paper 575C, 32–7.

Lambrick, H. T. 1964. *Sind, a general introduction.* Hyderabad.

Lamplugh, G. W. 1902. Calcrete. *Geological Magazine* **9**, 75.

Lancaster, N. 1978a. The pans of the southern Kalahari, Botswana. *Geographical Journal* **144**, 81–98.

Lancaster, N. 1978b. Composition and formation of south-ern Kalahari pan margin dunes. *Zeitschrift für Geomorphologie* **22**, 148–69.

Lancaster, N. 1979. *Quaternary environments in the arid zone of southern Africa.* Department of Geography and Environmental Studies, University of Witwatersrand, Occasional Paper 22.

Lancaster, N. 1980a. Dune systems and palaeoenviron-ments in southern Africa. *Palaeologica Africana* **23**, 185–9.

Lancaster, N. 1980b. The formation of seif dunes from barchans – supporting evidence for Bagnold's model from the Namib Desert. *Zeitschrift für Geomorphologie* **NF24**, 160–7.

Lancaster, N. 1981a. Palaeoenvironmental implication of fixed dune systems in southern Africa. *Palaeogeography, Palaeoclimatology, Palaeoecology* **33**, 327–46.

Lancaster, N. 1981b. Aspects of the morphometry of linear dunes of the Namib Desert. *South African Journal of Science* **77**, 366–8.

Lancaster, N. 1981c. Grain size characteristics of Namib Desert linear dunes. *Sedimentology* **28**, 115–22.

Lancaster, N. 1982a. Dunes on the Skeleton Coast, South-west Africa/Namibia: geomorphology and grain size relat-ionships. *Earth Surface Processes and Landforms* **7**, 575–87.

Lancaster, N. 1982b. Linear dunes. *Progress in Physical Geography* **6**, 476–504.

Lancaster, N. 1982c. Spatial variations in linear dune morphology and sediments in the Namib sand sea. *Palaeoecology of Africa* **15**, 173–82.

Lancaster, N. 1983a. Linear dunes of the Namib sand sea. *Zeitschrift für Geomorphologie*, Supplementband 45, 27–49.

Lancaster, N. 1983b. Controls on dune morphology in the Namib Sand Sea. See Brookfield & Ahlbrandt (1983), 261–90.

Lancaster, N. 1984a. Aridity in southern Africa: age, origins and expression in landforms and sediments. See Vogel (1984), 433–44.

Lancaster, N. 1984b. Characteristics and occurrence of wind erosion features in the Namib Desert. *Earth Surface Processes and Landforms* 9, 469–78.

Lancaster, N. 1984c. Aeolian sediments, processes and landforms. *Journal of Arid Environments* 7, 249–54.

Lancaster, N. 1985a. Variations in wind velocity and sand transport on the windward flanks of desert sand dunes. *Sedimentology* 32, 581–93.

Lancaster, N. 1985b. Winds and sand movements in the Namib sand sea. *Earth Surface Processes and Landforms* 10, 607–19.

Lancaster, N. 1986a. Grain-size characteristics of linear dunes in the southwest Kalahari. *Journal of Sedimentary Petrology* 56, 395–400.

Lancaster, N. 1986b. Dynamics of deflation hollows in the Elands Bay area, Cape Province, South Africa. *Catena* 13, 139–53.

Lancaster, N. 1987a. Variations in wind velocity and sand transport on the windward flanks of desert sand dunes – reply. *Sedimentology* 43, 516–20.

Lancaster, N. 1987b. Formation and reactivation of dunes in the southwestern Kalahari: palaeoclimatic implications. *Palaeoecology of Africa* 18, 103–10.

Lancaster, N. 1987c. *Variations in surface shear stress over a small aeolian dune*. Reports of Program for Planetary Geology and Geophysics.

Lancaster, N. 1987d. Dunes of the Gran Desierto sand sea, Sonora, Mexico. *Earth Surface Processes and Landforms* 12, 277–88.

Lancaster, N. 1988a. On desert sand seas. *Episodes* 11, 12–17.

Lancaster, N. 1988b. The development of large aeolian bedforms. *Sedimentary Geology* 55, 69–90.

Lancaster, N. 1988c. Controls of eolian dune size and spacing. *Geology* 16, 972–5.

Lancaster, N. 1989a. *The Namib sand sea – dune forms, processes and sediments*. Rotterdam: Balkema.

Lancaster, N. 1989b. The dynamics of star dunes: an example from the Gran Desierto, Mexico. *Sedimentology* 36, 273–90.

Lancaster, N. 1989c. Star dunes. *Progress in Physical Geography* 13, 67–91.

Lancaster, N. 1991. The orientation of dunes with respect to sand-transporting winds: a test of Rubin and Hunter's gross bedform-normal rule. *Acta Mechanica* 2, 89–102.

Lancaster, N. & R. Greeley 1990. Sediment volume in the north polar sand seas of Mars. *Journal of Geophysical Research* 95, 921–7.

Lancaster, N., R. Greeley, P. R. Christensen 1987. Dunes of the Gran Desierto sand sea, Sonora, Mexico. *Earth Surface Processes and Landforms* 12, 277–88.

Lancaster, N. & C. D. Ollier 1983. Sources of sand for the Namib Sand Sea. *Zeitschrift für Geomorphologie*, Supplementband 45, 71–83.

Lancaster, N. & J. T. Teller 1988. Interdune deposits of the Namib sand sea. *Sedimentary Geology* 55, 91–108.

Lancey-Forth, N. B. de 1930. More journeys in search of Zerzura. *Geographical Journal* 75, 49–64.

Land, L. S. 1964. Eolian cross-bedding in the beach-dune environment, Sapelo Island, Georgia. *Journal of Sedimentary Petrology* 34, 289–94.

Land, L. S. 1970. Phreatic versus vadose meteoric diagenesis of limestones: evidence from a fossil water table. *Sedimentology* 14, 175–85.

Landsberg, H. & N. A. Riley 1943. Wind influences on the transport of sand over a Michigan sand dune. *University of Iowa, Studies in Engineering Bulletin* 27, 342–52.

Langbein, W. B. 1961. *Salinity and hydrology of closed lakes*. USGS Professional Paper 412.

Langbein, W. B. & L. B. Leopold 1966. *River meanders and the theory of minimum variance*. USGS Professional Paper 442M.

Langbein, W. B. & S. A. Schumm 1958. Yield of sediment in relation to mean annual precipitation. *American Geophysical Union, Transactions* 39, 1076–84.

Langer, A. M. & P. F. Kerr 1966. Mojave playa crusts: physical properties and mineral content. *Journal of Sedimentary Petrology* 36, 372–96.

Langford, R. P. 1989. Fluvial–aeolian interaction, part I: modern systems. *Sedimentology* 36, 1023–36.

Langford-Smith, T. 1962. Riverine plains chronology. *Australian Journal of Science* 25, 96–7.

Langford-Smith, T. (ed.) 1978a. A select review of silcrete research in Australia. See Langford-Smith (1978b), 1–12.

Langford-Smith, T. (ed.) 1978b. *Silcrete in Australia*. Armidale: University of New South Wales Press.

Langford-Smith, T. 1982. The geomorphic history of the Australian deserts. *Striae* 17, 4–19.

Larson, M. K. & T. L. Péwé 1986. Origin of land subsidence and earth fissuring, northeast Phoenix, Arizona. *Bulletin of the Association of Engineering Geology* 23, 139–65.

Lartet, L. 1865. Sur la formation du bassin de la mer morte ou lac asphaltite, et sur les changements survenus dans le niveau de ce lac. *Académie des Sciences à Paris, Comptes Rendus* 60, 796–800.

Lattman, L. H. 1973. Calcium carbonate cementation of alluvial fans in southern Nevada. *Bulletin of the Geological Society of America* 84, 3013–28.

Lattman, L. H. 1977. Weathering of caliche in southern Nevada. See Doehring (1977), 221–31.

Lattman, L. H. & S. K. Lauffenburger 1974. Proposed role of gypsum in the formation of caliche. *Zeitschrift für Geomorphologie*, Supplementband 20, 140–9.

Lawson, A. C. 1915. The epigene profiles of the desert. *University of California Bulletin, Department of Geology* 9, 23–48.

Lawson, T. J. 1971. Haboob structure at Khartoum. *Weather* **26**, 105–12.

Lazarenko, A. A. 1984. The loess of Central Asia. See Velichko (1984), 125–31.

Lazarenko, A. A., N. S. Bolikhovskaya, V. V. Semenov 1981. An attempt at a detailed stratigraphic subdivision of the loess association of the Tashkent region. *International Geology Review* **23**, 1335–46.

Le Houérou, H. N. 1984. Rain use efficiency: a unifying concept in arid-land ecology. *Journal of Arid Environments* **7**, 213–49.

Le Ribault, L. 1978. The exoscopy of quartz sand grains. See Whalley (1978), 319–27.

Le Roux, J. S. & Z. N. Roos 1986a. The relationship between size of particles in surface wash sediment and rainfall characteristics on a low-angle slope in a semi-arid climate. *Zeitschrift für Geomorphologie* **30**, 357–62.

Le Roux. J. S. & Z. N. Roos 1986b. Wash erosion on a debris-covered slope in a semi-arid climate. *Zeitschrift für Geomorphologie* **30**, 477–83.

Leatherman, S. P. 1979. Barrier dune systems: a reassessment. *Sedimentary Geology* **24**, 1–16.

Lecce, S. A. 1988. *Influence of lithology on alluvial fan morphometry, White and Inyo Mountains, California and Nevada*. MA thesis, Arizona State University.

Lecce, S. A. 1991. Influence of lithologic erodibility in alluvial fan area, Western White Mountains, California and Nevada. *Earth Surface Processes and Landforms* **16**, 11–18.

Lee, K. E. 1981. Effects of biotic components on abiotic components. See *Arid-land ecosystems: structure, functioning and management*, vol. 2, D. W. Goodall, R. A. Perry, K. M. W. Howes (eds), 105–23. Cambridge: Cambridge University Press.

Leeder, M. R. 1975. Pedogenic carbonates and flood sediment accretion rates: a quantitative model for alluvial arid-zone lithofacies. *Geological Magazine* **112**, 257–70.

Leeder, M. R. 1977. Folk's bedform theory. *Sedimentology* **24**, 863–4.

Lees, B. 1989. Lake segmentation and lunette initiation. *Zeitschrift für Geomorphologie* **33**, 475–84.

LeGrand, H. E. 1953. Streamlining of the Carolina Bays. *Journal of Geology* **61**, 263–74.

Lehmann, O. 1933. Morphologische Theorie der Verwitterung von Steinschlagwänden. *Vierteljahrsschrift Naturforschung Gesellschaft Zürich* **78**, 83–126.

LeMone, M. A. 1973. The structure and dynamics of horizontal roll vortices in the planetary boundary layer. *Journal of Atmospheric Sciences* **30**, 1077–91.

Leonard, R. J. 1927. Pedestal rocks resulting from disintegration, Texas Canyon, Arizona. *Journal of Geology* **35**, 469–74.

Leopold, L. B. 1951. Rainfall frequency: an aspect of climatic variation. *American Geophysical Union, Transactions* **32**, 347–57.

Leopold, L. B. 1976. Reversal of erosion cycle and climatic change. *Quaternary Research* **6**, 557–62.

Leopold, L. B., W. W. Emmett, R. M. Myrick 1966.

Channel and hillslope processes in a semiarid area, New Mexico. USGS Professional Paper 352G.

Leopold, L. B. & W. B. Langbein 1962. *The concept of entropy in landscape evolution*. USGS Professional Paper 500A.

Leopold, L. B. & T. Maddock 1953. *The hydraulic geometry of stream channels and some physiographic implications*. USGS Professional Paper 252.

Leopold, L. B. & J. P. Miller 1956. *Ephemeral streams – hydraulic factors and their relation to the drainage net*. USGS Professional Paper 282A.

Leopold, L. B., M. G. Wolman, J. P. Miller 1964. *Fluvial processes in geomorphology*. San Francisco: W. H. Freeman.

Lettau, H. 1969. Note on aerodynamic roughness parameter estimation on the basis of roughness element description. *Journal of Applied Meteorology* **8**, 828–32.

Lettau, H. H. 1978. Explaining the world's driest climate. See Lettau & Lettau (1978b), 182–248.

Lettau, K. & H. H. Lettau 1969. Bulk transport of sand by the barchans of La Pampa La Hoja in southern Peru. *Zeitschrift für Geomorphologie* **13**, 182–95.

Lettau, K. & H. H. Lettau 1978a. Experimental and micrometeorological studies of dune migration. See Lettau & Lettau (1978b), 110–47.

Lettau, H. H. & K. Lettau 1978b. *Exploring the world's driest climate*. Madison, Wisconsin: Institute of Environmental Studies, University of Wisconsin.

Lever, A. & I. N. McCave 1983. Eolian components in Cretaceous and Tertiary North Atlantic sediments. *Journal of Sedimentary Petrology* **53**, 811–32.

Lewis, A. D. 1936a. Roaring sands of the Kalahari Desert. *South African Geographical Journal* **19**, 33–9.

Lewis, A. D. 1936b. Fulgurites from Witsands on the southeastern border of the Kalahari. *South African Geographical Journal* **19**, 50–7.

Lewis, P. F. 1960. Linear topography in southwestern Palouse, Washington–Oregon. *Annals of the AAG* **50**, 98–111.

Leys, J. F. 1990. The erodibility of a range of soils using a portable wind tunnel from south-west New South Wales, Australia. *Acta Mechanica* **2**, 103–112.

Li, Z. & P. D. Komar 1986. Laboratory measurements of pivoting angles for application to selective entrainment of gravel in cement. *Sedimentology* **33**, 413–23.

Lillesand, T. M. & R. W. Kiefer 1979. *Remote sensing and image interpretation*. New York: John Wiley.

Lindsey, J. F. 1973. Ventifact evolution in Wright Valley, Antarctica. *Geological Society of America, Bulletin* **84**, 1791–8.

Lindsey, J. F., D. R. Criswell, T. R. Criswell, B. S. Criswell 1976. Sound-producing dune and beach sands. *Geological Society of America, Bulletin* **87**, 463–73.

Lines, G. E. 1979. *Hydrology and surface morphology of the Bonneville Salt Flats and Pilot Valley Playa, Utah*. USGS Water Supply Paper 2057.

Lister, L. A. & C. D. Secrest 1985. Giant desiccation cracks and differential surface subsidence, Red Lake

Playa, Mohave County, Arizona. *Bulletin of the Association of Engineering Geology* 22, 299–314.

Little, I. P., T. M. Armitage, R. J. Gilkes 1978. Weathering of quartz in dune sands under subtropical conditions in eastern Australia. *Geoderma* 20, 225–37.

Liu Tung Sheng, Gu Xiong Fei, An Zhi Sheng, Fan Yong Xiang 1982. The dust fall in Beijing, China, on April 18, 1980. Geological Society of America Special Paper 186, 149–58.

Livingstone, D. A. 1954. On the orientation of lake basins. *American Journal of Science* 252, 547–54.

Livingstone, I. 1986. Geomorphological significance of wind flow patterns over a Namib linear dune. See Nickling (1986), 97–112.

Livingstone, I. 1987a. Using the response diagram to recognise zones of aeolian activity: a note on evidence from a Namib dune. *Journal of Arid Environments* 13, 25–30.

Livingstone, I. 1987b. Grain-size variation on a "complex" linear dune in the Namib Desert. See Frostick & Reid (1987), 281–92.

Livingstone, I. 1988. New models for the formation of linear sand dunes. *Geography* 73, 105–15.

Livingstone, I. 1989a. Monitoring surface change on a Namib linear dune. *Earth Surface Processes and Landforms* 14, 317–32.

Livingstone, I. 1989b. Applying Besler's response diagram: a comment. *Zeitschrift für Geomorphologie* 33, 499–502.

Livingstone, I. 1989c. Temporal trends in grain-size measures on a linear sand dune. *Sedimentology* 36, 1017–22.

Lofgren, B. E. & R. L. Klausing 1969. *Land subsidence due to groundwater withdrawal, Tulane, Wasco area, California*. USGS Professional Paper 437B.

Logan, R. F. 1960. *The Central Namib Desert, South West Africa*. Washington DC: National Academy of Sciences.

Logie, M. 1981. Wind-tunnel experiments on sand dunes. *Earth Surface Processes* 6, 365–74.

Logie, M. 1982. Influence of roughness elements and soil moisture on the resistance of sand to wind erosion. See Yaalon (1982), 161–73.

Long, J. T. & R. P. Sharp 1964. Barchan-dune movement in the Imperial Valley, California. *Geological Society of America, Bulletin* 75, 149–56.

Lonsdale, P. & F. N. Spiess 1977. Abyssal bedforms explored with a deeply towed instrument package. *Marine Geology* 23, 57–75.

Lovegrove, B. G. & W. R. Siegfried 1986. Distribution and formation of mima-like earth mounds in the western Cape Province of South Africa. *South African Journal of Science* 82, 432–6.

Lowdermilk, J. D. 1931. On the origin of desert varnish. *American Journal of Science*, series 5 21, 51–66.

Lowdermilk, J. D. & A. O. Woodruff 1932. Concerning rillensteine. *American Journal of Science*, series 5 23, 135–54.

Lowdermilk, W. C. & H. L. Sundling 1950. Erosion pavement, its formation and significance. *American Geophysical Union, Transactions* 31, 96–100.

Lowe, D. R. 1976. Grain flow and grain-flow deposits.

Journal of Sedimentary Petrology 46, 188–99.

Lu Yanchou 1981. Pleistocene climatic cycles and variations of $CaCO_3$ contents in a loess profile [in Chinese].

Lucchitta, B. K. 1982. Ice sculpture in Martian outflow channels. *Journal of Geophysical Research* 87, 9951–73.

Luk, S-H., A. D. Abrahams, A. J. Parsons 1986. A simple rainfall simulator and trickle system for hydrogeomorphological experiments. *Physical Geography* 7, 344–56.

Lustig, L. K. 1965. *Clastic sedimentation in Deep Springs Valley, California*. USGS Professional Paper 352F, 131–92.

Lustig, L. K. 1969a. *Trend surface analysis of the Basin and Range province and some geomorphic implications*. USGS Professional Paper 500D.

Lustig, L. K. 1969b. Quantitative analysis of desert topography. See McGinnies et al. (1969), 47–58.

Lyles, L. 1976. Wind patterns and soil erosion on the Great Plains. Paper presented at a *Symposium on shelter belts on the Great Plains, Denver, Colorado, April 20–22, 1976*, R. W. Tinns (ed.), 22–9. Lincoln, Nebraska: The Great Plains Agricultural Council.

Lyles, L. 1977. Wind erosion: processes and effects on soil productivity. *American Society of Agricultural Engineering, Transactions* 20, 880–4.

Lyles, L. & R. K. Krauss 1971. Threshold velocities and initial particle motion as influenced by air turbulence. *American Society of Agricultural Engineering, Transactions* 14, 563–6.

Lyles, L. & R. L. Schrandt 1972. Wind erodibility as influenced by rainfall and soil salinity. *Soil Science* 114, 367–72.

Ma Chengyuan 1982. Preliminary analysis of the sand driving wind in east Henan and north Henan. *Journal of Desert Research* 2, 17–75.

Mabbutt, J. A. 1963. Wanderrie banks: micro-relief patterns in semi-arid Australia. *Geological Society of America, Bulletin* 74, 529–40.

Mabbutt, J. A. 1965. Stone distribution in a Stony Tableland soil. *Australian Journal of Soil Research* 3, 131–42.

Mabbutt, J. A. 1966a. Mantle-controlled planation of pediments. *American Journal of Science* 264, 78–91.

Mabbutt, J. A. 1966b. Landforms of the western McDonnell Ranges. See *Essays in geomorphology*, G. H. Dury (ed.), 83–119. London: Heinemann.

Mabbutt, J. A. 1967. Denudation chronology in central Australia: structure, climate and landform inheritance in the Alice Springs area. See *Landform studies from Australia and New Guinea*, J. N. Jennings & J. A. Mabbutt (eds), 144–81. Cambridge: Cambridge University Press

Mabbutt, J. A. 1968. Aeolian landforms in central Australia. *Australian Geographical Studies* 6, 139–50.

Mabbutt, J. A. 1969. Landforms of arid Australia. See *Arid lands of Australia*, R. O. Slayter & R. A. Perry (eds). Canberra: ANU Press.

Mabbutt, J. A. 1971. The Australian arid zone as a prehistoric environment. See *Aboriginal man and environment in Australia*, D. J. Mulvaney & P. J. Golson (eds), 66–79. Canberra: ANU Press.

Mabbutt, J. A. 1977. *Desert landforms*. Cambridge, Mass.: MIT Press

Mabbutt, J. A. 1980. Some general characteristics of the aeolian landscapes. See Storrier & Stannard (1980), 1–16.

Mabbutt, J. A. 1984. Landforms of the Australian deserts. See El-Baz (1984b), 79–94.

Mabbutt, J. A. & M. E. Sullivan 1968. The formation of longitudinal dunes: evidence from the Simpson Desert. *Australian Geographer* 10, 483–7.

Mabbutt, J. A. & R. A. Wooding 1983. Analysis of longitudinal dune patterns in the northwestern Simpson Desert, central Australia. *Zeitschrift für Geomorphologie*, Supplementband 45, 51–69.

Mabbutt, J. A., J. N. Jennings, R. A. Wooding 1969. The asymmetry of Australian desert sand ridges. *Australian Journal of Science* 32, 159–60.

MacCarthy, G. R. 1935. Eolian sands: a comparison. *American Journal of Science* 30, 81–95.

MacCarthy, G. R. 1937. The Carolina Bays. *Geological Society of America, Bulletin* 48, 1211–26.

MacCarthy, G. R. & J. W. Huddle 1938. Shape-sorting of sand grains by wind action. *American Journal of Science* 35, 64–73.

MacGregor, A. N. & D. E. Johnson 1971. Capacity of desert algal crusts to fix atmospheric nitrogen. *Soil Science Society of America, Proceedings* 35, 843–4.

Machette, M. N. 1985. Calcic soils of the southwestern United Sates. See Weide (1985), 1–21.

MacKay, J. R. 1956. Notes on oriented lakes of the Liverpool Bay area, Northwestern Territories. *Revue Canadienne de Géographie* 10, 169–73.

MacKie, W. 1897. On the laws that govern the rounding of grains of sand. *Edinburgh Geological Society, Transactions* 7, 298–311.

Mackin, J. H. 1970. Origin of pediments in the western United States. See Pécsi (1970), 85–105.

Macumber, P. G. 1969. Inter-relationship between physiography, hydrology, sedimentation and salinization of the Loddon River Plains. Australia. *Journal of Hydrology* 7, 39–57.

MacVicar, C. N., J. L. Hutson, R. W. Fitzpatrick 1985. Soil formation in the coast aeolianites and sands of Natal. *Journal of Soil Science* 36, 373–88.

Madigan, C. T. 1936. *Central Australia*. London: Oxford University Press.

Madigan, C. T. 1946. The Simpson Desert expedition 1939; scientific reports, 6: geology – the sand formations. *Royal Society of South Australia, Transactions* 70, 45–63.

Magaritz, M. & G. A. Goodfriend 1987. Movement of the desert boundary in the Levant from the latest Pleistocene to early Holocene. See *Abrupt climatic change*, W. H. Berger & L. D. Labeyrie (eds), NATO ASI Series C216, 173–83. Dordrecht: Reidel.

Magura, L. M. & D. E. Wood 1980. Flood hazard identification and flood plain management on alluvial fans. *Water Resources Bulletin* 16, 56–62.

Mahmoudi, F. 1977. Les nebkhas de Lut, Iran. *Annales de Géographie* 86, 315–21.

Mainguet, M. 1968. Le Borkou, aspect d'une modèle éolienne. *Annales de Géographie* 77, 296–322.

Mainguet, M. 1970. Un étonnant paysage: les cannelures grèseuse du Bembéché (nord du Tchad): essai d'explication géomorphologique. *Annales de Géographie* 79, 58–66.

Mainguet, M. 1972. *Le modelé des grès: problèmes généreaux*. Paris: Institut Géographique National.

Mainguet, M. 1982. L'épaisseur des dépôts sableux éoliens, est-elle un indicateur d'aridité? L'aridité saharienne. *Association Géographes de France, Bulletin* 483–4, 64–7.

Mainguet, M. 1983a. Tentative mega-geomorphological study of the Sahara. See Gardner & Scoging (1983), 113–33.

Mainguet, M. 1983b. Dune vives, dunes fixées, dunes vêtues: une classification selon le bilan d'alimentation. Le regime éolien et la dynamique des édifices sableux. *Zeitschrift für Geomorphologie*, Supplementband 45, 265–86.

Mainguet, M. 1984a. A classification of dunes based on aeolian dynamics and the sand budget. See El-Baz (1984b), 31–58.

Mainguet, M. 1984b. Space observations of Saharan aeolian dynamics. See El-Baz (1984b), 59–77.

Mainguet, M. 1984c. Cordons longitudinaux (sand ridges); dunes allongées à ne plus confondre avec les sifs, autres dunes liniaires. *Institut de Géographie de Reims, Travaux* 59–60, 61–83.

Mainguet, M. 1985. Le Sahel, une laboratoire naturel pour l'étude du vent, mécanisme principal de la désertification. See Barndorff-Nielsen et al. (1985), 545–61.

Mainguet, M. 1986. The wind and desertification processes in the Saharo-Sahelian and Sahelian regions. See El-Baz & Hassan (1986), 210–40.

Mainguet, M. & Y. Callot 1978. L'erg de Fachi-Bilma (Tchad–Niger). Contribution à la connaissance de la dynamique des ergs et des dunes des zones arides chaudes. *Mémoires et Documents*, NS 18.

Mainguet, M., Y. Callot, M. Guy 1974. Systèmes crêtes – couloirs. *Photo-Interpretation* 13, 24–30.

Mainguet, M., L. Canon, M-C. Chemin 1980a. Le Sahara: géomorphologie et paléogéomorphologie éoliennes. See Williams & Faure (1980), 17–35.

Mainguet, M., L. Canon-Cossus, A. M. Chapelle 1980b. Utilisation des images Météosat pour préciser les trajectoires éoliennes au sol, au Sahara et sur les marges sahéliennes. *Société Française de Photogrammetrie et de Télédetection, Bulletin* 78, 1–12.

Mainguet, M. & M-C. Chemin 1981. Utilisation des images Landsat pour la cartographie de la dynamique éolienne et la définition de son influence dans la désertification en milieu Sahélien. *4ieme Colloque International du GDTA*, 135–49.

Mainguet, M. & M-C. Chemin 1983. Sand seas of the

Sahara and Sahel: an explanation of their thickness and sand dune type by the sand budget principle. See Brookfield & Ahlbrandt (1983), 353–63.

Mainguet, M. & M-C. Chemin 1984. Les dunes pyramidales du Grand Erg Oriental: un double dynamique pour un même édifice éolien. *Institut de Géographie de Reims, Travaux* **59–60**, 49–60.

Mainguet, M. & M-C. Chemin 1988. Wind system and sand dunes in the Taklamakan Desert. *Chinese Journal of Arid Land Research* **1**, 135–42.

Mainguet, M. & L. Cossus 1980. Sand circulation in the Sahara: geomorphological relations between the Sahara Desert and its margins. See Sarnthein et al. (1980), 69–78.

Mainguet, M., L. Cossus, A. M. Chapelle 1980. Utilisation des images Météostat pour préciser les trajectoires éoliennes au sol, au Sahara et sur les marges Saheliennes. *Société Française de Photogrammetrie et de Télédétection* **78**, 1–14.

Mainguet, M., M. Vimeux-Richeux, M-C. Chemin 1983. Autochthone et allochthonie des sables de la zone Saharo-Sahélienne du Niger. *Revue de Géographie Physique et de Géologie Dynamique* **24**, 167–75.

Maiti, D. & Y-F. Thomas 1975. *Interaction du vent et des plantes en zone dunaire littoral*. Mémoires du Laboratoire de Géomorphologie de l'École Practique des Hautes Études 28.

Maizels, J. K. 1981. *Freeze/thaw experiments in the simulation of sediment cracking patterns*. Bedford College London, Papers in Geography 13.

Maizels, J. K. 1987. Plio-Pleistocene raised channel systems of the western Sharqiya (Wahiba), Oman. See Frostick & Reid (1987), 31–50.

Maizels, J. K. 1988. Palaeochannels: Plio-Pleistocene raised channel systems of the western Sharkiyah. *Journal of Oman Studies*, Special Report 3, 95–112.

Maizels, J. K. 1990. Raised channel systems as indicators of paleohydrologic change: a case study from Oman. *Palaeogeography, Paleoclimatology, Paleoecology* **76**, 241–77.

Maley, J. 1977. Palaeoclimates of central Sahara during the Holocene. *Nature* **269**, 573–7.

Mammerickx, J, 1964a. Pédiments désertiques et pédiments tropicaux. *Acta Geographica Lovaniensia* **3**, 359–70.

Mammerickx, J. 1964b. Quantitative observations on pediments in the Mojave and Sonoran deserts (southwestern United States). *American Journal of Science* **262**, 417–35.

Manabe, S. & D. G. Hahn 1977. Simulation of the tropical climate of an ice age. *Journal of Geophysical Research* **82**, 3889–911.

Mangin, A. 1869. *The desert world*. London: Nelson.

Marathe, A. R., S. N. Rajayuri, V. S. Lele 1977. On the problem of the origin and age of the Miliolite rocks of the Hiran Valley, Saureshtra, western India. *Sedimentary Geology* **19**, 197–215.

Margolis, S. V. & D. H. Krinsley 1971. Submicroscopic frosting on eolian and subaqueous quartz sand grains. *Geological Society of America, Bulletin* **82**, 3395–406.

Marion, G. M., W. H. Schlesinger, P. J. Fonteyn 1985. CALDEP: a regional model for soil $CaCO_3$ (caliche) depos-ition in southwestern deserts. *Soil Science* **139**, 468–81.

Marsh, G. P. 1864. *Man and nature*. [D. Lowenthal (ed.)] Cambridge, Mass.: Belknap.

Marshall, J. K. 1970. Assessing the protective role of shrub-dominated rangeland vegetation against soil erosion by wind. *11th International Grassland Congress, Proceedings*, 19–22. Brisbane: University of Queensland Press.

Marshall, J. K. 1971. Drag measurements in roughness arrays of varying density and distribution. *Agricultural Meteorology* **8**, 269–92.

Marshall, J. K. 1973. Drought, land use and soil erosion. See *The environmental, economic and social significance of drought*, J. V. Lovett (ed.), 55–77. Sydney: Angus & Robertson.

Marshall, J. R. 1979. *Experimental abrasion of natural materials*. PhD dissertation, University College London.

Marshall, T. R. 1987. Morphotectonic analysis of the Wesselsborn panveld. *South African Journal of Geology* **90**, 209–18.

Marsland, J. & J. G. Woodruff 1937. A study of the effects of wind transportation on grains of several minerals. *Journal of Sedimentary Petrology* **7**, 18–30.

Martini, I. P. 1978. Tafoni weathering, with examples from Tuscany, Italy. *Zeitschrift für Geomorphologie* **22**, 44–67.

Martonne, E. 1927. Regions of interior-basin drainage. *Geographical Review* **17**, 397–414.

Mason, C. C. & R. L. Folk 1958. Differentiation of beach, dune and eolian flat environments by size analysis, Mustang Island, Texas. *Journal of Sedimentary Petrology* **28**, 211–26.

Mason, P. J. & B. R. Morton 1987. Trailing vortices in the wakes of surface-mounted obstacles. *Journal of Fluid Mechanics* **175**, 247–93.

Mason, P. J. & R. I. Sykes 1979. Flow over a single isolated hill of moderate slope. *Royal Meteorological Society, Quarterly Journal* **105**, 383–95.

Matchinski, M. 1951. Sur la structure du vent et sur les phenomènes secondaires ("wind marks" sur le sable et sur la neige, etc.) qu'elle provoque. *Academie des Sciences à Paris, Comptes Rendus* **233**, 580–2.

Matchinski, M. 1952. Sur les formations sableuses des environs de Beni-Abbès. *Société Géologique de la France, Comptes Rendus* **9–10**, 171–4.

Matchinski, M. 1955. La formation des dunes dans les déserts. *La Nature* **3241**, 169–75.

Matchinski, M. 1962a. Sur la distribution des petits mares de l'Île de France. *Academie des Sciences à Paris, Comptes Rendus* **254**, 331–4.

Matchinski, M. 1962b. Sur l'alignement des mares de Sucy-en-Brie (Seine et Oise). *Société Géologique de France, Comptes Rendus Sommaires* **7**, 173–4.

Mather, K. D. & G. S. Miller 1966. Wind drainage of the high plateau of Antarctica. *Nature* **209**, 281–4.

Mattox, R. B. 1955. Eolian shape sorting. *Journal of Sedimentary Petrology* **25**, 111–14.

Matter, A. & M. E. Tucker (eds) 1978. *Modern and ancient lake sediments*. Oxford: Blackwell Scientific.

Mattsson, J. O. 1976. Wind-tilted pebbles in sand – some field observations and simple experiments. *Nordic Hydrology* **7**, 181–208.

Maxson, J. H. 1940. Fluting and faceting of rock fragments. *Journal of Geology* **48**, 717–51.

Maxwell, T. A. & C. V. Haynes 1989. Large-scale, low amplitude bedforms (chevrons) in the Selima Sand Sheet. *Science* **243**, 1179–82.

Mayer, M., L. D. McFadden, J. W. Harden 1988. Distribution of calcium carbonate in desert soils: a model. *Geology* **16**, 303–6.

Mazullo, J. M. & S. Magenheimer 1987. The original shapes of quartz sand grains. *Journal of Sedimentary Petrology* **57**, 479–87.

McArthur, D. S. 1987. Distinctions between grain-size distributions of accretion and encroachment deposits in an inland dune. *Sedimentary Geology* **54**, 147–64.

McBride, E. F. & M. O. Hayes 1962. Dune cross-bedding on Mustang island, Texas. *American Association of Petroleum Geologists, Bulletin* **46**, 546–51.

McBurney, C. B. M. & R. W. Hey 1955. *Prehistory and Pleistocene Geology in Cyrenaican Libya*. Cambridge: Cambridge University Press.

McCauley, J. F., M. J. Grolier, C. S. Breed 1977a. *Yardangs of Peru and other desert regions*. USGS Interagency Report, Astrogeology 81.

McCauley, J. F., C. S. Breed, M. J. Grolier 1977b. Yardangs. See Doehring (1977), 233–69.

McCauley, J. F., G. G. Schaber, M. J. Grolier, C. V. Haynes, B. Issain, C. Elachi, R. Blom 1982. Subsurface valleys and geoarcheology of the eastern Sahara revealed by shuttle radar. *Science* **218**, 1004–20.

McCauley, J. F., G. G. Schaber, W. P. McHugh, B. Issain, C. V. Haynes, M. J. Grolier, A. Al-Kilani 1986. Paleodrainages of the eastern Sahara – the radar rivers revisited (SIR-A/B implications for a mid-Tertiary trans-African drainage system. *Transactions Geoscience and Remote Sensing* GE–24, 624–48.

McCleary, J. A. 1968. The biology of desert plants. See *Desert biology*, G. W. Brown Jr (ed.), 141–94. London: Academic Press.

McClure, H. A. 1976. Radiocarbon chronology of late-Quaternary lakes in the Arabian desert. *Nature* **263**, 755–6.

McClure, H. A. 1978. The Rub' al Khali. See Al-Sayari & Zötl (1978), 252–63.

McCoy Jr, F. W., W. G. Nokleberg, R. M. Norris 1967. Speculations on the origin of the Algodones Dunes, southern California. *Geological Society of America, Bulletin* **78**, 1039–44.

McCullagh, M., N. Hardy, W. O. Lockman 1972. Formation and migration of sand dunes: a simulation of their effect on the sedimentary environment. In *Mathematical models of sedimentary processes*, D. F. Merriam (ed.),

175–90. New York: Plenum.

McFadden, L. D. & J. C. Tinsley 1985. Rate and depth of pedogenic carbonate accumulation in soils: formulation and testing of a compartment model. See Weide (1985), 23–41.

McFadden, L. D., S. G. Wells, J. C. Dohrenwend 1986. Influence of Quaternary climatic changes on processes of soil development on desert loess deposits of the Cima volcanic field, California. *Catena* **13**, 361–89.

McFadden, L. D., S. G. Wells, M. J. Jercinovich 1987. Influences of eolian and pedogenic processes on the origin and evolution of desert pavements. *Geology* **15**, 504–8.

McFarlane M. F. & C. T. Twidale 1987. Karstic features associated with tropical weathering profiles. *Zeitschrift für Geomorphologie*, Supplementband 64, 73–95.

McGee, A. W., P. A. Bull, A. S. Goudie 1988. Chemical textures on quartz grains: an experimental approach. *Earth Surface Processes and Landforms* **13**, 665–70.

McGee, W. J. 1896. Expedition to Papagenia and Seriland. *American Author* **9**, 93–8.

McGee, W. J. 1897. Sheetflood erosion. *Geological Society of America, Bulletin* **8**, 87–112.

McGinnies, W. G. & B. J. Goldman (eds) 1969. *Arid lands in perspective*. Tucson: University of Arizona Press.

McGinnies, W. G., B. J. Goldman, P. Paylove (eds) 1969. *Deserts of the world*. Tucson: University of Arizona Press.

McGreevy, J. P. 1981. Some perspectives on frost shattering. *Progress in Physical Geography* **5**, 56–75.

McGreevy, J. P. 1982. "Frost and salt" weathering: further experimental results. *Earth Surface Processes and Landforms* **7**, 475–88.

McGreevy, J. P. 1985. Thermal properties as controls on rock surface temperature maxima and possible implications for rock weathering. *Earth Surface Processes and Landforms* **10**, 125–36.

McGreevy, J. P. & B. J. Smith 1984. The possible role of clay minerals in salt weathering. *Catena* **11**, 169–75.

McHugh, W. P., J. F. McCawley, C. S. Breed, G. G. Schaber, C. V. Haynes 1988. Paleorivers and geoarchaeology in the southern Egyptian Sahara. *Geoarchaeology* **3**, 1–40.

McKee, E. D. 1962. Origin of the Nubian and similar sandstones. *Sonderdruck aus der Geologische Rundschau* **52**, 551–87.

McKee, E. D. 1966. Structures of dunes at White Sands National Monument, New Mexico, (and a comparison with structures of dunes from other selected areas). *Sedimentology* **7**, 1–69.

McKee, E. D. (ed.) 1979a. *A study of global sand seas*. USGS Professional Paper 1052.

McKee, E. D. 1979b. Introduction to a study of global sand seas. See McKee (1979a), 1–19.

McKee, E. D. 1979c. Sedimentary structures in dunes. See McKee (1979a) 83–113.

McKee, E. D. 1979d. Ancient sandstones considered to be

eolian. See McKee (1979a), 187–233.

McKee, E. D. 1982. *Sedimentary structures in dunes of the Namib Desert, South West Africa.* Geological Society of America, Special Paper 188.

McKee, E. D. 1983. Eolian sand bodies of the world. See Brookfield & Ahlbrandt (1983), 1–25.

McKee, E. D. & J. R. Douglass 1971. Growth and movement of dunes at White Sands national Monument, New Mexico. USGS Professional Paper 750D, 108–14.

McKee, E. D., J. R. Douglass, S. Rittenhouse 1971. Deformation of lee-side laminae in eolian dunes. *Geological Society of America, Bulletin* 82, 359–78.

McKee, E. D. & R. J. Moiola 1975. Geometry and growth of the White Sands dune field, New Mexico. USGS *Journal of Research* 3, 59–66.

McKee, E. D. & G. C. Tibbits Jr 1964. Primary structures of a seif dune and associated deposits in Libya. *Journal of Sedimentary Petrology* 43, 5–17.

McKee, E. D. & W. C. Ward 1983. Eolian environments. See *Carbonate depositional environments*, P. A. Scholle, D. G. Bebont, C. H. Moore (eds), AAPG Memoir 33, 131–70.

McKenna-Neuman, C. 1989. Kinetic energy transfer through impact and its role in entrainment by wind of particles from frozen surfaces. *Sedimentology* 36, 1007–15.

McKenna-Neuman, C. & R. Gilbert 1986. Aeolian environments and landforms in glaciofluvial environments of southeastern Baffin Island, NWT. See Nickling (1986), 213–35.

McKenna-Neuman, C. & W. G. Nickling 1989. A theoretical and wind-tunnel investigation of the effect of capillary water on the entrainment of soil by wind. *Canadian Journal of Soil Science* 69, 79–96.

McLaren, P. 1981. An interpretation of trends in grain-size measures. *Journal of Sedimentary Petrology* 51, 611–24.

McLaren, P. & D. Bowles 1985. The effects of sediment transport on grain-size distributions. *Journal of Sedimentary Petrology* 55, 457–70.

McLean, S. R. & J. D. Smith 1986. A model for flow over two-dimensional bed forms. *Journal of Hydraulic Engineering* 112, 300–17.

McMahon, J. A. 1979. Hydrological characteristics of arid zones. IASH Publication 128, 105–23.

McTainsh, G. H. 1984. The nature and origin of aeolian mantles in central northern Nigeria. *Geoderma* 33, 13–38.

McTainsh, G. H. 1985. Dust processes in Australia and West Africa: a comparison. *Search* 16, 104–6.

McTainsh, G. H. 1987. Desert loess in northern Nigeria. *Zeitschrift für Geomorphologie* 31, 145–65.

McTainsh, G. H. & P. H. Walker 1982. Nature and distribution of Harmattan dust. *Zeitschrift für Geomorphologie* 26, 417–36.

Meckelein, W. 1959. *Forschungen in der zentralen Sahara.* Braunschweig: Westermann.

Meckelein, W. 1965. Beobachtungen und Gedanken zu geomorphologischen Konvergenzen in Polar- und Wärme-

wüsten. *Erdkunde* 19, 31–9.

Meckelein, W. 1974. Aride Verwitterung in Polargebieten im Vergleich zum subtropischen Wüstengürtel. *Zeitschrift für Geomorphologie*, Supplementband 20, 178–88.

Meddlicott, H. B. & W. T. Blanford 1879. *A manual of the geology of India* [5 volumes]. Calcutta: Government of India.

Meher-Homji, V. M. 1973. Is the Sind–Rajasthan Desert the result of a recent climatic change? *Geoforum* 15, 47–57.

Meigs, P. 1953. The world distribution of arid and semi-arid homoclimates. See *Reviews of research on arid zone hydrology,* 203–09. Paris: UNESCO.

Melhorn, W. N. & D. T. Trexler 1977. The Maria effect: equilibrium and activation of aeolian processes in the Great Basin of Nevada. See Doehring (1976), 271–2.

Melton, F. A. 1940. A tentative classification of sand dunes. Its application to dune history on the High Plains. *Journal of Geology* 48, 113–74.

Melton, F. A. & W. Schriever 1933. The Carolina Bays – are they meteorite craters? *Journal of Geology* 41, 52–66.

Melton, M. A. 1957. *An analysis of the relations among elements of climate, surface properties and geomorphology.* Office of Naval Research Technical Report II (Project NR 389–042).

Melton, M. A. 1965a. The geomorphic and palaeoclimatic significance of alluvial deposits in southern Arizona. *Journal of Geology* 73, 1–38.

Melton, M. A. 1965b. Debris-covered hillslopes of the southern Arizona desert – consideration of their stability and sediment contribution. *Journal of Geology* 73, 715–29.

Membery, D. A. 1983. Low-level wind profiles during the Gulf Shamal. *Weather* 38, 18–24.

Mensching, H. 1970. Planation in arid subtropic and tropic regions. See Pécsi (1970), 73–84.

Mensching, H. 1978. Inselberge, Pedimente und Rumpfflächen im Sudan (Republik). *Zeitschrift für Geomorphologie,* Supplementband 30, 1–19.

Mensching, H., K. Giessner, G. Stuckmann 1970. Die Hochwasserkatastrophe in Tunesien im Herbst 1969. *Geographisches Zeitschrift* 58, 81–94.

Mercer, J. H. 1976. Glacial history of southernmost South America. *Quaternary Research* 6, 125–66.

Merriam, R. 1969. Source of sand dunes of southeastern California and northwestern Sonora, Mexico. *Geological Society of America, Bulletin* 80, 531–3.

Michel, P. 1973. *Les bassins des fleuves Sénégal et Gambie, études géomorphologiques.* Mémoir ORSTOM, Paris [3 volumes].

Middleton, N. J. 1985. Effect of drought on dust production in the Sahel. *Nature* 316, 431–4.

Middleton, N. J. 1986. Dust storms in the Middle East. *Journal of Arid Environments* 10, 83–96.

Middleton, N. J., A. S. Goudie, G. L. Wells 1986. The frequency and source areas of dust storms. See Nickling (1986), 237–60.

Miller, K. J. (ed.) 1984. *The international Karakoram*

Project, [2 vols]. **Cambridge: Cambridge University Press.**

Miller, R. P. & P. D. Komar 1977. The development of sediment threshold curves for unusual environments (Mars) and for inadequately studied materials (foram sands). *Sedimentology* **24**, 709–21.

Miller, R. P. 1937. Drainage in bas-relief. *Journal of Geology* **45**, 432–8.

Millington, A. C., A. R. Jones, N. Quarmby, J. R. G. Townshend 1987. Remote sensing of sediment transfer processes in playa basins. See Frostick & Reid (1987), 369–81.

Millington, A. C. et al. 1988. Monitoring playas using thematic mapper data. *Proc. IGARSS '88 Symposium, Edinburgh, Scotland, 13–16 Sept, 1988*, 377–80.

Milnes, A. K. & N. H. Ludbrook 1986. Provenance of microfossils in aeolian calcarenites and calcretes in southern South Australia. *Australian Journal of Earth Science* **33**, 145–55.

Milnes, A. K. & C. R. Twidale 1983. An overview of silicification in Cainozoic landscapes of arid central and southern Australia. *Australian Journal of Soil Research* **21**, 387–410.

Mitchell, C. W. & J. A. Howard 1978. The application of LANDSAT imagery to soil degradation mapping. *UN/FAO Series AGLT*, **4/78**.

Mitchell, C. W. & R. M. S. Perrin 1967. The subdivision of hot deserts of the world into physiographic units. *Institut Français de Pétrole, Révue* **21**, 1855–72.

Mitchell, C. W., R. Webster, P. M. T. Beckett, B. Clifford 1979. An analysis of terrain classification for long-range prediction of conditions in deserts. *Geographical Journal* **145**, 72–85.

Mitchell, C. W. & S. G. Willimott 1974. Dayas of the Moroccan Sahara and other arid regions. *Geographical Journal* **140**, 441–53.

Mitha, S., M. Q. Tran, B. T. Werner, P. K. Haff 1986. The grain-bed impact process in aeolian saltation. *Acta Mechanica* **63**, 267–78.

Monod, Th. 1958. Le Majâbat al-Koubrâ. *Institut Français de l'Afrique Noire, Memoirs* **52**.

Monod, Th. 1961. Le Majâbat al-Koubrâ (Supplement). *Institut Français de l'Afrique Noire, Bulletin* **23**, 591–637.

Moore, D. O. 1968. Estimating runoff in ungauged semiarid areas. *Bulletin of the International Association of Scientific Hydrology* **13**, 29–39.

Morales, C. (ed.) 1979. *Saharan dust*. Chichester, England: John Wiley.

Morgan, R. P. C. & H. J. Finney 1987. Drag coefficients of single row crops and their implications for wind erosion control. See Gardiner (1987), 449–58.

Morisawa, M. (ed.) 1973 [reissued 1983]. *Fluvial geomorphology*. Boston: Allen & Unwin.

Morisawa, M. & J. T. Hack (eds) 1985. *Tectonic geomorphology*. Boston: Allen & Unwin.

Morris, W. J. 1957. Effects of sphericity, roundness and velocity on traction transportation of sand grains. *Journal of Sedimentary Petrology* **27**, 27–31.

Morrison, R. B. 1966. Predecessors of Great Salt Lake. In *The Great Salt Lake*, W. L. Stokes (ed.), 77–104. Salt Lake City: Utah Geological Society.

Morrison, R. B. 1968. Pluvial lakes. See Fairbridge (1968), 873–83.

Morrison, R. B. & H. E. Wright Jr (eds) 1968. *Means of correlation of Quaternary successions*. Salt Lake City: University of Utah Press.

Mortensen, H. 1927. *Der Formenschatz der nordchilenishen Wüste*. Berlin: Weidmannsche.

Mortensen, H. 1933. Die "Salzprengung" und ihre bedeutung für die regionalklimatische Gliederung der Wüsten. *Petermanns Geographische Mitteilungen* **79**, 130–5.

Mortimore, M. J. 1989. *Adapting to drought: farmers, famines and desertification in West Africa*. Cambridge: Cambridge University Press.

Moss, A. J. 1979. Thin-flow transportation of solids in arid and non-arid areas: a comparison of process. IASH Publication 128, 435–45.

Moss, J. H. 1977. The formation of pediments: scarp backwearing or surface downwearing. See Doehring (1977), 51–75.

Motts, W. S. 1965. Hydrologic types of playas and closed valleys and some relations of hydrology to playa geology. See *Geology, mineralogy and hydrology of US playas*, J. T. Neal (ed.), US Air Force Cambridge Research Laboratories, Environmental Research Papers 96, 73–104.

Motts, W. S. (ed.) 1970. *Geology and hydrology of selected playas in western United States*. Bedford, Mass.: US Air Force Cambridge Research Laboratories.

Mouat, D. A., J. A. Bale, K. E. Foster, B. D. Treadwell 1981. The use of remote sensing for an integrated inventory of a semi-arid area. *Journal of Arid Environments* **4**, 169–79.

Mueller, G. 1968. Genetic histories of nitrate deposits from Antarctica and Chile. *Nature* **219**, 113–16.

Mulcahy, M. J. & E. Bettenay 1972. Soil and landscape studies in Western Australia, 1: the major drainage divides. *Geological Society of Australia, Journal* **18**, 349–57.

Mulhearn, P. J. & E. F. Bradley 1977. Secondary flow in the lee of porous shelterbelts. *Boundary-layer Meteorology* **12**, 75–92.

Mulligan, K. R. 1987. Velocity profiles on the windward slope of a transverse dune. *Earth Surface Processes and Landforms* **13**, 573–82.

Murata, T. 1966. A theoretical study of the forms of alluvial fans. Tokyo Metropolitan University Geographical Report 1, 33–43.

Musick, H. B. 1975. Barrenness of desert pavement in Yuma County, Arizona. *Arizona–Nevada Academy of Sciences, Journal* **10**, 24–8.

Mustoe, G. E. 1982. The origin of honeycomb weathering. *Geological Society of America, Bulletin* **93**, 108–15.

Mustoe, G. E. 1983. Cavernous weathering in the Capitol

Reef Desert, Utah. *Earth Surface Processes and Landforms* **8**, 517–26.

Nahon, D. B. 1976. *Cuirasses ferrugineuses et encroûtements calcaires au Sénégal occidental et Mauritanie. Systèmes évolutifs, géochimie, structures, relais et coexistence.* Sciences Géologiques Mémoires 44, Université Louis Pasteur de Strasbourg, Institut de Géologie.

Nahon, D. B. 1986. Evolution of iron crusts in tropical landscapes. See *Rates of chemical weathering of rocks and minerals*, S. M. Colman & D. P. Dethier (eds), 169–91. London: Academic Press.

Nahon, D. B. & R. Trompette 1982. Origin of siltstone: glacial grinding versus weathering. *Sedimentology* **29**, 25–35.

Nakata, J. K., H. G. Wilshire, C. G. Barnes 1976. Origin of Mojave Desert dust plumes photographed from space. *Geology* **4**, 644–48.

Nalpanis, P. 1985. Saltating and suspended particles over flat and sloping surfaces, II: experiments and numerical simulations. See Barndorff-Nielsen et al. (1985), 37–66.

Nalpanis, P., J. C. R. Hunt, C. F. Barrett 1990. Saltating particles in turbulent air flow over flat erodible surfaces. [MS]

Nanson, G. C., B. R. Rust, G. Taylor 1986. Coexistent mud braids and anastomosing channels in an arid-zone river: Cooper's Creek, Central Australia. *Geology* **14**, 175–8.

Naranjo, J. A. & P. Cornejo 1989. Avalanchas multiples del Volcan Chaco en el norte de Chile: un mecanismo de degradacion de volcanes miocenos. *Revista Geologica Chile* **16**, 61–72.

Naranjo, J. A. & R. Paskoff 1985. Evolution Cenozoica del piedemonte Andino en la Pampa del Tamarugal, Norte de Chile (18°–21°S). *IV Congresso Geologico Chileno* **5**, 149–65.

Neal, J. T. 1965a. *Geology, mineralogy and hydrology of US playas.* US Air Force Cambridge Research Laboratories, Environmental Research Papers 96.

Neal, J. T. 1965b. *Giant desiccation polygons of Great Basin playas.* US Air Force Cambridge Research Laboratories Environmental, Research Papers 123.

Neal, J. T. 1968. Playa surface changes at Harper Lake, California, 1962–67. In *Playa surface morphology: miscellaneous investigations*, J. T. Neal (ed.), US Air Force Cambridge Research Laboratories, Environmental Research Papers 283.

Neal, J. T. 1969. Playa variation. See McGinnies et al. (1969), 13–44.

Neal, J. T. (ed.) 1975. *Playas and dried lakes.* Stroudsburg, Pennsylvania: Dowden, Hutchinson and Ross.

Neal, J. T., A. M. Langer, P. G. Kerr 1968. Giant desiccation polygons of Great Basin Playas. *Geological Society of America, Bulletin* **79**, 69–90.

Neal, J. T. & W. S. Motts 1967. Recent geomorphic changes in playas of western United States. *Journal of Geology* **75**, 511–25.

Nechaec, L. A. 1969. A study of duststorms at pale sandy sierozems of the piedmont valleys of the northern Tyan-Shan [in Russian]. *Vestnik Sel'skokhozyaystvennoy Nauki* **7**, 15–8.

Needham, C. E. 1937. Ventifacts in New Mexico. *Journal of Sedimentary Petrology* **7**, 31–3.

Nemec, J. & J. A. Rodier 1979. Streamflow characteristics in areas of low precipitation (with special reference to low and high flows). ISAH Publication 128, 125–40.

Netterberg, F. 1969. Ages of calcretes in southern Africa. *South African Archaeological Bulletin* **24**, 88–92.

Netterberg, F. 1978. Dating and correlation of calcretes and other pedocretes. *Geological Society of South Africa, Transactions* **81**, 379–91.

Netterberg, F. 1979. Dating and correlation of calcretes and other pedocretes. *Geological Society of South Africa, Transactions* **81**, 379–92.

Netterberg, F. & P. A. Loudon 1980. Simulation of salt damage to roads with laboratory model experiments. *7th Regional Conference for Africa on Soil Mechanical and Foundation Engineering*, 355–61.

Nettleton, W. D., B. R. Brasher, J. M. Yenter, T. W. Priest 1990. Geomorphological age and genesis of some San Luis Valley, Colorado, soils. *Soil Science Society of America, Journal* **53**, 165–70.

Nettleton, W. D., J. E. Witty, R. E. Nelson, J. W. Hawley 1975. Genesis of argillic horizons in soils of desert areas of the southwest United States. *Soil Science Society of America, Proceedings* **39**, 919–26.

Nevyazhskiy, I. I. & R. A. Bidzhiev 1960. Aeolian relief forms in central Yakutia [in Russian]. *Akademiia Nauk, SSR, Izvestiya, Serie Geograficheskaya* **3**, 90–5.

Newbold, D. 1924. A desert odyssey of a thousand miles. *Sudan Notes and Records* **7**, 43–92, 104–7.

Newell, R. E., S. Gould-Stewart, J. C. Chung 1981. A possible interpretation of palaeoclimatic reconstructions for 18,000 BP for the region 60°N to 60°S, 60°W to 100°E. *Palaeoecology of Africa* **13**, 1–19.

Nichol, J. E. 1991. The extent of desert dunes in northern Nigeria as shown by image enhancement. *Geographical Journal* **57**, 13–24.

Nicholas, R. M. & J. C. Dixon 1986. Sandstone scarp form and retreat of Standing Rocks, Canyonlands National Park, Utah. *Zeitschrift für Geomorphologie* **30**, 167–87.

Nicholson, S. E. 1981. Saharan climates in historic times. See *The Sahara: ecological change and early economic history*, J. A. Allan (ed.), 35–59. London: Menas Press.

Nicholson, S. E. & H. Flohn 1981. African climate changes in late Pleistocene and Holocene and the general atmospheric circulation. IASH Publication 131, 295–301.

Nickling, W. G. 1978. Eolian sediment transport during dust storms: Slims River Valley, Yukon Territory. *Canadian Journal of Earth Science* **15**, 1069–84.

Nickling, W. G. 1983. Grain size characteristics of sediment transported during dust storms. *Journal of Sedimentary Petrology* **53**, 1011–24.

Nickling, W. G. 1984. The stabilizing role of bonding agents on the entrainment of sediment by wind.

Sedimentology **31**, 111–8.

Nickling, W. G. (ed.) 1986. *Aeolian geomorphology.* **Boston: Allen & Unwin.**

Nickling, W. G. 1988. The initiation of particle movement by wind. *Sedimentology* **35**, 499–511.

Nickling, W. G. 1990. Potential dust emission from desert soils in Mali, West Africa. In *Extended abstracts,* NATO *Advanced Workshop on sand, dust and soil in their relation to aeolian and littoral processes.* 14–18 May, 1990, Sandbjerg, Denmark.

Nickling, W. G. & M. Ecclestone 1980. A technique for detecting grain motion in wind tunnels and flumes. *Journal of Sedimentary Petrology* **50**, 652–4.

Nicot, J. 1955. Remarques sur les peuplements de micromycetes des sables désertiques. *Academie des Sciences à Paris, Comptes Rendus* **240**, 2082–4.

Nielson, J. & G. Kocurek 1986. Climbing zibars of the Algodones. *Sedimentary Geology* **48**, 313–48.

Nielson, J. & G. Kocurek 1987. Surface processes, deposits, and development of star dunes, Dumont dune field, California. *Geological Society of America, Bulletin* **99**, 177–86.

Nilsen, T. H. & T. E. Moore 1984. *Bibliography of alluvial-fan deposits.* Norwich: Geobooks.

Nordstrom, K. F., J. M. McCluskey, P. S. Rosen 1986. Aeolian processes and dune characteristics of a developed shoreline: Westhampton Beach, New York. See Nickling (1986), 131–47.

Norin, E. 1941. *The Tarim Basin and its border regions.* Leipzig: Akademie.

Norris, R. M. 1956. Crescentic beach cusps and barchans. *AAPG Bulletin* **40**, 1681–6.

Norris, R. M. 1966. Barchan dunes of Imperial Valley, California. *Journal of Geology* **74**, 292–306.

Norris, R. M. 1969. Dune reddening and time. *Journal of Sedimentary Petrology* **39**, 7–11.

Norris, R. M. & K. S. Norris 1961. Algodones dunes of southeastern California. *Geological Society of America, Bulletin* **72**, 605–20.

Noy-Meir, I. 1973. Desert ecosystems: environment and producers. *Annual Review of Ecology and Systematics* **4**, 25–51.

Noy-Meir, I. & Y. Harpaz 1976. Nitrogen cycling in annual pastures and crops in a semi-arid region. See *Symposium on cycling of mineral nutrients in agricultural systems: agro-ecosystems,* 360–4. Amsterdam: Elsevier.

Nutting, W. L., M. I. Haverty, J. P. LaFage 1987. Physical and chemical alteration of soil by two subterranean termite species in Sonoran Desert grassland. *Journal of Arid Environments* **12**, 233–9.

Oakley, K. P. 1952. Dating the Libyan Desert silica glass. *Nature* **170**, 447–9.

Obeid, M. & A. Seif el Din 1971. Ecological studies of the vegetation of the Sudan, III: the effect of simulated rainfall distribution at different isohyets of the regeneration of *Acacia senegal* (L) Wild. on clay and sandy soils. *Journal of Applied Ecology* **8**, 203–16.

Oberlander, T. M. 1972. Morphogenesis of granitic boulder slopes in the Mojave Desert, California. *Journal of Geology* **80**, 1–20.

Oberlander, T. M. 1974. Landscape inheritance and the pediment problem in the Mojave Desert of southern California. *American Journal of Science* **274**, 849–75.

Oberlander, T. M. 1977. Origin of segmented cliffs in massive sandstones of southeastern Utah. See Doehring (1977), 79–114.

Obika, B., R. J. Freer-Hewish, P. G. Fookes 1989. Soluble salt damage to thin bituminous road and runway surfaces. *Quarterly Journal of Engineering Geology* **22**, 59–73.

Ochsenius, C. 1982. Atacama: the hologenesis of the Pacific coastal desert in the context of the tropical South American Quaternary. *Striae* **17**, 112–31.

Odum, H. T. 1952. Carolina Bays and the Pleistocene weather map. *American Journal of Science* **250**, 263–70.

Oertel, G. F. & M. Larsen 1976. Development sequences in Georgia coastal dunes and distribution of dune plants. *Georgia Academy of Sciences Bulletin* **34**, 35–48.

Offer, Z. I. & D. Goosens 1990. Airborne dust in the northern Negev Desert (January–December 1987); general occurrence and dust content measurement. *Journal of Arid Environments* **18**, 1–20.

Ohmori, H., K. Iwasaki, K. Takeuchi 1983. Relationships between the recent dune activities and the rainfall fluctuations in the southern part of Australia. *Geographical Review of Japan* **56**, 131–50.

Oliver, F. W. 1945. Dust storms in Egypt and their relation to the war period, as noted in Maryut, 1939–45. *Geographical Journal* **106**, 26–49.

Ollier, C. D. 1966. Desert gilgai. *Nature* **212**, 581–3.

Ollier, C. D. 1978. Silcrete and weathering. See Langford-Smith (1978b), 13–17.

Ollier, C. D. 1984. *Weathering,* 2nd edn. Harlow, England: Longman.

Ollier, C. D. 1988. The regolith in Australia. *Earth Science Reviews* **25**, 355–62.

Ollier, C. D. & J. E. Ash 1983. Fire and rock breakdown. *Zeitschrift für Geomorphologie* **27**, 363–74.

Ollier, C. D. & R. W. Galloway 1990. The laterite profile, ferricrete and unconformity. *Catena* **17**, 97–109.

Olson, J. S. 1958a. Lake Michigan dune development, 1: wind-velocity profiles. *Journal of Geology* **66**, 254–63.

Olson, J. S. 1958b. Lake Michigan dune development, 2: plants as agents and tools in geomorphology. *Journal of Geology* **66**, 345–51.

Olson, J. S. 1958c. Lake Michigan dune development, 3: lake level, beach and dune oscillations. *Journal of Geology* **66**, 473–83.

Olsson, L. 1985. *An integrated study of desertification: applications of remote sensing,* GIS *and spatial models in semiarid Sudan.* Meddelanden fron Lunds Universitets Geografiska Institutionen Arhandlingar 98.

Olsson-Seffer, P. 1908. Relation of wind to topography of coastal drift sands. *Journal of Geology* **16**, 549–64.

Orians, G. H. & O. T. Solberg 1977. A cost–income

model of leaves and roots with special reference to arid and semi-arid areas. *The American Naturalist* **111**, 677–90.

Orme, A. R. & V. P. Tchakerian 1986. Quaternary dunes of the Pacific coast of the Californias. See Nickling (1986), 149–75.

Osborn, G. 1985. Evolution of the late Cenozoic inselberg landscape of southwestern Jordan. *Palaeogeography, Palaeoclimatology, Palaeoecology* **49**, 1–23.

Osborn, H. B. & L. Lane 1969. Precipitation–runoff relations for very small semiarid rangeland watersheds. *Water Resources Research* **5**, 419–25.

Osterkamp, W. R. & W. W. Wood 1987. Playa-lake basins on the southern High Plains of Texas and New Mexico, part I: hydrologic, geomorphic and geologic evidence for their development. *Geological Society of America, Bulletin* **99**, 215–23.

Otterman, J. 1975. Observations of wind streaklines over the Red Sea from the ERTS-1 imagery. *Remote Sensing of the Environment* **4**, 79–94.

Otterman, J. & V. Gornitz 1983. Saltation versus soil stabilization: two processes determining the character of surfaces in arid regions. *Catena* **10**, 339–62.

Otterman, J., Y. Waisel, E. Rosenberg 1975. Western Negev and Sinai ecosystem: comparative study of vegetation, albedo and temperatures. *Agro-Ecosystems* **2**, 47–59.

Owen, P. R. 1964. Saltation of uniform grains in air. *Journal of Fluid Mechanics* **20**, 225–42.

Owen, P. R. 1985. Erosion of dust by a turbulent wind. See El-Baz & Hassan (1986).

Pachur, H. J. & S. Kröpelin 1987. Wadi Howar: paleo-climatic evidence from an extinct river system in the south-eastern Sahara. *Science* **237**, 298–300.

Pachur, H-J. & P. Hoelzmann 1991. Palaeoclimatic implications of late Quaternary lacustrine sediments in Western Nubia. *Sudan Quaternary Research* **36**, 257–76.

Packer, I. J. & R. L. Ison 1980. Stabilization of aeolian sand in the Braidwood area. *Journal of the Soil Conservation Service of New South Wales* **36**, 90–9.

Paige, S. 1912. Rock-cut surfaces in the desert ranges. *Journal of Geology* **20**, 442–50.

Pandey, S. & P. C. Chatterji 1970. Genesis of "Mitha Ranns", "Kharia Rann" and "Kanodurala Ranns" in the Great Indian Desert. *Annals of Arid Zone* **9**, 175–80.

Park, C. C. 1977. World-wide variations in hydraulic geometry exponents of stream channels: an analysis and some observations. *Journal of Hydrology* **33**, 133–46.

Parke, J. G. 1857. *Report of explorations for railroad routes*. US Congress House Executive Document 91, 33rd Congress, 2nd Session.

Parker, G. G. 1963. Piping, a geomorphic agent in landform development in drylands. IASH Publication 65, 103–13.

Parker, G. G. & E. A. Jenne 1967. The structural failure of US highways caused by piping. *USGS Water Resources Division* (Washington, DC) for 46th Annual Meeting of the Highways Research Board.

Parkin, D. W. & N. J. Shackleton 1973. Trade wind and temperature correlations down a deep-sea core off the Saharan coast. *Nature* **245**, 455–7.

Parmenter, C. & D. W. Folger 1974. Eolian biogenic detritus in deep-sea sediments: a possible index of equatorial ice age aridity. *Science* **185**, 695–8.

Parsons, A. J. & A. D. Abrahams 1984. Mountain mass denudation and piedmont formation in the Mojave and Sonoran deserts. *American Journal of Science* **284**, 255–71.

Parsons, A. J., A. D. Abrahams, S. Luk 1991. Size characteristics of sediment in interrill overland flow on a semiarid hillslope, southern Arizona. *Earth Surface Processes and Landforms* **16**, 143–52.

Partridge, J. & V. R. Baker 1987. Palaeoflood hydrology of the Salt River, Arizona. *Earth Surface Processes and Landforms* **12**, 109–25.

Paskoff, R. 1978–9. Sobre la evolucion geomorfologia del gran acantilado costero del Norte Grande de Chile. *Norte Grande* (Inst. Geogr. Univ. Catolica de Chile) **6**, 1–12.

Passarge, S. 1904. *Die Kalahari* [2 volumes]. Berlin: Reimer.

Passarge, S. 1911. Die pfannenformigen Hohlformen der Südafrikanischen Steppen. *Petermanns Mitteilungen* **57**, 130–5.

Passarge, S. 1930. Ergebnisse einer Studienreise nach Sud-tunisien im Jahre 1928. *Mitteilungen der Geographische Gesellschaft, Hamburg* **61**, 96–122.

Patton, P. C. & S. A. Schumm 1981. Ephemeral-stream processes: implications for studies of Quaternary valley fills. *Quaternary Research* **15**, 24–43.

Pearse, J. R., D. Lindley, D. C. Stevenson 1981. Wind flow over ridges in simulated atmospheric boundary layers. *Boundary-layer Meteorology* **21**, 77–92.

Pécsi, M. (ed.) 1970. *Problems of relief planation.* Budapest: Akademie Kiado.

Pedgeley, D. E. 1970. The climate of the interior of Oman. *Meteorological Magazine* **99**, 29–37.

Pedro, G. 1957. Nouvelles recherches sur l'influence des sels dans la désagregation des roches. *Academie des Sciences à Paris, Comptes Rendus* **244**, 2822–4.

Peel, R. F. 1941. Denudational landforms of the central Libyan desert. *Journal of Geomorphology* **4**, 3–23.

Peel, R. F. 1968. Landscape sculpture by wind. *21st International Geographical Congress, India, Papers I*, 99–104.

Peel, R. F. 1974. Insolation weathering: some measurements of diurnal temperature change in exposed rocks in the Tibesti region, central Sahara. *Zeitschrift für Geomorphologie*, Supplementband 21, 19–28.

Peignot, Lt. 1913. Notice sur la subdivision de Zigueï. *La Géographie* **27**, 322–3.

Penck, A. 1905. Climatic features of the land surface. *American Journal of Science* **19**, 165–74.

Penck, W. 1920. Der südrand der Puna de Atacama. Ak. Wissenschaften, Leipzig, Math–Phys. Klass. Abhandhungen 37.

Penck, W. 1953. *Morphological analysis of land forms.*

[Translated from the German by H. Czech & K. Boswell.] London: Macmillan.

Perrin, R. M. S. & C. W. Mitchell 1969/1971. *An appraisal of physiographic units for predicting site conditions in arid areas.* [UK] Military Engineering Experimental Establishment Report 1111 [2 volumes].

Perry, R. A. et al. 1962. *General report on the lands of the Alice Springs area, Northern Territory, 1956–57.* CSIRO Land Research Series 6.

Perry, R. S. & J. B. Adams 1978. Desert varnish: evidence for the cyclic deposition of manganese. *Nature* **276**, 489–91.

Perthouisot, P. & A. Jauzein 1975. Sebkhas et dunes d'argile; enclave endoreique du Pont du Fahs, Tunisie. *Revue de Géographie Physique et de Géologie Dynamique* **17**, 295–306.

Pesce, A. 1968. *Gemini space photographs of Libya and Tripoli.* Tripoli: Petroleum Exploration Society of Libya.

Peterson, F. F. 1980. Holocene desert soil formation under sodium salt influence in a playa-margin environment. *Quaternary Research* **13**, 172–86.

Peterson, F. F. 1981. *Landforms of the Basin and Range Province defined for soil survey.* Nevada Agricultural Experiment Station, University of Nevada, Reno, Technical Bulletin 28.

Peterson, G. L. 1980a. Sediment size reduction by salt weathering in arid environments. *Geological Society of America, Abstracts* **12**, 301.

Peterson, G. L. 1980b. Broken rocks in the desert environment. See *Geology and mineral wealth of the California Desert*, D. L. Fife & A. R. Brown (eds). South Coast Geological Society (field trip guidebook).

Petrie, W. M. Flinders 1889. Wind action in Egypt. *Royal Geographical Society, Proceedings* **11**, 646–50.

Petrov, M. P. 1962. Types de déserts de l'Asie centrale. *Annales de Géographie* **71**, 131–55.

Petrov, M. P. 1976. *Deserts of the world.* New York: John Wiley.

Péwé, T. L. 1974. Geomorphic processes in polar deserts. See *Polar deserts and modern man*, T. H. Smiley & J. H. Zumberge (eds), 33–69. Tucson: University of Arizona Press.

Péwé, T. L. 1981. Dust, an overview. Geological Society of America, Special Paper 186, 1–10.

Péwé, T. L. 1989. Environmental geology in Arizona. See *Geology of Arizona* [draft MS]. Tucson: Arizona Geological Society.

Phillips, J. A. 1882. The red sands of the Arabian desert. *Geological Society of London, Quarterly Journal* **38**, 110–3.

Phillips, S. E., A. R. Milnes, R. C. Foster 1987. Calcified filaments: an example of biogenic influence in the formation of calcrete in South Australia. *Australian Journal of Soil Research* **25**, 405–28.

Phillips, S. E. & P. G. Self 1987. Morphology, crystallography and origin of needle-fibre calcite in Quaternary pedogenic calcretes in South Australia. *Australian Journal of Soil Research* **25**, 429–44.

Picard, M. D. & L. H. High Jr 1973. *Sedimentary structures of ephemeral streams.* Amsterdam: Elsevier.

Pilgrim, G. E. 1908. *The geology of the Persian Gulf and the adjoining portions of Persia and Arabia.* Memoirs of the Geological Survey of India, 34.

Pillans, B. 1989. Lake shoulders – aeolian clay sheets associated with ephemeral lakes in basalt terrain, southern New South Wales. *Search* **87**, 313–5.

Pissart, A. 1960. Les dépressions fermées de la région Parisienne. *Revue de Géomorphologie Dynamique* **11**, 12.

Plafker, G. 1964. Oriented lakes and lineaments in northeastern Bolivia. *Geological Society of America, Bulletin* **75**, 503–22.

Poesen, J. 1986. Surface sealing as influenced by slope angle and position of simulated stones in the top layer of loose sediments. *Earth Surface Processes and Landforms* **11**, 1–10.

Pogue, J. E. 1911. A possible limiting effect of ground water on eolian action. *Journal of Geology* **19**, 270–1.

Porath, A. & A. P. Schick 1974. The use of remote sensing systems in monitoring desert floods. IASH Special Publication 112, 133–8.

Porter, M. L. 1986. Sedimentary record of erg migrations. *Geology* **14**, 497–500.

Potter, R. M. & G. R. Rossman 1977. Desert varnish: the importance of clay minerals. *Science* **196**, 1446–8.

Powell, J. W. 1875. *Exploration of the Colorado River of the West (1869–72)* (Washington DC). [Reprinted in 1957 by the University of Chicago Press.]

Powell, J. W. 1878. *Report on the lands of the arid region of the United States.* Washington DC: Department of the Interior.

Powers, M. C. 1953. A new roundness scale for sedimentary particles. *Journal of Sedimentary Petrology* **23**, 117–9.

Prell, A. J. 1984. Monsoonal climate of the Arabian Sea during the late Quaternary: a response to changing solar variations? In *Milankovitch and climate*, A. L. Berger, J. Imbrie, S. Hays, G. Kukla, B. Saltzman (eds) 349–66. Dordrecht: Reidel.

Prell, W. L., W. H. Hutson, D. F. Williams, A. W. H. Be, K. Geitzenauer, B. Molfino 1980. Surface circulation of the Indian Ocean during the last glacial maximum, approximately 18,000 BP. *Quaternary Research* **14**, 309–36.

Prell, W. L. & E. Van Campo 1986. Coherent response of Arabian Sea upwelling and pollen transport to late Quaternary monsoonal winds. *Nature* **323**, 526–8.

Price, W. A. 1944. Greater American deserts. *Texas Academy of Sciences, Proceedings and Transactions* **27**, 163–70.

Price, W. A. 1949. Pocket gophers as architects of mima (pimple) mounds of the western United States. *Texas Journal of Science* **1**, 1–17.

Price, W. A. 1950. Saharan sand dunes and the origin of the longitudinal dune: a review. *Geographical Review* **40**, 462–5.

Price, W. A. 1963. Physico-chemical and environmental

factors in clay-dune genesis. *Journal of Sedimentary Petrology* **33**, 766–78.

Price-Williams, D., A. Watson, A. S. Goudie 1982. Quaternary colluvial stratigraphy, archaeological sequences and palaeoenvironments in Swaziland, southern Africa. *Geographical Journal* **148**, 50–67.

Prince, H. C. 1961. Some reflections on the origin of hollows in Norfolk compared to those in the Paris region. *Revue de Géomorphologie Dynamique* **12**, 109–17.

Prince, H. C. 1964. The origin of pits and depressions in Norfolk. *Geography* **49**, 15–32.

Prohaska, F. J. 1973. New evidence on the climatic controls along the Peruvian coast. See *Coastal deserts: their natural and human environments*, D. H. K. Amiran & A. W. Wilson (eds), 91–107. Tucson: University of Arizona Press.

Prospero, J. M., R. A. Glaccum, R. T. Nees 1981. Atmospheric transport of soil dust from Africa to South America. *Nature* **289**, 570–2.

Prouty, W. F. 1952. Carolina Bays and their origin. *Geological Society of America, Bulletin* **63**, 167–224.

Psuty, N. P. 1989. An application of science to the management of coastal sand dunes along the Atlantic coast of the USA. See Gimmingham et al. (1989), 289–307.

Pullen, R. A. 1979. Termite hills in Africa: their characteristics and evolution. *Catena* **6**, 267–91.

Purser, B. H. (ed.) 1973. *The Persian Gulf*. **Berlin: Springer.**

Purser, B. H. & G. Evans 1973. Regional sedimentation along the Trucial Coast, Persian Gulf. See Purser (1973), 211–31.

Pye, K. 1980. Beach salcrete and eolian sand transport: evidence from North Queensland. *Journal of Sedimentary Petrology* **50**, 257–61.

Pye, K. 1982a. Morphological development of coastal dunes in a humid tropical environment, Cape Bedford and Cape Flattery, North Queensland. *Geografiska Annaler* **64A**, 212–27.

Pye, K. 1982b. Morphology and sediments of the Ramsay Bay sand dunes, Hinchinbrook Island, North Queensland. *Royal Society of Queensland, Proceedings* **93**, 31–47.

Pye, K. 1982c. SEM observations on some sand fulgerites from northern Australia. *Journal of Sedimentary Petrology* **52**, 991–8.

Pye, K. 1983a. Coastal dunes. *Progress in Physical Geography* **7**, 531–46.

Pye, K. 1983b. Dune formation on the humid tropical sector of the North Queensland Coast, Australia. *Earth Surface Processes and Landforms* **8**, 371–81.

Pye, K. 1983c. Early post-depositional modification of aeolian sands. See Brookfield & Ahlbrandt (1983), 197–221.

Pye, K. 1983d. Red beds. See Goudie & Pye (1983), 227–63.

Pye, K. 1984a. Loess. *Progress in Physical Geography* **8**, 176–217.

Pye, K. 1984b. Models of transgressive dune building episodes and their relationship to Quaternary sea-level

changes: a discussion with reference to evidence from eastern Australia. See *Coastal research: UK perspectives*, M. Clark (ed.), 81–104. Norwich: Geobooks.

Pye, K. 1985. Controls on fluid threshold velocity, rates of aeolian sand transport and dune grain-size parameters along the Queensland coast. See Barndorff-Nielsen et al. (1985), 483–509.

Pye, K. 1987. *Aeolian dust and dust deposits*. London: Academic Press.

Pye, K., A. S. Goudie, D. S. G. Thomas 1984. A test of petrological control in the development of bornhardts and kopples on the Matopos batholith, Zimbabwe. *Earth Surface Processes and Landforms* **9**, 455–67.

Pye, K. & C. H. B. Sperling 1983. Experimental investigation of silt formation by static breakage processes: the effect of temperature, moisture and salt on quartz dune and granitic regolith. *Sedimentology* **30**, 49–62.

Pye, K. & H. Tsoar 1987. The mechanics and geological implications of dust transport and deposition in deserts, with particular reference to loess formation and dune sand diagenesis in the northern Negev, Israel. See Frostick & Reid (1987), 139–56.

Queiroz, J., A. R. Southard, G. L. Woolridge 1982. Characteristics of soils in a stone-free surficial deposit in central Utah. *Soil Science Society of America, Journal* **46**, 777–81.

Queney, P. 1953. Classification des rides de sable et théorie ondulatoire de leur formation. See *Actions éoliennes* (CNRS). *Colloques Internationaux* **35**, 179–95.

Rabenhorst, M. C. & L. P. Wilding 1986a. Pedogenesis on the Edwards Plateau, Texas, II: formation and occurrence of diagnostic subsurface horizons in a climosequence. *Soil Science Society of America, Journal* **50**, 687–92.

Rabenhorst, M. C. & L. P. Wilding 1986b. Pedogenesis on the Edwards Plateau, Texas, III: New model for the formation of petrocalcic horizons. *Soil Science Society of America, Journal* **50**, 693–9.

Rachocki, A. 1981. *Alluvial fans*. Chichester, England: John Wiley.

Rahn, P. H. 1966. Inselbergs and ruckpoints in southwestern Arizona. *Zeitschrift für Geomorphologie* **10**, 217–25.

Rahn, P. H. 1967. Sheetfloods, streamfloods and the formation of pediments. *Annals of the AAG* **57**, 593–604.

Ranwell, D. S. 1960. Newborough Warren, Anglesey, III: changes in the vegetation on parts of the dune system after the loss of rabbits by myxamatosis. *Journal of Ecology* **48**, 385–95.

Ranwell, D. S. 1972. *The ecology of salt marshes and sand dunes*. London: Chapman & Hall.

Rapp, A. 1975. Some views on the Ordovician palaeoglaciation in Saharan Africa. *Geologiska Föreningens i Stockholm, Förhandlingar* **97**, 142–50.

Rapp, A., L. Berry, P. H. Temple 1972. Soil erosion and sedimentation in Tanzania – the project. *Geografiska Annaler* **54**, ser. 9A, 3–4.

Rasmussen, K. 1989. Some aspects of flow over coastal

dunes. See Gimmingham et al. (1989), 129–47.

Rasmussen, K., M. Sorensen, B. B. Willetts 1985. Measurement of saltation and wind strength on beaches. See Barndorff-Nielsen et al. (1985), 301–26.

Raudkivi, A. J. 1966. Bedforms in alluvial channels. *Journal of Fluid Mechanics* **22**, 113–33.

Raudkivi, A. J. 1976. *Loose boundary hydraulics*, 2nd edn. Oxford: Pergamon.

Raupach, M. R. 1990. Gusts and turbulence in saltation layers. *Acta Mechanica*, **2**, 83–96.

Raupach, M. R., A. S. Thom, I. Edwards 1980. A wind-tunnel study of turbulent flow close to regularly arrayed rough surfaces. *Boundary-layer Meteorology* **18**, 373–97.

Raverty, H. G. 1878. The Mihran of Sind. *Asiatic Society of Bengal, Journal* **61**.

Ravina, I. & D. Zaslavsky 1974. The electrical double layer as a possible factor in desert weathering. *Zeitschrift für Geomorphologie* **21**, 13–18.

Rea, D. K. & M. Leinen 1988. Asian aridity and the zonal westerlies: late Pleistocene and Holocene record of aeolian deposition in the northwest Pacific. *Palaeogeography, Palaeoclimatology, Palaeoecology* **66**, 1–8.

Read, D. J. 1989. Mycorrhizas and nutrient cycling in sand dune ecosystems. See Gimmingham et al. (1989), 89–110.

Rebillard, P. & J. L. Ballais 1984. Surficial deposits of two Algerian playas as seen on SIR-A, Seasat and Landsat coregistered data. *Zeitschrift für Geomorphologie* **28**, 483–98.

Reed, R. D. 1930. Recent sands of California. *Journal of Geology* **38**, 223–45.

Reeves Jr, C. C. 1966. Pluvial lake basins of western Texas. *Journal of Geology* **74**, 269–91.

Reeves Jr, C. C. 1983. Pliocene channel calcrete and suspenparallel drainage in west Texas and New Mexico. See R. C. L. Wilson (1983), 179–83.

Reheis, M. C. 1987a. Climate implications of alternating clay and carbonate formation in semi-arid soils of south-central Montana. *Quaternary Research* **27**, 270–82.

Reheis, M. C. 1987b. *Gypsic soils on the Kane alluvial fans, Bighorn County, Wyoming*. USGS Bulletin 1590C.

Reheis, M. C. 1990. Influence of climate and eolian dust on the major-element chemistry and clay mineralogy of soils in the western Bighorn Basin, USA. *Catena* **17**, 219–48.

Reid, D. G. 1985. Wind statistics and the shape of sand dunes. See Barndorff-Nielsen et al. (1985), 393–419.

Reid, I. & L. E. Frostick 1987. Flow dynamics and suspended sediment properties in arid zone flash floods. *Hydrological Processes* **1**, 239–53.

Reiter, E. 1969. Tropospheric circulation and jet streams. See *Climate of the free atmosphere*, D. F. Rex (ed.), 85–203. Amsterdam: Elsevier.

Renard, K. G. 1970. Hydrology of semi-arid rangeland watersheds. *US Department of Agriculture Research Service Publication* ARS, 41–162.

Renard, K. G. 1972. Sediment problems in the arid and semi-arid Southwest. *27th Annual Meeting of the Soil Conservation Society America* (Portland, Oregon), 225–32.

Renard, K. G. & R. V. Keppel 1966. Hydrographs of ephemeral streams in the Southwest. *American Society of Civil Engineers, Proceedings; Journal of the Hydrology Division* **92**, 33–52.

Rendell, H. & D. Alexander 1979. Note on some spatial and temporal variations in ephemeral channel form. *Geological Society of America, Bulletin* **90**, 761–72.

Rex, R. W. 1961. Hydrodynamic analysis of circulation and orientation of lakes in northern Alaska. See *Geology of the Arctic*, G. O. Raasch (ed.), *1st International Symposium on Arctic Geology* 2, 1021–43. Toronto: University of Toronto Press.

Reynolds, A. J. 1965. Waves on the erodible bed of an open channel. *Journal of Fluid Mechanics* **22**, 113–33.

Rhoads, B. L. 1986. Flood hazard assessment for land-use planning near desert mountains. *Environmental Management* **10**, 97–106.

Rhoads, B. L. 1988. Mutual adjustments between processes and form in a desert mountain fluvial system. *Annals of the AAG* **78**, 271–87.

Rhodes, D. D. & G. P. Williams (eds) 1979. *Adjustments of the fluvial system*. Boston: Allen & Unwin.

Rhodes, E. G. 1982. Depositional model for a chenier plain, Gulf of Carpentaria, Australia. *Sedimentary Geology* **29**, 201–21.

Rice, A. 1976. Insolation warmed over. *Geology* **4**, 61–2.

Rich, J. L. 1935. Origin and evolution of rock fans and pediments. *Geological Society of America, Bulletin* **46**, 999–1024.

Richards, K. J. 1980. The formation of ripples and dunes on an erodible bed. *Journal of Fluid Mechanics* **99**, 597–618.

Richards, K. J. 1982. *Rivers – form and process in alluvial channels*. London: Methuen.

Richards, K. J. 1986. Turbulent flow over topography, with applications to sand wave development. See El-Baz & Hassan (1986), 435–61.

Richardson, P. D. 1968. The generation of scour marks near obstacles. *Journal of Sedimentary Petrology* **38**, 965–70.

Riche, G., D. Rambaud, M. Riera 1982. Étude morphologique d'un encroûtement calcaire, région d'Irece, Bahia, Brésil. *Cahiers de l'ORSTOM, Serie Pédologie* **19**, Paris, 257–70.

Ridgway, K. & R. Rupp 1970. Whistling sand of Porth Oer, Caernarfon. *Nature* **226**, 158.

Ridgway, K. & J. B. Scotton 1972. Whistling sands and seabed sand transport. *Nature* **238**, 212–3.

Riggs, H. C. 1979. Inventorying streamflows in areas of low precipitation. IASH Publication 128, 151–7.

Rim, M. 1952. The collection of sand and dust carried by the atmosphere. *Research Council of Israel, Bulletin* **2** 195–7.

Roberts, C. R. & C. W. Mitchell 1987. Spring mounds in southern Tunisia. See Frostick & Reid (1987), 321–34.

Roberts, N. 1983. Age, palaeoenvironments, and climatic

significance of late Pleistocene Konya Lake, Turkey. *Quaternary Research* **19**, 154–71.

Roberts, N. 1990. Ups and downs of African lakes. *Nature* **346**, 107.

Robertson, E. C. 1962. The Carolina Bays and emergence of the coastal plain of the Carolinas and Georgia. USGS Professional Paper 450C, 87–90.

Rockwell, T. K., E. A. Keller, D. L. Johnson 1985. Tectonic geomorphology of alluvial fans and mountain fronts near Ventura, California. See Morisawa & Hack (1985), 183–207.

Rögner, K. 1987. Temperature measurements of rock surfaces in hot deserts (Negev, Israel). See Gardiner (1987), 1271–86.

Rognon, P. 1976. Essai d'interpretation des variations climatiques au Sahara depuis 40,000 ans. *Revue Géographie Physique et Géologique Dynamique* **18**, 251–82.

Rognon, P. 1990. Field measurements of dust near the ground, correlated with surrounding soils in the Sahara and Sahel. In *Extended abstracts, NATO Advanced Workshop on sand, dust and soil in their relation to aeolian and littoral processes*. 14–18 May, 1990, Sandbjerg, Denmark.

Rognon, P. & G. Coudé-Gaussen. 1987. Réconstruction paléoclimatique à partir des sédiments du Pléistocène superieur et de l'Holocène du nord de Fuerteventura (Canaries). *Zeitschrift für Geomorphologie* **31**, 1–19.

Rognon, P. & M. A. J. Williams 1977. Late Quaternary climatic changes in Australia ad North Africa, a preliminary interpretation. *Palaeogeography, Palaeoclimatology, Palaeoecology* **21**, 285–327.

Rosato, A., F. Prinz, K. J. Strandburg, R. H. Swendsen 1987. Monte Carlo simulation of particulate matter segregation. *Powder Technology* **49**, 59–69.

Ross, G. M. 1983. Bigbear Erg: a Proterozoic intermontane eolian sand sea in the Hornby Bay Group, Northwest Territories, Canada. See Brookfield & Ahlbrandt (1983), 483–511.

Rossignol-Strick, M. & D. Duzer 1980. Late Quaternary West African climate inferred from palynology of Atlantic deep-sea cores. *Palaeoecology of Africa* **12**, 227–8.

Rostagno, C. M. & H. F. del Valle 1988. Mounds associated with shrubs in the aridic soils of northeastern Patagonia: characteristics and probable genesis. *Catena* **15**, 347–59.

Rozycki, S. Z. 1968. The directions of winds carrying loess dust as shown by analysis of accumulative loess forms in Bulgaria. See *Loess and related deposits of the world*, C. B. Shultz & J. C. Frye (eds), 235–46. Lincoln, Nebraska: University of Nebraska Press.

Rubin, D. M. 1984. Factors determining desert dune type – discussion. *Nature* **309**, 91–2.

Rubin, D. M. 1990. Lateral migration of linear dunes in the Strzelecki Desert, Australia. *Earth Surface Processes and Landforms* **15**, 1–14.

Rubin, D. M. & R. E. Hunter 1982. Bedform climbing in theory and nature. *Sedimentology* **29**, 121–38.

Rubin, D. M. & R. E. Hunter 1983. Reconstructing bedform assemblages from compound crossbedding. See Brookfield & Ahlbrandt (1983), 407–27.

Rubin, D. M. & R. E. Hunter 1985. Why deposits of longitudinal dunes are rarely recognised in the geologic record. *Sedimentology* **32**, 147–57.

Rubin, D. M., & R. E. Hunter 1987. Bedform alignment in directionally varying flows. *Science* **237**, 276–8.

Rubin, D. M. & H. Ikeda 1990. Flume experiments on the alignment of transverse, oblique, and longitudinal dunes in directionally varying flows. *Sedimentology* **37**, 673–84.

Rubin, D. M. & D. S. McCulloch 1980. Single and superimposed bedforms: a synthesis of San Francisco Bay and flume observations. *Sedimentary Geology* **26**, 207–31.

Rude, A. 1959. Les galets éolisées de l'Isle Herd (Australie) et les galets éolisées fossiles de Pont-Le-Chateau (Limogne d'Auvergne). *Revue de Géomorphologie Dynamique* **10**, 33–4.

Ruellan, A., G. Beaudet, G. Millot, D. Nahon, H. Paquet, P. Rognon 1979. Rôle des encroûtements calcaires dans le façonnement des glacis d'ablation en régions arides et semi-arides du Maroc. *Academie des Sciences à Paris, Comptes Rendus* **289D**, 619–22.

Russell, I. C. 1883. Sketch of the geological history of Lake Lahontan. *Third Annual Report of the US Geological Survey*, 189–235.

Russell, W. L. 1927. Drainage alignment in the western Great Plains. *Journal of Geology* **37**, 249–55.

Rust, B. R. 1981. Sedimentation in an arid zone anastomosing fluvial system: Cooper's Creek, Central Australia. *Journal of Sedimentary Petrology* **51**, 745–55.

Rutin, J. 1983. *Erosional processes on a coastal sand dune, De Blink, Noordwijkerhiut, The Netherlands*, Dissertatie Universiteit van Amsterdam (Publicaties van het Fysische Geografisch en Bodenkundig Laboratorium van de Universiteit van Amsterdam, Kaal, B. V., Amsterdam).

Ruxton, B. P. & I. Berry 1961. Weathering profiles and geomorphic position on granite in two tropical regions. *Revue de Géomorphologie Dynamique* **12**, 16–31.

Ryan, J. A. 1972. Relation of dust devil frequency and diameter to atmospheric temperature. *Journal of Geophysical Research* **77**, 7133–7.

Ryan, J. A. & J. J. Carrol 1970. Dust devil wind velocities: mature state. *Journal of Geophysical Research* **75**, 531–41.

Saarinen, T. F., V. R. Baker, R. Durrenberger, T. Maddock 1984. *The Tucson, Arizona, flood of October, 1983*. Washington DC: National Academy of Sciences.

Sagan, C. & R. A. Bagnold 1975. Fluid transport on Earth and aeolian transport on Mars. *Icarus* **26**, 209–18.

Sagan, C., J. Veverka, P. Gierasch 1971. Observational consequences of Martian wind regions. *Icarus* **15**, 253–78.

Said, R. 1960. New light on the origin of the Qattara Depression. *Société Géographique d'Egypt, Bulletin* **33**,

37–44.

Said, R. 1962. *The geology of Egypt*. Amsterdam: Elsevier.

Said, R. 1982. The geological evolution of the River Nile in Egypt. *Zeitschrift für Geomorphologie* **26**, 305–14.

Sainsbury, C. L. 1956. Wind-induced stone tracks, Prince of Wales Island, Alaska. *Geological Society of America, Bulletin* **67**, 1659–60.

Sakamoto-Arnold, C. M. 1981. Eolian features produced by the December, 1977 windstorm, southern San Joaquin Valley, California. *Journal of Geology* **89**, 129–37.

Salem, M. J. & M. T. Busrewil (eds) 1980. *Geology of Libya*, vol. III. London: Academic Press.

Salomons, W., A. S. Goudie, W. G. Mook 1978. Isotopic composition of calcrete deposits from Europe, Africa and India. *Earth Surface Processes* **3**, 43–57.

Samojlov, I. W. 1956. *Die Flussmundungen*. [Translated from the Russian by F. Tutenberg & H. Taubert]. Gotha, Germany: Geographie, Kartographie Austalt.

Sandford, K. S. 1933. Geology and geomorphology of the southern Libyan Desert. *Geographical Journal* **82**, 213–19.

Santos, P. F., N. Z. Elkins, Y. Steinberger, W. G. Whitford 1984. A comparison of surface and buried *Larrea tridentata* leaf litter decomposition in North American hot deserts. *Ecology* **65**, 278–84.

Sarnthein, M. 1972. Sediments and history of the postglacial transgression in the Persian Gulf and the northwest Gulf of Oman. *Marine Geology* **12**, 245–66.

Sarnthein, M. 1978. Sand deserts during the last glacial maximum and climatic optimum. *Nature* **272**, 43–6.

Sarnthein, M. & L. Diester-Haass 1977. Eolian sand turbidites. *Journal of Sedimentary Petrology* **47**, 868–90.

Sarnthein, M. & B. Koopman 1980. Late Quaternary deep-sea record of northwest African dust supply and wind circulation. *Palaeoecology of Africa* **12**, 239–53. Rotterdam: Balkema.

Sarnthein, M., G. Tetzlaff, B. Koopmann, K. Wolter, U. Pflaumann 1981. Glacial and interglacial wind regimes over the eastern tropical Atlantic and North-West Africa. *Nature* **293**, 193–6.

Sarnthein, M. & K. Walger, K. 1974. Der äolische Sandstrom aus der W-Sahara zür Atlantikküste. *Geologische Rundschau* **63**, 1065–87.

Sarnthein, M., E. Siebold, P. Rognon (eds) 1980. *Sahara and surrounding seas*. Rotterdam: Balkema.

Sarre, R. D. 1987. Aeolian sand transport. *Progress in Physical Geography* **11**, 157–82.

Sarre, R. D. 1988a. The morphological significance of vegetation and relief on coastal foredune processes. *Zeitschrift für Geomorphologie*, Supplementband 73, 17–31.

Sarre, R. D. 1988b. Evaluation of aeolian sand transport equations using intertidal zone measurements, Saunton Sands, England. *Sedimentology* **35**, 671–9.

Sarre, R. D. 1989. Aeolian sand drift from the intertidal zone on a temperate beach: potential and actual rates. *Earth Surface Processes and Landforms* **14**, 247–58.

Sarre, R. D. 1990. Evaluation of aeolian sand transport

equations using intertidal-zone measurements, Saunton Sands, England – reply. *Sedimentology* **37**, 389–92.

Sarre, R. D. & C. C. Chancey 1990. Size segregation during aeolian saltation on sand dunes. *Sedimentology* **37**, 357–65.

Savat, J. 1982. Common and uncommon selectivity in the process of fluid transportation: field observations and laboratory experiments on bare surfaces. See Yaalon (1982), 139–59.

Saxena, S. K. & S. Singh 1976. Some observations on the sand dunes and vegetation of Bikaner District in western Rajasthan. *Annals of the Arid Zone* **15**, 313–22.

Sayward, J. M. 1984. *Salt action on concrete*. US Army Corps of Engineers, Cold Regions Research and Engineering Laboratory Special Report 84-25.

Scattergood, R. O. & J. L. Routbort 1983. Velocity exponent in solid particle erosion. *American Ceramic Society, Journal* **66**, C184–C186.

Schaber, G. G., G. L. Berlin, W. E. Brown 1976. Variations in surface roughness within Death Valley, California: geologic evaluation of 25cm wavelength radar images. *Geological Society of America, Bulletin* **87**, 29–41.

Schattner, I. 1961. Weathering phenomena in the crystalline of the Sinai in the light of current notions. *Bulletin of the Research Council of Israel* IDG, 247–66.

Scheffer, F. & E. Meyer 1963. Biologische Ursachen der Wüstenlackbildung. B. and Kalk. *Zeitschrift für Geomorphologie* **7**, 112–9.

Schenk, C. J. 1983. Textural and structural characteristics of some experimentally formed eolian strata. See Brookfield & Ahlbrandt (1983), 41–9.

Schenk, C. J. & S. G. Fryberger 1988. Early diagenesis of eolian dune and interdune sands at White Sands, New Mexico. *Sedimentary Geology* **55**, 109–20.

Schick, A. P. 1970. Desert floods: interim results of observations in the Nahal Yael Research Watershed, southern Israel, 1965–1970, *IASH–UNESCO Symposium* (Wellington, New Zealand), 478–93.

Schick, A. P. 1971. A desert flood: physical characteristics; effects on man, geomorphic significance, human adaptation – a case study of the southern Arava watershed. *Jerusalem Studies in Geography* **2**, 91–155.

Schick, A. P. 1974a. Formation and obliteration of desert stream terraces – a conceptual analysis. *Zeitschrift für Geomorphologie*, Supplementband 21, 88–105.

Schick, A. P. 1974b. Alluvial fans and desert roads – a problem in applied geomorphology. *Abhandlungen der Akademic Wissenshalten in Göttingen, Math.–Physik.*, Klasse II **29**, 418–25.

Schick, A. P. 1977. A tentative sediment budget for an extremely arid watershed in the southern Negev. See Doehring (1987), 139–61.

Schick, A. P. 1987. Hydrologic aspects of floods in extreme arid environments. See *Flood geomorphology*, V. R. Baker, R. C. Kochel, P. C. Patton (eds), 189–203. Chichester, England: John Wiley.

Schick, A. P. & J. Lekach 1987. A high magnitude flood

in the Sinai Desert. See *Catastrophic flooding*, L. Mayer & D. Nash (eds), 381–410. Boston: Allen & Unwin.

Schick, A. P., J. Lekach, M. A. Hassan 1987. Vertical exchange of coarse bedload in desert streams. See Frostick & Reid (1987), 7–16.

Schick, A. P. & D. Sharon 1974. *Geomorphology and climatology of arid watersheds*. Mimeo. Report, Department of Geography, Hebrew University, Jerusalem.

Schipull, K. von 1978. Waterpockets (Opferkessel) in Sandsteinen des zentralen Colorado-Plateaus. *Zeitschrift für Geomorphologie* 22, 426–38.

Schipull, K. von 1980. Die Cedar Mesa – Schichtstufe auf dem Colorado-Plateau – ein Beispiel für die Morphodynamik arider Schichtstufen. *Zeitschrift für Geomorphologie* 24, 318–31.

Schlesinger, W. H. 1985. The formation of caliche in soils of the Mojave Desert, California. *Geochimica et Cosmochimica Acta* 49, 57–66.

Schlichting, E. 1987. Anhydromorphic and hydromorphic saline soils. *Catena* 14, 325–31.

Schmidt, K-H. 1985. Regional variations of mechanical and chemical denudation, Upper Colorado River Basin, USA. *Earth Surface Processes and Landforms* 10, 497–508.

Schoewe, W. H. 1932. Experiments on the formation of wind-faceted pebbles. *American Journal of Science* 5th Series 24, 111–34.

Schreiber, H. A. & D. R. Kincaid 1967. Regression models for predicting on-site runoff from short-duration convective storms. *Water Resources Research* 3, 389–95.

Schumm, S. A. 1956. The role of creep and rainwash on the retreat of badland slopes. *American Journal of Science* 254, 693–706.

Schumm, S. A. 1960. The shape of alluvial channels in relation to sediment type. USGS Professional Paper 352B.

Schumm, S. A. 1962. Erosion of miniature pediments in Badlands National Monument, South Dakota. *Geological Society of America, Bulletin* 73, 719–24.

Schumm, S. A. 1965. Quaternary palaeohydrology. See *The Quaternary of the United States*, H. E. Wright Jr & D. G. Frey (eds), 783–94. Princeton, New Jersey: Princeton University Press.

Schumm, S. A. 1967. Rates of surficial rock creep on hillslopes in western Colorado. *Science* 155, 560–1.

Schumm, S. A. 1968. *River adjustment to altered hydrologic regimen – Murrumbidgee River and paleochannels, Australia*. US Geological Survey Professional Paper 598.

Schumm, S. A. 1969. River metamorphosis. *American Society of Civil Engineers, Proceedings; Journal of the Hydraulics Division* 95, 255–73.

Schumm, S. A. 1977. *The fluvial system*. New York: John Wiley.

Schumm, S. A. 1985. Patterns of alluvial rivers. *Annual Review of Earth and Planetary Sciences* 13, 5–27.

Schumm, S. A. & R. J. Chorley 1964. The fall of Threatening Rock. *American Journal of Science* 262, 1041–54.

Schumm, S. A. & R. J. Chorley 1966. Talus weathering and scarp recession in the Colorado plateaus. *Zeitschrift*

für *Geomorphologie* 10, 11–36.

Schumm, S. A. & R. F. Hadley 1957. Arroyos and the semiarid cycle of erosion. *American Journal of Science* 255, 161–74.

Schumm, S. A. & R. W. Lichty 1965. Time space and causality in geomorphology. *American Journal of Science* 263, 110–9.

Schütz, L., R. Jänicke, H. Pietrek 1981. Saharan dust transport over the North Atlantic Ocean. See *Desert dust; origin, characteristics and effects on man*, T. L. Péwé (ed.), Geological Society of America, Special Paper 186, 87–100.

Schwan, J. 1988. The structure and genesis of Weischelian to early Holocene sand sheets in western Europe. *Sedimentary Geology* 55, 197–232.

Schwarz, E. L. 1906. Natural mounds in Cape Colony. *Geographical Journal* 27, 67–9.

Scoging, H. M. 1988. *A theoretical and empirical investigation of soil erosion in a semi-arid environment*. PhD thesis, University of London.

Scoging, H. M. & J. B. Thornes 1979. Infiltration characteristics in a semiarid environment. *Publications of the International Association for Scientific Hydrology* 128, 159–68.

Scorer, R. S. 1978. *Environmental aerodynamics*. Chichester, England: Ellis Horwood

Seevers, P. M., D. T. Lewis, J. V. Drew 1975. Use of ERTS-1 imagery to interpret the wind erosion in Nebraska's Sand Hills. *Journal of Soil and Water Conservation* 30, 181–4.

Segerstrom, K. 1963. Matureland of northern Chile and its relationship to ore deposits. *Geological Society of America, Bulletin* 74, 513–8

Seginer, I. 1975. Flow round a windbreak in an oblique wind. *Boundary-layer Meteorology* 9, 133–41.

Sehgal, J. L. & G. Stoops 1972. Pedogenic calcite accumulation in arid and semiarid regions of the Indo-Gangetic plain of erstwhile Punjab (India), their morphology and origin. *Geoderma* 8, 59–72.

Seisser, W. G. 1978. Aridification of the Namib Desert, evidence from oceanic cores. See *Antarctic glacial history and world palaeoenvironments*, E. M. van Zinderen Bakker (ed.), 105–13. Rotterdam: Balkema.

Seisser, W. G. 1980. Late Miocene origin of the Benguela upwelling system off northern Namibia. *Science* 208, 283–85.

Selby, M. J. 1977a. Bornhardts of the Namib Desert. *Zeitschrift für Geomorphologie* 21, 1–13.

Selby, M. J. 1977b. Palaeowind directions in the central Namib Desert as indicated by ventifacts. *Madoqua* 10, 195–8.

Selby, M. J. 1982a. *Hillslope materials and processes*, Oxford: Oxford University Press.

Selby, M. J. 1982b. Form and origin of some bornhardts of the Namib Desert. *Zeitschrift für Geomorphologie* 26, 1–115.

Selby, M. J. 1982c. Rock mass strength and the form of some inselbergs in the central Namib Desert. *Earth*

Surface Processes and Landforms **7**, 489–97.

Selby, M. J. 1982d. Engineering geology of collapsing soils in south Australia. *Proceedings, IVth International Congress, Association of Engineering Geologists* 1, 469–75.

Selby, M. J. 1985. *Earth's changing surface: an introduction to geomorphology.* Oxford: Oxford University Press.

Selby, M. J., C. H. Hendy, M. K. Seely 1979. A late Quaternary lake in the central Namib Desert, southern Africa and some implications. *Palaeogeography, Palaeoclimatology, Palaeoecology* **26**, 27–41.

Selby, M. J., R. B. Raines, P. W. Palmer 1973. Eolian deposits of the ice-free Victoria Valley, southern Victoria Land, Antarctica. *New Zealand Journal of Geology and Geophysics* **17**, 543–62.

Selivanov, E. I. 1982. Dashte-Lut Desert of Iran. *Problems of desert development* [in Russian], (1) 14–19.

Semeniuk, V. 1986. Holocene climatic history of coastal southwestern Australia using calcrete as an indicator. *Palaeogeography, Palaeoecology, Palaeoclimatology* **53**, 289–308.

Semeniuk, V. & D. K. Glassford 1988. Significance of aeolian limestone lenses in quartz sand formations: an interdigitation of coastal and continental facies, Perth Basin, southwestern Australia. *Sedimentary Geology* **57**, 199–210.

Semeniuk, V. & T. D. Meagher 1981. Calcrete in Quaternary coastal dunes in southwestern Australia: a capillary rise phenomenon associated with plants. *Journal of Sedimentary Petrology* **51**, 47–68.

Semenuik, V. & D. J. Searle 1985. Distribution of calcrete in coastal sands in relation to climate, southwestern Australia. *Journal of Sedimentary Petrology* **55**, 86–95.

Seppälä, M. 1971. Evolution of eolian relief of the Kaamasjoki–Kiellajoki River basin in Finnish Lapland. *Fennia* **104**.

Seppälä, M. 1972. Location, morphology and orientation of inland dunes in northern Sweden. *Geografiska Annaler* **54A**, 85–104.

Seppälä, M. 1974. Some quantitative measurements of present-day deflation of Hietatievat, Finnish Lapland. *Abhandlungen der Akademie der Wissenschaften in Göttingen, Mathematisch–Physikalisch* Klasse **III, 29**, 208–20.

Seppälä, M. 1981. Forest fires as activator of geomorphic processes in Kuttanen esker-dune region, northernmost Finland. *Fennia* **159**, 221–88.

Seppälä, M. 1986. The origin of palsas. *Geografiska Annaler* **A68**, 141–47.

Seppälä, M. & K. Lindé 1978. Wind tunnel studies of ripple formation. *Geografiska Annaler* **60A**, 29–40.

Servant, M. & S. Servant-Vildary 1980. L'environnement Quaternaire du bassin du Tchad. See Williams & Faure (1980), 133–62.

Servant-Vildary, S. 1973. Le Plio-Quaternaire ancien du Tchad: evolution des associations de diatomées, stratigraphie, paléogéographie. *Cahiers ORSTOM, Série Géolo-*

gique **5**, 169–217.

Seth, S. K. 1963. A review of evidence concerning changes of climate in India during the protohistorical and historical periods. *Arid Zone Research* **20**, 443–54.

Sevenet, Lt. 1943. Étude sur le Djouf (Sahara occidental). *Institut Français de l'Afrique Noire, Bulletin* **5**, 1–26.

Shackleton, N. J. & N. D. Opdyke 1977. Oxygen isotope and palaeomagnetic evidence for early northern hemisphere glaciation. *Nature* **270**, 216–18.

Shand, S. J. 1946. Dust devils. Parallelism between the South African salt pans and the Carolina Bays. *Science Monthly* **62**, 95.

Sharon, D. 1962. On the nature of hamadas in Israel. *Zeitschrift für Geomorphologie* **6**, 129–47.

Sharp, R. P. 1940. Geomorphology of the Ruby–East Humboldt Range, Nevada. *Geological Society of America, Bulletin* **51**, 337–72.

Sharp, R. P. 1949. Pleistocene ventifacts east of the Bighorn Mountains, Wyoming. *Journal of Geology* **57**, 175–95.

Sharp, R. P. 1957. Geomorphology of Cima Dome, Mojave Desert, California. *Geological Society of America, Bulletin* **68**, 273–90.

Sharp, R. P. 1963. Wind ripples. *Journal of Geology* **71**, 617–36.

Sharp, R. P. 1964. Wind-driven sand in the Coachella valley, California. *Geological Society of America, Bulletin* **75**, 785–804.

Sharp, R. P. 1966. Kelso Dunes, Mojave Desert, California. *Geological Society of America, Bulletin* **77**, 1045–74.

Sharp, R. P. 1979. Interdune flats of the Algodones chain, Imperial Valley, California. *Geological Society of America, Bulletin* **90**, 908–16.

Sharp, R. P. 1980. Wind-driven sand in the Coachella valley, California: further data. *Geological Society of America, Bulletin* **91**, 724–30.

Sharp, R. P. & D. L. Carey 1976. Sliding stones, Racetrack Playa, California. *Geological Society of America, Bulletin* **87**, 1704–17.

Sharp, R. P. & R. S. Saunders 1978. Aeolian activity in westernmost Coachella Valley and at Garnet Hill. See *Aeolian features of southern California: a comparative planetary geology handbook*, R. Greeley, M. B. Womer, R. P. Papson, P. D. Spudis (eds), 9–27. Washington DC: Office of Planetary Geology, NASA.

Sharp, W. E. 1960. The movement of playa scrapers by the wind. *Journal of Geology* **68**, 567–72.

Shaw, P. A. & J. J. De Vries 1988. Duricrusts, groundwater and valley development in the Kalahari of southwest Botswana. *Journal of Arid Environments* **14**, 245–54.

Shawe, D. R. 1963. Possible wind-erosion origin of linear scarps on the Saga Plain, southwestern Colorado. USGS Professional Paper 475C, 138–42.

Shepard, F. P. & R. Young 1961. Distinguishing between beach and dune sands. *Journal of Sedimentary Petrology* **31**, 196–214.

Sherman, D. J. 1990. Evaluation of aeolian sand transport

equations using intertidal-zone measurements, Saunton Sands, England – discussion. *Sedimentology* 37, 385–89.

Shinn, E. A. 1973. Sedimentary accretion along the leeward SE coast of the Qatar peninsula, Persian Gulf. See Purser (1973), 199–209.

Shlemon, R. J. 1978. Quaternary soil–geomorphic relationships, southeastern Mojave desert, California and Arizona. See *Quaternary soils*, W. C. Mahaney (ed.), 187–207. Norwich: Geobooks.

Shmida, A., M. Evenari, I. Noy-Meir 1986. Hot desert ecosystems: an integrated view. See Evenari et al. (1986), 379–87.

Short, A. D. 1988. Holocene coastal dune formation in South Australia: a case study. *Sedimentary Geology* 55, 121–43.

Short, A. D. & P. A. Hesp 1982. Wave, beach and dune interactions in southeastern Australia. *Marine Geology* 48, 259–84.

Shotton, F. W. 1956. Some aspects of the New Red desert in Britain. *Liverpool and Manchester Geological Journal* 1, 450–65.

Sidwell, R. & W. F. Tanner 1939. Sand grain patterns of West Texas dunes (abstract). *American Journal of Science* 237, 181.

Simonett, D. S. 1949. Sand dunes near Castlereagh, New South Wales. *Australian Geographer* 5, 3–10.

Simonett, D. S. 1960. Development and grading of dunes in western Kansas. *Annals of the AAG* 50, 216–41.

Simons, D. B., E. V. Richardson, C. F. Nordin Jr 1965. *Bedload equation for ripples and dunes*. USGS Professional Paper 462H, H1–H9.

Simons, F. S. 1956. A note on Pur-Pur dune, Viru Valley, Peru. *Journal of Geology* 64, 517–21.

Simons, F. S. & G. E. Eriksen 1953. Some desert features of northwest central Peru. *Sociedad Geologica del Peru, Boletin* 26, 229–45.

Simonson, R. W. & C. E. Hutton 1954. Depositional curves for loess. *American Journal of Science* 252, 99–105.

Sinclair, P. C. 1969. General characteristics of dust devils. *Journal of Applied Meteorology* 8, 32–45.

Sinclair, P. C. 1973. The lower structure of dust devils. *Journal of Atmospheric Sciences* 30, 1599–1619.

Singer, A. 1979. Palygorskite in sediments: detrital diagenetic or neoformed: a critical review. *Geologische Rundschau* 68, 996–1008.

Singer, A. 1984. The palaeoclimatic interpretation of clay minerals in sediments – a review. *Earth Science Reviews* 21, 251–93.

Singer, A. 1989. Illite in the hot-aridic soil environment. *Soil Science* 147, 126–33.

Singer, A. & A. J. Amiel 1974. Characteristics of Nubian sandstone-derived soils. *Journal of Soil Science* 25, 310–19.

Singer, A. & K. Norrish 1974. Pedogenic palygorskite occurrences in Australia. *American Mineralogist* 59, 508–17.

Singh, G. 1969. Palaeobotanical features of the Thar Desert. *Annals of the Arid Zone* 8, 188–95.

Singh, G., R. D. Joshi, A. P. Singh 1972. Stratigraphic and radiocarbon evidence for the age and development of three salt lake deposits in Rajasthan, India. *Quaternary Research* 2, 496–501.

Sirocko, F. & H. Lange 1991. Clay mineral accumulation rates in the Arabian Sea during the later Quaternary. *Marine Geology* 97, 105–19.

Skidmore, E. L. 1986a. Wind erosion climatic erosivity. *Climatic Change* 9, 195–208.

Skidmore, E. L. 1986b. Soil erosion by wind: an overview. See El-Baz & Hassan (1986), 261–73.

Skocek, V. & A. A. Saadallah 1972. Heavy minerals in eolian sands, southern desert, Iraq. *Sedimentary Geology* 8, 29–43.

Slatyer, R. O. & J. A. Mabbutt 1964. Hydrology of arid and semiarid regions. See Chow (1964), 24–46.

Smale, D. 1978. Silcretes and associated silica diagenesis in southern Africa and Australia. See Langford-Smith (1978b), 261–80.

Smalley, I. J. 1970. Cohesion of soil particles and the intrinsic resistance of simple soil systems to wind erosion. *Journal of Soil Science* 21, 154–61.

Smalley, I. J. 1974. Fragmentation of granite quartz in water; discussion. *Sedimentology* 21, 633–5.

Smalley, I. J. & D. Krinsley 1979. Eolian sedimentation on Earth and Mars: some comparisons. *Icarus* 40, 276–88.

Smalley, I. J. & V. Smalley 1983. Loess material and loess deposits: formation, distribution and consequences. See Brookfield & Ahlbrandt (1983), 51–68.

Smalley, I. J. & C. Vita-Finzi 1968. The formation of fine particles in sandy deserts and the nature of "desert" loess. *Journal of Sedimentary Petrology* 38, 766–74.

Smettan, U. & H-P. Blume 1987. Salts in sandy soils, southwestern Egypt. *Catena* 14, 333–43.

Smith, B. J. 1977. Rock temperature measurements from northwest Sahara and their implications for rock weathering. *Catena* 4, 41–63.

Smith, B. J. 1978. The origin and geomorphic implications of cliff foot recesses and tafoni on limestone hamadas in the northwest Sahara. *Zeitschrift für Geomorphologie* 22, 21–43.

Smith, B. J. 1987. An integrated approach to the weathering of limestone in an arid area and its role in landscape evolution: a case study from southeast Morocco. See Gardiner (1987), 637–57.

Smith, B. J. & J. J. McAlister 1986. Observations on the occurrence and origins of salt weathering phenomena near Lake Magadi, southern Kenya. *Zeitschrift für Geomorphologie* 30, 445–60.

Smith, B. J., J. P. McGreevy, W. B. Whalley 1987. Silt production by weathering of a sandstone under hot, arid conditions: an experimental study. *Journal of Arid Environments* 12, 199–214.

Smith, B. J. & W. B. Whalley 1981. Late Quaternary drift deposits in north central Nigeria examined by scanning electron microscopy. *Catena* 8, 345–67.

Smith, B. J. & W. B. Whalley 1982. Observations on the

composition and mineralogy of an Algerian duricrust complex. *Geoderma* **28**, 285–311.

Smith, B. J. & W. B. Whalley 1988. A note on the characteristics and possible origins of desert varnish from southeastern Morocco. *Earth Surface Processes and Landforms* **13**, 251–8.

Smith, B. J. & W. B. Whalley 1989. A note on the characteristics and possible origins of desert varnish from southeastern Morocco. A reply to comments by R. I. Dorn. *Earth Surface Processes and Landforms* **14**, 171–2.

Smith, D. M., C. R. Twidale, J. A. Bourne 1975. Kappakoola Dunes – aeolian landforms induced by man. *Australian Geographer* **13**, 90–6.

Smith, G. I. 1968. Late Quaternary geologic and climatic history of Searles Lake, southeastern California. See *Means of correlation of Quaternary successions*, R. N. Morrison & H. E. Wright Jr (eds), 293–310. Salt Lake City: University of Utah Press.

Smith, G. I. & F. A. Street-Perrott 1983. Pluvial lakes of the western United States, in *Late-Quaternary environments of the United States, volume 1: the Late Pleistocene*, S. C. Porter (ed.), 190–212. Harlow, England: Longman.

Smith, H. T. U. 1938. Sand dune cycle in western Kansas (abstract). *Geological Society of America, Bulletin* **50**, 1934–35.

Smith, H. T. U. 1945. Giant grooves in northwest Canada [abstract]. *Geological Society of America, Bulletin* **56**, 1198.

Smith, H. T. U. 1954. Eolian sand on desert mountains. *Geological Society of America, Bulletin* **65**, 1036–7.

Smith, H. T. U. 1965. Dune morphology and chronology in central and western Nebraska. *Journal of Geology* **73**, 557–78.

Smith, H. T. U. 1966. Windeformte Geröllwellen in der Antarktis. *Umschau* **66**, 334.

Smith, H. T. U. 1969. *Photo interpretation studies of desert basins in northern Africa*. US Air Force Cambridge Research Labs. Final Report Contract AF19(628) 2486.

Smith, J. 1949. *Distribution of tree species in the Sudan in relation to rainfall and soil texture*. Sudan Government, Ministry of Agriculture Bulletin 4.

Smith, J. D. 1970. Stability of a sand wave subject to a shear flow of a low Froude number. *Journal of Geophysical Research* **75**, 5928–40.

Smith, J. D. & S. R. McLean 1977. Spatially averaged flow over a wavy surface. *Journal of Geophysical Research* **82**, 1735–46.

Smith, K. G. 1958. Erosional processes and landforms in Badlands National Monument, South Dakota. *Geological Society of America, Bulletin* **69**, 975–1008.

Smith, R. E. 1972. The infiltration envelope: results from a theoretical infiltrometer. *Journal of Hydrology* **17**, 1–21.

Smith, R. M., P. C. Twiss, R. K. Krauss, M. J. Brown 1970. Dust deposition in relation to site, season and climatic variables. *Soil Science Society of America, Proceedings* **34**, 112–7.

Smith, R. S. U. 1981. Birth and death of barchan dunes in southern Algodones dune chain, California and Mexico. *Geological Society of America, Abstracts with Programs* **13**, 107.

Smith, R. S. U. 1982. Sand dunes in North American deserts. See *Reference handbook on the deserts of North America*, G. L. Bender (ed.), 481–524. Westport, Connecticut: Greenwood Press.

Snead, R. E. 1966. *Physical geography reconnaissance, the Las Bela coastal plain, West Pakistan*. Coastal Studies Series 13, Louisiana State University.

Sneed, E. D. & R. L. Folk 1958. Pebbles in the Lower Colorado River, Texas. A study in particle morphogenesis. *Journal of Geology* **66**, 114–50.

Sneh, A. 1982. Drainage systems of the Quaternary in northern Sinai with emphasis on Wadi El Arish. *Zeitschrift für Geomorphologie* **26**, 179–97.

Sneh, A. & T. Weissbrod 1983. Size–frequency distribution on longitudinal dune ripple flank sands compared to that of slipface sands of various dune types. *Sedimentology* **30**, 717–26.

So, C. L. 1982. Wind-induced movements of beach sand at Port-Sea, Victoria. *Royal Society of Victoria, Proceedings* **94**, 61–8.

Sokolow, N. A. 1894. *Die Dünen, Bildung, Entwicklung und ihrer Bau* [translated by A. Arzruni]. Berlin: Springer.

Soleilharoup, R. E. 1972. The infiltration envelope: results from a theoretical infiltrometer. *Journal of Hydrology* **17**, 1–21.

Solger, F. 1910. Studien über Nordostdeutsche Inlanddünen. *Forschungen zur Deutschen Landes und Volkskunde* **19**, 1–99.

Soliman, M. M. 1982. Geochemical prospecting in Hamash area, southeastern desert, Egypt. *Geographical Magazine* **119**, 319–23.

Sonntag, C., R. J. Thorweite, E. P. Lohnert, C. Junghans, K. O. Munnick, E. Klitzsch, E. M. El Shazly, F. M. Swailem 1980. Isotopic identification of Saharan groundwater – groundwater formation in the past. *Palaeoecology of Africa* **12**, 159–71.

Sorensen, M. 1985. Estimation of some aeolian saltation transport parameters from transport rate profiles. See Barndorff-Nielsen et al. (1985), 141–90.

Sorensen, M. 1986. *Statistical analysis of data from a wind tunnel experiment using radioactive grains*. Research Report 141, Department of Theoretical Statistics, University of Aarhus.

Sorey, M. L. & W. G. Matlock 1969. Evaporation from an ephemeral streambed. *American Society of Civil Engineers, Proceedings; Journal of the Hydraulics Division* **95**, 423–38.

Spath, H. 1977. Rezente Verwitterung und Abtragung and Schicht- und Rumpfstufen im semiariden Westaustralien. *Zeitschrift für Geomorphologie*, Supplementband **28**, 81–100.

Spath, H. 1987. Landform development and laterites in northwestern Australia. *Zeitschrift für Geomorphologie*,

Supplementband 64, 163–80.

Spence, M. T. 1957. Soil blowing in the Fens in 1956. *Meteorological Magazine* **86**, 21–2.

Spencer, B. (ed.) 1896. *Report on the work of the Horn scientific expedition to Central Australia* (4 volumes). London: Dulan.

Sperling, C. H. B. & A. S. Goudie 1975. The miliolite of western India: a discussion of the aeolian and marine hypothesis. *Sedimentary Geology* **13**, 71–5.

Sperling, C. H. B., A. S. Goudie, D. R. Stoddart, G. G. Poole 1977. Dolines of the Dorset chalklands and other areas in southern Britain. *Institute of British Geographers, Transactions* **2**, 205–23.

Sperling, C. H. B. & R. U. Cooke 1985. Laboratory simulation of rock weathering by salt crystallization and hydration processes in hot, arid environments. *Earth Surface Processes and Landforms* **10**, 541–55.

Sprigg, R. C. 1959. Stranded sea beaches and associated sand accumulations of the upper South East. *Royal Society of South Australia, Transactions* **82**, 183–95.

Sprigg, R. C. 1963. Geology and petroleum prospects of the Simpson Desert. *Royal Society of South Australia, Transactions* **86**, 35–65.

Sprigg, R. C. 1979. Stranded and submerged sea-beach systems of southeastern South Australia and the aeolian desert cycle. *Sedimentary Geology* **22**, 53–97.

Springer, M. E. 1958. Desert pavement and vesicular layer of some desert soils in the desert of the Lahontan Basin. *Soil Science Society of America, Proceedings* **22**, 63–6.

Staal, P. 1958. Un exemple d'intense activité éolienne actuelle. *Revue de Géomorphologie Dynamique* **9**, 85–6.

Stahr, K., R. Jahn, A. Huth, J. Gauer 1989. Influence of eolian sedimentation and soil formation in Egypt and Canary Island Deserts. *Catena Supplement* **14**, 127–44.

Stalder, P. J. 1975. Cementation of Pliocene–Quaternary fluviatile clastic deposits in and along the Oman Mountains. *Geologie en Mijnbouw* **54**, 148–56.

Stanley, G. M. 1955. Origin of playa stone tracks, Racetrack Playa, Inyo County, California. *Geological Society of America, Bulletin* **66**, 1329–50.

Stapor, F. W. 1973. Heavy mineral concentrating processes and density/shape/size equilibria in marine and coastal dune sands of the Apalachicola, Florida region. *Journal of Sedimentary Petrology* **43**, 396–407.

Stapor, F. W., J. P. May, J. Barwis 1983. Eolian shape-sorting and aerodynamic traction equivalence in the coastal dunes of Hout Bay, Republic of South Africa. See Brookfield & Ahlbrandt (1983), 149–64.

Statham I. & S. C. Francis 1986. The influence of scree accumulation and weathering on the development of steep mountain slopes. See Abrahams (1986), 245–67.

Stearns, C. R. 1978. Micrometeorological instrumentation and data analysis. See Lettau & K. Lettau (1978b), 30–53.

Steele, R. P. 1983. Longitudinal draa in the Permian Yellow Sands of North-East England. See Brookfield & Ahlbrandt (1983), 543–50.

Steele, R. P. 1985. Early Permian (Rotliegendes) palaeo-

winds of the North Sea – comment. *Sedimentary Geology* **45**, 293–313.

Steensrup, K. J. V. 1894. Om Klitterns Vandrung. *Meddelelser Dansk Geol. Förening*, Copenhagen **1**, 1–14.

Steffan, E-M. 1983. *Untersuchunger zür Morphologie und Genese der äolischen Akkumulationsformen der Östsahara mit Hilfe der Fernerkundung.* Berliner Geowissenschaftliche Abhandlungen A45.

Steiner, E. A. 1976. Age of some ventifacts at Garnet Hill, California. *The Compass* **54**, 9–13.

Stembridge Jr, J. R. 1978. Vegetated coastal dunes: growth detected from aerial infrared photography. *Remote Sensing of the Environment* **7**, 73–6.

Stephen, I., E. Bellis, A. Muir 1956. Gilgai phenomena in tropical black clays of Kenya. *Journal of Soil Science* **7**, 1–9.

Stephens, C. G. 1971. Laterite and silcrete in Australia a study of the genetic relationships of laterite and silcrete and their companion materials, and their collective significance in the formation of the weathered mantle, soils, relief and drainage of the Australian continent. *Geoderma* **5**, 5–52.

Stephens, C. G. & R. L. Crocker 1946. Composition and genesis of lunettes. *Royal Society of South Australia, Transactions* **70**, 302–12.

Sterr, H. 1985. Rates of change and degradation of hillslopes formed in unconsolidated materials: a morphometric approach to date Quaternary fault scarps in western Utah, USA. *Zeitschrift für Geomorphologie* **29**, 315–33.

Stipho, A. S. 1985. On the engineering properties of salina soil. *Quarterly Journal of Engineering Geology* **18**, 129–37.

Stoertz, G. E. & G. E. Erickson 1974. *Geology of salars in northern Chile.* USGS Professional Paper 811.

Stokes, W. L. 1964. Incised wind-aligned stream patterns of the Colorado Plateau. *American Journal of Science* **262**, 808–16.

Stokes, W. L. 1968. Multiple parallel-truncation bedding planes – a feature of wind-deposited sandstone formations. *Journal of Sedimentary Petrology* **38**, 510–15.

Stolt, M. H. & M. C. Rabenhorst 1987. Carolina bays on the eastern shore of Maryland, I: soil characterization and classification. *Soil Science Society of America, Journal* **51**, 394–8.

Stone, R. O. 1967. A desert glossary. *Earth Science Reviews* **3**, 211–68.

Storrier, R. R. & M. E. Stannard (eds) 1980. *Aeolian landscapes in the semi-arid zone of south eastern Australia.* Riverina: Australian Society of Soil Science.

Story, R. 1982. *Notes on parabolic dunes, winds and vegetation in northern Australia.* CSIRO, Division of Water and Land Resources, Technical Paper 43,

Story, R., G. A. Yapp, A. T. Dunn 1976. LANDSAT patterns considered in relation to Australian resource surveys. *Remote Sensing of the Environment* **4**, 281–303.

Stow, C. D. 1969. Dust and sand storm electrification. *Weather* **24**, 134–40.

Street, A. & F. Gasse 1981. Recent developments in research into the Quaternary climatic history of the Sahara. See Allah (1981), 7–28.

Street, F. A. & A. T. Grove 1979. Global maps of lake level fluctuations since 30,000 years ago. *Quaternary Research* 12, 83–118

Street-Perrott F. A., N. Roberts, S. Metcalfe 1985. Geomorphic implications of late Quaternary hydrological and climatic changes in the Northern Hemisphere tropics. See Douglas & Spencer (1985), 164–83.

Striem, H. L. 1954. The seifs on the Israeli/Sinai border and the correlation of their alignment. *Research Council of Israel, Bulletin* 4, 195–8.

Suarez, D. L. 1981. Predicting Ca and Mg concentrations in arid land soils [abstract]. *Eos* 62, 285.

Subramaniam, A. R. & A. V. R. Kesava Rao 1981. Dew fall in sand dune areas of India. *9th International Congress of Biometeorology*, D. Overdiek, J. Muller, H. Lieth (eds), abstract volume, 197–200.

Sugden, W. 1964. Origin of faceted pebbles in some recent desert sediments of southern Iraq. *Sedimentology* 3, 65–74.

Sumer, B. M. 1985. The mechanics of sediment suspension in turbulent boundary-layer flows. See Barndorff-Nielsen et al. (1985), 191–224.

Summerfield, M. A. 1982. Distribution, nature and probable genesis of silcrete in arid and semi-arid southern Africa. *Catena*, Supplement 1, 37–66.

Summerfield, M. A. 1983a. Silcrete. See Goudie & Pye (1983), 59–91.

Summerfield, M. A. 1983b. Silcrete as a palaeoclimatic indicator: evidence from southern Africa. *Palaeogeography, Palaeoclimatology, Paleoecology* 41, 65–79.

Summerfield, M. A. 1986. Reply to discussion – silcrete as a palaeoenvironmental indicator evidence from southern Africa. *Palaeogeography, Palaeoclimatology, Palaeoecology* 52, 356–60.

Sundborg, A. 1986. Sedimentation processes. In *International symposium on erosion and sedimentation in Arab countries, Baghdad, Iraq, Feb. 15–19, 1986*, 1–27.

Suslov, S. P. 1961. *Physical geography of Asiatic Russia* [translated by N. D. Gershevsky]. London: Freeman.

Suzuki, T. & K. Takahashi 1981. An experimental study of wind abrasion. *Journal of Geology* 89, 23–36.

Svasek, J. N. & J. H. J. Terwindt 1974. Measurement of sand transport by wind on a natural beach. *Sedimentology* 21, 311–22.

Swart, H. 1986. Prediction of wind-driven transport rates. *20th Coastal Engineering Conference, Proceedings*, B. Edge (ed.), 1595–1611. New York: American Society of Civil Engineers.

Swarzendruber, D. & D. Hillel 1975. Infiltration and run-off for small field plots under constant intensity rainfall. *Journal of Water Resources Research* 11, 445–51.

Sweet, M. L., J. Nielson, K. Havholm, J. Farrelley 1988. Algodones dune field of southern California: case history of a migrating modern dune field. *Sedimentology* 35, 939–52.

Sweeting, M. M. & N. Lancaster, N. 1982. Solutional and wind erosion forms on limestone in the central Namib Desert. *Zeitschrift für Geomorphologie* 26, 197–207.

Swift, D. J. P., B. F. Molina, R. G. Jackson III 1978. Intermittent structure of the atmospheric boundary layer made visible by entrained sediment: example from the Copper River Delta, Alaska. *Journal of Sedimentary Petrology* 48, 897–900.

Syers, J. K. & T. W. Walker 1969. Phosphorus transformations in a chronosequence of soils developed on wind-blown sand in New Zealand, I: total and organic phosphorus. *Journal of Soil Science* 20, 57–64.

Symmons, P. M. & C. F. Hemming 1968. A note on wind-stable stone-mantles in the southern Sahara. *Geographical Journal* 134, 60–4.

Szabo, B. J., W. P. McHugh, G. G. Schaber, C. V. Haynes, C. S. Breed 1989. Uranium series dated authigenic carbonates and Acheulean sites in southern Egypt. *Science* 243, 1053–6.

Tabuteau, M. M. 1960. Étude graphique pour les conséquences hydro-érosives du climat Méditerrané. *Association Géographes Français, Bulletin* 295, 130–42.

Takahara, H. 1973. Sounding mechanism of singing sand. *Acoustical Society of America, Journal* 53, 634–9.

Talbot, M. R. 1980. Environmental responses to climatic change in the West African Sahel over the past 20,000 years. See Williams & Faure (1980), 37–62.

Talbot, M. R. 1984. Late Pleistocene rainfall and dune building in the Sahel. *Palaeoecology of Africa* 16, 203–14.

Talbot, M. R. 1985. Major bounding surfaces in aeolian sandstones: a climatic model. *Sedimentology* 32, 257–66.

Talbot, M. R. & M. A. J. Williams 1978. Erosion of fixed dunes in the Sahel, Central Niger. *Earth Surface Processes and Landforms* 3, 107–13.

Talbot, M. R. & M. A. J. Williams 1979. Cyclic alluvial fan sedimentation on the flank of fixed dunes, Janjari, central Niger. *Catena* 6, 43–62.

Talmage, S. B. 1932. The origin of the gypsum sands of the Tala Rosa River [abstract]. *Geological Society of America, Bulletin* 43, 185–6.

Tanner, C. B. & W. L. Pelton 1960. Potential evapotranspiration estimates by the approximate energy balance method of Penman. *Journal of Geophysical Research* 65, 3391–413.

Tanner, W. F. 1963. Spiral flows in rivers, shallow seas, dust devils and models. *Science* 139, 41–2.

Tanner, W. F. 1967. Ripple mark indices and their uses. *Sedimentology* 9, 89–104.

Tator, B. A. 1953. The climatic factor and pedimentation. *International Geological Congress Alger, C. R.* 7, 121–30.

Taylor, P. A. 1981. Model predictions of neutrally stratified planetary boundary-layer flow over ridges. *Royal Meteorological Society, Quarterly Journal* 107, 111–20.

Taylor, P. A., P. J. Mason, E. F. Bradley 1987.

Boundary-layer flow over low hills (a review). *Boundary-layer Meteorology* **39**, 107–32.

Taylor-George, S., F. E. Palmer, J. T. Staley, D. J. Borns, B. Curtiss, J. B. Adams 1983. Fungi and bacteria involved in desert varnish formation. *Microbial Ecology* **9**, 227–45.

Teichert, C. 1939. Corrasion by wind-blown snow in polar regions. *American Journal of Science* **237**, 146–8.

Teller, J. T. 1972. Aeolian deposits of clay sand. *Journal of Sedimentary Petrology* **42**, 648–86.

Terzaghi, K. & R. B. Peck 1948. *Soil mechanics in engineering practice*. New York: John Wiley.

Tetzlaff, G. & K. Wolter 1980. Meteorological patterns and the transport of mineral dust from the North African continent. See Sarnthein et al. (1980), 31–42.

Theakstone, W. H. 1982. Sediment fans and sediment flows generated by snowmelt: observations at Auster-dalsisan, Norway. *Journal of Geology* **90**, 583–8.

Thesiger, W. 1949. A further journey across the Empty Quarter. *Geographical Journal* **113**, 21–46.

Thom, B. G. 1970. Carolina Bays in Hory and Marion Counties, South Carolina. *Geological Society of America, Bulletin* **81**, 783–814.

Thomas, B. 1931. *Arabia Felix: across the empty quarter of Arabia*. London: Jonathan Cape.

Thomas, D. S. G. 1984a. Ancient ergs of the former arid zones of Zimbabwe, Zambia and Angola. *Institute of British Geographers, Transactions* **9**, 75–88.

Thomas, D. S. G. 1984b. Geomorphic evolution and river channel orientation in northwestern Zimbabwe. *Geographical Association of Zimbabwe, Proceedings* **14**, 45–55.

Thomas, D. S. G. 1984c. *Late Quaternary environmental change in central southern Africa with particular reference to extension of the arid zone*. DPhil thesis, University of Oxford.

Thomas, D. S. G. 1986a. Dune pattern statistics applied to the Kalahari dune desert, southern Africa. *Zeitschrift für Geomorphologie* **30**, 231–42.

Thomas, D. S. G. 1986b. The response diagram and ancient desert sands – a note. *Zeitschrift für Geomorphologie* **30**, 363–9.

Thomas, D. S. G. 1987a. Discrimination of depositional environments in the Mega-Kalahari. See Frostick & Reid (1987), 293–306.

Thomas, D. S. G. 1987b. The roundness of aeolian quartz sand grains. *Sedimentary Geology* **52**, 149–53.

Thomas, D. S. G. 1988a. The nature and depositional setting of arid and semi-arid Kalahari sediments. *Journal of Arid Environments* **14**, 17–26.

Thomas, D. S. G. 1988b. Analysis of linear dune sediment–form relationships in the Kalahari dune desert. *Earth Surface Processes and Landforms* **13**, 545–53.

Thomas, D. S. G. 1988c. The biogeomorphology of arid and semi-arid environments. See *Bio-geomorphology*, H. A. Viles (ed.), 193–221. Oxford: Basil Blackwell.

Thomas, D. S. G. 1988d. The geomorphological role of vegetation in the dune systems of the Kalahari. See *Geomorphological studies in South Africa*, G. F. Dardis

& B. P. Moon (eds), 145–58. Rotterdam: Balkema.

Thomas, D. S. G. 1989a. Arid geomorphology. *Progress in Physical Geography* **13**, 442–51.

Thomas, D. S. G. 1989b. Reconstructing ancient arid environments. See Thomas (1989c), 312–34.

Thomas, D. S. G. (ed) 1989c. *Arid zone geomorphology*. London: Pinter (Belhaven).

Thomas, D. S. G. & A. S. Goudie 1984. Ancient ergs of the southern hemisphere. See Vogel (1984), 407–18.

Thomas, D. S. G. & H. E. Martin 1987. Grain size characteristics of linear dunes in the southwestern Kalahari: discussion. *Journal of Sedimentary Petrology* **57**, 572–3.

Thomas, D. S. G. & P. A. Shaw 1991. *The Kalahari environment*. Cambridge: Cambridge University Press.

Thomas, M. F. 1966. Some geomorphological implications of deep weathering patterns in crystalline rocks in Nigeria. *Institute of British Geographers, Transactions* **40**, 173–93.

Thomas, M. F. 1974. *Tropical geomorphology*. London: Macmillan.

Thomas, Y-F. 1975. *Actions éoliennes en milieu littoral – la Pointe de la Courbe*. Mémoires de la Laboratoire de Géomorphologie de l'École des Hautes Études 29.

Thompson, C. H. & G. M. Bowman 1984. Subaerial denudation and weathering of vegetated coastal dunes in eastern Australia. See *Coastal geomorphology in Australia*, B. G. Thom (ed.), 263–90. London: Academic Press.

Thompson, R. D. 1975. *The climatology of the arid world*. University of Reading, Department of Geography, Geographical Papers 35.

Thornes, J. B. 1976. *Semi-arid erosional systems: case studies from Spain*. London: LSE Geographical Papers.

Thornes, J. B. 1977. Channel changes in ephemeral streams: observations, problems and models. See Gregory (1977), 317–35.

Thornes, J. B. 1980. Structural instability and ephemeral channel behaviour. *Zeitschrift für Geomorphologie* **36**, 233–44.

Thornes, J. B. 1985. The ecology of erosion. *Geography* **70**, 222–35.

Thorson, R. M. & G. Bender 1985. Eolian deflation by katabatic ancient winds: a late Quaternary example from the north Alaska Range. *Geological Society of America, Bulletin* **96**, 702–9.

Thortensen, D. C., F. T. McKenzie, B. L. Ristvet 1972. Experimental vadose and phreatic cementation of skeletal carbonate sand. *Journal of Sedimentary Petrology* **42**, 162–7.

Tilho, J. 1911. *Documents scientifiques de la mission Tilho* (Ministère des Colonies, Imprimeure Nationale) [3 volumes].

Tolman, C. F. 1909. Erosion and deposition in the southern Arizona bolson region. *Journal of Geology* **17**, 136–63.

Tolstead, W. C. 1942. Vegetation of the northern part of Cherry County, Nebraska. *Ecological Monographs* **12**, 257–92.

Torgersen, T., P. de Dekker, A. R. Chivas, J. M. Bowler

1986. Salt lakes: a discussion of processes influencing palaeoenvironmental interpretation and recommendations for future study. *Palaeogeography, Palaeoclimatology, Palaeoecology* **54**, 7–19.

Torrent, J. & U. Schwertmann 1987. Influence of hematite on the color of red beds. *Journal of Sedimentary Petrology* **57**, 682–94.

Townshend, J. R. G. et al. 1989. Monitoring playa sediment transport systems using thematic mapper data. *Advances in Space Research* **9**, 177–83.

Toy, T. J. 1977. Hillslope form and climate. *Geological Society of America, Bulletin* **88**, 16–22.

Tremblay, L. P. 1961. Wind striations in northern Alberta and Saskatchewan, Canada. *Geological Society of America, Bulletin* **72**, 1561–4.

Tricart, J. 1954a. Une forme de relief climatique: les sebkhas. *Revue de Géomorphologie Dynamique* **5**, 97–101.

Tricart, J. 1954b. Étude experimentale du problème de la gelivation. *Bulletin Périglaciation* **4**, 285–318.

Tricart, J. 1955. Nouvelles observations sur les sebkhas de l'Aftout es-Sahel Mauritanien et du delta du Senegal. *Revue de Géomorphologie Dynamique* **6**, 177–87.

Tricart, J. 1959. Géomorphologie dynamique de la moyenne vallée du Niger (Soudan). *Annales de Géographie* **368**, 333–43.

Tricart, J. 1969. Actions éoliennes dans la Pampa Deprimada (Republica Argentina). *Revue de Géomorphologie Dynamique* **19**, 178–89.

Tricart, J. 1985. Evidence of Upper Pleistocene dry climates in northern South America. See Douglas & Spencer (1985), 197–217.

Tricart, J. & M. Brochu 1955. Le grand erg ancien du Trarza et Cayor (sud-ouest de la Mauritanie et nord du Sénégal). *Revue de Géomorphologie Dynamique* **4**, 145–76.

Tricart, J. & A. Cailleux 1960. *Le modelé des régions sèches*. Paris: Le Cours de Sorbonne, Centre Documentation Universitaire.

Tricart, J. & A. Cailleux 1989. *Le modelé des régions sèches* [2 volumes]. Paris: SEDES.

Tricart, J. & M. Mainguet 1965. Caractéristiques granulométriques de quelques sables éoliens du désert Pérouvien: aspects de la dynamique des barchans. *Revue de Géomorphologie Dynamique* **15**, 110–21.

Tricart, J., R. Raynal, J. Besançon 1972. Cônes rocheux, pédiments, glacis. *Annales de Géographie* **81**, 1–24.

Trichet, J. 1963. Déscription d'une forme d'accumulation de gypse par voie éolienne dans le sud Tunisien. *Société Géologique de la France, Bulletin*, 7ᵉ Série **5**, 617–21.

Trichet, J. 1968. Étude des faciès d'une dune gypseuse, sud d'Oran, Algérie. *Société Géologique de la France, Bulletin*, Serie 7 **9**, 865–75.

Troeh, F. R. 1965. Landform equations fitted to contour maps. *American Journal of Science* **263**, 616–27.

Tseo, G. 1990. Reconnaissance of the dynamic characteristics of an active Strzelecki longitudinal dune, south-central Australia. *Zeitschrift für Geomophologie* **34**, 19–36.

Tsoar, H. 1974. Desert dunes, morphology and dynamics, El-Arish, northern Sinai. *Zeitschrift für Geomorphologie*, Supplementband 20, 41–61.

Tsoar, H. 1978. *The dynamics of longitudinal dunes*. Final Technical Report DA-ERO 76-G-072, European Research Office, US Army, London.

Tsoar, H. 1982. Internal structure and surface geometry of longitudinal (seif) dunes. *Journal of Sedimentary Petrology* **52**, 823–31.

Tsoar, H. 1983. Wind tunnel modelling of echo and climbing dunes. See Brookfield & Ahlbrandt (1983), 247–60.

Tsoar, H. 1984. The formation of seif dunes from barchans – a discussion. *Zeitschrift für Geomorphologie* **28**, 99–104.

Tsoar, H. 1985. Profiles analysis of sand dunes and their steady state signification. *Geografiska Annaler* **67A**, 47–59.

Tsoar, H. 1986. Two-dimensional analysis of dune profiles and the effect of grain size on sand dune morphology. See El-Baz & Hassan (1986), 94–108.

Tsoar, H. 1989. Linear dunes – forms and formation. *Progress in Physical Geography* **13**, 507–28.

Tsoar, H. 1990. Grain-size characteristics of wind ripples on a desert seif dune. *Geography Research Forum* **10**, 37–50.

Tsoar, H. & D. Blumberg 1990. The effect of sea cliffs on inland encroachment of aeolian sand. *Acta Mechanica* **2**, 131–46.

Tsoar, H., R. Greeley, A. R. Peterfreund 1979. Mars: the north polar sand sea and related wind patterns. *Journal of Geophysical Research* **84**, 8167–80.

Tsoar, H. & J-T. Moller 1986. The role of vegetation in the formation of linear sand dunes. See Nickling (1986), 75–97.

Tsoar, H. & K. Pye 1987. Dust transport and the question of desert loess formation. *Sedimentology* **34**, 139–54.

Tsoar, H., K. Rasmussen, M. Sorensen, B. B. Willetts 1985. Laboratory studies of flow over dunes. See Barndorff-Nielsen et al. (1985), 327–50.

Tsoar, H. & D. H. Yaalon 1983. Deflection of sand movement on a sinuous longitudinal (seif) dune: use of fluorescent dye as a tracer. *Sedimentary Geology* **36**, 25–40.

Tuan, Yi-Fu, 1959. *Pediments in southeastern Arizona*. University of California Publications in Geography 13.

Tuan, Yi-Fu, 1962. Structure, climate and basin land forms in Arizona and New Mexico. *Annals of the AAG* **52**, 51–68.

Tucker, M. E. 1978. Gypsum crusts (gypcrete) and patterned ground from northern Iraq. *Zeitschrift für Geomorphologie* **22**, 89–100.

Twidale, C. R. 1964. A contribution to the general theory of domed inselbergs: conclusions drawn from observations in South Australia. *Institute of British Geographers, Transactions* **34**, 91–113.

Twidale, C. R. 1967. Origin of the piedmont angle as evidenced in South Australia. *Journal of Geology* **75**, 393–411.

Twidale, C. R. 1968. *Geomorphology with special reference to Australia*. Melbourne: Nelson.

Twidale, C. R. 1972. Evolution of sand dunes in the Simpson Desert, Central Australia. *Institute of British Geographers, Transactions* **56**, 77–109.

Twidale, C. R. 1976. On the survival of palaeoforms. *American Journal of Science* **276**, 77–95.

Twidale, C. R. 1978. Singing sands. See Fairbridge (1968, 1978), 994–5.

Twidale, C. R. 1980. Origin of minor sandstone landforms. *Erdkunde* **34**, 219–24.

Twidale, C. R. 1981a. Age and origin of longitudinal dunes in the Simpson and other sand ridge deserts. *Die Erde* **112**, 231–41.

Twidale, C. R. 1981b. Granite inselbergs: domed, block-strewn and castellated. *Geographical Journal* **147**, 54–71.

Twidale, C. R. 1982a. *Granite landforms*. Amsterdam: Elsevier.

Twidale, C. R. 1982b. The evolution of bornhardts. *American Scientist* **70**, 268–76.

Twidale, C. R. 1983a. Australian laterites and silcretes: ages and significance. *Revue de Géographie Physique et de Géologie Dynamique* **24**, 35–45.

Twidale, C. R. 1983b. Pediments, peneplains and ulti-plains. *Revue de Géomorphologie Dynamique* **32**, 1–35.

Twidale, C. R. 1987. Etch and intracutaneous landforms and their implications. *Australian Journal of Earth Sciences* **34**, 367–86.

Twidale, C. R. & J. A. Bourne 1975. Episodic exposure of inselbergs. *Geological Society of America, Bulletin* **86**, 1473–81.

Twidale, C. R. & J. A. Bourne 1976. Origin and significance of pitting on granitic rocks. *Zeitschrift für Geomorphologie* **20**, 405–16.

Twidale, C. R. & E. M. Corbin 1963. Gnammas. *Revue de Géomorphologie Dynamique* **14**, 1–20.

Twidale, C. R. & J. T. Hutton 1986. Silcrete as a climatic indicator: discussion. *Palaeogeography, Palaeoclimatology, Palaeoecology* **52**, 351–60.

Tyson, P. D. 1964. Berg winds of South Africa. *Weather* **19**, 7–11.

Tyson, P. D. & M. K. Seely 1980. Local winds over the central Namib. *South African Geographical Journal* **62**, 135–50.

UNESCO, 1979. *Map of the world distribution of arid regions*. MAB Technical Note 7.

Ungar, J. E. & P. K. Haff 1987. Steady state saltation in air. *Sedimentology* **34**, 289–300.

Urvoy, Y. 1933a. Les formes dunaires à l'Ouest du Tchad. *Annales de Géographie* **42**, 506–15.

Urvoy, Y. 1933b, Modelé dunaire entre Zinder et le Tchad. *Association de Géographes Français, Bulletin* **10**, 79–82.

Urvoy, Y. 1942. *Les bassins du Niger: étude de géographie physique et de paleogéographie*. Institut Français de l'Afrique Noire, Mémoire 4.

US Department of Commerce 1978. *National Flash Flood Program development plan FY 1979–84*. Washington DC: USDC.

Vaché-Grandet, G. 1959. L'erg du Trarza, notes de géomorphologie dunaire. *Institut de Recherches Sahariennes, Traveaux* **18**, 161–72.

Van Berkalow, A. 1945. Angle of repose and angle of sliding friction: an experimental study. *Geological Society of America, Bulletin* **56**, 669–707.

Van der Graaf, W. J. E., R. W. A. Crowe, J. A. Bunting, M. J. Jackson 1977. Relict early Cenozoic drainages in arid Western Australia. *Zeitschrift für Geomorphologie* **21**, 379–400.

Van den Ancker, J. A. M., P. D. Jungerius, L. R. Mur 1985. The role of algae in the stabilization of coastal dune blowouts. *Earth Surface Processes and Landforms* **10**, 189–92.

Van der Meulen, F. & P. D. Jungerius 1989. Landscape development in Dutch coastal dunes: the breakdown and restoration of geomorphological and geohydrological processes. See Gimmingham et al. (1989), 219–29.

Van der Walt, C. F. J. 1940. Roaring sands. *South African Geographical Journal* **22**, 35–9.

Van Rooyen, T. H. & E. Verster 1983. Granulometric properties of the roaring sands in the southeastern Kalahari. *Journal of Arid Environments* **6**, 215–22.

van Zinderen Bakker, E. M. (1969). *Palaeoecology of Africa.* **Cape Town: Balkema.**

van Zinderen Bakker, E. M. & J. A. Coetzee (eds) 1979. *Palaeoecology of Africa and the surrounding islands***, volumes 10–11. Rotterdam: Balkema.**

van Zinderen Bakker Sr, E. M. 1980. Comparisons of late-Quaternary climatic evolutions in the Sahara and the Namib–Kalahari region. See Sarnthein et al. (1980), 381–94.

van Zinderen Bakker Sr, E. M. 1982. African palaeo-environments 18,000 BP. See Vogel et al. (1982), 77–99.

van Zinderen Bakker Sr, E. M. 1984. Elements for the chronology of late Cainozoic African climates. See *Correlation of Quaternary chronologies*, W. C. Mahaney (ed.), 23–37. Norwich: Geobooks.

Vanney, J. R. 1960. *Plue et crue dans le Sahara Nord-occidental*. Mémoire, Régionales Institut Recherches Sahariennes 4.

Vanoni, W. A. (ed.) 1975. *Sedimentation engineering*. New York: American Society of Civil Engineers.

Veblen, D. R., D. H. Krinsley, M. Thompson 1981. Transmission electron microscope study of quartz micro-structures produced by eolian bombardment. *Sedimentology* **28**, 853–8.

Veenstra, H. J. 1982. Size, shape and origin of sands of the East Frisian Islands, (North Sea, Germany). *Geologie en Mijnbouw* **61**, 141–6.

Velde, B. 1985. *Clay minerals: a physico-chemical explanation of their occurrence*. Amsterdam: Elsevier.

Verger, F. 1964. Mottereaux et gilgais. *Annales de Géographie* **73**, 413–30.

Verlaque, C. 1958. Les dunes d'In Salah. *Institut*

Recherches Sahariennes, Traveaux **17**, 12–58.

Velichko, A. A. (ed.) 1984. *Late Quaternary environments of the Soviet Union.* **Harlow, England: Longman.**

Verma, S. B. & J. E. Cermak 1974. Wind-tunnel investigation of mass transfer from soil corrugations. *Journal of Applied Meteorology* **13**, 578–87.

Verstappen, H. Th. 1968. On the origin of longitudinal (seif) dunes. *Zeitschrift für Geomorphologie* **12**, 200–20.

Verstappen, H. Th. 1970. Aeolian geomorphology of the Thar Desert and palaeoclimates. *Zeitschrift für Geomorphologie,* Supplementband 10, 104–20.

Verstappen, H. Th. 1972. On dune types, families and sequences in areas of unidirectional winds. *Göttinger Geographische Abhandelungen* **60**, 341–54.

Vincent, P. J. 1984. Particle size variation over a transverse dune in the Nefud as Sirr, central Saudi Arabia. *Journal of Arid Environments* **7**, 329–36.

Vincent, P. J. 1985. Some Saudi Arabian dune sands: a note on the use of the response diagram. *Zeitschrift für Geomorphologie* **29**, 117–22.

Vincent, P. J. 1988. The response diagram and sand mixtures. *Zeitschrift für Geomorphologie* **32**, 221–6.

Vine, H. 1987. Wind-blown materials and West African soils: an explanation of the "ferallitic soil over loose sandy sediments" profile. See Frostick & Reid (1987), 171–86.

Vogel, J. C. (ed.) 1984. *Late Cainozoic palaeoenvironments of the Southern Hemisphere.* **Rotterdam: Balkema.**

Vogel, J. C. 1989. Evidence of past climatic change in the Namib Desert. *Palaeogeography, Palaeoclimatology, Palaeoecology* **70**, 355–66.

Vogel, J. A., E. A. Voigt, T. C. Partridge (eds) 1982. *South African Society for Quaternary Research, Proceedings* (Coetzee et al. 1982, 53–76). Rotterdam: Balkema.

Vogt, T. 1982. Sur la présence d'algues d'eau douce dans une croûte calcaire Quaternaire d'Algèrie du nord. *Academie des Sciences à Paris, Comptes Rendus,* Série II **295**, 703–8.

Völkel, J. & J. Grunert 1990. To the problem of dune formation and dune weathering during the Late Pleistocene and Holocene in the southern Sahara and the Sahel. *Zeitschrift für Geomorphologie* **34**, 1–17.

Wade, A. 1911. Some observations on the Eastern desert of Egypt with considerations bearing upon the origin of the British Trias. *Quarterly Journal of the Geological Society of London* **67**, 238–62.

Waisel, Y. 1971. Patterns of distribution of some xerophytic species in the Negev. *Israel Journal of Botany* **20**, 101–10.

Walker, A. S. 1982. Deserts of China. *American Scientist* **70**, 366–76.

Walker, A. S. 1985. Large format camera photographs: a new tool for understanding arid environments. See *Global geomorphology,* NASA Publication 2312, 89–90.

Walker, A. S., J. W. Olsen, A. B. Bagen 1987. The Badain Jaran Desert: remote sensing investigations. *Geographical Journal* **153**, 205–10.

Walker, A. S. & C. J. Robinove 1981. *Annotated bibliography of remote sensing methods for monitoring desertification.* USGS Circular 851.

Walker, J. D. 1981. *An experimental study of wind ripples.* MS thesis, Massachusetts Institute of Technology.

Walker, J. D. & J. B. Southard 1982. Experimental study of wind ripples. *International Association of Sedimentologists, 11th Congress,* (Hamilton, Ontario), 65.

Walker, P. H. 1964. Sedimentary properties and processes on a sandstone hillside. *Journal of Sedimentary Petrology* **34**, 328–34.

Walker, T. R. 1979. Red color in dune sand. See McKee (1979a), 52–81.

Walker, T. R., B. Waugh, A. J. Crone 1978. Diagenesis of first-cycle desert alluvium of Cenozoic age, southwestern United States and northwestern Mexico. *Geological Society of America, Bulletin* **89**, 19–32.

Wallace, R. E. 1977. Profiles and ages of young fault scarps, north-central Nevada. *Geological Society of America, Bulletin* **88**, 1267–81.

Wallén, C. C. 1967. Aridity definitions and their applicability. *Geografiska Annaler* **49**, 367–84.

Walling, D. E. & A. H. A. Kleo 1979. Sediment yields of rivers in areas of low precipitation: a global view. IASH Publication 128, 479–93.

Walling, D. E. & B. W. Webb 1983. Patterns of sediment yield. See *Background to palaeohydrology: a perspective,* K. J. Gregory (ed.), 69–100. Chichester, England: John Wiley.

Walmsley, J. L. & A. D. Howard 1985. Application of a boundary-layer model to flow over an eolian dune. *Journal of Geophysical Research* **90**, D6, 10, 631–40.

Walmsley, J. L., J. R. Salmon, P. A. Taylor 1982. On the application of a model of boundary-layer flow over low hills to real terrain. *Boundary-layer Meteorology* **23**, 17–46.

Walter, W. 1951. Nouvelles recherches sur l'influence des facteurs physiques sur la morphologie des sables éoliens et des dunes. *Revue de Géomorphologie Dynamique* **2**, 242–58.

Walther, J. 1912. *Das Gesetz der Wüstenbildung in Gegenwart und Vorzeit.* Berlin: Quelle und Meyer.

Walther, J. 1915. Laterit in West Australien. *Zeitschrift der Deutsches Geologischen Gesellschaft* **67B**, 113–40.

Walther, J. 1924. *Das Gesetz der Wüstenbildung im Gegenwart und Vorzeit.* Berlin: Reimer.

Warburg, M. R. 1964. Observations on microclimate in habitats of some desert vipers in the Negev, Arava and Dead Sea Region. *Vie et Milieu* **15**, 1017–41.

Ward, A. W., K. B. Doyle, P. J. Helm, M. K. Weisman, N. E. Witbeck 1985. Global map of eolian features on Mars. *Journal of Geophysical Research* **90**, 2038–56.

Ward, A. W. & R. Greeley 1984. Evolution of the yardangs at Rogers Lake, California. *Geological Society of America, Bulletin* **95**, 829–37.

Warnke, D. E. 1969. Pediment evolution in the Halloran Hills, central Mojave Desert, California. *Zeitschrift für Geomorphologie* **13**, 357–89.

Warren, A. 1968. *The Qoz Region of Kordofan*. PhD thesis, University of Cambridge.

Warren, A. 1970. Dune trends and their implications in the central Sudan. *Zeitschrift für Geomorphologie*. Supplementband 10, 154–79.

Warren, A. 1971. The dunes of the Ténéré Desert. *Geographical Journal* **137**, 458–61.

Warren, A. 1972. Observations on dunes and bimodal sands in the Ténéré Desert. *Sedimentology* **19**, 37–44.

Warren, A. 1976a. Morphology and sediments of the Nebraska Sand Hills in relation to Pleistocene winds and the development of eolian bedforms. *Journal of Geology* **84**, 685–700.

Warren, A. 1976b. Dune trend and the Ekman spiral. *Nature* **259**, 653–4.

Warren, A. 1984. The moving desert margins. *Geographical Magazine* **56**, 457–62.

Warren, A. 1985. Arid geomorphology. *Progress in Physical Geography* **9**, 434–41.

Warren, A. 1988a. The dunes of the Wahiba Sands. *Journal of Oman Studies*, Special Report 3: *The scientific results of the Royal Geographical Society's Oman Wahiba Sands Project 1985-7*, 131–60.

Warren, A. 1988b. The dynamics of network dunes in the Wahiba Sands; a progress report. *Journal of Oman Studies*, Special Report 3: *The scientific results of the Royal Geographical Society's Oman Wahiba Sands Project 1985-7*, 169–81.

Warren, A. 1988c. A note on vegetation and sand movement in the Wahiba Sands. *Journal of Oman Studies*, Special Report 3: *The scientific results of the Royal Geographical Society's Oman Wahiba Sands Project 1985-7*, 251–5.

Warren, A. & C. T. Agnew 1988. *An assessment of desertification and land degradation in arid and semi-arid areas*. London: International Institute for Environment and Development.

Warren, A. & S. A. W. Kay 1987. The dynamics of dune networks. See Frostick & Reid (1987), 205–12.

Warren, A. & P. Knott 1983. Desert dunes: a short review of needs in desert dune research and a recent study of micrometeorological dune-initiation mechanisms. See Brookfield & Ahlbrandt (1983), 343–52.

Warren, J. K. 1983. On pedogenetic calcrete as it occurs in the vadose zone of Quaternary calcareous dunes in coastal South Australia. *Journal of Sedimentary Petrology* **53**, 787–96.

Washburn, A. L. 1979. *Geocryology*. London: Edward Arnold.

Wasson, R. J. 1974. Intersection point deposition on alluvial fans: an Australian example. *Geografiska Annaler* **56A**, 83–92.

Wasson, R. J. 1976. Holocene aeolian landforms of the Belarabon area SW of Cobar, NSW. *Royal Society of New South Wales, Journal and Proceedings* **109**, 91–101.

Wasson, R. J. 1977a. Catchment processes and the evolution of alluvial fans in the lower Derwent Valley, Tasmania. *Zeitschrift für Geomorphologie* **21**, 147–68.

Wasson, R. J. 1977b. Last-glacial alluvial fan sedimentation in the lower Derwent Valley, Tasmania. *Sedimentology* **24**, 781–99.

Wasson, R. J. 1983a. Dune sediment types, sand colour, sediment provenance and hydrology in the Strzelecki–Simpson Dunefield, Australia. See Brookfield & Ahlbrandt (1983), 165–95.

Wasson, R. J. 1983b. The Cainozoic history of the Strzelecki and Simpson dune fields (Australia), and the origin of the desert dunes. *Zeitschrift für Geomorphologie*, Supplementband 45, 85–115.

Wasson, R. J. 1984. Late Quaternary palaeo-environments in the desert dunefields of Australia. See Vogel (1984), 419–32.

Wasson, R. J. 1986. Geomorphology and Quaternary history of the Australian continental dunefields. *Geographical Review of Japan* **59**, Series B, 53–67.

Wasson, R. J. & R. Hyde 1983a. A test for the granulometric control of desert dune geometry. *Earth Surface Processes and Landforms* **8**, 301–12.

Wasson, R. J. & R. Hyde 1983b. Factors determining desert dune type. *Nature* **304**, 337–9.

Wasson, R. J. & R. Hyde 1984. Reply to D. M. Rubin on factors determining desert dune type. *Nature* **309**, 92.

Wasson, R. J. & P. M. Nanninga 1986. Estimating wind transport of sand on vegetated surfaces. *Earth Surface Processes and Landforms* **11**, 505–14.

Wasson, R. J., S. N. Rajaguru, V. N. Misra, D. P. Agarwal, R. P. Dhir, A. K. Singhoi, K. Kameswara Rao 1983. Geomorphology, late Quaternary stratigraphy and palaeoclimatology of the Thar dune field. *Zeitschrift für Geomorphologie*, Supplementband 45, 117–51.

Wasson, R. J., G. I. Smith, D. P. Agrawal 1984. Late Quaternary sediments, minerals, and inferred geochemical history of Didwana Lake, Thar Desert, India. *Palaeogeography, Palaeoclimatology, Palaeoecology* **46**, 345–72.

Watson, A. 1979. Gypsum crusts in deserts. *Journal of Arid Environments* **2**, 3–20.

Watson, A. 1983. Gypsum crusts. See Goudie & Pye (1983), 133–61.

Watson, A. 1985. Structure, chemistry and origins of gypsum crusts in southern Tunisia and the central Namib desert. *Sedimentology* **32**, 855–76.

Watson, A. 1986. Grain-size variations on a longitudinal dune and a barchan dune. *Sedimentary Geology* **46**, 49–66.

Watson, A. 1987. Discussion: variations in wind velocity and sand transport on the windward flanks of desert sand dunes. *Sedimentology* **34**, 511–20.

Watson, A. 1988. Desert gypsum crusts as palaeoenvironmental indicators: a micropetrographic study of crusts from southern Tunisia and the central Namib Desert. *Journal of Arid Environments* **15**, 19–42.

Watson, R. A. & H. E. Wright Jr 1969. The Saidmarreh

landslide, Iran. Geological Society of America Special Paper 123, 115–39.

Watt, A. S. 1937. On the origin and development of blowouts. *Journal of Ecology* **25**, 91–112.

Watts, N. L. 1977. Pseudo-anticlines and other structures in some calcretes of Botswana and South Africa. *Earth Surface Processes* **2**, 63–74.

Watts, N. L. 1978. Displacive calcite: evidence from recent and ancient calcretes. *Geology* **6**, 699–703.

Watts, N. L. 1980. Quaternary pedogenic calcretes from the Kalahari (southern Africa): mineralogy, genesis and diagenesis. *Sedimentology* **27**, 661–86.

Watts, S. H. 1975. Mound springs. *Australian Geographer* **13**, 52–3.

Watts, S. H. 1978a. A petrographic study of silcrete from inland Australia. *Journal of Sedimentary Petrology* **48**, 987–94.

Watts, S. H. 1978b. The nature and occurrence of silcrete in the Tibooburra area of northwestern New South Wales. See Langford-Smith (1978b), 167–86.

Webb, R. H. & H. G. Wilshire (eds) 1983. *Environmental effects of off-road vehicles*. New York: Springer.

Webb, R. H., P. T. Pringle, G. R. Rink 1987. *Debris flows from tributaries of the Colorado River, Grand Canyon National Park, Arizona*. USGS Open-file Report, 87–118.

Webster, R. 1988. The Bedouin of the Wahiba Sands: pastoral ecology and management. *Journal of Oman Studies*, Special Report **3**, 443–51.

Weedman, S. D. & R. Slingerland 1985. Experimental study of sand streaks formed in turbulent boundary layers. *Sedimentology* **32**, 133–46.

Wehmeier, E. 1986. Water-induced sliding of rocks on playas: Alkali Flat in Big Smoky Valley, Nevada. *Catena* **13**, 197–209.

Weide, D. L. (ed.) 1985. *Soils and Quaternary geology of the southwestern United States*. Geological Society of America, Special Paper 203, 23–41.

Weinert, H. H. 1965. Climatic factors affecting the weathering of igneous rock. *Agricultural Meteorology* **2**, 27–42.

Weischet, W. 1969. Zur Geomorphologie des Glatthang-Reliefs in der ariden Subtropenzone des Kleinen Nordens von Chile. *Zeitschrift für Geomorphologie*, **NF13**, 1–21.

Weise, O. R. 1970. Zur Morphodynamik der Pediplanation, *Zeitschrift für Geomorphologie*, Supplementband 10, 64–87.

Weise, O. R. 1978. Morphodynamics and morphogenesis of pediments in the deserts of Iran. *Geographical Journal* **144**, 450–62.

Weisrock, A. 1982. Signification paléoclimatique des dunes littorales d'Essaouira – Cap Sim (Maroc). *Revue de Géomorphologie Dynamique* **31**, 91–107.

Wellendorf, W. & D. H. Krinsley 1980. Wind velocities determined from the surface textures of sand grains. *Nature* **283**, 372–3.

Wellington, J. H. 1943. The Lake Chrissie problem. *South African Geographical Journal* **25**, 50–64.

Wells, G. L. 1983. Late-glacial circulation over central North America revealed by eolian features. See *Variations in the global water budget*, A. Street-Perrot, M. Beran, R. Ratcliffe (eds), 317–30. Dordrecht: Reidel.

Wells, G. L., T. F. Gullard, L. N. Smith 1982. Origin and evolution of deserts in the Basin and Range of the Colorado Plateau provinces of western North America. *Striae* **17**, 101–11.

Wells, L. A. 1902. *Journal of the Calvert scientific exploring expedition 1896-7*. Western Australia Parliamentary Paper 46.

Wells, P. V. 1976. Macrofossil analysis of woodrat (*Neotoma*) middens as a key to the Quaternary vegetational history of arid America. *Quaternary Research* **6**, 223–48.

Wells, S. G., J. C. Dohrenwend, I. D. McFadden, B. D. Turrin, K. D. Mahrer 1985. Late Cenozoic landscape evolution on lava flow surfaces of the Cima volcanic field, Mojave Desert, California. *Geological Society of America, Bulletin* **96**, 1518–29.

Wells, S. G., L. D. McFadden, J. C. Dohrenwend 1987. Influence of late-Quaternary climatic changes on geomorphic and pedogenic processes on a desert piedmont, eastern Mojave Desert, California. *Quaternary Research* **27**, 130–46.

Wells, S. G., L. D. McFadden, R. I. Dorn, M. J. DeNiro, H. O. Ajie 1987. Comment and reply on "Isotopic evidence for climatic influence on alluvial-fan development in Death Valley, California". *Geology* **15**, 1178–80.

Wendorf, F., R. Schild, R. Said, C. V. Haynes, A. Gautier, P. Kobusiewicz 1976. The prehistory of the Egyptian Sahara. *Science* **193**, 103–16.

Werner, B. T. 1990. A steady-state model of wind-blown sand transport. *Journal of Geology* **98**, 1–17.

Werner, B. T. & P. K. Haff 1988. The impact process in aeolian saltation: two-dimensional simulations. *Sedimentology* **35**, 189–96.

Werner, B. T., P. K. Haff, R. P. Livi, R. S. Anderson 1986. The measurement of eolian ripple cross-sectional shapes. *Geology* **14**, 743–5.

Western, S. 1972. The classification of arid-zone soils, I: an approach to the classification of arid-zone soils using depositional features. *Journal of Soil Science* **23**, 266–78.

Westin, F. C. 1982. Resource inventories of arid and semi-arid lands using LANDSAT. See *ERIM Proceedings International Symposium on Remote Sensing: remote sensing of arid and semi-arid lands*, 83–8.

Whalley, W. B. 1983. Desert varnish. See Goudie & Pye (1983), 179–226.

Whalley, W. B. 1984. High altitude rock weathering processes. See Miller (1984), 365–73.

Whalley, W. B. (ed.) 1978. *Scanning electron microscopy in the study of sediments*. Norwich, England: Geobooks.

Whalley, W. B., G. R. Douglas, J. P. McGreevy 1982a. Crack propagation and associated weathering in igneous rocks. *Zeitschrift für Geomorphologie* **26**, 33–54.

Whalley, W. B., J. R. Marshall, B. J. Smith 1982b. Origin

of desert loess from some experimental observations. *Nature* **300**, 433–5.

Whalley, W. B., B. J. Smith, J. J. McAlister, A. J. Edwards 1987. Aeolian abrasion of quartz particles and the production of silt-size fragments: preliminary results. See Frostick & Reid (1987), 129–38.

Whitaker, C. R. 1973. *A bibliography of pediments*. Norwich: Geobooks.

Whitaker, C. R. 1974. Split boulder. *Australian Geographer* **12**, 562–3.

Whitaker, C. R. 1979. The use of the term "pediment" and related terminology. *Zeitschrift für Geomorphologie* **23**, 427–39.

White, B. R. 1985. The dynamics of particle motion in saltation. See Barndorff-Nielsen et al. (1985), 101–40.

White, B. R. & J. C. Schulz 1977. Magnus effect in saltation. *Journal of Fluid Mechanics* **81**, 497–512.

White, E. M. 1961. Drainage alignment in western South Dakota. *American Journal of Science* **259**, 207–10.

White, E. M. & R. G. Bonestall 1960. Some gilgaied soils in South Dakota. *Soil Science Society of America, Proceedings* **24**, 305–9.

White, K. H. 1988. *The relationship between catchment structure and piedmont development in the Tunisian southern Atlas*. (MS)

White, L. P. 1971. The ancient erg of Hausaland in southwestern Niger. *Geographical Journal* **137**, 69–73.

White, L. P. & R. Law 1969. Channelling of alluvial depression soils in Iraq and Sudan. *Journal of Soil Science* **20**, 84–90.

Whitney, J. W. 1983. *Erosional history and surficial geology of western Saudi Arabia*. USGS Technical Record USGS-TR-04-1.

Whitney, J. W., D. J. Faulkender, M. Rubin 1983. *The environmental history and present condition of the northern sand seas of Saudi Arabia*. USGS, Open File Report USGS-OF-03-95.

Whitney, M. I. 1978. The role of vorticity in developing lineation by wind erosion. *Geological Society of America, Bulletin* **89**, 1–18.

Whitney, M. I. 1979. Electron micrography of mineral surfaces subject to wind-blast erosion. *Geological Society of America, Bulletin* **90**, 917–34.

Whitney, M. I. 1983. Eolian features shaped by aerodynamic and vorticity processes. See Brookfield & Ahlbrandt (1983), 223–45.

Whitney, M. I. & H. B. Brewer 1968. Discoveries in aerodynamic erosion with wind tunnel experiments. *Michigan Academy of Science Arts and Letters* **53**, 91–104.

Whitney, M. I. & R. V. Dietrich 1973. Ventifact sculpture by wind-blown dust. *Geological Society of America, Bulletin* **84**, 2561–82.

Whitney, M. I. & J. F. Splettstoesser 1982. Ventifacts and their formation; Darwin Mountains, Antarctica. See Yaalon (1982), 175–94.

Whittaker, R. H., S. W. Buol, W. A. Niering, Y. H. Havens 1968. A soil and vegetation pattern in the Santa Catalina Mountains, Arizona. *Soil Science* **105**, 440–51.

Wieder, M. & D. H. Yaalon 1982. Micro-morphological fabrics and development stages of carbonate nodular forms related to soil characteristics. *Geoderma* **28**, 203–20.

Wiegland, C. L., L. Lyles, D. L. Carter 1966. Interspersed salt affected and unaffected dryland soils of the lower Rio Grande Valley, II: occurrence of salinity in relation to infiltration rates and profile characteristics. *Soil Science Society of America, Proceedings* **30**, 106–9.

Wiggs, G. F. S. 1992. *Desert dune dynamics and the evaluation of shear velocity: an integrated approach*. Oxford: Blackwell Scientific.

Wilhelmy, H. 1969. Das Urstromtal am Ostrand der Industene und der Sarasvati-Problem. *Zeitschrift für Geomorphologie*, Supplementband 8, 76–93.

Willetts, B. B. 1983. Transportation by wind of granular materials of different grain shapes and densities. *Sedimentology* **30**, 669–80.

Willetts, B. B. 1989. Physics of sand movement in vegetated dune systems. See Gimmingham et al. (1989), 37–49.

Willetts, B. B., I. K. McEwan, M. A. Rice 1990. Initiation of motion of quartz sand grains. *Acta Mechanica* **1**, 123–34.

Willetts, B. B. & M. A. Rice 1983. Practical representation of characteristic grain shape in sands: a comparison of methods. *Sedimentology* **30**, 557–65.

Willetts, B. B. & M. A. Rice 1985a. Inter-saltation collisions. See Barndorff-Nielsen et al. (1985), 83–100.

Willetts, B. B. & M. A. Rice 1985b. Wind-tunnel tracer experiments using dyed sand. See Barndorff-Nielsen et al. (1985), 225–42.

Willetts, B. B. & M. A. Rice 1986. Collision in aeolian transport; the saltation/creep link. See Nickling (1986), 1–19.

Willetts, B. B. & M. A. Rice 1988. Particle dislodgement from a flat sand bed by wind. *Earth Surface Processes and Landforms* **13**, 717–28.

Willetts, B. B. & M. A. Rice 1989. Collisions of quartz grains with a sand bed: the influence of incident angle. *Earth Surface Processes and Landforms* **14**, 719–30.

Willetts, B. B., M. A. Rice, S. E. Swaine 1982. Shape effects in aeolian grain transport. *Sedimentology* **29**, 409–17.

Williams, C. B. 1923. A short bio-climatic study in the Egyptian desert. Ministry of Agriculture (Egypt) Technical and Scientific Service Bulletin 81, 567–72.

Williams, G. E. 1967. Characteristics and origin of a preCambrian pediment. *Journal of Geology* **77**, 183–207.

Williams, G. E. 1973. Late Quaternary piedmont sedimentation, soil formation and paleoclimates in arid South Australia. *Zeitschrift für Geomorphologie* **17**, 102–25.

Williams, G. P. 1964. Some aspects of the eolian saltation load. *Sedimentology* **3**, 257–87.

Williams, J. I. & R. M. Farias 1972. Utilization and taxonomy of the desert grass *Panicum turgidum*. *Economic Botany* **26**, 13–20.

Williams, J. J., G. R. Butterfield, D. G. Clark 1990. Aerodynamic entrainment thresholds and dislodgment rates on impervious and permeable beds. *Earth Surface Processes and Landforms* **15**, 255–64.

Williams, M. A. J. 1968. Soil salinity in the west central Gezira, Republic of the Sudan. *Soil Science* **105**, 451–64.

Williams, M. A. J. 1975. Late Pleistocene tropical aridity synchronous in both hemispheres? *Nature* **253**, 617–8.

Williams, M. 1984. Geology. See *Sahara Desert*, J. L. Cloudsley-Thompson (ed.), 31–9. Oxford: Pergamon Press.

Williams, M. A. J. 1985. Pleistocene aridity in tropical Africa, Australia and Asia. See Douglas & Spencer (1985), 219–33.

Williams, M. A. J. & D. A. Adamson 1974. Late Pleistocene desiccation along the White Nile. *Nature* **248**, 584–6.

Williams, M. A. J. & H. Faure (eds) 1980. *The Sahara and the Nile*. Rotterdam: Balkema.

Williams, M. A. J., D. A. Adamson, H. H. Abdulla 1981. Landforms and soils of the Gezira: a Quaternary legacy of the Blue and White Nile Rivers. See *A land between two Niles: Quaternary geology and biology of the central Sudan*, M. A. J. Williams & D. A. Adamson (1981), 111–42. Rotterdam: Balkema.

Williams, M. A. J. & M. F. Clarke 1984. Late Quaternary environments in north-central India. *Nature* **308**, 633–65.

Williams, P. B. & P. H. Kemp 1972. Initiation of ripples by artificial disturbances. *American Association of Civil Engineers, Journal of the Hydraulics Division* **98 HY6**, Proceedings Paper 8952, 1057–70.

Williams, R. G. B. 1975. The British climate during the last glaciation: an interpretation based on periglacial phenomena. See *Ice ages: ancient and modern*, A. E. Wright & F. Moseley, 95–119. Liverpool: Seel House Press.

Wilshire, H. G. 1980. Human causes of accelerated wind erosion in California's deserts. See *Thresholds in geomorphology*, D. R. Coates & J. D. Vitek (eds), 415–34. Boston: Allen & Unwin.

Wilshire, H. G. & J. K. Nakata 1976. Off-road vehicle effects on California's Mojave Desert. *California Geology*, June, 123–32.

Wilshire, H. G., J. K. Nakata, B. Hallett 1981. Field observations on the December 1977 wind storm, San Joaquin Valley, California. Geological Society of America, Special Paper 186, 233–52.

Wilson, A. T. 1979. Geochemical problems of the Antarctic dry areas. *Nature* **280**, 205–8.

Wilson, I. G. 1970. *The external morphology of wind-laid sand deposits*. PhD dissertation, University of Reading.

Wilson, I. G. 1971. Desert sandflow basins and a model for the development of ergs. *Geographical Journal* **137**, 180–99.

Wilson, I. G. 1972a. Sand waves. *New Scientist* **23**, 634–7.

Wilson, I. G. 1972b. Aeolian bedforms – their development and origins. *Sedimentology* **19**, 173–210.

Wilson, I. G. 1972c. Universal discontinuities in bedforms produced by the wind. *Journal of Sedimentary Petrology* **42**, 667–9.

Wilson, I. G. 1973a. Ergs. *Sedimentary Geology* **10**, 77–106.

Wilson, I. G. 1973b. Equilibrium cross-section of meandering and braided rivers. *Nature* **241**, 393–4.

Wilson, L. 1969. Les relations entre les processus géomorphologiques et le climat moderne comme méthode de paléoclimatologie. *Revue de Géographie Physique et Geologie Dynamique*, sér. 2 **11**, 303–14.

Wilson, L. 1973. Variations in mean sediment yield as a function of mean annual precipitation. *American Journal of Science* **273**, 335–49.

Wilson, P. 1979. Experimental investigation of etch-pit formation on quartz sand grains. *Geological Magazine* **116**, 477–82.

Wilson, R. C. L. (ed.) 1983. *Residual deposits, surface-related weathering and materials*. Oxford: Blackwell Scientific.

Wilson, S. J. & R. U. Cooke 1980. Wind erosion. See *Soil erosion*, M. J. Kirkby & R. P. C. Morgan (eds), 217–52. Chichester, England: John Wiley.

Winkelmolen, A. M. 1971. Rollability, a functional shape property of grains. *Journal of Sedimentary Petrology* **41**, 703–14.

Winkler, E. M. 1979. Role of salts in development of granitic tafoni, South Australia: a discussion. *Journal of Geology* **87**, 119–20.

Winkler, E. M. & A. Rice 1977. Insolation warmed over: comment and replies. *Geology* **5**, 188–90.

Winkler, E. M. & E. J. Wilhelm 1970. Salt burst by hydration pressures in architectural stone in urban atmosphere. *Geological Society of America, Bulletin* **81**, 567–72.

Winstanley, D. 1973. Rainfall patterns and general atmospheric circulation. *Nature* **245** 190–4.

Wippermann, F. K. 1969. The orientation of vortices due to instability of the Ekman boundary layer. *Beiträge zür Physik der Atmosphäre* **42**, 225–244. Also in *The planetary boundary layer*, F. Wippermann (1973). Offenbach am Main: Deutscher Wetterdienst.

Wippermann, F. K. & G. Gross 1986. The wind-induced shaping and migration of an isolated dune: a numerical experiment. *Boundary-layer Meteorology* **36**, 319–34.

Wirrman, D. & L. F. de O. Almeida 1987. Low Holocene level (7,700 to 3,650 years ago) of Lake Titicaca (Bolivia). *Palaeogeography, Palaeoclimatology, Palaeoecology* **59**, 315–23.

Wojciechowski, A. 1979. Reconstructions of the directions of near-ground air streams behind the lee slope of a dune on the basis of the distribution of eolian ripple marks (in Polish). *Badenia Fizjograficne nad Polska Zachodnia, Seria A, Geografia Fizyczna* **32**, 169–90.

Wolman, M. G. & R. Gerson 1978. Relative scales of time and effectiveness of climate in watershed geomorphology. *Earth Surface Processes* **3**, 189–208.

Wood, C. D. & P. W. Espenschade 1965. Mechanisms of dust erosion. *Society of Atomic Engineers* **73**, 515–23.

Wood, M. K., W. H. Blackburn, R. E. Eckert, F. F. Peterson 1978. Interrelations of the physical properties of coppice dunes and vesicular dune interspace soils with grass seedling emergence. *Journal of Range Management* **31**, 189–92.

Wood, W. H. 1970. The rectification of wind-blown sand. *Journal of Sedimentary Petrology* **40**, 29–37.

Woodruff, N. P. & F. H. Siddoway 1965. A wind erosion equation. *Soil Science Society of America, Proceedings* **29**, 602–8.

Woolnough, W. G. 1927. Presidential address, part I: The chemical criteria of peneplanation; part II: The duricrust of Australia. *Royal Society New South Wales, Journal and Proceedings* **61**, 1–53.

Wopfner, H. 1978. Silcretes of northern South Australia and adjacent regions. See Langford-Smith (1978b), 93–142.

Wopfner, H. & C. R. Twidale 1967. Geomorphological history of the Lake Eyre Basin. See *Landform studies from Australia and New Guinea*, J. N. Jennings & J. A. Mabbutt (eds), 119–43. Cambridge: Cambridge University Press.

Worral, G. A. 1974. Observations of some wind-formed features in the southern Sahara. *Zeitschrift für Geomorphologie* **18**, 291–302.

Wright, A. C. S. & H. Urzúa 1963. Meteorizacion en la region costera del desierto del Norte de Chile. *Comm. y Res. y Trabajos Conf. Latino-american para el estudio de las regiones Andas*, 26–8.

Wright, H. E. Jr 1958. An extinct wadi system in the Syrian Desert. *Bulletin of the Research Council of Israel* **76**, 53–9.

Wright, V. P., N. H. Platt, W. A. Wimbledon 1988. Biogenic laminar calcretes: evidence of calcified rootmat horizons in palaeosols. *Sedimentology* **35**, 603–20.

Wyrwoll, H. & G. K. Smyth 1985. On using the log normal distribution to describe the textural characteristics of eolian sediments. *Journal of Sedimentary Petrology* **55**, 471–8.

Xu Shuying & Xu Defu 1983. A primary observation of aeolian sand deposits on eastern shore of the Qinghai Lake. *Journal of Desert Research* **3**, 11–17.

Yaalon, D. H. 1967. Factors affecting the lithification of aeolianite and interpretation of its environmental significance in the coastal plain of Israel. *Journal of Sedimentary Petrology* **37**, 1189–99.

Yaalon, D. H. 1970. Parallel stone cracking, a weathering process on desert surfaces. *Geological Institute Bucharest, Technical & Economic Bulletin* **18**, 107–11.

Yaalon, D. H. (ed.) 1971. *Palaeopedology: origin, nature and dating of palaeosols*. Jerusalem: Israels University Press.

Yaalon, D. H. 1974. Note on some geomorphic effects of temperature changes on desert surfaces. *Zeitschrift für Geomorphologie*, Supplementband 21, 29–34.

Yaalon, D. H. 1975. Discussion of "Internal geometry of vegetated coastal dunes". *Journal of Sedimentary Petrology* **45**, 359.

Yaalon, D. H. 1981. Pedogenic carbonate in aridic soils –

magnitude of the pool and annual fluxes [abstract]. *International Conference on Aridic Soils*, Jerusalem.

Yaalon, D. H. (ed.) 1982. *Aridic soils and geomorphic processes*. Catena, Supplement 1.

Yaalon, D. H. & J. Dan 1974. Accumulation and distribution of loess-derived deposits in the semi-arid desert fringe area of Israel. *Zeitschrift für Geomorphologie*, Supplementband 20, 91–105.

Yaalon, D. H. & E. Ganor 1966. The climatic factor of wind erodibility and dust blowing in Israel. *Israel Journal of Earth Science* **15**, 27–32.

Yaalon, D. H. & E. Ganor 1973. The influence of dust on soils during the Quaternary. *Soil Science* **116**, 146–55.

Yaalon, D. H. & E. Ganor 1975. Rates of aeolian dust accretion in the Mediterranean and desert fringe environments of Israel. *9th International Sedimentological Congress, Nice, Proceedings* 2, 169–74.

Yaalon, D. H. & E. Ganor 1979. East Mediterranean trajectories of dust-carrying storms from the Sahara to Sinai. See Morales (1979), 187–93.

Yaalon, D. H. & D. Ginzbourg 1966. Sedimentary characteristics and climatic analysis of easterly dust storms in the Negev (Israel). *Sedimentology* **6**, 315–32.

Yaalon, D. H. & J. Laronne 1971. Internal structures in eolianites and palaeowinds, Mediterranean coast of Israel. *Journal of Sedimentary Petrology* **41**, 1059–64.

Yaalon, D. H. & S. Singer 1974. Vertical variation in strength and porosity of calcrete (nari) on chalk, Shefala, Israel, and interpretation of its origin. *Journal of Sedimentary Petrology* **44**, 1016–23.

Yaalon, D. H. & M. Wieder 1976. Pedogenic palygorskite in some arid brown (calciorthid) soils in Israel. *Clay Minerals* **11**, 73–80.

Yair, A. 1983. Hillslope hydrology water harvesting and areal distribution of some ancient agricultural systems in the northern Negev desert. *Journal of Arid Environments* **6**, 283–301.

Yair, A. 1990. Runoff generation in a sandy area – the Nizzana Sands, western Negev, Israel. *Earth Surface Processes and Landforms* **15**, 597–609.

Yair, A. & J. De Ploey 1979. Field observations and laboratory experiments concerning the creep process of rock blocks in an arid environment. *Catena* **6**, 245–58.

Yair, A. & Y. Enzel 1987. The relationship between annual rainfall and sediment yield in arid and semi-arid areas: the case of the northern Negev. *Catena* **10**, 121–35.

Yair, A. & M. Klein 1973. The influence of surface properties on flow and erosion processes on debris-covered slopes in an arid area. *Catena* **1**, 1–18.

Yair, A. & H. Lavee 1976. Runoff generative process and runoff yield from arid talus mantled slopes. *Earth Surface Processes* **1**, 235–47.

Yair, A. & H. Lavee 1981. An investigation of source areas of sediment and sediment transport by overland flow along arid hillslopes. See *Erosion and sediment transport measurement, symposium proceedings*, Florence, June 1981. IASH Publication 133, 433–46.

Yair A. & H. Lavee 1985. Runoff generation in arid and

semi-arid zones. See *Hydrological forecasting*, M. G. Anderson & T. P. Burt (eds), 183–20. Chichester, England: John Wiley.

Yair, A., H. Lavee, M. Shachak, Y. Enzel 1987. Sede Boger and Horav Plateau experimental sites, in *Workshop on erosion, transport and deposition processes with emphasis on semi-arid and arid areas*, field guidebook: main excursion (IASH/IGU, Israel, 1–4 April 1987), section 4.

Yair, A. & J. Rutin 1981. Some aspects of the regional variation in the amounts of available sediment produced by isopods and porcupines, northern Negev, Israel. *Earth Surface Processes and Landforms* 6, 221–34.

Yair, A., D. Sharon, H. Lavee 1980. Trends in runoff and erosion processes over an arid limestone hillside, northern Negev, Israel. *Hydrological Sciences Bulletin* 25, 245–55.

Yalin, M. S. 1964. Geometrical properties of sand waves. *American Society of Civil Engineers, Proceedings; Journal of the Hydraulics Division* 91 HY3, 343–66.

Yalin, M. S. 1977. *Mechanics of sediment transport*. Oxford: Pergamon.

Yarham, E. R. 1958. Singing sands: a strange concert heard in the desert. *Unesco Courier* 6, 26–7.

Young, A. 1972. *Slopes*. Edinburgh: Oliver & Boyd.

Young, A. R. M. 1987. Salt as an agent in the development of cavernous weathering. *Geology* 15, 962–66.

Young, J. A. & R. A. Evans 1986. Erosion and deposition of fine sediments from playas. *Journal of Arid Environments* 10, 103–15.

Young, R. W. 1985. Silcrete distribution in eastern Australia. *Zeitschrift für Geomorphologie* 29, 21–36.

Zaidberg, R., J. Dan, H. Koyumdjisky 1982. The influence of parent material, relief and exposure on soil formation in the arid region of eastern Samaria. *Catena Supplement* 1, 117–37.

Zakrewska, B. 1963. An analysis of landforms in a part of the central Great Plains. *Annals of the AAG* 53, 536–68.

Zenkovich, Z. P. 1967. Aeolian processes on sea coasts. See *Processes of coastal development*, J. A. Steers (ed.), 586–617. New York: Wiley-Interscience.

Zhu Zhenda 1984. Aeolian landforms of the Taklamakan desert. See El-Baz (1984b), 133–44.

Zhu Zhenda, Zou Bengong, Yang Youlin 1987. The characteristics of aeolian landforms and the control of mobile dunes in China. See Gardiner (1987), 1211–15.

Zimmerman, L. I. 1969. Atmospheric wake phenomena near the Canary Islands. *Journal of Applied Meteorology* 8, 896–907.

Zingg, A. W. 1952. Wind tunnel studies of the movement of sedimentary material. *5th Hydraulic Conference, Proceedings*, Iowa Institute of Hydraulics, Iowa City, 111–35.

Zingg, A. W. 1953. Speculations on climate as a factor in the wind erosion problem of the Great Plains. *Kansas Academy of Sciences, Transactions* 56, 371–7.

Zwolinski, Z. 1985. Depositional model for desert creek channels: Lake Eyre region, central Australia. *Zeitschrift für Geomorphologie*, Supplementband 55, 39–56.

GEOGRAPHICAL INDEX

SYSTEMATIC INDEX